ELSEVIER'S

DICTIONARY OF TELEVISION AND VIDEO RECORDING

ELSEVIER'S

DICTIONARY OF TELEVISION AND VIDEO RECORDING

IN SIX LANGUAGES

ENGLISH/AMERICAN-FRENCH-SPANISH-ITALIAN DUTCH AND GERMAN

Compiled and arranged on an English alphabetical basis by

W.E. CLASON
Geldrop, The Netherlands

ELSEVIER SCIENTIFIC PUBLISHING COMPANY
AMSTERDAM/OXFORD/NEW YORK
1975

ELSEVIER SCIENTIFIC PUBLISHING COMPANY
335 JAN VAN GALENSTRAAT
P.O. BOX 211, AMSTERDAM, THE NETHERLANDS

AMERICAN ELSEVIER PUBLISHING COMPANY, INC.
52 VANDERBILT AVENUE
NEW YORK, NEW YORK 10017

Library of Congress Card Number: 74-77577

ISBN: 0-444-41224-7

Printed in The Netherlands

PUBLISHER'S NOTE

By the most conservative estimates, there are at least two thousand basic languages in current use today. By some definitions, this figure could reach more than five thousand languages, not including a vast number of dialects. Added to this, the incredible current advances in engineering and science have rendered most existing technical dictionaries obsolete. The accumulated result has been a fast growing need for specialized multilingual dictionaries, particularly in the face of international scientific cooperation and exchange of knowledge.

To meet this need, Elsevier publishes a number of multilingual technical dictionaries relating to special fields of science and industry. Edited under authoritative auspices, they draw upon rich sources of knowledge.

In planning this dictionary, the author and publishers have been guided by the principles proposed by the United Nations Educational, Scientific and Cultural Organization (UNESCO). The aim is to ensure that each dictionary will fit into place in a pattern which may progressively extend over all interrelated fields of science and technology and cover all necessary languages.

For each language, there is an alphabetical list of words referring to corresponding numbers in the basic table. The system of thumb-indexing enables one to find any language at once. The binding, smooth paper and convenient size result in an enjoyable and valuable reference book.

PREFACE

The enormous expansion of the use of television throughout the world, the elopment and increasingly wide distribution of colour television, the im-tant impact of film on the compilation of programmes, and also the advent . rapid increase in video recording have drastically altered the terminology :his field since the early stage of 1954.

I have consulted the most recent literature and believe that the resulting ns reflect accurately the present situation with regard to the technical ninology. I have had the cooperation of official organizations in the :herlands, Germany, England, France and Italy, and for Spanish problems vas able to obtain the assistance of authorized technical translators in drid.

This dictionary is based on Elsevier's Dictionary of Television, Radar and tennas, which has been completely revised. The terms on radar and ennas that appeared in that dictionary are here deleted, with the excep-n of course of television antennas, but they will be incorporated in the re-:d edition of Elsevier's Telecommunication Dictionary.

As always, I have been assisted by my wife, who typed the complete nuscript and gave valuable assistance in correcting the proofs.

W.E. Clason

ABBREVATIONS

aea	aerials/antennas	opt	optics
ani	animation	rec	recording
aud	audio technique	rep	reproducing
cpl	coupling circuits and devices	stu	studios
crt	cathode-ray tubes	svs	servicing systems
ct	colo(u)r and light science	th	theatres
ctv	colo(u)r television	tv	television
fi	film technique	vr	video recording
med	miscellaneous applications		

adj	adjective	v	verb
f	feminine	f	Français
GB	English, British usage	e	Español
m	masculine	i	Italiano
n	neuter	n	Nederlands
pl	plural	d	Deutsch
US	English, American usage		

BASIC TABLE

A

1 A-B CUT MIXER rec
A vision mixer in which the output of the A and B banks are combined and returned to one input of a separate cut bank.
f mélangeur *m* vidéo à un étage
e mezclador *m* video de una etapa
i mescolatore *m* video ad uno stadio
n eentrapsbeeldmenger
d Einstufenbildmischer *m*

2 ABC, tv
AUTOMATIC BACKGROUND CONTROL,
AUTOMATIC BRIGHTNESS CONTROLL
A circuit used in a TV receiver to keep the average brightness of the reproduced image essentially constant.
f contrôle *m* automatique de la brillance, réglage *m* automatique de la luminosité
e control *m* automático del brillo, regulación *f* automática de la luminosidad
i controllo *m* automatico della luminosità
n automatische helderheidsregeling
d selbsttätige Helligkeitsregelung *f*

3 A-B-C-D CUT MIXER rec
A vision mixer which enables two sets of two sources to be mixed separately, and the resulting outputs to be themselves mixed.
f mélangeur *m* vidéo à deux étages
e mezclador *m* video de dos etapas
i mescolatore *m* video a due stadi
n tweetrapsbeeldmenger
d Zweistufenbildmischer *m*

4 ABERRATION crt/tv
An image defect that occurs when an optical lens or mirror does not bring all light rays to the same focus, or when an electron lens does not bring the electron beam to the same sharp focus at all points on the screen of a cathode-ray tube.
f aberration *f*
e aberración *f*
i aberrazione *f*
n aberratie, afwijking
d Aberration *f*

5 ABSOLUTE THRESHOLD OF LUMINANCE ct
Lowest luminance perceptible.
f seuil *m* absolu de luminance
e umbral *m* absoluto de luminancia
i soglia *f* assoluta di luminanza
n absolute luminantiedrempel
d absolute Luminanzschwelle *f*

6 ABSORBER CIRCUIT, cpl
ABSORPTION CIRCUIT
A series-resonant circuit used to absorb power at an unwanted signal frequency.
f circuit *m* absorbant, circuit *m* d'absorption
e circuito *m* absorbente, circuito *m* de compensación
i circuito *m* d'assorbimento, circuito *m* di compensazione
n zuigkring
d Absorptionskreis *m*

7 ABSORBING FILTER ge
f filtre *m* absorbant
e filtro *m* absorbente
i filtro *m* assorbente
n absorberend filter *n*
d absorbierendes Filter *n*

8 ABSORBING POWER ge
f pouvoir *m* absorbant
e poder *m* absorbente
i assorbità *f*, potere *m* assorbente
n absorptievermogen *n*
d Absorptionsvermögen *n*

9 ABSORPTION ge
f absorption *f*
e absorción *f*
i assorbimento *m*
n absorptie
d Absorption *f*

10 ABSORPTION LOSS cpl
Power loss in a transmission circuit caused by coupling to an adjacent circuit.
f perte *f* par absorption
e pérdida *f* por absorción
i perdita *f* per assorbimento
n absorptieverlies *n*
d Absorptionsverlust *m*

11 ABSORPTION MODULATION, cpl/tv
LOSS MODULATION
Amplitude modulation of the output of a transmitter modulated by means of a variable-impedance device inserted, or coupled to, the output circuit.
f modulation *f* par absorption
e modulación *f* por absorción
i modulazione *f* per assorbimento
n absorptiemodulatie
d Absorptionsmodulation *f*

12 ABSORPTION PEAK cpl
Abnormally high attenuation at a particular frequency as a result of absorption loss.
f crête *f* d'absorption
e cresta *f* de absorción
i cresta *f* d'assorbimento, punta *f* d'assorbimento
n absorptiepiek
d Absorptionsspitze *f*, Saugspitze *f*

13 ABSORPTION TRAP cpl
A parallel-tuned circuit used to absorb and thereby attenuate interfering signals.

f piège *m* à absorption
e trampa *f* de absorción
i trappola *f* ad assorbimento
n absorptieval
d Absorptionsfalle *f*

14 ABSTRACT SET (US), rec/tv
STUDIO DECORATION (GB)
The decorative means in a studio during the making of a picture.
f décoration *f* imaginative d'un studio
e adorno *m* de estudio,
decoración *f* de estudio
i allestimento *m* di studio
n studiodecor *n*
d Studiodekor *n*

15 ABSTRACT STUDIO DESIGN stu
f projet *m* abstract de décoration
e diseño *m* abstracto de decoración
i disegno *m* astratto d'allestimento
n abstract decorontwerp *n*
d abstrakter Dekorentwurf *m*

16 A.C. MAGNETIC BIASING, vr
ALTERNATING CURRENT MAGNETIC BIASING
Magnetic biasing with alternating current, usually well above the signal frequency range, in magnetic recording.
f polarisation *f* magnétique par courant alternatif
e polarización *f* magnética por corriente alterna
i polarizzazione *f* magnetica per corrente alternata
n wisselstroomvoormagnetisatie
d Wechselstromvormagnetisierung *f*

17 A.C. TRANSMISSION tv
A form of transmission in which the d.c. component is not directly represented.
f transmission *f* sans composante continue utile
e transmisión *f* sin componente continua
i trasmissione *f* senza componente continua
n transmissie zonder werkzame nul-component
d Übertragung *f* ohne Gleichstromanteil

18 ACC, ctv
AUTOMATIC CHROMINANCE CONTROL
A circuit which automatically controls the magnitude of the chrominance signal, such as in a colo(u)r receiver.
f réglage *m* automatique de la chrominance
e regulación *f* automática de la crominancia
i regolazione *f* automatica della crominanza
n automatische chrominantieregeling
d selbsttätige Chrominanzregelung *f*

19 ACCELERATING ELECTRODE, crt
ACCELERATOR
An electrode whose voltage provides the electric field to increase the velocity of the electrons in a beam.
f accélérateur *m*,
électrode *f* accélératrice
e acelerador *m*,
electrodo *m* acelerador
i elettrodo *m* acceleratore
n versnellingselektrode
d Beschleunigungselektrode *f*

20 ACCELERATION VOLTAGE crt
The voltage applied to the accelerating electrode.
f tension *f* d'accélération
e tensión *f* acelerante
i tensione *f* d'accelerazione
n versnellingsspanning
d Beschleunigungsspannung *f*

21 ACCENTUATION, rec
PRE-EMPHASIS
A method of placing emphasis on higher or lower studio frequencies in audio systems.
f accentuation *f*,
amplification *f* préférentielle,
amplification *f* sélective
e amplificación *f* selectiva,
preénfasis *f*
i preenfasi *f*
n opduw
d Anhebung *f*

22 ACCENTUATION FILTER, rec
PRE-EMPHASIS FILTER
f filtre *m* d'amplification préférentielle,
filtre *m* d'amplification sélective
e filtro *m* de amplificación selectiva
i filtro *m* per preenfasi
n opduwfilter *n*
d Anhebungsfilter *n*

23 ACCENTUATOR, rec
PRE-EMPHASIS CIRCUIT
A circuit or network inserted to provide accentuation of certain frequencies.
f amplificateur *m* préférentiel,
amplificateur *m* sélectif
e amplificador *m* selectivo
i amplificatore *m* di preenfasi
n opduwschakeling
d Anhebeschaltung *f*

24 ACCEPTABLE CONTRAST RATIO, rep/tv
ACR
Relation between highest and lowest illumination level at which a camera will display a satisfactory picture.
f plage *f* de contraste,
rapport *m* de contraste acceptable
e relación *f* de contraste admisible
i rapporto *m* di contrasto ammissibile
n aanvaardbare contrastverhouding
d annehmbares Kontrastverhältnis *n*,
Kontrastumfang *m*

25 ACCEPTANCE ANGLE opt
The solid angle within which all received light reaches the cathode of a phototube in its housing.
f angle *m* d'admission, angle *m* de capture

e ángulo *m* de admisión
i angolo *m* accettore
n invanghoek
d Einfangwinkel *m*

26 ACCEPTANCE TEST METHOD tv
Method for calculating the exact amplitude-frequency and phase-frequency responses of a TV circuit.
f essai *m* de réception
e prueba *f* de recepción
i collaudo *m* d'accettazione,
prova *f* d'accettazione
n aanvaardingskeuring,
afnemingstest
d Abnahmeprüfung *f*

27 ACCEPTOR CIRCUIT cpl/svs
1. A rejector circuit shunted across the signal circuit to offer a low-impedance path to the unwanted frequency.
2. A tuned circuit responding to a signal of one specific frequency.
f circuit *m* accepteur,
piège *m* série
e circuito *m* aceptador
i circuito *m* accettatore
n kortsluitcircuit *n*,
serieresonantiekring
d Saugkreis *m*

28 ACCESSORY cpt/svs
A part, subassembly, or assembly that contributes to the effectiveness of a piece of equipment without changing its basic function.
f accessoire *m*
e accesorio *m*
i accessorio *m*, pezzo *m* staccato
n onderdeel *n*
d Einzelteil *m*, Zubehör *n*

29 ACCESSORY KIT, svs/tv
TEST KIT
A collection of tools and component parts for servicing TV sets.
f équipement *m* d'accessoires,
trousse *f* de dépannage
e equipo *m* de accesorios
i equipaggiamento *m* d'accessori
n onderdelentas, servicetas
d Kompendium *n*

30 ACCOMMODATION, opt
OCULAR ACCOMMODATION
The ability of the eye to adjust for clear vision at different distances.
f accommodation *f*
e acomodación *f*
i accomodazione *f*
n accommodatie
d Akkommodation *f*

31 ACCOMPANYING SOUND tv
A sound which is generated by interference between the sound and picture carriers.
f bruit *m* propre
e sonido *m* propio
i suono *m* proprio
n bijgeluid *n*
d Eigenton *m*, mitkommender Ton *m*

32 ACCOMPANYING SOUND tv
SUPPRESSION
f suppression *f* du bruit propre
e supressión *f* del sonido propio
i soppressione *f* del suono proprio
n bijgeluidonderdrukking
d Eigentonunterdrückung *f*

33 ACCOMPANYING SOUND TRAP tv
f piège *m* de bruit propre
e trampa *f* para el sonido propio
i trappola *f* per il suono proprio
n bijgeluidvanger
d Falle *f* für den mitkommenden Ton

34 ACHROMATIC ct
Without colo(u)r; capable of transmitting light without breaking it up into constituent colo(u)rs.
f achromatique adj,
sans tonalité chromatique
e acromático adj
i acromatico adj
n achromatisch adj,
kleurloos adj
d achromatisch adj,
unbunt adj

35 ACHROMATIC AERIAL, aea
ACHROMATIC ANTENNA
An aerial (antenna) whose characteristics are uniform in a specified frequency band.
f antenne *f* achromatique
e antena *f* acromática
i antenna *f* acromatica
n achromatische antenne
d achromatische Antenne *f*

36 ACHROMATIC LENS, opt
ACHROMATIC OBJECTIVE
A lens combination that gives correction for achromatic aberration.
f objectif *m* achromatique
e objetivo *m* acromático
i obiettivo *m* acromatico
n achromaat, achromatische lens
d Achromat *n*

37 ACHROMATIC LOCUS, ct/ctv
ACHROMATIC REGION
The area in a chromaticity diagram embracing all colo(u)rs capable of being accepted as white under given conditions of observation.
f domaine *m* achromatique,
région *f* achromatique
e lugar *m* acromático,
región *f* acromática
i luogo *m* acromatico
n achromatisch gebied *n*
d achromatischer Bereich *m*,
Unbuntbereich *m*

38 ACHROMATIC POINT, ct/ctv
WHITE POINT

A point on a chromaticity diagram that represents an acceptable reference white standard.
f point *m* blanc
e punto *m* blanco
i punto *m* bianco
n achromatisch punt *n*, kleurpunt *n* van het wit
d Weisspunkt *m*

39 ACHROMATIC SENSATIONS ct
f sensations *pl* achromatiques
e sensaciones *pl* acromáticas
i sensazioni *pl* acromatiche
n achromatische gewaarwordingen *pl*, achromatische indrukken *pl*
d achromatische Empfindungen *pl*, achromatische Wahrnehmungen *pl*

40 ACHROMATIC STIMULUS ct/ctv
A visual stimulus that gives the sensation of white light, having no hue.
f stimulus *m* achromatique
e estímulo *m* acromático
i stimolo *m* acromatico
n achromatische prikkel, achromatische stimulus
d achromatischer Reiz *m*

41 ACHROMATIC THRESHOLD ct/ctv
The smallest light stimulus detectable by the adapted eye.
f seuil *m* achromatique
e umbral *m* acromático
i soglia *f* acromatica
n achromatische drempel
d achromatische Schwelle *f*

42 ACOUSTIC BACKING, ACOUSTIC LINING rec
Materials which absorb sound energy used to control the acoustics of studios and theatres.
f revêtement *m* acoustique
e revestimiento *m* acústico
i rivestimento *m* acustico
n akoestische bekleding
d akustische Verkleidung *f*

43 ACOUSTIC FEEDBACK, HOWL ROUND aud/rep
A positive feedback loop, usually involving a loudspeaker and a microphone in which sound is fed back and amplified until the system overloads and a loud howl is set up.
f réaction *f* acoustique
e reacción *f* acústica
i reazione *f* acustica
n akoestische terugkoppeling
d akustische Rückkopplung *f*, Schallrückkopplung *f*

44 ACOUSTIC MATERIAL aud/rec
f matériaux *pl* acoustiques
e materiales *pl* acústicos
i materiali *pl* acustici
n geluidabsorberend materiaal *n*
d Schluckmaterial *n*, Schluckstoff *m*

45 ACOUSTIC MONITORING aud/stu
f contrôle *m* acoustique
e control *m* acústico
i controllo *m* acustico
n akoestische controle
d akustische Kontrolle *f*

46 ACOUSTICAL SWITCH vr
f commutateur *m* acoustique
e conmutador *m* acústico
i commutatore *m* acustico
n akoestische schakelaar
d akustischer Schalter *m*

47 ACOUSTICS aud/rec
The science of sound.
f acoustique *f*
e acústica *f*
i acustica *f*
n akoestiek
d Akustik *f*

48 ACOUSTICS OF THE OPEN AIR rec
f acoustique *f* à l'air libre
e acústica *f* al aire libre
i acustica *f* all'aria aperta
n akoestiek in de open lucht
d Akustik *f* im Freien

ACR
see: ACCEPTABLE CONTRAST RATIO

49 ACTIVATION (US), SENSITIZATION (GB) crt
The process of improving the properties of luminescent materials.
f activation *f*
e activación *f*, sensitización *f*
i attivazione *f*, sensibilizzazione *f*
n activering
d Aktivierung *f*

50 ACTIVATOR (US), SENSITIZER (GB) crt
An impurity which increases the efficiency of luminescence of a material.
f activateur *m*
e activador *m*
i attivatore *m*
n activator
d Aktivator *m*, Aktivierungsmittel *n*

51 ACTIVE AERIAL (GB), DRIVEN RADIATOR, EXCITER (US), PRIMARY RADIATOR aea
That portion of an aerial (antenna) system which is energized by a transmitter either directly or through a feeder.
f antenne *f* active
e antena *f* activa
i radiatore *m* attiva, radiatore *m* primario
n actieve antenne
d aktiver Strahler *m*, wirksame Antenne *f*

2 ACTIVE COMPONENT, IN-PHASE COMPONENT ctv
That portion of the chrominance signal having the same phase as or opposite phase of the subcarrier modulated by the I signal.
composante *f* active,
composante *f* en phase,
composante *f* wattée
componente *f* en fase,
componente *f* vatada
componente *f* attiva,
componente *f* in fase
actieve component,
in-faze-component
In-Phase-Komponente *f*

3 ACTIVE DIPOLE aea
dipôle *m* actif
dipolo *m* activo
dipolo *m* attivo
actieve dipool
aktiver Dipol *m*

4 ACTIVE FIELD PERIOD, SIGNAL FIELD PERIOD cpl
That part of a TV signal field period that carries the image information, the field period less the field blanking period.
temps *m* utile de trame
tiempo *m* útil de cuadro
tempo *m* utile di trama
werkzame rastertijd
wirksame Halbbildzeit *f*

5 ACTIVE LINE tv
A horizontal line that carries picture information in TV.
ligne *f* d'analyse,
ligne *f* utile
línea *f* activa,
línea *f* de exploración,
línea *f* útil
linea *f* attiva,
linea *f* d'analisi
aftastlijn
Abtastzeile *f*

6 ACTIVE LINE LENGTH tv
longueur *f* de la ligne d'analyse
longitud *f* de la línea de exploración
lunghezza *f* della linea d'analisi
aftastlijnlengte
Abtastzeilenlänge *f*

7 ACTIVE LINE NUMBER tv
nombre *m* de lignes d'analyse
número *m* de líneas de exploración
numero *m* di linee d'analisi
aantal *n* aftastlijnen,
lijnenaantal *n*
Abtastzeilenzahl *f*,
Zeilenzahl *f*

8 ACTIVE MATERIAL crt
A fluorescent material used in screens for cathode-ray tubes.
matière *f* active
materia *f* activa
i materia *f* attiva
n actief materiaal *n*
d aktives Material *n*

59 ACTIVE SATELLITE tv
f satellite *m* actif
e satélite *m* activo
i satellite *m* attivo
n actieve satelliet
d aktiver Satellit *m*

60 ACTIVE SATELLITE REPEATER tv
A repeater, which contains electronic equipment, receives the signal from the ground transmitter and, after the amplification process and carrier-frequency change, retransmits the signal to the ground receiver.
f répéteur *m* de satellite actif
e repetidor *m* de satélite activo
i ripetitore *m* di satellite attivo
n versterker voor actieve satelliet
d Verstärker *m* für aktiven Satellit

61 ACTIVE TIME tv
f période *f* active,
temps *m* d'analyse
e período *m* activo,
tiempo *m* de exploración
i periodo *m* attivo,
tempo *m* d'analisi
n aftasttijd
d Abtastzeit *f*

62 ACTIVE TUNING UNIT tv
f unité *f* d'accord active
e bloque *m* de sintonía activo
i unità *f* di sintonia attiva
n werkzame afstemeenheid
d wirksame Abstimmeinheit *f*

63 ACTOR fi/th/tv
f acteur *m*, artiste *m*
e actor *m*
i attore *m*
n acteur, toneelspeler
d Darsteller *m*, Schauspieler *m*

64 ACTRICE fi/th/tv
f actrice *f*, artiste *f*
e actriz *f*
i attrice *f*
n actrice, toneelspeelster
d Darstellerin *f*, Schauspielerin *f*

65 ACTUAL MONITOR, OUTPUT MONITOR tv
The control apparatus which controls the last stage of the TV picture before being broadcast.
f moniteur *m* de sortie
e monitor *m* de salida
i monitore *m* d'uscita
n uitgangsmonitor
d Ausgangsmonitor *m*

66 ACUITY, crt/opt
SHARPNESS
f finesse *f*, netteté *f*
e nitidez *f*
i finezza *f*, nitidezza *f*
n scherpte
d Schärfe *f*

67 ACUITY MATCHING, ctv
ACUTANCE MATCHING
The principle wherein the bandwidth of the signals (luminance, chrominance) are chosen to match the acuity of the eye.
f adaptation *f* de l'acuité visuelle
e adaptación *f* de la agudeza visual
i adattamento *m* dell'acutezza visiva
n aanpassing van de gezichtsscherpte
d Anpassung *f* der Sehschärfe

68 ACUITY OF COLOUR DEFINITION (GB), ct
ACUTENESS OF COLOR RESOLUTION (US)
f pouvoir *m* de résolution chromatique
e poder *m* de resolución cromática
i potere *m* risolutivo cromatico
n kleuronderscheidingsvermogen *n*
d Farbauflösungsvermögen *n*

69 ACUITY OF HEARING (GB), aud
HEARING ACUTENESS (US)
f acuité *f* auditive
e agudeza *f* auditiva
i acutezza *f* auditiva
n gehoorscherpte
d Hörschärfe *f*

70 ACUITY OF THE EYE (GB), opt
VISUAL ACUTENESS (US)
f acuité *f* visuelle
e agudeza *f* visual
i acutezza *f* visiva
n gezichtsscherpte
d Sehschärfe *f*

71 ADAPTABILITY opt
f adaptativité *f*
e adaptabilidad *f*
i adattabilità *f*
n aanpassingsvermogen *n*
d Anpassungsfähigkeit *f*

72 ADAPTATION opt
Physiological process whereby the eye adjusts its sensitivity for different levels of illumination.
f adaptation *f*
e adaptación *f*
i adattamento *m*
n aanpassing
d Anpassung *f*

73 ADAPTER ge
f adaptateur *m*
e adaptador *m*
i adattatore *m*
n aanpasstuk *n*
d Anpassglied *n*

74 ADAPTIVE COLO(U)R SHIFT c
The change in the perceived colo(u)r of a object caused solely by change of chromatic adaptation.
f change *m* adaptatif de couleur
e cambio *m* adaptativo de color
i cambio *m* adattativo di colore
n adaptatieve kleurverandering
d adaptative Farbänderung *f*

75 ADDER cpl/ctv
In a colo(u)r TV receiver, the circuit tha combines the chrominance and luminance signals.
f circuit *m* mélangeur
e circuito *m* aditivo,
circuito *m* de adición
i circuito *m* combinatore
n mengkring, optelschakeling
d Beimischer *m*

76 ADDITIONAL REFERENCE TRANSMISSION, t
ART METHOD
f transmission *f* avec signal de référence additionnel
e transmisión *f* con señal de referencia adicional
i trasmissione *f* con segnale di riferiment addizionale
n transmissie met behulp van toegevoegd referentiesignaal
d Art-Methode *f*,
Übertragung *f* mit zusätzlichem Bezugssignal

77 ADDITIVE COLO(U)R MIXTURE ctv
Superposition or other non-destructive combination of light of different chromaticities.
f mixage *m* additif de couleurs
e mezclado *m* aditivo de colores
i mescolanza *f* additiva di colori
n additieve kleurenmenging
d additive Farbmischung *f*

78 ADDITIVE COLO(U)R PROCESS, ct/ctv
ADDITIVE PROCESS
Process of colo(u)r reproduction based on the addition of light of the three primary colo(u)rs on a single surface.
f procédé *m* additif de la photographie en couleurs
e proceso *m* aditivo de la fotografía en colores
i processo *m* additivo della fotografia a colori
n additief procédé *n*
d additives Verfahren *n*

79 ADDITIVE COLO(U)R SYNTHESIS c
f synthèse *f* additive de couleurs
e síntesis *f* aditiva de colores
i sintesi *f* additiva di colori
n additieve kleurensynthese
d additive Farbensynthese *f*

ADDITIVE COLO(U)R SYSTEM ct/ctv
A system that adds two colo(u)rs to form a third.
système *m* additif de couleurs
sistema *m* aditivo de colores
sistema *m* additivo di colori
additief kleurensysteem *n*
additive Farbfernsehmethode *f*,
additives Farbensystem *n*

ADDITIVE COMPLEMENTARY ct
COLO(U)RS
couleurs *pl* complémentaires additives
colores *pl* complementarios aditivos
colori *pl* complementari additivi
additieve complementaire kleuren *pl*
additive Komplementarfarben *pl*

ADDITIVE MIXING tv
The addition of signals in such a way that the mixed signal is equal or proportional to the sum of the separate signal components.
mixage *m* additif
mezclado *m* aditivo
mescolanza *f* additiva
additieve menging
additive Mischung *f*

ADDITIVE PRIMARIES, ct/ctv
PRIMARY ADDITIVE COLO(U)RS
The colo(u)rs red, green and blue.
couleurs *pl* primaires additives,
primaires *pl* additives
colores *pl* primarios aditivos
colori *pl* primari additivi
additieve primaire kleuren *pl*
additive Primärfarben *pl*

4 ADDITIVE TRICHROMY ct
trichromie *f* additive
tricromía *f* aditiva
tricromia *f* additiva
additieve trichromie
additive Trichromie *f*

5 ADDITIVE TRICHROMY BY ct
APPROACHING
trichromie *f* additive par rapprochement
tricromía *f* aditiva por aproximación
tricromia *f* additiva per approssimazione
additieve trichromie met benadering
additive Trichromie *f* mit Annäherung

6 ADDITIVE TRICHROMY BY ct
SUPERPOSITION
trichromie *f* additive superposée
tricromía *f* aditiva superpuesta
tricromia *f* additiva per sovrapposizione
gesuperponeerde additieve trichromie
überlagerte additive Trichromie *f*

7 ADJACENT AUDIO CARRIER, cpl
ADJACENT SOUND CARRIER
The nearest audio carrier in an adjacent channel.
porteuse *f* son adjacente
portadora *f* de sonido adyacente
i portante *f* audio di canale adiacente
n buurgeluidsdraaggolf
d Nachbartonträger *m*

88 ADJACENT CHANNEL cpl
A channel of which the frequency is immediately above or below that of the required signal.
f canal *m* adjacent, canal *m* voisin
e canal *m* adyacente
i canale *m* adiacente, canale *m* vicino
n buurkanaal *n*
d Nachbarkanal *m*

89 ADJACENT CHANNEL cpl
ATTENUATION,
SECOND CHANNEL ATTENUATION
f atténuation *f* de canaux adjacents
e atenuación *f* de canales adyacentes
i attenuazione *f* nel canale adiacente
n buurkanaaldemping
d Nachbarkanaldämpfung *f*

90 ADJACENT CHANNEL dis
INTERFERENCE,
MONKEY CHATTER,
SIDEBAND INTERFERENCE,
SIDEBAND SPLASH
Interference in a circuit by a transmitter which is assigned for operation in an adjacent channel.
f interférence *f* adjacente
e interferencia *f* del canal adyacente,
interferencia *f* entre canales adyacentes
i interferenza *f* del canale adiacente
n buurkanaalstoring
d Nachbarkanalstörung *f*

91 ADJACENT CHANNEL REJECTOR cpl
f piège *m* son du canal adjacent
e trampa *f* de sonido del canal adyacente
i trappola *f* di suono del canale adiacente
n geluidsval van het buurkanaal
d Nachbartonfalle *f*

92 ADJACENT CHANNEL cpl
SELECTIVITY
The ability of a circuit when using an adjacent channel to discriminate between the desired signal and signals at other frequencies.
f sélectivité *f* adjacente
e selectividad *f* contra canales adyacentes,
selectividad *f* de canal adyacente
i selettività *f* contro canali adiacenti
n buurkanaalselectiviteit
d Nahselektion *f*,
Trennschärfe *f* gegen Nachbarkanal.

93 ADJACENT CHROMINANCE TRAP ctv
Filter circuit included in the intermediate frequency selectivity arrangement of a TV receiver to prevent interference with the wanted vision signal caused by the adjacent-channel chrominance signal.
f filtre *m* de suppression du signal de
chrominance du canal adjacent

e filtro *m* de supresión de la señal de crominancia del canal adyacente
i filtro *m* di soppressione del segnale di crominanza del canale adiacente
n onderdrukkingsfilter *n* voor het chrominantiesignaal van het buurkanaal
d Unterdrückungsfilter *n* für das Chrominanzsignal des Nachbarkanals

94 ADJACENT CIRCUIT cpl
f circuit *m* adjacent
e circuito *m* adyacente
i circuito *m* adiacente
n buurkring
d Nachbarkreis *m*

95 ADJACENT IMAGE POINTS tv
f spots *pl* d'image adjacents
e puntos *pl* de imagen adyacentes
i punti *pl* d'immagine adiacenti
n aan elkaar grenzende beeldpunten *pl*
d aneinandergrenzende Bildpunkte *pl*

96 ADJACENT LINES cpl
f lignes *pl* adjacentes
e líneas *pl* adyacentes
i linee *pl* adiacenti
n buurleidingen *pl*
d Nachbarleitungen *pl*

97 ADJACENT PICTURE CARRIER, ADJACENT VIDEO CARRIER, ADJACENT VISION CARRIER cpl
The nearest video carrier in an adjacent channel.
f porteuse *f* image adjacente
e portadora *f* de imagen adyacente
i portante *f* video di canale adiacente
n buurbeelddraaggolf
d Nachbarbildträger *m*

98 ADJACENT PICTURE CARRIER SPACING, ADJACENT VISION CARRIER SPACING, VISION CARRIER SPACING tv
The frequency distance between the picture carriers of two directly adjacent TV channels.
f écart *m* entre porteuses
e distancia *f* entre portadoras
i distanza *f* tra portanti
n beelddragerafstand
d Bildträgerabstand *m*

99 ADJACENT PICTURE CARRIER TRAP, PICTURE CARRIER TRAP tv
An absorption circuit for suppressing the disturbance which can be generated by the picture carrier.
f piège *m* (sur la porteuse image) du canal voisin
e trampa *f* del canal adyacente
i trappola *f* del canale adiacente
n buurkanaalval
d Bildträgersperre *f*, Nachbarbildfalle *f*

100 ADJUSTABLE INTERCONNECTION BLOCKS s
f blocs *pl* amovibles de mutation
e bloques *pl* interconectadores ajustables
i elementi *pl* di circuito intercambiabili
n uitwisselbare schakelelementen *pl*
d austauschbare Schalteinheiten *pl*

101 ADJUSTABLE TIMER v
f minuterie *f* ajustable
e minutero *m* ajustable
i interruttorio *m* orario
n automatische in- en uitschakelklok, instelklok
d Schaltuhr *f*

102 ADJUSTMENT g
f ajustage *m*, réglage *m*
e ajuste *m*, arreglo *m*, regulación *f*
i aggiustaggio *m*, regolazione *f*
n instelling, regeling
d Einstellung *f*, Regelung *f*

103 ADVERTISING FILM rec / re
f film *m* de publicité, film *m* de réclame
e film *m* de publicidad, película *f* de publicidad
i film *m* pubblicitario, pellicola *f* pubblicitaria
n reclamefilm
d Werbefilm *m*

104 AERIAL (GB), ANTENNA (US) t
A device for the transmission or recepti of electromagnetic waves.
f antenne *f*
e antena *f*
i antenna *f*
n antenne
d Antenne *f*

105 AERIAL AMPLIFIER, ANTENNA AMPLIFIER aea / cp
f amplificateur *m* d'antenne
e amplificador *m* de antena
i amplificatore *m* d'antenna
n antenneversterker
d Antennenverstärker *m*

106 AERIAL ASSEMBLY, ANTENNA ASSEMBLY aea / tv
f ensemble *m* d'antennes
e conjunto *m* de antenas
i complesso *m* d'antenne
n antennesamenstel *n*
d Antennenanordnung *f*

107 AERIAL CASING, ANTENNA CASING aea / svs
f boîtier *m* antenne
e caja *f* de antena
i scatola *f* d'antenna
n antennedoos
d Antennendose *f*

8 AERIAL COIL, aea
ANTENNA COIL
The first coil in a receiver through which aerial (antenna) current flows.
bobine *f* d'antenne
bobina *f* de antena
bobina *f* d'antenna
antennespoel
Antennenankopplung *f*,
Antennenspule *f*

9 AERIAL DIRECTIVITY, aea
ANTENNA DIRECTIVITY
The property of an aerial (antenna) by virtue of which it radiates more strongly in some direction than in others.
directivité *f* d'antenne
efecto *m* direccional de antena
direttività *f* d'antenna
richteffect *n* van de antenne
Antennenrichtwirkung *f*

0 AERIAL EFFICIENCY, aea
ANTENNA EFFICIENCY,
RADIATION EFFICIENCY
The ratio of the power radiated to the power supplied to an aerial (antenna) at a specified frequency.
coefficient *m* de rayonnement de l'antenne
rendimiento *m* de radiación de la antena
rendimento *m* d'antenna
antennerendement *n*
Antennenwirkungsgrad *m*

1 AERIAL ELEMENT, aea
ANTENNA ELEMENT
A primary or secondary radiator which may be regarded as a unit for the purpose of design or construction.
élément *m* d'antenne
elemento *m* de antena
elemento *m* d'antenna
antenne-element *n*
Antennenelement *n*

12 AERIAL GAIN, aea
ANTENNA GAIN
The effectiveness of a directional aerial (antenna) as compared to a standard non-directional aerial (antenna).
gain *m* d'antenne
ganancia *f* en la antena
guadagno *m* d'antenna
antennewinst
Antennengewinn *m*

13 AERIAL IMAGE, opt
VIRTUAL IMAGE
An image formed in space and not received on a surface.
image *f* aérienne,
image *f* virtuelle
imagen *f* real aérea,
imagen *f* virtual
immagine *f* virtuale
virtueel beeld *n*
virtuelles Bild *n*

114 AERIAL INPUT, aea
ANTENNA INPUT
f entrée *f* d'antenne
e entrada *f* de antena
i entrata *f* d'antenna
n antenne-invoer
d Antenneneingang *m*

115 AERIAL LEAD, aea
ANTENNA LEAD
f descente *f* d'antenne
e bajada *f* de antena
i discesa *f* d'antenna
n antenne-invoerleiding
d Antennenzuleitung *f*

116 AERIAL MAINTENANCE, aea
ANTENNA MAINTENANCE
f entretien *m* de l'antenne
e conservación *f* de la antena,
mantenimiento *m* de la antena
i manutenzione *f* dell'antenna
n antenneonderhoud *n*
d Antenneninstandhaltung *f*,
Antennenpflege *f*

117 AERIAL MAST, aea
AERIAL TOWER,
ANTENNA MAST,
ANTENNA TOWER
f mât *m* d'antenne,
poteau *m* d'antenne,
pylone *m* d'antenne
e mástil *m* de antena,
torre *f* de antena
i pilone *m* d'antenna,
torre *f* d'antenna
n antennemast
d Antennenmast *m*,
Antennenturm *m*

118 AERIAL MATCHING DEVICE, aea
ANTENNA MATCHING DEVICE
f adaptation *f* d'antenne,
dispositif *m* à bouchon d'antenne
e adaptador *m* de antena
i adattatore *m* d'antenna
n antenneaanpasstuk *n*
d Antennenanpassglied *n*

119 AERIAL OUTPUT, aea
ANTENNA OUTPUT
f sortie *f* antenne
e salida *f* de antena
i uscita *f* d'antenna
n antenne-uitgang
d Antennenausgang *m*

120 AERIAL POWER GAIN, aea
ANTENNA POWER GAIN,
DIRECTIVE GAIN
In a given direction 4 π times the ratio of the radiation intensity in that direction to the total power delivered to the aerial (antenna).
f facteur *m* de directivité
e coeficiente *m* de directividad

i guadagno *m* d'antenna direttiva
n richtversterkingsfactor
d Richtverstärkungsfaktor *m*

121 AERIAL SIGNAL, ANTENNA SIGNAL — aea
f signal *m* d'antenne
e señal *f* de antena
i segnale *m* d'antenna
n antennesignaal *n*
d Antennensignal *n*

122 AERIAL SIGNAL DISTRIBUTION, ANTENNA SIGNAL DISTRIBUTION — aea
f distribution *f* des signaux d'une antenne
e distribución *f* de las señales de una antena
i distribuzione *f* dei segnali d'un'antenna
n verspreiding van antennesignalen
d Verteilung *f* von Antennensignalen

123 AERIAL SYSTEM FOR SATELLITES, ANTENNA SYSTEM FOR SATELLITES — aea
f système *m* d'antenne pour satellites
e sistema *m* de antena para satélites
i sistema *m* d'antenna per satelliti
n antennesysteem *n* voor satellieten
d Antennensystem *n* für Satelliten

124 AERIAL TERMINAL (GB), ANTENNA CONNECTION (US) — aea
The terminal which is introduced into the socket of the TV apparatus.
f borne *f* d'antenne
e conexión *f* de antena, terminal *m* de antena
i morsetto *m* d'antenna
n antenneklem
d Antennenklemme *f*

125 AEROPLANE FLUTTER, AIRCRAFT FLUTTER, AIRPLANE FLUTTER — dis/tv
Interference caused by reflection from moving aircraft.
f interférence *f* d'avion
e interferencia *f* de avión
i interferenza *f* d'aereo
n vliegtuigflikkering, vliegtuiginterferentie
d Flugzeugflattern *n*

126 AF, AUDIO FREQUENCY — cpl
A frequency that can be detected as a sound by the human ear.
f basse fréquence *f*, fréquence *f* acoustique
e audiofrecuencia *f*, baja frecuencia *f*
i audiofrequenza *f*, bassa frequenza *f*, frequenza *f* udibile
n audiofrequentie, hoorfrequentie, lage frequentie
d Hörfrequenz *f*, Niederfrequenz *f*, Tonfrequenz *f*

127 AF AMPLIFIER NOISE, AF NOISE — c
f perturbation *f* de basse fréquence, perturbation *f* dans l'amplificateur à basse fréquence
e ruido *m* de audiofrecuencia
i disturbo *m* a bassa frequenza
n hoorfrequentieruis
d Niederfrequenzstörung *f*, Tonfrequenzrauschen *n*

128 AF HUM PICK-UP — dis/s
A disturbance due to floating grid or unearthed screened lead.
f pénétration *f* du ronflement dans le domaine de fréquences acoustiques
e penetración *f* del zumbido en el campo audiofrecuencias
i penetrazione *f* del rumore nel campo d frequenze udibili
n brombesmetting in hoorfrequentiegebied
d Tonfrequenz-Brummeinkopplung *f*

129 AFC, AUTOMATIC FREQUENCY CONTRO — c
Electronic or mechanical means for compensating small variations in the frequencies of the incoming signal and/or local oscillator.
f réglage *m* automatique de fréquence
e control *m* automático de frecuencia
i regolazione *f* automatica di frequenza
n automatische frequentieregeling
d AFN, automatische Frequenznachstimmung *f*, automatische Scharfabstimmung *f*, automatischer Frequenznachlauf *m*

130 AFC CORRECTION RATIO — c
f taux *m* de correction pour réglage automatique de fréquence
e relación *f* de corrección con control automático de frecuencia
i rapporto *m* di correzione con regolazion automatica di frequenza
n verbeteringsgraad bij automatische frequentieregeling
d Verbesserungsgrad *m* bei automatischer Frequenznachstimmung

131 AFC DISCRIMINATOR — cpl/sv
In automatic frequency control a circuit in which magnitude and polarity of the output voltage depend on how an input signal differs from a standard or from another signal.
f discriminateur *m* en réglage automatique de fréquence
e discriminador *m* en control automático de frecuencia
i discriminatore *m* in regolazione automati di frequenza
n discriminator in automatische frequentieregeling
d Diskriminator *m* bei automatischer Frequenznachstimmung

2 AFC LINE CIRCUITS cpl/svs
circuits *pl* de ligne à réglage automatique de fréquence
circuitos *pl* de lînea con control automático de frecuencia
circuiti *pl* di linea a regolazione automatica di frequenza
leidingsketen met automatische frequentieregeling
Leitungskreis *m* mit automatischer Frequenznachstimmung

3 AFC PULL-IN ADJUSTMENT cpl/svs
ajustage *m* de l'effet d'insertion du réglage automatique de fréquence
ajuste *m* del efecto de inserción del control automático de frecuencia
aggiustaggio *m* del campo di funzionamento della regolazione automatica di frequenza
correctie voor het inlaseffect van de automatische frequentieregeling
Korrektur *f* des Einfügungseffekts der selbsttätigen Frequenzregelung

4 AFTER-ACCELERATION, crt
CATHODE-RAY TUBE POST-DEFLECTION ACCELERATION,
POST-ACCELERATION
In a cathode-ray tube, a technique in which the beam is accelerated, in part, after it has passed through the deflecting plates.
postaccélération *f*
postaceleración *f*
postaccelerazione *f*
naversnelling
Nachbeschleunigung *f*

5 AFTER-ACCELERATION CATHODE-RAY TUBE, crt
POST-ACCELERATION CATHODE-RAY TUBE,
POST-DEFLECTION CATHODE-RAY TUBE
An electrostatic cathode-ray tube in which the electron beam is accelerated to its final high velocity after passing through the deflection electrodes.
tube *m* cathodique à postaccélération
tubo *m* catódico de postaceleración
tubo *m* catodico a postaccelerazione
katodestraalbuis met naversnelling
Nachbeschleunigungselektronenstrahlröhre *f*

36 AFTER-DEFLECTION FOCUSING, crt
POST-DEFLECTION FOCUSING
Focusing of electron beam after deflection.
focalisation *f* après déviation
enfoque *m* después de desviación
focalizzazione *f* dopo deflessione,
focalizzazione *f* dopo deviazione
focussering na afbuiging
Fokussierung *f* nach Ablenkung

137 AFTERGLOW, crt
PERSISTENCE
Output of light from a phosphor after the excitation has been removed.
f effet *m* de phosphorescence, persistance *f*
e persistencia *f*
i persistenza *f*, postluminescenza *f*
n nalichten *n*
d Nachleuchteffekt *m*, Nachleuchten *n*

138 AFTERGLOW CORRECTION crt/dis
f correction *f* de persistance
e corrección *f* de persistencia
i correzione *f* di persistenza
n nalichtcorrectie
d Nachleuchtkorrektur *f*

139 AFTER-IMAGE opt/rep/tv
f traînage *m*
e imagen *f* persistente
i immagine *f* persistente, trascinamento
n nabeeld *n*
d Nachbild *n*, Nachziehen *n*

140 AGC, cpl
AUTOMATIC GAIN CONTROL,
AUTOMATIC VOLUME CONTROL
The process whereby the received signal causes the gain of the receiver to vary substantially-inversely as the magnitude of the radio-frequency input so as to maintain the output level substantially constant for a given modulation depth.
f commande *f* automatique de gain, réglage *m* automatique de sensibilité
e regulación *f* automática de ganancia, regulación *f* automática de volumen
i regolazione *f* automatica di sensibilità, regolazione *f* automatica di volume
n automatische versterkingsregeling
d Schwundregelung *f*, selbsttätige Verstärkungsregelung *f*, selbsttätiger Schwundausgleich *m*

141 AGC AMPLIFIER cpl
f amplificateur *m* de réglage
e amplificador *m* de regulación
i amplificatore *m* di regolazione
n regelversterker
d Regelverstärker *m*

142 AGC DETECTOR cpl
f diode *f* de réglage
e diodo *m* de regulación
i diodo *m* di regolazione
n regeldiode
d Regeldiode *f*

143 AIR ABSORPTION ac/tv
Sound absorption caused by the surrounding air.
f absorption *f* par l'air ambiant
e absorción *f* por aire ambiente

i assorbimento *m* per aria ambiente
n absorptie door omgevingslucht
d Absorption *f* durch Umgebungsluft

144 AIR FILM vr
A thin film of air between the tape and the guide posts.
f couche *f* intermédiaire d'air
e estrato *m* de aire
i strato *m* atmosferico
n luchtkussen *n*, luchtlaag
d Luftkissen *n*, Luftschicht *f*

145 AIR-GLASS INTERFACE opt
f zone *f* de séparation air-verre
e zona *f* interfacial aire-vidrio
i superficie *f* di separazione aria-vetro
n lucht-glas-grensvlak *n*
d Luft-Glas-Grenzfläche *f*

146 AIR TRAFFIC CONTROL tv
f réglage *m* du mouvement aérien
e regulación *f* de tránsito aéreo
i regolazione *f* di traffico aereo
n luchtverkeersregeling
d Flugverkehrsleitung *f*

147 AIRBORNE TELEVISION RECEIVER tv
f téléviseur *m* de bord
e televisor *m* aerotransportable
i televisore *m* di bordo
n televisieontvanger aan boord van vliegtuig
d Bordfernsehempfänger *m*

148 AIRBORNE TELEVISION RECEIVER AND RADAR TRANSPONDER tv
f téléviseur *m* et émetteur-récepteur *m* radar de bord
e televisor *m* y radiofaro *m* de respuesta aerotransportable
i televisore *m* e radiofaro *m* a risposta di bordo
n televisieontvanger en antwoordbaken *n* aan boord van vliegtuig
d Bordfernsehempfänger *m* und Bordimpulswiederholer *m*

AIRCRAFT FLUTTER,
AIRPLANE FLUTTER
see: AEROPLANE FLUTTER

149 ALIGNMENT cpl/tv
Usually refers to tuning the circuit of a radio frequency or intermediate frequency amplifier to one frequency or a band of frequencies.
f ajustage *m*, alignement *m*
e ajuste *m*, alineación *f*
i aggiustaggio *m*, allineamento *m*
n afregeling, afstemming
d Abgleichung *f*, Justierung *f*

150 ALIGNMENT OF HEADS, HEADS ALIGNMENT vr
f étalonnage *m* de têtes
e ajuste *m* de cabezas
i aggiustaggio *m* di testine, taratura *f* di testine
n instellen *n* van de koppen, kopafregeling
d Abgleich *m* der Köpfe, Einmessen *n* der Köpfe

151 ALL-BAND AMPLIFIER, WIDE-BAND AMPLIFIER aea/
f amplificateur *m* à large bande
e amplificador *m* de banda ancha
i amplificatore *m* a banda larga
n breedbandversterker
d Breitbandverstärker *m*

152 ALL-BAND TUNER
A channel selector for all channels used in TV broadcasting, containing common electrical construction elements for the frequency ranges I, III and IV/V, which are specific for tuning and functioning.
f sélecteur *m* pour tous les canaux
e selector *m* para todos los canales
i selettore *m* per tutti i canali
n kiezer voor alle kanalen
d Allbereichkanalwähler *m*

153 ALL-GLASS CINESCOPE t
f cinescope *m* tout-verre
e cinescopio *m* todo vidrio
i cinescopio *m* tutto vetro
n geheel glazen beeldbuis
d Allglasbildröhre *f*

154 ALL-PURPOSE CAMERA, IRON MAN t
Multi-purpose studio camera capable of being moved from its normal position for televising captions, photographs and diagrams.
f caméra *f* universelle
e cámara *f* universal
i camera *f* universale
n universele camera
d Allzweckkamera *f*

155 ALL-SULPHIDE PHOSPHOR cr
f sulfure *m* fluorescent
e sulfuro *m* fluorescente
i sulfuro *m* fluorescente
n fluorescerend sulfide *n*
d Sulfidleuchtstoff *m*

156 ALLOCATED FREQUENCY, ASSIGNED FREQUENCY cpl/tv
Of a transmitting station, a frequency located in some specified manner in the frequency band in which the station is authorized to work.
f fréquence *f* assignée
e frecuencia *f* asignada
i frequenza *f* assegnata
n toegewezen frequentie
d Verfügungsfrequenz *f*, zugeteilte Sollfrequenz *f*

157 ALPHA LOOP, ALPHA WRAP vr
f guidage *m* alpha
e bucle *m* alfa

i doppino *m* alfa
n alfalus
d Alpha-Bandführung *f*

158 ALTERNATE SCANNING fi/tv
Method of scanning developed to compensate for the difference in frame frequencies when televising from cinematograph film.
f analyse *f* de film alternée
e exploración *f* de película alterna
i analisi *f* di pellicola alternata
n sprongaftasting van films
d Sprungabtastung *f* von Filmen

159 ALTERNATING BURST, ctv
SWINGING BURST
Chrominance-carrier phase reference in the PAL colo(u)r TV system.
f salve *f* alternante,
salve *f* de référence à décalage de phase correcte
e sobreimpulso *m* alterno,
sobreimpulsión *f* de color con desplazamiento de fase correcto
i salva *f* alternata,
salva *f* di colore a spostamento corretto
n salvo *n* met juiste fazeverschuiving,
springend salvo *n*,
wisselend salvo *n*
d alternierender Burst *m*,
Farbsynchronsignal *n* mit richtiger Phasenverschiebung,
wechselndes Farbsynchronsignal *n*

ALTERNATING CURRENT MAGNETIC BIASING
see: A.C. MAGNETIC BIASING

160 ALTERNATION, ctv
COLO(U)R PHASE ALTERNATION
In colo(u)r TV, the alternation of the phase of one or more of the sidebands of the colo(u)r subcarrier signal with respect to the colo(u)r carrier reference.
f alternation *f* de phase colorimétrique
e alternación *f* de fase colorimétrica
i alternazione *f* di fase colorimetrica
n kleurfazewisseling
d Farbphasenwechslung *f*

161 ALUMINIZED PICTURE-TUBE crt
SCREEN
A screen with a coating of aluminum.
f écran *m* aluminisé
e pantalla *f* aluminiada
i schermo *m* alluminizzato
n gealuminiseerd scherm *n*
d aluminisierter Bildschirm *m*,
Schirm *m* mit aufgedampfter Al-Haut

162 ALUMINIZING tv
f aluminisation *f*
e aluminización *f*
i alluminatura *f*
n alumineren *n*
d Aluminierung *f*

163 ALYCHNE, ct
ZERO LUMINANCE PLANE
A plane in colo(u)r space representing the locus of colo(u)rs of zero luminance.
f plan *m* de luminance zéro
e plano *m* de luminancia cero
i piano *m* di luminanza zero
n nul-luminantievlak *n*
d Null-Luminanzebene *f*

164 AM, tv
AMPLITUDE MODULATION
System of modulation in which the amplitude of the transmitted carrier wave is varied in accordance with the impressed signals.
f modulation *f* d'amplitude
e modulación *f* de amplitud
i modulazione *f* d'ampiezza
n AM,
amplitudemodulatie
d AM,
Amplitudenmodelung *f*,
Amplitudenmodulation *f*

165 AMBIENT ILLUMINATION, stu
AMBIENT LIGHTING
In viewing, normal room lighting which illuminates the picture screen from the outside causing reflections from the front of the cathode-ray tube and illumination of the outer side of the fluorescent screen.
f éclairage *m* de l'ambiance
e iluminación *f* del ambiente
i illuminazione *f* ambientale
n ruimteverlichting
d Raumbeleuchtung *f*

166 AMBIENT LIGHT stu
f lumière *f* ambiante
e luz *f* ambiente
i luce *f* ambiente
n omgevingslicht *n*
d Umlicht *n*

167 AMBIENT LIGHT FILTER (GB), tv
GREY GLASS FILTER,
OPTIC LIGHT FILTER (US)
A screen of e.g. grey colo(u)r which decreases the reflection caused by ambient light and/or by the cathode-ray tube itself.
f filtre *m* optique
e filtro *m* de luz
i filtro *m* ottico
n filterplaat, grijsfilter *n*
d Grauglasscheibe *f*, Grauscheibe *f*

168 AMBIENT TEMPERATURE ge
f température *f* ambiante
e temperatura *f* ambiente
i temperatura *f* ambiente
n omgevingstemperatuur
d Umgebungstemperatur *f*

169 AMBIT (US), ge
CONTOUR (GB)
The line forming the circumference of an object.

f contour *m*
e contorno *m*
i contorno *m*
n contour, omtreklijn
d Kontur *f*, Umrisslinie *f*

170 AMPLIFICATION, GAIN — ge
The process of increasing the strength (current, voltage or power) of a signal.
f amplification *f*, gain *m*
e amplificación *f*, ganancia *f*
i amplificazione *f*, guadagno *m*
n versterking
d Verstärkung *f*

171 AMPLIFIER — cpl
A device in which an input signal controls a local source of power in such a way as to produce an output which bears some desired relationship to, and is generally greater than, the input signal.
f amplificateur *m*
e amplificador *m*
i amplificatore *m*
n versterker
d Verstärker *m*

172 AMPLIFIER BAY — cpl/stu
f baie *f* d'amplificateurs
e fila *f* de amplificadores
i sezione *f* d'amplificatori
n versterkerkolom
d Verstärkerbucht *f*

173 AMPLIFIER RACK — cpl
f bâti *m* d'amplificateurs
e bastidor *m* de amplificadores
i telaio *m* d'amplificatori
n versterkerrek *n*
d Verstärkergestell *n*

174 AMPLIFIER UNIT IN VIDEO RECORDING, CARTRIDGE AMPLIFIER, CASSETTE AMPLIFIER — vr
f amplificateur *m* en tiroir
e amplificador *m* para caseta
i amplificatore *m* a cassetta
n cassetteversterker
d Kassettenverstärker *m*

175 AMPLITUDE CHARACTERISTIC, RESPONSE CHARACTERISTIC — tv
The dependence of amplitude upon frequency.
f caractéristique *f* d'amplitude, courbe *f* de réponse
e respuesta *f* de amplitud
i curva *f* di risposta
n responsiekarakteristiek
d Durchlasscharakteristik *f*, Wiedergabecharakteristik *f*

176 AMPLITUDE CONTROL — cpl
f réglage *m* de l'amplitude
e regulación *f* de la amplitud
i regolazione *f* dell'ampiezza
n amplituderegeling
d Amplitudenregelung *f*

177 AMPLITUDE CORRECTION, CORRECTION OF AMPLITUDE — cpl
f correction *f* d'amplitude
e corrección *f* de amplitud
i correzione *f* d'ampiezza
n amplitudecorrectie
d Amplitudenentzerrung *f*

178 AMPLITUDE CORRECTOR — cpl
f circuit *m* correcteur d'amplitude
e circuito *m* corrector de amplitud
i circuito *m* correttore d'ampiezza
n amplitudecorrectieschakeling
d Amplitudenausgleichskreis *m*

179 AMPLITUDE DISTORTION — cpl/dis
f distorsion *f* d'amplitude
e distorsión *f* de amplitud
i distorsione *f* d'ampiezza
n van amplitude afhankelijke dempschakeling
d amplitudenabhängige Dämpfungsänderung

180 AMPLITUDE EQUALIZATION — cpl
f égalisation *f* d'amplitude
e igualación *f* de amplitud
i uguagliamento *m* d'ampiezza
n amplitudevereffening
d Amplitudenausgleich *m*

181 AMPLITUDE-ERROR CORRECTION — ctv/dis
f correction *f* de défauts d'amplitude
e corrección *f* de errores de amplitud
i correzione *f* d'errori d'ampiezza
n correctieschakeling voor amplitudefouten
d Amplitudenfehlerkorrektur *f*

182 AMPLITUDE EXCURSION — tv
In TV, the power level of a carrier wave affected by the components of the amplitude modulation.
f excursion *f* d'amplitude
e excursión *f* de amplitud
i escursione *f* d'ampiezza
n amplitude-excursie
d Amplitudenausflug *m*

183 AMPLITUDE FILTER, AMPLITUDE SEPARATOR, SYNC SEPARATOR — tv
A circuit for separating the sync signal out of the signal mixture.
f séparateur *m* d'amplitudes, séparateur *m* du signal de synchronisation
e separador *m* de amplitudes, separador *m* de la señal de sincronización
i separatore *m* del segnale di sincronizzazione, separatore *m* d'ampiezza
n afscheidingschakeling van het synchronisatiesignaal, amplitudezeef voor televisie
d Amplitudensieb *n* beim Fernsehen

84 AMPLITUDE-FREQUENCY RESPONSE — cpl
The variation of gain, loss, amplification or attenuation as a function of frequency.
réponse *f* amplitude-fréquence, réponse *f* de fréquence
respuesta *f* a amplitud y frecuencia, respuesta *f* de **frecuencia**
risposta *f* di frequenza
amplitude-frequentieverhouding
Frequenzgang *m*

85 AMPLITUDE INCREASE — tv
The rise in amplitude of a vibration.
gain *m* d'amplitude
ganancia *f* de amplitud
guadagno *m* d'ampiezza, innesco *m* d'ampiezza
amplitudeopslingering
Amplitudenaufschaukelung *f*

.86 AMPLITUDE LIMITER CIRCUIT (GB), AMPLITUDE LOPPER (US), AMPLITUDE SEPARATION CIRCUIT (GB), CLIPPER CIRCUIT (US) — cpl/tv
A circuit by means of which the signal is clipped for so far as its amplitude is above or below a certain level.
circuit *m* d'écrêtage, circuit *m* limiteur d'amplitude
circuito *m* limitador de amplitud
circuito *m* tosatore, limitatore *m* d'ampiezza, limitatore *m* di cresta, limitatore *m* di picco
n amplitudebegrenzer, amplitudekopper
d Amplitudenabkapper *m*, Amplitudenbegrenzer *m*

187 AMPLITUDE LINEARITY — cpl
Extent to which the transfer characteristic relating input to output of a circuit or device is a straight line function and thus is free from amplitude distortion.
f linéarité *f* d'amplitude
e linealidad *f* de amplitud
i linearità *f* d'ampiezza
n amplitudelineariteit
d Amplitudenlinearität *f*

188 AMPLITUDE-MODULATED SOUND TRANSMITTER — rep
f émetteur *m* son modulé en amplitude
e transmisor *m* de señal audible modulado en amplitud
i trasmettitore *m* audio modulato in ampiezza
n geluidszender met amplitudemodulatie
d AM-Tonsender *m*, amplitudenmodulierter Tonsender *m*

189 AMPLITUDE-MODULATED VHF TRANSMITTER — tv
Used in general only for picture transmission.
f émetteur *m* pour hyperfréquences à modulation d'amplitude
e emisor *m* para muy altas frecuencias modulado en amplitud
i emettitore *m* per altissime frequenze a modulazione d'ampiezza
n VHF-zender met amplitudemodulatie, zender voor zeer hoge frequentie met amplitudemodulatie
d amplitudenmodulierter VHF-Sender *m*, VHF-Sender *m* mit Amplitudenmodulation

190 AMPLITUDE-MODULATION MONITOR — tv
A monitor designed to check the characteristics of an amplitude-modulated broadcast transmission.
f moniteur *m* pour modulation d'amplitude
e monitor *m* para modulación de amplitud
i monitore *m* per modulazione d'ampiezza
n AM-monitor, monitor voor amplitudemodulatie
d AM-Monitor *m*, Monitor *m* für Amplitudenmodulation

191 AMPLITUDE-MODULATION SUPPRESSION — tv
In a TV receiver, the ratio of undesired output to standard test output, expressed in decibels.
f suppression *f* de modulation d'amplitude
e supresión *f* de modulación de amplitud
i soppressione *f* di modulazione d'ampiezza
n AM-onderdrukking
d AM-Unterdrückung *f*

192 AMPLITUDE NON-LINEARITY — cpl/dis
f nonlinéarité *f* d'amplitude
e no linealidad *f* de amplitud
i nonlinearità *f* d'ampiezza
n niet-lineariteit van een amplitude
d Nichtlinearität *f* einer Amplitude

193 AMPLITUDE OF VIDEO SIGNAL — tv
The peak-to-peak value of the signal.
f amplitude *f* du signal vidéo
e amplitud *f* de la señal video
i ampiezza *f* del segnale video
n amplitude van het videosignaal
d Amplitude *f* des Videosignals

194 AMPLITUDE PEAK — cpl/tv
Maximum positive or negative excursion from zero of any periodic disturbance.
f crête *f* d'amplitude
e cresta *f* de amplitud
i cresta *f* d'ampiezza
n amplitudetop
d Amplitudenspitze *f*

195 AMPLITUDE RANGE OF VIDEO SIGNAL — tv
f gamme *m* des amplitudes image
e gama *f* de las amplitudes de la señal video
i gamma *f* delle ampiezze del segnale video
n amplitudegebied *n* van het videosignaal, videospan
d Amplitudenbereich *m* des Videosignals, Bildamplitudenbereich *m*

196 AMPLITUDE SEPARATION ctv
Separation of the components of a signal using amplitude range multiplexing, e.g. the separation of the sync pulses from the picture information in a composite video signal by means of a synchronizing separator.
f séparation *f* réalisée par l'amplitude
e separación *f* realizada por la amplitud
i separazione *f* realizzata per l'ampiezza
n door de amplitude bepaalde scheiding
d amplitudengesteuerte Trennung *f*

197 ANALYSIS FILTER, ANALYZING FILTER ct/ctv
A filter placed in front of a single camera to intercept, field by field, the scene to be televised.
f filtre *m* analyseur de couleurs
e filtro *m* analizador de colores
i filtro *m* analizzatore di colori
n kleurenanalyserend filter *n*
d farbenanalysierendes Filter *n*

198 ANASTIGMAT opt
A lens with correction for astigmatism.
f anastigmat *m*, objectif *m* anastigmatique
e objetivo *m* anastigmático
i obiettivo *m* anastigmatico
n anastigmaat
d Anastigmat *m*

199 ANASTIGMATIC DEFLECTION YOKE crt
f bobine *f* de balayage anastigmatique, bobine *f* de déviation anastigmatique
e bobina *f* de barrido anastigmático, bobina *f* de desviación anastigmática
i bobina *f* di deflessione anastigmatica, bobina *f* di deviazione anastigmatica
n anastigmatische afbuigspoel
d anastigmatische Ablenkspule *f*

200 ANECHOIC ROOM aud
A room whose walls are made as absorbent as possible to obtain a sound field free of reflections.
f chambre *f* sourde
e cámara *f* anecoica
i camera *f* anecoica, camera *f* assorbente
n dode kamer, nagalmvrije kamer
d schalltoter Raum *m*

201 ANGLE MODULATION cpl
Modulation in which the angle of a sine-wave carrier is the characteristic varied.
f modulation *f* angulaire
e modulación *f* angular
i modulazione *f* angolare
n hoekmodulatie
d Winkelmodulation *f*

202 ANGLE OF BEAM TILT, BEAM TILT ANGLE aea
f angle *m* d'inclinaison du diagramme
e ángulo *m* de inclinación del diagrama
i angolo *m* d'inclinazione del diagramma
n neigingshoek van het stralingsdiagram
d Diagrammneigungswinkel *m*

203 ANGLE OF DIVERGENCE, DIVERGENCE ANGLE cr
The angle formed by the edge of the beam in a cathode-ray tube and a longitudinal line through the center(re).
f angle *m* de divergence
e ángulo *m* de divergencia
i angolo *m* di divergenza
n spreidingshoek
d Streuwinkel *m*

204 ANGLE OF EMERGENCE, EMERGENCE ANGLE opt
f angle *m* d'émergence
e ángulo *m* de salida
i angolo *m* d'uscita
n uitvalshoek
d Austrittswinkel *m*

205 ANGLE OF INCIDENCE, INCIDENCE ANGLE ge/opt
Angle between incidence ray or beam of radiation and the perpendicular to the surface at the incidence point.
f angle *m* d'incidence
e ángulo *m* de incidencia
i angolo *m* d'incidenza
n invalshoek
d Einfallswinkel *m*

206 ANGLE OF REFLECTION, REFLECTION ANGLE opt
The angle between a wave or beam leaving a surface and the perpendicular to the surface.
f angle *m* de réflexion
e ángulo *m* de reflexión
i angolo *m* di riflessione
n reflectiehoek
d Ausfallwinkel *m*, Reflexionswinkel *m*

207 ANGULAR APERTURE, ANGULAR FIELD opt
f angle *m* de champ
e abertura *f* angular
i apertura *f* angolare
n beeldhoek
d Bildwinkel *m*

208 ANGULAR DIVERGENCE crt
f divergence *f* angulaire
e divergencia *f* angular
i divergenza *f* angolare
n hoekspreiding
d Winkelstreuung *f*

209 ANGULAR WIDTH OF SPOT, APERTURE OF THE BEAM (US), BEAM WIDTH (GB) crt
f largeur *f* du faisceau
e anchura *f* del haz
i apertura *f* del fascio, larghezza *f* del fascio
n bundelbreedte
d Bündelbreite *f*, Strahlbreite *f*

0 ANIMATED CAPTION ani
titre *m* animé
título *m* animado
titolo *m* animato
tructitel
Tricktitel *m*

1 ANIMATED CARTOON ani
animation *f* dessinée,
dessin *m* animé
dibujo *m* animado
cartone *m* animato,
disegno *m* animato
tructekening
Zeichentrick *m*

2 ANIMATED FILM, ani
ANIMATED PICTURE
film *m* animé,
film *m* d'animation
película *f* animada
pellicola *f* animata
tekenfilm
Trickfilm *m*,
Zeichentrickfilm *m*

3 ANIMATING MOVEMENT ani
mouvement *m* de film de dessins animés
movimiento *m* de animación
movimento *m* d'animazione,
movimento *m* d'immagini ad effetti speciali
tekenfilmbeweging
Trickfilmbewegung *f*

14 ANIMATION, ani
ANIMATION TECHNIQUE
animation *f*
animación *f*
animazione *f*
tekenfilmtechniek
Zeichentrickfilmtechnik *f*

15 ANIMATION COMBINATIONS ani
The result of frame by frame recording of the painted background and celluloids.
combinaisons *pl* d'animations
combinaciones *pl* animadas
combinazioni *pl* d'animazioni
tekenfilmcombinaties *pl*
Trickfilmkombinationen *pl*

16 ANIMATION DESK ani
table *f* de mixage de films à dessins animés
pupitre *m* de mezclado de películas animadas
mescolatore *m* d'immagini ad effetti speciali
tekenfilmmengtafel
Trickmischpult *n*

17 ANIMATION DISSOLVE ani
fondu *m* enchaîné d'animations
desvanecimiento *m* gradual de animaciones
dissolvenza *f* graduale delle animazioni
tekenfilmoverlapping
Trickfilmüberblendung *f*

218 ANIMATION EQUIPMENT ani
f équipement *m* d'animation
e equipo *m* de animación
i attrezzatura *f* d'animazione
n tekenfilmapparatuur
d Zeichentrickfilmapparatur *f*

219 ANIMATION STAND ani
f banc *m* d'animation,
support *m* d'équipement d'animation
e soporte *m* de equipo de animación
i sopporto *m* d'attrezzatura d'animazione
n tekenfilmstelling
d Zeichentrickfilmgerät *n*

220 ANIMATION TIMING, ani
TIMING IN ANIMATION
f synchronisation *f* de l'animation
e sincronización *f* de la animación
i sincronizzazione *f* dell'animazione
n tekenfilmsynchronisatie
d Trickfilmsynchronisierung *f*

221 ANIMATOR, ani
CARTOONIST
f animateur *m*,
dessinateur *m* de dessins animés,
intervalliste *m*
e animador *m*,
dibujante *m* de películas animadas
i animatore *m*,
disegnatore *m* di cartoni animati
n tekenfilmtekenaar
d Phasenzeichner *m*,
Trickfilmzeichner *m*,
Trickzeichner *m*

222 ANISOTROPIC COMA opt
Distortion in a TV image arising from inclination of the objective, which is the first electric lens adjacent to the electron-emitting cathode.
f coma *m* anisotrope
e coma *m* anisótropo
i coma *m* anisotropo
n anisotrope coma
d anisotropes Koma *n*

223 ANNOUNCER, tv
NEWS READER,
NEWSCASTER
f annonceur *m* du journal parlé,
speaker *m*
e anunciador *m* del diario hablado,
locutor *m*
i annunziatore *m* del telegiornale
n nieuwslezer, omroeper
d Nachrichtenansager *m*,
Sprecher *m*

224 ANNOUNCER STUDIO rep
f studio *m* d'annonceurs,
studio *m* de speakers
e estudio *m* de locutores
i studio *m* d'annunziatori
n omroepersstudio
d Ansageraum *m*, Ansagerstudio *n*

225 ANNULAR COLLECTOR ELECTRODE, crt
COLLECTOR RING
Annular electrode in an iconoscope on the inner surface of the bulb which collects the electrons.
f électrode *f* collectrice annulaire
e electrodo *m* colector anular
i elettrodo *m* collettore anulare
n ringvormige verzamelelektrode
d ringförmige Sammelelektrode *f*

226 ANNULAR MAGNET, crt
RING MAGNET
f bague-aimant *m* circulaire
e imán *m* anular
i magnete *m* anulare
n ringvormige magneet
d Ringmagnet *m*

227 ANODE crt
An electrode of e.g. a cathode-ray tube.
f anode *f*
e ánodo *m*
i anodo *m*
n anode
d Anode *f*

228 ANODE FOLLOWER, cpl
GROUNDED CATHODE AMPLIFIER
An amplifier with cathode at zero alternating potential, drive on the grid and power taken from the anode.
f amplificateur *m* à cathode mise à la masse
e amplificador *m* con cátodo a la massa
i amplificatore *m* con catodo a massa
n katodebasisversterker
d Anodenverstärker *m*,
Katodenbasisverstärker *m*

229 ANODE-VOLTAGE-STABILIZED CAMERA TUBE (GB), tv
HIGH-ELECTRON-VELOCITY CAMERA TUBE,
PLATE-VOLTAGE-STABILIZED CAMERA TUBE (US)
A camera tube operating with a beam of electrons having velocities such that the average target voltage stabilizes at a value near to that of the electron gun anode.
f tube *m* analyseur à électrons rapides
e tubo *m* de cámara de electrones rápidos
i tubo *m* da ripresa a elettroni veloci
n iconoscoopcamerabuis
d Aufnahmeröhre *f* mit schnellen Elektronen

230 ANOPTIC SYSTEM tv
A system that does not employ optical devices.
f système *m* anoptique
e sistema *m* anóptico
i sistema *m* anottico
n anoptisch systeem *n*
d anoptisches System *n*

ANTENNA (US)
see: AERIAL

ANTENNA AMPLIFIER
see: AERIAL AMPLIFIER

ANTENNA ASSEMBLY
see: AERIAL ASSEMBLY

ANTENNA CASING
see: AERIAL CASING

ANTENNA COIL
see: AERIAL COIL

ANTENNA CONNECTION (US)
see: AERIAL TERMINAL

ANTENNA DIRECTIVITY
see: AERIAL DIRECTIVITY

ANTENNA EFFICIENCY
see: AERIAL EFFICIENCY

ANTENNA ELEMENT
see: AERIAL ELEMENT

ANTENNA GAIN
see: AERIAL GAIN

ANTENNA INPUT
see: AERIAL INPUT

ANTENNA LEAD
see: AERIAL LEAD

ANTENNA MAINTENANCE
see: AERIAL MAINTENANCE

ANTENNA MAST,
ANTENNA TOWER
see: AERIAL MAST

ANTENNA MATCHING DEVICE
see: AERIAL MATCHING DEVICE

ANTENNA OUTPUT
see: AERIAL OUTPUT

ANTENNA POWER GAIN
see: AERIAL POWER GAIN

ANTENNA SIGNAL
see: AERIAL SIGNAL

ANTENNA SIGNAL DISTRIBUTION
see: AERIAL SIGNAL DISTRIBUTION

ANTENNA SYSTEM FOR SATELLITES
see: AERIAL SYSTEM FOR SATELLITES

231 ANTICIPATING SIGNAL tv
f signal *m* d'anticipation,
signal *m* d'avance
e señal *f* anticipadora
i segnale *m* in anticipo
n voorsignaal *n*
d Vorläufer *m*

2 ANTI-CLOCHE CIRCUIT, ctv
CHROMINANCE CARRIER
ATTENUATING CIRCUIT,
COMPLEMENTARY GAUSSIAN
CIRCUIT
A circuit in the SECAM system to attenuate the colo(u)r carrier and the adjacent channels.
circuit *m* anticloche,
circuit *m* cloche complémentaire
circuito *m* atenuador de la portadora de crominancia
circuito *m* attenuatore della portante di crominanza
antiklokschakeling,
complementaire klokschakeling
Anticloche-Schaltung *f*,
Farbträgerschwächer *m*,
komplementärer Glockenkreis *m*

3 ANTIMONY TRISULPHIDE crt/rec/tv
VIDICON
vidicon *m* à couche en trisulfure d'antimoine
vidicón *m* con capa de trisulfuro de antimonio
vidicon *m* con strato a trisulfuro d'antimonio
vidicon *n* met antimoontrisulfidelaag
Vidikon *n* mit Antimontrisulfidschicht

4 ANTINODAL POINT, aud
ANTINODE,
LOOP
A point, line, or surface in a standing-wave system at which some characteristic of the wave has maximum amplitude.
ventre *m* d'une oscillation
antinodo *m* de oscilación
vientre *m* de oscilación
antinodo *m* d'oscillazione,
ventre *m* d'oscillazione
trillingsbuik
Schwingungsbauch *m*

35 ANTI-REFLECTION COATING opt
Thin coating of a transparent substance applied to the surface of a lens.
couche *f* antireflet,
revêtement *m* antireflet
capa *f* antirreflexión,
revestimiento *m* antirreflexión
rivestimento *m* antiriflessione,
strato *m* antiriflessione
antireflectielaag
Antireflexbelag *m*,
Blauschicht *f*

36 APERTURE, opt
LENS STOP
The space through which light passes in any optical instrument.
ouverture *f*
abertura *f*
apertura *f*
apertuur, lensopening
Apertur *f*, Blende *f*, Blendenöffnung *f*,
Öffnung *f*

237 APERTURE COLO(U)R, ct
NON-OBJECT PERCEIVED
COLO(U)R
Colo(u)r perceived as non-located in depth, such as that perceived as filling a hole in a screen.
f couleur *f* immatérielle
e color *m* inmaterial
i colore *m* immateriale
n immateriële kleur
d nichtphysikalische Farbe *f*

238 APERTURE COMPENSATION crt/tv
(GB),
APERTURE CORRECTION (GB),
APERTURE EQUALIZATION (US)
Method of compensating for loss of higher picture frequencies caused by the scanning spot in a camera tube having a finite size and thus failing to respond sharply to sudden vertical boundaries between dark and light areas.
f correction *f* de l'ouverture
e corrección *f* de la abertura
i correzione *f* dell'apertura
n apertuurcorrectie
d Aperturkorrektur *f*

239 APERTURE CORRECTOR (GB), opt
APERTURE EQUALIZER (US),
f correcteur *m* d'ouverture
e corrector *m* de abertura
i correttore *m* d'apertura
n apertuurcorrector
d Aperturkorrektor *m*

240 APERTURE DISK, tv
APERTURED DISK
A mechanical scanning device of the earlier days of TV.
f disque *m* perforé
e disco *m* agujereado,
disco *m* de orificios
i disco *m* forato
n gatenschijf
d Lochscheibe *f*

241 APERTURE DISTORTION crt/tv
Distortion produced by the finite size of the scanning spot, when covering more elements of the mosaic simultaneously.
f distorsion *f* d'ouverture
e distorsión *f* de abertura
i distorsione *f* d'apertura,
errore *m* d'apertura
n apertuurvervorming
d Aperturverzerrung *f*

242 APERTURE LENS crt
In electron optics, a hole in a plate electrode which separates two electric fields.
f lentille *f* électronique
e lente *f* electrónica
i lente *f* elettronica
n elektronenlens
d Elektronenlinse *f*

243 APERTURE MASK rep
A mask used in a projector to make the aperture correspond to the reduced dimensions of the safe action area.
f masque *m* d'ouverture
e máscara *f* de abertura
i maschera *f* d'apertura
n apertuurmasker *n*
d Aperturmaske *f*

244 APERTURE MASK, ctv
COLO(U)R TELEVISION MASK,
PLANAR MASK,
SHADOW MASK
In three-gun colo(u)r TV tubes, a thin sheet of metal mounted in the back of a phosphor dot plate, containing small holes through which each beam must pass.
f masque *m* d'ombre
e máscara *f* de sombra
i maschera *f* d'ombra
n schaduwmasker *n*
d Lochmaske *f*

APERTURE OF THE BEAM (US)
see: ANGULAR WIDTH OF SPOT

245 APERTURE RATIO, tv
ASPECT RATIO,
PICTURE RATIO
The ratio of the width to the height of the picture image projected on the screen, usually 3:4.
f rapport *m* de format,
rapport *m* largeur/hauteur de l'image
e formato *m* del cuadro,
relación *f* de aspecto,
relación *f* del ancho al alto del cuadro
i formato *m* del quadro,
rapporto *m* di formato
n breedte-hoogteverhouding
d Seitenverhältnis *n*

246 APERTURE STOP, crt/tv
DIAPHRAGM
A device to control the diameter of the light beam between the object and image.
f diaphragme *m*
e diafragma *m*
i diaframma *m*
n diafragma *n*
d Blende *f*

247 APL, tv
AVERAGE PICTURE LEVEL
f luminosité *f* moyenne de l'image
e luminosidad *f* media de la imagen
i luminosità *f* media dell'immagine
n gemiddelde beeldhelderheid
d mittlere Bildhelligkeit *f*

248 APOCHROMAT opt
A lens that has been corrected for chromatic aberration for three colo(u)rs.
f apochromat *m*,
objectif *m* apochromatique
e objetivo *m* apocromático
i obiettivo *m* apocromatico
n apochromaat
d Apochromat *m*

249 APPARATUS ROOM,
CONTROL ROOM,
OPERATIONS ROOM
A room containing audio-frequency and video-frequency equipment, mainly for amplifying, switching and distributing program(me) outputs.
f salle *f* de commande,
salle *f* des équipements
e sala *f* de equipos,
sala *f* de mando
i sala *f* di comando,
sala *f* d'equipaggiamenti
n bedieningsruimte,
controlekamer,
regelkamer
d Geräteraum *m*

250 APPARENT BLACK
f noir *m* apparent
e negro *m* aparente
i nero *m* apparente
n schijnbaar zwart *n*
d scheinbares Schwarz *n*

251 APPARENT BRIGHTNESS,
LUMINOSITY,
SUBJECTIVE BRIGHTNESS
Attribute of visual sensation according t which an area appears to emit more or less light.
f luminosité *f*
e luminosidad *f*
i luminosità *f*
n helderheid
d Helligkeit *f*

252 APPLE TUBE ct
Colo(u)r display tube with spaced vertica red, green and blue stripes.
f tube *m* dit pomme
e tubo *m* tipo manzana
i tubo *m* cosidetto mela
n appelbuis
d Apfelröhre *f*

253 AQUADAG COATING cr
A coating on the inside of cathode-ray tubes.
f enduit *m* aquadag
e revestimiento *m* aquadag
i rivestimento *m* aquadag
n aquadaglaag
d Aquadagbelag *m*, Graphitschicht *f*

254 ARBITRARY FUNCTION t
GENERATOR
A function generator in which a mask having the desired waveform outline is inserted in a photoelectric scanning system.
f générateur *m* de fonctions arbitraires
e generador *m* de funciones arbitrarias
i generatore *m* di funzioni arbitrarie
n generator voor willekeurige functies
d Generator *m* für beliebige Funktionen

5 ARCOM, ARCTIC COMMUNICATION SATELLITE EARTH STATION — tv
station *f* terrestre arctique pour satellite de communication
estación *f* terrestre ártica para satélite de comunicación
stazione *f* terrestre artica per satellite di comunicazione
noordpoolstation *n* voor communicatie-satelliet
Nordpolstation *f* für Kommunikations-satellit

6 ART DIRECTOR — fi/th/tv
architecte-décorateur *m*, chef-décorateur *m*
escenógrafo *m*
architetto *m*, scenografo *m*
artistiek leider
Bühnenbildner *m*, Filmarchitekt *m*, Szenenbildner *m*

RT METHOD
see: ADDITIONAL REFERENCE TRANSMISSION

7 ARTICULATION — aud
Of a system used for transmitting or reproducing speech, the percentage number, or fraction, of speech components correctly recognized over the system.
articulation *f*
articulación *f*
articolazione *f*
articulatie, spraakklankvorming
Deutlichkeit *f*, Sprachverständlichkeit *f*, Verständlichkeit *f*

8 ARTIFICIAL BLACK SIGNAL, NOMINAL BLACK SIGNAL — cpl
Electronic signal equivalent to black level inserted into the composite video signal waveform for measurement purposes.
signal *m* d'essai pour le noir
señal *f* de prueba para el negro
segnale *m* di prova per il nero
nominaal-zwartsignaal *n*
Testsignal *n* für Schwarz

9 ARTIFICIAL EARTH SATELLITE — tv
Circulating vehicle at a height above the earth to reflect or transmit back radio waves as a means of communication, using computer-controlled terrestrial aerials (antennas).
satellite *m* artificiel de la terre
satélite *m* artificial de la tierra, **satélite *m* terrestre artificial**
satellite *m* terrestre artificiale
kunstmatige aardsatelliet
künstlicher Erdsatellit *m*

260 ARTIFICIAL ECHO — aud
An echo added to the output from a program(me) source to stimulate reverberation.
f écho *m* artificiel
e eco *m* artificial
i eco *m* artificiale
n kunstmatige echo
d künstliches Echo *n*

261 ARTIFICIAL ECHO PRODUCTION — aud
f production *f* d'échos artificiels
e producción *f* de ecos artificiales
i produzione *f* d'echi artificiali
n opwekken *n* van kunstmatige echo's
d Erzeugung *f* künstlicher Echos

262 ARTIFICIAL ECHO UNIT — aud
A device for producing an artificial echo.
f générateur *m* d'échos artificiels
e generador *m* de ecos artificiales
i generatore *m* d'echi artificiali
n kunstmatige-echo-opwekker
d künstlicher Echoerzeuger *m*

263 ARTIFICIAL LIGHT — ct/stu
f lumière *f* artificielle
e luz *f* artificial
i luce *f* artificiale
n kunstlicht *n*
d Kunstlicht *n*

264 ARTIFICIAL LIGHTING — rec/rep/stu
f éclairage *m* artificiel
e alumbrado *m* artificial
i illuminazione *f* artificiale
n kunstverlichting
d künstliche Beleuchtung *f*

265 ARTIFICIAL REVERBERATION — aud
f réverbération *f* artificielle
e reverberación *f* artificial
i riverberazione *f* artificiale
n kunstmatige nagalm
d künstlicher Nachhall *m*

266 ARTIFICIAL WHITE SIGNAL, NOMINAL WHITE SIGNAL — cpl
Electronic signal equivalent to white level inserted into the composite video signal waveform for measurement purposes.
f signal *m* d'essai pour le blanc
e señal *f* de prueba para el blanco
i segnale *m* di prova per il bianco
n nominaal-witsignaal *n*
d Testsignal *n* für Weiss

ASPECT RATIO
see: APERTURE RATIO

267 ASPECT-RATIO ADJUSTMENT — rep/tv
f ajustage *m* du format
e ajuste *m* del aspecto
i aggiustaggio *m* del formato
n vastlegging van de breedte-hoogte-verhouding
d Seitenverhältnisjustierung *f*

268 ASPHERIC CORRECTOR PLATE opt
Lens, one surface of which is specially shaped and is not part of the surface of a sphere. Used in large screen TV projectors.
f lentille *f* asphérique correctrice
e lente *f* asférica correctora
i lente *f* asferica correttrice
n asferische correctielens
d asphärische Korrekturlinse *f*

269 ASSEMBLY ge
A number of parts or subassemblies joined together to perform a specific function.
f ensemble *m*, montage *m*
e conjunto *m*, montaje *m*
i complesso *m*, montaggio *m*
n opstelling, samenstel *n*
d Anordnung *f*, Montage *f*, Zusammenbau *m*

270 ASSEMBLY MACHINE ge
A machine used for inserting components automatically in printed wire boards, such as on an in-line assembly machine or on a single-station assembly machine.
f machine *f* de montage
e máquina *f* de montaje
i macchina *f* di montaggio
n montagemachine
d Montagemaschine *f*

271 ASSEMBLY OF NEWS FILM rec/rep
f montage *m* d'actualités filmées
e montaje *m* de película de actualidades
i montaggio *m* di film di telegiornale
n nieuwsfilmopmaak
d Nachrichtenfilmzusammenschnitt *m*

ASSIGNED FREQUENCY
see: ALLOCATED FREQUENCY

272 ASSISTANT-CAMERAMAN, fi/tv
CAMERA ASSISTANT
f assistant-cadreur *m*,
assistant-opérateur *m*
e operador-ayudante *m*
i aiuto-operatore *m*
n assistentoperateur
d Kameraassistent *m*,
Schärfezieher *m*

273 ASSISTANT-DIRECTOR fi/th/tv
f assistant(e) *m*, *f* de réalisation,
assistant(e) *m*, *f* du metteur en scène
e asistenta(e) *f*, *m* del director de escena,
ayudante *m*, *f* del director de escena
i aiuto-regista *m*, *f*,
assistente *m*, *f* del(la) regista
n regieassistent(e)
d Regieassistent(e) *m*, *f*

274 ASTIGMATISM opt
An electron-beam tube defect in which electrons in the beam come to a focus in different axial planes as the beam is deflected so that the spot on the screen is distorted in shape and the image is blurred.
f astigmatisme *m*
e astigmatismo *m*
i astigmatismo *m*
n astigmatisme *n*
d Astigmatismus *m*

275 ASYMMETRICAL AERIAL INPUT, ae
ASYMMETRICAL ANTENNA INPUT
f entrée *f* d'antenne asymétrique
e entrada *f* de antena asimétrica
i entrata *f* d'antenna asimmetrica
n asymmetrische antenne-invoer
d asymmetrischer Antenneneingang *m*

276 ASYMMETRICAL DEFLECTION c
f balayage *m* asymétrique,
déviation *f* asymétrique
e barrido *m* asimétrico,
desviación *f* asimétrica
i deflessione *f* asimmetrica,
deviazione *f* asimmetrica
n asymmetrische afbuiging
d asymmetrische Ablenkung *f*

277 ASYNCHRONOUS OPERATION t
Generation of the sync signals of a TV system such that the field scanning frequency is not synchronous with the local electricity supply frequency.
f opération *f* sans enclenchement au rése
e operación *f* sin enclavamiento a la red
i operazione *f* senza agganciamento alla frequenza di rete
n werken *n* zonder netvergrendeling
d Arbeiten *n* ohne Netzverriegelung

278 ATMOSPHERIC INTERFERENCE, di
ATMOSPHERIC NOISE,
ATMOSPHERICS
Noise arising from natural phenomena within the atmosphere.
f parasites *pl* atmosphériques
e parásitos *pl* atmosféricos,
perturbaciónes *pl* atmosféricas
i parassiti *pl* atmosferici,
perturbazioni *pl* atmosferiche
n atmosferische storingen *pl*
d atmosphärische Störungen *pl*

279 ATTACK au
The production of a tone quality for the initial part of the sound differing from that of the remaining or continuing part, due to the fact that the higher overtones last for only a short time.
f attaque *f*, intonation *f*
e entonación *f*
i intonazione *f*
n inzet
d Ansatz *m*, Einsatz *m*

280 ATTACK TIME c
Operating time under defined conditions, of a piece of apparatus such as a sound limiter, for the gain to change from one value to another when the input signal is suddenly increased in level.
f temps *m* de mise au point
e tiempo *m* de puesta en punto

tempo *m* di messa a punto
inregeltijd
Einregelzeit *f*

1 ATTENDED STATION tv
A station which is permanently manned when in use.
station *f* normalement exploitée
estación *f* atendida
stazione *f* sorvegliata
bemand station *n*, bewaakt station *n*
bemannte Sendestation *f*

2 ATTENUATION cpl
The reduction in current, voltage, or power along the transmission path of a signal when transmitted through a network, along a line or waveguide or by radio propagation.
affaiblissement *m*, atténuation *f*
atenuación *f*, debilitamiento *m*
attenuazione *f*, smorzamento *m*
demping, verzwakking
Dämpfung *f*, Schwächung *f*

3 ATTENUATION DISTORTION aud
distorsion *f* d'atténuation
distorsión *f* de atenuación
distorsione *f* d'attenuazione
dempingsvervorming
Dämpfungsverzerrung *f*

4 ATTENUATION OF THE ADJACENT SOUND aud
atténuation *f* du son adjacent
atenuación *f* del sonido adyacente
attenuazione *f* del suono adiacente
omgevingsgeluiddemping
Umgebungsschalldämpfung *f*

5 ATTENUATION OF THE PROPER SOUND aud
atténuation *f* du son propre
atenuación *f* del sonido propio
attenuazione *f* del suono proprio
eigen-geluiddemping
Eigenschalldämpfung *f*

86 ATTENUATOR PROBE svs
sonde *f* atténuatrice
sonda *f* atenuadora
sonda *f* attenuatrice
verzwakkerkop
Dämpfungskopf *m*

87 AUDIBILITY aud
audibilité *f*
audibilidad *f*
udibilità *f*
hoorbaarheid
Hörbarkeit *f*

88 AUDIBILITY THRESHOLD, THRESHOLD OF AUDIBILITY aud
seuil *m* d'audibilité
umbral *m* de audibilidad
soglia *f* d'udibilità
hoorbaarheidsgrens
Hörbarkeitsgrenze *f*

289 AUDIBLE DISTORTION aud/dis/vr
f distorsion *f* audible
e distorsión *f* audible
i distorsione *f* udibile
n hoorbare vervorming
d hörbare Verzerrung *f*

290 AUDIENCE ABSORPTION rec
f absorption *f* du son par l'audience
e absorción *f* del sonido por los espectadores
i assorbimento *m* del suono per gli spettatori
n geluidsabsorptie door de toeschouwers
d Schallabsorption *f* durch die Zuschauer

291 AUDIENCE SEATING rec/rep
f accommodation *f* des spectateurs
e asientos *pl* para los espectadores
i posti *pl* a sedere per gli spettatori
n zitplaatsen *pl* voor de toeschouwers
d Sitzgelegenheit *f* der Zuschauer

292 AUDIENCE STUDIO stu
f salle *f* publique, studio *m* d'émissions publiques
e sala *f* de emisiones públicas
i studio *m* d'emissioni pubbliche
n gehoorzaal
d Sendesaal *m*, Unterhaltungsstudio *n*

293 AUDIENCE TELEVISION TRANSMISSION tv
f émission *f* publique de télévision
e emisión *f* pública de televisión
i emissione *f* pubblica di televisione
n openbare televisie-uitzending
d öffentliche Fernsehsendung *f*

294 AUDIO AMPLIFIER, LOW-FREQUENCY AMPLIFIER aud/cpl
f amplificateur *m* audio, amplificateur *m* basse fréquence
e amplificador *m* audio, amplificador *m* baja frecuencia
i amplificatore *m* audio, amplificatore *m* bassa frequenza
n audioversterker, laagfrequentversterker
d Niederfrequenzverstärker *m*, Tonfrequenzverstärker *m*

295 AUDIO CARRIER, SOUND CARRIER tv
The carrier wave modulated by the audio (sound) signal.
f porteuse *f* son
e portadora *f* audio, portadora *f* de sonido
i portante *f* audio
n geluidsdraaggolf
d Tonträger *m*

296 AUDIO CENTER(RE) FREQUENCY, aud/rec/tv
CENTER(RE) CARRIER FREQUENCY,
SOUND CARRIER FREQUENCY,
SOUND CENTER(RE) FREQUENCY
The position in the channel frequency spectrum for TV broadcast of the carrier (known as the aural carrier) frequency for the frequency-modulated audio signal which transmits the sound portion of a TV program(me).
f fréquence *f* nominale de la porteuse son
e frecuencia *f* nominal de la portadora sonido
i frequenza *f* nominale della portante audio
n nominale frequentie van de geluidsdraaggolf
d mittlere Tonzwischenfrequenz *f*

297 AUDIO CONTROL ENGINEER, ac/tv
AUDIO ENGINEER,
SOUND CONTROL ENGINEER,
SOUND ENGINEER
The engineer who controls the quality of the sound.
f ingénieur *m* du son
e ingeniero *m* de sonido
i tecnico *m* del suono,
tecnico *m* dell'audio
n geluidstechnicus
d Toningenieur *m*,
Tonmeister *m*

298 AUDIO FEEDBACK CIRCUIT aud/cpl
f retour *m* d'écoute
e retorno *m* de escucha
i ritorno *m* d'ascolto
n meeluistercontrole
d Mithörkontrolle *f*,
Studioausgangston *m*

299 AUDIO-FOLLOW-VIDEO OPERATION, rep
MARRIED OPERATION
An operation in which sound and vision are preselected on the same switching elements.
f opération *f* à présélection consécutive d'image et son
e funcionamiento *m* con preselección consecutiva de imagen y sonido
i operazione *f* a preselezione consecutiva d'immagine e suono
n opeenvolgende voorselectie van beeld en geluid
d aufeinanderfolgende Vorselektion *f* von Bild und Ton

AUDIO-FREQUENCY
see: AF

300 AUDIO-FREQUENCY PEAK LIMITER aud/cpl
f limiteur *m* de crêtes basse fréquence
e limitador *m* de crestas baja frecuencia
i limitatore *m* di creste bassa frequenza
n piekbegrenzer voor lage frequentie
d Niederfrequenzverstärkungsbegrenzer *m*,
Tonfrequenzspitzenbegrenzer *m*

301 AUDIO-LEVEL INDICATOR
f indicateur *m* du niveau audio
e indicador *m* del nivel audio
i indicatore *m* del livello audio
n geluidsniveau-aanwijzer
d Tonpegelanzeiger *m*

302 AUDIO MASKING, d
MASKING
The amount by which the threshold of audibility of a sound is raised by the presence of another masking sound.
f masquage *m*
e enmascaramiento *m*
i mascheramento *m*
n maskering
d Verdeckung *f*

303 AUDIO MIXER,
SOUND MIXER
The desk at which the sound tracks are combined.
f mélangeur *m* de son,
table *f* de mixage son
e mezclador *m* audio,
mezclador *m* de sonido
i mescolatore *m* di suono,
tavolo *m* di dosaggio
n geluidsmengpaneel *n*,
geluidsmengtafel
d Mischpult *n*,
Mischtafel *f*,
Tonmischpult *n*

304 AUDIO MONITORING t
The operation of controlling, by keeping a listening watch, the technical or artisti quality of audio signals.
f contrôle *m* d'écoute
e control *m* de escucha
i ascolto *m* di controllo
n geluidscontrole
d Abhörkontrolle *f*

305 AUDIO NUISANCE VALUE re
A value obtained by inserting a filter in a set for measuring the noise in audio-frequency work.
f valeur *f* de la perturbation produite par le bruit
e valor *m* de la interferencia del ruido,
valor *m* de la perturbación producida por el ruido
i valore *m* della potenza del rumore nella banda audio
n waarde van de ruis in het hoorfrequentiegebied
d Wert *m* des Rauschens im Hörfrequenzgebiet

306 AUDIO RECORDING aud/rec
The converting of audio-frequency electric variations into magnetic variations.
f enregistrement *m* d'impulsions sonores

registro *m* de variaciones de audiofrecuencia
registrazione *f* d'impulsi a frequenza vocale
optekening van geluidsimpulsen
Aufzeichnung *f* von Tonimpulsen

7 AUDIO RECORDING LEVEL CONTROL — vr
régleur *m* manuel du niveau d'enregistrement
regulador *m* manual de la profundidad de modulación
regolatore *m* manuale del livello di registrazione
audio-opneemniveauregelaar, modulatiediepteregelaar
Audioaussteuerungsregler *m*

8 AUDIO SIGNAL, AURAL SIGNAL, SOUND SIGNAL — tv
In TV, the signal corresponding to the sound portion of the TV program(me).
signal *m* audio, signal *m* basse fréquence
audioseñal *f*, señal *f* acústica
segnale *m* acustico, segnale *m* audio
geluidssignaal *n*
Tonsignal *n*

9 AUDIO SOCKET — vr
Used for connection of a microphone, gramophone or tape recorder to a video recorder.
connexion *f* audio, connexion *f* son
conexión *f* audio, conexión *f* sonido
connessione *f* audio, connessione *f* suono
audioaansluiting
Audioanschluss *m*

0 AUDIO SPECTRUM — ge
Band of sound frequencies to which the normal ear responds.
spectre *m* de fréquences acoustiques
espectro *m* de frecuencias acústicas
spettro *m* di frequenze acustiche
hoorfrequentiespectrum *n*
Tonfrequenzspektrum *n*

1 AUDIO TAPE RECORDER, SOUND RECORDER, TAPE RECORDER — aud/vr
A mechanical electronic device for recording voice, music and other audio frequencies on magnetic tape.
appareil *m* d'enregistrement sur bande magnétique, enregistreur *m* sur bande magnétique, magnétophone *m*
magnetófono *m*, registrador *m* de cinta magnetofónica
i registratore *m* a nastro magnetico, registratore *m* magnetico audio
n audiobandrecorder, bandrecorder, magnetofoon
d Bandgerät *n*, Magnettonaufnahmegerät *n*, Tonbandgerät *n*

312 AUDIO TAPE RECORDING, TAPE RECORDING — aud
f enregistrement *m* sur bande magnétique
e registro *m* en cinta magnética
i registrazione *f* su nastro magnetico
n audiobandrecording, bandopname, bandrecording
d Bandaufnahme *f*, Tonbandaufnahme *f*

313 AUDIO TRANSMITTER, SOUND TRANSMITTER — tv
The radio equipment used for the transmission of the audio (sound) signals from a TV broadcast station.
f émetteur *m* son pour télévision
e transmisor *m* de señal audible para televisión
i trasmettitore *m* audio per televisione
n geluidszender
d Tonsender *m*

314 AUDIO-VISUAL MEANS — tv
f moyens *pl* audiovisuels
e ayudas *pl* audiovisuales, medios *pl* auditivo-visuales
i mezzi *pl* ausiliari audiovisuali
n audiovisuele hulpmiddelen *pl*
d audiovisuelle Hilfsmittel *pl*

315 AUDIO-VISUAL SCREEN — rep
f écran *m* audiovisuel
e pantalla *f* audiovisual
i schermo *m* audiovisuale
n geluiddoorlatend beeldscherm *n*
d schalldurchlässiger Bildschirm *m*

316 AUDIO-VISUAL UNIT — rep
f unité *f* audiovisuelle
e unidad *f* audiovisual
i unità *f* audiovisuale
n geluid-beeldeenheid
d Schall-Bild-Einheit *f*

317 AUDITION, TRIAL PERFORMANCE — tv
A preliminary studio test of a performer, act, or complete program(me) for a TV or radio show.
f séance *f* d'essai
e representación *f* de prueba
i rappresentazione *f* di prova
n proefuitvoering, studioproefuitvoering
d Probeaufführung *f*

318 AUDITORIUM — rep
The room in which lectures, etc. are given.
f auditorium *n*

e auditorio *m*
i auditorio *m*
n auditorium *n*, gehoorzaal
d Hörsaal *m*, Zuhörerraum *m*

319 AUDITORIUM rep
The persons listening to a lecture, etc.
f audience *f*, auditoire *m*
e auditorio *m*
i auditorio *m*
n auditorium *n*, gehoor *n*, toehoorders *pl*
d Auditorium *n*, Hörer *pl*, Zuhörerschaft *f*

320 AUDITORIUM NOISE rep
f bruit *m* d'auditorium,
bruit *m* de salle
e ruido *m* de auditorio,
ruido *m* de sala
i rumore d'auditorio,
rumore *m* di sala
n zaalruis
d Saalgeräusch *n*

321 AUDITORY FATIGUE rep
Caused, when looking at TV program(me)s by a flat and oppressive wall of sound.
f fatigue *f* auditive
e fatiga *f* de escuchar
i stanchezza *f* uditiva
n luistermoeheid
d Gehörermüdung *f*

322 AURAL SENSITIVITY rep
The region between the levels of minimum and maximum intensity of sound.
f sensibilité *f* auditive
e sensibilidad *f* auditiva
i sensibilità *f* uditiva
n hoorgevoeligheid
d Gehörempfindlichkeit *f*

323 AUTHOR fi/th/tv
f auteur *m*
e autor *m*
i autore *m*
n auteur, schrijver
d Autor *m*, Schriftsteller *m*

324 AUTHORESS fi/th/tv
f auteur *f*
e autora *f*
i autrice *f*
n auteur, schrijfster
d Autor *f*, Schriftstellerin *f*

325 AUTOMATED ANIMATION ani
f animation *f* automatisée
e animación *f* automatizada
i animazione *f* automatizzata
n geautomatiseerde vervaardiging van tekenfilms
d automatisierte Herstellung *f* von Trickfilmen

326 AUTOMATED MASTER SWITCHING, cpl/rec
MASTER CONTROL AUTOMATION
f commutation *f* automatisée d'un programme
e conmutación *f* automatizado de un programa
i commutazione *f* automatizzata d'un programma
n geautomatiseerde programmaschakeling
d automatische Programmschaltung *f*

327 AUTOMATED VIDEOTAPE EDITING
f montage *m* automatique de la bande magnétoscopique
e montaje *m* automático de la cinta magnetoscópica,
montaje *m* automático de la cinta video
i montaggio *m* automatico del nastro magnetoscopico
n automatische montage van de videomagneetband
d selbsttätiger Schnitt *m* des Magnetbildbandes

328 AUTOMATIC AMBIENT-LIGHT ADAPTION s
Circuits in a TV receiver adjusting contrast or contrast and black-level settings in response to changes in ambient lighting conditions.
f adaptation *f* automatique à la lumière ambiante
e adaptación *f* automática a la luz de ambiente
i adattamento *m* automatico alla luce ambiente
n automatische aanpassing aan het omgevingslicht
d selbsttätige Anpassung *f* an das Umlicht

AUTOMATIC BACKGROUND CONTROL,
AUTOMATIC BRIGHTNESS CONTROL
see: ABC

329 AUTOMATIC BALANCE OF CONTRAST AND BRIGHTNESS rep/t
f équilibre *m* automatique contraste-luminosité
e equilibrio *m* automático contraste-luminosidad
i equilibrio *m* automatico contrasto-luminosità
n automatische contrast-helderheidsbalans
d automatisches Gleichgewicht *n* zwischen Kontrast und Helligkeit

330 AUTOMATIC BALANCE OF CONTRAST AND SATURATION rep/t
f équilibre *m* automatique contraste-saturation
e equilibrio *m* automático contraste-saturación
i equilibrio *m* automatico contrasto-saturazione
n automatische contrast-verzadigingsbalans
d automatisches Gleichgewicht *n* zwischen Kontrast und Sättigung

331 AUTOMATIC BLACK-AND-WHITE AND COLO(U)R SYSTEM CIRCUITS ctv
f montages *pl* automatiques pour la commutation des systèmes "noir et blanc" et "couleurs"
e circuitos *pl* automáticos para la conmutación de los sistemas "blanco y negro" y "colores"
i circuiti *pl* automatici per la conmutazione dei sistemi "bianco e nero" e "colori"
n automatische zwart-wit en kleuren-systeemschakelingen *pl*
d automatische Schwarz-Weiss- und Farb-normschaltungen *pl*

332 AUTOMATIC BLACK LEVEL CIRCUIT ctv
f montage *m* automatique du niveau du noir
e circuito *m* automático del nivel del negro
i circuito *m* automatico del livello del nero
n automatische zwartniveauschakeling
d automatische Schwarzwertschaltung *f*

333 AUTOMATIC BLACK LEVEL CONTROL tv
f commande *m* automatique du niveau du noir
e regulación *f* automática del nivel del negro
i regolazione *f* automatica del livello del nero
n automatische regeling van het zwartniveau
d automatische Schwarzwerthaltung *f*

334 AUTOMATIC CHROMA CONTROL, AUTOMATIC CHROMINANCE CONTROL, AUTOMATIC COLO(U)R GAIN CONTROL, AUTOMATIC SATURATION CONTROL ctv
réglage *m* automatique de la saturation
regulación *f* automática de la saturación
regolazione *f* automatica della saturazione
automatische verzadigingsregeling
Autochroma *n*

AUTOMATIC CHROMINANCE CONTROL
see: ACC

335 AUTOMATIC COMPRESSION, AUTOMATIC VOLUME COMPRESSION aud
A circuit for reducing the volume range of speech signals.
compression *f* automatique de volume
compresión *f* automática de volumen
compressione *f* automatica di volume
automatische volumecompressie
selbsttätige Volumenverdichtung *f*

336 AUTOMATIC CONTRAST CONTROL, AUTOMATIC PICTURE CONTROL tv
A circuit that varies the gain of the radio-frequency and video-intermediate-frequency amplifier in such a way that the contrast of the TV picture is maintained at a constant average level.
commande *f* automatique du contraste
e regulación *f* automática del contraste
i regolazione *f* automatica del contrasto
n automatische contrastregeling
d selbsttätige Kontrastregelung *f*

337 AUTOMATIC DEGAUSSING, AUTOMATIC DEMAGNETIZATION crt
A method of ensuring that the receiver remains unaffected by magnetic fields that could distort the picture.
f désaimantation *f* automatique
e desmagnetización *f* automática
i demagnetizzazione *f* automatica, smagnetizzazione *f* automatica
n automatische demagnetisering, ontmagnetisering
d automatische Entmagnetisierung *f*

338 AUTOMATIC DISSOLVE ani/tv
f caméra *m* fade, fondu *m* automatique
e desvanecimiento *m* automático
i dissolvenza *f* automatica
n automatische verdwijning van het beeld
d Kamerablende *f*

339 AUTOMATIC FIELD-PHASING rec/rep
f mise *f* en phase automatique des trames
e puesta *f* en fase automática de los cuadros
i messa *f* in fase automatica delle trame
n automatische rastersynchronisatie
d selbsttätige Teilbildsynchronisierung *f*

340 AUTOMATIC FLARE CORRECTION opt
A means for automatically counteracting the effect of light flares in optical systems.
f correction *f* automatique de taches lumineuses
e corrección *f* de manchas luminosas
i correzione *f* di macchie luminose
n automatische lichtvlekcorrectie
d selbsttätige Lichtfleckkorrektur *f*

341 AUTOMATIC FOCUSING tv
Electrostatic focusing in which the focusing anode of a TV picture tube is internally connected through a resistor to the cathode so that no external focusing voltage is required.
f focalisation *f* automatique
e enfoque *m* automático
i focalizzazione *f* automatica
n automatische focussering
d selbsttätige Fokussierung *f*

AUTOMATIC FREQUENCY CONTROL
see: AFC

AUTOMATIC GAIN CONTROL
see: AGC

342 AUTOMATIC HUE CONTROL, AUTOMATIC TINT CONTROL ctv
f contrôle *m* automatique de tonalité
e control *m* automático de tonalidad
i controllo *m* automatico di tonalità

n automatische kleurtoonregeling
d selbsttätige Farbwertregelung *f*

343 AUTOMATIC IRIS CONTROL cpl/opt
f réglage *m* automatique de diaphragme
e regulación *f* automática de diafragma
i regolazione *f* automatica di diaframma
n automatische diafragma-instelling
d selbsttätige Blendeneinstellung *f*

344 AUTOMATIC LIGHT CONTROL rec
f réglage *m* automatique de la lumière
e regulación *f* automática de la luz
i regolazione *f* automatica della luce
n automatische lichtregeling
d selbsttätige Lichtregelung *f*

345 AUTOMATIC LINE-PHASING rec
f mise *f* en phase automatique des lignes
e puesta *f* en fase automática de las líneas
i messa *f* in fase automatica delle linee
n automatische lijnsynchronisatie
d selbsttätige Zeilensynchronisierung *f*

346 AUTOMATIC MONITOR tv
Apparatus which compares the quality of transmission at different parts of a system, and raises an alarm if there is appreciable variation.
f moniteur *m* automatique
e monitor *m* automático
i monitore *m* automatico
n automatische monitor
d selbsttätiger Monitor *m*

347 AUTOMATIC PEAK LIMITER cpl
A limiter, the capacitor of which acquires a charge through the back resistance of a metal rectifier approximately equal to peak white.
f limiteur *m* automatique de la crête du blanc
e limitador *m* automático de la cresta del blanco
i limitatore *m* automatico della cresta del bianco
n automatische begrenzer van het helderste wit
d selbsttätiger Weissgipfelbegrenzer *m*

348 AUTOMATIC PHASE CONTROL, cpl
AUTOMATIC PHASE REGULATION
Synchronization method in which the pulses supplied by the sync pulse separator are led to a phase detector instead of directly to the timebase generator.
f réglage *m* automatique de phase
e regulación *f* automática de fase
i regolazione *f* automatica di fase
n automatische fazeregeling
d selbsttätige Phasenregelung *f*

349 AUTOMATIC PICTURE rep/tv
STABILIZATION
f stabilisation *f* automatique de l'image
e estabilización *f* automática de la imagen
i stabilizzazione *f* automatica dell'immagine
n automatisch geregelde beeldstabilisati
d selbsttätige Bildstabilitätsregelung *f*

350 AUTOMATIC STANDARDS
RECOGNITION
A standards switching system in development, in which detection circui respond to change of line frequency in the video signal and switch the timebas and detector polarity accordingly.
f commutation *f* automatique de systèm
e conmutación *f* automática de sistema
i commutazione *f* automatica di sistema
n automatische systeemomschakeling
d selbsttätige Systemumschaltung *f*

351 AUTOMATIC STATION
A station controlled by local automatic equipment.
f station *f* autocommandée
e estación *f* automática
i stazione *f* automatizzata
n automatisch station *n*, onbemand station *n*
d automatische Station *f*

352 AUTOMATIC SUBCARRIER-
BALANCE CONTROL
Process of maintaining carrier balance automatically in the modulators of a colo(u)r system coder for systems usin suppressed-carrier modulation.
f équilibre *m* automatique du signal de la porteuse de chrominance
e equilibrio *m* automático de la señal de l portadora de crominancia
i equilibrio *m* automatico del segnale dell portante di crominanza
n automatische balancering van het chrominantiedraaggolfsignaal
d selbsttätiger Ausgleich *m* des Chrominanzträgersignals

353 AUTOMATIC SWITCHOVER TO c
BLACK-AND-WHITE AFTER
COLO(U)R TRANSMISSION
f commutation *f* automatique de "noir et blanc" sur "couleurs"
e conmutación *f* automática de "blanco-negro" hacía "colores"
i commutazione *f* automatica di "bianco-nero" a "colori"
n automatische omschakeling van zwart/v naar kleurenuitzending
d automatische Umschaltung *f* Schwarzweiss/Farbe

354 AUTOMATIC TEMPERATURE re
CONTROL
f réglage *m* automatique de température
e regulación *f* automática de temperatura
i regolazione *f* automatica di temperatur
n automatische temperatuurregeling
d selbsttätige Temperaturregelung *f*

355 AUTOMATIC TRANSFER re
A means for switching from one event to another when using computer memories having sufficient capacity for storing

sources and duration times.
commutation *f* automatique
conmutación *f* automática
commutazione *f* automatica
automatische overschakeling
selbsttätige Umschaltung *f*

6 AUTOMATIC TRANSMITTER tv
A transmitter controlled by local automatic equipment.
émetteur *m* autocommandé
transmisor *m* automático
trasmettitore *m* automatizzato
automatisch bediende zender
automatische Station *f*

7 AUTOMATIC VIDEO NOISE LIMITER cpl
limiteur *m* automatique du bruit vidéo
limitador *m* automático del ruido video
limitatore *m* automatico del rumore video
automatische videoruisbegrenzer
automatischer Videorauschbegrenzer *m*

58 AUTOMATIC VOLTAGE REGULATOR cpl
Device for keeping the output voltage automatically at a predetermined level.
régulateur *m* automatique de tension
regulador *m* automático de tensión
regolatore *m* automatico di tensione
automatische spanningsregelaar
selbsttätiger Spannungsregler *m*

59 AUTOMATIC WHITE CONTROL ctv
A control which gives the correct white in black-and-white and colo(u)r transmissions.
correction *f* automatique du blanc, obtention *f* automatique du blanc préféré
corrección *f* automática del blanco
correzione *f* automatica del bianco
automatisch geprefereerd wit *n*
selbsttätige Weisskorrektur *f*

60 AUTO-PREVIEW rep
A preview of an event source by using the off-air channel for presetting and preview.
image *f* préalable automatique, preview *m* automatique
imagen *f* previa automática
immagine *f* previa automatica
automatische voorbezichtiging
selbsttätige Vorschau *f*

61 AUTOTRANSFORMER cpl
autotransformateur *m*
autotransformador *m*
autotrasformatore *m*
spaartransformator
Spartransformator *m*

62 AUXILIARY TRANSMITTER, GAP FILLER tv
Low-power transmitter filling in areas of insufficient field strength.
émetteur *m* d'appoint
emisor *m* auxiliar
emettitore *m* ausiliario
steunzender

d Hilfssender *m*

363 AVAILABLE LINE rec/rep/tv
The portion of the scanning line which can be used for picture signals.
f partie *f* de la ligne utilisable
e parte *f* útil de la línea
i parte *f* utile della linea
n beschikbaar lijngedeelte *n*
d nutzbarer Bildzeilenteil *m*

364 AVERAGE BRIGHTNESS tv
The ratio between the number of dark picture elements to that of the light ones.
f luminosité *f* moyenne
e luminosidad *f* media
i luminosità *f* media
n gemiddelde helderheid
d mittlere Helligkeit *f*

365 AVERAGE BRIGHTNESS STABILITY tv
f stabilité *f* de la luminosité moyenne
e estabilidad *f* de la luminosidad media
i stabilità *f* della luminosità media
n gemiddelde helderheidsstabiliteit
d mittlere Helligkeitsstabilität *f*

AVERAGE PICTURE LEVEL
see: APL

366 AVERAGING ctv
Effect appearing in PAL-system equipment using a one-line delay line for decoding or error correction.
f établissement *m* de la valeur moyenne
e establecimiento *m* del valor medio
i fissazione *f* del valore medio
n middelen *n*
d Mittelwertbildung *f*

367 AXIAL MAGNETIC FIELD crt
f champ *m* magnétique axial
e campo *m* magnético axial
i campo *m* magnetico assiale
n axiaal magnetisch veld *n*
d axiales Magnetfeld *n*, Axialmagnetfeld *n*

368 AXIAL MAGNIFICATION opt/tv
The ratio of the interval between two adjacent image points on the axis of an optical instrument to the interval between the object conjugate points.
f agrandissement *m* axial
e aumento *m* axial
i ingrandimento *m* longitudinale
n asvergroting
d Achsenvergrösserung *f*

369 AXIAL MODE aud
Resonance in a room due to each of the axes of that room:length, width, and height.
f mode *m* axial
e modo *m* axial
i modo *m* assiale
n axiale modus
d axialer Modus *m*

370 AXIALLY SYMMETRICAL ELECTRON GUN crt
An electron gun having an electron beam symmetrically placed around the axis of the tube.
f canon *m* à faisceau à symétrie axiale
e cañón *m* de haz de simetría axial
i cannone *m* a fascio a simmetria assiale
n axiaal-symmetrisch elektronenkanon *n*
d achsensymmetrische Elektronenkanone

371 AXIS OF THE TUBE, TUBE AXIS c
f axe *m* du tube cathodique
e eje *m* de tubo catódico
i asse *m* di tubo catodico
n buisas
d Röhrenachse *f*

B

72 B, ct/ctv
BLUE PRIMARY
bleu *m* primaire,
couleur *f* primaire bleue
azul *m* primario,
color *m* primario azul
blu *m* primario,
colore *m* fondamentale blu
blauwe grondkleur,
primaire kleur blauw
blaue Grundfarbe *f*,
blaue Primärfarbe *f*

73 B BLACK LEVEL, ctv
BLUE BLACK LEVEL
The minimum permissible level of the B signal.
niveau *m* minimal pour le signal bleu
nivel *m* mínimo para la señal azul
livello *m* minimo per il segnale blu
minimumniveau *n* voor het signaal blauw
Mindestpegel *m* für das Blausignal

74 B PEAK LEVEL, ctv
BLUE PEAK LEVEL
The maximum permissible level of the B signal.
niveau *m* maximal pour le signal bleu
nivel *m* máximo para la señal azul
livello *m* massimo per il segnale blu
maximumniveau *n* voor het signaal blauw
Höchstpegel *m* für das Blausignal

75 B-Y AXIS, ctv
BLUE COLO(U)R DIFFERENCE AXIS
axe *m* B-Y
eje *m* B-Y
asse *m* B-Y
as van het blauwe-kleurverschilsignaal,
B-Y-as
B-Y-Achse *f*

376 B-Y MATRIX, crt
BLUE COLO(U)R-DIFFERENCE MATRIX
f matrice *f* B-Y
e matriz *f* B-Y
i matrice *f* B-Y
n B-Y-matrix
d B-Y-Matrize *f*

377 B-Y MODULATOR, ctv
BLUE COLO(U)R DIFFERENCE MODULATOR
The modulator for the B-Y signal.
f modulateur *m* B-Y
e modulador *m* B-Y
i modulatore *m* B-Y
n B-Y-modulator,
modulator voor het blauwe-kleurverschil-signaal
d B-Y-Modulator *m*

378 B-Y SIGNAL, ctv
BLUE COLO(U)R DIFFERENCE SIGNAL
A blue-minus-luminance colo(u)r-difference signal used in colo(u)r TV.
f signal *m* B-Y
e señal *f* B-Y
i segnale *m* B-Y
n B-Y-signaal *n*,
blauwe-kleurverschilsignaal *n*
d B-Y-Signal *n*

379 BABY CAN, stu
SINGLE BROADSIDE
f petit réflecteur *m* diffuseur
e pequeño reflector *m* difusor
i piccolo riflettore *m* diffusore
n kleine reflector met diffuus licht
d kleiner Streulichtscheinwerfer *m*

380 BACK BIAS, crt
BACK LIGHTING,
BIAS LIGHTING,
ICONOSCOPE BACK LIGHTING
In a TV camera tube, the illumination of the rear surface of the mosaic resulting in greater sensitivity.
f éclairage *m* de l'arrière côté,
éclairage *m* du côté dorsal
e alumbrado *m* de la cara posterior,
alumbrado *m* del dorso
i illuminazione *f* del lato posteriore,
illuminazione *f* del rovescio
n achterkantverlichting
d Rückseitenbeleuchtung *f*

381 BACK LIGHT, stu
REAR LIGHT
f contre-jour *m*
e luz *f* de reverso
i controluce *f*
n tegenlicht *n*
d Gegenlicht *n*

382 BACK-PACK rec/tv
The camera facility housing the electronic components and carried on the back of the cameraman.
f boîtier *m* dorsal
e recipiente *m* de espalda
i cassa *f* dorsale
n rugpak *n*
d Rückentrage *f*, Rücksack *m*

383 BACK PORCH tv
The interval of time immediately following the line sync pulse during which the video signal is maintained at blanking level.
f palier *m* arrière,
palier *m* de suppression
e rellano *m* posterior,
umbral *m* posterior

i gradino *m* posteriore, pianerottolo *m* posteriore
n achterstoep
d hintere Austastschulter *f*, hintere Schwarzschulter *f*

384 BACK-PORCH CLAMPING cpl
f clamping *m* sur le palier du noir
e circuito *m* enclavador sobre el rellano posterior
i bloccaggio *m* sul gradino posteriore
n achterstoepvergrendeling
d Schwarzschulterklemmung *f*

385 BACK PROJECTION, REAR PROJECTION fi/rep/tv
A projection in which the picture is projected from behind on a transparent screen.
f projection *f* par transparence
e proyección *f* por transparencia
i proiezione *f* per trasparenza
n doorzichtprojectie
d Durchprojektion *f*, Rückprojektion *f*

386 BACK PROJECTOR, REAR PROJECTOR fi/rep/tv
f projecteur *m* pour transparence
e proyector *m* para transparencia
i proiettore *m* per trasparenza
n doorzichtprojector
d Rückprojektor *m*

387 BACK RUN fi
f marche *f* arrière
e marcha *f* hacia atrás
i marcia *f* indietro
n terugloop
d Rücklauf *m*

388 BACKGROUND ge
Ever-present effects in physical apparatus above which a phenomenon must manifest itself in order to be measured.
f fond *m*
e fondo *m*
i fondo *m*
n achtergrond
d Hintergrund *m*

389 BACKGROUND BRIGHTNESS stu
f luminosité *f* de fond
e luminosidad *f* media
i luminosità *f* media
n gemiddelde helderheid
d Grundhelligkeit *f*

390 BACKGROUND COLO(U)R ctv/svs
f couleur *m* de fond
e color *m* de fondo
i colore *m* di fondo
n achtergrondkleur
d Hintergrundfarbe *f*

391 BACKGROUND CONTROL ctv
In colo(u)r TV, a potentiometer which is used as a means of controlling the d.c. level of a colo(u)r signal at the inp[ut] of a tricolo(u)r picture tube.
f contrôle *m* de la luminosité de fond
e control *m* de fondo, regulador *m* de la luminosidad de fondo
i controllo *m* della tinta di fondo del cinescopio
n regeling van de achtergrondhelderheid
d Hintergrundhelligkeitsregelung *f*

392 BACKGROUND EFFECT SOURCE r
Noise records which are kept in store a[nd] used during TV recordings.
f bande *f* de bruitage
e registro *m* de ruido
i registrazione *f* di rumore
n geruisband, geruisopname
d Geräuschband *n*

393 BACKGROUND FADE-IN rec/
f apparition *f* graduelle d'un fond
e aparición *f* gradual de un fondo
i dissolvenza *f* in apertura d'un fondo
n geleidelijk opkomen *n* van een achtergro[nd]
d Hintergrundeinblendung *f*

394 BACKGROUND FADE-OUT rec/t
f disparition *f* graduelle d'un fond
e desaparición *f* gradual de un fondo
i dissolvenza *f* in chiusura d'un fondo
n geleidelijke verdwijning van een achtergrond
d Hintergrundausblendung *f*

395 BACKGROUND FILM re
f film *m* animé de fond
e película *f* animada de fondo
i pellicola *f* animata di fondo
n achtergrondtekenfilm
d Hintergrundtrickfilm *m*

396 BACKGROUND LOUDSPEAKER (GB), PLAYBACK LOUDSPEAKER (US) rec/re
A loudspeaker for producing the accompanying sound in the studio.
f hautparleur *m* de fond
e altavoz *m* de fondo
i altoparlante *m* di fondo
n decorluidspreker
d Hintergrundlautsprecher *m*

397 BACKGROUND MUSIC re
Music accompanying e.g. televised films, etc.
f musique *f* de fond
e música *f* de fondo
i musica *f* di fondo
n achtergrondmuziek
d Hintergrundmusik *f*

398 BACKGROUND NOISE, RANDOM NOISE dis/rec
Noise due to the aggregate of a large number of elementary disturbances with random occurrence in time.
f bruit *m* de fond
e ruido *m* de fondo

rumore *m* di fondo
achtergrondruis
Grundgeräusch *n*,
Hintergrundgeräusch *n*

399 BACKGROUND NOISE LEVEL rec
niveau *m* du bruit de fond
nivel *m* del ruido de fondo
livello *m* del rumore di fondo
niveau *n* van de achtergrondruis
Grundgeräuschpegel *m*

400 BACKGROUND PROJECTION rec
In televising studio scenes, the projection of pictures to provide appropriate background scenery.
projection *f* de fond
proyección *f* de fondo
proiezione *f* di fondo
achtergrondprojectie
Hintergrundprojektion *f*

401 BACKGROUND SCENERY, rec
BACKING
décor *m* de fond
decoración *f* de fondo
fondale *m*
achtergronddecor *n*
Hintergrunddekor *n*

402 BACKGROUND SOUND EFFECTS rec
Effects usually recorded on endless magnetic tape loops for being introduced in a performance.
décor *m* sonore
efectos *pl* sonoros
effetti *pl* sonori
geluidseffecten *pl*
Geräuschkulisse *f*

403 BACKPLATE crt
In a TV camera tube, the plate behind, and insulated from, the mosaic, which transmits the video signal by capacitance.
contre-électrode *f*,
électrode *f* opposée
contraelectrodo *m*,
electrodo *m* opuesto
controelettrodo *m*,
elettrodo *m* opposto
achterelektrode
Gegenelektrode *f*,
Hinterelektrode *f*

404 BACKWARD RADIATION aea
rayonnement *m* en arrière
radiación *f* retrocesa
radiazione *f* retrograda
achterwaartse straling
Rückwärtsstrahlung *f*

405 BALANCED MODULATOR cpl
Circuit containing a pair of matched tubes (valves) acting as modulators, used to suppress the carrier in suppressed carrier transmission.
f modulateur *m* équilibré
e modulador *m* equilibrado
i modulatore *m* bilanciato
n uitgebalanceerde modulator
d Gegentaktmodulator *m*

406 BALANCING MAGNETIC STRIPE aud
f piste *f* magnétique de compensation
e pista *f* magnética de compensación
i pista *f* magnetica di compensazione
n magnetisch compensatiespoor *n*
d Magnetausgleichsspur *f*

407 BALANCING SOUND LEVELS aud/rec
One of the control measures carried out by the sound control room.
f équilibrage *m* de niveaux sonores
e equilibrado *m* de niveles de sonidos
i equilibratura *f* di livelli di suoni
n uitbalanceren *n* van geluidsniveaus
d Ausgleichung *f* von Schallpegeln

408 BALL RECEIVER, rep
RELAY RECEIVER
Relay station in TV transmission.
f récepteur *m* en balle
e receptor *m* retransmisor
i relè *m* ricetrasmettitore,
ricettore *m* ritrasmettitore
n relaisontvanger
d Ballempfänger *m*

409 BALL RECEPTION, rep
RELAY TELEVISION
TV system in which the transmitted program(me) is received by a relay station and broadcast to a defined region.
f réception *f* en balle
e recepción *f* y retransmisión *f*
i ricezione *f* a rimbalzo
n gerelayeerde ontvangst
d Ballempfang *m*

410 BANANA TUBE ctv
A colo(u)r display device so called from the shape of the tube in which a single electron gun scans the sidewall of a cylindrical tube on which a single phosphor strip is coated in the black and white version, and three colo(u)r-emitting strips in the colo(u)r version.
f tube *m* banane,
tube *m* cathodique analyseur à tambour à lentilles
e tubo *m* en forma de plátano,
tubo *m* catódico explorador con tambor de lentes
i tubo *m* catodico analizzatore a corona di lenti,
tubo *m* in forma di banana
n banaanbuis,
katodestraalbuis met draaiende lenzentrommel
d Bananenröhre *f*,
Katodenstrahlröhre *f* mit Linsenkranzabtaster

411 BAND, tv
FREQUENCY BAND
A range of frequencies between the limits

41 MHz and 860 MHz, subdivided in 5 groups of bands for TV.
f bande *f*, bande *f* de fréquences
e banda *f*, banda *f* de frecuencias
i banda *f*, banda *f* di frequenze
n band, frequentieband
d Band *n*, Frequenzband *n*

412 BAND LIMITS tv
The five subdivisions of TV frequency bands:
I 41-68 MHz
II 87,5-100 MHz
III 162-230 MHz
IV 470-550 MHz
V 582-860 MHz.
f limites *pl* de bande
e límites *pl* de banda
i limiti *pl* di banda
n bandgrenzen *pl*
d Bandgrenzen *pl*

413 BAND-PASS, BAND-PASS FILTER cpl
A filter having a single transmission band neither of the cut-off frequencies being zero or infinite.
f filtre *m* passe-bande
e filtro *m* pasabanda
i filtro *m* di banda, filtro *m* passabanda
n bandfilter *n*
d Bandfilter *n*

414 BAND-PASS AMPLIFIER cpl
An amplifier designed to pass a definite band of frequencies with essentially uniform response.
f amplificateur *m* passe-bande
e amplificador *m* pasabanda
i amplificatore *m* passabanda
n bandfilterversterker
d Bandfilterverstärker *m*

415 BAND-PASS RESPONSE rep/tv
A response characteristic in which a definite band of frequencies is transmitted with essentially uniform response.
f courbe *f* de réponse d'un filtre passe-bande
e curva *f* de respuesta de un filtro pasabanda
i curva *f* di risposta di un filtro passabanda
n bandfilterdoorlaatkromme
d Bandfilterresonanzkurve *f*, Bandresonanzkurve *f*

416 BAND REJECTION FILTER, BAND-STOP, BAND-STOP FILTER cpl
f filtre *m* éliminateur de bande
e filtro *m* eliminador de banda
i filtro *m* a soppressione di banda, filtro *m* soppressore
n bandstopfilter *n*
d Bandsperre *f*, Sperrfilter *n*

417 BAND SHARING
Use of a common band of frequencies by two signal channels.
f utilisation *f* de bande commune de fréquence
e utilización *f* de banda común de frecuencia
i utilizzazione *f* di banda comune di frequenza
n frequentiebandscharing
d gemeinsame Frequenzbandbenutzung *f*

418 BAND-SHARING SIGNALS c
f signaux *pl* à bande commune
e señales *pl* de banda común
i segnali *pl* a banda comune
n geschaarde signalen *pl*
d Signale *pl* auf gemeinsamem Frequenzband

419 BANDWIDTH ge/t
The width of a band of frequencies which for a TV station is 6 mc.
f largeur *f* de bande
e ancho *m* de banda, anchura *f* de banda
i larghezza *f* di banda
n bandbreedte
d Bandbreite *f*

420 BANDWIDTH COMPRESSION cpl/t
A technique in TV employing two channe[ls] for picture information to obtain high definition.
f compression *f* de largeur de bande
e compresión *f* de anchura de banda
i compressione *f* di larghezza di banda
n bandbreedtecompressie
d Bandbreitekompression *f*

421 BANK rec/t
A row of push-button switches, one for each source corresponding to a particula[r] function of a switching equipment or vision mixer, such as cutting, preview, or A and B bands of an A-B-mixer.
f portant *m*
e banco *m*
i banco *m*
n schakelbank
d Schaltbahn *f*

422 BANK OF LAMPS, OVERHEAD LIGHTING BATTENS stu/t[v]
f ambiance *f*, herses *pl*
e caja *f* para el alumbrado superior
i luci *pl* superiori della ribalta
n bovenlichten *pl*
d Flächenleuchte *f*, Hängegitter *n*

23 BAR (GB), tv
FLAGPOLE (US)
Black vertical or horizontal line on the screen of a TV set, used for testing, but also appearing as interference during broadcasts.
barre *f*
barra *f*
barra *f*
balk
Balken *m*

24 BAR GENERATOR (GB), tv
FLAGPOLE GENERATOR (US)
générateur *m* de barres
generador *m* de barras
generatore *m* di barre
balkengenerator
Balkengeber *m*, Balkengenerator *m*

25 BAR PATTERN (GB), tv
FLAGPOLE PATTERN (US)
mire *f* en barres
patrón *m* en barras
figura *f* in barre
balkentoetsbeeld *n*
Balkenmuster *n*,
Strichrastertestbild *n*

26 BAR SIGNAL (GB), tv
FLAGPOLE SIGNAL (US)
signal *m* en forme de barre
señal *f* en forma de barra
segnale *m* in forma di barra
balksignaal *n*
Balkensignal *n*

27 BARREL DISTORTION opt/tv
Distortion of the picture in which the sides bulge out like a barrel.
distorsion *f* en barillet
distorsión *f* en tonel
distorsione *f* a bariletto
tonvervorming
Tonnenverzeichnung *f*

28 BASE crt
The part of the electronic tube (valve) which is attached to the envelope and carries the pins or contacts to connect the electrodes to the external circuit.
base *f*, culot *m*
base *f*, casquillo *m*, zócalo *m*
base *f*, zoccolo *m*
buisbodem, buisvoet
Fuss *m*, Teller *m*

29 BASE BAND cpl
In the process of modulation, the frequency band occupied by the aggregate of the transmitted signals when first used to modulate a carrier, used e.g. for the transmission of picture and sync signals in TV.
bande *f* de base,
bande *f* de fréquence de modulation
banda *f* de base,
banda *f* de frecuencia de modulación
i banda *f* di base,
banda *f* di frequenza di modulazione
n basisband,
modulatiefrequentieband
d Basisband *n*,
Modulationsfrequenzband *n*

430 BASE BAND rec/tv
Narrow band of frequencies at the low-frequency end of a broad band channel.
f bande *f* des plus basses fréquences
e banda *f* de las más bajas frecuencias
i banda *f* delle più basse frequenze
n band van de laagste frequenties
d Band *n* der niedrigsten Frequenzen,
frequenzniedrigstes Band *n*

431 BASE FILM vr
f pellicule *f* de base
e película *f* de base
i pellicola *f* di base
n basisfilm
d Basisfilm *m*

432 BASE SIGNAL cpl
A signal of substantially square waveform which is an auxiliary to scanning both for transmission and reception.
f signal *m* de base
e señal *f* de base
i segnale *m* di base
n basissignaal *n*
d Basissignal *n*

433 BASE SIGNAL GENERATOR cpl
f générateur *m* de signaux de base
e generador *m* de señales de base
i generatore *m* di segnali di base
n basissignaalgenerator
d Basissignalgeber *m*

434 BASHER stu
f bol *m*, réflecteur *m*
e lámpara *f* proyectante
i lampada *f* per proiezione
n slaglichtlamp
d Fluter *m*

435 BASIC NOISE ge
At a point in a system, all noise present under test when the system is not carrying any wanted signals.
f bruit *m* de base
e ruido *m* de base
i rumore *m* di base
n basisruis
d Basisrauschen *n*

436 BASIC SIGNALS cpl
f impulsions *pl* formatrices du signal
e impulsiones *pl* para formación de la señal
i impulsi *pl* per la costruzione del segnale
n impulsen *pl* voor signaalopbouw
d Signalaufbauimpulse *pl*

437 BASIC STIMULUS ct
f stimulus *m* basique
e estímulo *m* básico

i stimolo *m* basico
n basisstimulus
d Basisreiz *m*

438 BATWING AERIAL (GB), SUPERTURNSTILE ANTENNA (US) aea/tv
A modified turnstile aerial (antenna) having wingshaped dipole elements used in pairs mounted at right angles about a common vertical axis.
f antenne *f* croisée multiple
e antena *f* de molinete múltiple
i antenna *f* incrociata multipla
n meervoudige vlinderantenne
d mehrfache Schmetterlingantenne *f*

439 BEAM aea/crt
A radiation of electromagnetic energy restricted to a small solid angle.
f faisceau *m*
e haz *m*
i fascio *m*
n bundel, straal
d Bündel *n*, Strahl *m*

440 BEAM ALIGNMENT, CAMERA ALIGNMENT crt
An adjustment of the electron beam in a camera tube, performed on tubes using low-velocity scanning, to cause the beam to be perpendicular to the target at the target surface.
f alignement *m* du faisceau, centrage *m* du faisceau
e alineación *f* del haz, alineamiento *m* del haz
i allineamento *m* del fascio, centratura *m* del fascio
n bundelcentrering, bundelinstelling
d Strahlausrichtung *f*, Strahlzentrierung *f*

441 BEAM ANGLE crt
The solid angle of the cone of electrons emerging from the crossover.
f angle *m* du faisceau
e ángulo *m* del haz
i angolo *m* del fascio
n bundelhoek
d Strahlwinkel *m*

442 BEAM BENDER, ION TRAP MAGNET crt
One or more small permanent magnets with pole pieces, placed around the neck of a TV picture tube to provide a magnetic field for ion-trap action in the electron gun.
f aimant *m* de piège d'ions
e imán *m* de trampa de iones
i magnete *m* di trappola ionica
n ionenvalmagneet
d Ionenfallemagnet *m*

443 BEAM BENDING crt
Deflection of the scanning beam by the electrostatic field of the charges on the target of a camera tube.
f déplacement *m* du faisceau
e desplazamiento *m* del haz
i spostamento *m* del fascio
n bundelverbuiging, straalafbuiging
d Strahlversetzung *f*

444 BEAM BLACK-OUT (US), BEAM BLANKING (GB), BEAM CURRENT BLANKING (GB) c
f suppression *f* du courant du faisceau, suppression *f* du faisceau
e borrado *m* del haz, supresión *f* de la corriente del haz
i cancellazione *f* del fascio, soppressione *f* della corrente del fascio
n bundelonderdrukking, bundelstroomonderdrukking
d Strahlaustastung *f*, Strahlstromaustastung *f*

445 BEAM CONVERGENCE crt/c
The adjustment that makes the three electron beams of a three-gun colo(u)r picture tube meet or cross at a shadow-mask hole.
f convergence *f* des faisceaux
e convergencia *f* de los haces
i convergenza *f* dei fasci
n bundelconvergentie
d Strahldeckung *f*, Strahlenkonvergenz *f*

446 BEAM COUPLING c
The production of an alternating current in a circuit connected between two electrodes through or by which a density-modulated electron beam is passed.
f couplage *m* électronique
e acoplamiento *m* electrónico
i accoppiamento *m* elettronico
n elektronenkoppeling
d Elektronenkopplung *f*

447 BEAM-COUPLING COEFFICIENT c
The ratio of alternating driving current produced in a resonator to the alternatin component of the initiating beam current
f coefficient *m* de couplage électronique
e coeficiente *m* de acoplamiento electrónic
i coefficiente *m* d'accoppiamento elettronico
n elektronenkoppelingscoëfficiënt
d Elektronenkopplungskoeffizient *m*

448 BEAM CROSS SECTION cr
f section *f* du faisceau
e sección *f* del haz
i sezione *f* del fascio
n bundeldoorsnede
d Strahlquerschnitt *m*

449 BEAM CURRENT cr
The electron current of the beam arrivin at the screen.
f courant *m* de faisceau
e corriente *f* de haz

corrente *f* di fascio
bundelstroom
Strahlstrom *m*

0 BEAM CURRENT ADJUSTMENT crt
alignement *m* du courant du faisceau
alineación *f* de la corriente del haz
allineamento *m* della corrente del fascio
bundelstroominstelling
Strahlstromjustierung *f*

1 BEAM CURRENT CONTROL crt
potentiomètre *m* de réglage du courant du faisceau
potenciómetro *m* de regulación de la corriente del haz
potenziometro *m* di regolazione della corrente del fascio
bundelstroompotentiometer
Strahlstromsteller *m*

2 BEAM CURRENT FOCUSING, crt
BEAM FOCUSING
concentration *f* du faisceau
concentración *f* del haz
concentrazione *f* del fascio
concentratie van de bundelstroom
Strahlstrombündelung *f*

3 BEAM CURRENT LIMITATION crt
limitation *f* du courant du faisceau
limitación *f* de la corriente del haz
limitazione *f* della corrente del fascio
bundelstroombegrenzing
Strahlstrombegrenzung *f*

4 BEAM CURRENT MODULATION crt
The process of varying with time the magnitude of the beam current passing through a surface.
modulation *f* du courant du faisceau
modulación *f* de la corriente del haz
modulazione *f* della corrente del fascio
bundelstroommodulatie
Strahlstrommodulation *f*

5 BEAM CURRENT NOISE, crt/dis
BEAM NOISE
bruit *m* du faisceau
ruido *m* del haz
rumore *m* del fascio
bundelstroomruis
Strahlstromrauschen *n*

6 BEAM CUT-OFF crt
The grid-to-cathode negative voltage in a TV picture tube at which electron flow ceases.
tension *f* de coupure du faisceau
tensión *f* de corte del haz
tensione *f* di taglio del fascio
bundelafsnijspanning
Strahleinsatzspannung *f*

7 BEAM FOCUS crt
finesse *f* du faisceau
nitidez *f* del haz
finezza *f* del fascio
n bundelscherpte
d Strahlschärfe *f*

458 BEAM-FORMING ELECTRODE crt
An electrode so located and maintained at such potentials that it causes the electron stream to be concentrated into one or more beams.
f électrode *f* formatrice du faisceau
e electrodo *m* formador del haz
i elettrodo *m* formatore del fascio
n bundelvormende elektrode
d Strahlelektrode *f*

459 BEAM GATE crt
f blocage *m* du faisceau
e bloqueo *m* del haz
i bloccaggio *m* del fascio
n bundelblokkering
d Strahlsperre *f*

460 BEAM IMPACT ERRORS, crt
BEAM LANDING ERRORS
Errors which occur when the electron beam does not strike the target correctly.
f erreurs *pl* d'impact du faisceau
e errores *pl* de impacto del haz
i errori *pl* d'urto del fascio
n treffouten *pl* van de bundel
d Auftreffehler *pl* des Strahls

461 BEAM IMPACT POINT, crt
BEAM LANDING,
BEAM TARGET
The spot where the beam strikes the target.
f point *m* d'impact du faisceau
e punto *m* de impacto del haz
i punto *m* d'urto del fascio
n bundeltrefplaats
d Strahlauftreffstelle *f*

462 BEAM-INDEXING COLO(U)R TUBE, crt/ctv
BEAM-INDEXING COLO(U)R TV TUBE,
BEAM-INDEXING TUBE
A colo(u)r picture tube in which a signal, generated by an electron beam after deflection, is fed back to a control device or element in such a way as to produce an image in colo(u)r.
f tube *m* image en couleurs à alimentation inverse du signal
e cromoscopio *m* con alimentación inversa de la señal
i cromoscopio *m* con alimentazione inversa del segnale
n bundelindexbuis,
kleurenbeeldbuis met signaalterugvoeding
d Farbfernsehbildröhre *f* mit Signalrückspeisung

463 BEAM JITTER crt
Random movements of electron beam in cathode-ray tubes due to electronic noise.
f fluctuation *f* du faisceau
e tambaleo *m* del haz

i fluttuazione *f* del fascio
n bundeltrilling
d Strahlschwankung *f*

464 BEAM LOADING crt
The production of an electronic admittance between two grids when an initially unmodulated beam of electrons is shot across the gap between them.
f charge *f* du faisceau
e carga *f* del haz
i carico *m* del fascio
n bundelbelasting
d Strahlbelastung *f*

465 BEAM MAGNET, crt
BEAM POSITIONING MAGNET,
CONVERGENCE MAGNET
A magnet assembly whose magnetic field converges two or more electron beams, used in three-gun picture tubes.
f aimant *m* de convergence
e imán *m* de convergencia
i magnete *m* di convergenza
n convergentiemagneet
d Konvergenzmagnet *m*

466 BEAM MODULATION cpl/crt
Deliberate variation of the beam current in a cathode-ray tube by the application of a varying signal voltage between control grid and cathode.
f modulation *f* du faisceau
e modulación *f* del haz
i modulazione *f* del fascio
n bundelmodulatie
d Strahlenmodulation *f*

467 BEAM MODULATION PERCENTAGE, crt
PERCENTAGE BEAM MODULATION
f facteur *m* de modulation de faisceau
e factor *m* de modulación de haz
i fattore *m* di modulazione di fascio
n bundelmodulatiefactor
d Strahlenmodulationsfaktor *m*

468 BEAM POSITIONING crt/tv
f apposition *f* du faisceau
e posicionación *f* del haz
i impostazione *f* del fascio
n bundelinstelling
d Strahleinstellung *f*

469 BEAM POSITIONING SYSTEM crt
f système *m* déflecteur du faisceau
e sistema *m* deflector del haz
i sistema *m* deflettore del fascio
n bundelafbuigsysteem *n*
d Strahlablenksystem *n*

470 BEAM RETURN crt
f retour *m* du faisceau
e retroceso *m* del haz
i ritorno *m* del fascio
n bundelterugloop
d Strahlrücklauf *m*

471 BEAM REVERSING LENS opt/tv
f optique *f* inverseuse de faisceau
e óptica *f* inversora de haz
i ottica *f* invertitrice di fascio
n bundelomkeringsoptiek
d Strahlumkehroptik *f*

472 BEAM SPLITTER, ctv/
BEAM SPLITTING PRISM,
OPTICAL SPLITTER
f diviseur *m* optique du faisceau, prisme *m* diviseur du faisceau
e prisma *m* divisor del haz
i prisma *m* divisore del fascio
n optische bundeldeler
d optischer Strahlenteiler *m*

473 BEAM SPLITTING SYSTEM,
OPTICAL SPLITTING SYSTEM
System used in colo(u)r TV camera to form two or more separate images fro a single lens.
f système *m* de division du faisceau
e sistema *m* de división del haz
i sistema *m* di divisione del fascio
n bundeldelingssysteem *n*
d Strahlenteilungssystem *n*

474 BEAM SWITCHING TUBE crt/
A single gun colo(u)r picture tube in which the beam is switched to an appropriate colo(u)r phosphor correspo ing to the signal applied to the electron gun.
f tube *m* commutateur du faisceau
e tubo *m* conmutador del haz
i tubo *m* commutatore del fascio
n bundelschakelbuis
d Strahlschaltröhre *f*

475 BEAM TILT a
Tilt above or below the horizontal axis of the pattern of radiation from a directional aerial (antenna) array.
f inclinaison *f* du diagramme
e inclinación *f* del diagrama
i inclinazione *f* del diagramma
n neiging van het stralingsdiagram
d Diagrammneigung *f*

BEAM TILT ANGLE
see: ANGLE OF BEAM TILT

476 BEAM TRAP c
Bucket-shaped electrode mounted in a cathode-ray tube, to catch the electron beam when it is not required to excite fluorescence in the screen.
f électrode *f* collectrice du faisceau
e electrodo *m* colector del haz
i elettrodo *m* collettore del fascio
n bundelopvangelektrode, bundelvanger
d Auffängerelektrode *f*, Strahlverschlucker *m*

477 BEAM WIDENING
f élargissement *m* du faisceau
e ensanchamiento *m* del haz
i allargamento *m* del fascio

bundelverbreding
Strahlverbreiterung *f*

EAM WIDTH (GB)
see: APERTURE OF THE BEAM

8 BEAT PATTERN, dis
INTERFERENCE FIGURE,
INTERFERENCE PATTERN
The resulting space distribution when progressive waves of the same frequency and kind are superimposed.
figure *f* d'interférence
figura *f* de interferencia
figura *f* d'interferenza,
immagine *f* d'interferenza
interferogram *n*,
storingspatroon *n*
Interferenzbild *n*

79 BEATING OSCILLATOR, ge
LOCAL OSCILLATOR
Oscillator in a superheterodyne circuit which beats with the carrier frequency of the received signal to produce a constant IF carrier.
oscillateur *m* local
oscilador *m* local
oscillatore *m* locale
lokale oscillator
Überlagerungsoszillator *m*

80 BELT SCANNER, tv
DRUM SCANNER
One of the earlier mechanical scanning devices.
analyseur *m* à ruban,
analyseur *m* à tambour
analizador *m* de cinta,
explorador *m* de tambor
analizzatore *m* a tamburo
trommelaftaster
Trommelabtaster *m*

81 BENDING crt
In TV picture tubes, distortion of the vertical straight lines into curved lines.
courbure *f*, flexion *f*
curvatura *f*, flexión *f*
curvatura *f*, flessione *f*
afbuiging, kromming, verbuiging
Abbiegung *f*, Krümmung *f*, Verbiegung *f*

482 BENT-GUN ION TRAP crt
An ion trap that consists of a bend in the electron gun of a cathode-ray tube.
piège *m* d'ions en courbure du canon électronique
e trampa *f* de iones en curvatura del cañón electrónico
i trappola *f* ionica in curvatura del cannone elettronico
n ionenvangende bocht in elektronenkanon
d ionenfangende Krümmung *f* in Elektronenkanone

483 BIAS cpl
f polarisation *f*
e polarización *f*
i polarizzazione *f*
n voorspanning
d Vorspannung *f*

484 BIAS vr
f prémagnétisation *f*
e premagnetización *f*
i premagnetizzazione *f*
n voormagnetisatie
d Vormagnetisierung *f*

485 BIAS BOX tv
f appareil *m* de la tension de polarisation
e aparato *m* de la tensión de polarización
i apparecchio *m* della tensione di polarizzazione
n voorspanningsapparaat *n*
d Vorspannungsgerät *n*

486 BIAS FREQUENCY rec/vr
Supersonic frequency of the bias current applied to the recording head of a magnetic recorder to improve the linearity of the recording characteristic.
f fréquence *f* de polarisation
e frecuencia *f* de polarización
i frequenza *f* di polarizzazione
n voormagnetisatiefrequentie
d Vormagnetisierungsfrequenz *f*

BIAS LIGHTING
see: BACK BIAS

487 BIASING vr
Polarization of a recording head in magnetic-tape recording, using d.c. or a.c. much higher than the maximum audio frequency to be reproduced.
f polarisation *f*,
prémagnétisation *f*
e polarización *f*,
premagnetización *f*
i polarizzazione *f*,
premagnetizzazione *f*
n voormagnetisatie
d Vormagnetisierung *f*

488 BIASING CONTROL VOLTAGE vr
f tension *f* de commande de polarisation
e tensión *f* de mando de polarización
i tensione *f* di comando di polarizzazione
n voormagnetisatiestuurspanning
d Vormagnetisierungssteuerspannung *f*

489 BIASING CURRENT vr
f courant *m* de polarisation,
courant *m* de prémagnétisation
e corriente *f* de polarización,
corriente *f* de premagnetización
i corrente *f* di polarizzazione,
corrente *f* di premagnetizzazione
n voormagnetisatiestroom
d Vormagnetisierungsstrom *m*

490 BIG CLOSE SHOT, opt
BIG-CLOSE-UP
f très gros plan

e foto *m* muy de cerca
i primissimo piano
n detailopname
d ganz gross, ganz nah

491 BINDER crt
A material used to bind the particles of a fluorescent material before applying it to the face of the cathode-ray tube.
f liant *m*
e adhesivo *m*
i legante *m*
n bindmiddel *n*
d Bindemittel *n*

492 BIPOST LAMP rec
Studio spotlamp in which the connections are brought out to two posts on the base of the bulb.
f lampe *f* à deux bornes
e lámpara *f* de dos bornes
i lampada *f* a due piedini di contatto
n lamp met twee aansluitpennen
d Lampe *f* mit zwei Kontaktstiften

493 BISTABLE CIRCUIT rec/tv
A circuit which has two stable states which can be decided by input signals.
f circuit *m* bistable
e circuito *m* biestable
i circuito *m* bistabile
n bistabiele kring
d bistabiler Kreis *m*

494 BLACK ct
f noir adj, *m*
e negro adj, *m*
i nero adj, *m*
n zwart adj, *n*
d schwarz adj, Schwarz *n*

495 BLACK AFTER WHITE rep
A black line defect which follows a white object on a TV picture tube.
f noir après blanc
e negro detrás de blanco
i nero dopo bianco
n zwart na wit
d schwarz hinter weiss

496 BLACK-AND-WHITE CHANNEL, MONOCHROME CHANNEL tv
f canal *m* pour monochrome, canal *m* pour noir et blanc
e canal *m* para blanco y negro, canal *m* para monocromo
i canale *m* per bianco e nero, canale *m* per monocromo
n monochroomkanaal *n*, zwart-witkanaal *n*
d Monochromkanal *m*, Schwarz-Weiss-Kanal *m*

497 BLACK AND WHITE PHOSPHOR crt
f substance *f* luminescente pour télévision en noir et blanc
e substancia *f* luminiscente para televisión en blanco y negro
i sostanza *f* luminescente per televisione in bianco e nero
n fosfor *n* voor zwart-wittelevisie, luminescerende stof voor zwart-wittelevisie
d Leuchtstoff *m* für Schwarzweiss-Fernse[hen], Phosphor *m* für Schwarzweiss-Fernseh[en]

498 BLACK-AND-WHITE PICTURE, MONOCHROME PICTURE
f image *f* monochrome, image *f* noire et blanche
e imagen *f* blanca y negra, imagen *f* monocroma
i immagine *f* bianca e nera, immagine *f* monocroma
n monochroom beeld *n*, zwart-witbeeld *n*
d Schwarz-Weiss-Bild *n*

499 BLACK-AND-WHITE TELEVISION, MONOCHROME TELEVISION
TV in which only the luminance of a sce[ne] is reproduced and not the colo(u)r.
f télévision *f* en noir et blanc, télévision *f* monochrome
e televisión *f* en blanco y negro, televisión *f* monocroma
i televisione *f* in bianco e nero, televisione *f* monocroma
n monochroomtelevisie, zwart-wittelevisie
d Monochromfernsehen *n*, Schwarz-Weiss-Fernsehen *n*

500 BLACK-AND-WHITE TRANSMISSION, MONOCHROME TRANSMISSION t[v]
f transmission *f* en noir et blanc, transmission *f* monochrome
e transmisión *f* en blanco y negro, transmisión *f* monocroma
i trasmissione *f* in bianco e nero, trasmissione *f* monocroma
n monochroomtransmissie, zwart-wittransmissie
d Monochromübertragung *f*, Schwarz-Weiss-Übertragung *f*

501 BLACK BODY, FULL BODY c
f corps *m* noir
e cuerpo *m* negro
i corpo *m* nero
n zwart lichaam *n*, zwartstraler
d Planckscher Strahler *m*, schwarzer Körper *m*

502 BLACK-BODY RADIATION, FULL-BODY RADIATION ip[c]
The thermal radiation from a black body at a given temperature.
f rayonnement *m* des corps noirs
e radiación *f* de los cuerpos negros
i radiazione *f* dei corpi neri
n zwarte straling
d Hohlraumstrahlung *f*,

schwarze Strahlung *f*,
Schwarzkörperstrahlung *f*

503 BLACK CLIPPER cpl
Provided to prevent unwanted signal excursions below peak-black.
f écrêteur *m* du noir,
limiteur *m* du noir
e limitador *m* del negro,
recortador *m* del negro
i limitatore *m* del nero,
limitatore *m* del valore di soglia del nero
n drempelwaardebegrenzer van het zwart
d Schwellenwertbegrenzer *m* für Schwarz

504 BLACK CLIPPING tv
f limitation *f* du noir
e limitación *f* del negro
i limitazione *f* del nero
n zwartbegrenzing
d Schwarzwertbegrenzung *f*

505 BLACK COMPRESSION rep
The reduction in gain applied to a TV picture signal at those levels corresponding to dark areas in a picture with respect to the gain at that level corresponding to the mid-range light value in the picture.
f écrasement *m* des noirs,
tassement *m* des noirs
e compresión *f* del negro
i compressione *f* del nero
n compressie in het zwart
d Schwarzkompression *f*

506 BLACK CONTENT ct
The subjectively estimated amount of blackness seen in the visual sensation arising from a surface colo(u)r.
f contenu *m* en noir
e contenido *m* en negro
i contenuto *m* in nero
n zwartgehalte *n*
d Schwarzanteil *m*

507 BLACK CRUSHING dis/rep/tv
Compression of low values of signals, i.e. black, resulting in loss of significant detail in the darker picture areas.
f noircissement *m* des gris
e ennegrecimiento *m* de los grises
i annerimento *m* dei grigi
n zwarten *n* van het grijs
d Schattenschwärzung *f*

508 BLACK-FACE TUBE crt
A type of TV picture tube employing unilluminated phosphors of a dark grey shade, used to reduce reflections from light sources.
f tube *m* image à filtre optique
e cinescopio *m* con filtro óptico
i cinescopio *m* con filtro ottico,
tubo *m* di ricezione con filtro ottico
n beeldbuis met grijsfilter
d Bildröhre *f* mit Graufilter

509 BLACK INSTABILITY tv
f instabilité *f* du noir
e inestabilidad *f* del negro
i instabilità *f* del nero
n onstabiliteit van het zwart
d Schwarzinstabilität *f*

510 BLACK LEVEL, tv
Y LEVEL
In positive transmission, the minimum permissible level of the effective picture signal.
f niveau *m* du noir
e nivel *m* del negro
i livello *m* del nero
n zwartniveau *n*
d Schwarzpegel *m*, Schwarzwert *m*

511 BLACK-LEVEL ALIGNMENT tv
Restoration of the d.c. component by maintaining the blanking level at a reference value.
f alignement *m* du niveau du noir,
calage *m* du niveau du noir
e ajuste *m* del nivel del negro
i allineamento *m* del livello del nero
n herstel *n* van de nulcomponent met onderdrukkingsniveaureferentie
d Schwarzpegelsteuerung *f*,
Schwarzwertsteuerung *f*

512 BLACK-LEVEL CLAMPING cpl/rec/tv
Alignment of the black level by adjustment of the signal to its reference value at a given instant during the line-blanking interval.
f clamping *m* sur le niveau du noir,
verrouillage *m* du niveau du noir
e fijación *f* del nivel del negro
i allineamento *m* al livello del nero
n vergrendeling van het zwartniveau,
zwartniveauklem
d Schwarzpegelklemmung *f*,
Schwarzschulterklemmung *f*,
Schwarzwertklemmung *f*

513 BLACK-LEVEL CONTROL rep/tv
A control on a TV receiver for varying the amplitude of that portion of the video signal which produces the black part of the picture.
f réglage *m* du niveau du noir
e regulación *f* del nivel del negro
i regolazione *f* del livello del nero
n zwartniveauregeling
d Schwarzpegelregelung *f*,
Schwarzwertregelung *f*

514 BLACK-LEVEL PERIOD tv
f période *f* du niveau du noir
e período *m* del nivel del negro
i periodo *m* del livello del nero
n zwartniveauperiode
d Schwarzpegelperiode *f*,
Schwarzwertperiode *f*

515 BLACK-LEVEL SHIFT tv
A video or picture signal fault due to an incorrectly transmitted or reconstituted d.c. component.
f défoncement *m* du niveau du noir
e desplazamiento *m* del nivel del negro
i slittamento *m* del livello del nero
n zwartniveauverschuiving
d Überlagerung *f* von Stossspannungen

516 BLACK LIFT, tv
LIFT,
SET-UP
The ratio between reference black level and reference white level both measured from blanking level.
f décollement *m* du noir
e sobreelevación *f* del negro
i alzata *f* del nero
n zwart-witverhouding
d Schwarzabhebung *f*

517 BLACK-LIFT CONTROL, dis
LIFT CONTROL,
SET-UP CONTROL
f réglage *m* du décollement du noir
e regulación *f* de la sobreelevación del negro
i regolazione *f* dell'alzata del nero
n regeling van de zwart-witverhouding
d Schwarzabhebungsregelung *f*

518 BLACK-LIFT LEVEL, dis/tv
LIFT LEVEL,
SET-UP LEVEL
f niveau *m* du décollement du noir
e nivel *m* de la sobreelevación del negro
i livello *m* dell'alzata del nero
n niveau *n* van de zwart-witverhouding
d Schwarzabhebungspegel *m*,
Schwarzabhebungswert *m*

519 BLACK NEGATIVE tv
TV picture signals in which the voltage corresponding to black is negative in relation to the voltage corresponding to white.
f tension *f* négative pour le noir
e tensión *f* negativa para el negro
i tensione *f* negativa per il nero
n negatieve spanning voor zwart
d Negativspannung *f* für Schwarz

520 BLACK-OUT (US), rec/tv
BLANKING (GB)
f suppression *f*
e supresión *f*
i cancellazione *f*, soppressione *f*
n onderdrukking
d Austastung *f*

521 BLACK-OUT AMPLIFIER, tv
BLANKING AMPLIFIER
f amplificateur *m* de suppression
e amplificador *m* de borrado,
amplificador *m* de supresión
i amplificatore *m* di cancellazione,
amplificatore *m* di soppressione
n onderdrukkingsversterker
d Austastverstärker *m*

522 BLACK-OUT AND SYNC LEVEL STABILIZATION, tv
BLANKING AND SYNC LEVEL STABILIZATION
f stabilisation *f* du niveau
e estabilización *f* del nivel
i stabilizzazione *f* del livello
n niveaustabilisatie
d Pegelhaltung *f*

523 BLACK-OUT AND SYNC SIGNAL, tv
BLANKING AND SYNC SIGNAL
f signal *m* de suppression et de synchronisation
e señal de borrado y de sincronización,
señal *f* de supresión y de sincronización
i segnale *m* di cancellazione e di sincronizzazione,
segnale *m* di soppressione e di sincronizzazione
n onderdrukkings- en synchronisatiesignaal *n*
d AS-Signal *n*,
Austast-Synchronsignal *n*

524 BLACK-OUT CIRCUIT, tv
BLANKING CIRCUIT
f circuit *m* de suppression
e circuito *m* de borrado,
circuito *m* de supresión
i circuito *m* di cancellazione,
circuito *m* di soppressione
n onderdrukkingscircuit *n*
d Austastkreis *m*

525 BLACK-OUT INTERVALS, tv
BLANKING INTERVALS
f intervalles *pl* de suppression
e intervalos *pl* de borrado,
intervalos *pl* de supresión
i intervalli *pl* di cancellazione,
intervalli *pl* di soppressione
n onderdrukkingspauzes *pl*
d Austastlücken *pl*

526 BLACK-OUT LEVEL, tv
BLANKING LEVEL,
PEDESTAL LEVEL
The level in a composite TV picture signal that separates the range of the composite picture signal containing the picture information from the range containing synchronizing information.
f niveau *m* de suppression
e nivel *m* de borrado,
nivel *m* de supresión
i livello *m* di cancellazione,
livello *m* di soppressione
n onderdrukkingsniveau *n*
d Austastpegel *m*,
Austastwert *m*

527 BLACK-OUT OF VIDEO SIGNAL, tv
BLANKING OF VIDEO SIGNAL,
VIDEO SIGNAL BLACK-OUT
f suppression *f* du signal vidéo

e borrado *m* de la señal video,
supresión *f* de la señal video
i cancellazione *f* del segnale video,
soppressione *f* del segnale video
n onderdrukking van het beeldsignaal,
onderdrukking van het videosignaal
d Austastung *f* des Bildsignals

528 BLACK-OUT PULSE, tv
BLANKING PEDESTAL,
BLANKING PULSE
One of the pulses that make up the blanking signal.
f impulsion *f* de suppression
e impulso *m* de borrado,
impulso *m* de supresión
i impulso *m* di cancellazione,
impulso *m* di soppressione
n onderdrukkingsimpuls
d Austastimpuls *m*

529 BLACK-OUT PULSE INSERTION, tv
BLANKING PULSE INSERTION
f insertion *f* du signal de suppression
e inserción *f* de la señal de borrado
i inserzione *f* del segnale di cancellazione
n invoeging van het onderdrukkingssignaal
d Einfügung *f* des Austastsignals

530 BLACK-OUT SIGNAL, tv
BLANKING SIGNAL
A wave constituted of recurrent pulses, related in time to the scanning process, used to effect blanking.
f signal *m* de suppression
e señal *f* de borrado,
señal *f* de supresión
i segnale *m* di cancellazione,
segnale *m* di soppressione
n onderdrukkingssignaal *n*
d Austastsignal *n*

531 BLACK-OUT VOLTAGE, tv
BLANKING VOLTAGE
f tension *f* de suppression
e tensión *f* de borrado,
tensión *f* de supresión
i tensione *f* di cancellazione,
tensione *f* di soppressione
n onderdrukkingsspanning
d Austastspannung *f*

532 BLACK PEAK (US), tv
PEAK BLACK (GB)
A peak excursion of the TV picture signal in the black direction.
f crête *f* du noir
e cresta *f* del negro
i cresta *f* del nero
n diepste zwart *n* in het signaal,
maximaal zwart *n*
d Maximum *n* an Schwarz

533 BLACK PORCH tv
f palier *m* du noir
e rellano *m* del negro
i gradino *m* del nero
n zwartstoep
d Schwarzschulter *f*

534 BLACK POSITIVE tv
TV picture signal in which the voltage corresponding to black is positive in relation to the voltage corresponding to white.
f tension *f* positive pour le noir
e tensión *f* positiva para el negro
i tensione *f* positiva per il nero
n positieve spanning voor zwart
d Positivspannung *f* für Schwarz

535 BLACK REFERENCE LEVEL rec
The TV picture signal level corresponding to a specified maximum limit for black peaks.
f niveau *m* de référence pour le noir
e nivel *m* de referencia para el negro
i livello *m* di riferimento per il nero
n referentieniveau *n* voor zwart
d Bezugspegel *m* für Schwarz

536 BLACK SATURATION tv
f saturation *f* du noir
e saturación *f* del negro
i saturazione *f* del nero
n verzadiging in het zwart
d Schwarzsättigung *f*

537 BLACK SCREEN rec/tv
A plastic filter interposed between a TV picture and the viewer to reduce reflected light and so increase contrast.
f filtre *m* optique plastique
e filtro *m* óptico plástico
i filtro *m* ottico plastico
n lichtfilter *n* uit kunststof
d Grauscheibe *f* aus Kunststoff

538 BLACK SCREEN TELEVISION SET rep
System whereby i.a. the contrast of the picture image on the screen is increased by mounting in front of the cathode-ray tube a black or grey glass or plastic filter which reduces the light reflected by the tube itself.
f téléviseur *m* à filtre optique
e televisor *m* de filtro óptico
i televisore *m* a filtro ottico
n televisietoestel *n* met lichtfilter
d Fernseher *m* mit Grauscheibe

539 BLACK SHADED dis
Process of correcting black shading by introducing a signal the inverse of the error component.
f à compensation de voile
e con compensación de velo
i a compensazione di velo
n voor sluier gecompenseerd
d mit Schleierkompensation

540 BLACK SHADING dis
f voile *m*
e velo *m*
i velo *m*
n sluier
d Schleier *m*

541 BLACK SPOTTER, dis
INTERFERENCE INVERTER
A special diode, inserted in the TV circuit to counteract interference of short duration.
f diode *f* antibrouilleuse
e diodo *m* antiparasitario,
diodo *m* antiparásito
i diodo *m* antiparassita
n ontstoordiode, storingsompoler
d Entstördiode *f*

542 BLACK SPOTTING, dis
INTERFERENCE INVERSION
A method of counteracting interference by reversing the drive to the grid of the cathode-ray tube during a large pulse of white interference.
f inversion *f* d'interférence
e inversión *f* de interferencia
i inversione *f* d'interferenza
n storingsinversie
d Störungsinversion *f*

543 BLACK STRETCH rep/tv
f étalement *m* des noirs
e expansión *f* del negro
i espansione *f* del nero
n vergroting van het zwartgedeelte van het signaal,
zwartexpansie
d Schwarzdehnung *f*

544 BLACK-TO-WHITE AMPLITUDE RANGE, cpl
WHITE-TO-BLACK AMPLITUDE RANGE
f amplitude *f* totale blanc-noir,
amplitude *f* totale noir-blanc
e rango *m* de amplitud blanco a negro,
rango *m* de amplitud negro a blanco
i ampiezza *f* totale bianco-nero,
ampiezza *f* totale nero-bianco
n wit-zwartspan,
zwart-witspan
d Schwingungsbreite *f* Schwarz-Weiss,
Schwingungsbreite *f* Weiss-Schwarz

545 BLACK-TO-WHITE FREQUENCY SWING, cpl
WHITE-TO-BLACK FREQUENCY SWING
f excursion *f* de fréquence blanc-noir,
excursion *f* de fréquence noir-blanc
e excursión *f* de frecuencia blanco-negro,
excursión *f* de frecuencia negro-blanco
i deviazione *f* di frequenza bianco-nero,
deviazione *f* di frequenza nero-bianco
n frequentiezwaai voor wit-zwartspan,
frequentiezwaai voor zwart-witspan
d Frequenzhub *m* Schwarz-Weiss,
Frequenzhub *m* Weiss-Schwarz

546 BLACK-TO-WHITE TRANSITION, tv
BRIGHTNESS TRANSITION
f transition *f* de la luminosité,
transition *f* noir-blanc,
variation *f* de la luminosité
e transición *f* de la luminosidad,
transición *f* negro-blanco
i transizione *f* della luminosità,
transizione *f* nero-bianco
n helderheidsovergang,
zwart-witovergang
d Helligkeitsübergang *m*,
Schwarz-Weiss-Sprung *m*

547 BLACK-WHITE CONTROL t
f commande *m* clair-obscur
e mando *m* claro-obscuro
i comando *m* chiaro-oscuro
n helder-donkersturing
d Hell-Dunkelsteuerung *f*

548 BLACK-WHITE MONITORING t
f contrôle *m* noir-blanc
e control *m* blanco-negro
i controllo *m* bianco-nero
n zwart-witcontrole
d Schwarz-Weisskontrolle *f*

549 BLACKER-THAN-BLACK, rep/tv
INFRABLACK
f infranoir adj,
ultranoir adj
e infranegro adj,
ultranegro adj
i infranero adj,
ultranero adj
n infrazwart adj,
ultrazwart adj,
zwarter dan zwart
d schwärzer als schwarz,
ultraschwarz adj

550 BLACKER-THAN-BLACK LEVEL, rep
INFRABLACK LEVEL
In TV, a level of greater instantaneous amplitude than the black level, used for sync and control signals.
f niveau *m* de l'ultranoir
e nivel *m* del ultranegro
i livello *m* dell'infranero
n ultrazwartniveau *n*
d Ultraschwarzpegel *m*

551 BLACKER-THAN-BLACK REGION, rep
INFRABLACK REGION
That part of the video signal adjoining the black level, used for sync signals.
f zone *f* de l'ultranoir
e región *f* del ultranegro
i zona *f* dell'infranero
n ultrazwartgebied *n*
d Ultraschwarzgebiet *n*

552 BLANKED PICTURE SIGNAL, cpl/rep
BLANKED VIDEO SIGNAL
The signal resulting from blanking a picture (video) signal.
f signal *m* vidéo muni du signal de suppression
e señal *f* video borrada,
señal *f* video suprimada
i segnale *m* video cancellato,
segnale *m* video soppresso
n onderdrukt beeldsignaal *n*,
onderdrukt videosignaal *n*

d ausgetastetes Bildsignal *n*,
ausgetastetes Videosignal *n*,
Bildsignal *n* mit Austastung

BLANKING (GB)
see: BLACK-OUT

BLANKING AMPLIFIER
see: BLACK-OUT AMPLIFIER

BLANKING AND SYNC LEVEL STABILIZATION
see: BLACK-OUT AND SYNC LEVEL STABILIZATION

BLANKING AND SYNC SIGNAL
see: BLACK-OUT AND SYNC SIGNAL

BLANKING CIRCUIT
see: BLACK-OUT CIRCUIT

BLANKING INTERVALS
see: BLACK-OUT INTERVALS

BLANKING LEVEL
see: BLACK-OUT LEVEL

BLANKING OF VIDEO SIGNAL
see: BLACK-OUT OF VIDEO SIGNAL

BLANKING PEDESTAL,
BLANKING PULSE
see: BLACK-OUT PULSE

BLANKING PULSE INSERTION
see: BLACK-OUT PULSE INSERTION

BLANKING SIGNAL
see: BLACK-OUT SIGNAL

BLANKING VOLTAGE
see: BLACK-OUT VOLTAGE

553 BLEARY opt/tv
f confus adj
e confuso adj
i confuso adj
n diffuus adj, vaag adj
d verwaschen adj

554 BLEMISH dis/rec
A mosaic imperfection which affects the transmission of an image in a camera tube.
f défaut *m* de la mosaïque,
tache *f* de la mosaïque
e falta *f* del mosaico,
mancha *f* del mosaico
i difetto *m* del mosaico,
macchia *f* del mosaico
n mozaïekfout,
mozaïekvlek
d Mosaikfehler *m*,
Mosaikfleck *m*

555 BLENDING OF SOUNDS aud
f fusion *f* de sons,
mixage *m* de sons
e fusión *f* de sonidos,
mezcladura *f* de sonidos
i fusione *f* di suoni,
mescolanza *f* di suoni
n geluidsvermenging,
klankvermenging
d Klangverschmelzung *f*

556 BLIMP aud
f caisson *m* insonore
e blindaje *m* insonoro
i cuffia *f* afonica
n geluiddichte omhulling
d schalldichte Hülle *f*,
Schallschutzhaube *f*

557 BLIND AREA, rep/tv
SHADOW REGION
Region within the normal service range of a TV transmitter where the field strength is reduced below the useful level by local obstructions in the path.
f zone *f* de silence
e zona *f* de sombra
i zona *f* d'ombra
n schaduwgebied *n*
d Schattengebiet *n*

558 BLIND MONITORING rec
Control of microphone outputs in broadcasting when the operator cannot see the performers of the transmission.
f contrôle *m* du volume des microphones positionnés hors du champ de vision
e control *m* sin visión de la escena
i controllo *m* dei segnali microfonici con gli esecutori fuori del campo d'osservazione
n volumecontrole van buiten het gezichtsveld opgestelde microfonen
d Mikrophonausregelung *f* wenn die Mikrophone nicht im Sichtfeld sind

559 BLOCK ANTENNA, aea/tv
COLLECTIVE ANTENNA,
COMMON AERIAL,
COMMUNITY ANTENNA
Aerial (antenna) system to be used by a number of persons living in the same block of houses, etc.
f antenne *f* collective,
antenne *f* commune
e antena *f* colectiva
i antenna *f* collettiva
n blokantenne,
gemeenschappelijke antenne
d Gemeinschaftsantenne *f*

560 BLOCK LEVEL tv
The TV signal level which is constant at or closely approaching pedestal level.
f niveau *m* constant du signal
e nivel *m* constante de la señal
i livello *m* costante del segnale
n constant signaalniveau *n*
d konstanter Signalpegel *m*

561 BLOCKING BIAS crt
That voltage of a beam system of a cathode-ray tube at which the beam current is exactly zero.

f tension *f* de blocage
e tensión *f* de bloqueo
i tensione *f* di bloccaggio
n afknijpspanning, blokkeerspanning
d Sperrspannung *f*

562 BLOCKING OSCILLATOR cpl/tv
A circuit using a single tube (valve) with positive feedback through a transformer.
f oscillateur *m* de blocage, oscillateur *m* surcouplé
e oscilador *m* autoamortiguado, oscilador *m* de bloqueo
i oscillatore *m* bloccato periodicamente, oscillatore *m* di rilassamento a bloccaggio
n blokkeeroscillator
d Sperrschwinger *m*

563 BLOCKING SIGNAL tv
Basic signal interrupting momentaneously the scanning process in order to suppress the scanning spot during specified time intervals.
f signal *m* de blocage
e señal *f* de bloqueo
i segnale *m* di bloccaggio
n afknijpsignaal *n*, blokkeersignaal *n*
d Sperrsignal *n*

564 BLOOM, BLOOMING rep/tv
A defect in a TV receiver which causes the picture to lose brightness and at the same time to expand like a flower blooming.
f étouffement *m*, flou *m* d'image, hyperluminosité *f* de l'image
e expansión *f* excesiva de la imagen, hiperfluorescencia *f* de la imagen
i sfuocatura *f*, sopraluminosità *f* dell'immagine
n bloemig beeld *n*
d Bildweichheit *f*, Überstrahlung *f*

565 BLOOMED LENS, COATED LENS opt
f objectif *m* traité
e objetivo *m* con revestimiento antirreflectivo
i obiettivo *m* trattato
n gekote lens, reflectievrij objectief *n*
d vergütetes Objektiv *n*

566 BLOOP, SPLICE BUMP fi/vr
f bruit *m* de collage
e ruido *m* de empalme
i rumore *m* d'incollatura
n lasruis
d Klebstellengeräusch *n*

567 BLUE ct
A colo(u)r in which the spectral components are confined mainly to the 400-500 nm band of the spectrum.
f bleu *m*
e azul *m*
i blu *m*
n blauw *n*
d Blau *n*

568 BLUE ADDER ctv
f circuit *m* mélangeur pour le bleu
e circuito *m* aditivo para el azul
i circuito *m* combinatore per il blu
n mengkring voor blauw, optelschakeling voor blauw,
d Blau-Beimischer *m*

569 BLUE APEX ctv
In a chromaticity diagram or a Maxwell colo(u)r triangle, the apex corresponding to the blue primary.
f point *m* de couleur de la primaire bleu
e punto *m* de color del primario azul
i punto *m* di colore del primario blu
n kleurpunt *n* van de blauwe primaire kleur
d Farbort *m* der blauen Primärfarbe

570 BLUE BEAM crt
f faisceau *m* pour le bleu
e haz *m* para el azul
i fascio *m* per il blu
n bundel voor blauw
d Strahl *m* für Blau

571 BLUE-BEAM MAGNET crt/ctv
A small permanent magnet used as a convergence adjustment to change the direction of the electron beam for blue phosphor dots in a three-gun colo(u)r TV picture tube.
f aimant *m* du faisceau pour le bleu
e imãn *m* del haz para el azul
i magnete *m* del fascio per il blu
n blauwe-bundelmagneet
d Blaustrahlmagnet *m*

BLUE BLACK LEVEL
see: B BLACK LEVEL

572 BLUE CATHODE crt
f cathode *f* bleue
e cátodo *m* azul
i catodo *m* blu
n blauwe katode
d blaue Katode *f*

BLUE COLO(U)R DIFFERENCE AXIS
see: B-Y AXIS

BLUE COLO(U)R DIFFERENCE MATRIX
see: B-Y MATRIX

BLUE COLO(U)R DIFFERENCE MODULATOR
see: B-Y MODULATOR

BLUE COLO(U)R DIFFERENCE SIGNAL
see: B-Y SIGNAL

573 BLUE CONVERGENCE CIRCUIT crt
f circuit *m* de convergence pour le bleu
e circuito *m* de convergencia para el azul

i circuito *m* di convergenza per il blu
n blauwe-convergentieschakeling
d Blaukonvergenzschaltung *f*

574 BLUE ELECTRON GUN, ctv
BLUE GUN
The electron gun whose beam strikes phosphor dots emitting the primary colo(u)r in a three-gun colo(u)r TV picture tube.
f canon *m* du bleu
e cañón *m* del azul
i cannone *m* del blu
n kanon *n* voor het blauw
d Blaustrahlsystem *n*

575 BLUE GAIN CONTROL crt/ctv
A variable resistor used in the matrix of a three-gun colo(u)r picture TV receiver to adjust the intensity of the blue primary signal.
f régulateur *m* d'intensité pour le bleu
e regulador *m* de intensidad para el azul
i regolatore *m* d'intensità per il blu
n intensiteitsregelaar voor blauw
d Intensitätsregler *m* für Blau

576 BLUE GRID crt
The grid of the blue electron gun in a multigun cathode-ray tube.
f grille *f* bleue
e rejilla *f* azul
i griglia *f* blu
n blauw rooster *n*
d blaues Gitter *n*

577 BLUE HIGHS crt/ctv
f hautes fréquences *pl* pour le bleu
e altas frecuencias *pl* para el azul
i componenti *pl* d'alta frequenza per il blu
n hoge frequenties *pl* voor blauw
d Blauhöhe *f*,
höhe Frequenzen *pl* für Blau

578 BLUE HORIZONTAL SHIFT MAGNET, crt
BLUE LATERAL SHIFT MAGNET,
BLUE POSITIONING MAGNET
f aimant *m* latéral pour le bleu
e imán *m* lateral para el azul
i magnete *m* laterale per il blu
n lateraalmagneet voor blauw
d Blauschiebemagnet *m*

579 BLUE LOWS crt/ctv
f basses fréquences *pl* pour le bleu
e bajas frecuencias *pl* para el azul
i componenti *pl* di bassa frequenza per il blu
n lage frequenties *pl* voor blauw
d Blautiefe *f*,
niedrige Frequenzen *pl* für Blau

580 BLUE MODULATOR crt
A modulator having the blue primary-colo(u)r signal as the base-band input signal.
f modulateur *m* pour le signal bleu
e modulador *m* para la señal azul
i modulatore *m* per il segnale blu
n blauwsignaalmodulator
d Blausignalmodulator *m*

581 BLUE OVERCAST, crt
BLUE SHADING
The gradual change in the black level of the blue component of a colo(u)r image or the signal giving rise to it.
f domination *f* du bleu
e dominación *f* del azul
i dominazione *f* del blu
n blauwzweem
d Stich *m* ins Blaue

BLUE PEAK LEVEL
see: B PEAK LEVEL

BLUE PRIMARY
see: B

582 BLUE PRIMARY INFORMATION ctv
f information *f* du bleu primaire
e información *f* del azul primario
i informazione *f* del blu primario
n informatie van de blauwe primaire kleur
d Information *f* der blauen Primärfarbe

583 BLUE PRIMARY SIGNAL crt/ctv
f signal *m* du bleu primaire
e señal *f* del azul primario
i segnale *m* del blu primario
n signaal *n* van de blauwe primaire kleur
d Signal *n* der blauen Primärfarbe

584 BLUE RESTORER crt/ctv
The d.c. restorer for the blue channel of a three-gun colo(u)r TV picture circuit.
f restauration *f* de la composante de courant continu du niveau du bleu
e restauración *f* de la componente de corriente continua del nivel del azul
i reinserzione *f* della componente di corrente continua del livello del blu
n herstel *n* van het blauwniveau
d Wiederherstellung *f* des Blaupegels

585 BLUE SATURATION SCALER, crt
BLUE SCALE
f échelle *f* de bleus
e escala *f* de azules
i scala *f* di blu
n blauwtrap
d Blauskala *f*

586 BLUE SCREEN-GRID crt
f grille-écran *m* bleu
e rejilla-pantalla *f* azul
i griglia-schermo *m* blu
n blauw schermrooster *n*
d blaues Schirmgitter *n*

587 BLUE SHIFT crt
f décalage *m* du canevas bleu
e desplazamiento *m* de la cuadrícula azul
i spostamento *m* del quadro rigato blu
n verschuiving van het blauwe raster
d Verschiebung *f* des blauen Rasters

588 BLUE VIDEO VOLTAGE crt/ctv
The signal voltage output from the blue section of a colo(u)r TV camera, or the signal voltage between the receiver matrix and the blue gun grid of a three-gun colo(u)r TV picture tube.
f tension *f* vidéo pour le bleu
e tensión *f* video para el azul
i tensione *f* video per il blu
n beeldspanning voor blauw
d Bildspannung *f* für Blau

589 BLURRING rep/tv
The reduction of the apparent sharpness of definition of objects in the reproduced scene.
f brouillard *m* du fond,
flou *m* d'image
e borrosidad *f*,
imagen *f* borrosa,
pérdida *f* aparente de definición
i brulichio *m*,
immagine *f* a scarsa definizione,
indefinitezza *f*,
scarsa definizione *f*
n wazigheid
d Unschärfe *f*,
Verschmierung *f*,
Verschwimmung *f*

590 BOARDS rec
Lamps for studio lighting, fixed horizontally and longer than they are high.
f herse *f* lumineuse
e caja *f* horizontal de lámparas
i cassa *f* orizzontale di lampade
n horizontale lampenbalk
d horizontale Lichtleiste *f*

591 BOMBARDMENT-INDUCED CONDUCTIVITY TYPE OF COLO(U)R PICKUP TUBE crt/ctv
f tube *m* image en couleurs à conductivité induite par bombardement d'électrons
e tubo *m* imagen de colores con conductividad inducida por bombardeo de electrones
i tubo *m* di ripresa a colori con conduttività indotta per bombardamento d'elettroni
n kleurenbeeldbuis met door elektronenbombardement geïnduceerde geleidendheid
d Farbfernsehröhre *f* mit durch Elektronenbombardement induzierter Leitfähigkeit

592 BOOM rec
Mechanical arrangement for swinging the microphone clear of artists and cameras in sound film and TV studios.
f girafe *f*, perche *f*
e árbol *m*, jirafa *f*, pértiga *f*
i giraffa *f*, pertica *f*
n hengel
d Galgen *m*, Giraffe *f*, Schwenkarm *m*

593 BOOM-MOUNTED MICROPHONE rec
f microphone *m* monté sur girafe
e micrófono *m* montado su árbol
i microfono *m* montato su giraffa
n hengelmicrofoon
d Galgenmikrophon *n*,
Schwenkarmmikrophon *n*

594 BOOM OPERATOR rec
f opérateur *m* de girafe
e operador *m* de árbol
i operatore *m* di giraffa
n hengelaar
d Schwenkarmbediener *m*

595 BOOMINESS aud/rec/tv
f son *m* de tonneau
e sonido *m* hueco
i rimbombo *m*
n dreun
d Dröhn *m*

596 BOOSTER CIRCUIT, ENERGY REGENERATION, POWER FEEDBACK cpl/tv
f récupération *f* d'énergie
e recuperación *f* de energía
i ricupero *m* d'energia
n energieterugwinning
d Energierückgewinnung *f*

597 BOOSTER DIODE, EFFICIENCY DIODE, ENERGY RECOVERY DIODE, SERIES-EFFICIENCY DIODE cpl
A diode used in a deflection generator to return to the amplifier tube (valve), through a capacitor, the energy from the deflection field, so that the anode of the tube (valve) is virtually fed from an external source in series with an internal source.
f diode *f* de récupération,
diode *f* élévatrice
e diodo *m* de ganancia en serie,
diodo *m* reforzador
i diodo *m* di guadagno,
diodo *m* di ricupero
n spaardiode
d Boosterdiode *f*,
Schalterdiode *f*,
Spardiode *f*

598 BOOSTER LIGHT stu
f lumière *f* d'appoint
e luz *f* de refuerzo
i luce *f* di rinforzo
n aanvullingslicht *n*
d Zusatzlicht *n*

599 BOOSTER STATION rep/tv
A station which rebroadcasts a transmission received directly on the same wavelength.
f station-relais *m* de diffusion
e estación *f* retransmisa
i stazione *f* relè,
stazione *f* ripetitrice
n relaisstation *n*
d Relaisstation *f*,
Relaisstelle *f*

600 BOOSTER TENSION, BOOSTER VOLTAGE cpl/tv
Additional voltage supplied to the damper tube (valve) to the horizontal output, and oscillator tubes (valves) and to the vertical output tubes (valves) to attain a higher sawtooth current.
f tension *f* de récupération, tension *f* élévatrice
e tensión *f* de ganancia, tensión *f* reforzadora
i tensione *f* di guadagno, tensione *f* di ricupero
n boostspanning
d Boosterspannung *f*, Zusatzspannung *f*

601 BOOTSTRAP cpl
Circuit in which the output is directly connected to the input as where the output load is connected between the cathode and negative high tension and where the input is applied between cathode and grid.
f circuit *m* autoélévateur
e circuito *m* autoelevador
i circuito *m* autoelevatore
n diabolocircuit *n*, laadspanningsregeling
d Bootstrapschaltung *f*, Schaltung *f* mit mitlaufender Ladespannung

602 BOTH-WAY BROADBAND CHANNEL cpl/tv
Channel capable of carrying broadband signals in both directions at the same time.
f canal *m* à large bande à trafic simultané dans deux directions
e canal *m* de banda ancha con tráfico simultáneo en dos direcciones
i canale *m* di banda larga con traffico simultaneo in due direzioni
n breedbandkanaal *n* met gelijktijdig verkeer in beide richtingen
d Breitbandkanal *m* mit gleichzeitigem Verkehr in beiden Richtungen

603 BOUNCING (US), JUMPING (US), VERTICAL HUNTING rep/tv
Up and down movement of the picture due to faulty synchronization.
f instabilité *f* verticale de l'image, saut *m* de l'image, sautillement *m* vertical
e efecto *m* de bailoteo vertical
i instabilità *f* verticale del quadro, saltello *m* verticale, scatti *pl* verticali del quadro
n danseffect *n*, springen *n*, verticale beweging van het beeld
d Pumpen *n*, Tanzeffekt *m*

604 BOUNDARY CONTRAST, CONTRAST THRESHOLD, LUMINANCE DIFFERENCE THRESHOLD ct
Smallest difference of luminance perceptible.
f valeur *f* de seuil de luminance
e valor *m* de umbral de luminancia
i valore *m* di soglia di luminanza
n drempelcontrast *n*, luminantiedrempelwaarde
d Helligkeitsschwellenwert *m*

605 BRAUN TUBE crt
f tube *m* de Braun
e tubo *m* de Braun
i tubo *m* di Braun
n braunbuis
d Braunsche Röhre *f*

606 BREAK tv
An accidental interruption of a broadcast program(me).
f coupure *f*, panne *f*
e avería *f*, corte *m*
i interruzione *f* per avaria, interruzione *f* per guasto
n onderbreking
d Unterbrechung *f*

607 BREAK-BEFORE-MAKE SWITCH, LAPPING SWITCH cpl
f commutateur *m* de séquence repos-travail
e conmutador *m* de secuencia reposo-trabajo
i commutatore *m* riposo-lavoro ad azionamento successivo
n verbreek-voor-maak-schakelaar
d Folge-Ruhe-Arbeit-Schalter *m*

608 BREAK-BEFORE-MAKE SWITCHING, LAP SWITCHING aud/rec/rep/tv
f commutation *f* de séquence repos-travail
e conmutación *f* de secuencia reposo-trabajo
i commutazione *f* riposo-lavoro ad azionamento successivo
n verbreek-voor-maak-schakeling
d Folge-Ruhe-Arbeit-Schaltung *f*

609 BREAK POINT cpl
An abrupt change in a gamma correction circuit.
f point *m* de flexion
e punto *m* de flexión
i punto *m* di flessione
n knikpunt *n*
d Knickpunkt *m*

610 BREEZEWAY (US) ctv
The time interval between the trailing edge of the horizontal sync pulse and the start of the colo(u)r burst.
f palier *m* intermédiaire
e rellano *m* intermedio
i pianerottolo *m* intermedio
n tussenstoep
d Zwischenschulter *f*

611 BRIGHT STUDIO RESPONSE stu
f résonance *f* claire du studio
e resonancia *f* clara en el estudio

i risonanza *f* chiara nello studio
n heldere studioresonantie
d helle Studioresonanz *f*

612 BRIGHTNESS ct
That colo(u)r quality, a decrease in which is associated with the residual degradation which would result from the addition of a small quantity of neutral grey to the colo(u)ring material when the strength of the mixture has been readjusted to the original strength.
f vivacité *f*
e vivacidad *f*
i vivacità *f*
n levendigheid
d Helligkeit *f*

613 BRIGHTNESS tv
f luminosité *f*
e luminosidad *f*
i luminosità *f*
n helderheid
d Helligkeit *f*

614 BRIGHTNESS CONSTANCY opt
f constance *f* de luminosité
e constancia *f* de luminosidad
i costanza *f* di luminosità
n constantheid van de helderheid
d Konstanz *f* der Helligkeit

615 BRIGHTNESS CONTRAST tv
f contraste *m* de la luminosité
e contraste *m* de la luminosidad
i contrasto *m* della luminosità
n helderheidscontrast *n*
d Helligkeitskontrast *m*

616 BRIGHTNESS CONTROL rep
Means for correcting the overall brightness of the TV picture.
f commande *f* de luminosité
e mando *m* de luminosidad
i comando *m* di luminosità
n helderheidsregeling
d Helligkeitssteuerung *f*

617 BRIGHTNESS CURVE rep/tv
A curve built up by superimposing a set of basic patterns characterized by amplitude A and frequency 1/a.
f courbe *f* de luminosité
e curva *f* de luminosidad
i curva *f* di luminosità
n helderheidskromme
d Helligkeitscharakteristik *f*

618 BRIGHTNESS DECAY ctv
f affaiblissement *m* de luminosité
e decaimiento *m* de luminosidad
i abbassamento *m* di luminosità
n helderheidsvermindering
d Helligkeitsrückgang *m*

619 BRIGHTNESS DISTRIBUTION tv
f distribution *f* de la luminosité
e distribución *f* de la luminosidad
i distribuzione *f* della luminosità
n helderheidsverdeling
d Helligkeitsverteilung *f*

620 BRIGHTNESS FLICKER, IMAGE FLICKER rep/tv
f papillotement *m* de l'image, scintillement *m* de la luminosité
e centelleo *m* de la imagen, parpadeo *m* de la luminosidad
i sfarfallio *m* dell'immagine, sfarfallio *m* della luminosità
n flikkeren *n* van het beeld, helderheidsflikkeren *n*
d Bildflimmern *n*, Helligkeitsflimmern *n*

621 BRIGHTNESS MODULATION rep/tv
Modulation, e.g. of a high-frequency voltage caused by and related to luminosity variations.
f modulation *f* par la luminosité
e modulación *f* por la luminosidad
i modulazione *f* per la luminosità
n helderheidsmodulatie
d Helligkeitsmodulation *f*

622 BRIGHTNESS OF IMAGE, BRIGHTNESS OF PICTURE tv
f luminosité *f* de l'image
e luminosidad *f* de la imagen
i luminosità *f* dell'immagine
n beeldhelderheid
d Bildhelligkeit *f*

623 BRIGHTNESS OF THE SPOT, SPOT BRIGHTNESS tv
f luminosité *f* du spot
e luminosidad *f* del punto
i luminosità *f* del punto
n helderheid van de lichtstip
d Lichtpunkthelligkeit *f*

624 BRIGHTNESS RANGE tv
f plage *f* des luminosités
e campo *m* de las luminosidades
i campo *m* delle luminosità
n helderheidsgebied *n*
d Helligkeitsbereich *m*, Helligkeitsumfang *m*

625 BRIGHTNESS RATIO rep/tv
This ratio of the composed picture should, wherever detail is required, be kept to a value which can be tolerably reproduced in good home viewing conditions.
f rapport *m* de luminosité
e relación *f* de luminosidad
i rapporto *m* di luminosità
n helderheidsverhouding
d Helligkeitsverhältnis *n*

626 BRIGHTNESS SENSATION tv
f sensation *f* de luminosité
e sensación *f* de luminosidad
i sensazione *f* di luminosità
n helderheidsgewaarwording
d Helligkeitsempfindung *f*

627 BRIGHTNESS SEPARATION rep/tv
A method to give the maximum separation between the subject to be keyed-in and the background in overlay processes.
f délimitation *f* optimale de la luminosité
e delimitación *f* óptima de luminosidad
i delimitazione *f* ottima della luminosità
n optimale helderheidsafgrenzing
d optimale Helligkeitsabgrenzung *f*

628 BRIGHTNESS SIGNAL tv
f signal *m* de luminosité
e señal *f* de luminosidad
i segnale *m* di luminosità
n helderheidssignaal *n*
d Helligkeitssignal *n*

BRIGHTNESS TRANSITION
see: BLACK-TO-WHITE TRANSITION

629 BRIGHTNESS VALUE tv
f facteur *m* de luminosité,
valeur *f* de la luminosité
e factor *m* de luminosidad,
valor *m* de la luminosidad
i fattore *m* di luminosità,
valore *m* della luminosità
n helderheidswaarde
d Helligkeitswert *m*

630 BRIGHTNESS VARIATIONS rep/tv
f variations *pl* de la luminosité
e variaciones *pl* de la luminosidad
i variazioni *pl* della luminosità
n helderheidsschommelingen *pl*,
helderheidsvariaties *pl*
d Helligkeitsschwankungen *pl*

631 BRILLIANCE ct
That attribute of any colo(u)r in respect of which it may be classed as equivalent to some member of a series of grays ranging between black and white.
f brillance *f*
e brillantez *f*
i splendore *m*
n glans
d Glanz *m*

632 BROAD stu
f banc *m* d'éclairage,
rampe *f* d'ambiance
e banco *m* de lámparas
i banco *m* di lampade
n lichtkoof
d Lichtwanne *f*

633 BROAD BEAM crt
f faisceau *m* large
e haz *m* ancho
i fascio *m* largo
n brede bundel
d breiter Strahl *m*

634 BROAD LIGHT, stu
DIFFUSE LIGHT,
SPREAD LIGHT
f lumière *f* diffuse,
lumière *f* diffusée
e luz *f* difusa
i luce *f* diffusa
n diffuus licht *n*
d Streulicht *n*

635 BROAD PULSE, cpl/tv
HALF-LINE PULSE
A type of pulse usually repetitive at twice the line frequency and having a duration greater than that of the line sync pulse, a number of which collectively form the field sync pulse of a TV waveform.
f impulsion *f* de demiligne
e impulso *m* de media línea
i impulso *m* di mezza linea
n impuls met dubbele lijnfrequentie
d Halbzeilenimpuls *m*

636 BROADBAND tv
Band of frequencies exceeding the audio range and usually of several MHz.
f large bande *f*
e banda *f* ancha
i banda *f* larga
n brede band
d Breitband *n*, breites Band *n*

637 BROADBAND AERIAL, aea/tv
BROADBAND ANTENNA
f antenne *f* à large bande
e antena *f* de banda ancha
i antenna *f* a banda larga
n brede-bandantenne
d Breitbandantenne *f*

638 BROADBAND AMPLIFIER cpl
f amplificateur *m* à large bande
e amplificador *m* de banda ancha
i amplificatore *m* a banda larga
n brede-bandversterker
d Breitbandverstärker *m*

639 BROADBAND PROTECTION tv
CHANNEL
Used preferably for giving radio-relay systems a high degree of reliability.
f canal *m* de protection à large bande
e canal *m* de protección de banda ancha
i canale *m* di protezione a banda larga
n beschermd kanaal *n* voor brede band
d geschützter Kanal *m* für Breitband

640 BROADBAND SIGNAL tv
f signal *m* de large bande
e señal *f* de banda ancha
i segnale *m* di banda larga
n brede-bandsignaal *n*
d Breitbandsignal *n*

641 BROADCAST ge
f radiodiffusion *f*
e radiodifusión *f*
i radiodiffusione *f*
n omroep
d Rundfunk *m*

642 BROADCAST tv
f émission *f*, transmission *f*
e emisión *f*, transmisión *f*
i emissione *f*, trasmissione *f*

n transmissie, uitzending
d Sendung *f*, Übertragung *f*

643 BROADCAST NETWORK rep/tv
The combined installations required to ensure broadcasting of sound or TV program(me)s and based, e.g. on a single system, or a single country.
f réseau *m* de radiodiffusion
e red *f* de radiodifusión
i rete *f* di radiodiffusione
n omroepnet *n*
d Rundfunknetz *n*

644 BROADCAST NETWORK DESIGN rep/tv
f projet *m* de réseau de radiodiffusion
e proyecto *m* de red de radiodifusión
i progetto *m* di rete di radiodiffusione
n omroepnetplanning
d Rundfunknetzplanung *f*

645 BROADCAST RECEPTION STATION tv
A station designed to receive broadcast transmissions, for monitoring, retransmitting or recording the program(me)s.
f centre *m* de réception,
station *f* de réception
e centro *m* de recepción
i centro *m* ricevente
n omroepontvangststation *n*
d Rundfunkempfangsanlage *f*

646 BROADCAST STUDIO tv
f studio *m* de radiodiffusion
e estudio *m* de radiodifusión
i studio *m* di radiodiffusione
n omroepstudio
d Rundfunkstudio *n*,
Senderaum *m*

647 BROADCAST TRANSMITTING STATION tv
A broadcasting center(re) for sound and/or TV, comprising one or more transmitters, permanently installed with their subsidiary apparatus and associated aerials (antennae).
f centre *m* d'émission,
station *f* d'émission
e emisora *f* de radiodifusión
i stazione *f* di radiodiffusione,
stazione *f* radiofonica
n omroepzendstation *n*
d Rundfunksendeanlage *f*

648 BROADCASTING HOUSE tv
f maison *f* de la radio
e edificio *m* de radiodifusión
i edificio *m* di radiodiffusione
n omroepgebouw *n*
d Funkhaus *n*

649 BROADSIDE stu
f projecteur *m* diffuseur
e proyector *m* difusor
i proiettore *m* diffusore
n schijnwerper met diffuus licht
d Streulichtscheinwerfer *m*

650 BROADSIDE DIPOLE ARRAY aea/tv
f série *f* de dipôles
e antena *f* colineal
i antenna *f* collineale,
fila *f* di dipoli
n dipoolrij
d Dipolreihe *f*

651 BROKEN WHITE, ct
OFF-WHITE,
TINT
The weak colo(u)r resulting from the addition to white of a small amount of colo(u)ring matter.
f blanc *m* cassé
e blanco *m* mate, blancuzco *m*
i biancastro *m*
n gebroken wit *n*
d gebrochenes Weiss *n*

652 BUFFER AMPLIFIER, cpl
ISOLATION AMPLIFIER
An amplifier used to minimize the effects of a following circuit on the preceding circuit.
f amplificateur *m* tampon
e amplificador *m* tampón
i amplificatore *m* di blocco
n bufferversterker
d Pufferverstärker *m*

653 BUILD-UP TIME, cpl/tv
BUILDING-UP TIME,
RISE TIME
The interval between the instants at which the instantaneous value of a pulse or of its envelope first reaches specified lower and upper limits, namely 10 per cent and 90 per cent of the peak values unless otherwise stated.
f durée *f* d'établissement,
temps *m* de montée
e tiempo *m* de crecimiento,
tiempo *m* de establecimiento
i tempo *m* di formazione
n stijgtijd
d Anstiegzeit *f*,
d Einschwingzeit *f*,
Steigzeit *f*

654 BUILDING-UP OF IMAGE, rec/tv
IMAGE SYNTHESIS
f synthèse *f* de l'image
e síntesis *f* de la imagen
i sintesi *f* dell'immagine
n beeldsynthese
d Bildsynthese *f*

655 BUILDING-UP TIME, cpl/tv
TRANSIENT TIME
The time interval between the moment of switching on and the end of the building-up time.
f période *f* transitoire initiale
e período *m* transitorio inicial
i periodo *m* transitorio iniziale
n opslingertijd
d Einschwingzeit *f*

656 BUILT-IN AERIAL, aea/tv
BUILT-IN ANTENNA
f antenne *f* incorporée
e antena *f* incorporada
i antenna *f* incorporata
n ingebouwde antenne
d Einbauantenne *f*

657 BUILT-IN DEGAUSSING COIL, crt/ctv
BUILT-IN DEMAGNETIZING COIL
f bobine *f* de démagnétisation incorporée
e bobina *f* de desmagnetización incorporada
i bobina *f* di demagnetizzazione incorporata
n ingebouwde ontmagnetiseringsspoel
d eingebaute Entmagnetisierungsspule *f*

658 BUILT-IN TELESCOPIC AERIAL, aea
BUILT-IN TELESCOPIC ANTENNA
f antenne *f* téléscopique incorporée
e antena *f* telescópica incorporada
i antenna *f* telescopica incorporata
n ingebouwde telescoopantenne
d eingebaute Teleskopantenne *f*

659 BULK ERASER rec/vr
A device used to erase an entire reel of magnetic tape at once.
f four *m* à effacer
e borrador *m* total
i cancellatore *m* totale
n totaalwisser
d Löschdrossel *f*

660 BURN, rep
IMAGE RETENTION,
STICKING
In a cathode-ray tube the image is sometimes retained by the screen and appears again in subsequent recordings.
f rémanence *f* d'image
e retención *f* de imagen
i rimanenza *f* d'immagine
n vasthouden *n* van het beeld
d Bildkonservierung *f*

661 BURN-IN crt
A screen defect in cathode-ray tubes with magnetic deflection.
f brûlure *f*
e contaminación *f*, envenenamiento *m*
i bruciatura *f*
n inbranden *n*
d Einbrennen *n*

662 BURN-OUT crt
Loss of tonal gradation in the white and near-white portion of the TV image.
f perte *f* de gradation
e pérdida *f* de gradación
i perdita *f* di gradazione
n gradatieverlies *n*
d Gradationsverlust *m*

663 BURNED-IN IMAGE, crt
BURNED-IN PICTURE,
RETAINED IMAGE,
STICKING PICTURE
An image that persists in a fixed position in the output signal of a TV camera tube after the camera has been turned to a different scene.
f image *f* rémanente,
image *f* retenue
e imagen *f* retenida
i immagine *f* ritenuta
n vastgehouden beeld *n*
d konserviertes Bild *n*,
Nachwirkungsbild *n*

664 BURST, ctv
BURST SIGNAL,
COLO(U)R BURST
Component of transmitted signal which acts as reference for chrominance components.
f salve *f* de référence,
signal *m* de synchronisation de couleur
e ráfaga *f* de información de color,
señal *f* de sincronismo de color,
sobreimpulso *m* de color
i salva *f* di colori,
segnale *m* di sincronismo di colore,
sincronismo *m* di colore
n kleursalvo *n*
d Burst *m*,
Farbsynchronsignal *n*

665 BURST AMPLIFIER ctv
f amplificateur *m* de salve
e amplificador *m* de sobreimpulso de color
i amplificatore *m* di salva
n salvoversterker
d Burstverstärker *m*

666 BURST-CONTROLLED OSCILLATOR ctv
f oscillateur *m* contrôlé par la sousporteuse
e oscilador *m* controlado por la subportadora
i oscillatore *m* controllato dalla sottoportante
n door de hulpdraaggolf gecontroleerde oscillator
d Farbträgerregenerator *m*

667 BURST-ENERGY-RECOVERY EFFICIENCY ctv
f rendement *m* de salve
e rendimiento *m* de sobreimpulso de color
i rendimento *m* di salva
n salvorendement *n*
d Burstausbeute *f*

668 BURST FLAG GENERATOR, ctv
BURST-GATE GENERATOR
A switching pulse used for inserting a colo(u)r burst in the process of forming a colo(u)r video signal, or for extracting the colo(u)r burst from an existing colo(u)r video signal.
f générateur *m* de l'impulsion déclenchant le signal de synchronisation de la sousporteuse de chrominance
e generador *m* del impulso de desacoplo de la señal de sincronización de la subportadora de crominancia

i generatore *m* dell'impulso per l'inserzione del sincronismo di colore
n generator voor de salvosleutel, kleursalvopoortsignaalgenerator
d Burstkennimpulsgeber *m*

669 BURST GATE ctv
A keying or gating device used to extract the colo(u)r burst from a colo(u)r picture signal.
f porte *f* déclenchant le signal de synchronisation de la sousporteuse de chrominance
e puerta *f* de desacoplo de la señal de sincronización de la subportadora de crominancia
i porta *f* per l'inserzione del sincronismo di colore
n kleursalvopoort, salvosleutel
d Burstkennimpulsgeber *m*

670 BURST GATING PULSE ctv
A gating signal used to extract the colo(u)r burst from a colo(u)r picture signal.
f impulsion *f* déclenchant le signal de synchronisation de la sousporteuse de chrominance
e impulso *m* de desacoplo de la señal de sincronización de la subportadora de crominancia
i impulso *m* per l'inserzione del sincronismo di colore
n kleursalvopoortsignaal *n*, salvosleutelimpuls
d Burstkennimpuls *m*

671 BURST KEYING PULSE ctv
f impulsion *f* en phase avec le signal de synchronisation de couleur
e impulso *m* en fase con el sincronismo de color, impulso *m* manipulador del sincronismo de color
i impulso *m* in fase con il sincronismo di colore
n impuls in faze met het kleursalvo
d Burstaustastimpuls *m*, Impuls *m* in Phase mit dem Farbsynchronsignal

672 BURST-LOCKED OSCILLATOR ctv
Oscillator, e.g. in a colo(u)r receiver, locked to the colo(u)r burst for subsequent application to later circuits.
f oscillateur *m* verrouillé au signal de synchronisation de couleur
e oscilador *m* bloqueado por el sincronismo de color
i oscillatore *m* agganciato alla sottoportante del sincronismo di colore
n met het kleursalvo vergrendelde oscillator
d mit dem Farbsynchronsignal verriegelter Oszillator *m*

673 BURST PEDESTAL, COLO(U)R BURST PEDESTAL ctv
Part of a colo(u)r burst in a colo(u)r TV signal.
f piédestal *m* du signal de synchronisation de couleur
e pedestal *m* del sincronismo de color
i piedistallo *m* del sincronismo di colore
n kleursalvomengsel *n*, salvostoep
d Burstbasis *f*, Farbsynchronsignalgemisch *n*

674 BURST PHASE ctv
Phase of subcarrier signal forming the colo(u)r burst with which reference the modulated subcarrier is compared.
f phase *f* de la sousporteuse du signal de synchronisation de couleur
e fase *f* de la subportadora del sincronismo de color
i fase *f* della sottoportante del sincronismo di colore
n faze van de hulpdraaggolf van het kleursalvo
d Hilfsträgerphase *f* des Farbsynchronsignals

675 BURST SEPARATOR ctv
The circuit in a colo(u)r TV receiver that separates the colo(u)r burst from the composite video signal.
f circuit *m* séparateur du signal de synchronisation de couleur
e separador *m* de la señal de sincronismo de color
i circuito *m* separatore del sincronismo di colore
n kleursalvoafscheiding
d Farbsynchronsignalabtrennung *f*

676 BUTT SPLICE fi/tv
f collage *m* de bout
e empalme *m* a tope
i incollatura *f* di testa
n stomplas
d Stumpfklebestelle *f*

677 BYPASS MIXED HIGHS tv
The mixed-highs signal containing frequencies between 2 and 4 mc, that is shunted around the chrominance-subcarrier modulator or demodulator in a colo(u)r TV system.
f signal *m* de hautes fréquences mixtes en dérivation
e señal *f* de altas frecuencias mezcladas derivada
i segnale *m* di mescolanza delle componenti d'alta frequenza in derivazione
n signaal *n* voor gemengde hoge frequenties in parallelschakeling
d Nebenschlussignal *n* für gemischte Höhen

678 BYPASS MONOCHROME SIGNAL, SHUNTED MONOCHROME SIGNAL cpl/tv
A monochrome signal shunted around the chrominance modulator or demodulator.
f signal *m* noir-blanc shunté

e señal *f* blanco-negro derivada,
señal *f* monocroma derivada
i segnale *m* bianco-nero in derivazione

n zwart-witsignaal *n* in parallelschakeling
d Schwarz-Weiss-Signal *n* in Nebenschluss

C

679 CABINET, tv
CASE
The housing for a TV receiver.
f boîtier *m*, coffret *m*, meuble *m*
e caja *f*, gabinete *m*
i cassetta *f*, custodia *f*
n kast
d Gehäuse *n*, Kasten *m*, Truhe *f*

680 CABLE CAPACITANCE aud/dis
A capacitance caused by the cable across the microphone output terminals, causing a serious drop in the signal passed on to the apparatus.
f réactance *f* capacitive de câble
e capacitancia *f* de cable
i reattanza *f* capacitiva di cavo
n kabelcapacitantie
d Kabelkapazitanz *f*

681 CABLE COMPENSATION CIRCUITS cpl/dis
Circuits designed to compensate for losses in TV signal links and not designed for a wide frequency range.
f circuits *pl* de compensation de câble
e circuitos *pl* de compensación de cable
i circuiti *pl* di compensazione di cavo
n kabelcompensatiekringen *pl*
d Kabelkompensationskreise *pl*

682 CABLE CORRECTION cpl
A method for counteracting losses in decibels.
f correction *f* d'atténuation de câble
e corrección *f* de atenuación de cable
i correzione *f* d'attenuazione di cavo
n kabeldempingscorrectie
d Kabeldämpfungskorrektur *f*

683 CABLE FILM rep/tv
Substandard TV film, scanned at low speed for signal transmission up to 4.5 KHz through cables, for reconstruction and video transmission.
f film *m* pour transmission par câble, répiquage *m* par câble
e película *f* para transmisión por cable
i pellicola *f* per trasmissione per cavo
n film voor kabeloverdracht
d Film *m* für Übertragung auf Kabeln, Kabelüberspielung *f*

684 CABLE LENGTH COMPENSATION, cpl/tv
CABLE LENGTH EQUALIZATION
Necessary when a picture monitor receives its signal over a long length of video cable.
f compensation *f* de longueur de câble
e compensación *f* de longitud de cable
i compensazione *f* di lunghezza di cavo
n kabellengtecompensatie
d Kabellängenentzerrung *f*, Kabellängenkompensation *f*

685 CABLE LINK rec/tv
A cable connection which is designed to bring a signal mixture from outside to the recording apparatus.
f câble *m* de raccord, liaison *f* par câble
e conexión *f* de cable, enlace *m* de toma directa, línea *f* tributaria
i connessione *f* di cavo, linea *f* entrante
n invoerkabel, kabelverbinding
d Kabelbrücke *f*, Kabelzubringerlinie *f*

686 CABLE TRANSMISSION tv
f transmission *f* par câble
e transmisión *f* por cable
i trasmissione *f* per cavo
n kabeltransmissie
d Übertragung *f* auf Kabeln

687 CALIBRATION PULSE rec/tv
A pulse in the electronic viewfinder used by the cameraman to locally adjust the iris diaphragm.
f impulsion *f* de calibrage
e impulsión *f* de calibración
i impulso *m* di taratura
n ijkimpuls
d Eichimpuls *m*

688 CAMERA, rec/tv
TELECAMERA,
TELEVISION CAMERA
Consisting of an optical lens, a camera tube and a video amplifier to pass on the pulses from the camera tube to the transmitting apparatus.
f caméra *f* de télévision
e cámara *f* de televisión
i camera *f* di televisione, telecamera *f*
n televisiecamera
d Fernsehkamera *f*

689 CAMERA ACCESSORIES, rec/tv
CAMERA AUXILIARIES
f accessoires *pl* de caméra
e acesorios *pl* de cámara
i accessori *pl* di camera
n camera-accessoires *pl*, cameraonderdelen *pl*
d Kamerazubehörteile *pl*

690 CAMERA ALIGNMENT fi
f cadrage *m* de la caméra

e alineación *f* de la cámara
i allineamento *m* della camera
n camera-instelling
d Kameraeinstellung *f*

CAMERA ALIGNMENT
see: BEAM ALIGNMENT

691 CAMERA AMPLIFIER rec/tv
f amplificateur *m* de caméra
e amplificador *m* de cámara
i amplificatore *m* di camera
n cameraversterker
d Kameraverstärker *m*

692 CAMERA APERTURE, fi
FILM GATE,
PICTURE GATE
f fenêtre *f* d'exposition
e puerta *f* de película,
ventanilla *f* de película
i finestra *f* di pellicola
n beeldvenster *n*
d Bildfenster *n*

693 CAMERA APERTURE, rec/tv
CAMERA LENS APERTURE
f ouverture *f* du diaphragme
e abertura *f* del diafragma
i apertura *f* del diaframma
n diafragmaopening
d Blendenöffnung *f*

CAMERA ASSISTANT
see: ASSISTANT-CAMERAMAN

694 CAMERA BALANCING, fi/tv
CAMERA LEVEL(L)ING,
CAMERA WEIGHT ADJUSTMENT
f compensation *f* du poids de la caméra
e compensación *f* del peso de la cámara
i compensazione *f* del peso della camera
n compensatie van het cameragewicht
d Auswiegen *n* der Kamera,
Kameragewichtsausgleich *m*

695 CAMERA CABLES rec/tv
The cables which unite the camera with the monitors.
f câbles *pl* de caméra
e cables *pl* de cámara
i cavi *pl* di collegamento della telecamera
n camerakabels *pl*
d Kamerakabel *pl*

696 CAMERA CAR fi/th/tv
f voiture *f* caméra,
voiture-travelling *m*
e carro *m* de cámara
i carrello *m* di camera
n camerawagen
d Filmwagen *m*,
Kamerawagen *m*

697 CAMERA CHAIN rec/tv
Camera, camera control unit and the power supply.
f chaîne *f* de caméra
e cadena *f* de cámara
i catena *f* della ripresa,
catena *f* di telecamera
n camera-aggregaat *n*
d Kamerakette *f*

698 CAMERA CHANNEL, rec/tv
In a TV studio, the camera, with all its supplies, monitor, control position, and communication to the operator which forms a unit, with others, for supplying video signal to the control room.
f appareillage *m* de prise de vues,
voie *f* de caméra
e conjunto *m* de aparatos de toma de vistas
i apparecchiatura *f* di ripresa
n opneemapparatuur
d Aufnahmeapparatur *f*,
Kamerazug *m*

699 CAMERA CIRCUITRY, rec/tv
f montage *m* de caméra
e circuitería *f* de cámara,
conjunto *m* de circuitos de cámara
i schema *m* di distribuzione
n cameraschakelplan *n*
d Kameraschaltungsaufbau *m*

700 CAMERA CONTROL fi/tv
f contrôle *m* des voies de caméra
e control *m* de conjunto de aparatos de toma de vistas
i controllo *m* dell'apparecchiatura di ripresa
n opneemapparatuurcontrole
d Kamerazugkontrolle *f*

701 CAMERA CONTROL, rec/tv
CAMERA CONTROL UNIT
A control unit designed to provide each camera in the course of a TV transmission with all the adjustments needed to bring the picture output of the individual camera up to an adequate standard.
f unité *f* de contrôle des voies de caméra,
unité *f* de réglage de caméra
e unidad *f* de control de los aparatos de toma de vistas,
unidad *f* de regulación de cámara
i unità *f* di controllo dell'apparecchiatura di ripresa,
unità *f* di regolazione di camera
n cameracontrole-inrichting,
cameraregelinrichting
d Kamerakontrollgerät *n*,
Kameraregler *m*

702 CAMERA CONTROL OPERATOR rec/tv
Operator of the main variable controls of a camera channel which are brought out in the control room.
f contrôleur *m* de voies de caméra,
opérateur *m* de réglage
e operador *m* de control de cámara,
operador *m* de regulación
i operatore *m* di controllo di camera,
operatore *m* di regolazione

n cameracontroleur, regeltechnicus
d Bildoperateur *m*, Kamerabedienung *f*, Regeltechniker *m*

703 CAMERA CONTROL ROOM, VIDEO CONTROL ROOM, VISION CONTROL ROOM rec/tv
That part of the studio where adjustments to the electronics of the camera are carried out to suit the particular scene.
f salle *f* de contrôle d'image, salle *f* de contrôle de vidéo
e cuarto *m* de control de imagen, cuarto *m* de control de video
i sala *f* di controllo d'immagine, sala *f* di controllo di video
n beeldcontrolekamer
d Bildregieraum *m*

704 CAMERA COVERAGE, CAMERA FIELD OF VIEW rec/tv
The area covered by the camera.
f champ *m* de prise de vues de la caméra
e campo *m* de toma de vistas de la cámara
i campo *m* di ripresa della telecamera
n cameraveld *n*, gezichtsveld *n*
d Gesichtsfeld *n*, Kamerabereich *m*, Sehfeld *n*, Sichtfeld *n*

705 CAMERA CRANE stu
f grue *f* de caméra
e grúa *f* de cámara
i gru *f* di camera
n camerakraan
d Kamerakran *m*

706 CAMERA DESIGN ctv/rec/tv
f projet *m* de caméra
e proyecto *m* de cámara
i progetto *m* di camera
n cameraontwerp *n*
d Kameraentwurf *m*

707 CAMERA DOLLY rec/tv
Wheeled support for carrying and moving the camera.
f travelling *m* de caméra
e carro *m* portacámara, pie *m* rodante
i carrello *m* della telecamera
n cameralorrie
d Dolly *m*, fahrbares Kamerastativ *n*, Kamerafahrgestell *n*

708 CAMERA EFFECTS ani
Effects realized by the cameraman during the production of animated films.
f effets *pl* de caméra
e efectos *pl* de cámara
i effetti *pl* di camera
n camera-effecten *pl*
d Kameraeffekte *pl*

709 CAMERA FILM SCANNER rec
f caméra *f* pour analyse de film
e cámara *f* exploradora de película
i camera *f* analizzatrice di pellicola
n filmaftastcamera
d Filmabtastkamera *f*

710 CAMERA FOR COLO(U)R TELEVISION, COLO(U)R TELEVISION CAMERA ctv
f caméra *f* pour télévision en couleur
e cámara *f* para televisión en colores
i camera *f* per televisione a colori
n kleurentelevisiecamera
d Farbfernsehkamera *f*

711 CAMERA FOR INDIRECT LARYNGOSCOPY, TELEVISION CAMERA FOR INDIRECT LARYNGOSCOPY med
f caméra *f* de télévision pour laryngoscopie indirecte
e cámara *f* de televisión para laringoscopía indirecta
i camera *f* di televisione per laringoscopia indiretta
n televisiecamera voor indirecte laryngoscopie
d Fernsehkamera *f* für indirekte Laryngoskopie

712 CAMERA HOOD (US), LENS HOOD (GB) rec/tv
A shield to keep undesired light out.
f paresoleil *m*
e parasol *m*, visera *f* de cámara
i paraluce *m*
n camerakap, lenskap, zonnekap
d Kameralichtkappe *f*, Sonnenblende *f*

713 CAMERA LENS, TELEVISION CAMERA LENS rec/tv
f lentille *f* de caméra de télévision
e lente *f* de cámara televisiva
i lente *f* di telecamera
n televisiecameralens
d Fernsehkameralinse *f*

714 CAMERA LIFTING PLATFORM (GB), CAMERA STACKER (US) rec/tv
Camera platform which can be raised.
f chariot *m* élévateur, plateau *m* élévateur
e plataforma *f* elevable
i carrello *m* elevatore, piattaforma *f* elevatrice
n cameraheftruck
d Hebebühne *f*, Hebetisch *m*, Kamerabühne *f*

715 CAMERA LINE-UP (GB), LINING-UP THE CAMERA (US) rec
Consisting of alignment, aspect-ratio adjustment, shading, s-distortion and high-peaker test.
f ajustage *m* de la caméra

e ajuste *m* de la cámara,
puesta *f* a punto de la cámara
i aggiustaggio *m* della camera
n camera-instelling
d Justierung *f*

716 CAMERA MONITOR (GB), PREVIEW MONITOR (US) rec/tv
Monitor for the control of the picture of a camera in the studio.
f moniteur *m* de caméra
e monitor *m* de primera visión,
monitor *m* previo
i monitore *m* d'attesa,
monitore *m* di telecamera
n cameramonitor
d Kamerakontrollgerät *n*,
Kameramonitor *m*

717 CAMERA NOISE aud/tv
f bruit *m* de caméra
e ruido *m* de cámara
i rumore *m* di camera
n cameraruis
d Kamerageräusch *n*

718 CAMERA OPERATION rec
f opération *f* de caméra
e operación *f* de cámara
i operazione *f* di ripresa
n camerabediening
d Kamerabedienung *f*

719 CAMERA OPERATOR, CAMERAMAN rec/tv
f cadreur *m*,
cameraman *m*,
opérateur *m* de la caméra
e operador *m* de la cámara
i cameraman *m*,
operatore *m* della telecamera,
operatore *m* di ripresa
n cameraman,
cameraoperateur
d Kameraman *m*,
Kameratechniker *m*

720 CAMERA OPTICS, TELEVISION CAMERA OPTICS opt/tv
f optique *f* de la caméra
e óptica *f* de la cámara
i ottica *f* della camera
n cameraoptiek
d Kameraoptik *f*

721 CAMERA PEDESTAL tv
f support *m* de caméra
e soporte *m* de cámara
i telaio *m* di camera
n cameradrager
d Kameragestell *n*, Kameraträger *m*

722 CAMERA POSITION, CAMERA STAND rec/tv
f point *m* de prise de vues
e posición *f* de toma de vistas
i posizione *f* di ripresa
n camerapositie
d Kamerastandort *m*

723 CAMERA PREAMPLIFIER, HEAD AMPLIFIER rec
f préamplificateur *m* de caméra
e preamplificador *m* de cámara
i preamplificatore *m* di telecamera
n cameravoorversterker
d Kameravorverstärker *m*

724 CAMERA SCANNING AMPLITUDE rec/tv
The amplitude of the scanning signals which determine the area of the TV camera mosaic interrogated for signal content.
f amplitude *f* d'analyse de la caméra
e amplitud *f* de exploración de la cámara
i ampiezza *f* d'analisi della camera
n camera-aftastamplitude
d Kameraabtastamplitude *f*

725 CAMERA SCRIPT rec/tv
f scénario *m* de caméra,
script *m* caméra
e scenario *m* de cámara
i copione *m* di camera,
sceneggiatura *f* di camera
n cameradraaiboek *n*
d Kameradrehbuch *n*,
Kamerascript *n*

726 CAMERA SHIFTING (GB), ZOOM (US) rec/tv
The act of rapidly changing the physical position of the TV camera or camera lens with reference to the fixed subject being televised, or the simulation of the effect of this act by other means.
f panoramique *m* rapide
e cambio *m* rápido de plano,
panorámica *f* rápida
i panoramica *f* rapida
n snelle camerazwenking
d Schnellschwenkung *f* einer Kamera

727 CAMERA SHOOTING (US), RECORDING (GB) rec/tv
f prise *f* de vues
e toma *f* de vistas
i ripresa *f* con telecamera
n cameraopname
d Kameraschuss *m*

728 CAMERA SIGNAL rec/tv
The video output signal from a TV camera.
f signal *m* de caméra
e señal *f* de cámara
i segnale *m* di telecamera
n camerasignaal *n*
d Kamerasignal *n*

729 CAMERA SIGNAL CHARACTERISTIC, CAMERA SPECTRAL CHARACTERISTIC ctv
The sensitivity with reference to wavelength, in colo(u)r TV, of each of the colo(u)r separation channels.
f caractéristique *f* du signal de caméra
e característica *f* de la señal de cámara

i caratteristica *f* del segnale di camera
n karakteristiek van het camerasignaal
d Charakteristik *f* des Kamerasignals

730 CAMERA SPECTRAL RESPONSE, rec/tv
CAMERA SPECTRAL SENSITIVITY
CHARACTERISTIC
f sensibilité *f* chromatique de la caméra
e sensibilidad *f* cromática de la cámara
i sensibilità *f* cromatica della camera
n kleurgevoeligheid van de camera
d Farbempfindlichkeit *f* der Kamera

731 CAMERA STAND rec/tv
f trépied *m* transportable
e trípode *m* transportable
i cavalletto *m* di camera,
treppiede *m* di camera
n camerastatief *n*
d Kamerastativ *n*

732 CAMERA SWITCHING rec
f substitution *f* de caméras
e conmutación *f* de cámaras
i commutazione *f* di telecamere
n cameraomschakeling
d Kameraumschaltung *f*

733 CAMERA TALKBACK rec/tv
A two-way system linking each cameraman with his control operator in the vision control room.
f circuit *m* d'interphone pour la caméra
e circuito *m* de intérfono para la cámara
i impianto *m* interfonico per la camera
n cameraruggespraakschakeling
d Kameragegensprechschaltung *f*

734 CAMERA TUBE (GB), crt/rec/tv
IMAGE TUBE (GB),
PICKUP TUBE (US)
f tube *m* analyseur,
tube *m* de prise de vues
e tubo *m* de cámara,
tubo *m* de toma de vistas
i tubo *m* di ripresa,
tubo *m* per telecamera
n beeldopneembuis,
camerabuis
d Aufnahmeröhre *f*,
Bildaufnahmeröhre *f*,
Fernsehaufnahmeröhre *f*

735 CAMERA TUBE CHARACTERISTICS crt
f caractéristiques *pl* du tube de prise de vues
e características *pl* del tubo de toma de vistas
i caratteristiche *pl* del tubo di ripresa
n beeldopneembuiskarakteristieken *pl*
d Bildaufnahmeröhrecharakteristiken *pl*

736 CAMERA TUBE SENSITIVITY crt/tv
The signal current developed per unit incident radiation density.
f sensibilité *f* du tube de prise de vues
e sensibilidad *f* del tubo de toma de vistas
i sensibilità *f* del tubo di ripresa
n beeldopneembuisgevoeligheid
d Bildaufnahmeröhre-Empfindlichkeit *f*

737 CANNED SOUND rep
Sound recorded on disk or tape.
f son *m* conservé
e sonido *m* registrado
i suono *m* registrato
n ingeblikt geluid *n*
d konservierter Schall *m*,
konservierter Ton *m*

738 CANOE vr
Configuration of the tape between the input and output guides due to the head assembly.
f canal *m* de bande
e canoa *f*
i canoa *f* del nastro
n canoa
d Canoe *n*

739 CANTED SHOT rec/tv
Angle shot obtained by canting the camera.
f bascule *f*, prise *f* de vues inclinée
e toma *f* de vistas inclinada
i ripresa *f* inclinata
n schuine opname
d verkantete Aufnahme *f*

740 CAPACITIVE SAWTOOTH cpl/tv
GENERATOR
f générateur *m* capacitif de dents de scie
e generador *m* capacitivo de dientes de sierra
i generatore *m* capacitivo d'oscillazioni a denti di sega
n capacitieve zaagtandgenerator
d kapazitiver Sägezahngenerator *m*

741 CAPSTAN rec/tv
Enlarged shaft or cylinder, sometimes an extension of the driving motor spindle, against which magnetic tape is pressed by a larger pinch roller to impart a steady movement to the tape.
f arbre *m* de commande,
cabestan *m*,
galet *m* d'entraînement
e árbol *m* de arrastre,
cabrestante *m*
i albero *m* motore,
argano *m*
n aandrijfas,
aandrijfrol,
bandtransportrol
d Antriebrolle *f*,
Bandantriebachse *f*,
Bandtransportrolle *f*

742 CAPSTAN DRIVE vr
f commande *f* à cabestan
e mando *m* con cabrestante
i comando *m* ad argano
n aandrijving met bandtransportrol
d Antrieb *m* mit Bandtransportrolle

743 CAPTION tv
Graphic material prepared for TV presentation.
f carton *m* de générique
e inserto *m* de título
i inserto *m* di titolo
n titelinlas
d Einblendtitel, Titelinsert *m*

744 CAPTION SCANNER ani
f analyseur *m* de titre
e explorador *m* de título
i analizzatore *m* di titolo
n titelaftaster
d Titelabtaster *m*

745 CAPTION STAND stu
f banc-titre *m*
e banco *m* para títulos
i banco *m* per didascalia, banco *m* per titoli, titolatrice *f*
n titellessenaar
d Graphikständer *m*, Insertpult *n*, Titelständer *m*

746 CAPTION VIDEO SIGNAL ani
f signal *m* vidéo pour titres
e señal *f* video para títulos
i segnale *m* video per titoli
n videosignaal *n* voor titels
d Videosignal *n* für Titel

747 CAPTIONS fi/tv
f légendes *pl*
e títulos *pl*
i didascalia *f*, titoli *pl*
n titels *pl*
d Graphik *f*

748 CAPTURE RANGE, COLLECTING ZONE, LOCK-IN RANGE, PULL-IN RANGE cpl/tv
The frequency band over which the oscillator will come into lock with the input frequency when initially unlocked.
f plage *f* de rattrapage
e margen *m* de captación
i intervallo *m* d'agganciamento
n vanggebied *n*
d Einspringbereich *m*, Fangbereich *m*, Mitnahmebereich *m*

749 CARDINAL STIMULI ct
Four standard stimuli by means of which the three reference stimuli and the basic stimulus of any trichromatic system may be defined.
f stimuli *pl* normaux
e estímulos *pl* normales
i stimoli *pl* normali
n standaardstimuli *pl*
d Normalreize *pl*

750 CARRIER, CARRIER WAVE cpl/ge
That component of a modulated wave which is independent of the modulation.
f onde *f* porteuse, porteuse *f*
e onda *f* portadora, portadora *f*
i onda *f* portante, portante *f*
n draaggolf
d Träger *m*, Trägerwelle *f*

751 CARRIER AMPLITUDE cpl
f amplitude *f* de la porteuse
e amplitud *f* de la portadora
i ampiezza *f* della portante
n draaggolfamplitude
d Trägeramplitude *f*

752 CARRIER-AMPLITUDE REGULATION cpl
The change in amplitude of the carrier wave in an amplitude-modulated transmitter when modulation is applied symmetrically.
f réglage *m* de l'amplitude de la porteuse
e regulación *f* de la amplitud de la portadora
i regolazione *f* dell'ampiezza della portante
n regeling van de draaggolfamplitude
d Regelung *f* der Trägeramplitude

753 CARRIER BALANCE ctv
In the NTSC and PAL system, the process which ensures that the colo(u)r subcarrier level is zero, for zero chrominance input.
f équilibrage *m* de la sousporteuse de couleur
e equilibrado *m* de la subportadora de color
i bilanciamento *m* della sottoportante di colore
n evenwicht *n* van de kleurhulpdraaggolf
d Farbhilfsträgergleichgewicht *n*

754 CARRIER-BALANCE GATING PULSE ctv
A switching pulse used in maintaining the balance of the chrominance modulators.
f impulsion *f* d'équilibrage de la sous-porteuse de couleur
e impulso *m* equilibrador de la subportadora de color
i impulso *m* di controllo del bilanciamento della sottoportante di colore
n evenwichtsimpuls van de kleurhulpdraaggolf
d Gleichgewichtsimpuls *m* des Farbhilfsträgers

755 CARRIER CHROMINANCE SIGNAL, CHROMINANCE SIGNAL ctv
In TV, the sidebands of the modulated chrominance subcarrier (plus the chrominance subcarrier if not suppressed) which are added to the monochrome signal to convey colo(u)r information.

f signal *m* de chrominance
e señal *f* de crominancia
i segnale *m* di crominanza
n chrominantiesignaal *n*
d Chrominanzsignal *n*

756 CARRIER DIFFERENCE SYSTEM, cpl/tv
INTERCARRIER SOUND SYSTEM
System of amplification of vision and sound signals together employed in TV receivers for use with certain TV receivers including the British 625-line system.
f réception *f* par battements, système *m* son à porteuse intermédiaire
e sistema *m* de sonido por interportadora, sistema *m* sonoro de portadora múltiple
i sistema *m* della portante differenziale
n interdraaggolfgeluidssysteem *n*
d Differenzträgerverfahren *n*

757 CARRIER OFFSET cpl
Frequency relationship between one carrier and another or between a carrier of some other signal parameter such as a harmonic of the scanning frequency or frequencies.
f décalage *m* de la porteuse
e desplazamiento *m* de la portadora
i spostamento *m* della portante
n draaggolfverzet *n*
d Trägerversetzung *f*

758 CARRIER REFERENCE WHITE LEVEL, tv
PICTURE WHITE
The carrier amplitude corresponding to reference white level.
f niveau *m* de blanc de la porteuse, niveau *m* du blanc maximal
e nivel *m* de blanco de la portadora, ultrablanco *m* de la imagen
i livello *m* di bianco della portante, ultrabianco *m* dell'immagine
n draaggolfamplitude voor maximaal wit, niveau *n* van het helderste wit
d Bildweiss *n*, Weisspegel *m*

759 CARRIER SPACING cpl
f intervalle *m* d'espacement des fréquences porteuses
e intervalo *m* de espaciación de las frecuencias portadoras
i intervallo *m* di spaziatura delle frequenze portanti
n draaggolfafstand
d Trägerabstand *m*

760 CARRIER TRANSMISSION tv
f transmission *f* par modulation de la porteuse
e transmisión *f* por portadora
i trasmissione *f* per modulazione della portante
n draaggolftransmissie
d Trägerfrequenzübertragung *f*

CARTOONIST
see: ANIMATOR

761 CARTRIDGE (US), vr
CASSETTE (GB)
Holder for reels of magnetic tape.
f cassette *f*
e caseta *f*
i cassetta *f*
n cassette
d Kassette *f*

CARTRIDGE AMPLIFIER,
CASSETTE AMPLIFIER
see: AMPLIFIER UNIT IN VIDEO RECORDING

762 CARTRIDGE DECK (US), aud/vr
CASSETTE DECK (GB)
f platine *m* à cassette
e tocacasetas *m*
i apparecchio *m* a cassette
n cassettespeler
d Kassettenplatine *f*, Kassettenspieler *m*

763 CARTRIDGE HOLDER (US), vr
CASSETTE HOLDER (GB)
f logement *m* de la cassette
e recipiente *m* de la caseta
i contenitore *m* della cassetta
n cassettehouder, cassettelift
d Kassettenhalter *m*

764 CASCODE AMPLIFIER, cpl/tv
WALLMAN AMPLIFIER
Circuit in which a grounded-cathode triode followed by a grounded-grid triode provides a low-noise amplifier for very high frequencies.
f amplificateur *m* cascode
e amplificador *m* cascode
i amplificatore *m* cascode
n cascodeversterker
d Kaskodenverstärker *m*

CASE
see: CABINET

765 CASING ge
f boîtier *m*
e caja *f*
i scatola *f*
n huis *n*
d Gehäuse *n*

766 CATADIOPTRIC opt/tv
Said of an optical system in projection TV receivers which uses reflecting mirrors as well as lenses.
f catadioptrique adj
e catadióptrico adj
i catadiottrico adj
n katadioptrisch adj
d katadioptrisch adj

767 CATHODE crt
An electrode of e.g. a cathode-ray tube.
f cathode *f*
e cátodo *m*
i catodo *m*
n katode
d Kathode *f*, Katode *f*

768 CATHODE COMPENSATION cpl/tv
Video frequency stage in which negative feedback sets in as a result of a small cathode bypass capacitor.
f compensation *f* cathodique
e compensación *f* catódica
i compensazione *f* catodica
n katodecompensatie
d Katodenkompensation *f*

769 CATHODE FOLLOWER, GROUNDED ANODE CIRCUIT cpl/tv
A network in which the output load is in the cathode circuit, the input being applied between the grid and the end of the load remote from the cathode.
f circuit *m* à charge cathodique
e circuito *m* seguidor de cátodo
i circuito *m* ad accoppiamento catodico
n anodebasisschakeling, katodevolger
d Anodenbasisschaltung *f*, Katodenfolger *m*

770 CATHODE-POTENTIAL STABILIZED CAMERA TUBE, CATHODE-VOLTAGE STABILIZED CAMERA TUBE, LOW-ELECTRON-VELOCITY CAMERA TUBE crt
A camera tube operating with a beam of electrons having velocities such that the average target voltage stabilizes at a value near that of the cathode.
f tube *m* analyseur à électrons lents
e tubo *m* de cámara de electrones lentos
i tubo *m* di ripresa ad elettroni lenti
n orthiconcamerabuis
d Aufnahmeröhre *f* mit langsamen Elektronen

771 CATHODE-RAY BEAM, CATHODE-RAY PENCIL crt
f faisceau *m* de rayons cathodiques
e haz *m* de rayos catódicos
i fascio *m* di raggi catodici, pennello *m* di raggi catodici
n katodestraalbundel
d Katodenstrahlbündel *n*

772 CATHODE-RAY BEAM INTENSITY MODULATION crt
f contrôle *m* de l'intensité du faisceau
e control *m* de la intensidad del haz
i controllo *m* dell'intensità del fascio
n controle van de bundelintensiteit
d Intensitätskontrolle *f* eines Katodenstrahls

773 CATHODE-RAY CAMERA tv
Combination of a cathode-ray tube and a moving-film camera.
f caméra *f* à tube cathodique
e cámara *f* de tubo catódico
i camera *f* a tubo catodico
n camera met katodestraalbuis
d Kamera *f* mit Katodenstrahlröhre

774 CATHODE-RAY CURRENT crt
f courant *m* du faisceau
e corriente *f* del haz
i corrente *f* del fascio
n bundelstroom
d Strahlstrom *m*

775 CATHODE-RAY SCREEN, CATHODE-RAY TUBE SCREEN, FLUORESCENT SCREEN, LUMINESCENT SCREEN crt
The luminescent screen of a cathode-ray tube which receives the cathode-ray beam.
f écran *m* fluorescent
e pantalla *f* fluorescente
i schermo *m* fluorescente
n buisscherm *n*, fluorescentiescherm *n*
d Fluoreszenzschirm *m*

776 CATHODE-RAY TUBE, CRT crt
An electronic tube in which a well-defined and controllable beam of electrons is produced and directed on to a surface to give a visible or otherwise detectable display or effect.
f tube *m* à rayons cathodiques, tube *m* cathodique
e tubo *m* catódico, tubo *m* de rayos catódicos
i tubo *m* a raggi catodici, tubo *m* catodico
n katodestraalbuis
d Katodenstrahlröhre *f*

777 CATHODE-RAY TUBE CONTROL CHARACTERISTIC crt
A series of plots of the beam current versus the grid voltage for various values of voltage on the third anode of a cathode-ray tube.
f caractéristique *f* de commande d'un tube cathodique, caractéristique *f* de réglage d'un tube cathodique
e característica *f* de mando de un tubo catódico, característica *f* de regulación de un tubo catódico
i caratteristica *f* di comando d'un tubo catodico, caratteristica *f* di regolazione d'un tubo catodico
n besturingskarakteristiek van een katodestraalbuis, regelkarakteristiek van een katodestraalbuis
d Regelcharakteristik *f* einer Katodenstrahlröhre, Steuercharakteristik *f* einer Katodenstrahlröhre

778 CATHODE-RAY TUBE DISPLAY crt
Luminous pattern produced on the screen of a cathode-ray tube by the controlled movements of an electron beam impinging on the phosphor of the screen.

f image *f* d'écran
e imagen *f* de pantalla
i immagine *f* di schermo
n schermbeeld *n*
d Schirmbild *n*

779 CATHODE-RAY TUBE PERSISTENCE CHARACTERISTIC — crt
The relation between the illumination of relative brightness of an excited phosphor on a cathode-ray tube screen and time, after the excitation is removed.
f caractéristique *f* de persistance d'un tube cathodique
e característica *f* de persistencia de un tubo catódico
i caratteristica *f* di persistenza d'un tubo catodico
n nalichtkarakteristiek van een katodestraalbuis
d Nachleuchtcharakteristik *f* einer Katodenstrahlröhre

CATHODE-RAY TUBE POST-DEFLECTION ACCELERATION
see: AFTER-ACCELERATION

780 CATHODE-RAY TUBE PROJECTOR, CRT PROJECTOR, PROJECTION CATHODE-RAY TUBE — crt
A special tube which can be operated at high voltage and beam current, and so produces a bright picture which can be projected with a suitable optical system.
f tube *m* cathodique de projection
e tubo *m* catódico de proyección
i tubo *m* catodico per proiezione
n projectiekatodestraalbuis
d Projektionskatodenstrahlröhre *f*

781 CATHODE RAYS — crt
Stream of negatively charged particles emitted normally from the surface of the cathode in a rarefied gas.
f rayons *pl* cathodiques
e rayos *pl* catódicos
i raggi *pl* catodici
n katodestralen *pl*
d Katodenstrahlen *pl*

782 CATHODOELECTROLUMINESCENCE, ELECTROCATHODOLUMINESCENCE — crt
Effect obtained on a fluorescent screen excited by cathode rays.
f électroluminescence *f* cathodique
e electroluminiscencia *f* catódica
i elettroluminescenza *f* catodica
n elektrokatodoluminescentie
d Elektrokatodolumineszenz *f*

783 CATHODOLUMINESCENCE — crt
Luminescence produced by high-velocity electrons.
f cathodoluminescence *f*
e catodoluminiscencia *f*
i catodoluminescenza *f*
n katodoluminescentie
d Katodolumineszenz *f*

784 CATOPTRIC — opt/tv
In optics, a projection system using only mirrors.
f catoptrique adj
e catóptrico adj
i catottrico adj
n katoptrisch adj
d katoptrisch adj

785 CATV, CENTRAL AERIAL TELEVISION, CENTRAL ANTENNA TELEVISION — aea/tv
Method of distributing TV by means of multiple outlets in one building.
f télévision *f* à antenne centrale
e televisión *f* de antena central
i televisione *f* ad antenna centrale
n centrale-antenne-televisie
d Zentralantennenfernsehen *n*

786 CATWALK, GANTRY — fi/th/tv
f passerelle *f* de projecteurs
e puente *m* de proyectores
i ponte *m* delle luci
n lichtbrug
d Beleuchterbrücke *f*

787 CAVITY RESONATOR — cpl
Type of tuned circuit used at ultra-high frequencies (Bands IV and V).
f cavité *f* résonnante
e cavidad *f* resonante
i cavità *f* risonante
n trilholte
d Hohlraumresonator *m*, Schwingtopf *m*, Topfkreis *m*

788 CCTV, CLOSED-CIRCUIT TELEVISION, RADIOVISION — cpl/tv
Essentially comprises a pick-up device, a signal distribution system and one or more displays.
f télévision *f* à circuit fermé
e televisión *f* por circuito cerrado
i televisione *f* a circuito chiuso
n televisie met gesloten kring
d Fernsehübertragung *f* im Kurzschlussverfahren

789 CEL — ani
f bande *f* en celluloid, cel *m*
e cinta *f* de celuloide
i nastro *m* celluloide
n celluloidstrook
d Zelluloidstreifen *m*

790 CEL ANIMATION — ani
The superimposition of the moving parts of the picture on to the static background by drawing or painting them on to transparent celluloid sheets.
f application *f* des animations sur bande en celluloid
e aplicación *f* de animaciones sobre cinta de celuloide

i applicazione *f* d'animazioni su nastro celluloide
n aanbrengen *n* van tekenfilmbeelden op celluloidstrook
d Anbringen *n* von Zeichentrickfilmelementen auf Zelluloidstreifen

791 CENSOR KEY, ctv
CENSOR SWITCH,
CUT KEY
A device used in sound or TV broadcasting to cut off a channel or change to another channel.
f interrupteur *m* de canal
e interruptor *m* de canal
i interruttore *m* di canale
n kanaalonderbreker
d Kanalunterbrecher *m*

792 CENTER FEED (US), aea
CENTRAL FEED (GB)
f alimentation *f* médiane
e alimentación *f* central
i alimentazione *f* a presa centrale
n middelpuntvoeding
d Mittelpunktsspeisung *f*

CENTER(RE) CARRIER FREQUENCY
see: AUDIO CENTER(RE) FREQUENCY

793 CENTER(RE) FREQUENCY, ge
RESTING FREQUENCY
As applied to frequency modulation, the frequency of the carrier wave in the absence of modulation.
f fréquence *f* nominale
e frecuencia *f* nominal
i frequenza *f* di riposo
n centrale frequentie
d Mittenfrequenz *f*,
Ruhefrequenz *f*,
Ruheträgerfrequenz *f*

794 CENTER(RE) FREQUENCY and/tv
STABILITY
The stability of the transmitter to maintain an assigned center(re) frequency in the absence of modulation.
f stabilité *f* de la fréquence nominale
e estabilidad *f* de la frecuencia nominal
i stabilità *f* di riposo nominale
n stabiliteit van de centrale frequentie
d Stabilität *f* der Ruhefrequenz

795 CENT(E)RING crt
The process of adjusting the position of the image on the screen of the cathode-ray tube.
f cadrage *m*, centrage *m*
e centrado *m*
i centratura *f*
n centrering
d Zentrierung *f*

796 CENT(E)RING CONTROL rep/tv
Method of cent(e)ring the picture on the screen.
f réglage *m* du cadrage,
réglage *m* du centrage
e regulación *f* del centrado
i regolazione *f* della centratura
n centreringsregeling
d Bildstandsregelung *f*,
Zentrierungsregelung *f*

797 CENT(E)RING CONTROL rep/tv
DEVICE
f dispositif *m* de cadrage,
dispositif *m* de centrage
e dispositivo *m* de centrado
i dispositivo *m* di centratura
n centreerapparaat *n*
d Zentriergerät *n*

CENTRAL AERIAL TELEVISION,
CENTRAL ANTENNA TELEVISION
see: CATV

798 CENTRAL APPARATUS ROOM tv
Room containing the pulse generators, test signal generators, and other common signal-handling equipment.
f salle *f* centrale des équipements
e sala *f* central de aparatos
i sala *f* centrale d'apparecchi
n centrale apparatenkamer
d Bild-und Tonschaltraum *m*,
Zentralgeräteraum *m*

799 CENTRAL CONTROL, rec/tv
CONTINUITY PROGRAM(ME) CONTROL,
MASTER CONTROL,
PRESENTATION SUITE
The point where program(me)s originating from studios, outside broadcasts, network, film or video tape are combined into a complete smooth presentation.
f régie *f* centrale
e control *m* central
i regia *f* centrale
n centrale regie
d zentrale Regie *f*

800 CENTRAL CONTROL DESK, rec/tv
MASTER CONTROL DESK
f pupitre *m* de régie centrale
e pupitre *m* de control central
i banco *m* di regia centrale
n centrale regielessenaar
d Zentralpult *n*,
Zentralregiepult *n*

801 CENTRAL CONTROL ROOM rec/tv
A room into which the modulating signals from different studios or control rooms are channel(l)ed with a view to routing these signals to continuity rooms, transmitters, recording rooms or to other control rooms.
f C.D.M.,
centre *m* de commutation,
centre *m* distributeur de modulation
e puesto *m* central de control
i sala *f* di controllo centrale
n centrale schakelkamer,
schakelcentrum *n*

d Hauptschaltraum *m*,
Schaltzentrale *f*

802 CENTRAL MONITORING POSITION, CMP — tv
f centrale *f* de moniteurs
e central *f* de monitores
i centrale *f* di monitori
n monitorcentrale
d Überwachungszentrale *f*

803 CENTRAL PROGRAM(ME) CONTROL ROOM, MASTER PROGRAM(ME) CONTROL ROOM — rec/tv
f salle *f* de régie centrale
e sala *f* de control central
i sala *f* di regia centrale
n centrale regieruimte
d Regieraum *m*, Zentralregieraum *m*

804 CHAIN — tv
A network of TV stations connected by special telephone lines, coaxial cables or relay links so all can operate as a group for broadcast purposes.
f chaîne *f* de stations de télévision
e cadena *f* de estaciones de televisión
i catena *f* di ripetitori televisivi
n keten van televisiestations
d Fernsehkette *f*

805 CHAIN BROADCASTING — tv
f diffusion *f* à relais
e retransmisión *f* de televisión
i ritrasmissione *f* di televisione
n relaisomroep
d Relaisrundfunk *m*

806 CHANGE-OVER — fi/tv
f enchaînement *m*
e transición *f*
i passaggio *m*
n overgang
d Überblendung *f*

807 CHANGE-OVER CUE, CUE MARK — fi/tv
f marque *f* de passage,
marque *f* de répère d'enchaînement
e marca *f* de transición
i marca *f* di passaggio
n overgangsteken *n*
d Stanze *f*,
Überblendungszeichen *n*

808 CHANGING OF PRESET CHANNELS — rep/tv
f modification *f* des canaux choisis
e modificación *f* de los canales seleccionados
i modificazione *f* dei canali selezionati
n wijzigen *n* van de ingestelde kanalen
d Ändern *n* der eingestellten Kanäle

809 CHANNEL — tv
A band of radio frequencies allocated for a particular purpose.
f canal *m*
e canal *m*
i canale *m*
n kanaal *n*
d Kanal *m*,
Übertragungskanal *m*

810 CHANNEL ALLOCATION — tv
f allocation *f* des canaux
e asignación *f* de los canales
i allogazione *f* dei canali
n kanaaltoewijzing
d Kanalzuteilung *f*

811 CHANNEL AMPLIFIER — cpl
f amplificateur *m* de canal
e amplificador *m* de canal
i amplificatore *m* di canali
n kanaalversterker
d Kanalverstärker *m*

812 CHANNEL BALANCING — cpl
f équilibrage *m* de canal
e equilibración *f* de canal
i bilanciamento *m* di canale
n kanaalaanpassing
d Kanalanpassung *f*

813 CHANNEL BANDWIDTH, CHANNEL WIDTH — tv
f largeur *f* de bande du canal
e anchura *f* de banda del canal
i larghezza *f* di banda del canale
n kanaalbandbreedte
d Kanalbandbreite *f*, Kanalbreite *f*

814 CHANNEL CAPACITY, — tv
In a specific channel, the maximum number of symbols per second that may be transmitted.
f capacité *f* de canal
e capacidad *f* de canal
i capacità *f* di canale
n kanaalcapaciteit
d Kanalkapazität *f*

815 CHANNEL COMBINING UNIT DIPLEXER — cpl
f coupleur *m* sélectif de canaux
e acoplador *m* selectivo de canales
i accoppiatore *m* selettivo di canali
n kanaalscheidingsfilter *n*
d Kanalweiche *f*

816 CHANNEL CONVERSION — rep/tv
The conversion from one standard of 625-line reception to another one.
f conversion *f* de canal
e conversión *f* de canal
i conversione *f* di canale
n kanaalomschakeling,
kanaalomzetting
d Kanalumschaltung *f*,
Kanalumsetzung *f*

817 CHANNEL-GAIN STABILITY — tv
Constancy of gain of a channel in the presence of changes due to ageing,

supply voltage, temperature, etc.
f stabilité *f* de gain de canal
e estabilidad *f* de ganancia de canal
i stabilità *f* di guadagno di canale
n stabiliteit van de kanaalversterking
d Stabilität *f* der Kanalverstärkung

818 CHANNEL PRESELECTION rep/tv
f présélection *f* des canaux
e preselección *f* de los canales
i preselezione *f* dei canali
n instelling vooraf van de kanalen
d Kanalvorwahl *f*

819 CHANNEL REJECTOR CIRCUIT cpl
f réjecteur *m* de canal
e circuito *m* supresor de canal
i circuito *m* soppressore di canale
n kanaalsperkring
d Kanalsperrkreis *m*

820 CHANNEL SELECTOR, rep
MULTICHANNEL SELECTOR
A switch or other control used to tune in the desired channel in a TV receiver.
f sélecteur *m* de canaux
e selector *m* de canales
i selettore *m* di canali
n kanalenkiezer
d Kanalwähler *m*,
Kanalwahlschalter *m*

821 CHANNEL SPACING tv
f écartement *m* entre canaux
e separación *f* de canales
i spaziatura *f* dei canali
n kanaalafstand
d Kanalabstand *m*

822 CHARACTERISTIC, ge
CHARACTERISTIC CURVE
A curve plotted on graph paper to show the relation between two changing values.
f caractéristique *f*,
courbe *f* caractéristique
e característica *f*,
curva *f* característica
i caratteristica *f*,
curva *f* caratteristica
n karakteristiek,
kromme
d Charakteristik *f*,
Kennlinie *f*

823 CHARGE IMAGE (GB), crt
IMAGE PATTERN (US)
The total of charged particles of the fluorescent substance on an insulating surface in a cathode-ray tube.
f image *f* de potentiel,
image *f* électronique de charge
e imagen *f* de potencial
i immagine *f* di potenziale,
immagine *f* elettronica di carica
n ladingsbeeld *n*
d Ladungsbild *n*

824 CHARGE STORAGE crt/tv
Principle whereby image information is stored on a mosaic, especially in camera tubes for TV.
f emmagasinage *m* de charge
e almacenamiento *m* de carga
i immagazzinamento *m* di carica
n ladingsopslag
d Ladungsspeicherung *f*

825 CHARGE-STORAGE LEVEL crt
f niveau *m* de l'emmagasinage de charge
e nivel *m* de almacenamiento de carga
i livello *m* d'immagazzinamento di carica
n niveau *n* van de ladingsopslag
d Pegel *m* der Ladungsspeicherung

826 CHARGE-STORAGE TUBE crt
A storage tube which retains the information on its active surface in the form of electric charges.
f tube *m* d'emmagasinage de charge
e tubo *m* de almacenamiento de carga
i tubo *m* d'immagazzinamento di carica
n ladingsopslagbuis
d Ladungsspeicherröhre *f*,
Speicherröhre *f*

827 CHARGED PARTICLE crt
A particle which carries a positive or negative electric charge.
f particule *f* chargée
e partícula *f* cargada
i particella *f* caricata
n geladen deeltje *n*
d geladenes Teilchen *n*

828 CHASSIS ge
The metal frame on which circuit components are mounted.
f châssis *m*,
plaque *f* de montage
e bastidor *m*,
chasis *m*
i chassis *m*,
piastra *f* di montaggio
n chassis *n*,
montageplaat
d Aufbauplatte *f*,
Chassis *n*,
Montageplatte *f*

829 CHECKERBOARD FREQUENCY (US), tv
CHEQUERBOARD FREQUENCY (GB),
CHESSBOARD FREQUENCY
f fréquence *f* de damier
e frecuencia *f* de ajedrez
i frequenza *f* di scacchiera
n schaakbordfrequentie
d Schachbrettfrequenz *f*

830 CHECKERBOARD PATTERN rep/tv
(US),
CHEQUERBOARD PATTERN (GB),
CHESSBOARD PATTERN
Pattern consisting of at right angles intersecting horizontal and vertical bars.

f mire *f* à damier
e imagen *m* patrón en forma de ajedrez
i immagine *f* campione a scacchiera
n schaakbordtoetsbeeld *n*
d Schachbrettestbild *n*

831 CHEESE CLOTH, SCRIM fi/tv
f écran *m* diffuseur, tarlatane *f*
e gasa *f* difusora
i velatino *m*
n gaasscherm *n*
d Gazeschirm *m*, Softscheibe *f*

832 CHEMICAL FADE fi/tv
f fondu *m* chimique
e desaparición *f* química
i dissolvenza *f* chimica
n chemische vloeier
d chemische Blende *f*

833 CHEST MICROPHONE, NECKLACE MICROPHONE tv
f microphone *m* cravate, microphone *m* plastron
e micrófono *m* de pecho, microplastrón *m*
i microfono *m* pettorale
n borstmicrofoon
d Brustmikrophon *n*, Umhängemikrophon *n*

834 CHICKEN dis
A term used for the signal corresponding to a line of scanning when the amplitude of the signal wanders spuriously towards black, due to accumulation of charges on the mosaic of the camera tube.
f signal *m* errant vers le noir
e señal *f* errante hacia el negro
i segnale *m* errante verso il nero
n zwerfsignaal *n* naar zwart
d Wandersignal *n* nach Schwarz

835 CHOIR fi/th/tv
f choeur *m*
e coro *m*
i coro *m*
n koor *n*
d Chor *m*

836 CHROMA ct
That quality which characterizes a colo(u)r without reference to its brightness; it embraces hue and saturation.
f chroma *m*
e croma *m*
i croma *m*
n chroma *n*
d Chroma *n*

837 CHROMA, MUNSELL CHROMA, SATURATION ct/ctv
The dimension of the Munsell system of colo(u)r which corresponds most closely to saturation.
f chroma *m* de Munsell, saturation *f*
e croma *m* de Munsell, saturación *f*
i croma *m* da Munsell, saturazione *f*
n chroma *n* volgens Munsell, verzadiging
d Chroma *n* nach Munsell, Sättigung *f*

838 CHROMA CODER ctv
A device for producing simultaneous colo(u)r signals from field sequential colo(u)r signals.
f convertisseur *m* de système
e convertidor *m* de sistema
i convertitore *m* di sistema
n systeemomzetter
d Systemumsetzer *m*

839 CHROMA CONTROL, COLO(U)R INTENSITY CONTROL, SATURATION CONTROL ctv
In a colo(u)r display apparatus, a gain control in the chrominance channel.
f réglage *m* de saturation de la couleur
e regulación *f* de saturación del color
i regolazione *f* di saturazione del colore
n kleurverzadigingsregeling
d Buntregler *m*, Buntspannungsregler *m*

840 CHROMA KEY (US), COLOUR SEPARATION OVERLAY (GB) ani/ctv
Insertion of one picture signal into another electronically triggered by the colo(u)r difference between these signals.
f mixage *m* de signaux d'image à activation électronique par différence de couleur
e entremezclado *m* de señales de imagen con activación electrónica por diferencia de color
i tecnica *f* di composizione di differenti segnali d'immagini, basata sul loro differente contenuto cromatico
n kleursleutel
d Blaustanzverfahren *n*, Farbschablonentrick *m*, Stanztrick *m*

841 CHROMATIC ct
Relating to colo(u)r.
f chromatique adj
e cromático adj
i cromatico adj
n chromatisch adj, kleur-
d chromatisch adj, farb-

842 CHROMATIC ABERRATION crt
An image defect due to the variations in the initial velocities of the electrons as they leave the cathode of the tube.
f aberration *f* chromatique
e aberración *f* cromática
i aberrazione *f* cromatica
n chromatische aberratie

d chromatische Aberration *f*,
chromatischer Fehler *m*

843 CHROMATIC ABERRATION opt
An optical lens defect causing colo(u)r fringes, because the lens material brings different colo(u)rs of light to a focus at different points.
f erreur *f* de teinte
e error *m* de tinta
i errore *m* di tinta
n kleurafwijking
d Farbabweichung *f*

844 CHROMATIC ADAPTATION opt
The adaptation of the eye for colo(u)r.
f adaptation *f* chromatique
e adaptación *f* cromática
i adattamento *m* cromatico
n kleuraanpassing
d Farbanpassung *f*,
farbige Verstimmung *f*

845 CHROMATIC CHAIN, CHROMATIC CHANNEL ctv
f chaîne *f* de chromatisme,
voie *f* de chrominance
e canal *m* de crominancia
i canale *m* di crominanza
n kleursoortkanaal *n*
d Farbartkanal *m*

846 CHROMATIC COEFFICIENT ct
f coefficient *m* chromatique
e coeficiente *m* cromático
i coefficiente *m* cromatico
n kleurcoëfficiënt
d Farbkoeffizient *m*

847 CHROMATIC COLO(U)RS ct
Colo(u)rs having a hue.
f couleurs *pl* chromatiques
e colores *pl* cromáticos
i colori *pl* cromatici
n chromatische kleuren *pl*
d chromatische Farben *pl*

848 CHROMATIC COMPONENT ct
f composante *f* chromatique
e componente *f* cromática
i componente *f* cromatica
n kleurcomponent
d Farbauszug *m*

849 CHROMATIC DEFECT, COLO(U)R DEFECT ct
f dénaturation *f* de couleur
e desnaturalización *f* de color
i denaturazione *f* di colore
n kleurontaarding
d Farbfehler *m*

850 CHROMATIC DEFINITION (GB), CHROMATIC RESOLUTION (US) ct
f netteté *f* de la transition colorée
e nitidez *f* de la transición colorada
i finezza *f* della transizione colorata
n chromatische contourscherpte
d Farbkonturschärfe *f*

851 CHROMATIC SENSATIONS ct
f sensations *pl* chromatiques
e sensaciones *pl* cromáticas
i percezioni *pl* dei colori,
sensazioni *pl* cromatiche
n kleurgewaarwordingen *pl*,
kleurindrukken *pl*
d Farbempfindungen *pl*,
Farbwahrnehmungen *pl*

852 CHROMATIC SPLITTING ct
f division *f* chromatique
e división *f* cromática
i divisione *f* cromatica
n kleursplitsing
d Farbaufteilung *f*

853 CHROMATICITY ct
The colo(u)r quality of a light definable by its chromaticity co-ordinates, or by its dominant (or complementary) wavelength and its purity taken together.
f chromaticité *f*
e cromaticidad *f*
i cromaticità *f*
n kleursoort
d Farbart *f*

854 CHROMATICITY ABERRATION opt
f erreur *f* de teinte
e error *m* de tinta
i errore *m* di tinta
n kleurafwijking, kleurfout
d Farbabweichung *f*,
Farbartabweichung *f*

855 CHROMATICITY ACUITY ct
f sensibilité *f* de chromaticité
e sensibilidad *f* de cromaticidad
i sensibilità *f* di cromaticità
n kleursoortgevoeligheid
d Farbartempfindlichkeit *f*

856 CHROMATICITY COEFFICIENT ct
Measure of the purity of a colo(u)r.
f coefficient *m* de chromaticité
e coeficiente *m* de cromaticidad
i coefficiente *m* di cromaticità
n kleursoortcoëfficiënt
d Farbartkoeffizient *m*

857 CHROMATICITY CO-ORDINATES ct
The ratios of each of the tristimulus values of a light to their sum.
f coordonnées *pl* de chromaticité,
coordonnées *pl* trichromatiques
e coordenadas *pl* de cromaticidad
i coordinate *pl* di cromaticità
n kleursoortcoördinaten *pl*
d Farbartkoordinate *pl*,
Farbwertkoordinate *pl*

858 CHROMATICITY DIAGRAM ct
A plane diagram formed by plotting one of the three chromaticity co-ordinates against another.
f diagramme *m* chromatique
e diagrama *m* de cromaticidad
i diagramma *m* di cromaticità

n kleurendiagram *n*,
volledige kleurendriehoek
d Farbtafel *f*

859 CHROMATICITY FLICKER ctv
Flicker which results from fluctuation of chromaticity only.
f papillotement *m* chromatique
e parpadeo *m* cromático
i sfarfallio *m* cromatico
n kleursoortgeflikker *n*
d Farbartflimmern *n*

860 CHROMATICNESS ct
The attribute of a visual sensation combining the hue and the saturation.
f sensation *f* de chroma
e sensación *f* de croma
i sensazione *f* di croma
n chromagewaarwording
d Chromaempfindung *f*

861 CHROMATICS ct
f chromatique *f*
e cromática *f*
i cromatica *f*
n kleurenleer
d Farbenlehre *f*

862 CHROMATON crt
A simplified modification of a chromatron.
f chromaton *m*
e cromatón *m*
i cromatone *m*
n chromaton *n*
d Chromaton *n*

863 CHROMATRON, ctv
CHROMOSCOPE,
FOCUS-MASK TUBE,
LAWRENCE TUBE,
POST-DEFLECTION FOCUS TUBE
A cathode-ray tube for colo(u)r TV, the screen of which is composed of four layers of phosphors, three of them for the primary colo(u)rs red, blue and green, the fourth one being connected to a high potential.
f chromatron *m*,
tube *m* tricolore à canon unique
e cromatrón *m*,
cromoscopio *m*
i cromatrone *m*,
cromoscopio *m*
n chromatron *n*,
chromoscoop
d Chromatron *n*,
Chromoskop *n*,
Gitterablenkröhre *f*

864 CHROME-DIOXIDE TAPE aud/vr
f bande *f* à enduit en bioxyde de chrome
e cinta *f* de capa en bióxido de cromo
i nastro *m* a strato in biossido di cromo
n band met chroomdioxydelaag
d Band *n* mit Chromdioxydschicht

865 CHROMINANCE ct/ctv
The colorimetric difference between the colo(u)r stimulus and a reference stimulus of the same luminance and given chromaticity, usually that of a visual grey in the particular viewing conditions.
f chrominance *f*
e crominancia *f*
i crominanza *f*
n chrominantie, kleurverschil *n*
d Chrominanz *f*, Farbdifferenz *f*

866 CHROMINANCE ADJUSTMENT svs
f ajustage *m* de la chrominance
e ajuste *m* de la crominancia
i aggiustaggio *m* della crominanza
n chrominantie-instelling
d Chrominanzeinstellung *f*

867 CHROMINANCE AXIS ct/ctv
One of a pair of axes in colo(u)r space along which the chrominance components are plotted.
f axe *m* de chrominance
e eje *m* de crominancia
i asse *m* di crominanza
n chrominantie-as
d Chrominanzachse *f*

868 CHROMINANCE BAND ctv
The frequency band which contains the colo(u)r information.
f bande *f* occupée par le signal de chrominance
e banda *f* de la señal de crominancia
i banda *f* del segnale di crominanza
n chrominantieband
d Chrominanzband *n*,
Farbfrequenzband *n*

869 CHROMINANCE BANDWIDTH, ctv
CHROMINANCE CHANNEL BANDWIDTH
The bandwidth of the path intended to carry the chrominance signal in a colo(u)r TV system.
f largeur *f* de la bande de chrominance
e anchura *f* de la banda de crominancia
i larghezza *f* della banda di crominanza
n chrominantiebandbreedte
d Chrominanzbandbreite *f*,
Farbfrequenzbandbreite *f*

870 CHROMINANCE CANCELLATION ctv
A cancellation of the brightness variations produced by the chrominance signal on the screen of a monochrome picture tube.
f suppression *f* de chrominance
e supresión *f* de crominancia
i soppressione *f* di crominanza
n chrominantieonderdrukking
d Chrominanzaustastung *f*

871 CHROMINANCE CARRIER, ctv
CHROMINANCE SUBCARRIER,
COLO(U)R SUBCARRIER
The carrier which is modulated to form

the chrominance signal.
f porteuse *f* couleur,
sousporteuse *f* couleur
e portadora *f* color,
subportadora *f* color
i portante *f* colore,
sottoportante *f* colore
n kleurdraaggolf,
kleurhulpdraaggolf
d Farbhilfsträger *m*,
Farbträger *m*

CHROMINANCE CARRIER ATTENUATING CIRCUIT
see: ANTI-CLOCHE CIRCUIT

872 CHROMINANCE CARRIER FREQUENCY, ctv
COLO(U)R CARRIER FREQUENCY
f fréquence *f* de la porteuse couleur
e frecuencia *f* de la portadora color
i frequenza *f* della portante colore
n kleurdraaggolffrequentie
d Farbträgerfrequenz *f*

873 CHROMINANCE CARRIER MODIFIER, ctv
COLO(U)R CARRIER MODIFIER
f modificateur *m* de la porteuse couleur
e modificador *m* de la portadora color
i modificatore *m* della portante colore
n kleurdraaggolfmodificator
d Farbträgermodifikator *m*

874 CHROMINANCE CARRIER REFERENCE, ctv
CHROMINANCE SUBCARRIER REFERENCE
A continuous signal having the same frequency as the chrominance subcarrier in a colo(u)r TV system and having fixed phase with respect to the colo(u)r burst.
f signal *m* de référence pour la sous-porteuse couleur
e señal *f* de referencia para la subportadora color
i segnale *m* di riferimento per la sotto-portante colore
n referentiesignaal *n* voor de hulpdraaggolf
d Farbträgerbezugssignal *n*

875 CHROMINANCE CARRIER REINSERTION OSCILLATOR, ctv
COLO(U)R CARRIER REINSERTION OSCILLATOR
f oscillateur *m* de réinsertion de la porteuse couleur
e oscilador *m* de reinserción de la portadora color
i oscillatore *m* di reinserzione della portante colore
n wederinvoeringsoscillator van de kleurdraaggolf
d Farbträgeroszillator *m*,
Farbträgerwiedereinführungsoszillator *m*

876 CHROMINANCE CARRIER SYNC PULSE, ctv
COLO(U)R CARRIER SYNC PULSE
f signal *m* de synchronisation de la porteuse couleur
e señal *f* de sincronización de la portadora color
i segnale *m* di sincronizzazione della portante colore
n synchronisatiesignaal *n* voor de kleurdraaggolf
d Farbträgersynchronsignal *n*

877 CHROMINANCE CHANNEL ctv
In a colo(u)r TV system, any path which is intended to carry the chrominance signal.
f canal *m* de chrominance
e canal *m* de crominancia
i canale *m* di crominanza
n chrominantiekanaal *n*
d Chrominanzkanal *m*

878 CHROMINANCE CIRCUITS ctv
Circuits containing all the spectral circuitry for extracting and processing the colo(u)r information encoded on the composite signal into a form suitable for applying to the shadow mask tube.
f circuits *pl* de chrominance
e circuitos *pl* de crominancia
i circuiti *pl* di crominanza
n chrominantieketens *pl*
d Chrominanzkreise *pl*

879 CHROMINANCE COMPONENTS ct/ctv
The two components of the chrominance vector relative to a pair of axes in a constant luminous plane having common origin on the grey axis of a colo(u)r space system.
f composantes *pl* de chrominance,
composantes *pl* du vecteur de chrominance
e componentes *pl* de crominancia,
componentes *pl* del vector de crominancia
i componenti *pl* del vettore di crominanza,
componenti *pl* di crominanza
n chrominantiecomponenten *pl*,
vectorcomponenten *pl* van de chrominantie
d Chrominanzkomponente *pl*,
Vektorkomponente *pl* der Chrominanz

880 CHROMINANCE DEFINITION (GB), ct
CHROMINANCE RESOLUTION (US)
f résolution *f* chromatique
e resolución *f* cromática
i risolvenza *f* cromatica
n chrominantiescheiding
d Farbauflösung *f*

881 CHROMINANCE DEMODULATOR, rep/ctv
CHROMINANCE DETECTOR,
CHROMINANCE SUBCARRIER DEMODULATOR
A demodulator used in colo(u)r TV reception for deriving video-frequency components from the chrominance signal.

f démodulateur *m* couleur
e desmodulador *m* color
i demodulatore *m* colore
n kleurdemodulator
d Farbdemodulator *m*

882 CHROMINANCE INFORMATION ct/ctv
f information *f* couleur
e información *f* color
i informazione *f* colore
n kleurinformatie
d Farbinformation *f*

883 CHROMINANCE/LUMINANCE DELAY INEQUALITY, ctv
DELAY INEQUALITY
The delay of the chrominance channel relative to that of the luminance channel, measured under specified conditions and expressed in nanoseconds.
f inégalité *f* de retard,
retard *m* chrominance-luminance
e desigualdad *f* de retardo,
retardo *m* crominancia-luminancia
i disuguagliamento *m* di ritardo,
ritardo *m* crominanza-luminanza
n chrominantie-luminantie-vertraging,
vertragingsongelijkheid
d Chrominanz-Leuchtdichteverzögerung *f*,
Verzögerungsungleichheit *f*

884 CHROMINANCE/LUMINANCE FACTOR ctv
The ratio of the measured chrominance/luminance ratio at any point in the system to that specified for the system.
f facteur *m* chrominance-luminance
e factor *m* crominancia-luminancia
i fattore *m* crominanza-luminanza
n chrominantie-luminantie-factor
d Chrominanz-Leuchtdichtefaktor *m*

885 CHROMINANCE/LUMINANCE GAIN INEQUALITY, ctv
GAIN INEQUALITY
The voltage gain of the chrominance channel relative to that of the luminance channel, measured under specified conditions and expressed as a percentage reference.
f gain *m* chrominance-luminance,
inégalité *f* de gain
e desigualdad *f* de ganancia,
ganancia *f* crominancia-luminancia
i disuguagliamento *m* di guadagno,
guadagno *m* crominanza-luminanza
n chrominantie-luminantie-versterking.
versterkingsongelijkheid
d Chrominanz-Leuchtdichteverstärkung *f*,
Verstärkungsungleichheit *f*

886 CHROMINANCE/LUMINANCE RATIO ctv
For a specified colo(u)r TV system, the ratio of the amplitude of the chrominance signal to the amplitude of the luminance signal under specified conditions.
f rapport *m* chrominance-luminance
e relación *f* crominancia-luminancia
i rapporto *m* crominanza-luminanza
n chrominantie-luminantie-verhouding
d Chrominanz-Leuchtdichteverhältnis *n*

887 CHROMINANCE MATRIXING ctv
Matrixing operation using modulated chrominance signals.
f matriçage *m* des signaux de chrominance
e representación *f* por matrices de las señales de crominancia
i rappresentazione *f* per matrici dei segnali di crominanza
n matricering van de chrominantiesignalen
d Matrizierung *f* der Chrominanzsignale

888 CHROMINANCE MODULATOR, rep/ctv
CHROMINANCE SUBCARRIER MODULATOR
A modulator used in colo(u)r TV transmission for generating the chrominance signal from the video-frequency components and the chrominance subcarrier.
f modulateur *m* couleur
e modulador *m* color
i modulatore *m* colore
n kleurmodulator
d Farbmodulator *m*

889 CHROMINANCE PHASE ctv
f phase *f* de chrominance
e fase *f* de crominancia
i fase *f* di crominanza
n chrominantiefaze
d Chrominanzphase *f*

890 CHROMINANCE PRIMARY ct/ctv
One of the pair of colo(u)r stimuli defined by the directions of the chrominance axes in the plane of zero luminance.
f primaire *m* de chrominance
e primario *m* de crominancia
i primario *m* di crominanza
n primaire chrominantiestimulus
d primärer Farbreiz *m*

891 CHROMINANCE PRIMARY SIGNAL ctv
f signal *m* du primaire de chrominance
e señal *f* del primario de crominancia
i segnale *m* del primario di crominanza
n signaal *n* van de primaire chrominantie-stimulus
d primäres Farbreizsignal *n*

892 CHROMINANCE SECTION rep/svs/tv
f section *f* chrominance
e sección *f* crominancia
i sezione *f* crominanza
n chrominantiedeel *n*
d Chrominanzteil *m*

CHROMINANCE SIGNAL
see: CARRIER CHROMINANCE SIGNAL

893 CHROMINANCE SIGNAL COMPONENT ctv
A component of the chrominance signal,

produced when a chrominance primary signal suitably modulates a subcarrier voltage of the proper phase.
f composante *f* du signal de chrominance
e componente *f* de la señal de crominancia
i componente *f* del segnale di crominanza
n component van het chrominantiesignaal
d Komponenten *f* des Farbartsignals

894 CHROMINANCE SIGNAL VECTOR DIAGRAM, ctv
VECTOR DIAGRAM OF THE CHROMINANCE SIGNAL
f diagramme *m* vectoriel du signal de chrominance
e diagrama *m* vectorial de la señal de crominancia
i diagramma *m* vettoriale del segnale di crominanza
n vectordiagram *n* van het chrominantiesignaal
d Vektordiagramm *n* des Farbsignals

895 CHROMINANCE SUBCARRIER LEVEL, ctv
COLO(U)R SUBCARRIER LEVEL
f niveau *m* de la sousporteuse couleur
e nivel *m* de la subportadora color
i livello *m* della sottoportante colore
n niveau *n* van de kleurhulpdraaggolf
d Farbhilfsträgerpegel *m*

896 CHROMINANCE SUBCARRIER REFERENCE, ctv
COLO(U)R SUBCARRIER REFERENCE,
REFERENCE SUBCARRIER
In the NTSC system, a continuous wave having the same frequency as the chrominance subcarrier and having a fixed phase with respect to the colo(u)r burst.
f sousporteuse *f* couleur de référence
e subportadora *f* color de referencia
i sottoportante *f* colore di riferimento
n referentiekleurhulpdraaggolf
d Bezugsfarbhilfsträger *m*,
Bezugshilfsträger *m*

897 CHROMINANCE SUBCARRIER REGENERATOR ctv
In a NTSC colo(u)r TV receiver, the circuit which performs the function of generating a local subcarrier locked in phase with the transmitter burst.
f générateur *m* de la sousporteuse
e generador *m* de la subportadora
i generatore *m* della sottoportante
n generator van de hulpdraaggolf
d Hilfsträgergenerator *m*

898 CHROMINANCE SUBCARRIER SIGNAL, ctv
COLO(U)R SUBCARRIER SIGNAL
f signal *m* de la sousporteuse couleur
e señal *f* de la subportadora color
i segnale *m* della sottoportante colore
n kleurhulpdraaggolfsignaal *n*
d Farbhilfsträgersignal *n*

899 CHROMINANCE UNIT ctv
Set of components of a colo(u)r TV set.
f bloc *m* de chrominance
e bloque *m* de crominancia
i blocco *m* di crominanza
n chrominantie-eenheid
d Bunteinheit *f*,
Chrominanzeinheit *f*

900 CHROMINANCE VECTOR ctv
In a colo(u)r TV signal, the finite mathematical vector whose angle represents the hue and whose length represents saturation.
f vecteur *m* de chrominance
e vector *m* de crominancia
i vettore *m* di crominanza
n chrominantievector
d Chrominanzvektor *m*

901 CHROMINANCE VIDEO SIGNAL ctv
f **signal *m* d'une image en couleur**
e señal *f* de una imagen en colores
i segnale *m* d'un'immagine a colori
n kleurenbeeldsignaal *n*
d Farbbildsignal *n*

CHROMOSCOPE
see: CHROMATRON

902 CINCH, dis/vr
CINCHING
f glissement *m* des spires
e cinchado *m*
i nastro *m* arricciato
n cinch, cinching
d Fensterbildung *f* im Bandwickel

903 CIRCLE ANIMATION ani
f animation *f* à cercle
e animación *f* con círculo
i animazione *f* a cerchio
n cirkelbeweging in tekenfilm
d Kreisbewegung *f* in Trickfilm

904 CIRCLE OF CONFUSION, opt
CIRCLE OF LEAST CONFUSION
The circular image of a distant point object as formed in a focal plane by a lens.
f cercle *m* de confusion
e círculo *m* de confusión
i cerchio *m* di confusione
n uitvloeiingscirkel
d Unschärfering *m*,
Zerstreuungskreis *m*

905 CIRCUIT DIAGRAM ge
f schéma *m* de montage
e esquema *m* de conexiones
i schema *m* di connessioni
n schakelschema *n*
d Schaltbild *n*, Schaltschema *n*, Stromlauf *n*

906 CIRCUIT MODULE cpl
f module *m* de circuit
e módulo *m* de circuito
i modulo *m* di circuito

n schakelmoduul *n*
d Schaltmodul *n*

907 CIRCUITRY ge
f montage *m*
e conjunto *m* de circuitos
i schema di distribuzione
n schakelplan *n*
d Schaltungsaufbau *m*

908 CIRCULAR DIAGRAM AERIAL (GB), aea
CIRCULAR PATTERN ANTENNA (US)
f antenne *f* à diagramme de rayonnement circulaire
e antena *f* de diagrama de radiación circular
i antenna *f* a diagramma di radiazione circolare
n antenne met cirkelvormig stralingsdiagram
d Antenne *f* mit kreisförmigem Strahlungsdiagramm

909 CIRCULAR SWEEP crt/tv
A sweep circuit which provides a circular timebase for cathode-ray tube presentations.
f analyse *f* spirale
e exploración *f* en espiral
i analisi *f* a spirale
n spiraalaftasting
d Spiralabtastung *f*

910 CIRCULAR TIMEBASE cpl
Circuit for causing the spot on the screen of a cathode-ray tube to traverse a circular path at constant angular velocity.
f base *f* de temps circulaire
e base *f* de tiempo circular
i base *f* di tempi circolare
n cirkelvormige tijdbasis
d kreisförmige Zeitbasis *f*

911 CIRCULAR TRACE crt
f trace *f* circulaire
e traza *f* circular
i traccia *f* circolare
n cirkelspoor *n*
d kreisförmige Spur *f*

912 CIRCULAR WAVEGUIDE cpl
f guide *m* d'ondes circulaire
e guía *f* de ondas anular
i guida *f* d'onde circolare
n ronde golfgeleider
d runder Hohlleiter *m*,
Rundhohlleiter *m*

913 CLADDING MATERIAL crt
Material used in the manufacture of cathode-ray tubes.
f matériel *m* de revêtement
e material *m* de revestimiento
i materiale *m* di rivestimento
n bekledingsmateriaal *n*
d Verkleidungsmaterial *n*

914 CLAMP, cpl/ge
CLAMPING CIRCUIT
A circuit, usually electronic, designed to hold a particular part of a waveform at a predetermined voltage level.
f circuit *m* de clamp,
circuit *m* de verrouillage
e circuito *m* de bloqueo,
circuito *m* de fijación de amplitud
i circuito *m* allineatore,
circuito *m* di bloccaggio,
circuito *m* di fissazione d'ampiezza
n blokkeerschakeling,
klemschakeling
d Klemmschaltung *f*

915 CLAMP AMPLIFIER cpl
Video amplifier which re-establishes the d.c. level of the signal with respect to black on a line-by-line basis.
f amplificateur *m* restaurateur du niveau du courant continu
e amplificador *m* restaurador del nivel de la corriente continua
i amplificatore *m* ristoratore del livello della corrente continua
n blokkeerversterker
d Klemmverstärker *m*

916 CLAMP PULSE GENERATION, cpl
CLAMPING PULSE GENERATION
f génération *f* d'impulsions de clamp,
génération *f* d'impulsions de verrouillage
e generación *f* de impulsos de bloqueo,
generación *f* de impulsos de estabilización
i generazione *f* d'impulsi di bloccaggio
n opwekking van blokkeerimpulsen
d Klemmimpulserzeugung *f*

917 CLAMP PULSES, cpl
CLAMPING PULSES
Accurately timed pulses which render a clamp circuit operative, usually causing it to become conductive.
f impulsions *pl* de clamp,
impulsions *pl* de verrouillage
e impulsos *pl* de bloqueo,
impulsos *pl* de fijación
i impulsi *pl* di bloccaggio
n blokkeerimpulsen *pl*,
grendelimpulsen *pl*
d Klemmimpulse *pl*

918 CLAMPED D.C.RESTORATION cpl
f restauration *f* de la composante continue sur niveaux préfixés
e restauración *f* de la componente continua sobre niveles prefijados
i reinserzione *f* della componente continua su livelli prefissati
n herstellen *n* van de werkzame nulcomponent op vooraf vastgestelde niveaus
d getastete Schwarzwerthaltung *f*

919 CLAMPING cpl/tv
The process that establishes a fixed level

for the repetitive components of the video signal.
f clamp *m*,
verrouillage *m* du niveau
e bloqueo *m* del nivel,
fijación *f* del nivel
i bloccaggio *m* del livello,
fissazione *f* del livello
n niveaustabilisatie
d Klemmung *f*,
Pegelhaltung *f*

920 CLAMPING DIODE cpl/tv
f diode *f* de restitution du niveau de noir,
diode *f* de verrouillage
e diodo *m* de bloqueo,
diodo *m* de restauración del nivel de negro,
diodo *m* enclavador
i diodo *m* allineatore,
diodo *m* di bloccaggio,
diodo *m* di restaurazione del livello di nero
n blokkeerdiode,
klemdiode,
zwartniveaudiode
d Klemmdiode *f*,
Schwarzwertdiode *f*

921 CLEAN EFFECTS rec/tv
Background sound from a broadcast without added commentary.
f musique *f* de fond sans commentaire
e música *f* de fondo sin comento
i musica *f* di fondo senza commentario
n achtergrondmuziek zonder commentaar
d Hintergrundmusik *f* ohne Kommentar

922 CLEAN EFFECTS MIXER rec/tv
f mélangeur *m* de musique de fond sans commentaire
e mezclador *m* de música de fondo sin comento
i mescolatore *m* di musica di fondo senza commentario
n menger van achtergrondmuziek zonder commentaar
d Mischer *m* von Hintergrundmusik ohne Kommentar

923 CLEAN EFFECTS MIXING rec/tv
f mixage *m* de musique de fond sans commentaire
e mezclado *m* de música de fondo sin comento
i mescolanza *f* di musica di fondo senza commentario
n mengen *n* van achtergrondmuziek zonder commentaar
d Mischung *f* von Hintergrundmusik ohne Kommentar

924 CLEAN START, rep/tv
READY FOR TRANSMISSION
Commencement of transmission of a unit of program(me) when the gain has been previously adjusted to transmit normal volume or level, and without subsequent adjustment.
f bon pour diffusion,
prêt à diffuser
e listo para la transmisión
i pronto per la trasmissione
n startklaar voor uitzending
d sendefertig

925 CLEANING OF BASE FILM vr
One of the steps in the manufacture of magnetic tape.
f nettoyage *m* de la bande de base
e limpieza *f* de la cinta de base
i pulitura *f* del nastro di base
n schoonmaken *n* van de basisband
d Reinigung *f* des Basisbandes

926 CLEAR BINDER, vr
CLEAR MEDIUM
Substance used in the manufacture of magnetic tape.
f liant *m* clair
e adhesivo *m* claro
i legante *m* chiaro
n helder bindmiddel *n*
d klares Bindemittel *n*

927 CLINICAL TELEVISION RECORDING med
f enregistrement *m* de télévision pour buts médicaux
e registro *m* televisual en la clínica
i registrazione *f* televisiva per scopi ospedalieri
n televisieopname voor medische doeleinden
d Fernsehaufnahme *f* von klinischen Vorgängen

928 CLIP, fi/tv
TRIMS
f chute *f*
e recortes *pl*
i ritagli *pl*
n snijdsel *n*
d Verschnitt *m*

929 CLIPPER cpl
A limiter used specifically for producing waveforms of desired shapes.
f écrêteur *m*
e recortador *m*
i limitatore *m* del valore di soglia,
tosatore *m*
n drempelwaardebegrenzer
d Abschneidestufe *f*,
Schwellenwertbegrenzer *m*

930 CLIPPER AMPLIFIER cpl
An amplifier designed to limit the instantaneous value of its output to a predetermined maximum.
f amplificateur *m* limiteur
e amplificador *m* limitador
i amplificatore *m* limitatore
n begrenzende versterker
d begrenzender Verstärker *m*

CLIPPER CIRCUIT (US)
see: AMPLITUDE LIMITER CIRCUIT

931 CLIPPER TUBE (US), cpl
THRESHOLD TUBE (GB)
Electronic tube (valve) in a circuit such that the tube (valve) will pass a signal only in part or not at all, according to the level of the input signal.
f tube *m* de seuil
e válvula *f* de umbral
i valvola *f* di soglia
n drempelbuis
d Schwellenröhre *f*

932 CLIPPING LEVEL cpl
The amplitude level at which a waveform is clipped.
f niveau *m* d'écrêtage
e nivel *m* de recorte
i livello *m* di tosatura
n afsnijniveau *n*
d Abschneidepegel *m*

933 CLIPPING TIME cpl
The time constant of a clipping circuit.
f temps *m* d'écrêtage
e tiempo *m* de recorte
i tempo *m* di tosatura
n afsnijtijd
d Abschneidezeit *f*

934 CLOCHE CIRCUIT, ctv
GAUSSIAN CIRCUIT
In the SECAM system a circuit to accentuate within a frequency the transmission and/or the gain.
f circuit *m* cloche
e circuito *m* elevador de la ganancia
i circuito *m* elevatore del guadagno
n klokschakeling
d Glockenkreis *m*

935 CLOSE MEDIUM SHOT, rec/tv
MEDIUM CLOSE SHOT,
MEDIUM SHOT
f plan *m* moyen
e plano *m* medio
i piano *m* medio
n half-groot-opname
d Halbnahe *f*

936 CLOSE-MESH SPACING crt
Target-mesh spacing in an image orthicon tube.
f espace *m* minime entre cible et grille
e espacio *m* mínimo entre blanco y rejilla
i spazio *m* minimo tra bersaglio e griglia
n minimale ruimte tussen trefplaat en rooster
d Minimumabstand *m* zwischen Treffplatte und Gitter

937 CLOSE OF TRANSMISSION, tv
CLOSING DOWN OF BROADCASTING
f fin *f* de transmission
e clausurado *m* de transmisión
i fine *f* di trasmissione
n einde *n* van de uitzending
d Sendeschluss *m*

938 CLOSE SCANNING (US), rec/tv
FINE SCANNING (GB)
Scanning system which uses a beam of very small size and a large number of lines.
f analyse *f* à haute définition
e exploración *f* de alta definición
i analisi *f* ad alta definizione
n fijnaftasting
d Feinabtastung *f*

939 CLOSE SHOT, rec/tv
CLOSE-UP (GB),
CLOSE-UP VIEW (US),
CLOSE VIEW,
MUG SHOT (US)
f gros plan *m*,
plan *m* très rapproché
e primer plano *m*,
toma *f* cercana
i ripresa *f* di primi piani
n close-up,
opname van vlakbij
d Grossaufnahme *f*,
Nahaufnahme *f*

940 CLOSE-TALKING MICROPHONE, aud
LIP MICROPHONE,
NOISE-CANCEL(L)ING MICROPHONE
f microphone *m* bouche
e micrófono *m* de anteboca,
micrófono *m* para hablar de cerca
i microfono *m* labiale
n lipmicrofoon
d Lippenmikrophon *n*,
Mikrophon *n* für Nahbesprechung

941 CLOSE-TALKING RESPONSE, aud
CLOSE-TALKING SENSITIVITY
f efficacité *f* paraphonique,
réponse *f* paraphonique
e rendimiento *m* parafónico,
respuesta *f* parafónica
i sensibilità *f* a distanza ravvicinata
n gevoeligheid op korte afstand,
gevoeligheid bij nabijheidsbespreking
d Empfindlichkeit *f* bei Nahbesprechung,
Übertragungsfaktor *m*

942 CLOSED CIRCUIT tv
A channel between various parts of the technical equipment in broadcasting for testing, rehearsal, recording or other purposes not involving broadcast transmission.
f circuit *m* local
e circuito *m* cerrado
i circuito *m* chiuso,
circuito *m* locale
n gesloten controleketen
d Vorabhörweg *m*

943 CLOSED-CIRCUIT NETWORK cpl/tv
DESIGN
f projet *m* de réseau pour circuit fermé
e proyecto *m* de red para circuito cerrado
i progetto *m* di rete per circuito chiuso

n netontwerp *n* voor gesloten kring
d Netzentwurf *m* für geschlossenen Kreis

944 CLOSED-CIRCUIT RECORDING cpl/rec/tv
A recording for subsequent dissemination in broadcasting and not of a performance being broadcast.
f enregistrement *m* à circuit fermé
e registro *m* de circuito cerrado
i registrazione *f* a circuito chiuso
n opname met gesloten circuit
d Aufnahme *f* mit geschlossenem Kreis

CLOSED-CIRCUIT TELEVISION
see: CCTV

945 CLOSED LOOP DRIVE SYSTEM aud/vr
f système *m* de commande en boucle fermée
e sistema *m* de mando en anillo cerrado
i sistema *m* di comando ad anello chiuso
n stuursysteem *n* met gesloten lus
d Regelsystem *n*

946 CLOUD (US), DARK AND LIGHT SPOTS dis/rep/tv
The result of unequal boundary potential caused by differences in the field between various parts of the mosaic of an iconoscope and the parts themselves.
f effets *pl* d'ombre
e efectos *pl* de sombra, nublado *m*
i effetti *pl* d'ombra
n schaduwvlekken *pl*
d Schattenflecke *pl*

CMP
see: CENTRAL MONITORING POSITION

947 COARSE CHROMINANCE PRIMARY ctv
In the colo(u)r TV system for broadcasting in the USA, that one of the two chrominance primaries which is associated with the lower transmission bandwidth.
f primaire *m* de chrominance à largeur de bande mineure
e primario *m* de crominancia tosco
i segnale *m* di crominanza con minore larghezza di banda
n chrominantiesignaal *n* met de kleinste bandbreedte
d Farbartsignal *n* mit kleinster Bandbreite

948 COARSE SCANNING rec/tv
In TV, scanning with a light spot whose diameter is comparable to the detail of the image, resulting in a severe limitation of the number of lines.
f analyse *f* approximative
e exploración *f* basta
i analisi *f* grossolana
n grofaftasting
d Grobabtastung *f*

COATED LENS
see: BLOOMED LENS

949 COATED MAGNETIC TAPE, MAGNETIC-POWDER COATED TAPE vr
A magnetic tape coated with magnetic particles.
f bande *f* à couche magnétique
e cinta *f* de capa magnética
i nastro *m* a strato magnetico
n band met magnetische laag
d Zweischichtband *n*

950 COATING ge
Application of emulsion or magnetic material to base film.
f couche *f*, revêtement *m*
e capa *f*, revestimiento *m*
i rivestimento *m*, strato *m*
n laag
d Belag *m*, Schicht *f*, Überzug *m*

951 COATING OF LENSES opt
f traitement *m* de la surface des lentilles
e revestimiento *m* de lentes
i trattamento *m* superficiale di lenti
n behandeling van het lensoppervlak
d Beschichtung *f* der Linsenoberfläche, Vergütung *f* der Linsenoberfläche

952 COAXIAL AERIAL, COAXIAL ANTENNA, COAXIAL DIPOLE AERIAL, COAXIAL DIPOLE ANTENNA aea
Exposed $\lambda/4$ length of coaxial line, with reversed metal cover, acting as a dipole.
f antenne *f* coaxiale
e antena *f* coaxial
i antenna *f* coassiale
n coaxiale antenne
d Koaxialantenne *f*

953 COAXIAL CABLE cpl
A cable consisting principally of one or more coaxial pairs.
f câble *m* coaxial
e cable *m* coaxial
i cavo *m* coassiale
n coaxiale kabel
d koaxiales Kabel *n*, Koaxialkabel *n*

954 COAXIAL-CABLE TRANSMISSION tv
f transmission *f* sur câble coaxial
e transmisión *f* sobre cable coaxial
i teletrasmissione *f* su cavo coassiale
n transmissie over coaxiale kabel
d Koaxialkabelübertragung *f*

955 COAXIAL FEEDER aea
f ligne *f* d'alimentation coaxiale
e alimentador *m* coaxial
i alimentatore *m* coassiale
n coaxiale voedingsleiding
d konzentrische Speiseleitung *f*

956 COAXIAL LINE cpl
A transmission line formed from inner and outer cylindrical conductors with a common axis.
f ligne *f* coaxiale
e línea *f* coaxial

i linea *f* coassiale
n coaxiale leiding
d koaxiale Leitung *f*

957 COAXIAL CONNECTOR, COAXIAL TERMINAL cpl
A terminal to which a coaxial line may be connected.
f connexion *f* coaxiale
e conexión *f* coaxial
i raccordo *m* coassiale
n coaxiale aansluiting
d Koaxialanschluss *m*

958 CO-CHANNEL INTERFERENCE tv
Interference between TV stations operating on the same channel.
f interférence *f* par canal commun
e interferencia *f* de co-canales
i interferenza *f* per canale comune
n zelfde-kanaalstoring
d Gleichkanalstörung *f*

959 CO-CHANNEL STATIONS tv
f stations *pl* à canal commun
e estaciones *pl* de co-canales
i stazioni *pl* a canale comune
n zendstations *pl* in het zelfde kanaal
d Sendestellen *pl* im gleichen Kanal

960 CODER, COLO(U)R CODER, COLO(U)R ENCODER, ENCODER ctv
In colo(u)r TV transmission an apparatus for producing the colo(u)r picture signal and possibly colo(u)r burst from e.g., the camera signals and the chrominance subcarrier.
f codeur *m*
e codificador *m*
i codificatore *m*
n codeur, kleurcodeur
d Farbkoder *m*, Farbkodierer *m*, Koder *m*, Kodierer *m*

961 COERCIVE FORCE, COERCIVITY cpl
In a ferromagnetic material, the magnetizing force which must be applied to that material to reduce the magnetic flux density to zero.
f force *f* coercitive
e fuerza *f* coercitiva
i forza *f* coercitiva
n coërcitie, coërcitieve kracht
d Koerzitivkraft *f*

962 COGGING dis/rep/tv
Break-up of vertical edges caused by a displacement of the raster from one field to the next.
f engrenage *m*
e engranaje *m*
i ingranaggio *m*
n ineengrijpen *n*, vertanding
d Verkämmung *f*

963 COHERENT LIGHT cpt/vr
The powerful light provided by a laser and used in holography.
f lumière *f* cohérente
e luz *f* coherente
i luce *f* coerente
n coherent licht *n*
d kohärentes Licht *n*

964 COIN-FREED TELEVISION, FEE TELEVISION, PAY-AS-YOU-SEE TELEVISION, SUBSCRIPTION TELEVISION tv
f télévision *f* à prépaiement
e televisión *f* con pago previo
i televisione *f* a gettone, televisione *f* a moneta
n munttelevisiesysteem *n*
d Münzfernsehen *n*

965 COLD-LIGHT MIRROR tv
Lighting device used in the Schlieren large screen TV system.
f miroir *m* à lumière froide
e espejo *m* de luz fría
i specchio *m* a luce fredda
n koudlichtspiegel
d Kaltlichtspiegel *m*

966 COLD-MIRROR REFLECTOR stu/tv
Type of studio lamp in which the standard interval aluminized reflector is replaced by a special dichroic coating which reflects the visible light with little loss but transmits infrared wavelengths.
f réflecteur *m* à revêtement dichroïque
e reflector *m* de capa dicroica
i riflettore *m* a rivestimento dicroico
n reflector met dichroïsche laag
d Reflektor *m* mit dichroischer Schicht

COLLECTING ZONE
see: CAPTURE RANGE

COLLECTIVE ANTENNA
see: BLOCK ANTENNA

967 COLLECTOR crt
Anode which collects the secondary electrons emitted from the mosaic of an iconoscope or similar device.
f électrode *f* collectrice
e electrodo *m* colector
i elettrodo *m* collettore
n verzamelelektrode
d Sammelelektrode *f*

COLLECTOR RING
see: ANNULAR COLLECTOR ELECTRODE

968 COLLIMATION opt
Process of setting up a lens system to produce a parallel beam from a light source.
f collimation *f*
e colimación *f*
i collimazione *f*

n collimatie
d Kollimation *f*

969 COLLIMATOR opt
Instrument for the virtual production of an object at infinity, consisting of a graticule or other transparent object in the focal plane of a corrected lens.
f collimateur *m*
e colimador *m*
i collimatore *m*
n collimator
d Kollimator *m*

970 COLLINEAR AERIAL, COLLINEAR ANTENNA, COLLINEAR ARRAY aea
f nappe *f* de dipôles horizontaux, série *f* de dipôles
e antena *f* colineal
i antenna *f* collineale, fila *f* di dipoli
n dipoolrij
d Dipolreihe *f*, Dipolzeile *f*, lineare Dipolgruppe *f*

971 COLOR (US), COLOUR (GB) ct
A characteristic of light that can be specified in terms of luminance, dominant wavelength, and purity.
f couleur *f*
e color *m*
i colore *m*
n kleur
d Farbe *f*

972 COLO(U)R ANALYSIS ct/opt
f analyse *f* de couleur
e análisis *f* de color
i analisi *f* di colore
n kleuranalyse
d Farbanalyse *f*

973 COLO(U)R AND SOUND MONITOR ctv
f moniteur *m* couleur et son
e monitor *m* color y sonido
i monitore *m* colore e suono
n kleur- en geluidsmonitor
d Farb- und Tonkontrolle *f*

974 COLO(U)R ATLAS ct
Any collection of colo(u)r samples arranged systematically for evaluating colo(u)rs by visual matching.
f atlas *m* chromatique, atlas *m* de couleurs
e atlas *m* cromático, atlas *m* de colores
i atlante *m* cromatico, atlante *m* di colori
n kleurenatlas
d Farbenatlas *m*

975 COLO(U)R AXIS ct
A line in any three-dimensional representation of a colo(u)r space which denotes the direction of one of the reference stimuli.
f axe *m* de couleur
e eje *m* de color
i asse *m* di colore
n kleurenas
d Farbachse *f*

976 COLO(U)R BALANCE rec
Proper proportioning of phosphor efficiences and/or electrical and optical characteristics so that the scale of neutral greys in the picture is reproduced achromatically.
f équilibrage *m* de couleurs
e equilibrio *m* de colores
i equilibratura *f* di colori
n kleurvereffening
d Farbabgleich *m*, Farbabstimmung *f*

977 COLO(U)R BALANCE CONTROL ctv
f contrôle *m* de l'équilibrage de couleurs
e control *m* del equilibrio de colores
i controllo *m* dell'equilibratura di colori
n kleurvereffeningscontrole
d Farbabgleichkontrolle *f*, Farbabstimmungskontrolle *f*

978 COLO(U)R BALANCE CONTROL PANEL crt/ctv
f pupitre *m* de contrôle de l'équilibrage de couleurs
e pupitre *m* de control del equilibrio de colores
i tavolo *m* di controllo dell'equilibratura di colori
n kleurvereffeningslessenaar
d Farbabgleichpult *n*, Farbabstimmungspult *n*

979 COLO(U)R BAR ctv
Bar of colo(u)r produced on a colo(u)r TV display.
f barre *f* colorée
e barra *f* colorada
i barra *f* colorata
n kleurbalk
d Farbbalken *m*

980 COLO(U)R BAR GENERATOR, COLO(U)R BAR SIGNAL GENERATOR ctv
An electronic device which generates a test signal corresponding to a picture of bars of specified colo(u)rs.
f générateur *m* du signal de barre colorée
e generador *m* de la señal de barra colorada
i generatore *m* del segnale di barra colorata
n kleurbalkgenerator, kleurbalksignaalgenerator
d Farbbalkengenerator *m*, Farbbalkensignalgenerator *m*

981 COLO(U)R BAR PATTERN, COLO(U)R BAR TEST PATTERN ctv
f mire *f* à barres colorées
e patrón *m* de barras coloradas

i figura *f* di prova a barre colorate
n kleurbalkentoetsbeeld *n*
d Farbbalkenmuster *n*

982 COLO(U)R BAR SIGNAL ctv
The signal producing a colo(u)r bar.
f signal *m* de barre colorée
e señal *f* de barra colorada
i segnale *m* di barra colorata
n kleurbalksignaal *n*
d Farbbalkensignal *n*

983 COLO(U)R/BLACK-WHITE SWITCH ctv
f commutateur *m* couleurs/noir et blanc
e conmutador *m* colores/blanco y negro
i commutatore *m* colori/bianco e nero
n kleuren/zwart-witschakelaar
d Farben/Schwarzweiss-Schalter *m*

984 COLO(U)R BLENDING ct
f mixage *m* de couleurs
e mezclado *m* de colores
i mescolanza *f* di colori
n kleurmenging
d Farbblendung *f*

985 COLO(U)R BREAK UP, ctv/dis
COLO(U)R SPLITTING
Transitory separation of colo(u)rs in an intermittent TV picture arisen from motion of the eye of the observer.
f décomposition *f* de couleurs
e descomposición *f* de colores
i scomposizione *f* di colori
n kleurontleding, kleursplitsing
d Farbzerlegung *f*

986 COLO(U)R BRIGHTNESS ct
f brillance *f* de couleur
e brillo *m* de color
i brillanza *f* di colore
n kleurhelderheid
d Farbhelligkeit *f*

COLO(U)R BURST
see: BURST

COLO(U)R BURST PEDESTAL
see: BURST PEDESTAL

987 COLO(U)R CAMERA SYSTEM ctv
f système *m* de caméra à couleur
e sistema *m* de cámara de colores
i sistema *m* di camera a colori
n kleurencamerasysteem *n*
d Farbkamerasystem *n*

COLO(U)R CARRIER FREQUENCY
see: CHROMINANCE-CARRIER FREQUENCY

COLO(U)R CARRIER MODIFIER
see: CHROMINANCE CARRIER MODIFIER

COLO(U)R CARRIER REINSERTION OSCILLATOR
see: CHROMINANCE CARRIER REINSERTION OSCILLATOR

COLO(U)R CARRIER SYNC PULSE
see: CHROMINANCE CARRIER SYNC PULSE

988 COLO(U)R CAST ctv/dis
Predominance of one-colo(u)r primary in an colo(u)r image resulting from lack of colo(u)r balance.
f dominante *f*,
prédominance *f* d'une couleur primaire
e predominancia *f* de uno color primario
i predominanza *f* di uno colore primario
n overheersing van één primaire kleur,
primaire kleurzweem
d Farbstich *m*,
Überherrschung *f* einer Primärfarbe

989 COLO(U)R CELL crt
Of a display tube, the smallest portion of the picture area which is capable of reproducing colo(u)r information.
f cellule *f* de couleur
e celda *f* de color
i cella *f* di colore
n kleurcel
d Farbzelle *f*

990 COLO(U)R CENTER(RE), crt/ctv
U CENTER(RE)
Unit region of cathode-ray tube phosphor, determined by the electron beam and perforated screen in establishing a colo(u)r TV image.
f centre *m* de couleur
e centro *m* de color
i centro *m* di colore
n kleurcentrum *n*
d Farbzentrum *n*

991 COLO(U)R CHANNEL ctv
f canal *m* couleur
e canal *m* color
i canale *m* colore
n kleurkanaal *n*
d Farbkanal *m*, Farbleitung *f*

992 COLO(U)R CHART ct
f charte *f* chromatique,
charte *f* de couleurs,
planche *f* de couleurs
e carta *f* cromática,
cuadro *m* de colores
i tavola *f* a colori
n kleurkaart
d Farbtafel *f*

COLO(U)R CODER,
COLO(U)R ENCODER
see: CODER

993 COLO(U)R COMPARISON ct
f comparaison *f* de couleurs
e comparación *f* de colores
i comparazione *f* di colori
n kleurvergelijking
d Farbvergleich *m*

994 COLO(U)R COMPENSATING FILTER ct
f filtre *m* compensateur de couleur

e filtro *m* compensador de color
i filtro *m* compensatore di colore
n kleurcompensatiefilter *n*
d Farbausgleichsfilter *n*

995 COLO(U)R COMPONENTS ct
f composantes *pl* de couleur
e componentes *pl* de color
i componenti *pl* di colore
n kleurcomponenten *pl*
d Farbkomponenten *pl*

996 COLO(U)R CONTAMINATION ctv/dis
An error of colo(u)r rendition due to incomplete separation of paths carrying different colo(u)r components of the picture.
f distorsion *f* colorée due à la diaphotie
e contaminación *f* de color
i contaminazione *f* di colore
n kleurverontreiniging
d Farbübersprechen *n*, Farbwertverschiebung *f*

997 COLO(U)R CONTENT ct
The subjectively estimated amount of colo(u)rfullness seen in the visual sensation arising from a surface colo(u)r.
f contenu *m* de couleur
e contenido *m* de color
i contenuto *m* di colore
n kleurgehalte *n*
d Farbgehalt *m*

998 COLO(U)R CONTRAST ct
The ratio of the intensities of the sensations caused by two colo(u)rs.
f contraste *m* de couleurs
e contraste *m* de colores
i contrasto *m* di colori
n kleurcontrast *n*
d Farbkontrast *m*

999 COLO(U)R CONTROL ctv
f régleur *m* de la tension couleur
e regulador *m* de la tensión color
i regolatore *m* della tensione colore
n kleurspanningsregelaar
d Buntregler *m*, Buntspannungsregler *m*

1000 COLO(U)R CONTROLS ADJUSTING ctv/svs
f retouche *f* des réglages des couleurs
e ajuste *m* de los reguladores de los colores
i aggiustaggio *m* dei regolatori dei colori
n bijstellen *n* van de kleurregelaars
d Nachstellung *f* der Bedienungsknöpfe

1001 COLO(U)R CONVERTER, COLO(U)R STANDARDS CONVERTER ctv
A device for converting a fully coded colo(u)r signal from one system to another one.
f convertisseur *m* du signal de couleur
e convertidor *m* de la señal de color
i convertitore *m* del segnale di colore
n kleursignaalomzetter
d Farbsignalumsetzer *m*

1002 COLO(U)R CO-ORDINATE TRANSFORMATION ct
The mathematical transformation of colo(u)r data from one system of axes of colo(u)r to another.
f changement *m* du système de coordonnées trichromatiques
e cambio *m* del sistema de coordenadas tricromáticas
i cambio *m* del sistema di coordinate tricromatiche
n transformatie van de kleurcoördinaten
d Farbkoordinatentransformation *f*

1003 COLO(U)R CORRECTION ctv
f correction *f* de couleur
e corrección *f* de color
i correzione *f* di colore
n kleurcorrectie
d Farbkorrektur *f*

1004 COLO(U)R CORRECTION FILTER ctv
A manually operated filter wheel inserted between the zoom unit and the prism assembly.
f filtre *m* correcteur de couleurs
e filtro *m* corrector de colores
i filtro *m* correttore di colori
n kleurencorrectiefilter *n*
d Farbkorrekturfilter *n*

1005 COLO(U)R CORRECTION MASK ct
f masque *m* de correction de couleur
e máscara *f* de corrección de color
i maschera *f* di correzione di colore
n kleurcorrectiemasker *n*
d Farbkorrekturmaske *f*

1006 COLO(U)R DECODER, DECODER ctv
A circuit for deriving the signals for the colo(u)r display device from the colo(u)r picture signal and the colo(u)r burst.
f décodeur *m* couleur
e descodificador *m* de color
i decodificatore *m* di colore
n decodeur, kleurdecodeur
d Dekoder *m*, Dekodierer *m*, Farbdekoder *m*, Farbdekodierer *m*

1007 COLO(U)R DECODING ctv
f décodage *m* de couleur
e descodificación *f* de color
i decodificazione *f* di colore
n kleurdecodering
d Farbdekodierung *f*

COLO(U)R DEFECT
see: CHROMATIC DEFECT

1008 COLO(U)R DEMODULATOR ctv
f démodulateur *m* couleur
e desmodulador *m* de color
i demodulatore *m* di colore
n kleurdemodulator
d Farbdemodulator *m*

1009 COLO(U)R DEVIATION, COLO(U)R DISTORTION ct
f erreur *f* de teinte
e error *m* de tinta
i errore *m* di tinta
n kleurafwijking
d Farbabweichung *f*

1010 COLO(U)R DENSITY ct
f facteur *m* de pureté colorimétrique
e factor *m* de pureza colorimétrica
i fattore *m* di purezza colorimetrica
n kleurzuiverheidsfactor
d Farbdichte *f*

1011 COLO(U)R DIFFERENCE ct
f différence *f* de couleur
e diferencia *f* de color
i differenza *f* di colore
n kleurverschil *n*
d Farbdifferenz *f*, Farbunterschied *m*

1012 COLO(U)R DIFFERENCE INFORMATION ctv
f information *f* dans le signal de différence de couleur
e información *f* en la señal de diferencia de color
i informazione *f* nel segnale di differenza di colore
n informatie in het kleurverschilsignaal
d Information *f* im Farbdifferenzsignal

1013 COLO(U)R DIFFERENCE SENSITIVITY, COLO(U)R DISCRIMINATION ct
f sensibilité *f* différentielle de chrominance
e sensibilidad *f* diferencial de crominancia
i sensibilità *f* differenziale di crominanza
n kleuronderscheidingsvermogen *n*
d Farbunterscheidungsvermögen *n*

1014 COLO(U)R DIFFERENCE SIGNAL ctv
An electrical signal, which when added to the luminance signal, produces a signal representative of one of the tristimulus values, with respect to a stated set of primaries, of the transmitted colo(u)r.
f signal *m* de différence de couleur
e señal *f* de diferencia de color
i segnale *m* di differenza di colore
n kleurverschilsignaal *n*
d Farbdifferenzsignal *n*

1015 COLO(U)R DILUTION ct
A condition brought about by the mixing of a colo(u)r with white light, reducing the saturation.
f réduction *f* de la saturation
e reducción *f* de la saturación
i riduzione *f* della saturazione
n verzadigingsvermindering
d Sättigungsabstufung *f*, Sättigungsreduktion *f*

1016 COLO(U)R DISCRIMINATION ct
Perception of differences between colo(u)rs.
f discernement *m* de couleurs
e discernimiento *m* de colores
i discernimento *m* di colori
n kleurverschilwaarneming
d Farbdifferenzwahrnehmung *f*

1017 COLO(U)R DISK, COLO(U)R WHEEL ct
A rotating disk with three radial sections composed of the three primary colo(u)rs which continue to make white light.
f disque *m* chromatique
e disco *m* cromático
i disco *m* cromatico
n kleurenschijf
d Farbenscheibe *f*

1018 COLO(U)R DISPLAY TUBE ctv
Picture tubes used in colo(u)r TV receivers to translate a standard colo(u)r signal, suitably decoded, into a picture.
f tube *m* à image couleur
e tubo *m* de imagen en colores
i tubo *m* d'immagine a colori
n kleurenbeeldbuis
d Farbbildröhre *f*

1019 COLO(U)R EDGING ctv/dis
Spurious colo(u)r at the boundaries of differently colo(u)red areas in the picture.
f couleur *f* fausse dans les bords
e color *m* espurio en los bordes
i colore *m* spurio nei bordi
n kleurgrensfout, kleurzoomfout
d Farbrandfehler *m*

1020 COLO(U)R EIDOPHOR ctv
f système *m* eidophor en couleurs
e sistema *m* eidophor en colores
i sistema *m* eidophor in colori
n eidophorsysteem *n* in kleuren
d Farb-Eidophorsystem *n*

1021 COLO(U)R ELEMENT ctv
f élément *m* de couleur
e elemento *m* de color
i elemento *m* di colore
n kleurelement *n*
d Farbelement *n*

1022 COLO(U)R EQUALIZING ASSEMBLY, FIELD EQUALIZER MAGNETS ctv
A number of magnets having variable strength and direction of field, mounted around the periphery of the screen of a shadow-mask colo(u)r display tube using a metal cone construction.
f aimants *pl* de pureté de couleur
e imanes *pl* de pureza de color
i magneti *pl* di purezza di colore
n egalisatiemagneten *pl*, kleurzuiverheidsgordel
d Farbreinheitsmagnete *pl*

1023 COLO(U)R ERRORS ctv
Errors which can be allotted to two distinct classes, viz. a loss of colo(u)r balance or a loss of saturation.
f erreurs *pl* chromatiques
e errores *pl* cromáticos
i errori *pl* cromatici
n kleurfouten *pl*
d chromatische Fehler *pl*, Farbfehler *pl*

1024 COLO(U)R FIDELITY ctv
The degree to which a colo(u)r TV system is capable of reproducing faithfully the colo(u)rs in the original scene.
f fidélité *f* de couleur
e fidelidad *f* de color
i fedeltà *f* di colore
n kleurgetrouwheid
d Farbtreue *f*, Farbwiedergabetreue *f*

1025 COLO(U)R FIELD, PRIMARY COLO(U)R FIELD ctv
A subdivision of the colo(u)r picture formed by scanning once in one of the constituent colo(u)rs of the system.
f trame *f* à couleurs
e cuadro *m* de colores
i quadro *m* a colori
n kleurraster *n*
d Farbteilbild *n*

1026 COLO(U)R FIELD CORRECTOR crt/ctv
A device used outside a colo(u)r picture tube to produce an electric or magnetic field that acts on the electron beam after deflection to produce more uniform colo(u)r fields.
f aimant *m* d'uniformisation de la trame à couleurs
e imán *m* uniformador del cuadro de colores
i magnete *m* uniformante del quadro a colori
n kleurzuiverheidsmagneet
d Farbreinheitsmagnet *m*

1027 COLO(U)R FILTER ct
Film of material selectively absorbing certain wavelengths, and hence changing spectral distribution of transmitted radiation.
f filtre *m* coloré
e filtro *m* colorado
i filtro *m* colorato
n kleurfilter *n*
d Farbfilter *n*

1028 COLO(U)R FILTER DISK ctv
A spinning circular disk having red, blue and green sectors of filters to produce the individual red, blue and green pictures in a field-sequential system.
f disque *m* à filtres colorés
e disco *m* con filtros colorados
i disco *m* a filtri colorati
n kleurfilterschijf
d Farbfilterscheibe *f*

1029 COLO(U)R FLASH ctv
f éclat *m* de couleur
e destello *m* de color
i sprizzo *m* di colore
n kleurflits
d Farbblitz *m*

1030 COLO(U)R FLICKER ctv
Flicker due to fluctuations of both chromaticity and luminance in a colo(u)r TV receiver.
f papillotement *m* de couleur
e parpadeo *m* de colores
i sfarfallamento *m* di colori
n kleurgeflikker *n*
d Farbvalenzflimmern *n*

1031 COLO(U)R FREQUENCY ctv
f fréquence *f* de couleur
e frecuencia *f* de color
i frequenza *f* di colore
n kleurfrequentie
d Farbfrequenz *f*

1032 COLO(U)R FRINGING ctv
Unnatural fringes of colo(u)r at the edges of field-sequential pictures of objects which are moving rapidly in the field of view across the line of sight.
f frange *f* de couleur
e franja *f* de colores
i frangia *f* di colori
n kleurslip
d Farbsaum *m*

1033 COLO(U)R GAMUT ct
Group or range of colo(u)rs, particularly those falling within a given area on a chromaticity diagram, which can be matched by a given set of primaries.
f gamme *f* de couleurs
e gama *f* de colores
i gamma *f* di colori
n kleurengamma *n*
d Farbskala *f*

1034 COLO(U)R GATE cpl/ctv
Circuit in colo(u)r TV receiver which allows only the primary colo(u)r signal corresponding to excited phosphor to reach the modulation electrode of the tube.
f porte *f* de signal couleur
e compuerta *f* de señal color
i porta *f* di segnale colore
n kleursignaalpoort
d Farbsignalgatter *m*

1035 COLO(U)R GENERATOR LOCK, COLO(U)R GENLOCK, COLO(U)R SUBCARRIER LOCK ctv
Provided by locking a subcarrier oscillator to the colo(u)r burst by means of a phase-locked loop.
f enchaînement *m* de la sousporteuse de couleur
e enganche *m* de la subportadora de color
i incatenamento *m* della sottoportante di colore
n koppeling van de kleurhulpdraaggolf
d Genlock-Betriebsweise *f*

1036 COLO(U)R GRADING ct
f étalonnage *m* des couleurs
e escalonamiento *m* de los colores
i gradazione *f* dei colori
n kleurgradatie
d Farblichtbestimmung *f*

1037 COLO(U)R GRID crt
A structure of wire screens placed at the viewing end of a chromatron single-gun colo(u)r TV tube.
f grille *f* de couleur
e rejilla *f* de color
i griglia *f* di colore
n kleurenkiezend rooster *n*, kleurrooster *n*
d Farbgitter *n*

1038 COLO(U)R IDENTIFICATION ct
f identification *f* couleur
e identificación *f* de color
i identificazione *f* di colore
n kleuridentificatie
d Farbkennung *f*

1039 COLO(U)R IMAGE SEPARATION, SEPARATION OF COLO(U)R IMAGE ctv
The separation of the red, green and blue images.
f séparation *f* des images en couleurs primaires
e separación *f* de las imágenes en colores primarios
i separazione *f* delle immagini in colori primari
n scheiding van de primaire kleurbeelden
d Trennung *f* der Primärfarbbilder

1040 COLO(U)R INFORMATION ctv
f information *f* couleur
e información *f* color
i informazione *f* colore
n kleurinformatie
d Farbinformation *f*

COLO(U)R INTENSITY CONTROL
see: CHROMA CONTROL

1041 COLO(U)R KILLER ctv
A device in colo(u)r TV receiver which makes the chrominance channel inoperative when a monochrome signal is being received.
f dispositif *m* de suppression de la couleur
e supresor *m* de transmisión del color
i dispositivo di soppressione di trasmissione del colore
n kleuronderdrukker
d Farbsperre *f*

1042 COLO(U)R LEVEL ctv
Modulation level corresponding to a fixed tone of the image.
f niveau *m* de couleur
e nivel *m* de color
i livello *m* di colore
n kleurniveau *n*
d Farbenlibelle *f*

1043 COLO(U)R LOCK ctv
f enchaînement *m* du signal couleur
e enganche *m* de la señal color
i incatenamento *m* del segnale colore
n koppeling van het kleursignaal
d Verkopplung *f* des Farbsignals

1044 COLO(U)R MASKING ct
f masquage *m* chromatique
e enmascaramiento *m* cromático
i mascheramento *m* cromatico
n kleurmaskering
d Farbmaskierung *f*

1045 COLO(U)R MATCH, COLO(U)R MATCHING ct
The condition in which two colo(u)rs are judged to have exactly the same hue, saturation and luminosity.
f équivalence *f* de couleurs
e equivalencia *f* de colores
i equivalenza *f* di colori
n kleurgelijkheid, kleurovereenstemming
d Farbanpassung *f*, Farbübereinstimmung *f*

1046 COLO(U)R MATCHING CURVES ct
f courbes *pl* d'équivalence de couleurs
e curvas *pl* de equivalencia de colores
i curve *pl* d'equivalenza di colori
n kleurgelijkheidskrommen *pl*, kleurovereenstemmingskrommen *pl*
d Farbanpassungskurven *pl*, Farbübereinstimmungskurven *pl*

1047 COLO(U)R MATCHING FUNCTIONS ct
f fonctions *pl* d'équivalence de couleurs
e funciones *pl* de equivalencia de colores
i funzioni *pl* d'equivalenza di colori
n kleurgelijkheidsfuncties *pl*, kleurovereenstemmingsfuncties *pl*
d Farbanpassungsfunktionen *pl*, Farbübereinstimmungsfunktionen *pl*

1048 COLO(U)R MATRIX ctv
f matrice *f* de chrominance
e matriz *f* de crominancia
i matrice *f* di crominanza
n kleurenmatrix
d Farbmatrix *f*

1049 COLO(U)R MATRIX UNIT ctv
f réseau *m* matriciel de transformation des signaux de couleur, unité *f* de la matrice de chrominance
e unidad *f* de la matriz de crominancia
i unità *f* della matrice di crominanza
n kleurenmatrixeenheid
d Farbmatrixschaltung *f*

1050 COLO(U)R MIXTURE ct
Colo(u)r produced by the combination of light of different colo(u)rs.
f mélange *m* de couleurs
e mezcla *f* de colores
i mescolanza *f* di colori
n kleurenmengsel *n*
d Farbmischung *f*

1051 COLO(U)R MIXTURE CURVE ct
Representation of the specified three colo(u)rs which match a given colo(u)r.
f courbe *f* de dosage des trois primaires, courbe *f* de mélange de couleurs
e curva *f* de mezcla de colores
i curva *f* di mescolanza dei colori
n kleurenmengselkromme
d Spektralwertkurve *f*

1052 COLO(U)R MIXTURE DATA, TRISTIMULUS SPECIFICATIONS, TRISTIMULUS VALUES ct
The amount of the primaries that must be combined to establish a match with the sample.
f composantes *pl* primaires du mélange de couleurs
e valores *pl* de triple estímulo
i componenti *pl* primari della mescolanza dei colori
n kleurenmengselgegevens *pl*
d Farbmischwerte *pl*, Farbwerte *pl*

1053 COLO(U)R MIXTURE FUNCTION ct
A function having as variables the amounts or proportions of three standard colo(u)r components, represented by the colo(u)r mixture curve.
f fonction *f* du mélange de couleurs
e función *f* de la mezcla de colores
i funzione *f* della mescolanza di colori
n functie van het kleurenmengsel
d Farbmischungsfunktion *f*

1054 COLO(U)R MODULATOR ctv
f modulateur *m* couleur
e modulador *m* de color
i modulatore *m* di colore
n kleurmodulator
d Farbmodulator *m*

1055 COLO(U)R MONITOR, COLO(U)R PICTURE MONITOR ctv
f moniteur *m* d'image couleur, récepteur *m* de contrôle couleur
e monitor *m* de imagen en colores
i monitore *m* d'immagine a colori
n kleurenbeeldmonitor
d Farbbildkontrollempfänger *m*

1056 COLO(U)R NOISE ctv/dis
f bruit *m* du signal couleur
e ruido *m* de la señal color
i rumore *m* del segnale colore
n kleursignaalruis
d Farbsignalrauschen *n*

1057 COLO(U)R OVERLOAD ctv
f sursaturation *f* de la couleur
e sobresaturación *f* del color
i soprassaturazione *f* del colore
n kleuroververzadiging
d Farbübersättigung *f*

1058 COLO(U)R PATTERN svs
f mire *f* en couleurs
e patrón *m* en colores
i immagine *f* di prova a colori
n kleurentoetsbeeld *n*
d Farbentestbild *n*

1059 COLO(U)R PATTERN GENERATOR svs
f générateur *m* de mire en couleurs
e generador *m* de patrón en colores
i generatore *m* d'immagine di prova a colori
n kleurentoetsbeeldgenerator
d Farbentestbildgenerator *m*

1060 COLO(U)R PERCEPTION ct
f perception *f* des couleurs
e percepción *f* de los colores
i percezione *f* dei colori
n kleurperceptie
d Farbempfindung *f*

1061 COLO(U)R PHASE ctv
The difference in phase between a chrominance signal (I or Q) and the chrominance-carrier reference in a colo(u)r TV receiver.
f angle *m* de phase du signal de chrominance
e ángulo *m* de fase de la señal de crominancia
i angolo *m* di fase del segnale di crominanza
n fazehoek van het chrominantiesignaal
d Phasenwinkel *m* des Chrominanzsignals

COLO(U)R PHASE ALTERNATION
see: ALTERNATION

1062 COLO(U)R PHASE DIAGRAM ctv
A vector diagram which denotes the phase difference between the colo(u)r burst signal and the chrominance signal for each of the three primary and three complementary colo(u)rs.
f diagramme *m* de phase de couleur
e diagrama *m* de fase de color
i diagramma *m* di fase di colore
n kleurfazediagram *n*
d Farbenphasendiagramm *n*

1063 COLO(U)R PHOSPHOR crt
f substance *f* luminescente pour télévision en couleurs
e fósforo *m* de colores, substancia *f* luminiscente para televisión en colores
i sostanza *f* luminescente per televisione a colori
n fosfor voor kleurentelevisie, luminescerende stof voor kleurentelevisie
d Farbphosphor *m*, Leuchtstoff *m* für Farbfernsehen, Phosphor *m* für Farbfernsehen

1064 COLO(U)R PICTURE ctv
f image *f* couleur
e imagen *f* en colores
i immagine *f* a colori
n kleurenbeeld *n*
d Buntbild *n*, Farbbild *n*

1065 COLO(U)R PICTURE AND WAVEFORM MONITOR cpl
f appareil *m* de contrôle de la couleur, récepteur *m* oscilloscope chromatique
e receptor *m* osciloscopio cromático
i ricevitore *m* oscilloscopio cromatico
n kleurbeeldmonitor
d Farbbildkontrollgerät *n*

1066 COLO(U)R PICTURE SCREEN, COLO(U)R SCREEN crt/ctv
The screen in a colo(u)r display tube.
f écran *m* de couleurs
e pantalla *f* cromática, pantalla *f* de colores
i schermo *m* a colori
n kleurenscherm *n*
d Farbschirm *m*

1067 COLO(U)R PICTURE SIGNAL ctv
The electrical signal representing complete colo(u)r picture information, excluding all sync signals.
f signal *m* d'image couleur
e señal *f* de imagen color
i segnale *m* d'immagine colore
n kleurenbeeldsignaal *n*
d Farbaustastsignal *n*, Farbbildsignal *n*

1068 COLO(U)R PICTURE TUBE, COLO(U)R TELEVISION PICTURE TUBE crt
An electron tube used to provide an image in colo(u)r by the scanning of a raster and by varying the intensity of excitation of phosphors to produce light of the chosen primary colo(u)rs.
f tube *m* image couleur
e tubo *m* cinescopio de televisión en colores
i cinescopio *m* per televisione a colori
n kleurenbeeldbuis
d Farbbildröhre *f*, Farbbildwiedergaberöhre *f*

1069 COLO(U)R PICTURE TUBE OF 110° crt
f tube *m* image extra-plat 110°
e cinescopio *m* de 110°
i cinescopio *m* supersquadrato a 110°
n 110° kleurenbeeldbuis
d 110°-Farbbildröhre *f*, 110°-Weitwinkelbildröhre *f*

1070 COLO(U)R PLANE crt
In a multibeam colo(u)r picture tube, the plane containing all the colo(u)r centers (centres).
f plan *m* de couleurs
e plano *m* de colores
i piano *m* di colori
n kleurenvlak *n*
d Farbleitfläche *f*

1071 COLO(U)R PRIMARIES, PRIMARY COLO(U)RS ct
Set of usually three colo(u)rs from which multicolo(u)r images are built up in printing, photography and TV.
f couleurs *pl* primaires, primaires *pl*
e colores *pl* fundamentales, colores *pl* primarios
i colori *pl* fondamentali, colori *pl* primari
n grondkleuren *pl*, primaire kleuren *pl*
d Grundfarben *pl*, Primärfarben *pl*

1072 COLO(U)R PROGRAM(ME) ctv
f émission *f* en couleur, programme *m* de télévision en couleur
e emisión *f* en color, programa *m* de televisión en colores
i emissione *f* in colore, programma *m* di televisione a colori
n televisieprogramma *n* in kleur, uitzending in kleur
d Farbfernsehprogramm *n*, Farbsendung *f*

1073 COLO(U)R PURITY, PURITY ct
f pureté *f* de la couleur
e pureza *f* del color
i purezza *f* del colore
n kleurzuiverheid
d Farbreinheit *f*

1074 COLO(U)R PURITY COIL, PURITY COIL ctv
A coil consisting of two current-carrying windings used in a colo(u)r TV receiver to produce a magnetic field which will alter the directions of the three electron beams so that each beam will strike only the proper set of phosphor dots.
f bobine *f* de pureté de couleur
e bobina *f* de pureza de color
i bobina *f* di purezza di colore
n kleurzuiverheidsspoel
d Farbreinheitsspule *f*

1075 COLO(U)R PURITY CONTROL, PURITY CONTROL ctv
A potentiometer or rheostat used to adjust the direct current through the purity coil.
f régleur *m* de la pureté de couleur
e regulador *m* de la pureza de color
i regolatore *m* della purezza di colore
n kleurzuiverheidsregelaar
d Farbreinheitsregler *m*

1076 COLO(U)R PURITY ERROR ctv
f impureté *f* de la couleur
e impureza *f* del color
i impurezza *f* del colore
n kleuronzuiverheid
d Farbunreinheit *f*, Farbverfälschung *f*

1077 COLO(U)R PURITY MAGNET, PURITY MAGNET crt/ctv
A magnet in the neck region of a colo(u)r picture tube to alter the electron beam

for the purpose of improving colo(u)r purity.
f aimant *m* de pureté de la couleur
e imán *m* de pureza del color
i magnete *m* di purezza del colore
n kleurzuiverheidsmagneet
d Farbreinheitsmagnet *m*

1078 COLO(U)R PURITY RING, PURITY RING ctv/svs
f anneau *m* de pureté de couleur
e anillo *m* de pureza de color
i anello *m* di purezza di colore
n kleurzuiverheidsring
d Farbreinheitsring *m*

1079 COLO(U)R QUALITY ctv
f qualité *f* de la couleur
e cualidad *f* del color
i qualitá *f* del colore
n kleurkwaliteit
d Farbgüte *f*

1080 COLO(U)R RANGE ct
f domaine *m* chromatique
e campo *m* cromático
i campo *m* cromatico
n kleurgebied *n*
d Farbbereich *m*

1081 COLO(U)R RECEIVER, COLO(U)R TELEVISION RECEIVER ctv
f récepteur *m* de télévision couleur, téléviseur *m* de couleur
e receptor *m* de televisión en colores, telerreceptor *m* en colores, televisor *m* en colores
i ricevitore *m* televisivo a colori, televisore *m* a colori
n kleurentelevisieontvanger
d Farbfernsehempfänger *m*

1082 COLO(U)R RECORDING ctv
A method of thermoplastic recording consisting in modulating a single electron beam simultaneously with three primary colo(u)rs, red, green and blue, received from a colo(u)r TV camera.
f enregistrement *m* mécanique en couleur
e grabación *f* mecánica en color
i registrazione *f* meccanica in colore
n mechanische kleuropneming
d mechanische Farbaufzeichnung *f*

1083 COLO(U)R REFERENCE ct
f référence *f* couleur
e referencia *f* color
i riferimento *m* colore
n kleurreferentie
d Farbbezugspunkt *m*

1084 COLO(U)R REFERENCE SIGNAL cpl
Continuous signal which determines the phase of the burst signal.
f signal *m* de référence couleur
e señal *f* de referencia color
i segnale *m* di riferimento colore
n kleurreferentiesignaal *n*
d Farbreferenzsignal *n*

1085 COLO(U)R REFLECTION TUBE crt
A colo(u)r picture tube which produces an image by means of electron reflection in the screen region.
f tube *m* image couleur à réflexion d'électrons
e cromoscopio *m* de reflexión de electrones
i cromoscopio *m* a riflessione d'elettroni
n kleurenbeeldbuis met elektronenreflectie
d Farbfernsehröhre *f* mit Elektronenreflexion

1086 COLOR REGISTRATION (US), COLOUR SUPERIMPOSITION (GB) ctv
The accurate registration (superimposition) of the red, green and blue pictures used to form a specific complete colo(u)r picture.
f superposition *f* de couleurs
e superposición *f* de colores
i sovrapposizione *f* di colori
n kleurensuperponering
d Farbdeckung *f*

1087 COLO(U)R RENDERING, COLO(U)R RENDITION, COLO(U)R REPRODUCTION ct
General expression for the effect of an illuminant on the colo(u)r appearance of objects in conscious or subconscious comparison with their colo(u)r appearance under a reference illuminant.
f rendu *m* des couleurs
e rendimiento *m* de los colores
i rendimento *m* dei colori
n kleurweergeving
d Farbwiedergabe *f*

1088 COLO(U)R RENDERING INDEX ct
Measure of the degree to which the psychophysical colo(u)rs of objects illuminated by the source conform to those of the same objects illuminated by a reference illuminance for specified conditions.
f indice *m* du rendu des couleurs
e índice *m* del rendimiento de los colores
i indice *m* del rendimento dei colori
n kleurweergevingsindex
d Farbwiedergabeindex *m*

1089 COLO(U)R RENDERING PROPERTIES ct
Effect of a light source on the colo(u)r appearance of objects in comparison with their colo(u)r appearance under a reference illuminant for specified conditions.
f caractéristiques *pl* du rendu des couleurs
e características *pl* del rendimiento de los colores
i caratteristiche *pl* del rendimento dei colori
n kleurweergevingskenmerken *pl*
d Farbwiedergabekennzeichen *pl*

1090 COLO(U)R RESOLUTION ct/ctv
f résolution *f* de couleur
e resolución *f* de color

i risolvenza *f* di colore
n kleurscheidend vermogen *n*
d Farbauflösungsvermögen *n*

1091 COLOR RESOLUTION PATTERN (US), COLOUR DEFINITION CHART (GB), COLOUR TEST CARD (GB) ctv
A standard pattern used for assessing the quality of a colo(u)r TV transmission.
f mire *f* en couleurs
e patrón *m* en colores
i figura *f* di prova a colori
n kleurentoetsbeeld *n*
d Farbtestbild *n*

1092 COLO(U)R RESPONSE crt
Output of a TV camera tube with reference to the colo(u)r of the light incident on it.
f réponse *f* de couleur
e respuesta *f* de color
i risposta *f* di colore
n kleurresponsie
d Ansprechen *n* einer Farbe

1093 COLO(U)R SAMPLER, ELECTRONIC COLO(U)R SAMPLER ct
f commutateur *m* électronique de couleurs
e conmutador *m* electrónico de colores
i commutatore *m* elettronico di colori
n elektronische kleurwisselaar
d elektronischer Farbschalter *m*

1094 COLO(U)R SAMPLING, ELECTRONIC COLO(U)R SAMPLING ct
f commutation *f* électronique de couleurs
e conmutación *f* electrónica de colores
i commutazione *f* elettronica di colori
n elektronische kleurwisseling
d elektronische Farbschaltung *f*

1095 COLOR SAMPLING RATE (US), COLOUR SAMPLING FREQUENCY (GB) ct
The number of times per second each primary colo(u)r is sampled.
f fréquence *f* de commutation de couleurs
e frecuencia *f* de conmutación de colores
i frequenza *f* di commutazione di colori
n kleurwisselfrequentie
d Farbschaltfrequenz *f*

1096 COLO(U)R SAMPLING SEQUENCE ct
The order in which the three primary colo(u)rs are sampled.
f séquence *f* de commutation de couleurs
e sucesión *f* de conmutación de colores
i sequenza *f* di commutazione di colori
n kleurwisselvolgorde
d Farbschaltfolge *f*

1097 COLO(U)R SATURATION ctv
f saturation *f* de couleur
e saturación *f* de color
i saturazione *f* di colore
n kleurverzadiging
d Farbsättigung *f*

1098 COLO(U)R SATURATION ADJUSTMENT ct
f réglage *m* de la saturation de couleur
e regulación *f* de la saturación de color
i regolazione *f* della saturazione di colore
n kleurverzadigingsregeling
d Farbsättigungsregelung *f*

1099 COLO(U)R SATURATION CONTROL ct
f régleur *m* de la saturation de couleur
e regulador *m* de la saturación de color
i regolatore *m* della saturazione di colore
n kleurverzadigingsregelaar
d Farbsättigungsregler *m*

1100 COLO(U)R SCREEN, GELATIN FILTER ctv
f écran *m* coloré
e pantalla *f* colorada
i schermo *m* colorato
n kleurenschijf
d Gelatinescheibe *f*

COLO(U)R SCREEN
see: COLO(U)R PICTURE SCREEN

1101 COLO(U)R-SELECTING-ELECTRODE SYSTEM crt
A structure containing a plurality of openings mounted in the vicinity of the screen of a colo(u)r picture tube, the function of this structure being to cause electron impingement on the proper screen area by using either masking, focusing, deflection, reflection, or a combination of these effects.
f système *m* d'électrode sélectionneur de couleur
e sistema *m* de electrodo selector de color
i sistema *m* d'elettrodo selettore di colore
n kleurselecterend elektrodesysteem *n*
d farbselektierendes Elektrodensystem *n*

1102 COLO(U)R SELECTIVE MIRRORS ctv
f miroirs *pl* sélectifs de couleurs
e espejos *pl* selectivos de colores
i specchi *pl* selettivi di colori
n kleurenselecterende spiegels *pl*
d farbselektierende Spiegel *pl*

1103 COLO(U)R SENSATION ct
A subjective experience constituting the primary conscious response to stimulation of the eye by radiant energy in the visible region.
f sensation *f* de couleur
e sensación *f* de color
i sensazione *f* di colore
n kleurgewaarwording
d Farbempfindung *f*

1104 COLO(U)R SENSATION INDUCED BY CONTRASTING ENVIRONMENT ct
f sensation *f* de couleur due au contraste de l'environnement
e sensación *f* de color debida a la cercania contrastante
i sensazione *f* di colore dovuta al dintorno contrastante

n kleurgewaarwording veroorzaakt door kontrasterende omgeving
d Farbempfindung *f* verursacht durch kontrastierende Umgebung

1105 COLO(U)R SENSITIVITY ct
f sensibilité *f* chromatique
e sensibilidad *f* cromática
i sensibilità cromatica
n kleurgevoeligheid
d Farbempfindlichkeit *f*

1106 COLO(U)R SEPARATION ct
f division *f* chromatique,
séparation *f* chromatique
e división *f* cromática,
separación *f* cromática
i divisione *f* cromatica,
separazione *f* cromatica
n kleurscheiding
d Farbteilung *f*

1107 COLO(U)R SEPARATION CHANNEL ctv
f canal *m* de séparation chromatique
e canal *m* de separación cromática
i canale *m* di separazione cromatica
n kleurscheidingskanaal *n*
d Farbteilungskanal *m*

1108 COLO(U)R SEPARATION FILTER ct
f filtre *m* de séparation chromatique
e filtro *m* de separación cromática
i filtro *m* di separazione cromatica
n kleurscheidingsfilter *n*
d Farbteilungsfilter *n*

COLOUR SEPARATION OVERLAY (GB)
see: CHROMA KEY

1109 COLO(U)R SEQUENCE ctv
The sequence red, green and blue with wavelengths 700, 546.1, and 435.8 nm.
f séquence *f* chromatique
e secuencia *f* cromática
i sequenza *f* cromatica
n kleurvolgorde
d Farbfolge *f*

1110 COLO(U)R SEQUENTIAL SYSTEM, ctv
COLO(U)R TELEVISION SEQUENTIAL SYSTEM
f système *m* séquentiel de couleurs,
système *m* séquentiel de télévision couleur
e sistema *m* de televisión por sucesión de colores,
sistema *m* secuencial de colores
i sistema *m* di televisione a colori sequenziali,
sistema *m* sequenziale di colori
n kleurwisselingsmethode,
volgrastersysteem *n* in kleurentelevisie
d Farbwechselverfahren *n*,
Zeitfolgeverfahren *n* beim Farbfernsehen

1111 COLO(U)R SHADING ctv
f virage *m* de teinte
e degradación *f* de tinta
i degradazione *f* di tinta
n omslaan *n* van de tint
d Farbtonverfälschung *f*

1112 COLO(U)R SIGNAL ctv
Any signal in a colo(u)r TV system, wholly or partially controlling the chromaticity of a colo(u)r TV picture.
f signal *m* chromatique,
signal *m* couleur
e señal *f* color,
señal *f* cromática
i segnale *m* colore,
segnale *m* cromatico
n kleursignaal *n*
d Farbsignal *n*,
Farbwertsignal *n*

1113 COLO(U)R SIGNAL EQUATION ctr
An equation specifying the colo(u)r picture signal in terms of the tristimulus signals.
f équation *f* du signal chromatique
e ecuación *f* de la señal cromática
i equazione *f* del segnale cromatico
n kleursignaalvergelijking
d Farbsignalgleichung *f*

1114 COLO(U)R SLIDE opt
f diapositive *f* colorée
e diapositiva *f* colorada
i diapositiva *f* colorata
n kleurendia
d Farbdia *n*

1115 COLO(U)R SLIDE CAMERA, ctv
COLO(U)R SLIDE SCANNER
f analyseur *m* de diapositives colorées
e explorador *m* de diapositivas coloradas
i analizzatore *m* di diapositive colorate
n kleurendia-aftaster
d Farbdiaabtaster *m*

1116 COLO(U)R SOLID ct
That part of colo(u)r space which is occupied by surface colo(u)rs.
f corps *m* chromatique
e cuerpo *m* cromático
i corpo *m* cromatico
n kleurenlichaam *n*
d Farbkörper *m*

1117 COLO(U)R SPACE ct
A manifold of three dimensions for the geometrical representation of colo(u)rs.
f espace *m* chromatique
e espacio *m* cromático
i spazio *m* cromatico
n kleurenruimte
d Farbraum *m*

COLO(U)R SPLITTING
see: COLO(U)R BREAK UP

COLO(U)R STANDARDS CONVERTER
see: COLO(U)R CONVERTER

1118 COLO(U)R STIMULUS ct
Physically defined radiation entering the eye and producing a sensation of colo(u)r.

f stimulus *m* de couleur
e estímulo *m* de color
i stimolo *m* di colore
n kleurprikkel, kleurstimulus
d Farbreiz *m*

1119 COLO(U)R STIMULUS FUNCTION ct
The relative spectral distribution of the colo(u)r stimulus.
f fonction *f* du stimulus de couleur
e función *f* de estímulo de color
i funzione *f* di stimolo di colore
n kleurprikkelfunctie
d Farbreizfunktion *f*

1120 COLO(U)R STIMULUS SPECIFICATION ct
The property of a colo(u)r stimulus which defines by simultaneous and spatially directly adjacent appearance of two colo(u)r stimuli the equality or difference of the stimulus.
f spécification *f* d'un stimulus de couleur, tristimulus *m*
e especificación *f* de un estímulo de color
i specificazione *f* d'uno stimolo di colore
n kleurvalentie
d Farbvalenz *f*

1121 COLO(U)R STRIPE SIGNAL ctv
Chrominance signal added to a monochrome signal to facilitate receiver installation during monochrome transmissions.
f signal *m* de bordure colorée
e señal *f* de tira colorada
i segnale *m* di striscia colorata
n kleurstripsignaal *n*
d Farbstreifensignal *n*

1122 COLO(U)R STUDIO TIMING ctv
f corrélation *f* du temps dans le studio de télévision couleur
e correlación *f* del tiempo en el estudio de televisión en colores
i correlazione *f* del tempo nello studio di televisione a colori
n tijdscorrelatie in kleurentelevisiestudio
d Zeitabstimmung *f* im Farbfernsehstudio

COLO(U)R SUBCARRIER
see: CHROMINANCE CARRIER

COLO(U)R SUBCARRIER LEVEL
see: CHROMINANCE SUBCARRIER LEVEL

COLO(U)R SUBCARRIER LOCK
see: COLO(U)R GENERATOR LOCK

COLO(U)R SUBCARRIER REFERENCE
see: CHROMINANCE SUBCARRIER REFERENCE

COLO(U)R SUBCARRIER SIGNAL
see: CHROMINANCE SUBCARRIER SIGNAL

1123 COLO(U)R SWITCH vr
Used to switch off the colo(u)r during both recording and reproduction.
f commutateur *m* couleur, interrupteur *m* couleur
e interruptor *m* color
i interruttore *m* colore
n kleurdover
d Farbausschalttaste *f*

1124 COLO(U)R SYMBOLISM ani/ctv
f symbolisme *m* de couleurs
e simbolismo *m* de colores
i simbolismo *m* di colori
n kleurensymboliek
d Farbensymbolik *f*

1125 COLO(U)R SYNC SIGNAL ctv
A sequence of colo(u)r bursts that is continuous except for a specified time interval during the vertical blanking period each burst occurring at a fixed time with respect to horizontal sync in a colo(u)r TV system.
f signal *m* de synchronisation couleur
e señal *f* de sincronización color, sincronismo *m* de color
i segnale *m* di sincronizzazione colore, sincronismo *m* di colore
n kleursynchronisatiesignaal *n*
d Farbsynchronsignal *n*

1126 COLO(U)R SYNCHRONIZING ctv
f synchronisation *f* de couleurs
e sincronización *f* de colores
i sincronizzazione *f* di colori
n kleursynchronisatie
d Farbsynchronisierung *f*

1127 COLO(U)R SYNTHESIS ct/ctv
f synthèse *f* de couleurs
e síntesis *f* de colores
i sintesi *f* di colori
n kleursynthese
d Farbsynthese *f*

1128 COLO(U)R SYSTEMS ctv
The modern compatible colo(u)r systems of which the NTSC, PAL and SECAM are the most important.
f systèmes *pl* de couleurs
e sistemas *pl* de colores
i sistemi *pl* di colori
n kleursystemen *pl*
d Farbsysteme *pl*

1129 COLO(U)R TELECINE, COLO(U)R TELECINE MACHINE, COLO(U)R TELEVISION FILM SCANNER ctv
A machine which converts the optical images previously recorded on standard colo(u)r cinematographic film into electrical TV signals suitable for broadcasting.
f appareil *m* d'analyse de film en couleur
e aparato *m* de exploración de película en colores
i apparecchio *m* d'analisi di pellicola a colori

n kleurenfilmaftastapparaat *n*
d Farbfilmabtastgerät *n*

1130 COLO(U)R TELERECORDING ctv
The recording of TV colo(u)r program(me) material on film, for editing and subsequent transmission.
f enregistrement *m* sur film en couleur d'images de télévision couleur
e registro *m* sobre película en color de imágenes de televisión en colores
i registrazione *f* su pellicola a colore d'immagini di televisione a colori
n op kleurenfilm optekenen *n* van kleurentelevisiebeelden
d Farbfernseh-Filmaufzeichnung *f*

1131 COLO(U)R TELEVISION tv
TV in which the reproduced picture simulates the colo(u)rs of the original scene.
f télévision *f* couleur
e televisión *f* en colores
i televisione *f* a colori
n kleurentelevisie
d Farbfernsehen *n*

COLO(U)R TELEVISION CAMERA
see: CAMERA FOR COLO(U)R TELEVISION

COLO(U)R TELEVISION MASK
see: APERTURE MASK

1132 COLO(U)R TELEVISION NETWORK cpl /ctv
f réseau *m* pour télévision couleur
e red *f* para televisión en colores
i rete *f* per televisione a colori
n kleurentelevisienetwerk *n*
d Farbfernsehnetz *n*

COLO(U)R TELEVISION PICTURE TUBE
see: COLO(U)R PICTURE TUBE

COLO(U)R TELEVISION SEQUENTIAL SYSTEM
see: COLO(U)R SEQUENTIAL SYSTEM

1133 COLO(U)R TEMPERATURE ct
That temperature of a black body which radiates with the same dominant wavelengths as those apparent from a source being described.
f température *f* de couleur
e temperatura *f* de color
i temperatura *f* di colore
n kleurtemperatuur
d Farbtemperatur *f*

1134 COLO(U)R TEMPERATURE METER ct
f appareil *m* de mesure de la température de couleur
e medidor *m* de la temperatura de color
i misuratore *m* della temperatura di colore
n kleurtemperatuurmeter
d Farbtemperaturmesser *m*

1135 COLO(U)R TEST CARD ctv
f mire *f* de télévision couleur
e patrón *m* de televisión en colores
i figura *f* di prova per televisione a colori
n kleurentelevisietoetsbeeld *n*
d Farbfernsehtestbild *n*

1136 COLO(U)R TEST SIGNAL ctv
A signal consisting of six vertical bars which have colo(u)rs that occur at points spread through the reproducible spectrum.
f signal *m* d'essai coloré
e señal *f* de prueba colorada
i segnale *m* di prova colorato
n kleurentoetssignaal *n*
d Farbtestsignal *n*

1137 COLO(U)R THRESHOLD ct
The luminance level below which colo(u)r differences are indiscernible.
f seuil *m* de différence de couleur
e umbral *m* de diferencia de color
i soglia *f* di differenza di colore
n kleurverschildrempel
d Farbdifferenzschwelle *f*

1138 COLO(U)R TRANSCODER, TRANSCODER ctv
f transcodeur *m* de couleur
e transcodificador *m* de color
i trascodificatore *m* di colore
n kleurtranscodeur
d Farbnormwandler *m*, Farbumkodierer *m*

1139 COLO(U)R TRANSITION ct
f transition *f* de couleur
e transición *f* de color
i transizione *f* di colore
n kleurovergang
d Farbübergang *m*

1140 COLO(U)R TRANSMISSION ctv
The transmission of a signal wave that represents both the luminance values and the chrominance values in the picture.
f transmission *f* de couleurs
e transmisión *f* de colores
i trasmissione *f* di colori
n kleurentransmissie
d Farbübertragung *f*

1141 COLO(U)R TRIAD, COLO(U)R TRIPLE crt
A colo(u)r cell of a three-colo(u)r phosphor-dot screen.
f triplet *m* de couleurs
e tríada *f* de colores
i tripla *f* di colori
n kleurendrieling
d Farbdrilling *m*, Farbtripel *m*

1142 COLO(U)R TRIANGLE, MAXWELL TRIANGLE ct
A projective transformation onto a two dimensional plane of the colo(u)r stimuli in any form of trichromatic colo(u)r space.
f triangle *m* des couleurs
e triángulo *m* de los colores
i triangolo *m* dei colori
n kleurendriehoek
d Farbdreieck *m*

1143 COLOUR VIDEO SIGNAL (GB), COMPOSITE COLOR SIGNAL (US) ctv
The combined colo(u)r picture signal and all sync signals.
f signal *m* complet en couleur, signal *m* de vidéo couleur
e señal *f* de video color
i segnale *m* di video colore
n kleurbeeldsignaalmengsel *n*, kleurbeeldvideosignaal *n*
d Farbbildsignalgemisch *n*, Farbbildaustastsynchronsignal *n*

1144 COLO(U)R VIDEO SWITCHER SYSTEM ctv
f système *m* mélangeur de couleurs
e sistema *m* mezclador de colores
i sistema *m* mescolatore di colori
n kleurmengersysteem *n*
d Farbmischersystem *n*

1145 COLO(U)R VISION ct
Ability of an observer to perceive chromatic colo(u)rs.
f vision *f* chromatique
e visión *f* cromática
i visione *f* cromatica
n kleurenzien *n*
d Farbensehen *n*, Farbenwahrnehmung *f*

COLO(U)R WHEEL
see: COLO(U)R DISK

1146 COLO(U)RED PATCHES dis/svs
f taches *pl* de couleur
e manchas *pl* de color
i macchie *pl* di colore
n kleurvlekken *pl*
d Farbflecken *pl*

1147 COLORANTS ct
Substances used to modify the colo(u)rs of objects.
f colorants *pl*
e colorantes *pl*
i coloranti *pl*
n kleurmiddelen *pl*
d Farbstoffe *pl*

1148 COLORATION, SOUND COLORATION aud/dis
An acoustic defect due to the considerable rising of reverberation time at particular frequencies.
f coloration *f* du son
e coloración *f* del sonido
i colorazione *f* del suono
n geluidskleuring
d Klangfärbung *f*

1149 COLORCAST (US), COLOUR BROADCASTING (GB) ctv
A TV broadcast on colo(u)r.
f émission *f* couleur
e emisión *f* en colores
i emissione *f* a colori
n uitzending in kleur
d Farbfernsehübertragung *f*

1150 COLORIMETER ct
f colorimètre *m*
e colorímetro *m*
i colorimetro *m*
n colorimeter
d Kolorimeter *n*

1151 COLORIMETRIC PURITY ct
f facteur *m* de pureté d'une couleur
e factor *m* de pureza de un color
i fattore *m* di purezza d'un colore
n verzadigingsgraad
d farbmetrische Echtheit *f*

1152 COLORIMETRIC SHIFT ct
The change in chromaticity and luminance factor of an object colo(u)r due to change of the illuminant.
f déplacement *m* colorimétrique
e desplazamiento *m* colorimétrico
i spostamento *m* colorimetrico
n colorimetrische verschuiving
d kolorimetrische Verschiebung *f*

1153 COLORIMETRIC SYSTEM ct
Quantative system of colo(u)r specification based on scales derived from either additive or subtractive colo(u)r mixture.
f système *m* colorimétrique
e sistema *m* colorimétrico
i sistema *m* colorimetrico
n colorimetrisch systeem *n*
d Kolorimetriesystem *n*

1154 COLORIMETRY ct
f colorimétrie *f*
e colorimetría *f*
i colorimetria *f*
n colorimetrie
d Farbmessung *f*, Kolorimetrie *f*

COLOUR DEFINITION CHART (GB), COLOUR TEST CARD (GB)
see: COLOR RESOLUTION PATTERN

COLOUR SAMPLING FREQUENCY (GB)
see: COLOR SAMPLING RATE

COLOUR SEPARATION OVERLAY (GB)
see: CHROMA KEY

COLOUR SUPERIMPOSITION (GB)
see: COLOR REGISTRATION

1155 COMA crt/opt
Of a cathode-ray tube, a comet-like appearance of the undeflected spot caused by the lack of alignment between the source of the beam and the electron-optical system.
f coma *f*
e coma *f*
i coma *f*
n coma *n*
d Koma *n*

1156 COMBINED CHROMINANCE SIGNAL $E_I + E_Q$

f signal *m* combiné de chrominance $V_I + V_Q$
e señal *f* combinada de crominancia $V_I + V_Q$
i segnale *m* combinato di crominanza $V_I + V_Q$
n gecombineerd chrominantiesignaal *n* $V_I + V_Q$
d kombiniertes Buntsignal *n* $U_I + U_Q$

1157 COMBINED MAGNETIC SOUND AND PICTURE RECORD, COMMAG — rep
Magnetic sound track striped on to a picture-carrying film.
f son *m* magnétique combiné
e sonido *m* magnético combinado
i suono *m* magnetico combinato
n gecombineerd magnetisch geluid *n*
d kombinierter Magnetton *m*

1158 COMBINED OPTICAL SOUND AND PICTURE RECORD, COMOPT — rep
Optical sound track printed onto a picture-carrying film.
f son *m* optique combiné
e sonido *m* óptico combinado
i suono *m* ottico combinato
n gecombineerd optisch geluid *n*
d kombinierter Lichtton *m*

1159 COMBINED TUNER — cpl
f sélecteur *m* O-dm/O-m, sélecteur *m* VHF-UHF
e selector *m* VHF-UHF
i selettore *m* VHF-UHF
n gecombineerde kanaalkiezer
d kombinierter Kanalwähler *m*

1160 COMBINER CIRCUIT, COMBINING AMPLIFIER — ctv
The mixing amplifier that combines the luminance and chrominance signals with the sync signals in a colo(u)r TV camera chain.
f circuit *m* combinateur
e circuito *m* combinador
i circuito *m* combinatore
n combineerkring
d Kombinatorkreis *m*

1161 COMBINING FILTER, COMBINING UNIT — aea
Passive device for feeding the outputs of two transmitters to a common aerial (antenna) system.
f coupleur *m*, multiplexeur *m* d'antenne
e multiplexor *m* de antena
i filtro *m* d'accoppiamento
n koppelfilter *n*
d Koppelweiche *f*, Senderweiche *f*

1162 COMMENTARY FACILITIES — rec/rep
f appareillage *m* de commentaire
e facilidades *pl* de comento
i apparecchiatura *f* di commentario
n commentaarapparatuur
d Kommentareinrichtungen *pl*

1163 COMMENTARY RECORDING — rec/rep
f enregistrement *m* du commentaire
e registro *m* del comento
i registrazione *f* del commentario
n commentaaropname
d Kommentaraufnahme *f*

1164 COMMENTARY VEHICLE — rec/rep
f voiture *f* de commentaire
e coche *m* de comento
i macchina *f* di commentario
n commentaarwagen
d Kommentarwagen *m*

1165 COMMENTATOR — tv
A person who edits and broadcasts news at a TV station, often interspersed with personal comments.
f commentateur *m*
e comentador *m*
i commentatore *m*
n commentator
d Kommentator *m*

1166 COMMERCIAL, SPOT — tv
An advertising message that is broadcast by TV.
f émission *f* de propaganda
e anuncio *m* comercial en televisión, comerciale *m* por televisión
i emissione *f* di propaganda, emissione *f* pubblicitaria
n reclamespotje *n*, reclame-uitzending
d Fernsehspot *m*, Werbesendung *f*

1167 COMMERCIAL TELEVISION (GB), SPONSORED TELEVISION (US) — ctv
TV in which advertising spots are produced during or between broadcasts.
f télévision *f* commanditée, télévision *f* publicitaire
e televisión *f* de propaganda, televisión *f* patrocinada
i televisione *f* patrocinata, televisione *f* pubblicitaria
n reclametelevisie
d kommerzielles Fernsehen *n*, Werbefernsehen *n*

COMMON AERIAL, COMMUNITY ANTENNA
see: BLOCK ANTENNA

1168 COMMON CHANNEL INTERFERENCE — dis
The interference caused by two broadcasting stations geographically separated,

using the same allocated frequency.
f interférence *f* de canal commun
e interferencia *f* de canal común
i interferenza *f* di canale comune
n storing in hetzelfde kanaal
d Gleichkanalstörung *f*

1169 COMMUNICATION CHANNELS rec/tv
Means by means of which the TV director is able to communicate within the studio with actors, cameraman, stage hands and the stage manager.
f canaux *pl* d'intercommunication
e canales *pl* de intercomunicación
i canali *pl* d'intercomunicazione
n ruggespraakkanalen *pl*
d Rücksprachekanäle *pl*

1170 COMMUNICATION SATELLITE, sat/tv
COMSAT
Artificial satellite projected into a predetermined orbit around the earth to act as a relay station.
f satellite *m* de télécommunication
e satélite *m* de telecomunicación
i satellite *m* di telecomunicazione
n telecommunicatiesatelliet
d Fernmeldesatellit *m*,
Kommunikationssatellit *m*

1171 COMMUNITY TELEVISION, tv
CTV
Distribution of sound and/or TV signals by means of cable in towns and large housing estates.
f télédistribution *f*,
télévision *f* collective
e televisión *f* colectiva
i televisione *f* collettiva
n groepstelevisie
d Gemeinschaftsfernsehen *n*

1172 COMPACT CARTRIDGE (US), aud/vr
COMPACT CASSETTE (GB)
f cassette *f* compacte
e caseta *f* compacta
i cassetta *f* compatta
n compacte cassette
d Kompaktkassette *f*

1173 COMPARISON OF COLO(U)R ctv
TELEVISION SYSTEMS
f comparaison *f* de systèmes de télévision couleur
e comparación *f* de sistemas de televisión en colores
i comparazione *f* di sistemi di televisione a colori
n vergelijking van kleurentelevisiesystemen
d Vergleich *m* von Farbfernsehsystemen

1174 COMPARISON SIGNALS vr
Auxiliary signals used in video recording.
f signaux *pl* de comparaison
e señales *pl* de comparación
i segnali *pl* di comparazione
n vergelijkingssignalen *pl*
d Vergleichssignale *pl*

1175 COMPATIBLE COLO(U)R ctv
TELEVISION
f télévision *f* couleur compatible
e televisión *f* de imágenes de color compatible,
televisión *f* en colores compatible
i televisione *f* a colori compatibile
n compatibile kleurentelevisie
d kompatibiles Farbfernsehen *n*

1176 COMPATIBLE COMPOSITION rec/rep
Composition of the action photographed in a motion picture frame in such a way that no important part of the action is lost if the image is projected at a different aspect ratio.
f composition *f* compatible
e composición *f* compatible
i composizione *f* compatibile
n compatibile compositie
d kompatibile Komposition *f*

1177 COMPATIBILITY ctv
The attribute of a colo(u)r TV system which permits monochrome receivers to produce a monochrome picture from a transmitted colo(u)r signal.
f compatibilité *f*
e compatibilidad *f*
i compatibilità *f*
n compatibiliteit
d Kompatibilität *f*

1178 COMPENSATED VIDEO tv
AMPLIFIER
f amplificateur *m* vidéo à compensation
e amplificador *m* video compensado
i amplificatore *m* video compensato
n gecompenseerde beeldversterker
d Kompensationsbildverstärker *m*

1179 COMPENSATION CIRCUIT cpl
A circuit in which the time constant of the coupling is made much smaller for the same ratio V_u/V_g.
f circuit *m* de compensation
e circuito *m* de compensación
i circuito *m* di compensazione
n compensatiestroomkring
d Ausgleichsstromkreis *m*

1180 COMPENSATOR tv
In TV, the inductance coils used in the grid and anode circuits of the video amplifier to compensate for the attenuation of the higher frequencies.
f compensateur *m* d'atténuation
e compensador *m* de atenuación
i compensatore *m* d'attenuazione
n dempingscompensator
d Dämpfungsausgleicher *m*

1181 COMPLEMENTARY ct
CHROMATICITY
The property of two light samples whereby they provide an achromatic visual stimulus when combined in suitable proportions.

f chromaticité *f* complémentaire
e cromaticidad *f* complementaria
i cromaticità *f* complementare
n complementaire chromaticiteit
d Komplementärfarbart *f*

1182 COMPLEMENTARY COLO(U)RS ct
Pairs of colo(u)rs which combine to give spectral white.
f couleurs *pl* complémentaires
e colores *pl* complementarios
i colori *pl* complementari
n complementaire kleuren *pl*
d Komplementärfarben *pl*

1183 COMPLEMENTARY FREQUENCY RESPONSE cpl
f réponse *f* de fréquence complémentaire
e respuesta *f* de frecuencia complementaria
i curva *f* di risposta complementare
n complementaire frequentieresponsie
d komplementärer Frequenzgang *m*

COMPLEMENTARY GAUSSIAN CIRCUIT
see: ANTI-CLOCHE CIRCUIT

1184 COMPLEMENTARY WAVELENGTH ct
The wavelength of light that, when combined with a sample colo(u)r in suitable proportions, matches a reference standard light.
f longueur *f* d'onde complémentaire
e longitud *f* de onda complementaria
i lunghezza *f* d'onda complementare
n complementaire golflengte
d kompensative Wellenlänge *f*

1185 COMPLETE COLO(U)R INFORMATION ctv
f information *f* complète couleur
e información *f* completa color
i informazione *f* completa colore
n volledige kleurinformatie
d vollständige Farbinformation *f*

1186 COMPLETELY TRANSISTORIZED TELEVISION RECEIVER rep/tv
f téléviseur *m* à transistorisation intégrale
e televisor *m* de transistorización integral
i televisore *m* a transistorizzazione integrale
n volledig getransistoriseerd televisietoestel *n*
d vollständig transistorisiertes Fernsehgerät *n*

COMPOSITE COLOR SIGNAL (US)
see: COLOUR VIDEO SIGNAL

1187 COMPOSITE COLO(U)R SYNC, COMPOSITE COLO(U)R SYNCHRONIZATION SIGNAL ctv
The signal comprising all the sync signals necessary for proper operation of a colo(u)r TV receiver.
f signal *m* de synchronisation couleur complet
e señal *f* de sincronización color completa
i segnale *m* di sincronizzazione colore completo
n volledig kleurensynchronisatiesignaal *n*
d vollständiges Farbsynchronsignal *n*

1188 COMPOSITE PICTURE SIGNAL (US), COMPOSITE VIDEO SIGNAL (GB) tv
The signal obtained by combining a monochrome TV picture signal with the horizontal and vertical blanking and sync signals.
f signal *m* d'image complet, signal *m* vidéo total
e señal *f* video compuesta
i segnale *m* video composto
n totaal beeldsignaal *n*
d BAS-Signal *n*, Signalgemisch *n*

1189 COMPOSITE SHOT tv
f prise *f* combinée, prise *f* à superposition
e toma *f* compuesta, toma *f* de superposición
i ripresa *f* composta, ripresa *f* di sovrapposizione
n gesuperponeerde opname
d Überlagerungsaufnahme *f*

1190 COMPOSITE SIGNALS cpl
Control signals, performing separate functions, added together to form a single waveform.
f signaux *pl* composés
e señales *pl* compuestas
i segnali *pl* composti
n samengestelde signalen *pl*
d zusammengesetzte Signale *pl*

1191 COMPOSITE SYNC GENERATOR tv
A generator which supplies the signal required to synchronize the scanning circuits in the display devices that are producing the picture signal.
f générateur *m* du signal de synchronisation complet
e generador *m* de la señal de sincronización completa
i generatore *m* del segnale di sincronizzazione completo
n generator van het volledige synchronisatiesignaal
d Generator *m* des vollständigen Synchronsignals

1192 COMPOSITE SYNC SIGNAL tv
A signal consisting of the horizontal and vertical sync pulses and the equalizing pulses.
f signal *m* de synchronisation complet
e señal *f* de sincronización completa
i segnale *m* di sincronizzazione completo
n volledig synchronisatiesignaal *n*
d S-Gemisch *n*, S-Signal *n*, vollständiges Synchronsignal *n*

1193 COMPRESSION, SOUND COMPRESSION, VOLUME COMPRESSION — aud
Function of a device which transfers a signal from its input to its output reducing at the same time the span of the amplitudes of the signal.
f compression *f* de volume
e compresión *f* de volumen
i compressione *f* della dinamica, compressione *f* di volume
n dynamiekcompressie
d Dynamikpressung *f*

1194 COMPRESSOR, SOUND COMPRESSOR, VOLUME COMPRESSOR — aud
Device used to reduce the volume range of sound without producing noticeable distortion.
f compresseur *m* de volume
e compresor *m* de volumen
i compressore *m* della dinamica
n dynamiekperser
d Dynamikpresser *m*

1195 COMPUTER ANIMATION — ani
f animation *f* par technique d'ordinateur
e animación *f* por técnica computadora
i animazione *f* per tecnica di calcolatore
n door computer vervaardigde tekenfilm
d mit Komputer hergestellter Zeichentrickfilm *m*

1196 COMPUTER-CONTROLLED TERRESTRIAL AERIAL, COMPUTER-CONTROLLED TERRESTRIAL ANTENNA — aea
f antenne *f* terrestre commandée par ordinateur
e antena *f* terrestre mandada por computadora
i antenna *f* terrestre comandata da calcolatore
n door computer bestuurde aarde-antenne
d maschinengesteuerte Erdantenne *f*

1197 COMPUTERIZED ELECTRONIC EDITING SYSTEM — fi/tv
f montage *m* à commande électronique par ordinateur
e montaje *m* de mando electrónico por computadora
i montaggio *m* a comando elettronico per calcolatore
n door computer gestuurde montage
d maschinengesteuertes Schnittsystem *n*

1198 CONCAVE MIRROR — opt
Component part of the eidophor system.
f miroir *m* concave
e espejo *m* cóncavo
i specchio *m* concavo
n concave spiegel, holle spiegel
d Hohlspiegel *m*, Konkavspiegel *m*

1199 CONCERT HALL ACOUSTICS — aud
f acoustique *f* de salle de concert
e acústica *f* de sala de conciertos
i acustica *f* di sala di concerto
n concertzaalakoestiek
d Musiksaalakustik *f*

1200 CONDENSER — opt
A lens for concentrating or bringing together rays of electromagnetic radiation.
f condenseur *m*
e condensador *m*
i condensatore *m*
n condensor
d Kondensor *m*

1201 CONDUCTOR — fi/th/tv
f chef *m* d'orchestre
e director *m* de orquesta
i direttore *m* d'orchestra
n dirigent
d Dirigent *m*

1202 CONE — crt
That part of the cathode-ray tube which lies between the neck and the screen.
f cône *m*, robe *f*
e cono *m*
i cono *m*
n conus
d Konus *m*

1203 CONES, RETINAL CONES — opt
One form of receptor cells in the retina of the eye.
f cônes *pl*
e conos *pl*
i coni *pl*
n kegeltjes *pl*
d Zäpfchen *pl*

1204 CONFETTI — dis
A disturbance in colo(u)r TV corresponding to snow in black and white reproduction.
f confetti *m*
e confeti *m*
i coriandoli *pl*
n confetti
d Konfetti *n*

1205 CONICAL MAGNETIC SHIELD — crt
f écran *m* magnétique conique
e pantalla *f* magnética cónica
i schermo *m* magnetico conico
n conisch magnetisch scherm *n*, kap voor magnetische afscherming
d Abschirmkonus *m*

1206 CONJUGATE FOCAL POINTS — opt
f foyers *pl* conjugués
e focos *pl* conjugados, lugares *pl* geométricos conjugados
i fuochi *pl* coniugati
n geconjugeerde brandpunten *pl*
d konjugierte Brennpunkte *pl*

1207 CONNECTION SOCKET FOR AERIALS, CONNECTION SOCKET FOR ANTENNAE — aea

f prise *f* pour antenne
e enchufe *m* de antena
i presa *f* di collegamento per antenna
n antenneaansluitbus
d Antennenbüchse *f*

1208 CONSECUTIVE FIELD rec/rep
f trame *f* consécutive
e cuadro *m* consecutivo
i trama *f* consecutiva
n aansluitend raster *n*
d anschliessendes Halbbild *n*

1209 CONSECUTIVE SCANNING rec/rep
f analyse *f* consécutive
e exploración *f* consecutiva
i analisi *f* consecutiva
n opeenvolgende aftasting
d Folgeabtastung *f*

1210 CONSOLE stu/tv
f pupitre *m*
e pupitre *m*
i tavolo *m*
n lessenaar
d Pult *n*

1211 CONSOLE tv
A large cabinet for a TV receiver, standing on the floor rather than on a table.
f boîtier *m*, coffret *m*
e caja *f*, gabinete *m*
i cassetta *f*, custodia *f*
n kast
d Kasten *m*, Truhe *f*

1212 CONSOLE RECEIVER tv
A TV receiver in a console.
f cabinet *m* de télévision
e gabinete *m* de televisión
i cassetta *f* di televisione
n televisiekast
d Fernsehtruhe *f*

1213 CONSTANT LINE NUMBER OPERATION, rec/tv
DRIFTLOCK
To overcome the phase jitter characteristic of field frequency locking, an additional phase-locked loop operating at the line frequency is arranged to take over control of twice the line frequency oscillator when field locking and phasing has been achieved.
f opération *f* à constance du nombre de lignes
e operación *f* de constancia del número de líneas
i operazione *f* a costanza del numero di linee
n werkwijze met constant aantal lijnen
d Zeilenzahlkonstanzverfahren *n*

1214 CONSTANT LUMINANCE ctv
The red, green and blue signals are added to form a monochrome signal of normal bandwidth whereas two narrow-band colo(u)r difference signals are formed in suitable matrices to realize constant luminance.
f luminance *f* constante
e luminancia *f* constante
i luminanza *f* costante
n constante luminantie
d konstante Luminanz *f*

1215 CONSTANT LUMINANCE COLO(U)R TRANSMISSION, ctv
CONSTANT LUMINANCE TRANSMISSION
1. A transmission in which the transmission primaries are a luminance primary and two chrominance primaries.
2. A transmission in which signals in the chrominance channel do not affect the luminance of the reproduced picture.
f système *m* de télévision couleur à luminance constante
e sistema *m* de televisión en colores de luminancia constante
i sistema *m* di televisione a colori a luminanza costante
n kleurtelevisiesysteem *n* met constante luminantie
d Farbfernsehsystem *n* mit konstanter Leuchtdichte

1216 CONSTANT LUMINANCE INDEX ctv
A measure of the fraction of the true luminance carried by the actual luminance signal in a colo(u)r TV system.
f quotient *m* de luminance
e cociente *m* de luminancia
i quoziente *m* di luminanza
n luminantiequotiënt *n*
d Luminanzquotient *m*

1217 CONSTANT-SPEED SCANNING rec
f analyse *f* à vitesse constante
e exploración *f* de velocidad constante
i analisi *f* a velocità costante
n aftasting met constante snelheid
d Abtastung *f* mit gleichbleibender Geschwindigkeit

1218 CONSTRUCTIVE INTERFERENCE ctv/vr
In the holographic process the ability of two light waves to add to one another.
f interférence *f* constructive
e interferencia *f* constructiva
i interferenza *f* costruttiva
n constructieve storing
d konstruktive Störung *f*

1219 CONTINUITY tv
Of a sequence of program(me)s or of the parts of a single program(me), the orderly procession from one part to the next according to a pre-arranged plan to produce a desired effect.
f déroulement *m* continu, enchaînement *m*
e continuidad *f*
i concatenazione *f*, montaggio *m*
n continuïteit,

programma-afwerking volgens plan
d Programmablauf *m*

1220 CONTINUITY APPARATUS ROOM rep
The room adjacent to a continuity studio containing the apparatus for the selection, switching, supervision and control of a sequency of program(me)s.
f régie *f* finale
e sala *f* de control central
i camera *f* delle apparecchiature annessa allo studio continuo
n centrale regelkamer
d Senderegieraum *m*

1221 CONTINUITY CONTROL rep
f contrôle *m* de déroulement, contrôle *m* d'enchaînement
e control *m* de continuidad
i controllo *m* di concatenazione
n continuïteitscontrole
d Einsatzkontrolle *f*, Programmablaufkontrolle *f*

1222 CONTINUITY GIRL, SCRIPT GIRL fi/tv
f script-girl *f*, secrétaire *f* de plateau
e adjunta *f* del realizador, script girl *f*, secretaria *f* de edición
i script girl *f*, segretaria *f* d'edizione
n regie-assistente, scriptgirl
d Ateliersekretarin *f*, Script-Girl *f*

1223 CONTINUITY PLANNING, PLANNING FOR CONTINUITY rep/tv
f projet *m* de déroulement, projet *m* d'enchaînement
e proyecto *m* de continuidad
i progetto *m* di concatenazione
n continuïteitsplanning
d Programmablaufplanung *f*

CONTINUITY PROGRAM(ME) CONTROL
see: CENTRAL CONTROL

1224 CONTINUITY STUDIO rep
A small studio from which an announcer supervises and controls the running of a sequence of program(me)s, makes opening and closing announcements, and interpolates interlude material when required.
f studio *m* tête de programme
e estudio *m* de continuidad
i studio *m* continuo, studio *m* per la continuità del programma
n hoofdomroepstudio
d Abwicklungsstudio *n*, Ansagestudio *n*

1225 CONTINUITY SUITE rep
A suite of rooms containing a continuity studio and a continuity apparatus room.
f bloc *m* studio
e complejo *m* de control de continuidad
i locali *pl* del complesso di continuità
n H-complex *n*, hoofdregelkamercomplex *n*
d Ablaufgruppe *f*

1226 CONTINUOUS BEAM MODULATION crt
Beam modulation in TV tubes to reproduce the gradation of light and shade in the picture.
f modulation *f* continue du faisceau
e modulación *f* continua del haz
i modulazione *f* continua del fascio
n continue bundelmodulatie
d kontinuierliche Strahlenmodulation *f*

1227 CONTINUOUS FILM SCANNER tv
A TV film scanner in which the motion-picture film moves continuously while being scanned by a flying spot kinescope.
f analyseur *m* continu de film
e explorador *m* continuo de película
i analizzatore *m* continuo di pellicola
n continue filmaftaster
d Durchlauffilmabtaster *m*

1228 CONTINUOUS NOISE
f parasite *m* récurrent
e perturbación *f* continua
i interferenza *f* continua
n permanente storing
d Dauerstörung *f*

1229 CONTINUOUS PROJECTION fi/tv
f projection *f* continue
e proyección *f* continua
i proiezione *f* continua
n doorlopende projectie
d fortlaufende Projektion *f*

1230 CONTINUOUS RANDOM NOISE dis
f parasites *pl* erratiques continus
e parásito *m* errático continuo
i disturbo *m* erratico continuo
n continue witte ruis
d kontinuierliches weisses Rauschen *n*

1231 CONTINUOUS WAVE, C.W. cpl
f onde *f* entretenue
e onda *f* continua, onda *f* sostenida
i onda *f* continua, onda *f* persistente
n ongedempte golf
d ungedämpfte Welle *f*

CONTOUR (GB)
see: AMBIT

1232 CONTOUR ACCENTUATION (GB), CONTOURING (GB), CRISPENING (US), EDGE ENHANCEMENT (GB) tv
Coupling in which the rising time of the signals is reduced in order to accentuate the contours of the objects in the image.
f accentuation *f* des contours, souligné *m*

e acentuación *f* de los contornos
i accentuazione *f* dei contorni
n aanzetten *n* van de contouren
d Konturenbetonung *f*,
Konturverstärkung *f*,
Konturversteilerung *f*

1233 CONTOUR CONVERGENCE ctv/rep
f convergence *f* des contours
e convergencia *f* de los contornos
i convergenza *f* dei contorni
n contourconvergentie
d Konturendeckung *f*

1234 CONTOUR CORRECTION rec/tv
f correction *f* de contours
e corrección *f* de contornos
i correzione *f* di contorni
n contourcorrectie
d Konturentzerrung *f*

1235 CONTOUR MAP tv
A map on which the boundaries of the range of a TV transmitter are indicated.
f carte *f* de la zone couverte
e mapa *f* de la zona útil
i carta *f* della zona utile
n kaart van het reikgebied
d Konturkarte *f*

1236 CONTOUR SHARPNESS ctv/rep
f netteté *f* du contour
e nitidez *f* del contorno
i finezza *f* del contorno
n contourscherpte
d Konturschärfe *f*

1237 CONTOURS OUT OF GREEN ctv
A system evolved to reduce registration errors in a three-tube camera by adding the green contour signal to the red and blue signals in addition to the green signal.
f addition *f* du signal du contour vert
e adición *f* de la señal del contorno verde
i addizione *f* del segnale del contorno verde
n toevoeging van het signaal voor de groene contour
d Beimischung *f* des Signals für die grüne Kontur

1238 CONTRAST ge
In TV, the ratio of two luminance levels in a scene or reproduced picture.
f contraste *m*
e contraste *m*
i contrasto *m*
n contrast *n*
d Kontrast *m*

1239 CONTRAST BALANCE rep/tv
f équilibre *m* des contrastes
e equilibrio *m* de los contrastes
i equilibrio *m* dei contrasti
n contrastevenwicht *n*
d Kontrastgleichgewicht *n*

1240 CONTRAST COMPRESSION (GB), tv
CONTRAST CRUSHING (US)
Distortion of the contrast gradient resulting in a loss of gradation, either at the white end of the range or at the black end.
f perte *f* de contraste
e pérdida *f* de contraste
i perdita *f* di contrasto
n contrastverlies *n*
d Kontrastverlust *m*

1241 CONTRAST CONTROL tv
Detailed control of high-lights and shadows in TV images.
f réglage *m* du contraste
e regulación *f* del contraste
i regolazione *f* del contrasto
n contrastregeling
d Kontrastregelung *f*

1242 CONTRAST CONTROL DEVICE tv
f dispositif *m* de réglage du contraste
e regulador *m* del contraste
i regolatore *m* del contrasto
n contrastregelaar
d Kontrastregler *m*

1243 CONTRAST DEGREE ct
f degré *m* de contraste
e grado *m* de contraste
i gradino *m* di contrasto
n contrastgraad
d Kontrastgrad *m*

1244 CONTRAST EFFECT rep/tv
f effet *m* de contraste
e efecto *m* de contraste
i effetto *m* di contrasto
n contrasteffect *n*
d Kontrasteffekt *m*

1245 CONTRAST EXPANSION tv
f expansion *f* du contraste
e expansión *f* del contraste
i espansione *f* del contrasto
n contrastverbetering
d Kontraststeigerung *f*

1246 CONTRAST GRADIENT, tv
POINT GAMMA
1. The instantaneous slope of the curve relating the logarithms of the intensity of the incident light and of the resulting output voltage.
2. The instantaneous slope of a curve relating the logarithms of the input voltage and of the intensity of the resulting light output.
f gradient *m* de contraste
e gradiente *m* de contraste
i gradiente *m* di contrasto
n contrastgradiënt
d Kontrastgradient *m*

1247 CONTRAST LIGHTING rec
f illumination *f* en contraste
e iluminación *f* contrastada
i illuminazione *f* in contrasto
n contrastverlichting
d Kontrastbeleuchtung *f*

1248 CONTRAST RANGE, DYNAMIC RANGE tv
The ratio of the luminance of the whitest portion of a picture to that of the blackest portion.
f domaine *m* de contraste, plage *f* de contraste
e **margen *m* de contraste, contraste *m* máximo**
i **contrasto *m* massimo, dinamica *f* di contrasto**
n contrastomvang, helderheidsdynamiek
d Helligkeitssprung *m*, Maximalkontrast *m*

1249 CONTRAST RANGE OF A CAMERA PICKUP EQUIPMENT rec/tv
The maximum ratio of the output voltage that can be supplied by a TV camera equipment.
f contraste *m* maximal d'un dispositif d'analyse, domaine *m* de contraste d'un dispositif d'analyse
e contraste *m* máximo de un dispositivo de exploración
i contrasto *m* massimo d'un tubo di ripresa
n uitgangsspanningsdynamiek van een beeldopneemtoestel
d Maximalkontrast *m* eines Bildaufnahmegerätes

1250 CONTRAST RANGE OF AN IMAGE REPRODUCTION EQUIPMENT rep/tv
The maximum ratio of the luminance that can be supplied by a receiving equipment.
f contraste *m* maximal d'un dispositif de synthèse, domaine *m* de contraste d'un dispositif de synthèse
e contraste *m* máximo de un dispositivo de exploración
i contrasto *m* massimo d'un tubo di riproduzione
n luminantiedynamiek van een beeldweergeeftoestel
d Maximalkontrast *m* eines Bildwiedergabegerätes

1251 CONTRAST RATIO tv
f rapport *m* de contrastes
e relación *f* de contrastes
i rapporto *m* di contrasti
n contrastverhouding
d Kontrastumfang *m*, Kontrastverhältnis *n*

1252 CONTRAST REDUCTION dis/tv
Adverse effect on the contrast in a picture tube caused i.a. by internal reflections.
f réduction *f* de contraste
e reducción *f* de contraste
i riduzione *f* di contrasto
n contrastvermindering
d Kontrastherabsetzung *f*

1253 CONTRAST SENSITIVITY ct
Liminal ratio of $\delta B/B$ of a small spot which differs from brightness B of the larger surroundings.
f sensibilité *f* de contraste
e sensibilidad *f* de contraste
i sensibilità *f* di contrasto
n contrastgevoeligheid
d Kontrastempfindlichkeit *f*

1254 CONTRAST TEST CARD, GRADATION TEST CARD tv
A test card for investigating the quality of reproduction of contrasts in a TV picture.
f mire *f* de demiteinte
e patrón *m* de contraste
i immagine *f* di prova per i grigi
n grijstraptoetsbeeld *n*
d Gradationstestbild *n*

CONTRAST THRESHOLD
see: BOUNDARY CONTRAST

1255 CONTRAST TRANSFER FUNCTION, TRANSMISSION GAMMA tv
f fonction *f* de transfert de contraste
e función *f* de transferencia de contraste
i funzione *f* di trasferimento di contrasto
n contrastoverdrachtsfunctie
d Kontrastübertragungsfunktion *f*, KÜF *f*

1256 CONTRIBUTION NETWORK cpl
Assembly of cables and radio links bringing to a central point the signals from various sources.
f circuit *m* de contribution
e circuito *m* tributario
i circuito *m* d'allacciamento
n koppelnetwerk *n*
d Zubringerleitung *f*

1257 CONTROL cpl
f commande *f*
e mando *m*
i comando *m*
n sturing
d Steuerung *f*

1258 CONTROL cpl
f contrôle *m*
e control *m*
i controllo *m*
n controle
d Kontrolle *f*, Überwachung *f*

1259 CONTROL cpl
f réglage *m*
e regulación *f*
i regolazione *f*
n regeling
d Regelung *f*

1260 CONTROL ASSEMBLY cpl
f bloc *m* de commande
e bloque *m* de mando
i blocco *m* di comando
n besturingsblok *n*
d Steuerfeld *n*

1261 CONTROL CUBICLE tv
f cabine *f* de commande,
cabine *f* technique,
contrôle *m* d'écoute
e cabina *f* de control,
puesto *m* de control
i cabina *f* di regia,
cabina *f* di smistamento
n bedieningscabine, regelcabine
d Abhörkabine *f*, Regieraum *m*

1262 CONTROL DESK (GB), tv
DIRECTOR'S CONSOLE (US)
A desk providing communicating, switching and monitoring facilities for the use of artistic and technical directors of TV shows.
f pupitre *m* de commande
e pupitre *m* de mando
i tavolo *m* di comando
n besturingslessenaar,
regeltafel
d Kontrollpult *n*,
Regiepult *n*

1263 CONTROL ELECTRODE crt
f cylindre *m* de commande
e cilindro *m* de mando
i cilindro *m* di comando
n stuurcilinder
d Steuerzylinder *m*

1264 CONTROL HEAD vr
f tête *f* de commande
e cabeza *f* de mando
i testina *f* di comando
n stuurkop
d Steuerkopf *m*

1265 CONTROL KNOB ge
f bouton *m* de commande
e botón *m* de mando
i bottone *m* di comando
n bedieningsknop
d Bedienungsknopf *m*

1266 CONTROL LINE tv
A telephone circuit used for technical conversations between points along a broadcast chain.
f ligne *f* de service
e línea *f* de servicio
i linea *f* di servizio
n dienstlijn
d Meldeleitung *f*

1267 CONTROL OF HORIZONTAL tv
SYNCHRONIZING
f réglage *m* de la synchronisation horizontale
e regulación *f* de la sincronización
horizontal
i regolazione *f* della sincronizzazione
orizzontale
n horizontale synchronisatieregeling
d Horizontalsynchronisationsregelung *f*

1268 CONTROL OF THE SPOT crt
LUMINOSITY
f commande *f* de la luminosité du spot
e mando *m* de la luminosidad del punto
i comando *m* della luminosità del punto
n sturing van de puntheIderheid
d Steuerung *f* der Punkthelligkeit

1269 CONTROL OF VERTICAL tv
SYNCHRONIZING
f réglage *m* de la synchronisation verticale
e regulación *f* de la sincronización vertical
i regolazione *f* della sincronizzazione
verticale
n verticale synchronisatieregeling
d Vertikalsynchronisationsregelung *f*

1270 CONTROL PANEL rep/svs/tv
f panneau *m* de réglage
e panel *m* de maniobra
i pannello *m* di manovra
n bedieningspaneel *n*
d Bedienungsfeld *n*

1271 CONTROL ROOM ge
f salle *f* de commande
e sala *f* de mando
i sala *f* di comando
n controlekamer
d Bedienungsraum *m*

CONTROL ROOM
see: APPARATUS ROOM

1272 CONTROL SIGNAL cpl
f signal *m* de commande
e señal *f* de mando
i segnale *m* di comando
n stuursignaal *n*
d Steuersignal *n*

1273 CONTROL SIGNALS IN SLAVE tv
LOCKING TECHNIQUE
D.c. signals appearing in the phase-locked loop between the phase detector and the voltage-controlled oscillator.
f signaux *pl* de commande dans le système
de verrouillage de l'oscillateur asservi
e señales *pl* de mando en el sistema de
enclavamiento del oscilador satélite
i segnali *pl* di comando nel sistema di
bloccaggio dell'oscillatore satellite
n stuursignalen *pl* in het systeem van
vergrendeling van de dochterzender
d Steuersignale *pl* im System der Tochter-
senderverriegelung

1274 CONTROL TRACK vr
f piste *f* d'asservicement,
piste *f* de pilotage
e pista *f* de control,
pista *f* piloto
i pista *f* controllo,
pista *f* pilota
n pilootspoor *n*,
stuurspoor *n*
d Steuerspur *f*

1275 CONTROLLED DELAY rec/rep/tv
LOCK
A method of bringing a remote signal into lock by introducing a controlled delay.

f verrouillage *m* retardé
e bloqueo *m* de señal por retardo regulado
i agganciamento *m* ritardato
n signaalvergrendeling door gestuurde vertraging
d Signalverriegelung *f* durch gesteuerte Verzögerung

1276 CONTROLLED RATE GENLOCK, SLOW GENLOCK cpl
f système *m* à générateur verrouillé à commande de vitesse
e sistema *m* de generador clavado de mando de velocidad
i sistema *m* a generatore bloccato a velocità comandata
n systeem *n* met vergrendelde generator en bestuurde snelheid
d geschwindigkeitsgesteuerte Genlock-Betriebsweise *f*

1277 CONVENTIONAL STUDIO stu
f studio *m* classique
e estudio *m* clásico
i studio *m* classico
n conventionele studio
d konventionnelles Studio *n*

1278 CONVERGENCE crt/ctv
A condition in which the electron beam of a multi-gun display tube intersects at a point so that the three primary colo(u)rs are correctly superposed.
f convergence *f*
e convergencia *f*
i convergenza *f*
n convergentie
d Konvergenz *f*

1279 CONVERGENCE ADJUSTMENTS
f ajustages *pl* de la convergence
e ajustes *pl* de la convergencia
i aggiustaggi *pl* della convergenza
n convergentie-instellingen *pl*
d Konvergenzeinstellungen *pl*

1280 CONVERGENCE ANODE crt/ctv
f anode *f* convergente
e ánodo *m* convergente
i anodo *m* convergente
n convergentieanode
d Konvergenzanode *f*

1281 CONVERGENCE ASSEMBLY crt
f bloc *m* de convergence
e bloque *m* de convergencia
i blocco *m* di convergenza
n convergentie-eenheid
d Konvergenzeinheit *f*

1282 CONVERGENCE CIRCUITS crt/ctv
Circuits provided for a multi-gun colo(u)r tube to align the electron beams accurately with their phosphor spots across the complete picture.
f circuits *pl* de convergence
e circuitos *pl* de convergencia
i circuiti *pl* di convergenza
n convergentiekringen *pl*
d Konvergenzkreise *pl*

1283 CONVERGENCE CONTROL crt/ctv
f réglage *m* de la convergence
e regulación *f* de la convergencia
i regolazione *f* della convergenza
n convergentieregeling
d Konvergenzregelung *f*

1284 CONVERGENCE ELECTRODE crt/ctv
f électrode *f* de convergence
e electrodo *m* de convergencia
i elettrodo *m* di convergenza
n convergentie-elektrode
d Konvergenzelektrode *f*

1285 CONVERGENCE ERRORS dis
f erreurs *pl* de convergence
e errores *pl* de convergencia
i errori *pl* di convergenza
n convergentiefouten *pl*
d Deckungsfehler *pl*, Konvergenzfehler *pl*

CONVERGENCE MAGNET
see: BEAM MAGNET

1286 CONVERGENCE PLANE crt/ctv
A plane containing the points at which the electron beams of a multibeam cathode-ray tube appear to experience a deflection applied for the purpose of obtaining convergence.
f plan *m* de convergence
e plano *m* de convergencia
i piano *m* di convergenza
n convergentievlak *n*
d Konvergenzebene *f*

1287 CONVERGENCE SYSTEM ctv
f système *m* de convergence
e sistema *m* de convergencia
i sistema *m* di convergenza
n convergentiesysteem *n*
d Konvergenzsystem *n*

1288 CONVERGENT BEAM crt
A beam focused at a point.
f faisceau *m* convergent
e haz *m* convergente
i fascio *m* convergente
n convergerende bundel
d konvergierende Strahlen *pl*

1289 CONVERSION fi/tv
f conversion *f*
e conversión *f*
i conversione *f*
n conversie, omzetting
d Konversion *f*, Umwandlung *f*

1290 CONVERSION FILTER fi/tv
f filtre *m* de conversion
e filtro *m* de conversión
i filtro *m* di conversione
n conversiefilter *n*
d Konversionsfilter *n*

1291 COOL COLO(U)RS ct
f couleurs *pl* froides
e colores *pl* fríos
i colori *pl* freddi

n koude kleuren *pl*
d kalte Farben *pl*

1292 COOLING-DOWN TIME tv
f temps *m* de refroidissement
e tiempo *m* de enfriamiento
i tempo *m* di raffreddamento
n afkoeltijd
d Abkühlungszeit *f*, Kühlzeit *f*

1293 CO-ORDINATE ge
One of a set of numbers specified in establishing the location of a point in space with respect to a co-ordinate system.
f coordonnée *f*
e coordenada *f*
i coordinata *f*
n coördinaat
d Koordinate *f*

1294 CO-ORDINATE SYSTEM ge
A set of numbers or surfaces which may be used to locate a point, line, or geometric element in space.
f système *m* de coordonnées
e sistema *m* de coordenadas
i sistema *m* di coordinate
n coördinatenstelsel *n*, coördinatensysteem *n*
d Koordinatenkreuz *n*, Koordinatensystem *n*

1295 COPLANAR CARTRIDGE (US), COPLANAR CASSETTE (GB) vr
f cassette *f* coplanaire
e caseta *f* coplanar
i cassetta *f* coplanare
n coplanaire cassette
d Koplanarkassette *f*

1296 COPY, DUB vr
Recording of signals obtained from another tape.
f copie *f*
e copia *f*
i copia *f*, riversamento *m*
n kopie
d Kopie *f*

1297 COPY LENS opt
f objectif *m* de reproduction, objectif *m* de tireuse optique
e objetivo *m* de reproducción
i obiettivo *m* di riproduzione
n reproduktie-objectief *n*
d Reproduktionsobjektiv *n*

1298 COPYING WORKSHOP tv
f atelier *m* de copie
e taller *m* de copia de registro
i officina *f* di copia di registrazione, officina *f* di riversamento
n kopieerwerkplaats
d Umschneideraum *m*, Umspielraum *m*

1299 CORKSCREW ANTENNA (US), HELICAL AERIAL (GB) aea
f antenne *f* hélicoïdale
e antena *f* helicoidal
i antenna *f* elicoidale
n spiraalantenne
d Wendelantenne *f*

1300 CORNEA opt
The transparent forward portion of the shell of the eye.
f cornée *f*
e córnea *f*
i cornea *f*
n hoornvlies *n*
d Hornhaut *f*

1301 CORNER CUTTING rep
The elimination of the corners of a TV picture.
f obscurcissement *m* des coins
e obscurecimiento *m* de los ángulos
i oscuramento *m* agli angoli
n hoekafscherming
d Abschattierung *f*

1302 CORNER DETAIL rep/tv
Perfect details at the corners of the TV picture.
f netteté *f* des coins
e nitidez *f* en los ángulos
i definizione *f* dell'immagine negli angoli
n hoekscherpte
d Eckenschärfe *f*

1303 CORNER INSERT tv
Picture insert in the corner of the main picture.
f insertion *f* au coin
e inserción *f* en el ángulo
i inserzione *f* nell'angolo
n hoekinlas
d Eckeinlage *f*

1304 CORRECT (TO) THE GAMMA, GAMMATE (TO) rep/tv
To apply a non-linear transfer characteristic to a TV signal in order to achieve a desired value of gamma.
f corriger v la gamme
e corregir v la gama
i correggere v la gamma
n de gamma verbeteren v
d das Gamma korrigieren v

1305 CORRECTING AMPLIFIER cpl
f amplificateur *m* correcteur
e amplificador *m* corrector
i amplificatore *m* correttore
n corrigerende versterker
d entzerrender Verstärker *m*

1306 CORRECTING COIL, PEAKING COIL cpl
A coil used in an amplifier to increase the gain in the higher frequencies.
f bobine *f* de crête
e bobina *f* correctora, bobina *f* de compensación

i bobina *f* correttrice
n ophaalspoel
d Versteilerungsspule *f*

1307 CORRECTING LENS, CORRECTION PLATE — opt
A lens in the Schmidt optical system to correct astigmatism.
f lentille *f* correctrice
e lente *f* correctora
i lente *f* correttrice
n correctielens
d Korrektionslinse *f*

1308 CORRECTING MAGNET — crt
f aimant *m* correcteur
e iman *m* corrector
i magnete *m* correttore
n correctiemagneet
d Gleichlaufmagnet *m*

1309 CORRECTING SEPARATING AMPLIFIER — cpl
f amplificateur *m* correcteur-séparateur
e amplificador *m* corrector-separador
i amplificatore *m* correttore-separatore
n scheidende corrigerende versterker
d entzerrender Trennverstärker *m*

1310 CORRECTION — ge
f correction *f*
e corrección *f*
i correzione *f*
n correctie
d Korrektur *f*

1311 CORRECTION FILTER — dis
f filtre *m* de correction
e filtro *m* de corrección
i filtro *m* di correzione
n correctiefilter *n*
d Korrekturfilter *n*

CORRECTION OF AMPLITUDE
see: AMPLITUDE CORRECTION

1312 CORRECTION SIGNAL — cpl
f signal *m* de correction
e señal *f* de corrección
i segnale *m* di correzione
n correctiesignaal *n*
d Korrektursignal *n*

1313 CORRECTOR CIRCUIT — tv
A circuit used in audio-frequency amplifiers, video stages in TV receivers, etc. to counteract the effect of deficiency in the amplifiers and/or transmission.
f circuit *m* compensateur,
circuit *m* correcteur
e circuito *m* compensador,
circuito *m* corrector
i circuito *m* compensatore,
circuito *m* correttore
n compensatiekring,
correctiekring,
d Ausgleichskreis *m*

1314 CORRECTOR CIRCUIT FOR COLO(U)R TELEVISION — ctv
f amplificateur *m* correcteur pour télévision couleur
e amplificador *m* corrector para televisión en colores
i amplificatore *m* correttore per televisione a colori
n correctieversterker voor kleurentelevisie
d Korrekturverstärker *m* für Farbfernsehen

1315 CORRELATED COLO(U)R TEMPERATURE — ct
The colo(u)r temperature corresponding to the point on the Planckian locus which is nearest to the point representing the chromaticity of the illuminant considered on an agreed uniform chromaticity scale diagram.
f température *f* de couleur corrélative,
température *f* de couleur proximale
e temperatura *f* de color correlativa
i temperatura *f* di colore correlativa
n correlatieve kleurtemperatuur
d korrelative Farbtemperatur *f*

1316 COSTUME ADVISER — fi/th/tv
f conseiller *m* pour les costumes
e consejero *m* para el vestuario
i consigliatore *m* per i costumi
n kostuumadviseur
d Kostumberater *m*

1317 COSTUME DESIGNER — fi/th/tv
f costumier *m*,
créateur *m* de costumes
e diseñador *m* de vestuario
i costumista *m*
n kostuumontwerper
d Kostümbildner *m*

1318 COVERAGE — tv
The size of, or the total number of inhabitants within, the service area of a broadcasting station.
f couverture *f*
e cobertura *f*
i copertura *f*
n bestreken gebied *n*
d Versorgung *f*

1319 CRAB DOLLY — rec
Type of movable camera platform with all wheels steerable.
f chariot *m* à galets d'orientation
e carretilla *f* con rodillos giratorios
i carrello *m* a rulli orientabili
n camerawagen met zwenkwielen
d Schwenkradkamerawagen *m*

1320 CRABBING — rec
Sideways movement of TV camera for taking an expended panning shot.
f travelling *m* latéral
e movimiento *m* lateral
i movimento *m* laterale
n zijwaartse beweging
d seitliche Kamerafahrt *f*

1321 CRAMPING rep/tv
Contraction of either side or the central section of a TV picture.
f contraction *f* d'image
e contracción *f* de imagen
i contratto *m* d'immagine
n beeldkrimp
d Bildeinschränkung *f*

1322 CRANE TRUCK rec
A power-driven mount for the TV camera.
f camion-grue *m*
e camión-grúa *m*
i autogru *m*
n kraanwagen
d Kranwagen *m*

1323 CRAWLING dis/tv
f filage *m*,
formation *f* de rayures
e formación *f* de rayas
i formazione *f* di coda,
formazione *f* di strie
n streepvorming
d Strichbildung *f*

1324 CREDIT TITLE, rec
CREDITS
Title which lists those creatively responsible for the film or TV program(me) to which it is attached.
f générique *m*
e lista *f* de nombres de colaboradores
i elenco *m* di nomi di collaboratori
n medewerkerslijst,
titels *pl*
d Namensvorspann *m*

1325 CREEPIE-PEEPIE, rec/tv
MOBILE CAMERA,
PORTABLE CAMERA
A camera which can be carried by the camera operator and functioning without cables.
f caméra *f* portative
e cámara *f* móvil,
cámara *f* portátil
i camera *f* portatile
n draagbare camera
d tragbare Kamera *f*

1326 CREW, stu
TEAM
f équipe *f*, personnel *m*
e equipo *m*, personal *m*
i equipo *m*, squadra *f*
n groep
d Gruppe *f*, Stab *m*, Team *n*

CRISPENING (US)
see: CONTOUR ACCENTUATION

1327 CRITICAL FLICKER FREQUENCY tv
The highest frequency at which flicker is just visible.
f fréquence *f* de papillotement critique
e frecuencia *f* de parpadeo crítica
i frequenza *f* di sfarfallio critica
n kritische flikkerfrequentie
d kritische Flimmerfrequenz *f*

1328 CROPPING rec
Restriction of the projected or transmitted image area of a film by making use of masks or apertures smaller than the image area available.
f réduction *f* du plan d'image
e reducción *f* del plano de imagen
i riduzione *f* del piano d'immagine
n beeldvlakverkleining
d Bildebeneverkleinerung *f*

1329 CROSS-COLO(U)R NOISE dis/ctv
An unwanted low frequency colo(u)r effect due to the introduction of noise from the luminous source into the chrominance channel because of band sharing.
f diachromie *f*
e diacromía *f*
i diacromia *f*
n diachromie,
kleuroverspreken *n*
d Diachromie *f*,
Farbkanalübersprechen *n*,
Leuchtdichte/Chrominanzübersprechen *n*

1330 CROSS-CUT, rec/tv
INTERCUT
Cutting so that sections of a scene are reproduced alternately.
f raccord *m*
e montaje *m* cruzado
i montaggio *m* incrociato
n kruismontage
d Kreuzschnitt *m*

1331 CROSS-CUTTING fi
f changement *m* rapide de cadre
e cambio *m* rápido de cuadro
i cambio *m* rapido di quadro
n snelle beeldwisseling
d schnelle Einstellungsfolge *f*

1332 CROSS-FADE tv
To fade in one channel while fading out another in order to substitute gradually the output of one for that of the other.
f fondu *m* enchaîné
e fundido *m* encadenado
i dissolvenza *f* incrociata
n langzame beeldwisseling
d Überblendung *f*,
Umblendung *f*

1333 CROSS-FIELD BIAS vr
A system of tape recording in which a separate head is used for the application of the high-frequency bias.
f polarisation *f* par tête à champ transversal
e polarización *f* por cabeza con campo transversal
i polarizzazione *f* per testina a campo trasversale
n voormagnetisatie door kruisveldkop
d Vormagnetisierung *f* durch Querfeldkopf

1334 CROSS-FIRE, tv
VISION CROSSTALK
f diaphotie *f*
e diafotía *f*
i diafotia *f*

n beeldoverspreken *n*,
diafotie
d Bildkanalübersprechen *n*,
Diaphotie *f*

1335 CROSS-GAMMA PATTERN opt/tv
An optical test pattern used as a signal source to adjust grey scale tracking of TV cameras.
f mire *f* de réglage du contraste dans le studio
e imagen *f* patrón de contraste en el estudio
i immagine *f* di prova per la correzione del gamma nello studio
n studiotoetsbeeld *n* voor de gamma-correctie
d Studiotestbild *n* für die Gammakorrektur

1336 CROSS-GAMMA SCALE cpl/tv
f échelles *pl* des gris en contre-sens
e escalas *pl* de grises en sentido inverso
i scale *pl* di grigi in senso inverso
n tegengesteld gerichte grijstrappen *pl*
d gegenläufige Grautreppen *pl*

1337 CROSS-HATCH PATTERN tv
A test pattern on a TV picture tube consisting of vertical and horizontal lines, produced by a special generator.
f mire *f* quadrillée
e mira *f* de cuadros,
patrón *m* de cuadros
i immagine *f* campione quadrettata
n geruit toetsbeeld *n*,
ruitpatroon *n*
d Rautenmuster *n*

1338 CROSS LIGHTS, stu
CROSSED SPOTS
Light sources symmetrically disposed in front of the set and at equal and opposite angles to the optical axis of the camera.
f lumières *pl* croisées
e luces *pl* cruzadas
i luci *pl* incrociate
n kruislichten *pl*
d Kreuzlichter *pl*

1339 CROSS MODULATION dis
The modulation of the carrier of the desired signal by an undesired signal.
f transmodulation *f*
e modulación *f* cruzada
i trasmodulazione *f*
n kruismodulatie
d Kreuzmodulation *f*

1340 CROSSOVER, crt
CROSSOVER AREA,
FOCAL SPOT
In an electron lens system, location where streams of electrons from the object pass through a very small area, substantially a point, before forming an image.
f convergence *f*,
point *m* de convergence
e punto *m* de cruce
i punto *m* d'incrociamento
n bundelknoop
d Bündelknoten *m*,
Überschneidungspunkt *m*

1341 CROSSTALK dis/tv
f diaphonie *f*,
introduction *f* furtive
e diafonía *f*,
introducción *f* furtiva
i diafonia *f*,
introduzione *f* furtiva
n insluiping,
overspreken *n*
d Einstreuung *f*,
Übersprechen *n*

1342 CROSSVIEW dis/rec/tv
Reception of an unwanted picture on a vision circuit analogous to crosstalk on sound.
f réception *f* de signal vidéo non-désiré due à l'intermodulation
e recepción *f* de señal video no deseada debida a la intermodulación
i ricezione *f* di segnale video non desiderato causata da intermodulazione
n door intermodulatie veroorzaakte niet-gewenste ontvangst van een video-signaal
d durch Zwischenmodulation verursachter Empfang *m* eines nichtgewünschten Videosignals

1343 CROWD NOISE dis
f bruit *m* de salle
e ruido *m* de sala
i rumore *m* di sala
n zaalruis *n*
d Saalgeräusch *n*

1344 CROWDING tv
f aplatissement *m*
e apiñamiento *m*
i affollamento *m*
n gedrongenheid
d Gedrängtheit *f*

CRT
see: CATHODE-RAY TUBE

CRT PROJECTOR
see: CATHODE-RAY TUBE PROJECTOR

1345 CRYSTAL TIMING OSCILLATOR cpl
A device and circuit used for the timing of the scanning functions in a TV system.
f oscillateur *m* de synchronisation à commande par cristal
e oscilador *m* de sincronización de mando por cristal
i oscillatore *m* di sincronizzazione a comando per cristallo
n door kristal gestuurde synchronisatie-oscillator
d kristallgesteuerter Synchronisations-oszillator *m*

1346 CRYSTALLINE LENS opt
A transparent moderately rigid body, deformation of which by the ciliary muscles produces the focusing action of the eye.
f cristallin *m*
e cristalino *m*
i lente *f* cristallina
n ooglens
d Augenlinse *f*

CTV
see: COMMUNITY TELEVISION

1347 CUE (GB), tv
TALLY SIGNAL (US)
A pre-arranged signal, for production purposes, to a studio or other program(me) source.
f signal *m* d'avertissement
e señal *f* de aviso
i segnale *m* d'avvertimento
n waarschuwingssignaal *n*
d Achtungssignal *n*

1348 CUE CARD rec/tv
Aid to the camera operator containing information about the various shots he will be making during a program(me).
f feuille *f* de service
e tarjeta *f* de apuntes
i foglietto *m* di servizio
n programma-instructies *pl*
d Programmanweisungen *pl*

1349 CUE CIRCUIT tv
A one-way communication circuit used to convey program(me) control information.
f ligne *f* de commande
e línea *f* de apunte,
línea *f* de mando
i linea *f* di comando
n instructielijn
d Kommandoleitung *f*

1350 CUE CLOCK, dis/tv
ELAPSED TIME CLOCK
A clock for registering the duration of station breaks.
f horloge *m* pour enregistrer les temps de panne
e reloj *m* registrador de los tiempos de avería
i orologio *m* per registrare i tempi d'avaria
n storingstijdklok
d Störungszeituhr *f*

1351 CUE DOTS, rep/tv
CUE MARKS
Small marks appearing in the corner of a picture image to indicate the approach of the end of a program(me).
f marques *pl* fin de programme
e marcaciones *pl* fin de programa
i marche *pl* fine di programma
n tekens *pl* einde programma
d Zeichen *pl* Programmende

1352 CUE HEAD vr
f tête *f* de repérage
e cabeza *f* de indicación
i testina *f* d'indicazione
n markeerkop
d Merkkopf *m*

1353 CUE-IN TIMES rec/tv
The times between the separate items of news in beforehand combined news films.
f intervalles *pl*
e intervalos *pl*
i intervalli *pl*
n tussentijden *pl*
d Zwischenzeiten *pl*

1354 CUE LIGHTS (GB), tv
TALLY LIGHTS (US)
f lampes *pl* de signalisation
e focos *pl* de apunte
i segnali *pl* luminosi
n signaallampen *pl*
d Studiosignallampen *pl*

CUE MARK
see: CHANGE-OVER CUE

1355 CUE TRACK vr
f piste *f* d'ordres,
piste *f* de repérage
e pista *f* de indicación,
pista *f* de órdenes
i pista *f* d'indicazione,
pista *f* ordini
n markeerspoor *n*,
d Merkspur *f*

1356 CUEING DEVICE rec/tv
A facility for presenting the script for an artist to read while appearing before the camera.
f portescénario *m*
e portascenario *m*
i portascenario *m*
n draaiboekhouder
d Drehbuchhalter *m*

1357 CURRENT AFFAIRS rec/tv
f actualités *pl*
e actualidades *pl*
i attualità *f*, notizie attuali
n actualiteiten *pl*
d Aktualitäten *pl*, Aktuelles *n*

1358 CURRENT AFFAIRS rec/tv
PROGRAM(ME)
f programme *m* d'actualités
e programa *m* de actualidades
i programma *m* d'attualità
n actualiteitenprogramma *n*
d Aktualitätenprogramm *n*

1359 CURRENT AMPLIFICATION, ge
CURRENT GAIN
The ratio of the output and input currents under specified conditions of impedance termination.
f amplification *f* de courant
e amplificación *f* de corriente

i amplificazione *f* di corrente
n stroomversterking
d Stromverstärkung *f*

1360 CURRENT COMMUNICATIONS SATELLITES rec/rep/tv
f satellites *pl* de communication actifs
e satélites *pl* de comunicación activos
i satelliti *pl* di comunicazione attivi
n actieve communicatiesatellieten *pl*
d aktive Kommunikationssatelliten *pl*

1361 CURRENT CONSUMPTION cpl
f courant *m* absorbé
e absorción *f* de corriente,
consumo *m* de corriente
i corrente *f* assorbita
n opgenomen stroom
d Stromaufnahme *f*

1362 CURVATURE OF THE FIELD rep/tv
The curved reproduction of a flat object.
f courbure *f* du champ
e curvatura *f* del campo
i curvatura *f* del campo
n beeldveldkromming
d Bildfeldwölbung *f*

1363 CUT vr
Physical severing of the tape.
f coupure *f*
e corte *m*
i taglio *m*
n knip, snede
d Schnitt *m*

1364 CUT, CUTTING rep/tv
In TV, the instantaneous transfer from one electronic picture signal to another.
f fondu *m* enchaîné rapide
e cambio *m* brusco de toma
i dissolvenza *f* istantanea,
sostituzione *f* istantanea
n sprongovergang
d harte Überblendung *f*

1365 CUT, CUTTING tv
1. Deletion of part of a program(me).
2. Abrupt termination of the output of a channel.
f coupure *f* brusque
e corte *m* brusco
i taglio *m* brusco
n afbreking, weglating
d Wegnahme *f*

1366 CUT BANK rep/tv
Facility of vision switching equipment used by a producer in the studio.
f table *f* de commutation vidéo
e mesa *f* de conmutación video
i tavola *f* di commutazione video
n beeldomschakeltafel
d Umblendetisch *m*

1367 CUT-IN BLANKING, INTERFIELD CUT rep/tv
An instantaneous transition between two vision signals which occurs during the field suppression period so that no disturbance to picture information occurs during the visible lines.
f raccord *m* pendant la suppression de trame
e transición *f* durante la supresión de cuadro
i transizione *f* durante la cancellazione di trama
n overgang gedurende de rasteronderdrukking
d Übergang *m* während der Teilbild-austastung

1368 CUT-IN SCENE tv
f plan *m* de raccord
e inserto *m*
i inserto *m*,
scena *f* di raccordo
n inlas
d Anschnitt *m*,
Einschnitt *m*,
Einschnittszene *f*

CUT KEY
see: CENSOR KEY

1369 CUT-OFF FREQUENCY cpl
f fréquence *f* de coupure
e frecuencia *f* de corte
i frequenza *f* di taglio
n afsnijfrequentie
d Grenzfrequenz *f*

1370 CUT-OFF POINT OF PICTURE TUBE crt/svs
f point *m* de coupure de tube image
e punto *m* de corte de tubo imagen
i punto *m* di taglio di tubo di ripresa
n afknijppunt *n* beeldbuis
d Bildröhrengrenzpunkt *m*

1371 CUTTER, EDITOR fi/tv
f chef-monteur *m*,
monteur *m*
e jefe *m* de corte,
montador
i montatore *m*
n cutter
d Cutter *m*,
Cutterin *f*,
Schnittmeister *m*

1372 CUTTING, EDITING fi/tv
f montage *m*
e montaje *m*
i montaggio *m*
n montage
d Schnitt *m*

1373 CUTTING DESK, CUTTING TABLE, EDITING DESK, EDITING TABLE fi/tv
f table *f* de montage
e mesa *f* de montaje
i moviola *f*, tavolo *m* di montaggio
n montagetafel
d Klebetisch *m*

1374 CUTTING ROOM, EDITING ROOM fi/tv
f atelier *m* de montage, salle *f* de montage
e sala *f* de montaje
i sala *f* di montaggio
n montageruimte
d Montageraum *m*, Schneideraum *m*

1375 CUTTING SOUND aud
f montage *m* son
e montaje *m* sonido
i montaggio *m* suono
n geluidsmontage
d Tonschnitt *m*

C.W.
see: CONTINUOUS WAVE

1376 C.W. REFERENCE SIGNAL ctv
In colo(u)r TV, a sinusoidal signal used to control the conduction time of a synchronous demodulator.
f signal *m* de référence entretenu
e señal *f* de referencia continua
i segnale *m* di riferimento continuo
n ongedempt referentiesignaal *n*
d ungedämpftes Bezugssignal *n*

1377 CYAN, MINUS RED ctv
A colo(u)r in which the spectral components are confined mainly to the 400-600 nm band of the spectrum, i.e. the 600-700 nm (red) band is missing.
f cyan *m*, moins rouge
e ciano *m*, menos rojo
i ciano *m*, meno rosso
n cyaan *n*, minus rood
d minus Rot, Zyan *n*

1378 CYLINDER ELECTRODE crt
f électrode *f* cylindrique
e electrodo *m* cilíndrico
i elettrodo *m* cilindrico
n cilinderelektrode
d Zylinderelektrode *f*

D

1379 DALTONISM, ct
DEFECTIVE COLO(U)R VISION
The condition in which colo(u)r discrimination is significantly reduced in comparison with the normal trichromat.
f daltonisme *m*
e daltonismo *m*
i daltonismo *m*
n kleurenblindheid
d Farbenblindheit *f*

1380 DAMAGE FROM WINDING, rec
TAPE DAMAGE FROM WINDING
f endommagement *m* de la bande par enroulement
e daño *m* de la cinta debido al arrollamiento
i danno *m* del nastro dovuto all'avvolgimento
n bandbeschadiging tijdens het opwikkelen
d Bandbeschädigung *f* während der Aufwicklung

1381 DAMPING cpl
A reduction of the amplitudes of signals caused in TV by stray capacitances in the wiring of a circuit or in the components.
f amortissement *m*
e amortiguamiento *m*
i smorzamento *m*
n demping
d Dämpfung *f*

1382 DAMPING DIODE, cpt/crt
DAMPING TUBE
Used to prevent highly undesirable oscillations in a circuit of a cathode-ray tube employing electromagnetic focusing.
f diode *f* d'amortissement
e diodo *m* amortiguador
i diodo *m* smorzatore
n dempdiode
d Dämpfungsdiode *f*, Zeilendiode *f*

1383 DARK ADAPTATION ct
Adaptation of the eye exposed to no light or only faint red light.
f adaptation *f* à l'obscurité
e adaptación *f* a la obscuridad
i adattamento *m* all'oscurità
n aanpassing aan het donker
d Dunkeladaptation *f*

DARK AND LIGHT SPOTS
see: CLOUD

1384 DARK BURN, crt
FATIGUE
The decrease in efficiency of a fluorescent material during excitation.
f fatigue *f* d'une substance fluorescente
e fatiga *f* de una substancia fluorescente
i fatica *f* d'una sostanza fluorescente
n vermoeidheid van een fluorescerende stof
d Ermüdung *f* eines Leuchtstoffes

1385 DARK CURRENT cpl
Current in a photoelectric or photoconductive device when the device is supplied with its normal operating voltages and there is no light on the target.
f courant *m* d'obscurité
e corriente *f* de obscuridad, corriente *f* de reposo, corriente *f* obscura
i corrente *f* d'oscurità, corrente *f* di riposo
n donkerstroom
d Dunkelstrom *m*

1386 DARK CURRENT CLIPPER cpl
f limiteur *m* du courant d'obscurité
e limitador *m* de la corrente obscura
i limitatore *m* della corriente d'oscurità
n donkerstroombegrenzer
d Dunkelstrombegrenzer *m*

1387 DARK DESATURATION ctv
Reduction of the saturation in the shadows of the picture.
f désaturation *f* dans les parties obscures
e desaturación *f* en las partes obscuras
i saturazione *f* inferiore nelle parti oscure
n verzadigingsreductie in de donkere gedeelten
d Farbverweisslichung *f* der Tiefen

1388 DARK PICTURE AREAS rep/tv
f zones *pl* obscures de l'image
e zonas *pl* obscuras de la imagen
i zone *pl* oscure dell'immagine
n donkere beeldgebieden *pl*
d Tiefen *pl*

1389 DARK SPOT crt
A spot sometimes seen in a TV image, due to an electron cloud formation in front of a portion of the mosaic in a TV camera tube.
f point *m* obscur
e mancha *f* obscura, punto *m* obscuro
i macchia *f* oscura, punto *m* oscuro
n donker punt *n*, donkere vlek
d Dunkelfleck *m*, Dunkelpunkt *m*

1390 DARK SPOT SIGNAL tv
f signal *m* de point obscur
e señal *f* de punto obscuro
i segnale *m* di punto oscuro
n donkervleksignaal *n*, zwartsignaal *n*
d Schwarzsignal *n*

1391 DARK TRACE TUBE, crt
SKIATRON
A type of cathode-ray tube in which the screen, coated with an alkali-halide, momentarily darkens under the action of an electron beam.
f skiatron *m*
e esquiatrón *m*,
tubo *m* catódico de pantalla absorbente
i tubo *m* catodico a schermo assorbente
n donkerspoorbuis
d Schwärzungsröhre *f*,
Skiatron *n*

1392 D.C. BIASING vr
f polarisation *f* en courant continu
e polarización *f* en corriente continua
i polarizzazione *f* in corrente continua
n gelijkstroomvoormagnetisatie
d Gleichstromvormagnetisierung *f*

1393 D.C. BLOCKING rep/tv
A method for eliminating unpleasant surges due to variations in the values of the d.c. component by placing a large capacitor between the incoming coaxial cable and the matrix.
f blocage *m* de la composante continue
e bloqueo *m* de la componente continua
i bloccaggio *m* della componente continua
n blokkering van de werkzame nulcomponent
d Blockierung *f* der Gleichstromkomponente

1394 D.C. CENT(E)RING crt
A method of cent(e)ring in which use is made of the passage of a direct current through the line and/or field deflection coils.
f centrage *m* par le courant continu
e centraje *m* por la corriente continua
i centratura *f* per la corrente continua
n gelijkstroomcentrering
d Gleichstromzentrierung *f*

1395 D.C. CLAMP DIODE, cpl
D.C. RESTORING DIODE
f diode *f* de niveau
e diodo *m* de nivel de la componente continua,
diodo *m* restaurador
i diodo *m* di ristabilimento della componente continua
n niveaudiode
d Schwarzsteuerdiode *f*

1396 D.C. COMPONENT ge
The component of the picture or video signal which represents the average luminance of the picture with respect to a certain level.
f composante *f* continue,
composante *f* utile
e componente *f* continua
i componente *f* continua
n werkzame nulcomponent
d Gleichstromanteil *m*,
Gleichstromkomponente *f*

1397 D.C. CONVERGENCE crt/ctv
STATIC CONVERGENCE
Convergence of the three electron beams at an opening in the center(re) of the shadow mask in the colo(u)r picture tube.
f convergence *f* statique
e convergencia *f* estática
i convergenza *f* statica
n statische convergentie
d statische Konvergenz *f*

1398 D.C. ELECTRON STREAM crt
RESISTANCE
The ratio of electron-stream potential to the d.c. component of stream current.
f résistance *f* en courant continu du flux électronique
e resistencia *f* en corriente continua del flujo electrónico
i resistenza *f* in corrente continua del flusso elettronico
n gelijkstroomweerstand van de elektronenflux
d Gleichstromwiderstand *m* der Elektronenströmung

1399 D.C. ERASING vr
The removal of signals in a tape by a unidirectional magnetic field using d.c. current.
f effacement *m* en courant continu
e borrado *m* en corriente continua
i cancellazione *f* in corrente continua
n gelijkstroomwissing
d Gleichstromlöschung *f*

1400 D.C. INSERTER cpl
A TV transmitter stage that adds to the video signal a d.c. component known as the pedestal level.
f circuit *m* d'insertion de la composante continue au signal vidéo
e circuito *m* de inserción de la componente continua en la señal video
i circuito *m* d'inserzione della componente continua nel segnale video
n toevoeging van de werkzame nulcomponent aan het videosignaal
d Einfügung *f* der Gleichstromkomponente in das Videosignal,
Gleichstromzuführungskreis *m*

1401 D.C. INSERTION rec/rep/tv
f insertion *f* de la composante continue
e inserción *f* de la componente continua
i inserzione *f* della componente continua
n toevoeging van de werkzame nulcomponent
d Gleichstromzuführung *f*

1402 D.C. LEVEL cpl
f niveau *m* de la composante continue
e nivel *m* de la componente continua
i livello *m* della componente continua
n niveau *n* van de nulcomponent
d Pegel *m* der Gleichstromkomponente

1403 D.C. LEVEL INSTABILITY cpl
f instabilité *f* du niveau de la composante continue

e inestabilidad *f* del nivel de la componente continua
i instabilità *f* del livello della componente continua
n onstabiliteit van het niveau van de nul-componente
d Instabilität *f* des Pegels der Gleichstrom-komponente

1404 D.C. PICTURE TRANSMISSION rep/tv
TV transmission in which the signal contains a d.c. component that represents the average illumination of the entire scene.
f transmission *f* de la composante continue du signal vidéo
e transmisión *f* de la componente continua de la señal video
i trasmissione *f* della componente continua del segnale d'immagine
n transmissie van de werkzame nulcomponent van het videosignaal
d Übertragung *f* der Gleichstromkomponente des Videosignals

1405 D.C. QUADRICORRELATOR ctv
An assembly of two synchronous detector circuits operating with reference signals of the same frequency but in phase quadrature.
f quadricorrélateur *m* à courant continu
e cuadricorrelador *m* de corriente continua
i quadricorrelatore *m* a corrente continua
n gelijkstroomquadricorrelator
d Gleichstromquadrikorrelator *m*

1406 D.C. REINSERTION, rec/ rep/tv
D.C. RESTORATION
The operation of reconstructing a picture signal or video signal, with its d.c. component, from a signal not carrying a d.c. component or from a signal influenced by some unwanted d.c. component.
f restauration *f* de la composante continue, restitution *f* de la composante continue
e restauración *f* de la componente continua
i restaurazione *f* della componente continua, ristabilimento *m* della componente continua
n herstellen *n* van de werkzame nulcomponent
d Schwarzwerthaltung *f*, Wiederherstellung *f* des Schwarzpegels

1407 D.C. RESTORER cpl
The mean of restoring a d.c. component of a signal after transmission in a line which only easily transmits high frequencies.
f restaurateur *m* de la composante continue
e restaurador *m* de la componente continua
i restauratore *m* della componente continua
n hersteller van de werkzame nulcomponent
d einfaches Schwarzwertsteuerglied *n*

1408 D.C. TRANSMISSION tv
A form of transmission in which the d.c. component is directly represented in the vision signal.
f transmission *f* avec composante continue utile
e transmisión *f* con componente continua
i trasmissione *f* con componente continua
n transmissie met werkzame nulcomponent
d Übertragung *f* mit Gleichstromanteil

1409 DE-ACCENTUATION, cpl
DE-EMPHASIS,
POST-EMPHASIS
Correction of frequency response distortion originated by pre-emphasis by introducing a complementary frequency response.
f désaccentuation *f*
e desacentuación *f*
i deenfasi *f*, disaccentuazione *f*
n nacorrectie, neerduw
d Nachentzerrung *f*

1410 DE-ACCENTUATION NETWORK, cpl
DE-EMPHASIS NETWORK,
POST-EMPHASIS NETWORK
An RC filter inserted in a system to restore pre-emphasized signals to their original form.
f réseau *m* de désaccentuation
e red *f* desacentuadora
i rete *f* correttrice di distorsione
n neerduwnetwerk *n*
d Entzerrungskreis *m*

1411 DE-ACCENTUATOR, cpl
DE-EMPHASIS CIRCUIT,
POST-EMPHASIS CIRCUIT
A device used in frequency-modulation receivers to de-emphasize the higher frequency in the received signal.
f circuit *m* de désaccentuation
e circuito *m* desacentuador
i correttore *m* di distorsione d'attenuazione
n neerduwketen
d Entzerrer *m*

1412 DEAD END aud
The end of a sound studio that has the greater sound-absorbing characteristics.
f paroi *f* sourde
e pared *f* anecóica
i parete *f* anecoica
n echovrije wand
d schalltote Wand *f*

1413 DEAD ROOM aud
A room characterized by an unusually large amount of sound absorption.
f chambre *f* sourde
e cámara *f* anecóica
i camera *f* anecoica
n echovrije kamer
d schalltoter Raum *m*

1414 DEAD SPACE, rep/tv
DEAD SPOT
A geographic location in which signals from a TV transmitter are received poorly or not at all.

f point *m* mort,
zone *f* de silence
e espacio *m* muerto,
zona *f* de silencio
i punto *m* morto,
zona *f* di silenzio
n dood gebied *n*
d Totzone *f*

1415 DEAD STUDIO rec/rep
A studio having a short reverberation time.
f studio *m* sourd
e estudio *m* muerto
i studio *m* assorbente
n studio met korte nagalm
d Studio *n* ohne Nachhall

1416 DEAD WINDOW aud/stu
f fenêtre *f* sourde
e ventana *f* muerta
i apertura *f* afonica
n echovrij venster *n*
d schalltotes Fenster *n*

1417 DECAY TIME cpl
The time taken by a quantity to decay to a stated fraction of its initial value.
f temps *m* de chute
e tiempo *m* de caída
i tempo *m* di caduta
n afvaltijd
d Abfallzeit *f*

1418 DECAY TIME, cpl/tv
FALL TIME,
PULSE DECAY TIME,
PULSE FALL TIME
The interval between the instants at which the instantaneous values of a pulse or of its envelope reaches specified upper and lower limits, namely 90 per cent and 10 per cent of the peak values unless otherwise stated.
f temps *m* de descente
e tiempo *m* de caída,
tiempo *m* de decaimiento
i tempo *m* di caduta,
tempo *m* di discesa
n uitslingertijd
d Abklingzeit *f*,
Ausschwingdauer *f*

1419 DECELERATING ELECTRODE, crt
RETARDING ELECTRODE
An auxiliary electrode in an image-orthicon at OV, to retard the photo-electrons on their way towards the screen.
f électrode *f* de freinage
e electrodo *m* de retardo,
electrodo *m* decelerador
i elettrodo *m* deceleratore,
elettrodo *m* ritardatore
n remelektrode
d Bremselektrode *f*

DECODER
see: COLO(U)R DECODER

1420 DECODING DECK ctv
f platine *f* de décodage
e placa *f* de descodificación
i placca *f* di decodificazione
n decodeerplaat
d Dekodierplatte *f*

1421 DECODING MATRIX ctv
f matrice *f* de décodage
e matriz *f* de descodificación
i matrice *f* di decodificazione
n decodeermatrijs
d Dekodiermatrix *f*, Dematrix *f*

1422 DEEPER ct
A difference in a colo(u)r apparently due to the presence of less white than in the original sample.
f plus foncé
e más subido
i più scuro
n dieper
d satter

DEFECTIVE COLO(U)R VISION
see: DALTONISM

1423 DEFINITION (GB), ge
RESOLUTION (US)
A measure of the sharpness of detail of a TV picture.
f définition *f*
e definición *f*
i definizione *f*
n definitie
d Auflösung *f*,
Bildschärfe *f*,
Zeichenschärfe *f*

1424 DEFINITION OF IMAGE crt
f définition *f* de l'image de l'écran
e definición *f* de la imagen de la pantalla
i definizione *f* dell'immagine dello schermo
n schermbeelddefinitie
d Rasterfeinheit *f*

1425 DEFINITION ON BORDER rec/tv
f définition *f* des bords
e definición *f* de los bordes
i definizione *f* dei bordi
n randscherpte
d Randschärfe *f*

1426 DEFINITION TEST CARD (GB), tv
RESOLUTION PATTERN (US)
A test card designed for checking the definition of a TV transmission.
f mire *f* de définition
e patrón *m* de definición
i immagine *f* di prova per la definizione
n definitietoetsbeeld *n*
d Auflösungstestbild *n*,
Probebild *n*

1427 DEFLECT (TO) crt
f balayer v, dévier v
e barrer v, desviar v

i deflettere v, deviare v
n afbuigen v
d ablenken v

1428 DEFLECTED BEAM crt
f faisceau de balayage,
faisceau *m* dévié
e haz *m* de barrido,
haz *m* desviado
i fascio *m* deflesso,
fascio *m* deviato
n afgebogen bundel
d abgelenkter Strahl *m*

1429 DEFLECTION crt
The bending of cathode rays from a straight line.
f balayage *m*, déviation *f*
e barrido *m*, desviación *f*
i deflessione *f*, deviazione *f*
n afbuiging
d Ablenkung *f*

1430 DEFLECTION AMPLIFIER crt
Output stage of a field or line timebase.
f amplificateur *m* de balayage,
amplificateur *m* de déviation
e amplificador *m* de barrido,
amplificador *m* de desviación
i amplificatore *m* di deflessione,
amplificatore *m* di deviazione
n afbuigversterker
d Ablenkverstärker *m*

1431 DEFLECTION ANGLE crt
f angle *m* de balayage,
angle *m* de déviation
e ángulo *m* de barrido,
ángulo *m* de desviación
i angolo *m* di deflessione,
angolo *m* di deviazione
n afbuighoek
d Ablenkwinkel *m*

1432 DEFLECTION CAPACITOR crt
f condensateur *m* de balayage,
condensateur *m* de déviation
e condensador *m* de barrido,
condensador *m* de desviación
i condensatore *m* di deflessione,
condensatore *m* di deviazione
n afbuigcondensator
d Ablenkkondensator *m*

1433 DEFLECTION CENTER(RE) crt
The intersection of the forward projection of the electron path prior to deflection and the backward projection of the electron path in the field-free space after deflection.
f centre *m* de déviation
e centro *m* de desviación
i centro *m* di deflessione,
centro *m* di deviazione
n afbuigmiddelpunt *n*
d Ablenkmittelpunkt *m*

1434 DEFLECTION COEFFICIENT, crt
DEFLECTION FACTOR
In cathode-ray tubes the reciprocal of deflection sensitivity.
f coefficient *m* de balayage,
coefficient *m* de déviation
e coeficiente *m* de barrido,
coeficiente *m* de desviación
i coefficiente *m* di deflessione,
coefficiente *m* di deviazione
n afbuigfactor
d Ablenkungsfaktor *m*

1435 DEFLECTION COIL (GB), crt
SCANNING COIL,
SWEEPING COIL (US)
f bobine *f* de balayage,
bobine *f* de déviation
e bobina *f* de barrido,
bobina *f* de desviación
i bobina *f* di deflessione,
bobina *f* di deviazione
n afbuigspoel
d Ablenkspule *f*

1436 DEFLECTION CURRENT crt
DEFLECTOR CURRENT
f courant *m* de la bobine de balayage,
courant *m* de la bobine de déviation
e corriente *f* de la bobina de barrido,
corriente *f* de la bobina de desviación
i corrente *f* della bobina di deflessione,
corrente *f* della bobina di deviazione
n afbuigspoelstroom
d Ablenkspulenstrom *m*

1437 DEFLECTION DEFOCUSING crt
The change in spot size or beam width due to the deflection process.
f élargissement *m* du spot
e desenfoque *m* del haz
i distorsione *f* della macchia,
sfocatura *f* del fuoco
n afbuigonscherpte
d Fleckunschärfe *f* bei Ablenkung

1438 DEFLECTION DISTORTION, crt
PATTERN DISTORTION
f distorsion *f* par déviation
e distorsión *f* por desviación
i distorsione *f* per deflessione,
distorsione *f* per deviazione
n afbuigvervorming
d Ablenkfehler *pl*

1439 DEFLECTION ELECTRODE crt
CONNECTION,
DEFLECTION PLATE CONNECTION
f couplage *m* des électrodes de balayage,
couplage *m* des électrodes de déviation
e acoplamiento *m* de los electrodos de barrido,
acoplamiento *m* de los electrodos de desviación
i accoppiamento *m* degli elettrodi di deflessione,
accoppiamento *m* degli elettrodi di deviazione

n aansluitschakeling voor de afbuigelektroden
d Ablenkplattenschaltung *f*

1440 DEFLECTION ELECTRODES, DEFLECTION PLATES crt
f électrodes *pl* de balayage, électrodes *pl* de déviation, plaques *pl* de balayage, plaques *pl* de déviation
e chapas *pl* de barrido, chapas *pl* desviadoras, electrodos *pl* de barrido, electrodos *pl* de desviación
i elettrodi *pl* di deflessione, elettrodi *pl* di deviazione, placche *pl* di deflessione, placche *pl* di deviazione
n afbuigelektroden *pl*, afbuigplaten *pl*
d Ablenkelektroden *pl*, Ablenkplatten *pl*

1441 DEFLECTION FIELD crt
f champ *m* de balayage, champ *m* de déviation
e campo *m* de barrido, campo *m* de desviación
i campo *m* di deflessione, campo *m* di deviazione
n afbuigingsveld *n*
d Ablenkungsfeld *n*

1442 DEFLECTION MAGNET crt
f aimant *m* de balayage, aimant *m* de déviation
e imán *m* de barrido, imán *m* de desviación
i magnete *m* di deflessione, magnete *m* di deviazione
n afbuigingsmagneet
d Ablenkungsmagnet *m*

1443 DEFLECTION NON-LINEARITY tv
The deviation of the scanning velocity of the deflected cathode-ray beam from a constant value.
f non-linéarité *f* de balayage, non-linéarité *f* de déflexion
e **no linealidad *f* de barrido,** no linealidad *f* de desviación
i non-linearità *f* di deflessione, non-linearità *f* di deviazione
n niet-lineariteit van de afbuiging
d Ablenknichtlinearität *f*

1444 DEFLECTION PLANE crt
A plane perpendicular to the tube axis containing the deflection center(re).
f plan *m* de déviation
e plano *m* de desviación
i piano *m* di deflessione, piano *m* di deviazione
n afbuigingsvlak *n*
d Ablenkungsebene *f*

1445 DEFLECTION SENSITIVITY, SENSITIVITY OF DEFLECTION crt
In a cathode-ray tube, the ratio of the displacement of the electron beam at the place of impact to the change in the deflection field.
f sensibilité *f* de balayage, sensibilité *f* de déviation
e sensibilidad *f* de barrido, sensibilidad *f* de desviación
i sensibilità *f* di deflessione, sensibilità *f* di deviazione
n afbuiggevoeligheid
d Ablenkempfindlichkeit *f*

1446 DEFLECTION SPACE crt
The space in which the generated fields act upon the electron beam.
f espace *m* de déviation
e espacio *m* de desviación
i spazio *m* di deflessione, spazio *m* di deviazione
n afbuigruimte
d Ablenkraum *m*

1447 DEFLECTION SPEED, DEFLECTION VELOCITY crt
The velocity of the light spot on the screen of cathode-ray tubes under the influence of deviation.
f vitesse *f* de balayage, vitesse *f* de déviation
e velocidad *f* de barrido, velocidad *f* de desviación
i velocità *f* di deflessione, velocità *f* di deviazione
n afbuigsnelheid
d Ablenkgeschwindigkeit *f*

1448 DEFLECTION SYSTEM, DEFLECTION YOKE, SCANNING YOKE crt
f bloc *m* de balayage, bloc *m* de déviation, culasse *f* de balayage, culasse *f* de déviation
e culata *f* de barrido, culata *f* de desviación, yugo *m* de barrido, yugo *m* de desviación
i blocco *m* di deflessione, blocco *m* di deviazione, giogo *m* di deflessione, giogo *m* di deviazione
n afbuigjuk *n*
d Ablenkjoch *n*

1449 DEFLECTION VOLTAGE crt
f tension *f* de balayage, tension *f* de déviation
e tensión *f* de barrido, tensión *f* de desviación
i tensione *f* di deflessione, tensione *f* di deviazione
n afbuigspanning
d Ablenkspannung *f*

1450 DEFLECTION YOKE PULLBACK crt/tv
1. The distance between the maximum possible forward position of the yoke and the position of the yoke to obtain

maximum colo(u)r purity.
2. The maximum distance the yoke can be moved along the tube axis without producing neck shadow.
f jeu *m* du bloc de balayage,
jeu *m* du bloc de déviation
e juego *m* del yugo de barrido,
juego *m* del yugo de desviación
i gioco *m* del giogo di deflessione,
gioco *m* del giogo di deviazione
n afbuigjukspeling
d Ablenkjochspiel *n*

1451 DEGAUSSING, DEMAGNETIZING crt/ctv
Removing residual magnetism from metal components in and around the colo(u)r picture tube.
f démagnétisation *f*
e desmagnetización *f*
i demagnetizzazione *f*,
smagnetizzazione *f*
n ontmagnetiseren *n*
d Entmagnetisierung *f*

1452 DEGAUSSING COIL, DEMAGNETIZING COIL crt/ctv
f bobine *f* de démagnétisation
e bobina *f* de desmagnetización
i bobina *f* di demagnetizzazione,
bobina *f* di smagnetizzazione
n ontmagnetiseringsspoel
d Achterspule *f*,
Entmagnetisierungsspule *f*

1453 DEGRADATION, IMAGE DETERIORATION rep/tv
In TV, deterioration of the image from the original scene to its immediate reproduction, or from the latter to some more removed image arrived at by transmission or duplication.
f détérioration *f* de l'image
e deterioro *m* de la imagen
i deterioramento *m* dell'immagine
n beeldverslechtering
d Bildverschlechterung *f*

1454 DEGRADED COLO(U)RS ct
Reproduced colo(u)rs which are dull (dark and desaturated) as compared with the original.
f couleurs *pl* dégradées
e colores *pl* degradados
i colori *pl* degradati
n verflauwde kleuren *pl*
d abgeflaute Farben *pl*

1455 DEGREE K, KELVIN DEGREE ct
Scale of basic temperature units, by which e.g. colo(u)r temperature is invariably measured.
f degré *m* absolu,
degré *m* Kelvin
e grado *m* absoluto,
grado *m* Kelvin
i grado *m* assoluto,
grado *m* Kelvin
n kelvingraad
d Kelvingrad *m*

1456 DELAY ge
In physical equipment or systems, the retardation of the time of arrival of a signal or impulse after transmission through that equipment or system.
f retard *m*,
temps *m* de transit
e retardo *m*,
tiempo *m* de tránsito
i ritardo *m*,
tempo *m* di transito
n looptijd,
vertraging
d Laufzeit *f*,
Verzögerung *f*

1457 DELAY CABLE cpl
Special coaxial cable designed to considerably delay the passage of signals; sometimes used to obtain the necessary time constants in sync generators.
f câble *m* à retard
e cable *m* de retardo
i cavo *m* a ritardatore
n vertragingskabel
d Verzögerungskabel *m*

1458 DELAY CIRCUIT cpl
A circuit which delays the wavefront of a signal.
f circuit *m* à retard
e circuito *m* de retardo
i circuito *m* ritardatore
n vertragingsketen
d Verzögerungskreis *m*

1459 DELAY DISTORTION ctv
In colo(u)r TV, a picture distortion occurring when the time for passage of signal channels is not the same at all frequencies.
f distorsion *f* du temps de transit
e distorsión *f* del tiempo de tránsito
i distorsione *f* del tempo di transito
n looptijdvervorming
d Laufzeitverzerrung *f*

DELAY INEQUALITY
see: CHROMINANCE/LUMINANCE DELAY INEQUALITY

1460 DELAY LINE ctv
A specially constructed cable used in the luminance channel of a colo(u)r receiver to provide a time delay to the luminance channel.
f ligne *f* de retard
e línea *f* de retardo
i linea *f* di ritardo
n vertragingslijn
d Verzögerungsleitung *f*

1461 DELAY SCREEN crt
A fluorescent screen in which the light intensity of any particular spot dies out gradually after the beam moves to a new position.

f écran *m* à retard
e pantalla *f* de retardo
i schermo *m* a ritardo
n vertragingsscherm *n*
d Verzögerungsschirm *m*

1462 DELAY SIGNAL cpl
f signal *m* de retard
e señal *f* de retardo
i segnale *m* ritardatore
n vertragingssignaal *n*
d Verzögerungssignal *n*

1463 DELAY TIME cpl
f temps *m* de retard
e tiempo *m* de retardo
i tempo *m* di ritardo
n vertragingstijd
d Verzögerungszeit *f*

1464 DELAYED BLANKING SIGNAL tv
Blanking signal which is delayed by artificial or coaxial transmission in TV circuits.
f signal *m* de suppression différé
e señal *f* de supresión retardada
i segnale *m* di cancellazione ritardato
n vertraagd onderdrukkingssignaal *n*
d verzögertes Austastsignal *n*

1465 DELAYED SCANNING (GB), DELAYED SWEEP (US) tv
A scanning (sweep) whose beginning is delayed for a definite time after the pulse that initiates the sweep.
f analyse *f* différée
e exploración *f* retardada
i analisi *f* ritardata
n vertraagde aftasting
d verzögerte Abtastung *f*

1466 DELAYED SYNC SIGNAL, DELAYED SYNCHRONIZATION SIGNAL, DELAYED SYNCHRONIZING SIGNAL cpl
f signal *m* de synchronisation retardé
e señal *f* de sincronización retardada
i segnale *m* di sincronizzazione ritardato
n vertraagd synchronisatiesignaal *n*
d verzögertes Synchronisierungssignal *n*

1467 DELTA-L CORRECTION, DELTA-L SYSTEM ctv
A technique for correcting the distinction between the monochrome signal transmitted and the gamma-corrected true luminance signal by forming a luminance difference signal delta-L from the RGB signals from the colo(u)r tubes by matrixing techniques.
f correction *f* delta-L
e corrección *f* delta-L
i correzione *f* delta-L
n delta-L-correctie
d Delta-L-Korrektur *f*

1468 DEMODULATION cpl
The process of reproduction of an original modulating signal from a modulated wave.
f démodulation *f*
e desmodulación *f*
i demodulazione *f*
n demodulatie
d Demodulation *f*, Entmodelung *f*

1469 DEMODULATION AXIS ctv
f axe *m* de démodulation
e eje *m* de desmodulación
i asse *m* di demodulazione
n demodulatieas
d Demodulationsachse *f*

1470 DENOISE (TO) dis
To reduce the noise in a circuit, electron beam, etc.
f éliminer v les perturbations
e eliminar v las perturbaciones
i eliminare v i disturbi
n ontstoren v
d entstören v

1471 DENSITY-MODULATED ELECTRON BEAM crt
f faisceau *m* électronique modulé en densité
e haz *m* electrónico modulado en densidad
i fascio *m* elettronico modulato in densità
n elektronenbundel met dichtheidsmodulatie
d Elektronenstrahl *m* mit Dichtemodulation

1472 DEPARTURE FROM NEUTRALITY OF THE GREY SCALE dis
A defect in colo(u)r reproduction caused by the input-output amplitude characteristics of each channel not being identical.
f décalage *m* chromatique des gris
e deslizamiento *m* cromático de los grises
i spostamento *m* cromatico dei grigi
n kleurverschuiving van de grijzen
d Grauwertverzerrung *f*

1473 DEPTH OF FIELD opt
Permissible range of distances of the object plane of an optical system such that the circle of light resulting from a point object and produced at the image plane is still acceptable as an image of the point object.
f profondeur *f* de champ
e profundidad *f* de campo
i profondità *f* di campo
n scherptediepte
d Schärfentiefe *f*, Tiefenschärfe *f*

1474 DEPTH OF FOCUS opt
In an optical system, the permissible range of distances of the image plane for a stationary object plane, over which the circle of light resulting from a point object is still acceptable as an image of the point object.
f différence *f* focale
e diferencia *f* focal
i differenza *f* focale
n scherpinstelgebied *n*

d Fokusdifferenz *f*,
Scharfeinstellbereich *m*

1475 DEPTH OF VELOCITY MODULATION WITH SMALL SIGNAL cpl/crt
The ratio of the peak amplitude of the velocity modulation of an electron beam, expressed in equivalent volts, to the electron beam voltage.
f profondeur *f* de modulation de vitesse pour signaux faibles
e profundidad *f* de modulación de velocidad para señales débiles
e profondità *f* di modulazione di velocità per piccoli segnali
n snelheidsmodulatiediepte voor zwakke signalen
d Geschwindigkeitsaussteuerung *f* bei kleinem Signal

1476 DEPTH PERCEPTION, PERCEPTION OF DEPTH cpt
f perception *f* de profondeur, vision *f* tridimensionnelle
e percepción *f* de profundidad, visión *f* tredimensional
i percezione *f* di profondità, visione *f* tridimensionale
n dieptewaarneming, driedimensionaal zien *n*
d plastisches Sehen *n*, Tiefenwahrnehmung *f*

1477 DERIVATIVE CORRECTOR, DERIVATIVE EQUALIZER tv
Equipment for introducing into the picture signal a correcting signal containing (a) derivative(s) of the original signal.
f correcteur *m* différentiateur
e corrector *m* diferenciador
i correttore *m* differenziatore
n disproportiecorrector
d differenzierender Entzerrer *m*

1478 DESATURATED COLO(U)RS ct
Reproduced colo(u)rs which are less saturated than the originals.
f couleurs *pl* désaturées
e colores *pl* desaturados
i colori *pl* a saturazione inferiore
n minder verzadigde kleuren *pl*, onverzadigde kleuren *pl*
d entsättigte Farben *pl*

1479 DESATURATION, PALING OUT ct
Reduction of colo(u)r saturation often as a result of distortion or defect in the TV transmission system.
f désaturation *f*
e desaturación *f*
i saturazione *f* inferiore
n mindere verzadiging, onverzadigdheid
d Entsättigung *f*

1480 DESIGNER, SET DESIGNER fi/th/tv
f décorateur *m*, dessinateur *m* de décors
e escenógrafo *m*
i disegnatore *m* teatrale
n decorateur, toneelschilder
d Bühnenbildner *m*, Szenenbildner *m*

1481 DESTRUCTIVE INTERFERENCE ctv
In the holographic process, the ability of two light waves to cancel each other.
f interférence *f* destructive
e interferencia *f* destructiva
i interferenza *f* distruttiva
n destructieve storing
d destruktive Störung *f*

1482 DETAIL tv
The extent to which image elements that are close together can be individually distinguished.
f détail *m*
e detalle *m*
i dettaglio *m*
n detail *n*
d Bildfeinheit *f*, Detail *n*, Einzelteil *m*

1483 DETAIL CONTRAST RATIO rep
f rapport *m* de contraste des détails
e relación *f* de contraste de los detalles
i rapporto *m* di contrasto dei dettagli
n detailcontrastverhouding
d Detailkontrastverhältnis *n*, Einzelheitenkontrastverhältnis *n*

1484 DETAIL RENDERING, DETAIL RENDITION rep/tv
f rendu *m* de détails
e rendición *f* de detalles
i riproduzione *f* di dettagli
n detailweergeving
d Detailwiedergabe *f*

1485 DETECTION UNIT PICTURE tv
f bloc *m* de détection vidéo
e bloque *m* de detección video
i blocco *m* di rivelazione video
n detectie-eenheid beeld
d Bildfrequenz-Gleichrichtereinheit *f*

1486 DETECTION UNIT SOUND tv
f bloc *m* de détection son
e bloque *m* de detección sonido
i blocco *m* di rivelazione suono
n detectie-eenheid geluid
d Tonfrequenz-Gleichrichtereinheit *f*

1487 DETECTOR CIRCUIT cpl
f circuit *m* détecteur
e circuito *m* detector
i circuito *m* rivelatore
n detectorkring
d Gleichrichterkreis *m*

1488 DEUTERANOMALOUS VISION, PARTIAL DEUTERONOPIA ct
A form of trichromatic vision in which more green is required in a mixture of red and green to match a spectral yellow than is the case for the normal trichromat.

f deutéranopie *f* anomale, deutéroanomalie *f*
e deuteranopía *f* anomal, deuteroanomalía *f*
i deuteranopia *f* anomale, deuteroanomalia *f*
n anomale deuteranopie, deuteranomalie
d anomale Deuteranopie *f*, Deuteranomalie *f*

1489 DEUTERANOPE ct
One who possesses deuteranopia.
f deutéranope *m*
e deuteranope *m*
i deuteranope *m*
n deuteranoop
d Deuteranop *m*

1490 DEUTERANOPIA ct
A form of dichromatic vision in which colo(u)rs can be matched by a mixture of yellow and blue stimuli, but in which the relative luminance efficiency does not differ from the normal.
f deutéranopie *f*
e deuteranopía *f*
i deuteranopia *f*
n deuteranopie
d Deuteranopie *f*

1491 DIAGONAL CLIPPING ctv
Distortion of a sinusoidal signal component due to transmitter overmodulation in the negative direction.
f distorsion *f* diagonale du signal
e distorsión *f* diagonal de la señal
i distorsione *f* diagonale del segnale
n diagonale signaalvervorming
d diagonale Signalverzerrung *f*

1492 DIAGONALIZING tv
Practice of radiating the same program(me) at different times, and on different wave-lengths.
f répétition *f* d'un programme sur longueurs différentes d'onde
e repetición *f* de un programa sobre longitudes diferentes de onda
i ripetizione *f* d'un programma su lunghezze differenti d'onda
n programmaherhaling op verschillende golflengten
d mehrfache Aussendung *f* desselben Programms auf verschiedenen Wellenlängen

1493 DIALOGUE CORRECTION aud
The removing of any irregularities in volume and frequency content from the dialogue track.
f correction *f* de la piste dialogue
e corrección *f* de la pista diálogo
i correzione *f* della pista dialogo
n bijwerking van het dialoogspoor
d Dialogspurkorrektur *f*

1494 DIALOGUE LEVEL aud
The volume level of a dialogue kept approximately 12 dB below 100% modulation.
f niveau *m* du dialogue
e nivel *m* del diálogo
i livello *m* del dialogo
n dialoogniveau *n*
d Dialogpegel *m*

1495 DIALOGUE RECORDING aud
f enregistrement *m* du dialogue
e registro *m* del diálogo
i registrazione *f* del dialogo
n dialoogopname
d Dialogaufzeichnung *f*

1496 DIALOGUE SYNCHRONIZATION aud
f doublage *m* du dialogue, synchronisation *f* du dialogue
e sincronización *f* del diálogo
i sincronizzazione *f* del dialogo
n dialoogsynchronisatie
d Dialogsynchronisierung *f*

1497 DIALOGUE TRACK aud
Sound track on which dialogue has been edited for dubbing.
f piste *f* du dialogue
e pista *f* del diálogo
i pista *f* del dialogo
n dialoogspoor *n*
d Dialogspur *f*

DIAPHRAGM
see: APERTURE STOP

1498 DICHROIC COATING opt
A substitute for the normal aluminized reflector of a studio lamp designed to reduce the heat radiation from the lamp.
f couche *f* dichroïque
e capa *f* dicroica
i strato *m* dicroico
n dichroïtische laag
d dichroitische Schicht *f*

1499 DICHROIC FILTER, DICHROIC MIRROR, DICHROIC REFLECTOR ct/opt
A mirror through which all of the light frequencies pass except those of the colo(u)r which the mirror is designed to reflect.
f miroir *m* dichroïque
e espejo *m* dicroico
i specchio *m* dicroico
n dichroïtische spiegel
d Zweifarbenfilterspiegel *m*

1500 DICHROISM ct/opt
A phenomenon in which a secondary source shows a marked change in hue with change in the observing condition.
f dichroïsme *m*
e dicroismo *m*
i dicroismo *m*
n dichroïsme *n*
d Dichroismus *m*, Zweifarbigkeit *f*

1501 DICHROMAT ct
One who possesses dichromatic vision.
f dichromate *m*
e dicrómato *m*
i dicromato *m*
n dichromaat
d Dichromat *m*

1502 DICHROMATIC VISION, DICHROMATISM ct
A form of defective vision in which all colo(u)rs can be matched by a mixture of only two suitably chosen stimuli.
f dichromasie *f*
e dicromasía *f*
i dicromasia *f*
n dichromasie
d Dichromasie *f*, Dichromatopsie *f*

1503 DIFFERENTIAL-CURRENT AMPLIFIER cpl/svs
f amplificateur *m* de courant différentiel
e amplificador *m* de corriente diferencial
i amplificatore *m* di corrente differenziale
n verschilstroomversterker
d Differenzstromverstärker *m*

1504 DIFFERENTIAL EQUALIZATION cpl
f égalisation *f* différentielle
e igualización *f* diferencial
i uguagliamento *f* differenziale
n differentiële egalisatie
d differentielle Egalisierung *f*

1505 DIFFERENTIAL FLUX RESPONSE FACTOR crt
f facteur *m* différentiel de flux du diaphragme
e factor *m* diferencial de flujo del diafragma
i fattore *m* differenziale di flusso del diaframma
n differentiële fluxfactor van het diafragma
d differentieller Fluxfaktor *m* der Blende

1506 DIFFERENTIAL GAIN cpl/ctv
Of a colo(u)r video channel, the voltage gain for a small chrominance subcarrier signal at a given luminance signal level expressed as a percentage difference relative to the gain at blanking or some specified level.
f amplification *f* différentielle, gain *m* différentiel
e amplificación *f* diferencial, ganancia *f* diferencial
i amplificazione *f* differenziale, guadagno *m* differenziale
n differentiële versterking
d differentielle Verstärkung *f*

1507 DIFFERENTIAL GAIN CONTROL cpl
f atténuation *f* sélective
e atenuación *f* selectiva
i attenuazione *f* selettiva
n selectieve demping
d Differentialverstärkungsregelung *f*

1508 DIFFERENTIAL GAIN DISTORTION ctv
Error produced in the signal being handled by a colo(u)r network when the luminance is varied, so producing an error in the amplitude of the subcarrier.
f distorsion *f* de l'amplification différentielle, distorsion *f* du gain différentiel
e distorsión *f* de la amplificación diferencial, distorsión *f* de la ganancia diferencial
i distorsione *f* del guadagno differenziale, distorsione *f* dell'amplificazione differenziale
n differentiële versterkingsvervorming
d differentielle Verstärkungsverzerrung *f*

1509 DIFFERENTIAL MICROPHONE aud
f microphone *m* différentiel
e micrófono *m* diferencial
i microfono *m* differenziale
n compensatiemicrofoon
d Kompensationsmikrophon *n*

1510 DIFFERENTIAL PHASE cpl/ctv
Of a colo(u)r video channel, the phase shift of a small chrominance subcarrier signal at a given luminance signal level, relative to the phase shift at blanking level.
f phase *f* différentielle
e fase *f* diferencial
i fase *f* differenziale
n differentiële faze
d differentielle Phase *f*

1511 DIFFERENTIAL PHASE DISTORTION ctv
A distortion which can arise in long links in the NTSC system.
f distorsion *f* de phase différentielle
e distorsión *f* de fase diferencial
i distorsione *f* di fase differenziale
n differentiële fazevervorming
d differentielle Phasenverzerrung *f*

1512 DIFFERENTIAL SIGNAL rep/tv
f signal *m* différentiel
e señal *f* diferencial
i segnale *m* differenziale
n differentiëel signaal *n*
d differentielles Signal *n*

1513 DIFFRACTION opt
Optical phenomenon, resulting from the wave structure of light, occurring at the edges of illuminated objects.
f diffraction *f*
e difracción *f*
i diffrazione *f*
n buiging, diffractie
d Beugung *f*, Diffraktion *f*

1514 DIFFRACTION EFFECT opt
An effect used in the Schlieren optical system.
f effet *m* de diffraction

e efecto *m* de difracción
i effetto *m* di diffrazione
n buigingseffect *n*
d Beugungseffekt *m*

DIFFUSE LIGHT
see: BROAD LIGHT

1515 DIFFUSE REFLECTION ge
f réflexion *f* diffuse
e reflexión *f* difusa
i riflessione *f* diffusa
n diffuse reflectie
d diffuse Reflexion *f*,
gestreute Reflexion *f*,
verstreute Reflexion *f*

1516 DIHEPTAL BASE crt
A base for cathode-ray tubes.
f culot *m* à 14 broches
e casquillo *m* de 14 clavijas
i zoccolo *m* a 14 contatti
n 14-pools-voet
d 14poliger Sockel *m*

1517 DIMENSIONS OF MAGNETIC SOUND TRACKS, aud
MAGNETIC SOUND TRACK DIMENSIONS
f dimensions *pl* de pistes sonores magnétiques
e dimensiones *pl* de pistas sonoras magnéticas
i dimensioni *pl* di piste sonore magnetiche
n afmetingen *pl* van magnetische geluidssporen
d Abmessungen *pl* von magnetischen Tonspuren

1518 DIODE CLIPPER cpl
A clipper circuit using a diode.
f circuit *m* de seuil à diode
e circuito *m* de umbral de diodo
i circuito *m* di soglia a diodo
n drempelschakeling met diode
d Schwellenwertbegrenzer *m* mit Diode

1519 DIODE RATE OF RISE NOISE LIMITER aud/dis
A limiter blocking the audio channel on noise peaks and thus improves the signal-to-noise ratio.
f limiteur *m* à diode à blocage du canal audio
e limitador *m* de diodo de bloqueo del canal audio
i limitatore *m* a diodo a bloccaggio del canale audio
n begrenzerdiode met audiokanaalblokkering
d Begrenzerdiode *f* mit Audiokanalsperrung

1520 DIODE SWITCH tv
f commutateur *m* à diode
e conmutador *m* de diodo
i commutatore *m* a diodo
n diodeschakelaar
d Diodenschalter *m*

1521 DIOPTER(RE) opt
Unit of power of a single lens, the reciprocal of its focal length in meter(re)s.
f dioptrie *f*
e dioptría *f*
i diottria *f*
n dioptrie
d Diopter *n*

1522 DIOPTER(RE) LENS opt
Simple supplementary lens used as an attachment in front of the main camera lens to allow the photography of extreme close-ups.
f bonnette *f*,
lentille *f* dioptrique
e lente *f* dióptrica
i lente *f* diottrica
n dioptrische lens
d dioptrische Linse *f*,
Gürtellinse *f*

1523 DIPLEXER aea
An apparatus used for preventing interference between the sound and video channels radiated from the TV aerial (antenna).
f diplexeur *m*,
séparateur *m* de fréquences
e diplexer *m*,
multiacoplador *m*
i separatore *m* di frequenze
n antennemenger
d Diplexer *m*,
Frequenzweiche *f*,
Trennweiche *f*

1524 DIPLEXER FILTER cpl
A device used in a TV transmitter combining the functions of a sideband attenuator filter and a diplexer of the sound and picture transmissions.
f filtre *m* diplexeur
e filtro *m* diplexer
i filtro *m* combinatore vestigiale
n filterdiplexer
d Restseitenbandfilter *n* mit Bild-Tonweiche

1525 DIPOLE, aea
DIPOLE AERIAL,
DIPOLE ANTENNA
An aerial(antenna) comprising a straight conductor, up to half- wavelength long, attached to a feed at its center(re).
f antenne *f* dipôle, dipôle *m*
e antena *f* dipolo, dipolo *m*
i antenna *f* dipolo, dipolo *m*
n dipool, dipoolantenne
d Dipol *m*, Dipolantenne *f*

1526 DIPOLE ARRAY, aea
GROUP AERIAL,
GROUP ANTENNA
f nappe *f* de dipôles verticaux
e grupo *m* de dipolos
i gruppo *m* di dipoli
n dipoolgroep
d Dipolgruppe *f*

1527 DIRECT COLORIMETRY ct
The science of measuring the intensity of colo(u)r.
f colorimétrie *f* directe
e colorimetría *f* directa
i colorimetria *f* diretta
n directe kleurmeting
d unmittelbare Farbmessung *f*

1528 DIRECT COMPATIBILITY tv
f compatibilité *f* directe
e compatibilidad *f* directa
i compatibilità *f* diretta
n directe compatibiliteit
d unmittelbare Kompatibilität *f*

1529 DIRECT GLARE opt
Glare due to a luminous object situated in the same or nearly the same direction as the object viewed.
f éblouissement *m* direct
e deslumbramiento *m* directo
i abbagliamento *m* diretto
n direct verblindend licht *n*
d direktes blendendes Licht *n*

1530 DIRECT IMAGING OPTICS ctv/opt
An optical system using three plumbicons and a prism beam splitter so designed as to reduce the angle of incidence to a low value in order to separate the red from green light as efficiently as possible.
f optique *f* formatrice d'images directes
e óptica *f* formadora de imágenes directas
i ottica *f* formatrice d'immagini dirette
n optiek voor directe beeldvorming
d Optik *f* für unmittelbare Bildformung

1531 DIRECT LIGHTING opt
f éclairage *m* direct
e alumbrado *m* directo
i illuminazione *f* diretta
n directe verlichting
d direkte Beleuchtung *f*

1532 DIRECT PICK-UP rec/tv
The transmission of TV images without intermediate photographic or magnetic recording.
f prise *f* de vues directe
e toma *f* de vistas directa
i ripresa *f* diretta
n directe beeldopname
d unmittelbare Bildaufnahme *f*

1533 DIRECT PICK-UP RECEIVER tv
A receiver designed to supply the modulating signals for a rebroadcasting transmitter.
f récepteur *m* pour l'alimentation d'un émetteur répétiteur
e receptor *m* para la alimentación de retransmisión en directo
i ricevitore *m* per alimentare un trasmettitore ripetitore
n ontvanger voor de voeding van een relaiszender
d Empfänger *m* für die Speisung eines Ballsenders

1534 DIRECT POSITIVE TELERECORDING vr
The system of obtaining a positive image from a negative displayed on the cathode-ray tube screen.
f enregistrement *m* positif de l'image
e telerregistro *m* positivo directo
i registrazione *f* dell'immagine in positivo
n directe positieve video-opname
d direkte Positivbildaufnahme *f*

1535 DIRECT RECORDING rec
Recording in which a visible record is produced immediately, without subsequent processing, in response to received signals.
f enregistrement *m* direct
e registro *m* directo
i registrazione *f* diretta
n directe opneming
d unmittelbare Aufzeichnung *f*

1536 DIRECT REFLECTION, MIRROR REFLECTION, REGULAR REFLECTION, SPECULAR REFLECTION opt
f réflexion *f* spéculaire
e reflexión *f* especular
i riflessione *f* regolare
n spiegelreflectie
d gerichtete Reflexion *f*

1537 DIRECT SCANNING tv
A scanning method in which the subject is illuminated at all times and only one elemental area of the subject is viewed at a time by the TV camera.
f analyse *f* directe
e exploración *f* directa
i analisi *f* diretta
n ---
d Abtastmethode *f* mit Dauerbeleuchtung der Bildvorlage und Ausblendung der jeweils abgetasteten Stelle im Bildfänger

1538 DIRECT SOUND aud
f son *m* direct
e sonido *m* directo
i suono *m* diretto
n direct geluid *n*
d unmittelbarer Schall *m*

1539 DIRECT SWITCHER tv
Vision switcher in which the push-buttons operate switches carrying the vision signal as opposed to sending a switching signal to a relay or matrix.
f commutateur *m* direct
e conmutador *m* directo
i commutatore *m* diretto
n directe schakelaar
d unmittelbarer Schalter *m*

1540 DIRECT TELEVISION TRANSMISSION tv
f émission *f* de télévision directe
e videodifusión *f* directe
i teleemissione *f* diretta

n directe televisie-uitzending
d direkte Fernsehsendung *f*

1541 DIRECT VIDEOTAPE RECORDING vr
f enregistrement *m* magnétoscopique temporaire
e grabación *f* directa en cinta magnetoscópica, grabación *f* directa en cinta video
i registrazione *f* videomagnetica temporanea
n tijdelijke videomagneetbandopname
d zeitweilige Bandaufnahme *f* von Fernsehsignalen

1542 DIRECT VIEW PICTURE TUBE crt
f tube *m* image à vision directe
e tubo *m* imagen de visión directa
i tubo *m* immagine a visione diretta
n directzichtbeeldbuis
d Direktsichtbildröhre *f*

1543 DIRECT VIEW SCREEN rep/tv
f écran *m* à vision directe
e pantalla *f* de visión directa
i schermo *m* a visione diretta
n directzichtscherm *n*
d Schirm *m* für direkte Betrachtung

1544 DIRECT VIEW SYSTEM tv
TV reception in which the received image is viewed directly on the screen of a cathode-ray tube without projection or reflecting device.
f système *m* à vision directe
e sistema *m* de visión directa
i sistema *m* a visione diretta
n directzichtsysteem *n*
d direktes Betrachtungsverfahren *n*

1545 DIRECT VISION CINESCOPE crt
f cinescope *m* à vision directe
e cinescopio *m* de visión directa
i cinescopio *m* a visione diretta
n directzichtcinescoop
d Direktsichtkineskop *n*

1546 DIRECT VISION TELEVISION RECEIVER rep
f récepteur *m* de télévision à vision directe
e televisor *m* de visión directa
i televisore *m* a visione diretta
n directzichttelevisieontvanger
d Direktsichtfernsehempfänger *m*

1547 DIRECTIONAL MICROPHONE aud
A microphone the response of which varied significantly with the direction of sound incidence.
f microphone *m* directionnel
e micrófono *m* direccional
i microfono *m* direttivo
n gerichte microfoon, microfoon met richtgevoeligheid
d gerichtetes Mikrophon *n*, Mikrophon *n* mit Richtwirkung, Richtmikrophon *n*

1548 DIRECTIONAL PATTERN, POLAR DIAGRAM, RADIATION PATTERN aea
A graphical representation of the radiation properties of the aerial (antenna) as a function of space co-ordinates.
f diagramme *m* polaire
e diagrama *m* polar
i diagramma *m* polare
n stralingsdiagram *n*
d Polardiagramm *n*

1549 DIRECTIONAL REFLECTANCE opt
The ratio of the radiance of a surface to the radiance of an ideal, non-absorbing, perfectly diffusing surface placed in the same position and similarly irradiated.
f albédo *m*
e albedo *m*
i albedo *m*
n albedo, reflectieversterkingsgraad
d Albedo *n*, Remissionsgrad *m*

1550 DIRECTIONAL RESPONSE PATTERN aud
Description of the response of an electroacoustic transducer as a function of the direction of propagation of the radiated or incident sound in a specified plane through the acoustical center(re) and at a specified frequency.
f diagramme *m* directionnel
e diagrama *m* direccional
i diagramma *m* direttivo
n richtkarakteristiek
d Richtcharakteristik *f*

1551 DIRECTIONAL SELECTIVITY aea/aud
Of an aerial (antenna) or microphone, the preferential sensitivity of the device to radiation from a particular direction.
f sensibilité *f* directionnelle
e selectividad *f* direccional, sensibilidad *f* direccional
i sensibilità *f* direzionale
n richtingsgevoeligheid
d Richtungsempfindlichkeit *f*, Richtungsselektivität *f*

1552 DIRECTIONS FOR USE, OPERATING INSTRUCTIONS ge
f mode *m* d'emploi
e modo *m* de empleo
i istruzioni *pl* per l'uso
n gebruiksaanwijzing
d Bedienungsanleitung *f*, Gebrauchsanweisung *f*

1553 DIRECTIVE AERIAL, DIRECTIVE ANTENNA aea
An aerial (antenna) that radiates or receives radio waves more effectively in some directions than others.
f antenne *f* directive
e antena *f* directiva, antena *f* dirigida

i antenna *f* direttiva
n richtantenne
d Richtantenne *f*,
Richtstrahler *m*

DIRECTIVE GAIN
see: AERIAL POWER GAIN

1554 DIRECTIVITY FUNCTION (US), GAIN FUNCTION (GB) aea
A function of direction relative to the aerial (antenna), the value of the function for any direction being the radiation power flux per unit solid angle in the given direction divided by the average value of the flux per unit solid angle for all directions.
f fonction *f* de directivité
e función *f* de directividad
i funzione *f* di direttività
n richtfunctie
d Richtfunktion *f*

1555 DIRECTOR aea
In an aerial (antenna) array, a parasitic element placed in front of a driven element.
f directeur *m*
e director *m*
i direttore *m*
n director, dirigator
d Direktor *m*, Wellenrichter *m*

1556 DIRECTOR fi/th/tv
f régisseur *m*
e director *m* de escena
i regista *m*
n regisseur
d Regisseur *m*

1557 DIRECTOR DIPOLE aea
A dipole combined with a parasitic element a fraction of a wavelength ahead of the receiving end of the aerial (antenna) to increase the gain in the direction of the major lobe.
f dipôle *m* directeur
e dipolo *m* director
i dipolo *m* direttore
n directordipool
d Direktordipol *m*

1558 DIRECTOR OF PHOTOGRAPHY, LIGHTING CAMERAMAN fi/th/tv
f chef-opérateur *m*,
directeur *m* de la photographie
e jefe operador *m*
i capo operatore *m*,
direttore *m* di fotografia
n chef-cameraman
d erster Kameramann *m*,
Lichtgestalter *m*

DIRECTOR'S CONSOLE (US)
see: CONTROL DESK

1559 DIRTIER ct
A difference apparently due to the presence of more black than in the original sample.
f plus sale
e más sucio
i più sudicio
n vuiler
d schmutziger

1560 DISABILITY GLARE opt
Glare which impairs the vision of objects without necessarily causing discomfort.
f éblouissement *m* diminuant la visibilité de l'objet
e deslumbramiento *m* menoscabador de la visibilidad del objeto
i abbagliamento *m* con indebolimento della visibilità dell'oggetto
n de zichtbaarheid van het object verminderende verblinding
d die Sichtbarkeit des Objekts herabsetzende Blendung *f*

1561 DISCOMFORT GLARE opt
Glare which causes discomfort without necessarily imparing the vision of objects.
f éblouissement *m* disturbant
e deslumbramiento *m* incómodo
i abbagliamento *m* disturbante
n hinderlijke verblinding
d unangenehme Blendung *f*

1562 DISCONTINUOUS RANDOM NOISE dis
f parasites *pl* erratiques discontinus
e ruido *m* errático discontinuo
i disturbo *m* erratico discontinuo
n niet-continue witte ruis
d nichtkontinuierliches weisses Rauschen *n*

1563 DISCRIMINATOR, FREQUENCY DISCRIMINATOR cpl
In a frequency of phase-modulation receiver, a device the output of which is substantially proportional to the deviation of the frequency of an alternating input from some predetermined value.
f discriminateur *m* de fréquence
e discriminador *m* de frecuencia
i discriminatore *m* di frequenza
n frequentiediscriminator
d Frequenzdiskriminator *m*

1564 DISK AERIAL, DISK ANTENNA aea
f antenne *f* en nappe
e antena *f* de disco
i antenna *f* a disco
n schijfantenne
d Scheibenantenne *f*

1565 DISK ANODE crt
Final anode in an electrostatically focused cathode-ray tube, usually a circular plate containing a central aperture.
f anode *f* à disque perforé
e ánodo *m* de disco perforado
i anodo *m* forato
n doorboorde schijfanode
d Lochscheibenanode *f*

1566 DISK PRISM tv
f disque *m* à prisme
e disco *m* de prisma
i disco *m* a prisma
n prisma-aftastschijf
d Prismenabtastscheibe *f*

1567 DISK SCANNER rec/tv
f disque *m* analyseur
e disco *m* explorador
i disco *m* analizzatore
n aftastschijf
d Abtastscheibe *f*

1568 DISPERSED-POWDER MAGNETIC TAPE, IMPREGNATED TAPE, MAGNETIC-POWDER IMPREGNATED TAPE aud/vr
A magnetic tape that consists of magnetic particles uniformly dispersed in a non-magnetic material.
f bande *f* magnétique homogène
e cinta *f* homogénea
i nastro *m* omogeneo
n homogene band
d Einschichtband *n*, Masseband *n*

1569 DISPERSION opt
The separation of electromagnetic radiation, compound sound waves, etc. with respect to some variable of the radiation concerned, such as energy, frequency or wavelength.
f dispersion *f*
e dispersión *f*
i dispersione *f*
n dispersie, kleurschifting
d Dispersion *f*

1570 DISPERSION vr
One of the steps in the manufacture of impregnated magnetic tape.
f dispersion *f*
e dispersión *f*
i dispersione *f*
n dispersie
d Dispersion *f*

1571 DISPLAY crt
f reproduction *f*
e reproducción *f*
i riproduzione *f*
n weergeefsysteem *n*, weergeving
d Wiedergabe *f*

1572 DISPLAY GAMMA crt
f gamme *f* de reproduction
e gama *f* de reproducción
i gamma *f* di riproduzione
n gamma van het weergeefsysteem
d Wiedergabebereich *m*

1573 DISPLAY PRIMARIES, RECEIVER PRIMARIES ct
The TV receiver primary colo(u)rs that, when mixed in proper proportions, serve to produce other desired colo(u)rs.
f primaires *pl* à la réception
e primarios *pl* a la presentación, primarios *pl* a la recepción
i primari *pl* alla ricezione
n primaire kleuren *pl* bij ontvangst
d Empfängerprimärvalenzen *pl*

1574 DISSECTOR, DISSECTOR TUBE, IMAGE DISSECTOR crt
A photo-emissive camera tube having a semi-transparent photocathode and a point anode in which an optical image gives rise to an electron image which is swept across the point anode to effect scanning.
f tube *m* dissecteur
e tubo *m* disector
i tubo *m* dissettore
n farnsworthbuis
d Elektronenbildzerleger *m*, Sondenröhre *f*, Zerleger *m*

1575 DISSOLVE, MIXES rep/tv
The merging of two TV signals in such a way, that as one scene disappears, another slowly appears.
f fondu *m* enchaîné
e disolución *f* superpuesta
i dissolvenza *f* incrociata
n overlapping
d Überblendung *f*

1576 DISTORSION EFFECTS ctv
f effets *pl* de distorsion
e efectos *pl* de distorsión
i effetti *pl* di distorsione
n vervormingseffecten *pl*
d Verzerrungseffekte *pl*

1577 DISTORSIONS dis
Errors in geometry of image due to imperfections in the focusing system.
f distorsions *pl*
e distorsiones *pl*
i distorsioni *pl*
n vervormingen *pl*
d Verzerrungen *pl*

1578 DISTRIBUTED AMPLIFIER, TRANSMISSION LINE AMPLIFIER cpl
An amplifier consisting of vacuum tubes (valves) distributed along two transmission lines.
f amplificateur *m* de ligne de transmission
e amplificador *m* de línea de transmisión
i amplificatore *m* di linea di trasmissione
n kettingversterker
d Kettenverstärker *m*

1579 DISTRIBUTING AMPLIFIER, DISTRIBUTION AMPLIFIER cpl
A radio-frequency amplifier used to feed TV signals to a number of receivers, as in an apartment house or hotel.
f amplificateur *m* de distribution

e amplificador *m* de distribución
i amplificatore *m* di distribuzione
n verdelerversterker
d Verteilerverstärker *m*

1580 DISTRIBUTION COEFFICIENTS ct
Chromaticity co-ordinates for spectral (monochromatic) radiations of equal power.
f coefficients *pl* de distribution
e coeficientes *pl* de distribución
i coefficienti *pl* di distribuzione
n verdelingscoëfficiënten *pl*
d Verteilungskoeffiziente *pl*

1581 DISTRIBUTION CONTROL tv
The means by which the distribution of scanning speeds is varied during the trace interval.
f commande *f* de base de temps
e mando *m* de base de tiempo
i comando *m* di base di tempi
n tijdbasisbesturing
d Einstellung *f* der Linearitäts-kompensation für die Ablenkschaltung

1582 DISTRIBUTION NOISE, dis
PARTITION NOISE
Noise resulting from random fluctuation in the division of cathode current between various electrodes in an electron tube (valve).
f bruit *m* de répartition
e ruido *m* de repartición,
ruido *m* por efecto de distribución
i rumore *m* di ripartizione
n verdelingsruis
d Stromverteilungsrauschen *n*

1583 DISTURBANCE ge
Any non-desired effect during transmission or reception of telecommunication.
f interférence *f*,
perturbation *f*
e interferencia *f*,
perturbación *f*
i disturbo *m*,
interferenza *f*,
perturbazione *f*
n storing
d Störung *f*

DIVERGENCE ANGLE
see: ANGLE OF DIVERGENCE

1584 DIVERGENT LENS opt
A lens with a negative focal length.
f lentille *f* divergente
e lente *f* divergente
i lente *f* divergente
n divergerende lens,
verspreidingslens
d Zerstreuungslinse *f*

1585 DOCUMENT TRANSMISSION tv
An application of closed circuit TV for transmitting documents or tabulated information between geographically separated points.
f transmission *f* par télévision de documents
e transmisión *f* televisora de documentos
i trasmissione *f* televisiva di documenti
n documenttransmissie met televisie-apparatuur
d Fernsehdokumentübertragung *f*

1586 DOCUMENTARY FILM tv
f documentaire *m*,
film *m* documentaire
e documental *m*,
película *f* documental
i documentario *m*,
pellicola *f* documentaria
n documentaire,
documentaire film
d Dokumentarfilm *m*

1587 DOLLY, rec/tv
TRUCK
f chariot *m*, dolly *m*
e carretilla *f*
i carrello *m*
n camerawagen, dolly
d Dolly *m*, Kamerawagen *m*

1588 DOLLY IN, rec/tv
TRACK IN
Decreasing the distance between the scene to be televised and the camera.
f travelling *m* en avant de la caméra
e acercamiento *m* de la cámara,
avance *m* de la cámara
i avanzamento *m* della telecamera
n voorwaartse beweging van de camera
d Kameravorschub *m*,
Vorfahrt *f* der Kamera

1589 DOLLY OUT, rec/tv
TRACK OUT
Increasing the distance between the scene to be televised and the camera.
f travelling *m* en arrière de la caméra
e alejamiento *m* de la cámara,
movimiento *m* separándose con la cámara
i arretramento *m* della telecamera
n achterwaartse beweging van de camera
d Rückfahrt *f* der Kamera,
Zurückfahren *n* der Kamera

1590 DOLLYING SHOT, rec/tv
FOLLOW SHOT,
TRAVEL(L)ING SHOT
A recording carried out by the travel(l)ing camera following the action of the scene.
f prise *f* de vues en mouvement,
travelling *m*
e toma *f* de vistas en movimiento
i carrellata *f*,
presa *f* a seguimento,
ripresa *f* in movimento
n travelling
d Fahraufnahme *f*

1591 DOMINANT WAVELENGTH ct
The single wavelength of light which predominates in any particular colo(u)r.

f longueur *f* d'onde dominante
e longitud *f* de onda dominante
i lunghezza *f* d'onda dominante
n dominerende golflengte
d dominierende Wellenlänge *f*,
farbtongleiche Wellenlänge *f*

1592 DO-NOT-MIX INDICATOR rec
A safeguard sign to the operator not to mix since a non-synchronous signal is selected.
f indicateur *m* de non-mixage
e aviso *m* de no mezclado
i attenzione *f*: non mescolatura
n niet-mengen waarschuwing
d Nichtmischen-Warnung *f*

1593 DOPE,
OXIDE PAINT vr
Intermediate product in the manufacture of magnetic tape.
f pâte *f* magnétique
e pasta *f* magnética
i pappa *f* magnetica
n magnetische brij
d magnetische Paste *f*

1594 DOPE SHEET fi/tv
f liste *f* des plans,
rapport *m* de tournage
e lista *f* de las tomas
i elenco *m* delle riprese
n opnamenlijst
d Aufnahmebericht *m*,
Filmbericht *m*,
Kamerabericht *m*,
Shotlist *f*

1595 DOPPLER SHIFT tv
The change in frequency and wavelength of electromagnetic radiation caused by translational motions of radiating atoms, molecules, or nuclei in the line of sight; of importance in satellite transmission.
f variation *f* de fréquence due à l'effet Doppler
e cambio *m* Doppler,
variación *f* de frecuencia por efecto Doppler
i spostamento *m* Doppler,
variazione *f* di frequenza dovuta all'effetto Doppler
n dopplerverschuiving
d Doppler-Verschiebung *f*

1596 DOT ctv
An element of the mosaic.
f point *m*
e punto *m*
i punto *m*
n punt *n*, stip, vlek
d Fleck *m*, Punkt *m*

1597 DOT-BAR GENERATOR tv
f générateur *m* de points et de barres
e generador *m* de puntos y de barras
i generatore *m* di punti e di barre
n punten- en balkengenerator
d Punkte-Balkengenerator *m*

1598 DOT FREQUENCY ctv/tv
The number of dots of the mosaic screen covered per second.
f fréquence *f* de points
e frecuencia *f* de puntos
i frequenza *f* di punti
n puntfrequentie,
stipfrequentie
d Punktfrequenz *f*

1599 DOT GENERATOR crt/tv
A generator which provides impulses for a test pattern for adjusting TV images on a cathode-ray tube screen.
f générateur *m* de points.
e generador *m* de puntos
i generatore *m* di punti
n puntengenerator,
stippengenerator
d Punktegenerator *m*

1600 DOT GRATING opt
f treillis *m* à points
e rejilla *f* de puntos
i graticcio *m* a punti
n puntenrooster *n*
d Punktgitter *n*

1601 DOT INTERLACE,
DOT INTERLACE SCANNING rec
A method of scanning in which the picture elements are explored in a regular but non-sequential order.
f analyse *f* entrelacée point par point
e exploración *f* entrelazada por puntos
i analisi *f* intercalata per punti
n interpunctering,
stipwisseling
d Punktsprungverfahren *n*,
Punktverflechtung *f*,
Zwischenpunktabtastung *f*

1602 DOT RECTIFICATION ctv
An increase in luminosity produced in the reproduced picture by the partial rectification of the chrominance signal resulting from the non-linear characteristics of the display device.
f augmentation *f* de la luminance des points
e aumento *m* de la luminancia de los puntos
i aumento *m* della luminanza dei punti
n verhoging van de puntluminantie
d Steigerung *f* der Punktleuchtdichte

1603 DOT SEQUENCE ctv/tv
f séquence *f* de points
e secuencia *f* de puntos,
sucesión *f* de puntos
i sequenza *f* di punti
n puntopeenvolging
d Punktfolge *f*,
Punktsequenz *f*

1604 DOT SEQUENTIAL COLO(U)R
TELEVISION ctv
Sequential colo(u)r TV in which signals derived from the primary colo(u)rs are transmitted in rotation, each sequence within a very small proportion of the duration of each line.

f système *m* à séquence de points
e sistema *m* de sucesión de puntos
i sistema *m* a sequenza di punti
n puntwisselsysteem *n*
d Punktfolgeverfahren *n*,
Punktsequentverfahren *n*

1605 DOUBLE-BEAM CATHODE-RAY TUBE, crt
DOUBLE-GUN TUBE,
DUAL-BEAM CATHODE-RAY TUBE
A cathode-ray tube containing two complete sets of beam-forming and beam-deflecting electrodes operated from the same cathode.
f tube *m* cathodique à deux faisceaux
e tubo *m* catódico de dos haces
i tubo *m* catodico a due fasci
n katodestraalbuis met twee bundels
d Zweistrahlkatodenröhre *f*

1606 DOUBLE BROAD stu
f projecteur *m* d'ambiance
e lámpara *f* proyectante
i lampada *f* di proiezione
n slaglichtschijnwerper
d Flutlichtscheinwerfer *m*

1607 DOUBLE CAMERA rec
One in which the film registering the outside scene is accompanied by a parallel film for optical or magnetic synchronous recording of the accompanying sound.
f caméra *f* pour image et son
e cámara *f* para imagen y sonido
i camera *f* per immagine e suono
n beeld- en geluidscamera
d Bild- und Tonkamera *f*

1608 DOUBLE-CONE LOUDSPEAKER aud
f hautparleur *m* à double cône
e altavoz *m* de doble cono
i altoparlante *m* a doppio cono
n dubbelconusluidspreker
d Doppelkonuslautsprecher *m*

1609 DOUBLE IMAGE, rep/tv
ECHO,
FOLD-OVER,
GHOST (GB),
MULTIPATH EFFECT (US)
f double image *f*,
image *f* fantôme,
image *f* multiple
e imagen *f* eco,
imagen *f* fantasma
i immagine *f* fantasma,
immagine *f* sdoppiata,
seconda immagine *f*
n echobeeld *n*
d Doppelbild *n*,
Echobild *n*,
Geisterbild *n*,
Schattenbild *n*

1610 DOUBLE-IMAGE SIGNALS, dis
ECHO SIGNALS,
GHOST SIGNALS
f signaux *pl* de double image,
signaux *pl* d'image fantôme
e señales *pl* de eco,
señales *pl* de fantasma
i segnali *pl* d'immagini fantasme
n echobeeldsignalen *pl*
d Doppelbildsignale *pl*,
Geisterbildsignale *pl*

1611 DOUBLE-LAYER SCREEN crt
Screen with separate layers of phosphors with differing properties.
f écran *m* à deux couches
e pantalla *f* de dos capas
i schermo *m* a due strati
n tweelagenscherm *n*
d Doppelschichtschirm *m*

1612 DOUBLE LIMITER, cpl/tv
GATE,
WINDOW
f circuit *m* à déclenchement périodique
e circuito *m* de desconexión periódica
i stadio *m* di sblocco periodico
n sleutelimpulsketen,
vensterketen
d Ausblendstufe *f*,
Einfallfeld *n*

1613 DOUBLE POLE-PIECE MAGNETIC HEAD, vr
DOUBLE POLE-TIP MAGNETIC HEAD
f tête *f* magnétique à deux pièces polaires
e cabeza *f* magnética de dos piezas polares
i testina *f* magnetica a due espansioni polari
n magneetkop met twee poolstukken
d magnetischer Kopf *m* mit Doppelpolstück

1614 DOUBLE-SIDEBAND TRANSMISSION tv
Transmission comprising the carrier frequency as well as both sidebands resulting from modulation of the carrier.
f transmission *f* à deux bandes latérales
e transmisión *f* de dos bandas laterales
i trasmissione *f* a due bande laterali
n transmissie met twee zijbanden
d Zweiseitenbandübertragung *f*

1615 DOUBLE-SIDED MOSAIC, crt
DOUBLE-SIDED SCREEN
An array of photosensitive elements insulated one from the other and mounted in a TV camera tube in such a way that an image can be projected optically on one side of the mosaic.
f mosaïque *m* à deux couches
e mosáico *m* de dos capas
i mosaico *m* a due strati
n tweelagenmozaïek *n*
d Zweischichtenbildschirm *m*

1616 DOUBLE-SIX ARRAY aea
A multiple aerial (antenna) system responsive to a number of channels.
f antenne *f* à deux fois six éléments
e antena *f* de dos veces seis elementos
i antenna *f* a due volte sei elementi
n antenne met tweemaal zes elementen
d Antenne *f* mit zweimal sechs Elementen

1617 DOUBLE-SUPER EFFECT dis
Interference occurring when the oscillator signal can penetrate to the grid of the radio-frequency tube (valve) where there is also present an interfering signal with such an amplitude that as a result of the linear characteristic of the radio-frequency tube (valve) a signal originates with a beat frequency which falls in the channel to which the receiver has been tuned.
f double action *f* hétérodyne
e doble acción *f* heterodina
i doppia azione *f* eterodina
n dubbel supereffect *n*
d Doppelsupereffekt *m*

1618 DOUBLE-SYSTEM SOUND RECORDING aud/rec
A system for taking picture and sound simultaneously possible by arranging special means for synchronizing camera and tape recorder.
f système *m* d'enregistrement simultané d'image et son
e sistema *m* de registro simultáneo de imagen y sonido
i sistema *m* di registrazione simultanea d'immagine e colore
n systeem *n* voor gelijktijdige opneming van beeld en geluid
d System *n* zur gleichzeitiger Aufnahme von Bild und Ton

1619 DOUBLE-TRACK RECORDER, DUAL-TRACK RECORDER, HALF-TRACK RECORDER vr
A tape recorder with a recording head that covers half the tape width, so two parallel tracks can be recorded on the tape.
f enregistreur *m* bipiste
e registrador *m* de dos pistas
i registratore *m* a due piste
n dubbelspoorgeluidsbandapparaat *n*
d Doppelspurtonbandgerät *n*

1620 DOUBLE-TRACK TAPE RECORDING, DUAL-TRACK TAPE RECORDING, HALF-TRACK TAPE RECORDING vr
Magnetic record in which two adjacent tracks are placed on the tape for doubling the recording capacity of the tape.
f enregistrement *m* bipiste
e registro *m* de dos pistas
i registrazione *f* a due piste
n dubbelspoorgeluidsopneming
d Doppelspurtonaufzeichnung *f*

1621 DOUBLE-TUNED CIRCUIT cpl
A circuit that is resonant to two adjacent frequencies.
f circuit *m* à double syntonisation
e circuito *m* de doble sintonización
i circuito *m* a doppia sintonizzazione
n dubbel afgestemde kring
d Zweikreisfilter *n*

1622 DOUBLING OF THE IMAGE tv
f doublage *m* d'images
e doblado *m* de imágenes, duplicación *f* de imágenes
i sdoppiamento *m* d'immagini
n beeldverdubbeling
d Bildverdopplung *f*, Verdopplung *f* des Bildes

1623 DOUBLY BALANCED MODULATOR cpl
A modulator circuit in which two class A amplifiers are supplied with modulating signals of equal amplitudes and opposite polarities and with carrier signals of equal amplitudes and opposite polarities.
f modulateur *m* à double équilibrage
e modulador *m* de doble equilibración
i modulatore *m* a doppio bilanciamento
n dubbelgebalanceerde modulator
d Ringmodulator *m*

1624 DOWN LEAD, LEAD IN aea
Feeder cable connecting an aerial (antenna) to a receiver.
f ligne *f* d'alimentation
e línea *f* de alimentación
i linea *f* d'alimentazione
n toevoerleiding
d Zuleitung *f*

1625 DOWN LEAD IMPEDANCE aea
f impédance *f* de la ligne d'alimentation
e impedancia *f* de la línea de alimentación
i impedenza *f* della linea d'alimentazione
n toevoerleidingsimpedantie
d Zuleitungsimpedanz *f*

1626 DOWNSTREAM vr
Pertaining to locations on the tape longitudinally displaced from a given reference point in the direction of tape motion.
f partie *f* en aval
e ---
i a valle
n stroomafwaarts
d in Bandlaufrichtung

1627 DOWNWARD CONVERSION tv
Standards conversion to a lower standard of lines.
f conversion *f* à un système de télévision à nombre de lignes plus bas
e conversión *f* hacia un sistema de televisión con número de líneas más bajo
i conversione *f* ad un sistema di televisione con numero di linee più basso
n overgang op een televisiesysteem met kleiner aantal lijnen
d Übergang *m* auf ein Fernsehsystem mit geringerer Zeilenzahl

1628 DRAMA STUDIO stu
f studio *m* de théâtre
e estudio *m* de teatro
i studio *m* teatrico
n toneelstudio
d Hörspielstudio *n*

1629 DRAWER TUNER rep/tv
f syntonisateur *m* à tiroir
e sintonizador *m* de estirador
i sintonizzatore *m* a cassetto
n schuiflade-kanalenkiezers *pl*
d Schubladenkanalwähler *pl*

1630 DRESS REHEARSAL rec/tv
f répétition *f* générale
e ensayo *m* general
i prova *f* generale
n generale repetitie
d Generalprobe *f*

1631 DRESSING THE SET rec/tv
The furnishing of the set with whatever is appropriate to the type of scene required by the story.
f décoration *f* du plateau
e decoración *f* del escenario
i decorazione *f* della scena
n aankleding van de planken
d Bühnenausstattung *f*

DRIFTLOCK
see: CONSTANT LINE NUMBER OPERATION

1632 DRIVE,
EXCITATION cpl
The signal voltage that is applied to the control electrode of an electron tube (valve).
f attaque *f* de grille,
excitation *f* de grille
e excitación *f* de rejilla
i eccitazione *f* di griglia
n roosterexcitatie
d Gitteranregung *f*

1633 DRIVE CONTROL,
HORIZONTAL DRIVE CONTROL rep/tv
The control in a TV receiver that adjusts the output of the horizontal deflection oscillator.
f réglage *m* du signal de commande horizontal
e regulación *f* de la señal de mando horizontal
i regolazione *f* del segnale di comando orizzontale
n regeling van het horizontale stuursignaal
d Horizontalsteuerung *f*

1634 DRIVE MOTOR vr
f moteur *m* de commande
e motor *m* de mando
i motore *m* di comando
n aandrijfmotor
d Antriebsmotor *m*

1635 DRIVE SPROCKET fi/tv
f tambour *m* d'entraînement
e rueda *f* dentada accionadora
i tamburo *m* dentato d'agganciamento
n aandrijftandrad *n*
d Schaltrolle *f*

1636 DRIVE UNIT tv
The unit which drives the large output stages in a high-powered TV transmitter.
f bloc *m* de commande
e bloque *m* de mando
i blocco *m* di comando
n stuureenheid
d Steuerungseinheit *f*

1637 DRIVE VOLTAGE crt
Of a cathode-ray tube, the difference between the cut-off voltage and the modulator voltage.
f tension *f* de signal de commande
e tensión *f* de señal de mando
i tensione *f* di segnale di comando
n stuursignaalspanning
d Steuersignalspannung *f*

1638 DRIVEN ELEMENTS aea
The elements in an aerial (antenna) which are fed by the transmitter.
f éléments *pl* alimentés
e elementos *pl* alimentados
i elementi *pl* alimentati
n gevoede elementen *pl*
d gespeiste Elemente *pl*

DRIVEN RADIATOR
see: ACTIVE AERIAL

1639 DRIVER,
DRIVING STAGE cpl
The amplifier stage preceding the output stage in a receiver or transmitter.
f étage *m* excitateur
e etapa *f* excitadora
i stadio *m* eccitatore
n stuurtrap
d Treiber *m*,
Treiberstufe *f*

1640 DRIVING PULSE cpl
f impulsion *f* excitatrice
e impulso *m* de excitación
i impulso *m* d'eccitazione
n stuurimpuls
d Treiberimpuls *m*

1641 DRIVING SIGNALS tv
The original line and field pulses which synchronize the whole TV system.
f signaux *pl* d'excitation
e señales *pl* de excitación
i segnali *pl* d'eccitazione
n stuursignalen *pl*
d Treibersignale *pl*

1642 DRIVING SYSTEM vr
f système *m* de commande

e sistema *m* de mando
i sistema *m* di comando
n aandrijfsysteem *n*
d Antriebssystem *n*

1643 DROP OUT rec/vr
In magnetic recording, sudden momentary losses of signal caused by impurities or other imperfections of the tape.
f manque *m* de signal
e falla *f* de señal
i mancanza *f* di segnale
n signaaluitval
d Signalausfall *m*

1644 DROP TIME tv
f temps *m* de descente
e tiempo *m* transitorio final
i tempo *m* di discesa
n uitslingertijd
d Ausschwingzeit *f*

1645 DROPPING OUT vr
f irrégularité *f* du son
e irregularidad *f* del sonido
i irregolarità *f* del suono
n brokkeligheid
d kurzzeitige Amplitudenschwankung *f*

1646 DRUM fi/tv
f tambour *m*, rouleau *m*
e tambor *m*
i tamburo *m*
n trommel
d Trommel *f*

DRUM SCANNER
see: BELT SCANNER

DUAL-BEAM CATHODE-RAY TUBE
see: DOUBLE-BEAM CATHODE-RAY TUBE

1647 DUAL CHANNEL SOUND, DUAL CHANNEL TELEVISION SOUND SYSTEM tv
A technique used in TV receivers in which a separate intermediary frequency stage is employed for sound and video signals after the common first detector stage.
f système *m* sonore à deux canaux
e sistema *m* sonoro de dos canales
i sistema *m* sonoro a due canali
n geluidssysteem *n* met twee kanalen
d Tonsystem *n* mit zwei Kanälen

1648 DUAL INPUT rec/rep
Two pairs of signal input sockets for picture monitors.
f entrée *f* double
e entrada *f* doble
i ingresso *m* doppio
n dubbele invoer
d doppelte Einfuhr *f*

1649 DUAL MASTER CONTROL rec/tv
The control of two transmissions from two identical presentation mixers fed from the same sources in those cases e.g. in which different main program(me)s are broadcast.
f contrôle *m* de mixage double
e control *m* principal doble
i controllo *m* principale doppio
n dubbele hoofdmengtafel
d doppelte Hauptmischstelle *f*

1650 DUAL NETWORKS, STRUCTURALLY DUAL NETWORKS cp
A pair of networks so arranged that any mesh of one corresponds to a cut-set of the other.
f réseaux *pl* réciproques
e redes *pl* estructuralmente doble
i strutture *pl* reciproche
n duale netwerken *pl*
d duale Netzwerke *pl*

1651 DUAL STANDARD RECEIVER rep/tv
Receiver capable of receiving TV pictures on more than one line standard.
f récepteur *m* de télévision pour systèmes différents de lignes
e receptor *m* de televisión para sistemas diferentes de líneas
i ricettore *m* televisivo per sistemi differenti di linee
n televisieontvanger voor verschillende lijnensystemen
d Zweinormen-Fernsehempfänger *m*

DUAL TRACK RECORDER
see: DOUBLE TRACK RECORDER

DUAL TRACK TAPE RECORDING
see: DOUBLE TRACK TAPE RECORDING

1652 DUAL WINDING CONVERGENCE MAGNET crt
f aimant *m* de convergence à enroulement bifilaire
e imán *m* de convergencia de arrollamiento bifilar
i magnete *m* di convergenza ad avvolgimento bifilare
n convergentiemagneet met bifilaire wikkeling
d bifilar gewickelter Konvergenzmagnet *m*

DUB
see: COPY

1653 DUBBED VERSION rec/tv
Foreign-language version of a film or TV recording in which the original dialogue has been replaced by a translation carefully synchronized to the movement of the actor's lip.
f film *m* doublé
e película *f* doblada
i pellicola *f* doppiata
n film met gesynchroniseerde dialoogvertaling
d Film *m* mit synchronisierter Dialogübersetzung

1654 DUBBER aud
Sound reproducer of the highest possible quality on which sound tracks are run in synchronism with other sound tracks on similar machines during the process of

dubbing or mixing.
f console *f* de mixage, doubleur *m*, mélangeur *m*
e combinador *m* de dos o más registros sonoros, mezclador *m* de dos o más registros sonoros
i doppiatore *m*, mescolatore *m*
n geluidskopieerinrichting, geluidsmenger
d Tonkopierer *m*, Tonmischer *m*

1655 DUBBING fi/tv
The adding of a new sound.
f mixage *m*, sonorisation *f*
e sonorización *f*
i sonorizzazione *f*
n sonorisatie
d Tonmischung *f*

1656 DUBBING fi/tv
The adding of a translated synchronized version.
f doublage *m*
e doblado *m*
i doppiaggio *m*
n synchronisatie
d Doubeln *n*, Synchronisierung *f*

1657 DUBBING rec/tv
Combination of several sound tracks into a single track.
f combinaison *f* de pistes
e combinación *f* de pistas
i combinazione *f* di piste
n geluidssporencombinatie
d Tonspurenkombination *f*

1658 DUBBING MIXER aud/rec/tv
f mélangeur *m* doubleur
e mezclador *m* doblador
i mescolatore *m* doppiatore
n verdubbelende menger
d verdopplender Mischer *m*

1659 DULLNESS ct
That colo(u)r quality, an increase in which is associated with the residual degradation which would result from the addition of a small quantity of neutral grey to the colo(u)ring material when the strength of the mixture has been readjusted to its original strength.
f manque *m* d'éclat
e falta *f* de claridad
i mancanza *f* di lustro
n saaiheid
d Glanzlosigkeit *f*

1660 DUPLEX SOUND TRACK aud/rec
Type of symmetrical variable-area optical sound track in the negative of the central axis and the surrounding modulated area are opaque.
f piste *f* à densité fixe duplex
e pista *f* sonora en duplex de área variable
i pista *f* sonora doppia a densità fissa
n duplexzaagtandspoor *n*
d Doppelzackenschrift *f*

1661 DYNAMIC CONVERGENCE ctv
The condition which exists when the three beams of a colo(u)r picture tube converge at the aperture mask as they are deflected both vertically and horizontally.
f convergence *f* dynamique
e convergencia *f* dinámica
i convergenza *f* dinamica
n dynamische convergentie
d dynamische Konvergenz *f*

1662 DYNAMIC CONVERGENCE CORRECTION crt/dis
f correction *f* de la convergence dynamique
e corrección *f* de la convergencia dinámica
i correzione *f* della convergenza dinamica
n dynamische convergentiecorrectie
d dynamische Konvergenzkorrektur *f*

1663 DYNAMIC DEMONSTRATOR, WORKING DIAGRAM tv
A large schematic circuit diagram that has been cemented to a board with all components mounted near or on their symbols and connected together to give a working circuit of a TV receiver.
f démonstration *f* pratique, diagramme *m* de fonctionnement
e diagrama *m* sinóptico
i schema *m* funzionale
n werkend schema *n*
d dynamisches Schaubild *n*

1664 DYNAMIC FOCUSING crt
The automatic process of varying the voltage on the focusing electrode of a colo(u)r TV picture tube so that the beam spots remain always in focus as they are swept across the flat screen.
f focalisation *f* dynamique
e enfoque *m* dinámico
i focalizzazione *f* dinamica
n dynamische focussering
d dynamische Fokussierung *f*

1665 DYNAMIC MISREGISTRATION ctv
A misregistration produced in a signal from a colo(u)r camera as a function of the picture content.
f défaut *m* de calage dynamique
e registro *m* dinámico falso
i registrazione errata *f* dipendente dal contenuto d'immagine
n door beeldinhoud veroorzaakte foutieve registratie
d durch Bildinhalt verursachte Fehlüberdeckung *f*

1666 DYNAMIC RANGE, VOLUME RANGE rep/tv
Of a program(me), the range within which its volume fluctuates over a specified period.
f dynamique *f* d'un signal de modulation

e margen *m* dinámico
i dinamica *f*
n dynamiek
d Dynamikumfang *m*,
  Lautstärkeumfang *m*

DYNAMIC RANGE
  see: CONTRAST RANGE

1667 DYNAMIC RANGE OF SPEECH aud
f gamme *f* dynamique de la parole
e gama *f* dinámica de las frecuencias audibles
i gamma *f* dinamica delle parole
n dynamisch gebied *n* van de spraak
d dynamischer Sprachbereich *m*

1668 DYNODE cpl
An electrode of an electron tube (valve) whose primary function is to supply secondary-electron emission.
f cathode *f* secondaire,
  dynode *f*
e dinodo *m*
i dinodo *m*
n dynode
d Dynode *f*,
  Sekundäremissionskatode *f*

1669 DYNODE EFFECT, crt
  DYNODE SPOTS
Spurious signals which may be produced in image orthicons.
f effet *m* de dynode
e efecto *m* de dinodo
i effetto *m* di dinodo
n dynode-effect *n*
d Dynodeneffekt *m*

E

1670 EAR MICROPHONE aud
A contact microphone shaped to fit into the ear where it picks up the voice of the bearer to be transmitted.
f microphone *m* d'oreille
e micrófono *m* para oído
i microfono *m* d'orecchio
n insteekmicrofoon, oormicrofoon
d Ohrmikrophon *n*

1671 EARPHONES aud
f casque *m*, écouteurs *pl*
e auriculares *pl*
i auricolari *pl*, cuffia *f*
n hoofdtelefoon
d Kopfhörer *pl*

1672 EAR-PLUG, RADIO CONTROL RECEIVER aud
f récepteur *m* d'ordres haute fréquence
e receptor *m* de instrucciones alta frecuencia
i ricettore *m* d'ordini alta frequenza
n draadloze bevelontvanger
d drahtloser Kommandoempfänger *m*

1673 EAR-SHOT aud
f distance *f* d'écoute, portée *f* de la voix
e distancia *f* máxima de audibilidad
i portata *f* della voce
n gehoorsafstand
d Hörweite *f*

1674 EARTH'S MAGNETIC FIELD, TERRESTRIAL MAGNETIC FIELD dis/ge
f champ *m* magnétique terrestre
e campo *m* magnético terrestre
i campo *m* magnetico terrestre
n magnetisch aardveld *n*
d Erdmagnetfeld *n*

1675 EARTH SATELLITE tv
f satellite *m* de la terre
e satélite *m* de la tierra
i satellite *m* della terra
n aardsatelliet
d Erdsatellit *m*

1676 EARTH STATION, GROUND STATION tv
A station which receives the very weak signals from a satellite and amplifies them for subsequent transmission.
f station *f* terrienne
e estación *f* terrestre
i stazione *f* terrestre
n aardstation *n*
d Bodenstation *f*, Erdefunktstelle *f*

1677 EAVES AERIAL, EAVES ANTENNA aea
f antenne *f* goutière
e antena *f* de canalón
i antenna *f* di grondaia
n dakgootantenne
d Dachrinnenantenne *f*

1678 ECHO aud
f écho *m*
e eco *m*
i eco *m*
n echo
d Echo *n*

ECHO
see: DOUBLE IMAGE

1679 ECHO ATTENUATION aud/vr
f affaiblissement *m* d'écho
e amortiguación *f* de eco
i smorzamento *m* d'eco
n echodemping
d Echodämpfung *f*

1680 ECHO CANCELLATION aea/tv
Elimination of the echo's (ghosts) in TV pictures by suitable disposal of the aerial (antenna).
f élimination *f* d'images fantômes
e eliminación *f* de imágenes fantasmas
i eliminazione *f* d'immagini fantasmi
n echo-onderdrukking
d Echokompensation *f*, Geisterbildbeseitigung *f*

1681 ECHO CHAMBER, ECHO ROOM aud
A reverberant room, containing only a microphone and a loudspeaker, through which an output from a studio or hall is passed in order to allow a variable degree of reverberation to be added to the direct output from the same source.
f chambre *f* d'écho
e cámara *f* de reverberación
i camera *f* riverberante
n echokamer, nagalmkamer
d Hallraum *m*, Nachhallraum *m*

1682 ECHO DELAY TIME dis
f temps *m* de transit d'un écho
e tiempo *m* de tránsito de un eco
i tempo *m* di transito d'un eco
n echolooptijd
d Echolaufzeit *f*

1683 ECHO EQUALIZER, ECHO KILLER, ECHO TRAP (GB), POWER EQUALIZER (US) tv
A device or circuit which suppresses

undesired echos.
f correcteur *m* d'écho, piège *m* d'écho, suppresseur *m* d'écho
e supresor *m* de eco, trampa *f* de eco
i soppressore *m* d'eco, trappola *f* d'eco
n echo-onderdrukker, echoval
d Echofalle *f*, Echosperre *f*

1684 ECHO INTENSITY dis
f intensité *f* d'un écho
e intensidad *f* de un eco
i valore *m* d'un eco
n echo-intensiteit
d Echointensität *f*

1685 ECHO SIGN dis
f signe *m* d'un écho
e signo *m* de un eco
i segno *m* d'un eco
n echoteken *n*
d Echozeichen *n*

ECHO SIGNALS
see: DOUBLE IMAGE SIGNALS

1686 ECHO WAVEFORM CORRECTOR dis/rep/tv
Corrector for linear phase and amplitude distortion on a TV signal, resulting from multipath signals or echos.
f correcteur *m* de la forme d'onde de l'écho
e corrector *m* de forma de onda del eco
i correttore *m* di forma d'onda dell'eco
n corrector van de echogolfvorm
d Echowellenformentzerrer *m*

1687 EDGE BUILD-UP, EDGE LIP vr
A fault occurring in magnetic tape due to winding up under too high a tension.
f bord *m* redressé, bord *m* relevé
e borde *m* ensortijado, borde *m* rizado
i bordo *m* rilevato
n omgekrulde kant, opstaande kant
d hochstehende Kante *f*

1688 EDGE CORRECTION tv/vr
f correction *f* de contour
e corrección *f* de contorno
i correzione *f* di contorno
n contourcorrectie
d Kantenanhebung *f*

1689 EDGE DETERIORATION vr
f endommagement *m* de bord
e daño *m* de borde
i danno *m* di bordo
n randbeschadiging
d Randbeschädigung *f*

1690 EDGE DISTORTION fi/tv
f distorsion *f* au bord de l'image
e distorsión *f* de borde de la imagen
i distorsione *f* al bordo dell'immagine
n randvervorming
d Randverzerrung *f*

1691 EDGE EFFECT dis
f effet *m* de souligné
e efecto *m* de borde
i effetto *m* di bordo
n randeffect *n*
d Kanteneffekt *m*

EDGE ENHANCEMENT (GB)
see: CONTOUR ACCENTUATION

1692 EDGE FLARE crt/tv
Unwanted luminosity occurring at the edges of the picture.
f éclaircissement *m* des bords
e sobreluminancia *f* marginal
i bordi *pl* luminosi
n randflits
d Randaufhellung *f*

1693 EDGE LUMINANCE ctv/tv
A luminance the intensity of which is slightly lower than that of the rest of the picture in order to evade flatness.
f luminance *f* de bord
e luminancia *f* de borde
i luminanza *f* di bordo
n randluminantie
d Randleuchtdichte *f*

1694 EDGE OF A PULSE, PULSE EDGE cpl
f flanc *m* d'impulsion
e flanco *m* de impulso
i fronte *f* d'impulso
n impulsflank
d Flanke *f*, Impulsflanke *f*

1695 EDGE OF TRACK BANDING vr
f effet *m* de bande de bord de piste
e efecto *m* de banda en entrada y salida de cabeza
i difetti *pl* vicino alla commutazione
n kopovergang banding
d Farbfehler *pl* der ersten Zeilen einer Kopfspur

1696 EDGED CAPTION rec/tv
A caption having a contrasting, electronically produced border on the vertical edges of each character.
f titre *m* à effet de profondeur
e título *m* con efecto de profundidad
i titolo *m* ad effetto di profondità
n titel met dieptewerking
d Titel *m* mit Tiefenwirkung

1697 EDIT vr
The point at which another recording, which has been substituted for a previous recording, begins.

f point *m* de montage
e montaje *m*
i giunta *f*, montaggio *m*
n montage
d Schnittstelle *f*

1698 EDIT PULSE vr
A magnetic pulse recorded on the control track of a transverse-scan video tape machine to mark the boundary between two successive pictures.
f impulsion *f* de montage
e impulso *m* de montaje
i impulso *m* di montaggio
n montage-impuls
d Schneideimpuls *m*,
Schnittmarke *f*

EDITING
see: CUTTING

1699 EDITING CUBICLE fi/stu/tv
f cellule *f* de montage
e cubículo *m* de montaje
i cabina *f* di montaggio
n montagecabine
d Schnittkabine *f*

EDITING DESK,
EDITING TABLE
see: CUTTING DESK

1700 EDITING OF THE TAPE, vr
TAPE EDITING
f montage *m* de bande
e montaje *m* de cinta
i montaggio *m* di nastro
n bandmontage
d Bandschnitt *m*

EDITING ROOM
see: CUTTING ROOM

EDITOR
see: CUTTER

1701 EDITORIAL NEWSROOM tv
f centre *m* de rédaction du téléjournal
e centro *m* editorial de noticias del diario
i centro *m* editoriale del telegiornale
n redactiekamer voor het journaal
d Redaktionsraum *m* der Tagesschau

1702 EDUCATIONAL BROADCAST tv
f émission *f* didactique
e emisión *f* educativa
i emissione *f* didattica
n onderwijsuitzending
d Lehrsendung *f*

1703 EDUCATIONAL FILM tv
f film *m* d'enseignement,
film *m* éducatif
e película *f* de enseñanza,
película *f* educativa
i pellicola *f* didattica
n leerfilm,
onderwijsfilm
d Lehrfilm *m*

1704 EDUCATIONAL TELEVISION, tv
SCHOOL AND COLLEGES
TELEVISION
f télévision *f* scolaire
e televisión *f* educativa
i telescuola *f*
n schooltelevisie
d Schulfernsehen *n*

1705 EFFECT AMPLIFIER aud
A device for realizing an adjustable distortion of a voice.
f correcteur *m* pour pièces radiophoniques
e corrector *m* de efectos acústicos
i filtro *m* d'effetti
n akoestische effectgever
d Hörspielverzerrer *m*

1706 EFFECT FILTER opt
f filtre *m* pour effets spéciaux
e filtro *m* para efectos especiales
i filtro *m* per effetti speciali
n effectfilter *n*
d Effektfilter *n*

1707 EFFECT LIGHT, fi/stu/tv
EFFECT LIGHTING
f éclairage *m* d'effet
e alumbrado *m* de efecto
i illuminazione *f* d'effetto
n effectverlichting
d Effektbeleuchtung *f*,
Effektlicht *n*,
Effektspitze *f*

1708 EFFECTIVE AERIAL AREA, aea
EFFECTIVE ANTENNA AREA
An area determined by dividing the maximum power which the aerial (antenna) could absorb from a plane wave by the incident power flux density, the load being matched to the aerial (antenna).
f portée *f* efficace d'antenne
e área *f* efectiva de antena
i area *f* effettiva d'antenna
n actieradius, effectief reikgebied *n*
d Antennenwirkfläche *f*

1709 EFFECTIVE APERTURE opt
f ouverture *f* photométrique
e abertura *f* efectiva
i apertura *f* effettiva
n fotometrische apertuur
d photometrische Öffnung *f*

1710 EFFECTIVE PICTURE SIGNAL cpl/tv
f signal *m* d'image effectif,
signal *m* vidéo effectif
e señal *f* video efectiva
i segnale *m* video effettivo
n effectief beeldsignaal *n*
d effektives Bildsignal *n*

1711 EFFECTIVE RADIATED POWER aea
Product of the power fed into a transmitter aerial (antenna) and the aerial (antenna) gain.
f puissance *f* effective émise

e potencia *f* efectiva de radiación
i potenza *f* effettiva irradiata
n effectief uitgestraald vermogen *n*
d äquivalente Strahlungsleistung *f*

1712 EFFECTS BANK tv
The part of a vision switching matrix used in conjunction with special effects equipment.
f unité *f* de commutation pour effets sonores
e unidad *f* de conmutación para efectos sonoros
i unità *f* di commutazione per effetti sonori
n schakelblok *n* voor geluidseffecten
d Schaltglied *n* für Toneffekte

1713 EFFECTS GENERATOR cpl
f générateur *m* d'effets
e generador *m* de efectos
i generatore *m* d'effetti
n effectengenerator
d Effektengeber *m*

1714 EFFECTS MIXER ctv
f mélangeur *m* d'effets
e mezclador *m* de efectos
i mescolatore *m* d'effetti
n effectenmenger
d Effektenmischer *m*

1715 EFFECTS TAPE vr
f bande *f* de bruitage
e cinta *f* de ruido
i nastro *m* di rumore
n geruisband
d Geräuschband *n*

1716 EFFECTS TRACK aud
A track containing sound effects not normally recorded when a film is being shot and usually available from the library.
f piste *f* à effets sonores,
piste *f* de bruitage
e pista *f* de efectos sonoros
i pista *f* ad effetti sonori
n geluidseffectenspoor *n*
d Geräuschspur *f*,
Toneffektespur *f*

EFFICIENCY DIODE
see: BOOSTER DIODE

1717 EFFICIENCY OF FLUORESCENT SCREEN crt
f rendement *m* de l'écran fluorescent
e rendimiento *m* de la pantalla fluorescente
i rendimento *m* dello schermo fluorescente
n rendement *n* van het fluorescerend scherm
d Leuchtschirmwirkungsgrad *m*

1718 EHT,
EXTRA HIGH TENSION cpl
f très haute tension *f*
e tensión *f* extraalta
i tensione *f* altissima
n extra hoge spanning
d Höchstspannung *f*

1719 EHT CAGE rep/svs
f cage *f* pour très haute tension
e caja *f* para tensión extraalta
i gabbia *f* per tensione altissima
n hoogspanningskooi
d EHT-Käfig *m*

1720 EHT RECTIFIER cpl
f redresseur *m* pour très haute tension
e rectaficador *m* para tensión extraalta
i raddrizzatore *m* per tensione altissima
n gelijkrichter voor extra hoge spanning
d Höchstspannungsgleichrichter *m*

1721 EHT SUPPLY cpl
f alimentation *f* en très haute tension
e alimentación *f* en tensión extraalta
i alimentazione *f* in tensione altissima
n extra-hoge-spanningsvoeding
d Höchstspannungsspeisung *f*

1722 EIDOPHOR SYSTEM tv
A projection TV system which employs diffraction effects to produce the modulation of light for the picture.
f système *m* "eidophor"
e sistema *m* "eidophor"
i sistema *m* "eidophor"
n eidophorsysteem *n*
d Eidophor-Verfahren *n*

1723 EIGHT-TRACK TAPE aud/vr
f bande *f* à huit pistes
e cinta *f* de ocho pistas
i nastro *m* ad otto piste
n achtsporenband
d Achtspurenband *n*

ELAPSED TIME CLOCK
see: CUE CLOCK

1724 ELECTRIC DEFLECTION crt
Deflection of the electron beam produced by an electric field across its path.
f balayage *m* électrique,
déviation *f* électrique
e barrido *m* eléctrico,
desviación *f* eléctrica
i deflessione *f* elettrica,
deviazione *f* elettrica
n elektrische afbuiging
d elektrische Ablenkung *f*

1725 ELECTRIC DEFLECTION SENSITIVITY crt
Of a cathode-ray tube, the quotient of the spot displacement and the change in voltage between the deflector plates for a given voltage on the final anode.
f sensibilité *f* de la déviation électrique,
sensibilité *f* du balayage électrique
e sensibilidad *f* del barrido eléctrico,
sensibilidad *f* de la desviación eléctrica
i sensibilità *f* della deflessione elettrica,
sensibilità *f* della deviazione elettrica
n elektrische afbuigingsgevoeligheid
d elektrische Ablenkempfindlichkeit *f*

1726 ELECTRIC IMAGE tv
An image consisting of charges on some insulating surface or other, the magnitude of each charge corresponding from point to point with the light intensity at the corresponding point in the optical picture.
f image *f* électrique de charge
e imagen *f* eléctrica de carga
i immagine *f* elettrica di carica
n elektrisch ladingsbeeld *n*
d elektrisches Ladungsbild *n*

1727 ELECTRIC INTERFERENCE dis
Interference caused by the operation of electrical apparatus other than radio stations.
f perturbation *f* par appareils électriques
e perturbación *f* por aparatos eléctricos
i disturbo *m* per apparecchiatura elettrica
n storing door elektrische apparaten
d Störung *f* durch elektrische Geräte

1728 ELECTRICAL CENT(E)RING rep/tv
The cent(e)ring of the picture on the screen by electrical means.
f centrage *m* électrique
e centraje *m* eléctrico
i centratura *f* elettrica
n elektrische centrering
d elektrische Zentrierung *f*

1729 ELECTRICAL MASKING, ctv
MASKING
1. The correction of inadequate characteristics in the colo(u)r analyzing equipment, by matrixing the colo(u)r signals.
2. In a film scanner, the adjustment of the signal to correct for photographic and/or processing errors in the film being scanned.
f masquage *m* électrique
e enmascaramiento *m* eléctrico
i mascheramento *m* elettrico
n elektrische maskering
d elektrische Maskierung *f*

1730 ELECTRICAL SPEECH LEVEL, aud
VOLUME LEVEL
At a point in a circuit, the level, relative to a specified level, of power or voltage due to sounds or speech transmitted by a microphone, measured at the point by means of a specified meter.
f niveau *m* vocal électrique
e nivel *m* eléctrico de las corrientes vocales
i livello *m* elettrico delle correnti vocali
n elektrisch spraakniveau *n*, elektrisch spreekniveau *n*
d elektrischer Sprachpegel *m*, elektrischer Sprechpegel *m*

ELECTROCATHODOLUMINESCENCE
see: CATHODOELECTROLUMINESCENCE

1731 ELECTRODE ASSEMBLY crt
f ensemble *m* d'électrodes
e conjunto *m* de electrodos
i castello *m* degli elettrodi, complesso *m* d'elettrodi
n elektrodensamenstel *n*
d Elektrodenanordnung *f*, Elektrodensatz *m*

1732 ELECTROMAGNETIC DEFLECTION, crt
MAGNETIC DEFLECTION
f balayage *m* électromagnétique, déviation *f* électromagnétique
e barrido *m* electromagnético, desviación *f* electromagnética
i deflessione *f* elettromagnetica, deviazione *f* elettromagnetica
n elektromagnetische afbuiging
d elektromagnetische Ablenkung *f*

1733 ELECTROMAGNETIC DEFLECTION SENSITIVITY, crt
MAGNETIC DEFLECTION SENSITIVITY
The quotient of the spot displacement and the change in deflection coil current for a given magnetic deflecting system, and for a given voltage on the final accelerator.
f sensibilité *f* de la déviation électromagnétique, sensibilité *f* du balayage électromagnétique
e sensibilidad *f* del barrido electromagnético, sensibilidad *f* de la desviación electromagnética
i sensibilità *f* della deflessione elettromagnetica, sensibilità *f* della deviazione elettromagnetica
n elektromagnetische-afbuigingsgevoeligheid, gevoeligheid voor elektromagnetische afbuiging
d elektromagnetische Ablenkempfindlichkeit *f*

1734 ELECTROMAGNETIC FOCUSING, crt
MAGNETIC FOCUSING
f focalisation *f* électromagnétique
e enfoque *m* electromagnético
i focalizzazione *f* elettromagnetica
n elektromagnetische focussering
d elektromagnetische Fokussierung *f*

1735 ELECTROMAGNETIC LENS, opt
MAGNETIC LENS
An electron lens employing electromagnetic focusing techniques.
f lentille *f* électronique à focalisation électromagnétique
e lente *f* electrónica de enfoque electromagnético
i lente *f* elettronica a focalizzazione elettromagnetica
n elektronenlens met elektromagnetische focussering
d Elektronenlinse *f* mit elektromagnetischer Fokussierung

1736 ELECTROMAGNETIC SPECTRUM cpl
The spectrum containing all wavelengths from 10^{-13}cm up to 10^{9}m.
f spectre *m* de longueurs d'ondes électromagnétiques

e espectro *m* de longitudes de ondas electromagnéticas
i spettro *m* di lunghezze d'onde elettromagnetiche
n golflengtespectrum *n* van elektromagnetische golven
d Wellenlängenspektrum *n* von elektromagnetischen Wellen

1737 ELECTROMECHANICAL AUTOMATED GRAM OPERATOR rep/stu/tv
A device that automatically carries out the work usually done by the human gram operator.
f appareil *m* électromécanique automatisé pour le service de disques et bandes
e aparato *m* electromecánico automatizado para el servicio de discos y cintas
i apparecchio *m* elettromeccanico automatizzato per il servizio di dischi e nastri
n elektromechanisch geautomatiseerd toestel *n* voor het bedienen van geluiddragers
d elektromechanisches automatisiertes Gerät *n* für die Bedienung der Schallträger

1738 ELECTRON BEAM crt
A focused flow of electrons.
f faisceau *m* électronique
e haz *m* electrónico
i fascio *m* elettronico, pennello *m* elettronico
n elektronenbundel
d Elektronenstrahl *m*

1739 ELECTRON-BEAM D.C. RESISTANCE crt
The quotient of electron-beam voltage and the d.c. component of the electron beam.
f résistance *f* équivalente du faisceau
e resistencia *f* equivalente del haz
i resistenza *f* equivalente del fascio
n bundelweerstand
d Elektronenstrahlwiderstand *m*

1740 ELECTRON-BEAM FILM SCANNING rec/vr
A scanning method in which a specially prepared film acts as its own source of light, the normal light-sensitive emulsion being coated with scintillators which emit 5 to 10 photons for every primary electron which strikes them.
f analyse *f* de films par faisceau électronique
e exploración *f* de películas por haz electrónico
i analisi *f* di pellicole per fascio elettronico
n elektronische filmaftasting
d elektronische Filmabtastung *f*

1741 ELECTRON-BEAM FOCUSING crt
f focalisation *f* de faisceau électronique
e enfoque *m* de haz electrónico
i focalizzazione *f* di fascio elettronico
n elektronenbundelfocussering
d Elektronenstrahlfokussierung *f*, Elektronenstrahlkonzentration *f*

1742 ELECTRON-BEAM RECORDING, ELECTRON-BEAM TELERECORDING vr
Telerecording in which the light-sensitive film passes into a vacuum chamber where it is directly exposed by an electron beam modulated by the video signal.
f enregistrement *m* par faisceau électronique
e registro *m* por haz electrónico
i registrazione *f* mediante fascio elettronico
n opname met elektronenbundel
d Elektronenstrahlaufnahme *f*

1743 ELECTRON-BEAM TRANSMISSION EFFICIENCY, ELECTRON-STREAM TRANSMISSION EFFICIENCY crt
At an electrode through which the electron beam passes, the ratio of the average current leaving the electrode to the average current density at the cathode surface.
f rendement *m* de transmission du faisceau électronique
e rendimiento *m* de transmisión del haz electrónico
i rendimento *m* di trasmissione del fascio elettronico
n transmissierendement *n* van de elektronenbundel
d Übertragungswirkungsgrad *m* des Elektronenstrahls

1744 ELECTRON-BEAM VOLTAGE, ELECTRON-STREAM POTENTIAL crt
At any point in an electron beam, the time average of the voltage between that point and the electron-emitting surface.
f tension *f* du faisceau électronique
e tensión *f* del haz electrónico
i tensione *f* del fascio elettronico
n elektronenbundelspanning
d Elektronenstrahlspannung *f*

1745 ELECTRON CAMERA, ELECTRONIC CAMERA, RADIO CAMERA rec/tv
A portable camera for recording news items.
f caméra *f* électronique
e cámara *f* electrónica
i camera *f* elettronica
n elektronische camera
d Elektronenkamera *f*, elektronischer Bildfänger *m*

1746 ELECTRON CLOUD crt
The total of negatively charged electrons emitted by the cathode and occupying the space between cathode and anode.
f nuage *m* électronique
e nube *f* de electrones
i nuvola *f* elettronica
n elektronenwolk
d Elektronenwolke *f*

1747 ELECTRON CURRENT crt
f courant *m* électronique
e corriente *f* electrónica

i corrente *f* elettronica
n elektronenstroom
d Elektronenstrom *m*

1748 ELECTRON-CURRENT DENSITY crt
f densité *f* du courant électronique
e densidad *f* de la corriente electrónica
i densità *f* della corrente elettronica
n elektronenstroomdichtheid
d Elektronenstromdichte *f*

1749 ELECTRON DENSITY crt
f densité *f* d'électrons
e densidad *f* de electrones
i densità *f* d'elettroni
n elektronendichtheid
d Elektronendichte *f*

1750 ELECTRON-EMITTING AREA, crt/ge
ELECTRON-EMITTING SURFACE
f surface *f* d'émission électronique
e superficie *f* emisora de electrones
i superficie *f* emettitrice d'elettroni
n elektronenemitterend oppervlak *n*
d Elektronenabgabefläche *f*

1751 ELECTRON-EMITTING CATHODE crt/ge
f cathode *f* à émission électronique
e cátodo *m* de emisión electrónica
i catodo *m* ad emissione elettronica
n elektronenemitterende katode
d Elektronenemissionskatode *f*

1752 ELECTRON GUN crt
An electrode structure which produces and may deflect, focus and control the position and intensity of an electron beam.
f canon *m* électronique
e cañón *m* electrónico
i cannone *m* elettronico
n elektronenkanon *n*
d Elektronenkanone *f*,
Elektronenschleuder *f*

1753 ELECTRON-GUN CONVERGENCE RATIO (GB), crt
ELECTRON-GUN DENSITY MULTIPLICATION (US)
The ratio of the average current density at any specified aperture through which the electron beam passes to the average current density at the cathode.
f rapport *m* de densité du faisceau électronique
e relación *f* de densidad del haz electrónico
i rapporto *m* di densità del fascio elettronico
n elektronenbundeldichtheidsverhouding
d Elektronenstrahldichteverhältnis *n*

1754 ELECTRON-GUN CURRENT crt
Total electronic current flowing to the anode, part of which forms the beam current.
f courant *m* global du faisceau
e corriente *f* total del haz
i corrente *f* totale del fascio
n totale bundelstroom
d gesamter Strahlstrom *m*

1755 ELECTRON IMAGE, opt
ELECTRONIC IMAGE
The virtual image formed in a beam of electrons in the cross-section of which the electron density is proportional to the luminance of the optical image which gave rise to the beam.
f image *f* électronique
e imagen *f* electrónica
i immagine *f* elettronica
n elektronenbeeld *n*,
ladingsbeeld *n*
d Elektronenabbildung *f*,
Elektronenbild *n*

1756 ELECTRON IMAGE AMPLIFIER crt
f amplificateur *m* de l'image électronique
e amplificador *m* de la imagen electrónica
i amplificatore *m* dell'immagine elettronica
n elektronenbeeldversterker
d Elektronenbildverstärker *m*

1757 ELECTRON IMAGE STABILIZATION tv
f stabilisation *f* de l'image électronique
e estabilización *f* de la imagen electrónica
i stabilizzazione *f* dell'immagine elettronica
n stabilisatie van het elektronenbeeld
d Elektronenbildstabilisierung *f*

1758 ELECTRON IMAGE TUBE crt
f tube *m* à image électronique
e tubo *m* de imagen electrónica
i tubo *m* ad immagine elettronica
n elektronenbeeldbuis
d Elektronenbildröhre *f*,
Elektronenbildwandler *m*

1759 ELECTRON IMPACT, crt
ELECTRON IMPINGEMENT
The impingement of the electron beam on the screen of the cathode-ray tube.
f impact *m* d'électrons
e impacto *m* de electrones
i rimbalzo *m* d'elettroni
n optreffen *n* van de elektronen
d Elektronenaufprall *m*,
Elektronenaufschlag *m*

1760 ELECTRON LENS opt
A device consisting of magnetic coils and electrodes used to reflect and focus electron beams in a manner closely resembling the action of an optical lens on beams.
f lentille *f* électronique
e lente *f* electrónica
i lente *f* elettronica
n elektronenlens
d Elektronenlinse *f*

1761 ELECTRON MULTIPLIER, cpl
SECONDARY EMISSION MULTIPLIER

A secondary emission tube used e.g. in telecine systems.
f multiplicateur *m* d'électrons secondaires
e multiplicador *m* de electrones secundarios
i moltiplicatore *m* d'elettroni secondari
n secundaire-elektronenvermenigvuldiger
d Sekundärelektronenvervielfacher *m*

1762 ELECTRON-OPTICAL ABERRATION dis/opt
Defect of an electron lens giving rise to distortions, analogous to optical aberration.
f aberration *f* électronoptique
e aberración *f* electronóptica
i aberrazione *f* elettronottica
n elektronenoptische aberratie
d elektronenoptische Aberration *f*

1763 ELECTRON-OPTICAL ANASTIGMATIC SYSTEM opt
An electron optical system that has been corrected for astigmatism.
f système *m* électronoptique anastigmatique
e sistema *m* electronóptico anastigmático
i sistema *m* elettronottico anastigmatico
n anastigmatisch elektronenoptisch systeem *n*
d anastigmatisches elektronenoptisches System *n*

1764 ELECTRON-OPTICAL ASTIGMATISM opt
In an electron-beam tube, a focus defect in which electrons in different axial planes come to focus at different points.
f astigmatisme *m* électronoptique
e astigmatismo *m* electronóptico
i astigmatismo *m* elettronottico
n elektronenoptisch astigmatisme *n*
d elektronenoptischer Astigmatismus *m*

1765 ELECTRON-OPTICAL IMAGE tv
Optical image produced by the electron spot on the luminous screen of the receiver.
f image *f* électronoptique
e imagen *f* electronóptica
i immagine *f* elettronottica
n elektronenoptisch beeld *n*
d elektronenoptisches Bild *n*

1766 ELECTRON OPTICS crt
Study of the behavio(u)r of electrons and electron beams under the influence of electric and magnetic fields.
f optique *f* électronique
e óptica *f* electrónica
i ottica *f* elettronica
n elektronenoptiek
d Elektronenoptik *f*

1767 ELECTRON PATH crt
f trajectoire *f* électronique
e trayectoria *f* de los electrones
i traiettoria *f* degli elettroni
n elektronenbaan
d Elektronenbahn *f*, Elektronenstrecke *f*

1768 ELECTRON SCANNING BEAM tv
f faisceau *m* électronique analyseur
e haz *m* electrónico explorador
i fascio *m* elettronico analizzatore
n elektronenaftastbundel
d Elektronenabtaststrahl *m*

1769 ELECTRON STREAM crt
f flux *m* électronique
e flujo *m* electrónico
i flusso *m* elettronico
n elektronenflux
d Elektronenströmung *f*

ELECTRON-STREAM POTENTIAL
see: ELECTRON-BEAM VOLTAGE

ELECTRON-STREAM TRANSMISSION EFFICIENCY
see: ELECTRON-BEAM TRANSMISSION EFFICIENCY

1770 ELECTRON SWITCH (GB), ELECTRONIC COMMUTATOR (US) tv
f commutateur *m* électronique
e conmutador *m* electrónico
i commutatore *m* elettronico
n elektronische schakelaar
d elektronischer Schalter *m*

1771 ELECTRON TELESCOPE crt
A telescope in which the infrared image of a distant object is focused on the photosensitive cathode of an image converter tube and the resulting electron image enlarged by electron lenses and displayed on a fluorescent screen.
f télescope *m* électronique
e telescopio *m* electrónico
i telescopio *m* elettronico
n elektronentelescoop
d Elektronenteleskop *n*

1772 ELECTRON TUBE, ELECTRONIC TUBE, ELECTRONIC VALVE cpl
f tube *m* électronique
e válvula *f* electrónica
i valvola *f* elettronica
n elektronenbuis
d Elektronenröhre *f*

1773 ELECTRONIC ADMITTANCE cpl
The admittance of a vacuum tube (valve) caused by the action of electron beams within the tube (valve).
f admittance *f* électronique
e admitancia *f* electrónica
i ammettenza *f* elettronica
n elektronische admittantie
d elektronischer Scheinleitwert *m*

ELECTRONIC COLO(U)R SAMPLER
see: COLO(U)R SAMPLER

ELECTRONIC COLO(U)R SAMPLING
see: COLO(U)R SAMPLING

1774 ELECTRONIC CONTROL rec/tv
Used in outside broadcasts by choosing the suitable gamma correction, control of exposure and black level and taking into account noise and definition.
f commande *f* électronique, réglage *m* électronique
e mando *m* electrónico, regulación *f* electrónica
i comando *m* elettronico, regolazione *f* elettronica
n elektronische besturing, elektronische regeling
d elektronische Regelung *f*, elektronische Steuerung *f*

1775 ELECTRONIC CONVERTER tv
A converter which carries out the line and/or field conversion entirely by electronic means.
f convertisseur *m* électronique
e convertidor *m* electrónico
i convertitore *m* elettronico
n elektronische omzetter
d elektronischer Wandler *m*

1776 ELECTRONIC CORRECTION OF COLO(U)R ERRORS ctv
f correction *f* électronique d'erreurs chromatiques
e corrección *f* electrónica de errores cromáticos
i correzione *f* elettronica d'errori cromatici
n elektronische correctie van kleur-afwijkingen
d elektronische Korrektur *f* von Farb-abweichungen

1777 ELECTRONIC EDITING vr
A method of adding to or replacing part of an existing record without cutting the tape, while maintaining the continuity of video, audio and control tracks.
f montage *m* électronique
e montaje *m* electrónico
i montaggio *m* elettronico
n elektronische montage
d elektronischer Bandschnitt *m*

1778 ELECTRONIC EDITOR vr
A device for assembling a master video-tape from studio signal sources or other tape segments with complete continuity of servo control signals, and without any physical cutting and splicing of the tape.
f appareil *m* de montage électronique
e aparato *m* de montaje electrónico
i apparecchio *m* di montaggio elettronico
n elektronisch montageapparaat *n*
d elektronischer Scheideapparat *m*

1779 ELECTRONIC FIELD FREQUENCY CONVERTER, ELECTRONIC FIELD RATE CONVERTER rec/rep/tv
An electronic digital device which converts the field frequency (rate) from 60 to 50 fields and then uses the line store method for line standard conversion.
f convertisseur *m* électronique de la fréquence de trame
e convertidor *m* electrónico de la frecuencia de cuadro
i convertitore *m* elettronico della frequenza di trama
n elektronische omzetter van de raster-frequentie
d elektronischer Teilbildfrequenzwandler *m*

1780 ELECTRONIC FLYING-SPOT SCANNER rec/tv
f analyseur *m* indirect électronique à spot mobile
e explorador *m* indirecto electrónico de punto móvil
i analizzatore *m* indiretto elettronico a punto mobile
n elektronische lichtpuntaftaster
d elektronischer Lichtpunktabtaster *m*

1781 ELECTRONIC IMAGE INSERTION, ELECTRONIC PICTURE INSERTION tv
Made possible by the sequential nature of the TV scanning process, together with the normal process of scan synchronization in cameras or other sources which contribute to a production.
f raccord *m* électronique d'une image
e inserción *f* electrónica de una imagen
i inserzione *f* elettronica di un'immagine
n elektronische beeldinlas
d elektronische Bildeinfügung *f*, elektronischer Zwischenschnitt *m*

1782 ELECTRONIC LINE SCANNING crt
That method of scanning which provides motion of the scanning spot along the scanning line by electronic means.
f analyse *f* électronique ligne
e exploración *f* electrónica línea
i analisi *f* elettronica linea
n elektronische lijnenaftasting
d elektronische Zeilenabtastung *f*

1783 ELECTRONIC MAGNIFICATION tv
The magnifying of the TV picture by electronic instead of by optical means.
f agrandissement *m* électronique
e agrandamiento *m* electrónico
i ingrandimento *m* elettronico
n elektronische vergroting
d elektronische Vergrösserung *f*

1784 ELECTRONIC MASKING ctv
Consists in adding to the obtained signal in colo(u)r telecine a mask-image signal drawn from all three channels.
f masquage *m* électronique
e enmascaramiento *m* electrónico
i mascheramento *m* elettronico
n elektronische maskering
d elektronische Verdeckung *f*

1785 ELECTRONIC NOISE ctv
Noise accompanying the colo(u)r difference signals.

f bruit *m* électronique
e ruido *m* electrónico
i rumore *m* elettronico
n electronische ruis
d elektronischer Rausch *m*

1786 ELECTRONIC POINTER, ELECTRONIC SPOTLIGHT rep/tv
A device to bring the viewer's attention to a special part of the picture by purely electronic means.
f projecteur *m* électronique d'index sur image
e aparato *m* electrónico de alumbrado intenso
i progettore *m* elettronico a fascio stretto di luce
n elektronisch spotlicht *n*, elektronische lichtpijl
d elektronischer Lichtzeiger *m*

1787 ELECTRONIC PROJECTION APPARATUS crt/opt/tv
Apparatus which represent an object or picture by means of free electrons, in the form of an electron image or a visible image.
f projecteur *m* électronoptique
e proyector *m* electronóptico
i progettore *m* elettronottico
n elektronenoptische projector
d elektronenoptisches Abbildungsgerät *n*

1788 ELECTRONIC RASTER SCANNING crt
That method of scanning in which motion of the scanning spot in both dimensions is accomplished by electronic means.
f analyse *f* électronique canevas
e exploración *f* electrónica cuadrícula
i analisi *f* elettronica quadro rigato
n elektronische rasteraftasting
d elektronische Teilbildabtastung *f*

1789 ELECTRONIC SCANNING tv
The dissection of the image at the transmission end or synthesis of the picture at the receiving end by means of a cathode-ray tube.
f analyse *f* électronique
e exploración *f* electrónica
i analisi *f* elettronica
n elektronische aftasting
d elektronische Abtastung *f*

1790 ELECTRONIC SPROCKET HOLES tv
A series of regular impulses generated in the camera and recorded on the tape in such a way that they do not interfere with the normal sound record; used for synchronization.
f marques *pl* d'entraînement enregistrées par voie électronique
e marcas *pl* de arrastre registradas per vía electrónica
i marche *pl* di trascinamento registrate per via elettronica
n elektronisch geproduceerde geleidetekens *pl*
d elektronisch hergestellte Transportzeichen *pl*

1791 ELECTRONIC TEST PATTERN cpl/tv
f mire *f* électronique
e imagen *f* de prueba electrónica
i figura *f* di prova elettronica
n elektronisch toetsbeeld *n*
d elektronisches Testbild *n*

1792 ELECTRONIC VIDEO RECORDING, EVR vr
f enregistrement *m* sur film par faisceau électronique
e registro *m* sobre película por haz electrónico
i registrazione *f* su pellicola per fascio elettronico
n elektronische filmopname
d elektronische Filmaufzeichnung *f*

1793 ELECTRONIC VIEWFINDER rep/tv
A TV camera viewfinder using a small cathode-ray picture tube to show the image being picked up.
f viseur *m* électronique
e visor *m* electrónico
i mirino *m* elettronico, visore *m* elettronico
n elektronische zoeker
d elektronischer Sucher *m*

1794 ELECTRONICALLY CONTROLLED KEYBOARD rep/tv
f clavier *m* à commande électronique
e teclado *m* de mando electrónico
i tastiera *f* a comando elettronico
n elektronisch bediend toetsenbord *n*
d elektronisch betätigtes Tastenbrett *n*

1795 ELECTRONICS OF THE CAMERA rec/tv
f électronique *f* de la caméra
e electrónica *f* de la cámara
i elettronica *f* della camera
n camera-elektronica
d Kamera-Elektronik *f*

1796 ELECTROSTATIC DEFLECTION crt
f balayage *m* électrostatique, déviation *f* électrostatique
e barrido *m* electroestático, desviación *f* electroestática
i deflessione *f* elettrostatica, deviazione *f* elettrostatica
n elektrostatische afbuiging
d elektrostatische Ablenkung *f*

1797 ELECTROSTATIC FOCUSING crt
Focusing an electron beam by the action of an electrostatic lens.
f focalisation *f* électrostatique
e enfoque *m* electroestático
i focalizzazione *f* elettrostatica
n elektrostatische focussering
d elektronische Fokussierung *f*, elektronische Strahlbündelung *f*

1798 ELECTROSTATIC LENS crt/opt
f lentille *f* électronique à focalisation électrostatique
e lente *f* electrónica de enfoque electroestático

i lente *f* elettronica a focalizzazione elettrostatica
n elektronenlens met elektrostatische focussering
d Elektronenlinse *f* mit elektrostatischer Fokussierung

1799 ELEMENTAL AREA crt
f élément *m* d'image, point *m* d'image
e elemento *m* de imagen, punto *m* de imagen
i elemento *m* d'immagine, punto *m* d'immagine
n beeldelement *n*, beeldpunt *n*
d Bildpunkt *m*

1800 ELEMENTS OF LINE SIGNAL WAVEFORM, SIGNAL WAVEFORM ELEMENTS cpl
f éléments *pl* de forme d'onde du signal
e elementos *pl* de forma de onda de la señal
i elementi *pl* di forma d'onda del segnale
n elementen *pl* van de golfvorm van het signaal
d Elemente *pl* der Wellenform des Signals

1801 ELLIPTICAL SUBCARRIER cpl/ctv
Chrominance subcarrier signal where the reproduced luminance depends upon subcarrier phase with an elliptical relationship.
f porteuse *f* elliptique de chrominance
e portadora *f* elíptica de crominancia
i portante *f* ellittica di crominanza
n elliptische chrominantiedraaggolf
d elliptischer Farbhilfsträger *m*

EMERGENCE ANGLE
see: ANGLE OF EMERGENCE

1802 EMERGENCY BYPASS tv
A switching system which, under emergency conditions routes an input to the output.
f dérivation *f* de secours
e derivación *f* de emergencia
i derivazione *f* d'emergenza
n noodomleiding
d Notumleitung *f*

1803 EMERGENT RAY opt
f rayon *m* émergent
e rayo *m* emergente
i raggio *m* emergente
n uittredende straal
d ausfallender Strahl *m*

1804 EMISSION OF LIGHT, LIGHT EMISSION opt
f émission *f* de lumière
e emisión *f* de luz
i emissione *f* di luce
n lichtemissie
d Lichtemission *f*

1805 ENCODED COLO(U)R CORRECTION SYSTEM ctv
f système *m* de correction de couleur
e sistema *m* de corrección de color
i sistema *m* di correzione di colore
n kleurcorrectiesysteem *n*
d FBAS-Farbkorrektursystem *n*

ENCODER
see: CODER

1806 ENCORE the
f bis *m*
e bis *m*
i bis *m*
n toegift
d Zugabe *f*

1807 END-FIRE AERIAL ARRAY, END-FIRE ANTENNA ARRAY aea
A linear array in which the direction of maximum radiation is along the axis of the array.
f réseau *m* d'antennes à rayonnement longitudinal
e red *f* de antenas de radiación longitudinal
i rete *f* d'antenne a radiazione longitudinale
n antennesysteem *n* met straling in de langsrichting
d Längsrichtstrahler *m*, Längsstrahler *m*

1808 END OF PROGRAM(ME) ge
f fin *f* d'émission, fin *f* de programme
e fin *f* de programa
i fine *f* di programma
n einde *n* van de uitzending, programma-einde *n*
d Ende *n* einer Sendung, Sendeschluss *m*

1809 END OF TAPE vr
f fin *f* de bande
e fin *f* de cinta
i fine *f* di nastro
n bandeinde *n*
d Bandende *n*

1810 END TITLE fi/tv
f générique *m* de fin
e título *m* de fin
i titolo *m* di fine
n eindtitel
d Schlusstitel *m*

1811 ENDOSCOPIC PICTURE med/tv
f image *f* endoscopique
e imagen *f* endoscópica
i immagine *f* endoscopica
n endoscopisch beeld *n*
d endoskopisches Bild *n*

ENERGY RECOVERY DIODE
see: BOOSTER DIODE

ENERGY REGENERATION
see: BOOSTER CIRCUIT

1812 ENVELOPE cpl
A graph defining the variations in amplitude of successive oscillations in an amplitude-modulated wave.
f enveloppante *f*
e envolvente *m*
i curva *f* inviluppo,
inviluppo *m*
n omhullende
d Hüllkurve *f*,
Umhüllende *f*

1813 ENVELOPE crt
The gas-tight container of the cathode-ray tube.
f ampoule *f*
e ampolla *f*
i bulbo *m*
n ballon
d Kolben *m*

1814 ENVELOPE DELAY, cpl/tv
ENVELOPE DELAY TIME,
GROUP DELAY TIME
The time of propagation of a certain feature of a wave envelope.
f temps *m* de transit de groupe
e tiempo *m* de tránsito de grupo
i tempo *m* di transito di gruppo
n groeplooptijd
d Gruppenlaufzeit *f*

1815 ENVELOPE DEMODULATION, tv
ENVELOPE DETECTION
f détection *f* de phase,
redressement *m* d'enveloppante
e detección *f* de fase,
rectificación *f* de envolvente
i modulazione *f* della curva inviluppo,
raddrizzamento *m* della curva inviluppo
n omhullendegelijkrichting
d Hüllkurvengleichrichtung *f*

1816 ENVELOPE DEMODULATOR, rep/tv
ENVELOPE DETECTOR
A diode detector whose output is shunted by a capacitor, so as to make the output proportional to the peaks of the rectified amplitude-modulated carrier.
f démodulateur *m* de l'enveloppante
e demodulador *m* del envolvente
i demodulatore *m* dell'inviluppo
n demodulator van de omhullende
d Hüllkurvendemodulator *m*,
Hüllkurvengleichrichter *m*

1817 ENVELOPE DISTORTION cpl
f distorsion *f* du temps de transit de groupe
e distorsión *f* del tiempo de tránsito de grupo
i distorsione *f* del tempo di transito di gruppo
n groeplooptijdvervorming
d Gruppenlaufzeitverzerrung *f*

1818 ENVELOPE VELOCITY, cpl/tv
GROUP VELOCITY
The velocity of propagation of a certain feature of the wave envelope.
f vitesse *f* de groupe
e velocidad *f* de grupo
i velocità *f* di gruppo
n groepsnelheid
d Gruppengeschwindigkeit *f*

1819 EQUAL-ENERGY SOURCE ct
A light source of which the colo(u)r is a theoretically perfect white.
f source *f* d'énergie égale,
source *f* équiénergie
e fuente *f* equienergética
i sorgente *f* d'energia uguale
n bron van gelijke energie
d energiegleiche Quelle *f*,
Gleichenergiequelle *f*

1820 EQUAL-ENERGY WHITE, tv
ILLUMINANT E
Uniform radiance on all wavelengths.
f blanc *m* d'énergie égale,
blanc *m* équiénergie
e blanco *m* equienergético
i bianco *m* d'energia uguale
n E-wit *n*,
wit *n* van gelijke energie
d energiegleiches Weiss *n*,
Gleichenergieweiss *n*

1821 EQUAL-SIGNAL WHITE tv
Total radiance when the signals have the same value in all channels used in colo(u)r TV.
f blanc *m* moyen
e blanco *m* medio
i bianco *m* medio
n middelwit *n*
d Mittelweiss *n*

1822 EQUALIZATION cpl/dis
In electronics, the reduction of distortion by the introduction of networks which compensate for the particular type of distortion over the required frequency band.
f correction *f* de distorsion,
réduction *f* de distorsion
e reducción *f* de distorsión
i riduzione *f* di distorsione
n vervormingscorrectie,
vervormingsreductie
d Entzerrung *f*

1823 EQUALIZATION dis
Technique of improving sound signal-to-noise ratio.
f amélioration *f* du rapport signal sonore-bruit
e mejoramiento *m* de la relación señal sonora-ruido
i miglioramento *m* del rapporto segnale sonoro-rumore
n verbetering van de geluidssignaal-ruisverhouding
d Verbesserung *f* Tonsignal-Rausch-verhältnis

1824 EQUALIZER cpl/dis
An electrical network designed to correct for unequal attenuation or phase shift in the transmission of signals over wires and cables.
f circuit *m* correcteur de distorsion
e circuito *m* corrector de distorsión
i circuito *m* antidistorcente
n egalisator, vervormingscorrectiekring
d Entzerrerschaltung *f*

1825 EQUALIZING AMPLIFIER cpl
f amplificateur *m* correcteur
e amplificador *m* corrector
i amplificatore *m* correttore
n vereffeningsversterker
d Entzerrerverstärker *m*

1826 EQUALIZING PULSES cpl
In some TV systems, pulses at twice the line frequency occurring just before and after the field synchronization signal.
f impulsions *pl* d'égalisation
e impulsos *pl* de igualación
i impulsi *pl* d'uguagliamento
n egalisatie-impulsen *pl*, vereffeningsimpulsen *pl*
d Ausgleichimpulse *pl*

1827 EQUI-BAND CODER ctv
Colo(u)r signal coder where both of the chrominance signal components have the same bandwidth.
f codeur *m* à largeurs de bande égales
e codificador *m* con anchos de banda iguales
i codificatore *m* con larghezze di banda uguali
n codeur met gelijke bandbreedten
d Äquibandkoder *m*, Koder *m* mit gleichen Bandbreiten

1828 EQUI-BAND COLO(U)R TELEVISION RECEIVER rep/ctv
A receiver in which the two chrominance signal channels have equal bandwidths.
f téléviseur *m* couleur à largeurs de bande égales
e televisor *m* en colores de anchos de banda iguales
i televisore *m* a colori a larghezze di banda uguali
n kleurentelevisieontvanger met gelijke bandbreedten
d Äquibandfarbfernsehempfänger *m*, Farbfernsehempfänger *m* mit gleichen Bandbreiten

1829 EQUI-BAND DEMODULATOR cpl
f démodulateur *m* à largeurs de bande égales pour les deux composantes du signal de chrominance
e desmodulador *m* con anchos de banda iguales para las dos componentes de la señal de crominancia
i demodulatore *m* a larghezze di banda uguali per le due componenti del segnale di crominanza
n demodulator met gelijke bandbreedten voor beide componenten van het kleursoortsignaal
d Äquibanddemodulator *m*

1830 EQUI-BAND OPERATION cpl
f opération *f* à largeurs de bande égales
e operación *f* con anchos de banda iguales
i operazione *f* a larghezze di banda uguali
n systeem *n* met gelijke bandbreedten
d Äquibandbetrieb *m*

1831 EQUI-ENERGY SPECTRUM ct
Spectrum in which the spectral concentration of energy evaluated on a wavelength basis is constant throughout the visible region.
f spectre *m* d'énergie égale
e espectro *m* de energía igual
i spettro *m* d'energia uguale
n spectrum *n* van gelijke energie
d energiegleiches Spektrum *n*, Gleichenergiespektrum *n*

1832 EQUIVALENT LIMIT FREQUENCY cpl
f fréquence *f* limite équivalente
e frecuencia *f* límite equivalente
i frequenza *f* limite effettiva
n equivalente grensfrequentie
d äquivalente Grenzfrequenz *f*

1833 EQUIVALENT LOUDNESS, LOUDNESS LEVEL aud
f niveau *m* de l'intensité sonore
e nivel *m* de la intensidad sonora
i livello *m* dell'intensità sonora
n geluidssterkteniveau *n*
d Lautstärkepegel *m*

1834 EQUIVALENT VEILING LUMINANCE ct
Luminance which has to be added, by superposition, to the luminance of both the adapting background and the object in order to make the luminance difference threshold in the absence of disability glare the same as that experienced in the presence of disability glare.
f luminance *f* équivalente de voile
e luminancia *f* equivalente de celado
i luminanza *f* equivalente di velatura
n equivalente luminantie bij sluiereffect
d Äquivalenzhelligkeit *f* bei Schleiereffekt

1835 ERASING, WIPING rec/rep
Process of demagnetizing a recording on magnetic tape.
f effacement *m*
e borrado *m*
i cancellazione *f*
n wissen *n*
d Löschung *f*

1836 ERASING CURRENT, WIPING CURRENT vr
f tension *f* d'effacement
e tensión *f* de borrado

i tensione *f* di cancellazione
n wisspanning
d Löschspannung *f*

1837 ERASING HEAD, WIPING HEAD aud/vr
The component in a magnetic recording equipment that obliterates previous recordings.
f tête *f* d'effacement
e cabeza *f* borradora
i testa *f* di cancellazione
n wiskop
d Löschkopf *m*

1838 EUROPIUM PICTURE TUBE crt
f tube *m* image à europium
e tubo *m* imagen de europio
i tubo *m* immagine ad europio
n europiumbeeldbuis
d Europiumbildröhre *f*

1839 EUROVISION tv
System for relays of TV program(me)s through the EBU between various countries.
f eurovision *f*
e eurovisión *f*
i eurovisione *f*
n eurovisie
d Eurovision *f*

1840 EVEN-LINE INTERLACE rec/tv
f analyse *f* entrelacée à nombre pair de lignes
e exploración *f* entrelazada de número par de líneas
i analisi *f* interlacciata a numero pari
n interliniëring met even aantal lijnen
d geradzahliger Zeilensprung *m*

1841 EVEN NUMBER OF LINES rec/tv
f nombre *m* de lignes pair
e número *m* de líneas par
i numero *m* di linee pari
n even aantal *n* lijnen
d gerade Zeilenzahl *f*

EVR
see: ELECTRONIC VIDEO RECORDING

1842 EXCESS NOISE dis
Noise produced e.g. in carbon resistors, carbon microphones, etc.
f bruit *m* de scintillation
e ruido *m* de centelleo
i rumore *m* di scintillazione
n gekraak *n*, knetterruis
d Funkelrauschen *n*

1843 EXCESSIVE SIBILANCE aud
A speech sound having an excessive hissing effect.
f effet *m* sibilant excessif
e efecto *m* sibilante excesivo
i effetto *m* sibilante eccessivo
n bovenmatige sisklank
d übermässiger Zischlaut *m*

1844 EXCITATION crt/ge
f excitation *f*
e excitación *f*
i eccitazione *f*
n aanslaan *n*, aanslag, excitatie
d Anregung *f*

EXCITATION
see: DRIVE

1845 EXCITATION PURITY ct
The ratio of the distances, measured on the C.I.E. chromaticity diagram, from the adopted achromatic stimulus to the sample stimulus and to the stimulus lying on the spectrum locus or the straight line joining its extremes which, by additive mixture with the adopted achromatic stimulus, can form a match with the sample stimulus.
f facteur *m* de pureté d'excitation
e factor *m* de pureza de excitación
i fattore *m* di purezza d'eccitazione
n verzadigingsgraad in afstandseenheden
d Farbe *f* maximaler Sättigung, spektrale Farbdichte *f*

1846 EXCITATION TIME OF A FLUORESCENT SCREEN crt
f temps *m* d'excitation d'un écran fluorescent
e tiempo *m* de excitación de una pantalla fluorescente
i tempo *m* d'eccitazione d'uno schermo fluorescente
n aanslagtijd van een fluorescentiescherm
d Anregungszeit *f* eines Leuchtschirms

1847 EXCITED PHOSPHOR crt
f substance *f* fluorescente excitée
e substancia *f* fluorescente excitada
i sostanza *f* fluorescente eccitata
n aangeslagen fluorescerende stof
d angeregter Leuchtstoff *m*

EXCITER (US)
see: ACTIVE AERIAL

1848 EXPANDER cpl
Circuit devised to expand the volume range in reception and compensate for compression in transmission.
f expanseur *m*
e expansor *m* dinámico
i espansore *m* di dinamica
n dynamiekverbreder
d Dynamikdehner *m*

1849 EXPANSION cpl
f expansion *f*
e expansión *f*
i espansione *f*
n dynamiekverbreding
d Dynamikdehnung *f*

1850 EXTENDED FIELD COVERAGE opt
f champ *m* de vue élargi
e campo *m* de visión extendido

i　campo *m* di visione esteso
n　vergroot gezichtsveld *n*,
　verruimd gezichtsveld *n*
d　erweiterter Blickfeld *n*

1851　EXTERNAL COMMUNICATIONS　rec tv
　Communication required by the studio personnel with external areas.
f　connexions *pl* vers l'extérieur
e　conexiones *pl* hacia el exterior
i　connessioni *pl* verso l'esteriore
n　verbindingen *pl* naar buiten
d　Verbindungen *pl* nach aussen

EXTRA HIGH TENSION
　see: EHT

1852　EXTREME LONG SHOT　opt
f　plan *m* de grand ensemble
e　plano *m* lejano extremo
i　campo *m* estremo lungo
n　opname op grote afstand
d　Weiteinstellung *f*

1853　EYE　ct
f　oeuil *m*
e　ojo *m*
i　occhio *m*
n　oog *n*
d　Auge *n*

1854　EYE LIGHT　stu
　Light source intended to give a reflection from the eyes of an artist without increasing the general light level.
f　lumière *f* éclairant les yeux de l'acteur
e　luz *f* para alumbrar los ojos del actor
i　luce *f* per illuminare gli occhi dell'attore
n　oogreflectielicht *n*
d　Augenlicht *n*

1855　EYESTRAIN　ct
f　effort *m* excessif des yeux
e　esfuerzo *m* excesivo de los ojos
i　fatica *f* eccessiva degli occhi
n　overinspanning van de ogen
d　Überanstrengung *f* der Augen

F

1856 FACE, crt
FACE-PLATE
Of a cathode-ray tube, a glass plate carrying the fluorescent screen and manufactured separately for attachment to the envelope.
f fenêtre *f*, fond *m*
e cara *f* frontal, fondo *m*, ventana *f*
i finestra *f*, piastra *f* frontale
n frontplaat, venster *n*
d Schirmträger *m*, Stirnfläche *f*

1857 FACE TONE, stu
FLESH TONE
f ton *m* chair
e color *m* del rostro
i carnagione *f*, incarnato *m*
n gelaatskleur
d Gesichtsfarbe *f*

1858 FACTOR OF MERIT, ge
FIGURE OF MERIT,
Q FACTOR
f coefficient *m* de qualité,
nombre *m* de mérite
e factor *m* de calidad,
factor *m* de mérito,
factor *m* Q
i fattore *m* di merito,
fattore *m* di qualità
n kwaliteitsfactor,
Q-factor
d Güteziffer *f*,
Q-Faktor *m*

1859 FADE-DOWN aud/rec/rep/tv
To reduce gradually to a lower level the output of a channel.
f abaissement *m* graduel du niveau
e variación *f* gradual del nivel
i variazione *f* graduale del livello
n neerregelen *n*
d Abschwächen *n*

1860 FADE-DOWN tv
The gradual disappearance of an image from top to bottom.
f disparition *f* graduelle en bas
e desaparición *f* gradual hacia abajo
i dissolvenza *f* in chiusura discendente
n langzaam verdwijnen *n* naar beneden
d Ausblendung *f* nach unten

1861 FADE-IN rec/rep/tv
The gradual appearance of an image.
f apparition *f* graduelle
e aparición *f* gradual
i dissolvenza *f* d'inizio,
dissolvenza *f* in apertura
n langzaam opkomen *n*
d Einblendung *f*

1862 FADE-IN AND -OUT rep/tv
The gradual appearance and disappearance of an image.
f apparition *f* et disparition graduelle
e aparición *f* y desaparición gradual
i dissolvenza *f* in apertura e chiusura
n langzaam opkomen *n* en verdwijnen *n*
d Überblendung *f*

1863 FADE-OUT rec/rep/tv
The gradual disappearance of an image.
f disparition *f* graduelle
e desaparición *f* gradual
i dissolvenza *f* di fine,
dissolvenza *f* in chiusura
n langzaam verdwijnen *n*
d Ausblendung *f*

1864 FADE-UP aud/rec/rep/tv
To increase gradually to a higher level the output of a channel.
f relèvement *m* graduel du niveau
e variación *f* gradual del nivel
i variazione *f* graduale del livello
n opregelen *n*
d Verstärkung *f*

1865 FADE-UP tv
The gradual appearance of an image from bottom to top.
f apparition *f* graduelle en haut
e aparición *f* gradual ascendente
i dissolvenza *f* in apertura ascendente
n langzaam opkomen *n* naar boven
d Einblendung *f* nach oben

1866 FADER rec/tv
Attenuation used in studio equipment for fading in and fading out a sound or vision program(me) or for introducing special effects.
f régleur *m* du niveau acoustique
e regulador *m* de nivel acústico
i regolatore *m* di livello acustico
n geluidsniveauregelaar
d Schallpegelregler *m*,
Tonpegelregler *m*

1867 FADING AMPLIFIER cpl/tv
Amplifier used in video mixing circuits to fade the vision signal up or down.
f amplificateur *m* régleur du niveau de sortie
e amplificador *m* regulador del nivel de salida
i amplificatore *m* regolatore del livello d'uscita
n regelversterker van het uitgangsniveau
d Regelverstärker *m* des Ausgangspegels

FALL TIME
see: DECAY TIME

1868 FALSE LINE LOCK dis
In a TV receiver a critical setting of the line-hold control, which results in a line lock with a part of the picture to the left and a part to the right of a dark and broad vertical bar.
f décalage *m* de phase du signal de synchronisation ligne dû à réglage erroné
e desplazamiento *m* de fase de la señal de sincronización línea debido a regulación errónea
i spostamento *m* di fase del sincronismo orizzontale dovuto a regolazione cattiva
n fazeverschuiving van het lijnsynchronisatiesignaal door foutieve regeling
d Phasenverschiebung *f* des Zeilensynchronisierungssignals durch fehlerhafte Regelung

1869 FAM, cpl
SIMULTANEOUS FREQUENCY AND AMPLITUDE MODULATION
f modulation *f* de fréquence et d'amplitude simultanée
e modulación *f* de frecuencia y de amplitud simultánea
i modulazione *f* di frequenza e d'ampiezza simultanea
n gelijktijdige frequentie- en amplitudemodulatie
d gleichzeitige Frequenz- und Amplitudenmodulation *f*

1870 FAST PULL-DOWN fi/tv
f escamotage *m* rapide
e empuje *m* rápido hacia abajo
i trasporto *m* rapido verso il basso
n snelle schoksgewijze filmvoortbeweging
d schnelle Filmfortschaltung *f*

FATIGUE
see: DARK BURN

1871 FAULT CAPTION dis
f diapositive *f* d'excuses
e diapositiva *f* de falta
i diapositiva *f* di guasto
n storingsdia
d Störungsdia *n*

1872 FAULT FINDING (GB), dis/svs
TROUBLE HUNTING (US),
TROUBLE SHOOTING (US)
f dépannage *m*,
recherche *f* des dérangements
e detección *f* de las faltas,
detección *f* de las fugas
i ricerca *f* dei guasti
n ontstoring
d Fehlersuche *f*,
Störungsbeseitigung *f*

FEE TELEVISION
see: COIN-FREED TELEVISION

1873 FEED (TO), aea/ge
INJECT (TO)
To supply a signal to a circuit, transmission line or aerial (antenna).
f alimenter v, contribuer v
e alimentar v, inyectar v
i alimentare v, immettere v
n invoeren v, toevoeren v
d einspeisen v, zuführen v

1874 FEED IN (TO) aud/vr
To bring a tape e.g. into a cassette.
f insérer v
e inserir v
i inserire v
n inleggen v
d einlegen v

1875 FEED REEL, vr
FEED SPOOL
The reel on a tape recorder that supplies the magnetic tape to the recording or playback head.
f bobine *f* débitrice,
bobine *f* dérouleuse
e bobina *f* desenrolladora
i bobina *f* svolgitrice
n afwikkelspoel
d Abwickelspule *f*

1876 FEEDBACK cpl
Return of a fraction of the voltage or current from the output circuit of an amplifier to the input circuit.
f réaction *f*
e reacción *f*
i reazione *f*
n terugkoppeling
d Rückkopplung *f*

1877 FEEDBACK ADMITTANCE (GB), cpl
SUSCEPTANCE (US)
A supplementary admittance in a circuit, due to the feedback effect of the anode of an electronic tube (valve).
f admittance *f* de réaction
e admitancia *f* de reacción
i ammettenza *f* di reazione
n terugkoppeladmittantie
d Rückwirkungsleitwert *m*

1878 FEEDBACK CIRCUIT rec/stu
A circuit by which the personnel at a program(me) source can hear a program(me) originally elsewhere.
f retour *m* d'écoute
e circuito *m* de escucha
i circuito *m* di ritorno acustico
n meeluisterverbinding
d Einspielkreis *m*

1879 FEEDBACK EQUALIZER cpl/vr
Negative feedback loop in an audio amplifier which incorporates combinations of resistance and capacitance to boost, attenuate or equalize selected frequencies within the range.
f circuit *m* d'égalisation à réaction négative
e circuito *m* igualador de reacción negativa
i circuito *m* uguagliatore a reazione negativa
n vereffeningsschakeling met negatieve terugkoppeling

d Ausgleichskreis *m* mit negativer Rückkopplung

1880 FEEDER aea
The means by which radio-frequency energy is supplied from the transmitter to the aerial (antenna).
f câble *m* d'alimentation
e cable *m* de alimentación, cable *m* de suministro
i alimentatore *m*, cavo *m* d'alimentazione
n voedingskabel
d Speisekabel *n*

1881 FIBER(RE) CORE crt
Used in cathode-ray tubes.
f noyau *m* à fibres
e núcleo *m* de fibras
i nucleo *m* di fibre
n vezelkern
d Faserkern *m*

1882 FICKLE COLO(U)R PATTERN svs
f image *f* en couleur irrégulière
e imagen *f* en color irregular
i immagine *f* in colore irregolare
n grillig kleurenpatroon *n*
d ungeordnetes Farbenbild *n*

1883 FICTITIOUS PRIMARIES, IMAGINARY PRIMARY COLO(U)RS, NON-PHYSICAL PRIMARIES tv
Primary colo(u)rs that are not generatable physically.
f couleurs *pl* primaires imaginaires
e colores *pl* primarios imaginarios
i colori *pl* primari immaginari
n denkbeeldige primaire kleuren *pl*
d imaginäre Primärfarben *pl*

1884 FIDELITY ge
The exactness of reproducibility of the input signal at the output end of a system transmitting information.
f fidélité *f*, qualité *f* de la reproduction
e calidad *f* de la reproducción, fidelidad *f*
i fedeltà *f*, qualità *f* della riproduzione
n natuurgetrouwheid
d Wiedergabegüte *f*

1885 FIDELITY OF FREQUENCY RESPONSE cpl
f précision *f* en fréquence
e fidelidad *f* de respuesta
i fedeltà *f* di risposta
n responsiegetrouwheid
d Frequenztreue *f*

1886 FIELD rec/rep/tv
In monochrome TV, a subdivision of the complete TV picture consisting of a series of sequentially scanned lines spaced equidistantly over the whole picture area, the repetition rate of the series being a multiple of that for the picture.
f trame *f*
e cuadro *m*
i trama *f*
n raster *n*
d Halbbild *n*, Teilbild *n*

1887 FIELD AMPLITUDE, VERTICAL AMPLITUDE tv
f amplitude *f* verticale
e amplitud *f* vertical
i ampiezza *f* verticale
n rasteramplitude, verticale amplitude
d Vertikalamplitude *f*

1888 FIELD AMPLITUDE CONTROL, VERTICAL AMPLITUDE CONTROL cpl
f réglage *m* de l'amplitude verticale
e regulación *f* de la amplitud vertical
i regolazione *f* dell'ampiezza verticale
n rasteramplituderegeling, verticale-amplituderegeling
d Teilbildhöhenregelung *f*, Vertikalamplitudenregelung *f*

1889 FIELD APPARATUS, OB APPARATUS, OUTSIDE BROADCAST APPARATUS tv
f appareillage *m* pour extérieurs
e aparatos *pl* para el exterior
i apparecchiatura *f* da campo
n buitenopnameapparatuur
d Apparatur *f* für Aussenreportage

1890 FIELD BEND, FIELD BEND CORRECTION rec/rep/tv
An approximately parabolic waveform, recurring at field frequency, introduced into the picture signal, to compensate for an inherent amplitude distortion in the camera output.
f signal *m* parabolique de correction de trame
e señal *f* parabólica de corrección de cuadro
i correzione *f* parabolica di luminosità della trama
n rasterparaboolcorrectie
d parabolische Vertikalsignale *pl*

1891 FIELD BLANKING, VERTICAL BLANKING rec/rep/tv
f suppression *f* de trame
e supresión *f* de cuadro
i soppressione *f* di trama
n rasteronderdrukking
d Teilbildaustastung *f*, Vertikalaustastung *f*

1892 FIELD BLANKING INTERVAL, VERTICAL BLANKING INTERVAL tv
The interval of time following the end of the field during which the field sync pulses and field suppression pulses are transmitted.
f intervalle *m* de suppression de trame
e intervalo *m* de supresión de cuadro
i intervallo *m* di cancellazione di trama
n rasteronderdrukkingstijd
d Teilbildaustastlücke *f*, Vertikalaustastlücke *f*

1893 FIELD BLANKING PULSE, tv
VERTICAL BLANKING PULSE
A negative pulse lasting for about six or eight lines during the field blanking interval.
f impulsion *f* de suppression de trame
e impulso *m* de supresión de cuadro
i impulso *m* di soppressione di trama
n rasteronderdrukkingsimpuls
d Teilbildaustastimpuls *m*, Vertikalaustastimpuls *m*

1894 FIELD CENT(E)RING CONTROL, cpl
VERTICAL CENT(E)RING CONTROL
f réglage *m* de centrage de trame, réglage *m* de décentrement vertical
e regulación *f* de centraje de cuadro, regulación *f* de centraje vertical
i regolazione *f* di centratura di trama, regolazione *f* di centratura verticale
n rastercentreringsregeling, verticale centreringsregeling
d Teilbildzentrierungsregelung *f*, Y-Lageregelung *f*

1895 FIELD CONVERGENCE, rec/rep/tv
FIELD-DERIVED CONVERGENCE,
VERTICAL CONVERGENCE
f convergence *f* de trame
e convergencia *f* de cuadro
i convergenza *f* di trama
n rasterconvergentie
d Teilbildkonvergenz *f*

1896 FIELD CONVERGENCE AMPLITUDE-CONTROL, rec/rep/tv
VERTICAL CONVERGENCE AMPLITUDE CONTROL
f réglage *m* de convergence et d'amplitude de trame
e regulación *f* de convergencia y de amplitud de cuadro
i regolazione *f* di convergenza e d'ampiezza di trama
n rasterconvergentie- en amplitude-regeling
d Teilbildkonvergenz- *f* und Amplitudenregelung *f*

1897 FIELD CONVERGENCE CONTROL, ctv
VERTICAL CONVERGENCE CONTROL
The control that adjusts the amplitude of the vertical convergent voltage in a colo(u)r TV receiver.
f réglage *m* de convergence de trame
e regulación *f* de convergencia de cuadro
i regolazione *f* di convergenza di trama
n rasterconvergentieregeling
d Teilbildkonvergenzregelung *f*

1898 FIELD CONVERGENCE SHAPE CONTROL, rec/rep/tv
VERTICAL CONVERGENCE SHAPE CONTROL
f réglage *m* de la configuration de la convergence de trame
e regulación *f* de la configuración de la convergencia de cuadro
i regolazione *f* de la configurazione della convergenza di cuadro
n regeling van de vorm van de rasterconvergentie
d Formregelung *f* der Teilbildkonvergenz

1899 FIELD CONVERGENCE TILT CONTROL, rec/rep/tv
VERTICAL CONVERGENCE TILT CONTROL
f réglage *m* linéaire de la convergence de trame
e regulación *f* lineal de la convergencia de cuadro
i regolazione *f* lineare della convergenza di trama
n rasterconvergentiezaagtandregeling
d Sägezahnregelung *f* der Teilbildkonvergenz

1900 FIELD CONVERGENCE YOKE, rec/rep/tv
VERTICAL CONVERGENCE YOKE
f bobine *f* de convergence de trame
e bobina *f* de convergencia de cuadro
i bobina *f* di convergenza di trama
n rasterconvergentiespoel
d Teilbildkonvergenzspule *f*

1901 FIELD COVERAGE, opt
FIELD OF VIEW,
FIELD OF VISION
The area or solid angle visible through an optical instrument.
f champ *m* de vue, champ *m* visuel
e campo *m* de visión
i campo *m* di visione
n gezichtsveld *n*
d Blickfeld *n*, Gesichtsfeld *n*

1902 FIELD DEFLECTION, rec/tv
VERTICAL DEFLECTION
f déviation *f* de la trame, déviation *f* verticale
e desviación *f* del cuadro, desviación *f* vertical
i deflessione *f* della trama, deviazione *f* verticale
n rasterafbuiging, verticale afbuiging
d Teilbildablenkung *f*, Vertikalablenkung *f*

1903 FIELD DEFLECTION ELECTRODES, crt
VERTICAL DEFLECTION ELECTRODES,
Y PLATES
The pair of electrodes that moves the electron beam up and down on the fluorescent screen of a cathode-ray tube employing electrostatic deflection.
f électrodes *pl* de déviation de trame
e electrodos *m* de desviación de cuadro
i elettrodi *pl* di deflessione di trama, elettrodi *pl* di deviazione di trama
n rasterafbuigingselektroden *pl*
d Teilbildablenkungselektroden *pl*

1904 FIELD DEFLECTION OSCILLATOR, crt
VERTICAL DEFLECTION OSCILLATOR,
VERTICAL OSCILLATOR
f oscillateur *m* de la déviation verticale
e oscilador *m* de la desviación vertical
i oscillatore *m* della deflessione verticale, oscillatore *m* della deviazione verticale
n oscillator voor de verticale afbuiging
d Oszillator *m* für die vertikale Ablenkung

1905 FIELD DIVIDER tv
Apparatus for reproducing signals at field frequency by subdivision from a reference frequency.
f diviseur *m* de fréquence de trame
e divisor *m* de frecuencia de cuadro
i divisore *m* per frequenza di trama
n rasterfrequentiedeler
d Frequenzteiler *m*, Teilbildfrequenzteiler *m*

1906 FIELD DRIVE SIGNAL tv
A signal used to establish field sync in studio systems, e.g. in non-composite working.
f signal *m* synchroniseur de trame
e señal *f* sincronizadora de cuadro
i segnale *m* sincronizzatore di trama
n rastersynchroniseersignaal *n*
d Teilbildsynchronisiersignal *n*

1907 FIELD DURATION cpl
f durée *f* de trame
e duración *f* de cuadro
i durata *f* di trama
n rasterduur
d Halbbilddauer *f*

1908 FIELD DYNAMIC FOCUS, rec/rep/tv
VERTICAL DYNAMIC FOCUS
f foyer *m* dynamique vertical
e foco *m* dinámico vertical
i fuoco *m* dinamico verticale
n verticaal-dynamisch focus *n*
d vertikal-dynamischer Brennpunkt *m*

1909 FIELD ELIMINATION GATE, rec/rep/tv
FIELD REMOVAL GATE
f porte *f* de trame
e compuerta *f* de cuadro
i porta *f* di trama
n rasterpoort
d Teilbildgatter *n*

FIELD EQUALIZER MAGNETS
see: COLO(U)R EQUALIZING ASSEMBLY

1910 FIELD FLYBACK, rec/tv
VERTICAL FLYBACK
f retour *m* de trame
e retorno *m* de cuadro
i ritorno *m* di trama
n verticale terugslag
d Teilbildrücklauf *m*, Vertikalrücklauf *m*

1911 FIELD FLYBACK PERIOD, rec/rep/tv
VERTICAL FLYBACK PERIOD
f temps *m* de retour de trame
e tiempo *m* de retorno de cuadro
i tempo *m* di ritorno di trama
n verticale terugslagtijd
d Teilbildrücklaufzeit *f*, Vertikalrücklaufzeit *f*

1912 FIELD FREQUENCY, tv
FIELD RATE,
FIELD REPETITION RATE,
VERTICAL FREQUENCY
The number of fields scanned per second.
f fréquence *f* de balayage vertical, fréquence *f* de trame
e frecuencia *f* de barrido vertical, frecuencia *f* de cuadro
i frequenza *f* di scansione verticale, frequenza *f* di trama
n rasterfrequentie
d Bildwechselfrequenz *f*, Teilbildfrequenz *f*, Vertikalfrequenz *f*

1913 FIELD FREQUENCY CONTROL, rec/tv
FIELD FREQUENCY TIMING,
VERTICAL FREQUENCY CONTROL
f réglage *m* de la fréquence de trame
e regulación *f* de la frecuencia de cuadro
i regolazione *f* della frequenza di trama
n rasterfrequentieregeling
d Bildwechselfrequenzregelung *f*

1914 FIELD FREQUENCY LOCKING, rec/tv
FIELD LOCKING,
VERTICAL FREQUENCY LOCKING
Achieved by locking the field sync pulses.
f enclenchement *m* de la trame
e enclavamiento *m* del cuadro
i bloccaggio *m* della trama
n rastervergrendeling
d Teilbildverriegelung *f*

1915 FIELD GATING CIRCUIT ctv/tv
Circuit used to extract or insert a signal occurring at field frequency.
f circuit *m* déclencheur de trame
e circuito *m* de desbloqueo periódico de cuadro
i circuito *m* di sblocco periodico di trama
n rasterpoortcircuit *n*
d Teilbildgatterkreis *m*

1916 FIELD HEIGHT, rec/rep/tv
VERTICAL HEIGHT
Vertical raster dimension of the active field.
f hauteur *f* de l'image
e altitud *f* de la imagen
i altezza *f* dell'immagine
n beeldhoogte
d Bildhöhe *f*

1917 FIELD HOLD CONTROL, rec/rep/tv
VERTICAL HOLD CONTROL,
VERTICAL SYNC CONTROL
f régleur *m* de la synchronisation de la trame
e regulador *m* de la sincronización del cuadro

i regolatore *m* della sincronizzazione della trama
n rastersynchronisatieregelaar
d Teilbildsynchronisationsregler *m*

1918 FIELD IDENTIFICATION SIGNAL, IDENT SIGNAL ctv
Signal incorporated in early PAL system to identify alternate fields.
f signal *m* d'identification de trame
e señal *f* de identificación de cuadro
i segnale *m* d'identificazione di trama
n rasteridentificeersignaal *n*
d Teilbildkennsignal *n*

1919 FIELD KEYSTONE CORRECTION, FIELD KEYSTONE WAVEFORM rec/rep/tv
A scanning waveform, occurring at field frequency, shaped to compensate for the geometrical distortion which would otherwise occur in tubes having electron gun inclined at an angle to the target or fluorescent screen.
f signal *m* de correction de trapèze-trames
e señal *f* trapezoidal de corrección de cuadro
i correzione *f* trapezoidale geometrica della trama
n trapeziumcorrectie in rasterrichting
d Trapezvertikalsignale *pl*

1920 FIELD LENS opt
Large diameter lens used in optical relay systems to increase the light transfer from the primary-input lens system to the output.
f lentille *f* de champ
e lente *f* de campo
i lente *f* di campo
n veldlens
d Feldlinse *f*

1921 FIELD LINEARITY CONTROL, GENERAL VERTICAL LINEARITY CONTROL rep/tv
The control which corrects the output waveform to bring about uniform velocity of scan in the vertical direction.
f réglage *m* de linéarité de la trame
e regulación *f* de linealidad del cuadro
i regolazione *f* di linearità della trama
n rasterlineariteitsregeling
d Teilbildlinearitätsregelung *f*

1922 FIELD MASKING ani
Used to obtain precise size of desired camera field and also to assess free space round titles.
f limitation *f* du champ
e limitación *f* del campo
i limitazione *f* del campo
n veldafbakening
d Feldbegrenzung *f*

1923 FIELD MONITORING TUBE rep/tv
Cathode-ray tube on the picture control desk, which represents the superposition of the vision signals of the field and the field or picture frequency.
f tube *m* moniteur de trame
e válvula-monitor *m* de cuadro
i valvola-monitore *m* di trama
n rastermonitorbuis
d Teilbildkontrollröhre *f*

1924 FIELD-NEUTRALIZING COIL ctv/rep
A device which encircles the perimeter of the face-plate of a colo(u)r picture tube to offset the effects that the earth's magnetic field and other stray magnetic fields may have on the electron beams.
f bobine *f* de blindage du champ magnétique
e bobina *f* de blindaje del campo magnético
i bobina *f* schermante del campo magnetico
n magneetveldafschermspoel
d Magnetfeldabschirmspule *f*

1925 FIELD-NEUTRALIZING MAGNET, RIM MAGNET ctv/rep
A permanent magnet mounted near the edge of the face plate of a colo(u)r picture tube to prevent stray magnetic fields from affecting the path of the electron beams.
f aimant *m* de blindage
e imán *m* de blindaje
i magnete *m* schermante
n afschermmagneet
d Abschirmmagnet *m*

1926 FIELD NON-LINEARITY rec/rep/tv
Distortion of a picture due to variation of the field sweep, in transmission or reception during the effective part of the field sweep.
f distorsion *f* de vitesse de balayage de trame
e no linealidad *f* de cuadro
i distorsione *f* della deflessione di trama
n rastervervorming
d Ablenknichtlinearität *f*

FIELD OF VIEW
see: FIELD COVERAGE

1927 FIELD OUTPUT TRANSFORMER, VERTICAL OUTPUT TRANSFORMER rec/rep/tv
f transformateur *m* de sortie de trame
e transformador *m* de salida de cuadro
i trasformatore *m* d'uscita di trama
n rasteruitgangstransformator
d Teilbildausgangstransformator *m*, Vertikalablenktransformator *m*

1928 FIELD PERIOD rec/rep/tv
f période *f* de trame
e período *m* de cuadro
i periodo *m* di trama
n rasterperiode
d Teilbildperiode *f*

1929 FIELD PHASING rec/tv
Action of setting the phase of the picture field with respect to a synchronizing source.

f mise *f* en phase de la trame
e puesta *f* en fase del cuadro
i messa *f* in fase della trama
n in faze brengen *n* van het raster
d Teilbildeinphasierung *f*

1930 FIELD PICKUP, tv
NEMO,
OB,
OUTSIDE BROADCAST,
REMOTE
f extérieur *m*,
prise *f* de vues en extérieur
e transmisión *f* de exteriores
i ripresa *f* esterna
n buitenopname
d Aussenaufnahme *f*,
Aussenreportage *f*,
Aussenübertragung *f*

1931 FIELD PULSE rec/tv
f impulsion *f* de trame,
impulsion *f* verticale
e impulso *m* de cuadro,
impulso *m* vertical
i impulso *m* di trama,
impulso *m* verticale
n rasterimpuls,
verticale impuls
d Teilbildimpuls *m*,
Vertikalimpuls *m*

1932 FIELD RATE CONVERTER, tv
A device designed to change the field rate of the TV signal from 50 to 60 Hz.
f convertisseur *m* du nombre de trames
e convertidor *m* del número de cuadros
i convertitore *m* del numero di trame
n rasteraantalomzetter
d Halbbildzahlumwandler *m*,
Teilbildzahlumwandler *m*

1933 FIELD RATE FLICKER rec/tv
f papillotement *m* en trames
e centelleo *m* de cuadro
i sfarfallamento *m* di trama
n rasterflikkering
d Teilbildflimmern *n*

1934 FIELD SAWTOOTH cpl
A waveform at the frequency of field deflection used for linearity checking.
f forme *f* d'onde en dents de scie
e forma *f* de onda en dientes de sierra
i forma *f* d'onda a denti di sega
n zaagtandgolfvorm
d Sägezahnwellenform *f*

1935 FIELD SCAN GENERATOR, rep/tv
FIELD TIME-BASE
TV receiver time-base which controls the field scanning of the picture tube.
f base *f* de temps de trame
e base *f* de tiempo de cuadro
i base *f* di tempi di trama
n rasterafbuigingsgenerator,
rastertijdbasis
d Teilbildzeitbasis *f*

1936 FIELD SCANNING, tv
VERTICAL SCANNING
f analyse *f* de trame,
analyse *f* verticale
e exploración *f* de cuadro,
exploración *f* vertical
i analisi *f* di trama,
analisi *f* verticale
n rasteraftasting,
verticale aftasting
d Teilbildabtastung *f*,
vertikale Abtastung *f*

1937 FIELD SEQUENCE, rec/rep/tv
INTERLACE SEQUENCE,
INTERLACING ORDER
f ordre *m* d'intercalage,
séquence *f* d'entrelacement
e orden *m* de intercalación,
sequenza *f* de interlazado
i ordine *m* d'intercalamento,
sequenza *f* d'interlacciamento
n interliniëringsvolgorde
d Teilbildfolge *f*,
Zeilensprungreihenfolge *f*

1938 FIELD-SEQUENTIAL ctv
COLO(U)R TELEVISION
Colo(u)r TV in which the camera sees red, blue and green images in turn through a rotating disk consisting of segments of colo(u)r filters in the correct sequence.
f télévision *f* couleur à séquence de trames
e televisión *f* en colores de secuencia de cuadros
i televisione *f* a colori a sequenza di trame
n kleurentelevisie volgens het raster-kleurwisselsysteem,
volgrasterkleurentelevisie
d Zeitfolgeverfahren *n* beim Farbfernsehen

1939 FIELD-SEQUENTIAL SCANNING ctv
f analyse *f* pour la télévision couleur à séquence de trames
e exploración *f* para la televisión en colores de secuencia de cuadros
i analisi *f* per la televisione a colori a sequenza di trame
n aftasting volgens het rasterkleurwissel-systeem
d Zeitfolgeverfahrenabtastung *f*

1940 FIELD SHIFT SWITCH ctv
f commutateur *m* de trame
e conmutador *m* de cuadro
i commutatore *m* di trama
n rasterschakelaar
d Halbbildschalter *m*,
Teilbildschalter *m*

1941 FIELD SIMULTANEOUS ctv
COLO(U)R TELEVISION
A colo(u)r TV system in which a complete colo(u)r field is presented simultaneously as a unit.
f système *m* simultané de télévision couleur

e sistema *m* simultáneo de televisión en colores
i sistema *m* simultaneo di televisione a colori
n simultaansysteem *n* voor kleurentelevisie
d Simultanverfahren *n* beim Farbfernsehen

1942 FIELD STORE CONVERTER ctv
A converter which can change the field rate by electronic means.
f convertisseur *m* électronique du nombre de trames par seconde
e convertidor *m* electrónico del número de cuadros per segundo
i convertitore *m* elettronico del numero di trame per secondo
n elektronische omzetter van het aantal rasters per seconde
d elektronischer Umwandler *m* der Teilbildzahl pro Sekunde

1943 FIELD STRENGTH tv
In TV the intensity, measured in microvolts per meter(re), of the carrier signal from a given transmitter which is present in a dipole 1 meter(re) long.
f intensité *f* de champ
e intensidad *f* de campo
i intensità *f* di campo
n veldsterkte
d Feldstärke *f*

1944 FIELD SWEEP, rec/tv
VERTICAL SWEEP
The vertical to-and-fro movement of the field frequency which, combined with the line sweep, causes the spot to trace out the lines which make up the field.
f balayage *m* de la trame
e barrido *m* del cuadro
i andata e ritorno della trama
n rasterheen- en terugslag
d Teilbildablenkung *f*

1945 FIELD SYNC PULSE, rep/tv
FIELD SYNCHRONIZING SIGNAL,
VERTICAL SYNC PULSE
A pulse or series of pulses transmitted at the end of each pulse to hold the receiver scanning process in synchronism with that of the transmitter.
f signal *m* de synchronisation de trame
e señal *f* de sincronización de cuadro
i segnale *m* di sincronizzazione di trama
n rastersynchronisatiesignaal *n*
d Bildsynchronisiersignal *n*,
Vertikalsynchronsignal *n*

1946 FIELD SYNCHRONIZATION, rec/tv
VERTICAL SYNCHRONIZATION
f synchronisation *f* de trames
e sincronización *f* de cuadros
i sincronizzazione *f* di trame
n rastersynchronisatie
d Teilbildsynchronisierung *f*

1947 FIELD TILT, rec/rep/tv
FIELD TILT CORRECTION
A sawtooth waveform, recurring at field frequency, introduced into the picture signal to compensate for an inherent amplitude distortion in the camera output.
f signal *m* de correction linéaire de trame
e señal *f* de corrección lineal de cuadro
i segnale *m* di correzione lineare di luminosità di trama
n rasterzaagtandcorrectie
d Sägezahnvertikalsignale *pl*

1948 FIGURE OF LINEARITY cpl
f facteur *m* de linéarité
e factor *m* de linealidad
i fattore *m* di linearità
n lineariteitsfactor
d Linearitätsmass *n*

FIGURE OF MERIT
see: FACTOR OF MERIT

1949 FILL-IN LIGHT, stu
FILL LIGHT,
FILLER,
KICKER LIGHT
Light used in studios directed to the shadow parts of the scene to prevent excessive contrast.
f lumière *f* complémentaire
e luz *f* complementaria
i luce *f* complementaria
n opvullicht *n*
d Aufheller *m*,
Füllicht *n*

1950 FILM ADAPTATION, fi/tv
FILM VERSION
f adaptation *f* cinématographique,
fabrication *f* film
e adaptación *f* cinematográfica
i adattamento *m* cinematografico
n filmbewerking
d Filmbearbeitung *f*

1951 FILM DEPARTMENT, stu
FILM UNIT
A department of the TV studio for handling all activities in which films are used.
f section *f* de films
e sección *f* de películas
i sezione *f* di pellicole
n filmafdeling
d Filmabteilung *f*

FILM GATE
see: CAMERA APERTURE

1952 FILM REPRODUCER, vr
MAGNETIC FILM REPRODUCER
Equipment in which sprocket-hole film is the medium from which a magnetic recording is reproduced.
f ensemble *m* à reproduction de film magnétique
e conjunto *m* de reproducción de película magnética
i complesso *m* per riproduzione di pellicola magnetica
n weergeefapparaat *n* voor magneetfilm
d Magnettonwiedergabegerät *n*

1953 FILM SCANNER, MOTION PICTURE PICKUP, TELECINE, TELECINE MACHINE, TELECINE PROJECTOR, TELEVISION FILM SCANNER — tv
A motion-picture projector adapted for use with a TV camera tube to televise 24-frame-per second motion-picture film at the 30-frame-per second rate required for TV.
f analyseur *m* de film, projecteur *m* de télécinéma, télécinéma *m*
e explorador *m* de película, proyector *m* de telecinema, telecinema *m*
i analizzatore *m* di pellicola, proiettore *m* di telecinema, telecinema *m*
n filmaftaster, televisiebeeldprojector
d Fernsehbildprojektor *m*, Filmabtaster *m*, Projektionsgerät *n* für Zwischenfilmfernsehbetrieb

1954 FILM SCANNING, MOTION PICTURE PICKUP — tv
The process of converting motion picture film into corresponding electric signals that can be transmitted by a TV system.
f analyse *f* de film
e exploración *f* de película
i analisi *f* di pellicola
n filmaftasting
d Filmabtastung *f*

1955 FILM SCHEDULE — fi
f plan *m* de tournage
e programa *m* de toma
i programma *m* di ripresa
n opnameplan *n*
d Aufnahmeplan *m*

1956 FILM SHOOTING TECHNIQUE — ctv/tv
f technique *f* de prise de vues
e técnica *f* de toma de vistas
i tecnica *f* di ripresa
n filmopneemtechniek
d Filmaufnahmetechnik *f*

1957 FILM SOUND RECORDER — vr
Equipment that uses coated sprocket-hole film as the medium for magnetic recording.
f enregistreur *m* magnétique de son
e registrador *m* magnético de sonido
i registratore *m* magnetico di suono
n magnetische geluidsopnemer
d Magnettongerät *n*

1958 FILM TITLE WITH MUSIC — fi
f composition *f* musicale
e composición *f* musical
i composizione *f* musicale
n muziektitel
d Musiktitel *m*

1959 FILM TRANSPORT — fi/tv
f avancement *m* du film, entraînement *m* du film, transport *m* du film
e transporte *m* de película
i trasporto *m* di pellicola
n filmtransport *n*
d Filmtransport *m*

1960 FILMING — fi/tv
f tournage *m*
e filmación *f*
i il filmare *m*
n filmen *n*
d Aufnahme *f*, Dreh *m*, Dreharbeiten *pl*

1961 FILTER FACTOR — opt
Numerical factor by which the length of an exposure must be increased to compensate for the light absorption of an optical filter through which the exposure is made.
f coefficient *m* du filtre
e coeficiente *m* del filtro
i coefficiente *m* del filtro
n filtercoëfficiënt
d Filterkoeffizient *m*

1962 FILTER FOR TELEVISION TRANSMITTER, VESTIGIAL SIDEBAND AERIAL (ANTENNA) FILTER — aea
f filtre *m* d'antenne pour émetteur de télévision à bande latérale restante
e filtro *m* de antena para emisor televisivo de banda lateral vestigial
i filtro *m* d'antenna per emettitore televisivo a banda laterale residua
n antennefilter *n* voor semi-eenzijbandtelevisiezender
d Antennenfilter *n* für Restseitenbandfernsehsender

1963 FILTERPLEXER — cpl
A device incorporating a vestigial-sideband filter and a diplexer.
f combinaison *f* diplexeur-filtre de bande latérale
e combinación *f* diplexador-filtro de banda lateral
i combinazione *f* separatore di frequenza-filtro di banda laterale
n zijbandfilter-antennemengercombinatie
d Frequenzweiche-Seitenbandfilterkombination *f*

1964 FINAL AMPLIFIER — aea/cpl
The transmitter stage that feeds the aerial (antenna).
f amplificateur *m* final
e amplificador *m* final
i amplificatore *m* finale
n eindversterker
d Endverstärker *m*

1965 FINAL ANODE — crt
In a cathode-ray tube the anode situated farthest from the cathode, to which the EHT is applied to accelerate the electrons.

f anode *f* finale,
anode *f* terminale
e ánodo *m* de salida,
ánodo *m* final
i anodo *m* finale
n eindanode
d Endanode *f*

1966 FINAL BLANKING tv
f postsuppression *f*
e postsupresión *f*
i postcancellazione *f*
n naonderdrukking
d Nachaustastung *f*

1967 FINAL CUT fi/tv
f montage *m* définitif
e montaje *m* final
i montaggio *m* finale
n eindmontage
d Feinschnitt *m*

1968 FINAL SHOT rec/tv
f prise *f* définitive
e toma *f* definitiva
i ripresa *f* definitiva
n eindopname
d heisse Probe *f*

1969 FINAL VERSION fi/tv
f version *f* définitive
e versión *f* definitiva
i versione *f* definitiva
n eindredactie
d Endfassung *f*

1970 FINDER, rec/tv
VIEWFINDER
An optical or electronic device that shows the field of action covered by a TV camera.
f viseur *m*
e visor *m*
i mirino *m*, visore *m*
n zoeker
d Sucher *m*

1971 FINDER MONITOR, stu
VIEWFINDER MONITOR
A picture monitor with a flatfaced black and white tube, usually mounted above the camera tube.
f moniteur *m* à viseur
e monitor *m* de visor
i monitore *m* a visore
n zoekermonitor
d Suchermonitor *m*

1972 FINE ADJUSTMENT cpl
f réglage *m* fin
e regulación *f* fina
i regolazione *f* fina
n fijnregeling
d Feinregelung *f*

1973 FINE CHROMINANCE PRIMARY ctv
Of a pair of chrominance primaries so chosen that one corresponds to the direction of maximum acuity of colo(u)r vision and the other to the direction of minimum acuity, that primary corresponding to the direction of maximum acuity of colo(u)r vision.
f primaire *f* principale de chrominance
e primario *m* principal de crominancia
i primario *m* principale di crominanza
n voornaamste primaire chrominantie-stimulus
d Hauptprimärfarbreiz *m*

1974 FINE DETAIL RESOLUTION rep/tv
f finesse *f* d'image
e nitidez *f* de imagen
i finezza *f* d'immagine
n detailscherpte
d Durchzeichnung *f*

FINE SCANNING (GB)
see: CLOSE SCANNING

1975 FINE STRUCTURE rep/tv
f structure *f* fine
e estructura *f* fina
i struttura *f* fina
n fijnstructuur
d Feinstruktur *f*

1976 FINENESS OF SCANNING rec/tv
f finesse *f* d'analyse,
finesse *f* du canevas
e fineza *f* de exploración,
nitidez *f* de la cuadrícula
i definizione *f* dell'immagine,
finezza *f* del quadro rigato
n rasterfijnheid
d Rasterfeinheit *f*

1977 FIRST ANODE crt
f première anode *f*
e primero ánodo *m*
i primo anodo *m*
n eerste anode
d Sauganode *f*, Voranode *f*

1978 FIRST GENERATION TAPE, vr
ORIGINAL
f bande *f* de première génération,
enregistrement *m* original
e grabación *f* original,
primera copia *f*
i prima generazione *f*,
registrazione *f* originale
n eerste generatie,
origineel *n*
d erste Generation *f*,
Originalaufzeichnung *f*

1979 FIRST REFLECTION tv
f première réflexion *f*
e primera reflexión *f*
i prima riflessione *f*
n eerste reflectie
d erste Reflexion *f*

1980 FISHPOLE rec/tv
A light pole, resembling a fishing rod carrying a microphone, the whole being projected out over a scene when a dialogue is being recorded.

f girafe *f* de microphone,
perche *f* de son
e pértiga *f* de micrófono
i giraffa *f* di microfono
n microfoonhengel
d Mikrophongalgen *m*

1981 FIVE-BEAM CATHODE-RAY TUBE crt
f tube *m* cathodique à cinq faisceaux
e tubo *m* catódico de cinco haces
i tubo *m* catodico a cinque fasci
n vijfbundelbuis
d Fünfstrahlröhre *f*

1982 FIXED POINT-TO-POINT TELEVISION SERVICE tv
f transmission *f* point à point pour la télévision
e equipo *m* de radiocomunicación para televisión entre puntos fijos
i trasmissione *f* a fasci direttivi per la televisione
n vaste televisieverbinding
d Richtfunk *m* für Fernsehen

1983 FLAG, tv
GOBO (US),
LENS SCREEN (GB)
Shield to protect lenses of TV cameras from unwanted light.
f coupe-flux *m*,
nègre *m*
e pantalla *f* a prueba de luz
i pannello *m* di mascheramento della luce,
schermo *m* paraluce
n lensscherm *n*
d Blende *f*,
Lichtblende *f*

FLAGPOLE (US)
see: BAR

FLAGPOLE GENERATOR (US)
see: BAR GENERATOR

FLAGPOLE PATTERN (US)
see: BAR PATTERN

FLAGPOLE SIGNAL (US)
see: BAR SIGNAL

1984 FLARE, rep/tv
FLARE SPOT (GB),
WOMP (US)
Bright area of light appearing usually near the center(re) of a TV picture, caused by internal reflections in the camera lens.
f diffusion *f* parasite,
tache *f* hyperlumineuse
e hiperluminosidad *f* del punto,
mancha *f* hiperluminosa
i macchia *f* di riflessione,
macchia *f* iperluminosa
n intensieve lichtvlek,
storende lichtvlek
d intensiver Lichtfleck *m*,
Überstrahlung *f*

1985 FLASH dis
A momentary superposition of a sound or picture, in a program(me).
f surimpression *f*
e transparencia *f*
i sovrapposizione *f*
n geluid-of-beeld-overheersing
d Einblendung *f*

1986 FLASHBACK rep/tv
Story-telling device in which the chronological unfolding of the plot is interrupted by a scene or sequence drawn from the past.
f reprise *f* d'une séquence antérieure,
rétrospective *f*,
scène *f* avec retour dans le temps
e escena *f* de recuerdo de lo pasado,
inserción *f* retrospectiva
i ripresa *f* d'una scena antecedente,
ripresa *f* retrospettiva
n terugblik,
terugblikonderbreking
d Rückblende *f*,
Rückblickunterbrechung *f*

1987 FLAT tv
Low contrast rendering of restricted density range in a photographic or TV reproduction.
f sans contraste
e sin contraste
i senza contrasto,
senza rilievo
n contrastarm adj
d kontrastarm adj

1988 FLAT-ENDED CATHODE-RAY TUBE, crt
FLAT-FACED CATHODE-RAY TUBE
f tube *m* cathodique à fond plat
e tubo *m* catódico de fondo plano
i tubo *m* catodico a piastra frontale piana
n katodestraalbuis met platte frontplaat,
platte beeldbuis
d Katodenstrahlröhre *f* mit ebener Sternfläche

1989 FLAT-FACED SCREEN crt
f écran *m* plat
e pantalla *f* plana
i schermo *m* piano
n vlak scherm *n*
d ebener Schirm *m*,
Planschirm *m*

1990 FLAT PICTURE, tv
f image *f* sans contraste
e imagen *f* sin contraste
i immagine *f* senza contrasto
n contrastarm beeld *n*
d kontrastarmes Bild *n*

1991 FLAT RANDOM NOISE, dis
WHITE NOISE
Random acoustic or electric noise having equal energy per cycle over a specified total frequency band.
f bruit *m* blanc

e ruido *m* blanco
i rumore *m* bianco
n witte ruis
d weisses Rauschen *n*

FLESH TONE
see: FACE TONE

1992 FLICKER dis
In the reproduced image, the unwanted rhythmic variations in the luminosity of the picture.
f papillotement *m*
e centelleo *m*
i sfarfallamento *m*, sfarfallio *m*
n flikker, flikkering
d Flimmern *n*

1993 FLICKER-FREE REPRODUCTION rep/tv
f reproduction *f* sans papillotement
e reproducción *f* sin centelleo
i riproduzione *f* senza sfarfallamento
n flikkerloze weergave
d flimmerfreie Wiedergabe *f*

1994 FLICKER FREQUENCY tv
f fréquence *f* de papillotement
e frecuencia *f* de centelleo
i frequenza *f* di sfarfallamento
n flikkerfrequentie
d Flimmerfrequenz *f*

1995 FLOAT fi/tv
f instabilité *f* d'image
e inestabilidad *f* de imagen
i instabilità *f* d'immagine
n beeldschommeling
d Bildzittern *n*

1996 FLOATING PEGS ani
An additional set of pegs with a north/south and east/west movement completely free of the table top, fixed either to the base or to the columns.
f chevilles *pl* à libre mouvement
e espigas *pl* flotantes
i carichi *pl* flottanti
n vrijstaande pennen *pl*
d freistehende Zapfen *pl*

1997 FLOOD stu
A lamp for providing a uniform illumination on the studio floor.
f lampe *f* à faisceau élargi, lampe *f* pour éclairage égal
e lámpara *f* para alumbrado igual
i lampada *f* per illuminazione uguale
n lamp voor gelijkmatige verlichting
d Lampe *f* für gleichmässige Beleuchtung

1998 FLOODLAMPS stu
f lampes *pl* pour éclairage par projection
e lámparas *pl* para alumbrado por proyección
i lampade *pl* per illuminazione per proiezione
n slaglichtlampen *pl*
d Flutlichtlampen *pl*

1999 FLOODLIGHT SCANNING rec/tv
f analyse *f* à éclairage par projection
e exploración *f* de alumbrado por proyección
i analisi *f* ad illuminazione per proiezione
n aftasting bij slaglicht
d Abtastung *f* bei Anstrahlung, Rampenabtastung *f*

2000 FLOOR stu
f plateau *m*
e escenario *m*, tablas *pl*
i palcoscenico *m*
n planken *pl*, toneel *n*
d Atelier *n*, Bretter *pl*, Bühne *f*

2001 FLOOR SHOOTING, STUDIO RECORDING rec/stu/tv
f prise *f* en studio
e toma *f* de estudio
i ripresa *f* di studio
n studio-opname
d Studioaufnahme *f*

2002 FLUCTUATION OF CHROMATICITY ct
f fluctuation *f* de la chromaticité
e fluctuación *f* de la cromaticidad
i fluttuazione *f* della cromaticità
n schommeling van de kleursoort
d Farbartschwankung *f*

2003 FLUORESCENCE tv
Emission of light or other electromagnetic radiation by a material exposed to another type of radiation or to a beam of particles.
f fluorescence *f*
e fluorescencia *f*
i fluorescenza *f*
n fluorescentie
d Fluoreszenz *f*

2004 FLUORESCENT LAMP (GB), FLUORESCENT TUBE (US) stu
Used sometimes in studios where a soft light is required.
f lampe *f* fluorescente
e lámpara *f* fluorescente
i lampada *f* fluorescente
n buisvormige fluorescentielamp, fluorescentielamp, TL-buis
d Fluoreszenzlampe *f*, Fluoreszenzröhre *f*, Leuchtstofflampe *f*, Leuchtstoffröhre *f*

2005 FLUORESCENT MATERIAL, LUMINESCENT MATERIAL, PHOSPHOR crt
f matière *f* fluorescente, substance *f* fluorescente
e materia *f* fluorescente, substancia *f* fluorescente
i materia *f* fluorescente, sostanza *f* fluorescente
n fluorescerende stof
d Leuchtstoff *m*

2006 FLUORESCENT SCANNING rec/tv
f analyse *f* de l'écran fluorescent
e exploración *f* de la pantalla fluorescente
i analisi *f* dello schermo fluorescente
n aftasting van het fluorescentiescherm
d Leuchtschirmabtastung *f*

FLUORESCENT SCREEN
see: CATHODE-RAY SCREEN

2007 FLUTTER aea
A spurious vibration of the dipole, e.g. caused by the wind which interferes with the quality of the TV picture.
f vibration *f* parasite
e vibración *f* parásita
i vibrazione *f* parassita
n parasitaire trilling
d Flackern *n*, Flattern *n*

2008 FLUTTER vr
In tape recording, rhytmic fluctuations of the pitch of sound caused by irregularities in the motion of e.g. a motor.
f pleurage *m*,
sautillement *m*
e variaciones *pl* rápidas de altura de sonido
i trillo *m*
n snelle toonhoogtevariaties *pl*
d schnelle Tonhöhenschwankungen *pl*

2009 FLUTTER BRIDGE dis
An apparatus used for measuring the flutter content of recording and reproducing devices by null balance methods.
f pont *m* de mesure pour le pleurage
e puente *m* de medida para variaciones de altura de sonido
i ponte *m* di misura per trillo
n meetbrug voor toonhoogtevariaties
d Messbrücke *f* für Tonhöhenschwankungen

2010 FLUTTER ECHOS aud/dis
f échos *pl* intermittants,
échos *pl* multiples
e écos *pl* múltiples
i echi *pl* intermittenti
n meervoudige echo's *pl*
d Flatterechos *pl*,
Mehrfachechos *pl*,
Schetterechos *pl*

2011 FLUTTER EFFECT aea/dis
f effet *m* vibratoire
e efecto *m* vibratorio
i effetto *m* vibratorio
n trillingseffect *n*
d Flattereffekt *m*

2012 FLYBACK, crt
KICKBACK,
RETRACE (GB),
RETURN TRACE (US)
In TV, the return of the electron beam from the end of a line or field to the commencement of the next.
f retour *m* du spot
e retorno *m* del punto
i ritorno *m* del punto
n terugslag
d Rücklauf *m*

2013 FLYBACK BLANKING, rep/tv
RETRACE BLANKING
At the receiver, the suppression of the spot during flyback.
f suppression *f* du retour
e supresión *f* del retorno
i cancellazione *f* del ritorno
n terugslagonderdrukking
d Rücklaufaustastung *f*

2014 FLYBACK EHT SUPPLY, tv
FLYBACK POWER SUPPLY,
KICKBACK POWER SUPPLY
A high-voltage power supply used to produce the d.c. voltage of about 10.000 to 25.000 volts required for the second anode of a cathode-ray tube in a TV receiver.
f alimentation *f* en très haute tension par retour du spot
e alimentación *f* de extra alta tensión por retorno
i alimentazione *f* d'altissima tensione per ritorno
n voeding met zeer hoge spanning uit de terugslag
d Rücklaufhöchstspannungsspeisung *f*

2015 FLYBACK TRANSFORMER, cpl
HORIZONTAL OUTPUT
TRANSFORMER,
LINE OUTPUT TRANSFORMER
A transformer used in a TV receiver to provide the horizontal deflection voltage, the high voltage for the second-anode power supply of the picture tube and the filament voltage for the high-voltage rectifier.
f transformateur *m* de retour du spot,
transformateur *m* de sortie lignes
e transformador *m* de retorno,
transformador *m* de salida líneas
i trasformatore *m* di ritorno,
trasformatore *m* d'uscita linee
n lijnuitgangstransformator,
terugslagtransformator
d Rücklauftransformator *m*,
Zeilenausgangstransformator *m*

2016 FLYING-SPOT ctv/rec/tv/vr
ANALYZER,
FLYING-SPOT SCANNER
f analyseur *m* à spot lumineux,
analyseur *m* à spot mobile
e explorador *m* de punto móvil
i analizzatore *m* a punto mobile
n lichtpuntaftaster
d Lichtpunktabtaster *m*

2017 FLYING-SPOT SCANNING, tv
LIGHT-SPOT SCANNING
A method of scanning in which a beam of

light, concentrated into a small spot, moves over the object or image in a series of lines, the reflected and transmitted light being converted into current variations by a photo-electric cell.
f analyse *f* à spot lumineux
e exploración *f* de punto móvil
i analisi *f* a punto mobile
n lichtpuntaftasting
d Lichtpunktabtastung *f*

2018 FLYING-SPOT TUBE SCANNING rec/tv
A cathode-ray tube in which the spot produced on a short-persistence screen by a scanning beam of electrons illuminates a physical object such as a transparency.
f tube *m* analyseur à spot lumineux, tube *m* analyseur à spot mobile, tube *m* analyseur d'image
e tubo *m* de exploración de punto móvil
i tubo *m* d'analisi a punto mobile
n lichtpuntaftastbuis
d Bildabtaströhre *f*, Lichtpunktabtaströhre *f*

2019 FLYWHEEL CIRCUIT cpl/ge
Tuned circuit of high Q which maintains oscillation for a relatively long time.
f circuit *m* à effet de volant
e circuito *m* con efecto de volante
i circuito *m* ad effetto di volano
n vliegwielschakeling
d Schwungradschaltung *f*

2020 FLYWHEEL SYNCHRONIZATION rep/tv
A form of synchronization in a TV receiver in which the scanning circuits are so designed that they will continue to operate at the system line frequency during a temporary disturbance of the sync signal.
f synchronisation *f* par effet de volant
e sincronización *f* por efecto de volante
i sincronizzazione *f* per effetto di volano
n vliegwielsynchronisatie
d Schwungradsynchronisierung *f*

2021 FLYWHEEL TIMEBASE rep/tv
In TV receivers, a method of controlling the picture frequency by the electrical inertia of the scanning circuits and not by synchronizing pulses.
f base *f* de temps à effet de volant
e base *f* de tiempo de efecto de volante
i base *f* di tempi ad effetto di volano
n vliegwieltijdbasis
d Schwungradzeitbasis *f*

2022 F.M., FREQUENCY MODULATION cpl
Angle modulation of a sine-wave carrier in which the instantaneous frequency of the modulated wave differs from the carrier frequency by an amount proportional to the instantaneous value of the modulating wave.
f modulation *f* de fréquence
e modulación *f* de frecuencia
i modulazione *f* di frequenza
n frequentiemodulatie
d Frequenzmodelung *f*, Frequenzmodulation *f*

2023 FOCAL DISTANCE, FOCAL LENGTH opt
In a thin lens, the distance from the center(re) of the lens to either principal focus.
f distance *f* focale
e distancia *f* focal
i distanza *f* focale, lunghezza *f* focale
n brandpuntsafstand
d Brennweite *f*

2024 FOCAL PLANE opt
Plane through the principal focus of a lens, perpendicular to its optical axis.
f plan *m* focal
e plano *m* focal
i piano *m* focale
n brandpuntsvlak *n*, brandvlak *n*
d Brennebene *f*, Brennpunktebene *f*

2025 FOCAL POINT, FOCUS opt
Point to which the converging rays of light converge.
f foyer *m*
e foco *m*
i fuoco *m*
n brandpunt *n*, focus
d Brennpunkt *m*, Fokus *m*

FOCAL SPOT
see: CROSSOVER

2026 FOCAL TIME crt/opt
In an aperture electron lens, the equivalent of the focal length in an optical lens.
f distance *f* focale de lentille électronique, temps *m* focal
e distancia *f* focal de lente electrónica, tiempo *m* focal
i distanza *f* focale di lente elettronica, tempo *m* focale
n brandpuntsafstand van een elektronenlens, brandpunttijd
d Brennpunktzeit *f*, Brennweite *f* einer Elektronenlinse

2027 FOCUS CONTROL, SCANNING SPOT CONTROL crt
A control that adjusts the spot size at the screen of a cathode-ray tube.
f réglage *m* du spot lumineux
e regulación *f* del punto luminoso
i regolazione *f* del punto luminoso
n regeling van de lichtpuntgrootte
d Lichtpunktgrösseregelung *f*

FOCUS-MASK TUBE
see: CHROMATRON

2028 FOCUS MODULATION crt
Variation of the focusing of a cathode-ray beam as it is deflected.
f modulation *f* de la convergence, modulation *f* de la focalisation
e modulación *f* de la convergencia, modulación *f* del enfoque
i modulazione *f* della convergenza, modulazione *f* della focalizzazione
n convergentiemodulatie, modulatie van de focussering
d Konvergenzmodulation *f*, Modulation *f* der Fokussierung

2029 FOCUS PULLER, FOLLOW CAMERAMAN, SECOND CAMERAMAN rec/tv
f assistant *m* de caméra, assistant *m* metteur au point
e ayudante *m* del operador de la cámara
i aiuto *m* operatore di ripresa
n camera-assistent
d Kameraassistent *m*, Schärfezieher *m*

2030 FOCUS SERVO-SYSTEM crt
System for remote adjustment of the focus of a TV camera.
f téléajustage *m* du foyer
e teleajuste *m* del foco
i teleaggiustaggio *m* del fuoco
n focusinstelling op afstand
d Fernfokuseinstellung *f*

2031 FOCUS UNIT, SHOT BOX rec/tv
Control unit for remotely operating a TV camera.
f appareil *m* de télécommande de la focalisation
e aparato *m* de telemando del enfoque
i apparecchio *m* di telecomando della focalizzazione
n verrebesturingsapparaat *n* van de televisiecamera
d Fernsteuergerät *n* der Fernsehkamera

2032 FOCUSING crt
The process of controlling the convergence of an electron beam.
f focalisation *f*
e enfoque *m*
i focalizzazione *f*
n focussering
d Bündelung *f*, Fokussierung *f*, Scharfeinstellung *f*

2033 FOCUSING AND SWITCHING GRILLE ctv
A colo(u)r selecting electrode system used in a special type of colo(u)r picture tube.
f grille *f* de Lawrence, sélecteur *m* de couleurs commutateur
e rejilla *f* de Lawrence, selector *m* de colores conmutador
i griglia *f* di Lawrence, selettore *m* di colori commutatore
n lawrencerooster *n*. schakelende kleurkiezer
d Gitter *n* nach Lawrence, schaltender Farbwähler *m*

2034 FOCUSING ANODE crt
An anode used in a cathode-ray tube to change the size of the electron beam at the screen.
f anode *f* de focalisation
e ánodo *m* de enfoque
i anodo *m* di focalizzazione
n focusseeranode
d Fokussieranode *f*

2035 FOCUSING COIL crt
f bobine *f* de focalisation
e bobina *f* de enfoque
i bobina *f* di focalizzazione
n focusseringsspoel
d Fokussierspule *f*, Fokussierungsspule *f*

2036 FOCUSING ELECTRODE crt
The component electrode of the electron gun which focuses the beam by means of an applied voltage.
f électrode *f* de focalisation
e electrodo *m* de enfoque
i elettrodo *m* di focalizzazione
n focusseringselektrode
d Fokussierelektrode *f*, Fokussierungselektrode *f*, Konzentrationselektrode *f*

2037 FOCUSING FIELD crt
The field produced by the focusing electrodes in the trajectory of an electron beam in a cathode-ray tube.
f champ *m* de concentration, champ *m* de focalisation
e campo *m* de concentración, campo *m* de enfoque
i campo *m* di concentrazione, campo *m* di focalizzazione
n focusseringsveld *n*
d Fokuseinstellungsfeld *n*, Fokussierungsfeld *n*

2038 FOCUSING MAGNET crt
f aimant *m* de focalisation
e imán *m* de enfoque
i magnete *m* di focalizzazione
n focusseringsmagneet
d Fokussiermagnet *m*, Fokussierungsmagnet *m*

2039 FOCUSING MICROSCOPE, VIEWFINDER TUBE opt
Apparatus used in rack-over viewfinding.
f viseur *m* à microscope focalisateur
e visor *m* de microscopio enfocador
i mirino *m* a microscopio focalizzatore
n zoeker met focusserende microscoop
d Sucher *m* mit Fokussierungsmikroskop

2040 FOCUSING VOLTAGE crt/svs
f tension *f* de focalisation
e tensión *f* de enfoque

i tensione *f* di focalizzazione
n focusseringsspanning
d Fokussierungsspannung *f*

2041 FOG opt
f voile *m*
e velo *m*
i velo *m*
n sluier
d Schleier *m*

2042 FOLD-OVER dis
Distortion in a TV picture showing itself as a white line on one of the edges.
f formation *f* de lignes pliées
e formación *f* de lĭneas plegadas
i formazione *f* di linee piegate
n omvouwen *n* van de lijnen
d Faltenbildung *f* der Zeilen

FOLD-OVER
see: DOUBLE IMAGE

2043 FOLDBACK LOUDSPEAKER aud/stu
High-quality loudspeaker system provided for reproducing music or effects on the studio floor.
f hautparleur *m* à membrane repliée
e altavoz *m* de membrana plegada
i altoparlante *m* a membrana piegata
n vouwluidspreker
d Faltenlautsprecher *m*

2044 FOLDED DIPOLE aea
An aerial (antenna) consisting of inter-connected parallel dipoles separated by a small fraction of the wavelength of operation, connected together at their outer ends, and fed at the center(re) of one of the dipoles.
f dipôle *m* replié
e dipolo *m* plegado
i dipolo *m* ripiegato
n gevouwen dipool
d Faltdipol *m*,
Schleifendipol *m*

2045 FOLDED LIGHT BEAM opt
A beam used in rear projection systems to greatly reduce the back-of-screen depth.
f faisceau *m* de lumière réfléchi
e haz *m* de luz reflejado
i fascio *m* di luce riflesso
n gereflecteerde lichtstraal
d reflektierter Lichtstrahl *m*

2046 FOLLOW-FOCUS SHOT fi/tv
f prise *f* de vues poursuite
e toma *f* de vistas siguiendo al ejecutante
i presa *f* a seguimento
n achtervolgingsopname
d Verfolgungsaufnahme *f*

2047 FOLLOW-FOCUS VIEWER rec/tv
f viseur *m* suivant le sujet
e visor *m* de enfocar constantemente el sujeto
i mirino *m* a costante messa a fuoco
n meelopende zoeker
d mitlaufender Sucher *m*

FOLLOW SHOT
see: DOLLYING SHOT

2048 FOOT,
STEM crt
That part of an electronic tube (valve) on which the electrodes are mounted.
f embase *f*
e base *f*
i base *f*
n buisbodem, buisvoet
d Quetschfuss *m*

2049 FOOTAGE fi/tv
f métrage *m*
e metraje *m*
i metraggio *m*
n lengte
d Meterlänge *f*

2050 FOOTLIGHT fi/tv
f feux *pl* de la rampe
e candilejas *pl* de escenario
i luci *pl* di proscenio
n voetlicht *n*
d Beleuchtungsrampe *f*,
Rampenlicht *n*

2051 FOREGROUND SIGNAL stu
f signal *m* du premier plan
e señal *f* del primer plano
i segnale *m* del primo piano
n voorgrondsignaal *n*
d Vordergrundsignal *n*

2052 FOREIGN COMMENTARY MIXER rec/rep
f mélangeur *m* de son pour commentateurs de l'étranger
e mezclador *m* de sonido para comentadores extranjeros
i mescolatore *m* di suono per commentatori stranieri
n geluidsmenger voor buitenlandse commentators
d Tonmischer *m* für Auslandskommentatore

2053 FOREIGN COMMENTATOR rec/tv
f commentateur *m* de l'étranger
e comentador *m* extranjero
i commentatore *m* straniero
n buitenlandse commentator
d Auslandskommentator *m*

2054 FOREIGN LANGUAGE CAPTIONS rec/rep/tv
f titres *pl* en langue étrangère
e tĭtulos *pl* en idioma extranjero
i titoli *pl* in idioma straniero
n vreemdtalige titels *pl*
d fremdsprachige Titel *pl*

2055 FOREIGN RELEASE,
FOREIGN VERSION fi/tv
f version *f* en langue étrangère
e versión *f* traducida

i versione *f* in altra lingua
n vertaalde versie
d übersetzte Fassung *f*

2056 FORM PERCEPTION ct
f perception *f* de forme
e percepción *f* de forma
i percezione *f* di forma
n vormperceptie
d Formempfindung *f*

2057 FORWARD RADIATION aea
f rayonnement *m* en avant
e radiación *f* hacia adelante
i radiazione *f* in avanti
n voorwaartse straling
d Vorwärtsstrahlung *f*

2058 FORWARD STROKE, rec/tv
TRACE
f aller *m*,
trace *f* de gauche à droite
e traza *f* de la izquierda a la derecha
i andata *f*,
traccia *f* da sinistra a destra
n heenslag
d Bildspur *f*,
Hinlauf *m*

2059 FORWARD-STROKE rec/rep/tv
INTERVAL,
TRACE INTERVAL
f durée *f* de la ligne,
temps *m* d'aller
e duración *f* de la línea,
tiempo *m* de traza
i durata *f* della linea,
tempo *m* d'andata
n heenslagtijd,
lijnaftastduur
d Hinlaufzeit *f*,
Zeilenabtastdauer *f*

2060 FORWARD WIND KEY vr
Used to wind at speed to the start of a recording or the end of the tape.
f touche *f* défilement rapide
e tecla *f* de devanado rápido
i pulsante *f* d'avvolgimento rapido
n snelspoeltoets,
vooruitspoeltoets
d Vorlauftaste *f*

2061 FOUR-TRACK SOUND aud
RECORDING
f enregistrement *m* sonore à quatre pistes
e registro *m* sonoro de cuatro pistas
i registrazione *f* sonora a quattro piste
n viersporengeluidsopname
d vierspurige Schallaufzeichnung *f*

2062 FOUR-TUBE CAMERA ctv
f caméra *f* à quatre tubes cathodiques
e cámara *f* de cuatro tubos catódicos
i camera *f* a quattro tubi catodici
n vierbuizencamera
d Vierröhrenkamera *f*

2063 FOVEA, ct
FOVEA CENTRALIS
Central part of the yellow spot, thinner and depressed, containing almost exclusively cones and forming the sight of most distinct vision.
f fovéa *f*
e fovea *f*
i fossetta *f*, fovea *f*
n netvlieskuiltje *n*
d Netzhautgrube *f*

2064 FRAME fi/tv
f cadre *m*, image *f*
e fotograma *m*
i fotogramma *m*
n beeld *n*
d Bild *n*, Einzelbild *n*

2065 FRAME (US), rec/rep/tv
PICTURE (GB)
f image *f*
e imagen *f*
i immagine *f*
n beeld *n*
d Bild *n*, Vollbild *n*

2066 FRAME AERIAL RECEPTION, rep/tv
FRAME ANTENNA RECEPTION
f réception *f* à cadre
e recepción *f* con antena de cuadro
i ricezione *f* con antenna a quadro
n raamantenneontvangst
d Rahmenempfang *m*

2067 FRAME AMPLITUDE (US), rec/rep/tv
FRAME HEIGHT (US),
PICTURE ALTITUDE (GB),
PICTURE HEIGHT (GB)
f hauteur *f* d'image
e altura *f* de imagen
i altezza *f* d'immagine
n beeldhoogte
d Bildhöhe *f*, Vollbildhöhe *f*

2068 FRAME-AMPLITUDE rep/tv
CONTROL (US),
PICTURE-ALTITUDE CONTROL (GB),
VERTICAL-POSITIONING CONTROL
f réglage *m* de la hauteur d'image
e regulación *f* de la altura de imagen
i regolazione *f* dell'altezza d'immagine
n beeldhoogteregelaar
d Bildamplitudenregler *m*,
Bildhöhenverschiebung *f*,
Vollbildhöhenverschiebung *f*

2069 FRAME BAR, fi/tv
FRAME LINE,
MASK LINE
f barrette *f*
e línea *f* de separación
i interlinea *f* d'immagine
n scheidingslijn
d Bildsteg *m*, Bildstrich *m*

2070 FRAME BORDER (US), rep/tv
PICTURE EDGE (GB)
f bord *m* de l'image

e borde *m* de la imagen
i bordo *m* dell'immagine
n beeldkant
d Bildkante *f*,
Vollbildkante *f*

2071 FRAME BY FRAME DISPLAY fi/tv
f marche *f* image par image
e marcha *f* imagen a imagen
i marcia *f* immagine ad immagine
n beeld-voor-beeldprojectie
d Einergang *m*

2072 FRAME COIL (US), rep/tv
PICTURE COIL (GB)
f bobine *f* d'image
e bobina *f* de imagen
i bobina *f* d'immagine
n beeldspoel
d Bildablenkspule *f*,
Bildeinstellungsspule *f*,
Vollbildablenkspule *f*

2073 FRAME DEFLECTION (US), rec/tv
PICTURE DEFLECTION (GB)
f déviation *f* de l'image
e desviación *f* de la imagen
i deflessione *f* dell'immagine,
deviazione *f* dell'immagine
n beeldafbuiging
d Bildablenkung *f*,
Vollbildablenkung *f*

2074 FRAME DISTORTION (US), rep/tv
PICTURE DISTORTION (GB)
f distorsion *f* d'image
e distorsión *f* de imagen
i distorsione *f* d'immagine
n beeldvervorming
d Bildverzerrung *f*,
Vollbildverzerrung *f*

2075 FRAME FREQUENCY (US), rec/tv
FRAME REPETITION RATE (US),
PICTURE FREQUENCY (GB),
PICTURE REPETITION RATE (GB)
f fréquence *f* d'image
e frecuencia *f* de imagen
i frequenza *f* d'immagine
n beeldfrequentie
d Bildfolgefrequenz *f*,
Bildwechselfrequenz *f*,
Vollbildfrequenz *f*

2076 FRAME INPUT TRANSFORMER rep/tv
(US),
PICTURE INPUT TRANSFORMER
(GB)
f transformateur *m* d'entrée d'image
e transformador *m* de entrada de imagen
i trasformatore *m* d'ingresso d'immagine
n beeldingangstransformator
d Bildeingangstransformator *m*,
Vollbildeingangstransformator *m*

2077 FRAME LINEARITY CONTROL rep/tv
(US),
PICTURE LINEARITY CONTROL (GB)
f réglage *m* de linéarité
e regulación *f* de linealidad
i regolazione *f* di linearità
n lineariteitsregeling
d Linearitätsregelung *f*

2078 FRAME OUTPUT rep/tv
TRANSFORMER (US),
PICTURE OUTPUT
TRANSFORMER (GB)
f transformateur *m* de sortie d'image
e transformador *m* de salida de imagen
i trasformatore *m* d'uscita d'immagine
n beelduitgangstransformator
d Bildausgangstransformator *m*

2079 FRAME PERIOD (US), tv
PICTURE PERIOD (GB)
A time interval equal to the reciprocal of the frame (picture) frequency.
f période *f* d'image
e período *m* de imagen
i periodo *m* d'immagine
n beeldperiode
d Bildperiode *f*,
Vollbildperiode *f*

2080 FRAME RATE (GB), tv
PICTURE DURATION (US)
f durée *f* de l'image
e duración *f* de la imagen
i durata *f* dell'immagine
n beeldduur
d Bilddauer *f*

2081 FRAME RETRACE TIME (US), rec/tv
PICTURE FLYBACK TIME (GB)
f temps *m* de retour d'image
e tiempo *m* de retorno de imagen
i tempo *m* di ritorno d'immagine
n beeldterugslagtijd
d Bildrücklaufzeit *f*

2082 FRAME SEQUENTIAL ctv
SYSTEM (US),
PICTURE SEQUENTIAL
SYSTEM (GB)
f système *m* à séquence d'images
e sistema *m* de secuencia de imágenes
i sistema *m* a sequenza d'immagini
n volgbeeldkleurensysteem *n*
d Zeitfolgeverfahren *n*

2083 FRAME SIMULTANEOUS ctv
SYSTEM (US),
PICTURE SIMULTANEOUS
SYSTEM (GB)
f système *m* additif de télévision couleur
e sistema *m* aditivo de televisión en colores
i sistema *m* additivo di televisione a colori
n televisiesysteem *n* met simultaan kleurenbeeld
d Simultanverfahren *n*

2084 FRAME SIZE (US), tv
PICTURE SIZE (GB)
f format *m* d'image
e formato *m* de imagen
i formato *m* d'immagine
n beeldformaat *n*

d Bildformat *n*,
Bildgrösse *f*,
Vollbildgrösse *f*

2085 FRAME SLIP (US), dis
PICTURE SLIP (GB),
VERTICAL SLIP
Lack of exact synchronization of the vertical scanning and the incoming signal whereby the reproduced picture progresses vertically.
f décalage *m* vertical,
glissement *m* vertical
e deslizamiento *m* vertical,
desplazamiento *m* vertical
i scorrimento *m* verticale,
spostamento *m* verticale
n verticale beeldverschuiving
d vertikale Bildverschiebung *f*

2086 FRAME SWEEP UNIT (US), tv
PICTURE SWEEP UNIT (GB)
f générateur *m* de déviation d'image
e generador *m* de desviación de imagen
i generatore *m* di deflessione d'immagine,
generatore *m* di deviazione d'immagine
n beeldafbuigingsgenerator
d Vollbildablenkgenerator *m*,
Vollbildkippgerät *n*

2087 FRAME SWEEP VOLTAGE (US), tv
PICTURE SWEEP VOLTAGE (GB)
f tension *f* de déviation d'image
e tensión *f* de desviación de imagen
i tensione *f* di deflessione d'immagine,
tensione *f* di deviazione d'immagine
n beeldafbuigingsspanning
d Vollbildkippspannung *f*

2088 FRAME SYNC PULSE (US), tv
PICTURE SYNC PULSE (GB)
f impulsion *f* de synchronisation d'image
e impulso *m* de sincronización de imagen
i impulso *m* di sincronizzazione d'immagine
n beeldsynchronisatie-impuls
d Bildgleichlaufimpuls *m*,
Bildsynchronisierimpuls *m*,
Vollbildsynchronisierimpuls *m*

2089 FRAME rec/rep/tv
SYNCHRONIZATION (US),
PICTURE SYNCHRONIZATION (GB)
f synchronisation *f* d'image
e sincronización *f* de imagen
i sincronizzazione *f* d'immagine
n beeldsynchronisatie
d Bildsynchronisierung *f*,
Vollbildsynchronisierung *f*

2090 FRAME SYNCHRONIZATION rep/tv
CONTROL (US),
PICTURE SYNCHRONIZATION
CONTROL (GB)
f dispositif *m* de réglage de la stabilité verticale
e regulador *m* de la estabilidad vertical
i regolatore *m* della stabilità verticale
n verticale stabiliteitsregelaar
d vertikaler Stabilitätsregler *m*

2091 FRAME TIMEBASE (US), tv
PICTURE TIMEBASE (GB)
f base *f* de temps d'image
e base *f* de tiempo de imagen
i base *f* di tempi d'immagine,
tempo *m* di durata d'immagine
n beeldtijdbasis
d Vollbildablenkgerät *n*,
Vollbildzeitbasis *f*

2092 FRAME TIMEBASE tv
OSCILLATION (US),
PICTURE TIMEBASE
OSCILLATION (GB)
f oscillation *f* de la base de temps d'image
e oscilación *f* de la base de tiempo de imagen
i oscillazione *f* a frequenza d'immagine,
oscillazione *f* della base di tempi d'immagine
n beeldtijdbasistrilling
d Bildkippschwingung *f*,
Vollbildkippschwingung *f*

2093 FRAMES PER SECOND fi/tv
f cadres *pl* par seconde
e fotogramas *pl* por segundo
i fotogrammi *pl* per secondo
n beelden *pl* per seconde
d Bilder *pl* je Sekunde

2094 FRAMING rec/rep/tv
The adjustment of the surface swept out by the spot into correct position in relation to the face of a camera or picture tube.
f cadrage *m*
e encuadre *m*
i centratura *f*
n beeldinraming,
beeldkader *n*
d Bildfang *m*,
Bildlageeinstellung *f*,
Höhe-Breite-Einstellung *f*

2095 FRAMING CONTROL rep/tv
A control that adjusts the cent(e)ring, width or height of the image on a TV receiver screen.
f réglage *m* du cadrage
e regulación *f* del encuadre
i regolazione *f* della centratura
n beeldinramingsregeling
d Bildfangregler *m*,
Bildregler *m*

2096 FRAMING MASK rec/rep/tv
Mask used in TV receiving sets to frame the picture.
f masque *m* de cadrage
e recuadro *m*
i inquadratura *f*
n beeldmasker *n*
d Bildmaske *f*

2097 FREE ECHOES aud/stu
f échos *pl* francs
e ecos *pl* de itineraria libre
i echi *pl* di propagazione libera
n ongehinderde echo's *pl*
d ungehinderte Echos *pl*

2098 FREE-FLOATING CAMERA MOUNT med
f montage *m* flottant de la caméra
e montaje *m* flotante de la cámara
i montaggio *m* flottante della camera
n zwevende cameramontage
d schwebende Kameramontage *f*

2099 FREE-RUNNING SWEEP tv
f balayage *m* non-synchronisé, déviation *f* non-synchronisée
e barrido *m* no sincronizado, desviación *f* no sincronizada
i deflessione *f* non sincronizzata, deviazione *f* non sincronizzata
n niet-gesynchroniseerde afbuiging
d nichtsynchronisierte Ablenkung *f*

2100 FREE-RUNNING SYSTEM tv
TV system in which the field frequency is independent of the main frequency and deduced from the line frequency which is kept constant.
f système *m* libre
e sistema *m* libre
i sistema *m* libero
n onvergrendeld systeem *n*
d Freilaufsystem *n*, unverkoppeltes System *n*

2101 FREE-SPACE PROPAGATION tv
f rayonnement *m* direct
e radiación *f* directa
i radiazione *f* diretta
n directe straling
d Freiraumausbreitung *f*

2102 FREEDOM FROM FLICKER tv
f sans papillotement
e sin parpadeo
i senza sfarfallio
n flikkerloosheid
d Flimmerfreiheit *f*

2103 FREEZE FRAME, HOLD FRAME, STOP FRAME rep/tv
A single frame of a moving shot arrested for as long as required electronically to concentrate on a phase of action.
f arrêt *m* d'image, tirage *m* en image fixe
e parada *f* de imagen
i arresto *m* d'immagine
n beeldstop
d Bildstopp *m*, Standbildverlängerung *f*

2104 FREQUENCY ge
The rate of repetition of a cycle.
f fréquence *f*
e frecuencia *f*
i frequenza *f*
n frequentie
d Frequenz *f*

FREQUENCY BAND
see: BAND

2105 FREQUENCY CHANGER, FREQUENCY CONVERTER ge
A circuit comprising a local oscillator and a mixer.
f changeur *m* de fréquence
e convertidor *m* de frecuencia
i convertitore *m* di frequenza
n frequentieomzetter
d Frequenzwandler *m*

FREQUENCY DISCRIMINATOR
see: DISCRIMINATOR

2106 FREQUENCY DIVIDER ge
A device delivering output power at a frequency which is an exact integral submultiple of the input frequency.
f diviseur *m* de fréquence
e divisor *m* de frecuencia
i divisore *m* di frequenza
n frequentiedeler
d Frequenzteiler *m*

2107 FREQUENCY FLUCTUATION aud
f fluctuation *f* de la fréquence
e variación *f* de la frecuencia
i fluttuazione *f* della frequenza
n frequentieschommeling
d Frequenzschwankung *f*

2108 FREQUENCY INTERLACE, FREQUENCY INTERLACING, FREQUENCY INTERLEAVING tv
Interlace of interfering signal frequencies with the spectrum of harmonics of scanning frequencies in TV, to minimize the effect of interfering signals by altering the appearance of their pattern in successive scans.
f entrelacement *m* de fréquences, entrelacement *m* spectral
e entrelazado *m* de frecuencias
i interlacciamento *m* di frequenze
n frequentievervlechting
d Frequenzverkämmung *f*, Spektralverkämmung *f*

2109 FREQUENCY JUMPING cpl
f saut *m* de fréquence
e salto *m* de frecuencia
i salto *m* di frequenza
n frequentieverspringing
d Frequenzsprung *m*

2110 FREQUENCY LIMITATIONS vr
As the bandwidth that can be reproduced on magnetic tape is limited to about 10 octaves by the nature of the reproducing system, there are limitations to the frequencies used.
f limitations *pl* de fréquence
e limitaciones *pl* de frecuencia
i limitazioni *pl* di frequenza
n frequentiebegrenzingen *pl*
d Frequenzbegrenzungen *pl*

2111 FREQUENCY LOCK cpl
f verrouillage *m* de fréquence
e enclavamiento *m* de frecuencia

i bloccaggio *m* di frequenza
n frequentievergrendeling
d Frequenzverriegelung *f*

FREQUENCY MODULATION
see: F.M.

2112 FREQUENCY OVERLAP cpl/ctv
In colo(u)r TV, that part of the frequency band common to both monochrome and chrominance channels.
f bande *f* commune
e recubrimiento *m* de frecuencias
i sovrapposizione *f* di frequenze
n frequentieoverlapping
d Frequenzüberlappung *f*

2113 FREQUENCY RANGE ge
The range of frequencies over which a device may be considered useful with various circuit and operating conditions.
f gamme *f* de fréquences
e gama *f* de frecuencias
i gamma *f* di frequenze
n frequentiegebied *n*
d Frequenzbereich *m*

2114 FREQUENCY RESPONSE CHARACTERISTIC ge
Of a system or device, the variation with frequency of its transmission gain or loss.
f réponse *f* en fréquence
e respuesta *f* de frecuencia
i curva *f* di risposta
n frequentieresponsie
d Frequenzgang *m*

2115 FREQUENCY RESPONSE DISTORTION cpl
f distorsion *f* de la réponse de fréquence
e distorsión *f* de la respuesta de frecuencia
i distorsione *f* della risposta di frequenza
n frequentieresponsievervorming
d Frequenzgangverzerrung *f*

2116 FREQUENCY SLIP cpl
f glissement *m* de fréquence
e deslizamiento *m* de frecuencia
i slittamento *m* di frequenza
n slippen *n* van de synchronisatie
d Frequenzabweichung *f*

2117 FREQUENCY SWING cpl
f excursion *f* de fréquence
e excursión *f* de frecuencia
i deviazione *f* di frequenza
n frequentiezwaai
d Frequenzhub *m*

2118 FREQUENCY TOLERANCE tv
The extent to which the carrier frequency of a TV transmitter may be permitted to depart from the frequency assigned by the licensing authorities.
f tolérance *f* de fréquence
e tolerancia *f* de frecuencia
i tolleranza *f* di frequenza
n frequentietolerantie
d Frequenztoleranz *f*

2119 FREQUENCY TRANSLATOR, TRANSLATOR cpl
A device for preventing that the amplified signal from the TV transmitter is fed back into the receiver.
f transpositeur *m* de fréquence
e traslador *m* de frecuencia
i traspositore *m* di frequenza
n frequentievertaler
d Frequenzumsetzer *m*

2120 FRESNEL LENS stu
Stepped-surface lens used in spot lamps.
f lentille *f* de Fresnel
e lente *f* de Fresnel
i lente *f* di Fresnel
n fresnellens
d Fresnelsche Linse *f*

2121 FRESNEL REGION, NEAR FIELD aea
The electromagnetic field that exists in the near region, within a distance of 1 wavelength from a transmitting aerial (antenna).
f champ *m* proche,
zone *f* de Fresnel
e campo *m* próximo,
zona *f* de Fresnel
i campo *m* a breve distanza,
zona *f* di Fresnel
n fresnelgebied *n*,
nabije veld *n*
d Fresnel-Bereich *m*,
Nahfeld *n*

2122 FRINGE AREA rep/tv
Area of reduced field strength beyond the useful service area of a transmitter where reception is not constantly reliable.
f zone *f* limite de propagation
e zona *f* límite de propagación
i zona *f* limite di propagazione
n randgebied *n*
d Randgebiet *n*

2123 FRINGE EFFECT dis
f effet *m* de bord
e efecto *m* de borde
i effetto *m* di bordo
n randeffect *n*
d Randeffekt *m*

2124 FRINGING crt
The outward bulging of the electrostatic field at the edges of the deflection plates of a cathode-ray tube.
f déformation *f* de champ,
distorsion *f* de champ
e deformación *f* de campo,
distorsión *f* de campo
i deformazione *f* di campo,
distorsione *f* di campo
n veldvervorming
d Feldverzerrung *f*

2125 FRINGING ctv/tv
Effect caused by imperfect registration in the superimposition of two or more images.

f effet *m* de défaut de calage
e efecto *m* de registración errónea
i effetto *m* di sovrapposizione erronea
n effect *n* van slechte dekking, gekleurde randen *pl*
d Fehlüberdeckungseffekt *m*

2126 FRONT CREDITS, OPENING CREDITS fi
f générique *m* de début
e cinta *f* inicial de nombres de colaboradores
i nastro *m* iniziale di nomi di collaboratori
n begintitels *pl*
d Namensvorspann *m*

2127 FRONT PANEL rep/tv
f panneau *m* frontal
e panel *m* frontal
i piastra *f* frontale
n frontpaneel *n*
d Frontplatte *f*

2128 FRONT PANEL GRILLE aud/rep/tv
f grille *f* de hautparleur
e rejilla *f* de altavoz
i griglia *f* d'altoparlante
n luidsprekerrooster *n*
d Lautsprechergitter *n*

2129 FRONT PORCH tv
The interval of time immediately proceeding the line sync pulse during which the video signal is maintained at blanking level.
f palier *m* avant
e rellano *m* anterior
i pianerottolo *m* anteriore
n voorstoep
d vordere Schwarzschulter *f*

2130 FRONT PROJECTION tv
A projection TV system that uses a non-translucent reflecting screen.
f projection *f* directe
e proyección *f* frontal
i proiezione *f* diretta, proiezione *f* frontale
n frontprojectie, opzichtprojectie
d Aufprojektion *f*

FULL BODY
see: BLACK BODY

FULL-BODY RADIATION
see: BLACK-BODY RADIATION

2131 FULL COLO(U)R ct
Surface colo(u)rs which are produced with the maximum colo(u)rfulness obtainable.
f couleur *f* intensive
e color *m* intensivo
i colore *m* intensivo
n intensieve kleur
d intensive Farbe *f*, kräftige Farbe *f*

2132 FULL COMPATIBILITY ctv/tv
Reception by a colo(u)r TV receiver of a colo(u)r TV signal when the chromaticity information is within the black-and-white video band of frequencies.
f compatibilité *f* réciproque
e compatibilidad *f* recíproca
i compatibilità *f* reciproca
n wederkerige compatibiliteit
d reziproke Kompatibilität *f*

2133 FUNCTION GENERATOR ge
A signal generator designed to produce test signals of various different wave-forms over a wide range of low frequencies.
f générateur *m* de fonction
e generador *m* de función
i generatore *m* di funzione
n functiegenerator
d Funktionsgeber *m*, Funktionsgenerator *m*

2134 FURNACE MONITOR cpl/tv
f moniteur *m* de four
e monitor *m* de horno
i monitore *m* di forno
n ovenmonitor
d Ofenmonitor *m*

2135 FUSION FREQUENCY ct
Frequency of succession of retinal images above which their differences of luminosity of colo(u)r are not longer perceptible.
f fusion *f* de sensations de couleurs
e fusión *f* de sensaciones de colores
i fusione *f* di sensazioni di colori
n versmeltingsfrequentie
d Verschmelzungsfrequenz *f*

G

2136 G, ct/ctv
GREEN PRIMARY
f couleur *f* primaire verte, vert *m* primaire
e color *m* primario verde, verde *m* primario
i colore *m* fondamentale verde, verde *m* primario
n groene grondkleur, primaire kleur groen
d grüne Primärfarbe *f*

2137 G BLACK LEVEL, ctv
GREEN BLACK LEVEL
The minimum permissible level of the G signal.
f niveau *m* minimal pour le signal vert
e nivel *m* mínimo para la señal verde
i livello *m* minimo per il segnale verde
n minimumniveau *n* voor het signaal groen, zwartniveau *n* in het groene signaal
d Mindestpegel *m* für das Grünsignal

2138 G PEAK LEVEL, ctv
GREEN PEAK LEVEL
The maximum permissible level of the G signal.
f niveau *m* maximal pour le signal vert
e nivel *m* máximo para la señal verde
i livello *m* massimo per il segnale verde
n maximumniveau *n* voor het signaal groen
d Höchstpegel *m* für das Grünsignal

2139 G-Y AXIS, ctv
GREEN COLO(U)R DIFFERENCE AXIS
f axe *m* V-Y
e eje *m* V-Y
i asse *m* V-Y
n as van het groene-kleurverschilsignaal, G-Y-as
d G-Y-Achse *f*

2140 G-Y MATRIX, svs
GREEN COLO(U)R DIFFERENCE MATRIX
f matrice V-Y
e matriz *f* V-Y
i matrice *f* V-Y
n G-Y-matrix
d G-Y-Matrize *f*

2141 G-Y MODULATOR, ctv
GREEN COLO(U)R DIFFERENCE MODULATOR
f modulateur *m* V-Y
e modulador *m* V-Y
i modulatore *m* V-Y
n G-Y-modulator, modulator voor het groene-kleurverschil-signaal
d G-Y-Modulator *m*

2142 G-Y SIGNAL, ctv
GREEN COLO(U)R DIFFERENCE SIGNAL
f signal *m* V-Y
e señal *f* V-Y
i segnale *m* V-Y
n G-Y-signaal *n*, groene-kleurverschilsignaal *n*
d G-Y-Signal *n*

2143 GABOR TUBE crt
A cathode-ray tube designed by Gabor developed to reduce tube length.
f tube *m* cathodique de Gabor
e tubo *m* catódico de Gabor
i tubo *m* catodico di Gabor
n gaborkatodestraalbuis
d Gabor-Katodenstrahlröhre *f*

2144 GAIN, ge
TRANSMISSION GAIN
The increase in power, usually expressed in decibels, in transmission from one point to another.
f amplification *f* de puissance, gain *m* de transmission
e amplificación *f* de potencia
i amplificazione *f* di potenza
n vermogensversterking
d Leistungsverstärkung *f*

GAIN
see: AMPLIFICATION

2145 GAIN-BANDWIDTH-PRODUCT rep/tv
The midband gain of an amplifier stage multiplied by the bandwidth in megacycles.
f produit *m* amplification/largeur de bande
e producto *m* amplificación/ancho de banda
i prodotto *m* amplificazione/larghezza di banda
n versterking/bandbreedte-produkt *n*
d Verstärkung/Bandbreite-Produkt *n*

2146 GAIN CONTROL, cpl
VOLUME CONTROL
A device for varying the amplification of an amplifier.
f réglage *m* de puissance
e regulación *f* de ganancia
i regolazione *f* di guadagno
n volumeregeling
d Lautstärkeregelung *f*, Verstärkungsregelung *f*

GAIN FUNCTION (GB)
see: DIRECTIVITY FUNCTION

GAIN INEQUALITY
see: CHROMINANCE/LUMINANCE GAIN INEQUALITY

2147 GAIN MARGIN cpl
The amount of increase in gain that would cause oscillation in a feedback control system.
f marge *f* de gain
e margen *m* de ganancia
i margine *m* di guadagno
n versterkingsmarge
d Verstärkungsgrenze *f*

2148 GALLOWS ARM stu
f potence *f* pour projecteur
e soporte *m* de proyector
i sostegno *m* di proiettore
n schijnwerperarm
d Scheinwerfergalgen *m*

2149 GAMMA, tv
OVERALL GAMMA
In TV, a relationship between the relative luminance of two points in the transmitted scene and the relative luminance of the corresponding points on the receiver screen.
f gamme *f*
e gama *f*
i gamma *f*
n gamma
d Gamma *n*

2150 GAMMA CONTROL tv
f réglage *m* de la gamme
e regulación *f* de la gama
i regolazione *f* della gamma
n gammaregeling
d Gammaregelung *f*

2151 GAMMA CORRECTION, tv
GAMMA CORRECTION CIRCUIT
Correction circuit of the effective value of gamma by introducing a non-linear output-input characteristic.
f correction *f* de la gamme
e corrección *f* de la gama
i correzione *f* della gamma
n gammacorrectie
d Gammakorrektur *f*

2152 GAMMA CORRECTOR tv
f correcteur *m* de gamme
e corrector *m* de gama
i correttore *m* di gamma
n gammacorrector
d Gammakorrektor *m*

2153 GAMMA ERROR, dis
GRADATION DISTORTION
f distorsion *f* des demiteintes,
erreur *f* de gradation
e error *m* de gradación
i errore *m* di gradazione
n gradatiefout
d Gradationsfehler *m*

2154 GAMMA OF PICTURE TUBE, crt
PICTURE TUBE GAMMA
The exponent of that power law which is used to approximate the curve of output magnitude versus input magnitude over the region of interest.
f gamme *f* du tube image
e gama *f* del tubo imagen
i gamma *f* del tubo immagine
n beeldbuisgamma
d Bildröhregamma *n*

GAMMATE (TO)
see: CORRECT (TO) THE GAMMA

2155 GAMUT OF CHROMATICITIES ctv
f gamme *f* chromatique
e gama *f* cromática
i gamma *f* cromatica
n kleursoortenschaal
d Farbartenskala *f*

GANTRY
see: CATWALK

2156 GAP vr
The small air space in an otherwise closed ferromagnetic circuit.
f entrefer *m*
e entrehierro *m*
i intraferro *m*
n spleet
d Spalt *m*

2157 GAP ADJUSTMENT, aud/vr
GAP SETTING
f équilibrage *m* d'entrefer
e ajuste *m* de entrehierro
i aggiustaggio *m* d'intraferro
n spleetinstelling
d Spalteinstellung *f*

2158 GAP EFFECT vr
Self-erasure occurring at high audio frequencies in tape reproduction when the gap of the playback head is equal to or greater than the wavelength of the recorded pattern on the tape.
f effet *m* d'entrefer
e efecto *m* de entrehierro
i effetto *m* d'intraferro
n spleeteffect *n*
d Spalteffekt *m*

GAP FILLER
see: AUXILIARY TRANSMITTER

2159 GAP LENGTH vr
In magnetic recording, the distance between adjacent surfaces of the poles of a magnetic head.
f largeur *f* d'entrefer
e distancia *f* de entrehierro
i larghezza *f* d'intraferro
n spleetlengte
d Spaltbreite *f*

2160 GAP LOSS aud/vr
f perte *f* d'entrefer
e pérdida *f* de entrehierro
i perdita *f* d'intraferro
n spleetdemping,
spleetverlies *n*
d Spaltdämpfung *f*,
Spaltverlust *m*

2161 GAP SWITCHING, aud/rec/rep/tv
MAKE-BEFORE-BREAK
SWITCHING
f commutation *f* à séquence travail-repos
e conmutación *f* de secuencia trabajo-reposo
i commutazione *f* lavoro-riposo ad azionamento successivo
n maak-voor-verbreek-schakeling
d Folge-Arbeits-Ruhe-Schaltung *f*, unterbrechungslose Schaltung *f*

2162 GAPPING SWITCH, tv
MAKE-BEFORE-BREAK SWITCH
f commutateur *m* à séquence travail-repos
e conmutador *m* de secuencia trabajo-reposo
i commutatore *m* lavoro-riposo ad azionamento successivo
n maak-voor-verbreek-schakelaar
d Folge-Arbeits-Ruhe-Schalter *m*, unterbrechungsloser Schalter *m*

2163 GAS FOCUSING crt
Focusing an electron beam by the presence of an ionized gas.
f concentration *f* du faisceau par gaz ionisé
e enfoque *m* del haz por gas
i concentrazione *f* per ionizzazione, focalizzazione *f* per ionizzazione
n gasfocussering
d Gasfokussierung *f*, Ionenfokussierung *f*

2164 GAS LASER vr
A laser using a low-energy discharge in a mixture of helium and neon gases to generate a continuous, very intense, and very sharply defined beam of infrared radiation.
f laser *m* à gaz
e laser *m* de gas
i laser *m* a gas
n gaslaser
d Gaslaser *m*

2165 GATE, cpl
GATE CIRCUIT
A network having one or more inputs which open or close a channel according to the combination of stimuli applied to the inputs.
f circuit *m* de porte
e circuito *m* de compuerta
i circuito *m* ad impulsi periodici di sblocco
n poort, poortschakeling
d Tor *n*, Torschaltung *f*

GATE
see: DOUBLE LIMITER

2166 GATE GENERATOR cpl
A circuit for generating gate pulses.
f générateur *m* d'impulsions de déblocage
e generador *m* de impulsos de desbloqueo
i generatore *m* d'impulsi di sblocco
n poortimpulsgenerator, sleutelimpulsgenerator
d Auftastgenerator *m*, Torimpulsgenerator *m*

2167 GATE WIDTH, cpl
GATING TIME
Time interval during which a gating circuit allows a signal to pass.
f temps *m* de déclenchement
e intervalo *m* de paso
i durata *f* dell'impulso di sblocco
n sleuteltijd
d Auftastzeit *f*, Torschaltzeit *f*

2168 GATED AUTOMATIC cpl
GAIN CONTROL
System of automatic gain control used in TV reception in which the biasing control voltage is made independent of the changing picture content by sampling the carrier during the black level period when the modulation remains constant.
f antifading *m* à déclenchement périodique
e regulación *f* automática de volumen de desbloqueo periódico
i regolazione *f* automatica di volume a sblocco periodico
n automatische volumeregeling met periodieke deblokkering, gesleutelde automatische versterkingsregeling
d torgesteuerte selbsttätige Verstärkungsregelung *f*

2169 GATED BEAM TUBE crt
f tube *m* de phase
e válvula *f* de compuerta
i valvola *f* di fase
n gesleutelde bundelbuis
d Doppelstromtor-Strahlsteuerungsröhre *f*

2170 GATING cpl
The operation or activation of a gate by selecting portions of a cyclic waveform between certain upper and lower limits so as to obtain a waveform of required magnitude and length.
f déclenchement *m* périodique
e desbloqueo *m* periódico
i sblocco *m* periodico
n poortbesturing, sleutelen *n*
d Torsteuerung *f*

2171 GATING PULSE cpl
A pulse that modifies the operation of a circuit for duration of a pulse.
f impulsion *f* de déclenchement
e impulso *m* de desbloqueo
i impulso *m* di sblocco
n deblokkeerimpuls, sleutelimpuls
d Auftastimpuls *m*

2172 GATING SIGNAL cpl
A periodic signal that keys on or

activates a circuit so as to permit the production or transmission of another signal.
f signal *m* de débloquage, signal *m* de porte
e señal *f* de compuerta, señal de desbloqueo
i segnale *m* di porta, segnale *m* di sblocco
n deblokkeersignaal *n*, poortsignaal *n*, sleutelsignaal *n*
d Auftastsignal *n*, Gattersignal *n*, Torsignal *n*

GAUSSIAN CIRCUIT
see: CLOCHE CIRCUIT

GELATIN FILTER
see: COLO(U)R SCREEN

2173 GENERAL COLO(U)R RENDERING INDEX — ct
A colo(u)r rendering index which refers to the quality of an illuminant in respect of any collection of objects which might be viewed.
f indice *m* général de reproduction de la couleur
e índice *m* general de reproducción del color
i indice *m* generale di riproduzione del colore
n algemene kleurweergevingsindex
d allgemeiner Farbwiedergabe-Index *m*

2174 GENERAL-PURPOSE TEST-SIGNAL GENERATOR — tv
f générateur *m* de mire universel
e generador *m* de imagen de prueba universal
i generatore *m* d'immagine di prova universale
n universele toetsbeeldgenerator
d Universaltestbildgeber *m*

GENERAL VERTICAL LINEARITY CONTROL
see: FIELD LINEARITY CONTROL

2175 GENLOCKING — tv
f système *m* à générateur synchroniseur, système *m* à générateur verrouillé
e sistema *m* de generador clavado, sistema *m* de generador sincronizador
i sistema *m* a generatore bloccato, sistema *m* a generatore sincronizzatore
n systeem *n* met synchroniserende generator, systeem *n* met vergrendelde generator
d Genlock-Betriebsweise *f*

2176 GEOMETRIC DISTORTION, GEOMETRICAL DISTORTION, PICTURE GEOMETRY FAULT — dis
A picture fault in which the geometrical proportions of forms transmitted are distorted.
f distorsion *f* géométrique
e distorsión *f* geométrica
i distorsione *f* geometrica
n geometrische vertekening, geometrische vervorming
d Geometriefehler *m*, geometrische Verzeichnung *f*, geometrische Verzerrung *f*

2177 GEOMETRIC INSTABILITY — rep/tv
f instabilité *f* géométrique
e inestabilidad *f* geométrica
i instabilità *f* geometrica
n geometrische instabiliteit
d geometrische Instabilität *f*

2178 GEOMETRY HUM, POSITIONAL HUM — dis
f ronflement *m* de géométrie
e imagen *f* de prueba de geometría
i ronzio *m* di geometria
n geometriebrom
d Geometriebrumm *m*

2179 GEOMETRY TEST CARD, LINEARITY TEST CARD — tv
A test card for investigating the geometric distortion of a picture transmitted by TV.
f mire *f* de géométrie
e mira *f* de geometría
i immagine *f* di prova per la geometria
n lineariteitstoetsbeeld *n*
d Geometrietestbild *n*

2180 GEO-STATIONARY SATELLITE — tv
f satellite *m* en orbite géostationnaire
e satélite *m* en órbita geoestacionaria
i satellite *m* in orbita geostazionaria
n in vaste baan om de aarde bewegende satelliet
d geostationärer Satellit *m*, Satellit *m* in geostationärer Umlaufbahn

2181 GETAWAY (US), MOVABLE SCENE (GB) — stu
f coulisse *f* de studio
e bastidor *m*
i quinta *f* di studio
n studiocoulisse
d Studiokulisse *f*

GHOST
see: DOUBLE IMAGE

2182 GHOST IMAGE — fi
f filage *m* image
e hilatura *f* de la imagen
i filatura *f* dell'immagine sullo schermo
n rafelend beeld *n*
d Ziehen *n* des Bildes

GHOST SIGNALS
see: DOUBLE IMAGE SIGNALS

2183 GLANCING LIGHT — stu
f lumière *f* frisante
e luz *f* de soslayo
i luce *f* di striscio
n schamplicht *n*
d Streiflicht *n*

2184 GLARE ct
Condition of vision in which there is discomfort or a reduction in the ability to see significant objects or both.
f éblouissement *m*
e deslumbramiento *m*
i abbagliamento *m*
n verblinding
d Blendung *f*

2185 GLARE INDEX ct
A numerical index which enables the discomfort glare from a light installation to be assessed and the permissible limits of discomfort glare from an installation to be prescribed.
f échelle *f* d'éblouissements
e escala *f* de deslumbramientos
i scala *f* d'abbagliamenti
n verblindingenschaal
d Blendungsskala *f*

2186 GO (TO) ON THE AIR stu/tv
f prendre v l'antenne
e andar v en el aire
i andare v in onda
n de uitzending beginnen v,
in de ether komen v
d auf Antenne kommen v

2187 GOBO aud
A screen used to absorb and/or attenuate the sound reaching the microphone.
f écran *m* sonore
e pantalla *f* sonora
i schermo *m* sonoro
n geluidsscherm *n*
d Schallschluckschirm *m*

GOBO (US)
see: FLAG

2188 GRADATION, rep/tv
KEY
Slope of the tangent to the brightness reproduction characteristic at each point.
f gradation *f*
e gradación *f*
i gradazione *f*
n gradatie
d Abstufung *f*,
Gradation *f*

GRADATION DISTORTION
see: GAMMA ERROR

GRADATION TEST CARD
see: CONTRAST TEST CARD

2189 GRADED NEUTRAL DENSITY crt
FILTER
Exposure corrector attached to the face of the cathode-ray tube.
f filtre *m* neutre à graduation
e filtro *m* neutro graduado
i filtro *m* grigio a graduazione
n getrapt neutraal filter *n*
d abgestuftes Graufilter *n*

2190 GRADUAL FILTER ct
f filtre *m* dégradé
e filtro *m* graduado
i filtro *m* graduato
n verloopfilter *n*
d Verlauffilter *m*

2191 GRAM OPERATOR rep/stu/tv
The man who produces at a signal selected fragments of music from disk or tape which he has rehearsed and selected beforehand.
f opérateur *m* de disque et bande
e operador *m* de disco y cinta
i operatore *m* di disco e nastro
n plaat- en bandoperateur
d Platten-und-Bandoperator *m*

2192 GRAPHIC DESIGN ani/tv
f graphique *m*
e diseño *m* gráfico
i disegno *m* grafico
n grafisch ontwerp *n*,
grafische vormgeving
d Graphik *f*

2193 GRAPHIC DESIGNER tv
The man who handles the typography suited to the proportions of the screen.
f dessinateur *m* graphique,
graphiste *m*
e diseñador *m* gráfico
i disegnatore *m* grafico
n grafische ontwerper
d Graphiker *m*,
graphischer Zeichner *m*

2194 GRAPHICAL DETERMINATION ct
OF THE COMPLEMENTARY
WAVELENGTH
The wavelength corresponding to the intersection with the spectrum locus of the straight line drawn from the point representing the sample and extended through the reference point is the complementary wavelength.
f détermination *f* graphique de la longueur d'onde complémentaire
e determinación *f* gráfica de la longitud de onda complementaria
i determinazione *f* grafica della lunghezza d'onda complementare
n grafische bepaling van de complementaire golflengte
d graphische Ermittlung *f* der kompensativen Wellenlänge

2195 GRAPHICAL DETERMINATION ct
OF THE DOMINANT WAVELENGTH
The wavelength corresponding to the intersection with the spectrum locus of the straight line drawn from the reference point and extended through the point representing the sample is the dominant wavelength.
f détermination *f* graphique de la longueur d'onde dominante
e determinación *f* gráfica de la longitud de onda dominante

i determinazione *f* grafica della lunghezza d'onda dominante
n grafische bepaling van de dominerende golflengte
d graphische Ermittlung *f* der dominierenden Wellenlänge, graphische Ermittlung *f* der farbtongleichen Wellenlänge

2196 GRAPHITE COATING crt
f couche *f* en graphite
e capa *f* de grafito
i rivestimento *m* di grafite
n grafietlaag
d Graphitschicht *f*

2197 GRATICULE crt
Transparency marked with limits or guide marks, placed over a signal trace in a cathode-ray tube.
f réticule *f*
e gratícula *f*
i graticola *f*
n rooster *n*
d Gitter *n*

2198 GRATING opt
f quadrillage *m*
e rejilla *f* de difracción
i quadrettato *m*
n raster *n*
d Gitter *n*

2199 GRATING GENERATOR opt
f générateur *m* de quadrillage
e generador *m* de rejilla de difracción
i generatore *m* di quadrettato
n rastergenerator
d Gittergenerator *m*

2200 GRATING REFLECTOR aea
An openwork metal structure designed to provide a good reflecting surface for microwave aerials (antennae).
f réflecteur *m* à quadrillage
e reflector *m* de rejilla
i riflettore *m* a quadrettato
n rasterreflector
d Gitterreflektor *m*

2201 GRAVURE COATING vr
Application of the dope by means of indented gravure rollers.
f revêtement *m* par rouleaux dentelés
e recubrimiento *m* por rodillos dentados
i ricoprimento *m* per rotoli dentati
n aanbrengen *n* van een laag door middel van getande rollen
d Beschichtung *f* mittels verzahnten Rollen

2202 GREEN ct
A colo(u)r in which the spectral components are confined mainly to the 500-600 nm band of the spectrum.
f vert *m*
e verde *m*
i verde *m*
n groen *n*
d Grün *n*

2203 GREEN ADDER ctv
f circuit *m* mélangeur pour le vert
e circuito *m* aditivo para el verde
i circuito *m* combinatore per il verde
n mengkring voor groen, optelschakeling voor groen
d Grün-Beimischer *m*

2204 GREEN APEX ctv
f point *m* de couleur de la primaire verte
e punto *m* de color del primario verde
i punto *m* di colore del primario verde
n kleurpunt *n* van de groene grondkleur
d Farbort *m* der grünen Primärfarbe

2205 GREEN BEAM crt/ctv
f faisceau *m* pour le vert
e haz *m* para el verde
i fascio *m* per il verde
n bundel voor groen
d Strahl *m* für Grün

2206 GREEN-BEAM MAGNET crt/ctv
f aimant *m* du faisceau pour le vert
e imán *m* del haz para el verde
i magnete *m* del fascio per il verde
n groene-bundelmagneet
d Grünstrahlmagnet *m*

GREEN BLACK LEVEL
see: G BLACK LEVEL

2207 GREEN CATHODE crt
f cathode *f* verte
e cátodo *m* verde
i catodo *m* verde
n groene katode
d grüne Katode *f*

GREEN COLO(U)R DIFFERENCE MATRIX
see: G-Y AXIS

GREEN COLO(U)R DIFFERENCE MATRIX
see: G-Y MATRIX

GREEN COLO(U)R DIFFERENCE MODULATOR
see: G-Y MODULATOR

GREEN COLO(U)R DIFFERENCE SIGNAL
see: G-Y SIGNAL

2208 GREEN CONVERGENCE CIRCUIT crt
f circuit *m* de convergence pour le vert
e circuito *m* de convergencia para el verde
i circuito *m* di convergenza per il verde
n groene-convergentieschakeling
d Grünkonvergenzschaltung *f*

2209 GREEN ELECTRON GUN, GREEN GUN crt/ctv
In a three-colo(u)r TV tube, the electron gun used to excite the green phosphor.
f canon *m* du vert
e cañón *m* del verde
i cannone *m* del verde
n kanon *n* voor het groen
d Grünstrahlsystem *n*

2210 GREEN GAIN CONTROL crt/ctv
A variable resistor used in the matrix of a three-gun colo(u)r picture TV receiver to adjust the intensity of the green primary signal.
f régleur *m* d'intensité pour le vert
e regulador *m* de intensidad para el verde
i regolatore *m* d'intensità per il verde
n intensiteitsregelaar voor groen
d Intensitätsregler *m* für Grün

2211 GREEN GRID crt
f grille *f* verte
e rejilla *f* verde
i griglia *f* verde
n groen rooster *n*
d grünes Gitter *n*

2212 GREEN HIGHS crt/ctv
f hautes fréquences *pl* pour le vert
e altas frecuencias *pl* para el verde
i componenti *pl* d'alta frequenza per il verde
n hoge frequenties *pl* voor groen
d Grünhöhe *f*,
höhe Frequenzen *pl* für Grün

2213 GREEN HORIZONTAL SHIFT MAGNET, crt
GREEN LATERAL SHIFT MAGNET,
GREEN POSITIONING MAGNET
f aimant *m* latéral pour le vert
e imán *m* lateral para el verde
i magnete *m* laterale per il verde
n lateraalmagneet voor het groen
d Grünschiebemagnet *m*

2214 GREEN LOWS crt/ctv
f basses fréquences *pl* pour le vert
e bajas frecuencias *pl* para el verde
i componenti *pl* di bassa frequenza per il verde
n lage frequenties voor groen
d Grüntiefe *f*,
niedrige Frequenzen *pl* für Grün

2215 GREEN-MAGENTA AXIS ctv
f axe *m* vert-magenta
e eje *m* verde-magenta
i asse *m* verde-magenta
n groen-magenta-as
d Grün-Magenta-Achse *f*

2216 GREEN MODULATOR ctv
f modulateur *m* pour le signal vert
e modulador *m* para la señal verde
i modulatore *m* per il segnale verde
n groensignaalmodulator
d Grünsignalmodulator *m*

2217 GREEN OVERCAST, crt
GREEN SHADING
The gradual change in the black level of the green component of a colo(u)r image or the signal giving rise to it.
f teinte *f* de vert
e tinta *f* de verde
i tinta *f* di verde
n groen zweem,
zweem naar het groen
d Stich *m* ins Grüne

GREEN PEAK LEVEL
see: G PEAK LEVEL

GREEN PRIMARY
see: G

2218 GREEN PRIMARY INFORMATION ctv
f information *f* du vert primaire
e información *f* del verde primario
i informazione *f* del verde primario
n informatie van de groene primaire kleur
d Information *f* der grünen Primärfarbe

2219 GREEN PRIMARY SIGNAL crt/ctv
f signal *m* du vert primaire
e señal *f* del verde primario
i segnale *m* del verde primario
n signaal *n* van de groene primaire kleur
d Signal *n* der grünen Primärfarbe

2220 GREEN RESTORER crt
The d.c. restorer for the green channel of a three-gun colo(u)r TV picture circuit.
f restauration *f* de la composante de courant continu du niveau du vert
e restauración *f* de la componente de corriente continua del nivel del verde
i reinserzione *f* della componente di corrente continua del livello del verde
n herstel *n* van het groenniveau
d Wiederherstellung *f* des Grünpegels

2221 GREEN SATURATION SCALE, crt
GREEN SCALE
f échelle *f* de verts
e escala *f* de verdes
i scala *f* di verdi
n groentrap
d Grünskala *f*

2222 GREEN SCREEN-GRID crt
f grille-écran *m* vert
e rejilla-pantalla *f* verde
i griglia-schermo *m* verde
n groen schermrooster *n*
d grünes Schirmgitter *n*

2223 GREEN SHIFT ctv
f décalage *m* du canevas vert
e desplazamiento *m* de la cuadrícula verde
i spostamento *m* del quadro rigato verde
n verschuiving van het groene raster
d Verschiebung *f* des grünen Rasters

2224 GREEN VIDEO VOLTAGE crt
The signal voltage output from the green section of a colo(u)r TV camera, or the signal voltage between the receiver matrix and the green-gun grid of a three-gun colo(u)r TV picture tube.
f tension *f* vidéo pour le vert
e tensión *f* video para el verde
i tensione *f* video per il verde
n beeldspanning voor groen
d Bildspannung *f* für Grün

2225 GREY ct
Any achromatic sensation of luminosity intermediate between black and white.
f gris *m*

e gris *m*
i grigio *m*
n grijs *n*
d Grau *n*

2226 GREY BODY, NON-SELECTIVE RADIATOR opt
A radiator whose spectral sensitivity is independent of wavelength within this region.
f corps *m* gris, radiateur *m* non-sélectif
e cuerpo *m* gris, radiador *m* no selectivo
i corpo *m* grigio, radiatore *m* non selettivo
n grijze straler
d grauer Körper *m*, grauer Strahl *m*, richtselektiver Strahler *m*

2227 GREY FILTER, NEUTRAL-DENSITY FILTER, NON-SELECTIVE ABSORBER ct
A body for which the spectral transmission factor is the same at all wavelengths.
f filtre *m* gris, filtre *m* neutre, filtre *m* non-sélectif
e filtro *m* absorbente no selectivo, filtro *m* neutro
i filtro *m* grigio, filtro *m* non selettivo
n dichtheidsfilter *n*, neutraal filter *n*
d Dichtefilter *n*, Graufilter *n*

GREY GLASS FILTER
see: AMBIENT LIGHT FILTER

2228 GREY SCALE ct
In a TV system, a scale of brightness values ranging from maximum to minimum brightness for the system.
f échelle *f* de gradations, échelle *f* de gris, mire *f* de gris
e escala *f* de gradaciones, escala *f* de grises
i scala *f* di gradazioni, scala *f* di grigi
n gradatieschaal, grijstrap
d Gradationstreppe *f*, Grauskala *f*, Graustufung *f*

2229 GREY SCALE SIGNAL tv
f signal *m* de gradation
e señal *f* de gradación
i segnale *m* di gradazione
n gradatiesignaal *n*, grijstrapsignaal *n*
d Treppensignal *n*

2230 GREY SCALE TRACKING ctv
The situation in which parallel channels are adjusted to reproduce an achromatic picture over the luminance range.
f reproduction *f* correcte de l'échelle de gris
e acuerdo *m* con la escala de grises
i riproduzione *f* corretta della scala di grigi
n correcte grijstrapreproduktie
d richtige Grauskalawiedergabe *f*

2231 GREY SCALE TRANSIENTS ctv
f transitions *pl* de l'échelle de gris
e transiciones *pl* de la escala de grises
i transizioni *pl* nella scala di grigi
n grijstrapovergangen *pl*
d Grauskala-Übergänge *pl*

2232 GREY SCALE VALUE ct
f luminosité *f* équivalente
e luminosidad *f* equivalente
i luminosità *f* equivalente
n grijswaarde
d Grauwert *m*

2233 GREY WEDGE, NEUTRAL WEDGE ctv
A wedge-shaped device containing a series of achromatic tones ranging from black to white.
f coin *m* des gris
e cuña *f* de los grises
i cuneo *m* dei grigi
n grijswig, trapverzwakker
d Graukeil *m*

2234 GRID MODULATION cpl
Modulation of a carrier wave effected by operating on the grid of an electron tube (valve) in any stage of the transmitter amplifier.
f modulation *f* dans la grille
e modulación *f* de rejilla
i modulazione *f* per griglia
n roostermodulatie
d Gittermodulation *f*

2235 GRID OF MIRRORS, MIRROR GRID tv
A component part of the eidophor system. The mirrors are arranged in parallel planes (so that no light from the lamp used can pass between them) and reflect light on to the face of a concave mirror, which carries the eidophor oil film.
f grille *f* à miroirs
e rejilla *f* de espejos
i griglia *f* di specchi
n spiegeltralie
d Spiegelgitter *n*

2236 GRID TUBE crt
A colo(u)r display tube of relatively simple design and potentially high efficiency still in the laboratory stage.
f tube *m* à grille de Lawrence
e tubo *m* de rejilla de Lawrence
i tubo *m* a griglia di Lawrence
n beeldbuis met lawrencerooster
d Bildröhre *f* mit Lawrence-Gitter, Gittermaskenröhre *f*

2237 GRIP (US), fi/th/tv
SCENE SHIFTER (GB),
STAGE HAND (GB)
f machiniste *m*
e obrero *m* escenógrafo,
tramoyista *m*
i macchinista *m*
n toneelknecht
d Atelierarbeiter *m*,
Baubühnenarbeiter *m*

2238 GROUND NOISE dis/vr
Residual system noise which in recording systems employing optical film or magnetic tape is principally due to the ultimate particle size of the recording medium.
f bruit *m* de fond
e ruido *m* de fondo
i rumore *m* di fondo
n eigen ruis
d Eigenrauschen *n*,
Grundrauschen *n*,
Wiedergabegeräusch *n*

GROUND STATION
see: EARTH STATION

GROUNDED ANODE CIRCUIT
see: CATHODE FOLLOWER

GROUNDED CATHODE AMPLIFIER
see: ANODE FOLLOWER

2239 GROUNDED GRID AMPLIFIER cpl
An amplifier circuit in which the control grid is at earth (ground) potential at the operating frequency.
f amplificateur *m* à grille à la masse
e amplificador *m* de rejilla a tierra
i amplificatore *m* con griglia a massa
n roosterbasisversterker
d Gitterbasisverstärker *m*

GROUP AERIAL,
GROUP ANTENNA
see: DIPOLE ARRAY

2240 GROUP DELAY EQUALIZATION cpl
f égalisation *f* du temps de transit de groupe
e igualamiento *m* del tiempo de tránsito de grupo
i uguagliamento *m* del tempo di transito di gruppo
n vereffening van de groeplooptijd
d Gruppenlaufzeitausgleich *m*

GROUP DELAY TIME
see: ENVELOPE DELAY

GROUP VELOCITY
see: ENVELOPE VELOCITY

2241 GUARD BAND ge/tv
A narrow frequency band left vacant between two channels to give a margin of safety against mutual interference.
f bande *f* de sécurité
e banda *f* de seguridad
i banda *f* di sicurezza
n veiligheidsfrequentieband
d Schutzband *n*,
Sicherheitsband *n*

2242 GUARD BAND vr
f piste *f* de sécurité
e pista *f* de seguridad
i pista *f* di sicurezza
n veiligheidsspoor *n*
d Sicherheitsspur *f*

2243 GUIDE RAILS, stu
PARALLELS
f rails *pl* pour caméra
e rieles *pl* de guía
i listelli *pl* di guida
n camerarails *pl*
d Laufschienen *pl*

2244 GUIDE SOUND (US), vr
GUIDE TRACK (GB)
An extra sound track recorded during the shooting of a film to assist in editing and postsynchronizing, but not used in the final version.
f piste *f* de contrôle,
piste *f* d'ordres
e pista *f* de guía
i pista *f* di guida
n geleidespoor *n*
d Hilfstonspur *f*,
Kontrollspur *f*

2245 GUN EFFICIENCY crt
The ratio between the beam current and the cathode current.
f rendement *m* de canon
e rendimiento *m* de cañón
i rendimento *m* di cannone
n kanonrendement *n*
d Strahlsystemausbeute *f*

2246 GUN MIKE, aud
RIFLE MIKE
Highly directional microphone that can be aimed like a rifle at the source of sound.
f microphone *m* canon
e micrófono *m* en forma de cañón,
micrófono *m* tubular
i microfono *m* direttivo in forma di cannone
n pistoolmicrofoon,
puntmicrofoon
d Punktrichtmikrophon *n*

H

2247 H AERIAL, aea
H ANTENNA
A dipole primary radiator with a dipole reflector.
f antenne *f* en H
e antena *f* en H
i antenna *f* in H
n H-antenne
d H-Antenne *f*

2248 H AMPLIFIER, crt
HORIZONTAL AMPLIFIER
f amplificateur *m* pour la déviation horizontale
e amplificador *m* horizontal,
amplificador *m* para la desviación horizontal
i amplificatore *m* per la deflessione orizzontale,
amplificatore *m* per la deviazione orizzontale
n versterker voor de horizontale afbuiging
d Horizontalverstärker *m*,
Verstärker *m* für horizontale Ablenkung

2249 H BAR CONTROL, tv
HORIZONTAL BAR CONTROL
f régleur *m* de barres horizontales
e regulador *m* de barras horizontales
i regolatore *m* di barre orizzontali
n horizontale-balkenregelaar
d Horizontalbalkenregler *m*

2250 HALATION, crt
HALO DISTURBANCES
Of a cathode-ray tube, an annular area surrounding a spot, due to the light emanating from the spot being reflected from the front and rear sides of the face.
f halo *m*
e halo *m*
i alone *m*
n halo,
overstraling
d Halo *n*,
Lichthof *m*,
Lichthofbildung *f*

HALF-LINE PULSE
see: BROAD PULSE

2251 HALF TONE opt/tv
A tone intermedian between the extreme lights and the extreme shades.
f demiteinte *f*
e media tinta *f*
i mezza tinta *f*
n halftint,
schakering,
tussentint
d Grauwert *m*,
Halbton *m*

2252 HALF-TONE DISTORTION dis/rep/tv
Distortion of the picture signal, the effect being that the values of the half tones are modified.
f distorsion *f* de demiteintes
e distorsión *f* de medias tintas
i distorsione *f* di grigi
n grijstrapvervorming
d Grauwertverzerrung *f*,
Halbtonverzerrung *f*

2253 HALF-TONE IMAGE, rep/tv
HALF-TONE PICTURE
An image in which there is a continuous range of contrast between black and white.
f image *f* nuancée
e imagen *f* matizada
i immagine *f* sfumata
n geschakeerd beeld *n*
d Halbtonbild *n*

2254 HALF-TONE RENDERING tv
f rendu *m* des demiteintes
e rendimiento *m* de las medias tintas
i riproduzione *f* dei grigi
n grijstrapweergeving
d Grauwertwiedergabe *f*

HALF-TRACK RECORDER
see: DOUBLE-TRACK RECORDER

HALF-TRACK TAPE RECORDING
see: DOUBLE-TRACK TAPE RECORDING

2255 HALF-WAVE DIPOLE aea
f dipôle *m* demi onde
e dipolo *m* de media onda
i dipolo *m* semionda
n halve-golflengtedipool
d Halbwellendipol *m*

2256 HAND-HELD MICROPHONE aud
f microphone *m* à main,
microphone *m* portatif
e micrófono *m* de mano
i microfono *m* a mano,
microfono *m* portatile
n handmicrofoon
d Handmikrophon *n*,
tragbares Mikrophon *n*

2257 HAND TEST, fi/tv
TEST STRIP
f bout *m* d'essai
e película *f* de prueba
i provino *m*
n filmproef
d Filmprobe *f*,
Probeaufnahme *f*,
Probestreifen *m*

2258 HANGOVER, ctv
TAILING
An effect in field-sequential colo(u)r TV produced by the pick-up tube whereby the signal produced by the scanning of one field persists during the scanning of a subsequent field.
f traînage *m*
e arrastre *m*
i persistenza *f* parassita
n nablijven *n*
d Fahnenbildung *f*,
Fahnenziehen *n*,
Überhangen *n*

2259 HARD IMAGE, crt/rep/tv
HARD PICTURE
A defect due to too high a voltage of the signal applied to the cathode-ray tube.
f image *f* dure
e imagen *f* dura
i immagine *f* dura
n hard beeld *n*
d hartes Bild *n*

2260 HARMONIC cpl
A sinusoidal oscillation having a frequency which is an integral multiple of a fundamental frequency.
f harmonique *f*
e harmónica *f*
i armonica *f*
n harmonische
d Harmonische *f*

2261 HARMONIC DISTORTION cpl/dis
A constituent of nonlinearity distortion, consisting of the production, in the response to a sinusoidal excitation, of sinusoidal components whose frequencies are integral multiples of the frequency of the excitation.
f distorsion *f* harmonique,
distorsion *f* nonlinéaire
e distorsión *f* harmónica,
distorsión *f* no lineal
i distorsione *f* d'armonica,
distorsione *f* non lineare
n niet-lineaire vervorming
d Klirrverzerrung *f*,
nichtlineare Verzerrung *f*

2262 HARMONIC DISTORTION dis
ATTENUATION
f atténuation *f* de distorsion non-linéaire
e atenuación *f* de distorsión no lineal
i attenuazione *f* di distorsione non lineare
n demping van de niet-lineaire vervorming
d Klirrdämpfung *f*

2263 HARMONIC DISTORTION FACTOR dis
f coefficient *m* de distorsion non-linéaire
e coeficiente *m* de distorsión no lineal
i coefficiente *m* di distorsione non lineare
n factor van de niet-lineaire vervorming
d Klirrfaktor *m*

2264 HARMONIC FILTER dis/cpl
A deviation for filtering out or removing unwanted harmonics or multiples of a signal which can be present at the output of a circuit.
f suppresseur *m* d'harmoniques
e supresor *m* de harmónicas
i soppressore *m* d'armoniche
n onderdrukker van de harmonischen
d Oberwellenfilter *n*

2265 HARMONIC GENERATOR cpl
A device delivering output power at one or more frequencies, each of which is an exact integral multiple of the input frequency.
f générateur *m* d'harmoniques
e generador *m* de harmónicas
i generatore *m* d'armoniche
n generator van harmonischen
d Oberwellengenerator *m*

2266 HAZINESS rep/tv
f manque *m* de netteté
e falta *f* de nitidez
i sfocatura *f*
n beeldonscherpte
d Bildunschärfe *f*

2267 HAZY IMAGE, rep/tv
HAZY PICTURE
f image *f* floue
e imagen *f* confusa
i immagine *f* confusa
n onscherp beeld *n*
d unscharfes Bild *n*

2268 HEAD, vr
MAGNETIC HEAD
The magnetic recording, reproducing, or erasing device of a tape recorder.
f tête *f*,
tête *f* magnétique
e cabeza *f*,
cabeza *f* magnética
i testina *f*,
testina *f* magnetica
n kop,
magneetkop
d Kopf *m*,
Magnetkopf *m*

2269 HEAD ADJUSTMENT, vr
HEAD GAP ADJUSTMENT
f équilibrage *m* de l'entrefer
e balanceo *m* del entrehierro
i bilanciamento *m* dell'intraferro
n instelling van de spleet
d Eintaumeln *n* des Kopfes

HEAD AMPLIFIER
see: CAMERA PREAMPLIFIER

2270 HEAD ASSEMBLY, vr
HEAD SUPPORT ASSEMBLY
f porte-têtes *m*
e portacabezas *m*
i portatestine *m*
n magneetkoppendrager
d Kopfträger *m*

2271 HEAD BANDING vr
f effet *m* de bande
e efecto *m* de bandas
i bandoni *pl*
n banding
d Kopfspuren *pl*

2272 HEAD CLOGGING vr
f colmatage *m* de tête
e suciedad *f* de cabeza
i testine *pl* intasate
n vuilslaan *n* van de koppen
d Kopfverschmutzung *f*

2273 HEAD DEMAGNETIZER vr
A device that eliminates residual magnetism built up in a magnetic-tape recording head.
f démagnétisateur *m* de tête
e desmagnetizador *m* de cabeza
i smagnetizzatore *m* di testina
n kopontmagnetiseringsapparaat *n*
d Kopfentmagnetisierungsgerät *n*

2274 HEAD DRUM vr
A drum used in video tape recording carrying the four heads with coils whose tips protrude and penetrate slightly into the magnetic coating of the tape.
f tambour *m* à têtes magnétiques
e tambor *m* de cabezas magnéticas
i tamburo *m* a testine magnetiche
n koptrommel, magneetkoptrommel
d Kopftrommel *f*, Magnetkopftrommel *f*

2275 HEAD DRUM MOTOR vr
f moteur *m* du tambour à têtes
e motor *m* del tambor de cabezas
i motore *m* del tamburo a testine
n koptrommelmotor
d Kopftrommelmotor *m*

2276 HEAD EFFICIENCY vr
f rendement *m* de tête
e rendimiento *m* de cabeza
i rendimento *m* di testina
n koprendement *n*
d Kopfwirkungsgrad *m*

2277 HEAD INTRUSION vr
f pénétration *f* de la tête dans la bande
e penetración *f* de la cabeza en la cinta
i penetrazione *f* della testina nel nastro
n indringen *n* van de kop in de band
d Eindringen *n* des Kopfes in das Band

2278 HEAD LEADER fi/tv
f amorce *f* de début, amorce *f* de départ
e tira *f* de película sin imágenes
i coda *f* iniziale
n beginstrook
d Startband *n*

2279 HEAD SET aud/stu
A pair of headphones with a microphone attached to them.
f casque *f* (téléphonique) à prise pour microphone
e receptor *m* de cabeza con micrófono acoplado
i cuffia *f* (telefonica) con microfono accoppiato
n koptelefoon met microfoonaansluiting
d Kopfhörer *m* mit Mikrophonanschluss

2280 HEAD SWITCH vr
f commutateur *m* de tête
e conmutador *m* de cabeza
i commutatore *m* di testina
n kopomschakelaar
d Kopfumschalter *m*

2281 HEAD TIPS vr
Small metallic protuberances on the magnetic head made from a hard material, a composition of aluminium, iron and silicon.
f protubérances *pl* de la tête magnétique
e protuberancias *pl* de la cabeza magnética
i puntine *pl* della testina magnetica
n magneetkopspitsen *pl*, magneetkoptippen *pl*
d Magnetkopfspitzen *pl*

2282 HEAD-TO-TAPE CONTACT vr
f contact *m* tête-bande
e contacto *m* cabeza-cinta
i contatto *m* testina-nastro
n contact *n* kop-band
d Kontakt *m* Kopf-Band

2283 HEAD-TO-TAPE VELOCITY vr
f rapport *m* vitesse tête-vitesse bande
e relación *f* velocidad cabeza-velocidad cinta
i rapporto *m* velocità testina-velocità nastro
n verhouding kop-en bandsnelheid
d Verhältnis *n* Kopf- und Bandgeschwindigkeit

2284 HEAD WEAR vr
Effect due to the tape abrading the magnetic head(s).
f usure *f* de tête
e desgaste *m* de cabeza
i usura *f* di testina
n slijtage van de kop
d Kopfabrieb *m*

2285 HEAD WHEEL vr
f disque *m* porte-tête, roue *f* porte-tête
e rueda *f* de portacabezas
i ruoto *f* di portatestine
n kopwiel *n*
d Kopfrad *n*

2286 HEAD WHEEL DRIVE vr
f commande *f* de roue de porte-têtes
e mando *m* de rueda de portacabezas
i comando *m* di ruota di portatestine
n kopwielaandrijving
d Kopfradantrieb *m*

2287 HEAD WHEEL MOTOR vr
f moteur *m* à roue de porte-têtes

e motor *m* de rueda de portacabezas
i motore *m* di ruoto di portatestine
n kopwielaandrijfmotor
d Kopfradantriebsmotor *m*

2288 HEAD WINDING vr
f bobinage *m* de tête
e devanado *m* de cabeza
i avvolgimento *m* di testina
n kopwikkeling
d Kopfwicklung *f*

2289 HEADPHONES aud/stu
f casque *m* (téléphonique)
e casco *m* (telefónico),
receptor *m* de cabeza
i cuffia *f* (telefonica)
n hoofdtelefoon,
koptelefoon
d Kopfhörer *pl*

HEADS ALIGNMENT
see: ALIGNMENT OF HEADS

HEARING ACUTENESS (US)
see: ACUITY OF HEARING

2290 HEAT RADIATION, ct
INFRARED RADIATION
Electromagnetic radiation within the wavelength range of about 7.500 to 100.000 ångstroms.
f rayonnement *m* infrarouge
e radiación *f* infrarroja
i radiazione *f* infrarossa
n infraroodstraling
d Infrarotstrahlung *f*

2291 HEAT-REDUCING FILTER stu
A filter placed in front of a carbon-arc lamp to reduce the colo(u)r temperature.
f filtre *m* absorbant la chaleur
e filtro *m* absorbente el calor
i filtro *m* assorbente il calore
n warmteabsorberend filter *n*
d wärmeabsorbierendes Filter *n*

2292 HEIGHT CONTROL crt/rep/tv
A device provided on a TV receiver to adjust the vertical amplitude of the image.
f régleur *m* de la hauteur d'image
e regulador *m* de la altura de imagen
i regolatore *m* dell'altezza d'immagine
n beeldhoogteregelaar
d Bildhöhenregler *m*

2293 HEIGHT CONTROL crt/rep/tv
Control provided on a TV receiver which adjusts the vertical amplitude of the image.
f réglage *m* de l'hauteur d'image
e regulación *f* de la altura de imagen
i regolazione *f* dell'altezza d'immagine
n beeldhoogte-instelling
d Bildhöhenregelung *f*,
Einstellung *f* der Bildhöhe

HELICAL AERIAL (GB)
see: CORKSCREW ANTENNA

2294 HELICAL RECORDING, vr
HELICAL SCANNING,
SPIRAL SCANNING
Type of video tape recording in which the tape is wrapped helically round a fixed drum of large diameter, while a rotation disk coaxial with the drum and mounted within it, carries the recording head(s).
f analyse *f* hélicoïdale,
enregistrement *m* hélicoïdal
e exploración *f* helicoidal,
grabación *f* helicoidal
i analisi *f* elicoidale,
registrazione *f* elicoidale
n spiraalaftasting,
spiraalopname
d Schrägspurverfahren *n*,
Spiralabtastung *f*

2295 HELICAL-SCAN VIDEO vr
RECORDER
f magnétoscope *m* à analyse hélicoïdale,
magnétoscope *m* à défilement hélicoïdal
e magnetoscopio *m* con exploración helicoidal,
magnetoscopio *m* helicoidal
i magnetoscopio *m* ad analisi elicoidale,
registratore *m* elicoidale
n beeldbandopnemer met spiraalaftasting
d Bildbandgerät *n* mit Spiralabtastung,
Schrägschriftaufzeichnungsgerät *n*,
Schrägspuraufzeichnungsanlage *f*

2296 HELIUM-NEON LASER vtr
Light source used in video long play system.
f laser *m* hélium-néon
e laser *m* helio-néon
i laser *m* elio-neon
n helium-neonlaser
d Helium-Neonlaser *m*

2297 HEMERALOPIA, ct
NIGHT-BLINDNESS
Anomaly of vision in which there is a pronounced inadequacy or complete absence of dark adaptation.
f héméralopie *f*
e ceguera *f* nocturna,
hemeralopía *f*
i emeralopia *f*
n nachtblindheid
d Nachtblindheit *f*

2298 HERRINGBONE PATTERN crt/dis
An interference pattern seen sometimes on a TV receiver screen consisting of a horizontal band of closely spaced V- or S-shaped lines.
f diagramme *m* de perturbation en boucles
e patrón *m* de disturbo en bucles
i figura *f* di disturbo ad attorcigliamenti
n kronkelstoringspatroon *n*
d Schlingenstörungsmuster *n*

2299 HETEROCHROMATIC STIMULI ct
Colo(u)r stimuli which, when acting simultaneously in adjacent fields, give rise to different colo(u)r sensations.

f stimuli *pl* hétérochromatiques
e estímulos *pl* heterocromáticos
i stimoli *pl* eterocromatici
n heterochromatische stimuli *pl*
d heterochromatische Reize *pl*

2300 HETERODYNING vr
f hétérodynation *f*
e heterodinación *f*
i eterodinazione *f*
n heterodyniseren *n*
d Überlagerung *f*

2301 H.F. BIAS, vr
HIGH-FREQUENCY BIAS
Bias current at a frequency of 30-100 kc/s applied to magnetic tape along with the signal during tape recording, to reduce the distortion that would otherwise result from the non-linear magnetic characteristics of the tape.
f polarisation *f* à haute fréquence
e polarización *f* de alta frecuencia
i polarizzazione *f* ad alta frequenza
n hoogfrequent-voormagnetisatie
d HF-Vormagnetisierung *f*

2302 H.F. CARRIER SYSTEM, cpl
HIGH-FREQUENCY CARRIER SYSTEM
System for conveying broadband signals over a cable using amplitude or frequency modulation of a sine wave carrier.
f système *m* à courants porteurs à haute fréquence
e sistema *m* de corrientes portadoras de alta frecuencia
i sistema *m* a correnti portanti ad alta frequenza
n hoogfrequent-draaggolfsysteem *n*
d HF-Trägerverfahren *n*

2303 H.F. COMPENSATION, cpl
HIGH-FREQUENCY COMPENSATION
f compensation *f* des hautes fréquences
e compensación *f* de las altas frecuencias
i compensazione *f* delle alte frequenze
n ophalen *n* van de hoge frequenties
d Hochfrequenzausgleich *m*

2304 H.F. EMPHASIS, cpl
HIGH-FREQUENCY EMPHASIS
f accentuation *f* des hautes fréquences, surcompensation *f* des aigus
e acentuación *f* de las altas frecuencias
i enfasi *f* delle alte frequenze
n opduw van de hoge frequenties
d Höhenanhebung *f*

2305 HI-FI, ge
HIGH FIDELITY
That property of reproducing speech, music and/or pictures which matches it as closely as possible to the original.
f fidélité *f* de reproduction
e fidelidad *f* de reproducción
i alta fedeltà *f* di riproduzione
n natuurgetrouwheid
d Wiedergabetreue *f*

2306 HIGH-ANGLE SHOT, opt
TOP SHOT
A shot taken with the axis of the camera nearly vertical.
f plongée *f*
e toma *f* angular a vista de pájaro
i inquadratura *f* dall'alto
n hoog hoekshot
d Vogelperspektive *f*

2307 HIGH BAND tv
The TV band extending from 174 to 216 mc, which includes channels 7 to 13.
f bande *f* des plus hautes fréquences
e banda *f* de las frecuencias más altas
i bande *f* delle frequenze più alte
n band van hogere frequenties
d frequenzhöheres Band *n*

2308 HIGH BEAM-VELOCITY VIDICON crt
f vidicon *m* à haute vitesse d'électrons d'analyse
e vidicón *m* de alta velocidad de electrones exploradores
i vidicon *m* ad alta velocità d'elettroni analizzatori
n vidicon *n* met snelle aftastelektronen
d Vidikon *n* mit schnellen Abtastelektronen

2309 HIGH BOOST, cpl
HIGH-FREQUENCY COMPENSATION
Increasing the amplification at high frequencies with respect to that at low and middle frequencies in a given band, such as in a video band or in an audio band.
f accentuation *f* des fréquences élevées, compensation *f* des fréquences élevées
e compensación *f* de las altas frecuencias
i compensazione *f* delle alte frequenze
n vereffening van de hoge frequenties
d Hochfrequenzausgleich *m*

2310 HIGH-CONTRAST IMAGE, rep/tv
HIGH-CONTRAST PICTURE
An image in which the contrast between black and white is great and intermediate tones are poor.
f image *f* contrastée
e imagen *f* de alto contraste
i immagine *f* ricca di contrasto
n contrastrijk beeld *n*
d kontrastreiches Bild *n*

2311 HIGH DEFINITION rep/tv
The TV equivalent of high fidelity, in which the reproduced image contains such a large number of accurately reproduced elements that picture details approximate those of the original scene.
f haute définition *f*
e alta definición *f*
i alta definizione *f*
n hoge definitie
d Bildschärfe *f*, Hochzeiligkeit *f*

2312 HIGH-DEFINITION IMAGE, rep/tv
HIGH-DEFINITION PICTURE

f image *f* à haute définition
e imagen *f* de alta definición
i immagine *f* ad alta definizione
n beeld *n* met hoge definitie
d hochzeiliges Bild *n*,
scharfes Bild *n*

2313 HIGH-DEFINITION SCANNING SYSTEM rec/rep/tv
f système *m* d'analyse à grand nombre de lignes
e sistema *m* de exploración de gran número de líneas
i sistema *m* d'analisi a gran numero di linee
n aftastsysteem *n* met groot aantal lijnen
d hochzeiliges Fernsehabtastsystem *n*

2314 HIGH-DEFINITION TELEVISION tv
TV system using more than 200 scanning lines per picture.
f télévision *f* à haute définition
e televisión *f* de alta definición
i televisione *f* ad alta definizione
n televisie met groot aantal lijnen
d hochzeiliges Fernsehen *n*

HIGH-ELECTRON VELOCITY CAMERA TUBE
see: ANODE-VOLTAGE-STABILIZED CAMERA TUBE

2315 HIGH-FREQUENCY ABSORPTION aud/stu
f absorption *f* des aigus
e absorción *f* de las altas frecuencias
i assorbimento *m* delle alte frequenze
n absorptie van de hoge frequenties
d Höhenabsorption *f*

2316 HIGH-GAIN AERIAL, HIGH-GAIN ANTENNA aea
Aerial (antenna) array which has a very high gain compared to a dipole and thus produces a high effective radiated power for a given input.
f antenne *f* à gain élevé
e antena *f* de alta ganancia
i antenna *f* ad alto guadagno
n antenne met hoge versterking
d Antenne *f* mit hoher Verstärkung

2317 HIGH-GAMMA TUBE crt
A TV picture tube in which the light intensity on the screen is directly proportional to the control-grid voltage.
f tube *m* image à gamme élevée
e tubo *m* imagen de alta gama
i tubo *m* immagine ad alta gamma
n beeldbuis met hoge gamma
d Röhre *f* mit harmonischer Gradation,
Röhre *f* mit hochwertiger Gradation

2318 HIGH-KEY IMAGE, HIGH-KEY PICTURE stu
Style of tonal rendering of a scene which emphasizes the middle and lighter tones.
f image *f* lumineuse
e imagen *f* luminosa
i immagine *f* molto chiara e luminosa
n helder en licht beeld *n*
d gradationsreiches Bild *n*

2319 HIGH-LEVEL MODULATION cpl/tv
f modulation *f* dans l'anode de l'étage final
e modulación *f* en alto nivel
i modulazione *f* di potenza
n eindtrapmodulatie
d Endstufenmodulation *f*,
Modulation *f* bei hohem Leistungspegel

2320 HIGH-LIGHT BRIGHTNESS tv
f luminosité *f* des blancs
e luminosidad *f* de los blancos
i luminosità *f* dei bianchi
n tophelderheid
d Spitzenhelligkeit *f*

2321 HIGH-PASS FILTER cpl
A filter having a single transmission band extending from the cut-off frequency up to infinite frequency.
f filtre *m* passe-haut
e filtro *m* paso alto
i filtro *m* passa-alto
n hoogdoorlatend filter *n*
d Hochpass *m*,
Hochpassfilter *n*

2322 HIGH-PEAKER TEST ctv/tv
f examen *m* de contrastes
e ensayo *m* de contrastes
i prova *f* di contrasti
n contrastonderzoek *n*
d Kontrastprüfung *f*

2323 HIGH-STABILITY UNLOCKED OPERATION ctv/tv
A system to provide approximate coincidence of the field pulses from a number of sources by using high-stability oscillators which drive the sync pulse generator.
f opération *f* à déclenchement de haute stabilité
e operación *f* de desenclavamiento de alta estabilidad
i operazione *f* sbloccata ad alta stabilità
n onvergrendelde werkwijze van grote stabiliteit
d unverriegelte Operation *f* hoher Stabilität

2324 HIGH-VACUUM CATHODE-RAY TUBE crt
f tube *m* cathodique à vide élevé
e tubo *m* catódico de alto vacío
i tubo *m* catodico a vuoto spinto
n hoogvacuümkatodestraalbuis
d Hochvakuumkatodenstrahlröhre *f*

2325 HIGH-VACUUM TELEVISION TUBE crt
f tube *m* image à vide élevé
e tubo *m* imagen de altovacío
i tubo *m* immagine a vuoto spinto
n hoogvacuümbeeldbuis
d Hochvakuumfernsehröhre *f*

2326 HIGH-VELOCITY SCANNING rec/tv
f analyse *f* par électrons de haute vitesse
e exploración *f* con electrones de alta velocidad

i analisi *f* con elettroni ad alta velocità
n aftasting met snelle elektronen
d Abtastung *f* mit schnellen Elektronen

2327 HIGHLIGHT crt
Very bright area in a screen or image.
f aire *f* de grande intensité,
plage *f* lumineuse
e área *f* clara,
claro *m*,
mancha *f* hiperluminosa
i macchia *f* iperluminosa
n hoge lichten *pl*,
intensieve lichtvlek
d hellste Stelle *f*

2328 HIGHLIGHT LUMINOSITY rec/rep/tv
The maximum luminosity in a TV image, corresponding to the area having highest luminosity in the original scene.
f luminosité *f* maximale
e luminosidad *f* máxima
i luminosità *f* massima
n maximale helderheid
d Spitzenhelligkeit *f*

2329 HOLD CONTROL (US), cpl
HOLDING CONTROL (US),
SYNC CONTROL (GB)
In TV, an adjustable resistor which varies the frequency of a synchronized time-base oscillator to keep it approximately equal to the frequency of the transmitted sync pulses.
f régleur *m* de synchronisation
e regulador *m* de enganche
i regolatore *m* di sincronismo
n synchronisatieregelaar
d Bildfangregler *m*

2330 HOLD CONTROL (US), cpl
HOLDING CONTROL (US),
SYNC CONTROL (GB)
f contrôle *m* de synchronisation
e control *m* de enganche
i controllo *m* di sincronismo
n synchronisatiecontrole
d Bildfang *m*

HOLD FRAME
see: FREEZE FRAME

2331 HOLD-IN RANGE tv
The frequency band over which the oscillator remains locked in the input frequency.
f zone *f* de l'enclenchement
e zona *f* del enganche
i zona *f* del bloccaggio
n vergrendelingsgebied *n*
d Sperrbereich *m*

2332 HOLD RANGE, rec/tv
RETAINING ZONE,
RETENTION RANGE (US)
The frequency range over which the oscillator will stay synchronized when the sync frequency is altered.
f bande *f* de synchronisation
e banda *f* de sincronización
i banda *f* di sincronizzazione
n hougebied *n*,
synchronisatiegebied *n*
d Haltebereich *m*,
Synchronisierungsbereich *m*

2333 HOLOGRAM tv
The product of holography.
f hologramme *m*
e holograma *m*
i ologramma *m*
n hologram *n*
d Hologramm *n*

2334 HOLOGRAM RECONSTRUCTION tv
A laser emits a coherent light beam which strikes developed emulsion of hologram; light, diffracted by interference pattern on hologram, results in a virtual and a real image, the virtual image is seen by the spectator.
f réconstruction *f* de hologramme
e reconstitución *f* de holograma
i ricostruzione *f* d'ologramma
n hologramreconstructie
d Hologrammrekonstruktion *f*

2335 HOLOGRAPHY tv
A photographic process applied to motion pictures and TV in which light from a laser is reflected from an object and recorded directly on an emulsion surface without passing through a lens so that all parts of the surface receive light from all visible parts of the object.
f holographie *f*
e holografía *f*
i olografia *f*
n holografie
d Holographie *f*

2336 HOME VIDEOTAPE vr
RECORDING
f magnétoscopie *f* d'amateur
e magnetoscopía *f* de aficionado
i registrazione *f* video con apparecchi non professionali
n videoregistratie in huis,
videoregistratie voor huiselijk gebruik
d Amateur-Videoaufzeichnung *f*

2337 HOOD crt
An opaque shield above the screen of a cathode-ray tube to eliminate extraneous light.
f calotte *f*, capot *m*
e visera *f*
i calotta *f*, cappa *f*, cappuccio *m*
n kap
d Haube *f*, Kappe *f*

2338 HOPPING FILM SCANNER, tv
HOPPING TELECINE MACHINE,
JUMP SCANNER
A machine with a twin-lens optical system, one lens for each field, or with a single lens and a raster which alternates at field frequency between two positions.

f analyseur *m* de film à saut
e explorador *m* de película de salto
i analizzatore *m* di pellicola a salto
n springende filmaftaster
d springender Filmabtaster *m*

2339 HOPPING PATCH tv
When producing an interlaced picture in film scanning, the raster on the face of a flying stop tube can be displaced between alternate fields so that the film frame can be scanned twice.
f spot *m* lumineux décalant
e punto *m* luminoso deslizante
i punto *m* luminoso spostante
n verspringende lichtvlek
d rückender Lichtfleck *m*

HORIZONTAL AMPLIFIER
see: H-AMPLIFIER

2340 HORIZONTAL AMPLITUDE, tv
LINE AMPLITUDE
f amplitude *f* de ligne
e amplitud *f* de línea
i ampiezza *f* di linea
n lijnamplitude
d Zeilenamplitude *f*

2341 HORIZONTAL AMPLITUDE cpl
ADJUSTMENT,
LINE AMPLITUDE ADJUSTMENT
f ajustage *m* de l'amplitude de ligne,
ajustage *m* de l'amplitude horizontale
e ajuste *m* de la amplitud de línea,
ajuste *m* de la amplitud horizontal
i aggiustaggio *m* dell'ampiezza di linea,
aggiustaggio *m* dell'ampiezza orizzontale
n instelling van de horizontale amplitude,
instelling van de lijnamplitude
d Einstellung *f* der Horizontalamplitude,
Einstellung *f* der Zeilenamplitude

2342 HORIZONTAL APERTURE crt
CORRECTION
f correction *f* de l'ouverture horizontale
e corrección *f* de la abertura horizontal
i correzione *f* dell'apertura orizzontale
n horizontale apertuurcorrectie
d HAK *f*,
horizontale Aperturkorrektion *f*

2343 HORIZONTAL BAR crt
f barre *f* horizontale
e barra *f* horizontal
i barra *f* orizzontale
n horizontale balk
d Horizontalbalken *m*

HORIZONTAL BAR CONTROL
see: H-BAR CONTROL

2344 HORIZONTAL BAR GENERATOR crt
f générateur *m* de barres horizontales
e generador *m* de barras horizontales
i generatore *m* di barre orizzontali
n horizontale-balkengenerator
d Horizontalbalkengenerator *m*

2345 HORIZONTAL BARREL cpl/dis
DISTORTION
f distorsion *f* en barillet horizontale
e distorsión *f* en tonel horizontal
i distorsione *f* a bariletto orizzontale
n horizontale tonvervorming
d horizontale Tonnenverzeichnung *f*

2346 HORIZONTAL BLANKING, rec/tv
LINE BLANKING
In TV, interruption of the electron beam of the cathode-ray tube during the flyback period between lines.
f suppression *f* de ligne
e supresión *f* de línea,
supresión *f* horizontal
i cancellazione *f* di linee,
cancellazione *f* orizzontale
n horizontale onderdrukking,
lijnonderdrukking
d zeilenfrequente Austastung *f*

2347 HORIZONTAL BLANKING rec/tv
INTERVALS,
LINE BLANKING INTERVALS
f intervalles *pl* de suppression horizontale
e intervalos *pl* de supresión horizontal
i intervalli *pl* di soppressione orizzontale
n horizontale-onderdrukkingspauzes *pl*
d Horizontalaustastlücken *pl*

2348 HORIZONTAL BLANKING LEVEL, crt
LINE BLANKING LEVEL
f niveau *m* de suppression de ligne
e nivel *m* de supresión de línea
i livello *m* di cancellazione di linea
n lijnonderdrukkingsniveau *n*
d Zeilenaustastpegel *m*

2349 HORIZONTAL BLANKING rec/tv
PULSE,
LINE-FREQUENCY BLANKING
PULSE
f impulsion *f* de suppression horizontale
e impulso *m* de supresión horizontal
i impulso *m* di soppressione orizzontale
n horizontale-onderdrukkingsimpuls
d Horizontalaustastimpuls *m*

2350 HORIZONTAL CENT(E)RING rep/tv
CONTROL
f dispositif *m* de décentrement,
dispositif *m* de réglage du centrage
horizontal
e regulador *m* del centrado horizontal
i regolatore *m* della centratura
orizzontale
n horizontale-centreringsregelaar
d Bildseitenverschiebungsregler *m*,
horizontale-Bildlage-Einstellungsgerät *n*,
Horizontalzentrierregler *m*

2351 HORIZONTAL CONVERGENCE rep/ctv
CONTROL,
HORIZONTAL CONVERGENCE
SHAPE CONTROL
The control that adjusts the amplitude of the horizontal dynamic convergence

voltage in a colo(u)r TV receiver.
f dispositif *m* de réglage de la convergence horizontale
e regulador *m* de la convergencia horizontal
i regolatore *m* della convergenza orizzontale
n horizontale-convergentieregelaar
d Horizontalkonvergenzregler *m*

2352 HORIZONTAL DEFINITION (GB), rep/tv
HORIZONTAL RESOLUTION (US),
LINE DEFINITION (GB),
LINE RESOLUTION (US)
In TV, the number of picture elements resolved along the scanning lines.
f définition *f* horizontale
e definición *f* horizontal
i definizione *f* orizzontale
n horizontale definitie
d Horizontalauflösung *f*

2353 HORIZONTAL DEFLECTION tv
f déviation *f* horizontale
e desviación *f* horizontal
i deviazione *f* orizzontale
n horizontale afbuiging
d horizontale Ablenkung *f*,
Zeilenablenkung *f*

2354 HORIZONTAL DEFLECTION rec/tv
CONTROL,
WIDTH CONTROL
f réglage *m* de la largeur d'image
e regulación *f* de la anchura de imagen
i regolazione *f* della larghezza d'immagine
n beeldbreedteregeling
d Bildbreiteregelung *f*

2355 HORIZONTAL DEFLECTION rec/tv
ELECTRODES,
HORIZONTAL DEFLECTION
PLATES,
X-PLATES
The pair of electrodes that moves the electron beam horizontally from side to side on the fluorescent screen of a cathode-ray tube employing electrostatic deflection.
f électrodes *pl* de déviation horizontale
e electrodos *pl* de desviación horizontal
i elettrodi *pl* di deflessione orizzontale,
elettrodi *pl* di deviazione orizzontale
n elektroden *pl* voor horizontale afbuiging,
X-afbuigingsplaten *pl*
d Horizontalablenkelektroden *pl*,
X-Ablenkplatten

2356 HORIZONTAL DEFLECTION rec/tv
OSCILLATOR,
HORIZONTAL OSCILLATOR
The oscillator that produces, under control of the horizontal sync signals, the sawtooth voltage waveform that is amplified to feed the horizontal deflection coils on the picture tube of a TV receiver.
f oscillateur *m* pour la déviation horizontale
e oscilador *m* para la desviación horizontal
i oscillatore *m* per la deflessione orizzontale,
oscillatore *m* per la deviazione orizzontale
n oscillator voor de horizontale afbuiging
d Oszillator *m* für die horizontale Ablenkung

2357 HORIZONTAL DEFLECTION rec/tv
UNIT
f unité *f* de déviation horizontale
e unidad *f* de desviación horizontal
i unità *f* di deflessione orizzontale,
unità *f* di deviazione orizzontale
n horizontale-afbuigeenheid
d Horizontalablenkgerät *n*

2358 HORIZONTAL DISPLAY rep/tv
MONITOR,
LINE DISPLAY MONITOR
f moniteur *m* de lignes
e monitor *m* de líneas
i monitore *m* di linee
n lijnenmonitor
d Zeilenmonitor *m*

HORIZONTAL DRIVE CONTROL
see: DRIVE CONTROL

2359 HORIZONTAL DYNAMIC ctv
CONVERGENCE
In colo(u)r TV, voltages added to the steady convergence voltage to keep the spot in register.
f convergence *f* dynamique horizontale
e convergencia *f* dinámica horizontal
i convergenza *f* dinamica orizzontale
n dynamische horizontale convergentie
d dynamische Horizontalkonvergenz *f*

2360 HORIZONTAL DYNAMIC ctv
FOCUSING
In colo(u)r TV, voltages added to the steady convergence voltage to keep the spot in focus.
f focalisation *f* dynamique horizontale
e enfoque *m* dinámico horizontal
i focalizzazione *f* dinamica orizzontale
n dynamische horizontale focussering
d dynamische Horizontalfokussierung *f*

2361 HORIZONTAL FLYBACK (GB), rec/tv
HORIZONTAL RETRACE (US),
LINE FLYBACK (GB),
LINE RETRACE (US)
Flyback in which the electron beam of a TV picture tube returns from the end of one scanning line to the beginning of the next line.
f retour *m* du spot de ligne,
retour *m* du spot horizontal
e retorno *m* del punto de línea,
retorno *m* del punto horizontal
i ritorno *m* di linea,
ritorno *m* orizzontale
n horizontale terugslag,
lijnterugslag
d horizontaler Rücklauf *m*,
Zeilenrücklauf *m*

2362 HORIZONTAL FREQUENCY, rec/rep/tv
HORIZONTAL SCANNING FREQUENCY,
LINE FREQUENCY,
LINE SCANNING FREQUENCY
The number of times per second that the scanning spot sweeps across the screen in a horizontal direction in a TV system.
f fréquence *f* d'analyse de ligne,
fréquence *f* d'analyse horizontale
e frecuencia *f* de exploración de línea,
frecuencia *f* de exploración horizontal
i frequenza *f* d'analisi di linea,
frequenza *f* d'analisi orizzontale
n horizontale aftastfrequentie,
lijnenaftastfrequentie
d Horizontalfrequenz *f*,
Zeilenabtastfrequenz *f*

2363 HORIZONTAL FREQUENCY rec/tv
DEVIATION,
LINE FREQUENCY DEVIATION
f déviation *f* de la fréquence de ligne
e desviación *f* de la frecuencia de línea
i deviazione *f* della frequenza di linea
n lijnfrequentieafwijking
d Zeilenfrequenzabweichung *f*

2364 HORIZONTAL FREQUENCY rec/tv
DIVIDER,
LINE DIVIDER,
LINE FREQUENCY DIVIDER
Apparatus for producing signals at line frequency by subdivisions from a constant reference frequency.
f diviseur *m* de fréquence de lignes
e divisor *m* de frecuencia de líneas
i divisore *m* per frequenza di linee
n lijnfrequentiedeler
d Frequenzteiler *m*

2365 HORIZONTAL HOLD cpl/rep/tv
CONTROL
Hold control of the line timebase in a TV receiver.
f réglage *m* de la fréquence de lignes
e regulación *f* de la frecuencia de líneas
i regolazione *f* della frequenza di linee
n lijnfrequentieregeling
d Zeilenfangregelung *f*,
Zeilenfrequenzeinstellung *f*

2366 HORIZONTAL HUNTING (GB), dis
JITTER (US)
Horizontal movement of the image, caused by faulty synchronization.
f écart *m* horizontal,
instabilité *f* horizontale de l'image,
sautillement *m* horizontal de l'image
e inestabilidad *f* horizontal de la imagen
i instabilità *f* orizzontale dell'immagine,
scatto *m* orizzontale
n bibbereffect *n*,
horizontaal trillen *n* van het beeld
d horizontale Bildstandschwankungen *pl*,
Schaukeleffekt *m*

2367 HORIZONTAL IMAGE SHIFT, rep/tv
HORIZONTAL PICTURE SHIFT
f décalage *m* horizontal de l'image
e deslizamiento *m* horizontal de la imagen
i spostamento *m* orizzontale dell'immagine
n horizontale beeldverschuiving
d horizontale Bildverschiebung *f*

2368 HORIZONTAL KEYSTONE ctv/dis/rep
CORRECTION,
LINE KEYSTONE CORRECTION,
LINE KEYSTONE WAVEFORM
A scanning waveform recurring at line frequency, shaped to compensate for the geometrical distortion which would otherwise occur in TV tubes having electron guns inclined at an angle to the fluorescent screen.
f signal *m* de correction de trapèze-lignes
e señal *f* trapezoidal de corrección de línea
i correzione *f* trapezoidale geometrica della linea
n lijnzaagtandcorrectie
d Trapezzeilensignale *pl*

2369 HORIZONTAL LINEARITY rec/rep/tv
CONTROL,
LINE LINEARITY CONTROL
A linearity control that permits narrowing or expanding the width of the left-hand half of a TV receiver image, to give linearity in the horizontal direction so that circular objects appear as true circles.
f réglage *m* de la linéarité horizontale
e regulación *f* de la linealidad horizontal
i regolazione *f* della linearità orizzontale
n horizontale-lineariteitsregeling
d Horizontallinearitätsregelung *f*

2370 HORIZONTAL rec/rep/tv
NONLINEARITY,
LINE NONLINEARITY
Distortion of the picture due to a variation in the line sweep speed, in transmission or reception, during the effective part of the line sweep.
f distorsion *f* de vitesse de balayage de ligne
e no linealidad *f* de línea
i distorsione *f* della deflessione di linea
n lijnvervorming
d Ablenknichtlinearität *f*

2371 HORIZONTAL OUTPUT, rec/rep/tv
LINE OUTPUT
The timebase which belongs to the scanning circuit of the line.
f base *f* de temps de ligne
e base *f* de tiempo de línea
i base *f* di tempi di linea
n lijntijdbasis
d Zeilenzeitbasis *f*

2372 HORIZONTAL OUTPUT rep/tv
PENTODE,
LINE OUTPUT PENTODE
In a TV receiver, the final amplifying tube (valve) in the line timebase generator.

f pentode *f* de sortie pour la base de temps de ligne
e pentodo *m* de salida para la base de tiempo de línea
i pentodo *m* d'uscita per la base di tempi di linea
n eindpentode voor de lijntijdbasis
d Horizontalendpentode *f*

2373 HORIZONTAL OUTPUT STAGE, rep/tv
LINE OUTPUT STAGE
The TV receiver stage that feeds the horizontal deflection coils of the picture tube through the horizontal output transformer.
f étage *m* de sortie de ligne
e etapa *f* final de línea
i stadio *m* finale di linea
n lijnuitgangstrap
d Zeilenendstufe *f*

HORIZONTAL OUTPUT TRANSFORMER
see: FLYBACK TRANSFORMER

2374 HORIZONTAL OUTPUT VOLTAGE, rep/tv
LINE OUTPUT VOLTAGE
f tension *f* de la base de temps de ligne
e tensión *f* de la base de tiempo de línea
i tensione *f* della base di tempi di linea
n lijntijdbasisspanning
d zeilenfrequente Kippspannung *f*

2375 HORIZONTAL PINCUSHION DISTORTION crt/dis
f distorsion *f* en coussin horizontale
e distorsión *f* en cojín horizontal
i distorsione *f* a cuscinetto orizzontale
n horizontale kussenvervorming
d horizontale Kissenverzeichnung *f*

2376 HORIZONTAL POLARIZATION cpl/dis
Combined with vertical polarization of a transmitted radiofrequency signal, horizontal polarization will minimize interference by adjacent channels when using microwave links.
f polarisation *f* horizontale
e polarización *f* horizontal
i polarizzazione *f* orizzontale
n horizontale polarisatie
d horizontale Polarisation *f*

2377 HORIZONTAL PULSE, rec/rep/tv
LINE PULSE
f impulsion *f* d'analyse,
impulsion *f* lignes
e impulso *m* líneas
i impulso *m* linee
n lijnimpuls
d Horizontalimpuls *m*

2378 HORIZONTAL SCANNING, tv
LINE SCANNING
f analyse *f* à lignes,
analyse *f* horizontale
e exploración *f* de líneas,
exploración *f* horizontal
i analisi *f* orizzontale,
analisi *f* per linee
n horizontale aftasting,
lijnaftasting
d horizontale Abtastung *f*,
Zeilenabtastung *f*

2379 HORIZONTAL SHADING dis
A type of raster shading occurring when the brightness differences appear from side to side.
f trame *f* à luminosité inégale horizontale
e cuadro *m* de luminosidad dispareja horizontal
i trama *f* a luminosità disuguale orizzontale
n horizontale ongelijke rasterhelderheid
d horizontale ungleiche Rasterhelligkeit *f*

2380 HORIZONTAL SIZE CONTROL (US), rec/rep/tv
LINE AMPLITUDE CONTROL (GB)
f dispositif *m* de réglage de la largeur de ligne
e regulador *m* de la anchura de línea
i regolatore *m* della larghezza di linea
n lijnbreedteregelaar
d Zeilenbreiteregler *m*

2381 HORIZONTAL SLIP dis
Horizontal displacement of sections of a TV picture to one side, usually due to faulty synchronization of the line timebase.
f décalage *m* horizontal,
glissement horizontal
e deslizamiento *m* horizontal,
desplazamiento *m* horizontal
i scorrimento *m* orizzontale,
spostamento *m* orizzontale
n horizontale verschuiving
d horizontale Verschiebung *f*

2382 HORIZONTAL SWEEP tv
The sweep of the electron beam from left to right and back across the screen of a cathode-ray tube.
f aller et retour horizontal
e barrido *m* de línea
i andata e ritorno orizzontale
n horizontale heen- en terugslag
d horizontale Ablenkung *f*,
horizontaler Hin- und Rücklauf *m*

2383 HORIZONTAL SYNC CONTROL, cpl/rec/rep/tv
LINE SYNC CONTROL
f dispositif *m* de réglage de synchronisation de lignes
e regulador *m* de sincronización de líneas
i regolatore *m* di sincronizzazione di linee
n horizontale-stabiliteitsregelaar
d Zeilensynchronregler *m*

2384 HORIZONTAL SYNC GENERATOR, rec/rep/tv
LINE SYNC GENERATOR
f générateur *m* de l'impulsion de synchronisation horizontale
e generador *m* del impulso de sincronización horizontal
i generatore *m* dell'impulso di sincronizzazione orizzontale

n horizontale-synchronisatieimpulsgenerator
d Generator *m* für die Horizontal-synchronisierimpulse

2385 HORIZONTAL SYNC PULSE, LINE SYNC PULSE rec/tv
In TV, a pulse transmitted at the end of each line to synchronize the start of the scan for the next line.
f impulsion *f* de synchronisation de ligne, impulsion *f* de synchronisation horizontale
e impulso *m* de sincronización de línea, impulso *m* de sincronización horizontal
i impulso *m* di sincronizzazione di linea, impulso *m* di sincronizzazione orizzontale
n horizontale synchronisatie-impuls, lijnsynchronisatie-impuls
d Horizontalsynchronimpuls *m*, Zeilengleichlaufimpuls *m*, Zeilenimpuls *m*, Zeilensynchronisierimpuls *m*

2386 HORIZONTAL TIMEBASE, LINE TIMEBASE, LINEAR TIMEBASE rep/tv
Circuits of a TV receiver which generate a current of sawtooth waveform for line scanning.
f base *f* de temps de lignes
e base *f* de tiempo de líneas
i base *f* di tempi di linee, generatore *m* di frequenza d'analisi
n aftastfrequentiegenerator, lijntijdbasis
d Zeilenkipper *m*, Zeilenkippgerät *n*, Zeilenzeitbasis *f*

2387 HORIZONTAL TIMEBASE GENERATOR, LINE TIMEBASE GENERATOR cpl/tv
f générateur *m* de base de temps horizontale
e generador *m* de base de tiempo horizontal
i generatore *m* di base di tempi orizzontale
n horizontale tijdbasisgenerator
d horizontale Zeitbasisgenerator *m*, Zeilenkipper *m*, Zeilenkippgenerator *m*, Zeilenkippschaltung *f*

2388 HOT LIGHTING, KEY LIGHTING stu
f éclairage *m* principal
e alumbrado *m* principal
i illuminazione *f* essenziale
n hoofdverlichting
d Führungslicht *n*, Hauptbeleuchtung *f*

2389 HOT LIGHTS, KEY LIGHTS stu
The most important lamps in the studio.
f lampes *pl* pour l'éclairage principal
e lámparas *pl* para el alumbrado principal
i lampade *pl* per l'illuminazione principale
n hoofdlampen *pl*, hoofdverlichtingslampen *pl*
d Hauptbeleuchtungskörper *pl*

HOWL ROUND
see: ACOUSTIC FEEDBACK

2390 HUE ct
The attribute of visual perception denoted by blue, green, yellow, red, purple, etc.
f teinte *f*, tonalité *f* chromatique
e matiz *m*, tonalidad *f* cromática
i tinta *f*, tonalità *f* cromatica
n kleurtoon
d Farbton *m*

2391 HUE CONTROL, PHASE CONTROL ctv
In colo(u)r picture display apparatus, a control for adjusting the hue of the picture.
f contrôle *m* de tonalité
e control *m* de tonalidad
i controllo *m* di tonalità
n kleurtoonregeling
d Farbwertregelung *f*

2392 HUE DIFFERENCE ct
f différence *f* de tonalité
e diferencia *f* de tonalidad
i differenza *f* di tonalità
n kleurtoonverschil *n*
d Farbtondifferenz *f*

2393 HUE ERRORS ct
f erreurs *pl* de tonalité
e errores *pl* de tonalidad
i errori *pl* di tonalità
n kleurtoonfouten *pl*
d Farbtonfehler *pl*

2394 HUM dis
f ronflement *m*
e zumbido *m*
i ronzio *m*
n brom
d Brumm *m*, Brummen *n*

2395 HUM BALANCER dis
f égalisateur *m* de ronflement
e igualador *m* de zumbido
i uguagliatore *m* di ronzio
n bromvereffenaar
d Brummausgleicher *m*

2396 HUM BAR dis/rep/tv
A dark horizontal bar extending across a TV picture, due to excessive hum in the video signal applied to the input of the picture tube.
f barre *f* due au ronflement excessif
e barra *f* obscura de zumbido
i barra *f* di ronzio, barra *f* orizzontale
n brombalk, bromband
d Brummstreifen *m*

2397 HUM DISPLACEMENT ctv
f décalage *m* du ronflement
e deslizamiento *m* del zumbido
i spostamento *m* del ronzio

n bromverschuiving
d Brummverschiebung *f*

2398 HUM REDUCTION FACTOR dis
f facteur *m* de filtrage
e factor *m* de filtración
i fattore *m* di filtrazione
n uitzeeffactor
d Siebfaktor *m*

2399 HUSH-HUSH dis
Variation in background noise when a noise-reduction system is in use.
f diminution *f* du bruit de fond
e diminución *f* del ruido de fondo
i diminuzione *f* del rumore di fondo
n achtergrondruisverzachting
d Herabsetzung *f* des Hintergrundrauschens

2400 HYBRID TELEVISION RECEIVER rep/tv
A receiver in which transistors are used as well as electron tubes (valves).
f téléviseur *m* à tubes électroniques et transisteurs
e televisor *m* con válvulas electrónicas y transistores
i televisore *m* con valvole elettroniche e transistori
n televisieontvanger met elektronenbuizen en transistors
d Fernsehempfänger *m* mit Elektronenröhren und Transistoren

2401 HYPOTHETICAL REFERENCE CIRCUIT cpl
f circuit *m* hypothétique de référence
e circuito *m* hipotético de referencia
i circuito *m* ipotetico di riferimento
n hypothetische referentieketen
d hypothetischer Bezugskreis *m*

I

2402 I AXIS, ctv
WIDE-BAND AXIS
In a system of the NTSC type, the chrominance axis along which the chrominance component corresponding to the fine chrominance primary is plotted.
f axe *m* de la large bande
e eje *m* de la banda ancha
i asse *m* della banda larga
n brede-bandas
d Breitbandachse *f*

2403 I COMPONENT ctv
In the NTSC system, that component of the chrominance signal formed by modulating the chrominance subcarrier with the I signal.
f composante *f* I
e componente *f* I
i componente *f* I
n I-component
d I-Komponente *f*

2404 I DEMODULATOR ctv
The demodulator in which the chrominance signal and the colo(u)r burst signal are combined to recover the I signal in a colo(u)r TV receiver.
f démodulateur *m* I
e desmodulador *m* I
i demodulatore *m* I
n I-demodulator
d I-Demodulator *m*

2405 I SIGNAL ctv
In the NTSC system, the signal corresponding to the component of chrominance information along the orange/cyan axis of the colo(u)r diagram.
f signal *m* I
e señal *f* I
i segnale *m* I
n I-signaal *n*
d I-Signal *n*

2406 ICONOSCOPE crt
TV camera tube in which the scene to be transmitted is optically focused on to a photo-mosaic formed by a large number of minute photo-emissive granules or cells, insulated from one another and backed by a metal(l)ized plate.
f iconoscope *m*
e iconoscopio *m*
i iconoscopio *m*
n iconoscoop
d Ikonoskop *n*

ICONOSCOPE BACK LIGHTING
see: BACK BIAS

2407 ICONOSCOPE MOSAIC, crt
MOSAIC
A device used in iconoscopes for electrical storage of the optical image to be televised.
f mosaïque *m*
e mosaico *m*
i mosaico *m*
n mozaïek *n*
d Mosaik *n*

2408 IDEAL MAGNETIC MEDIUM aud/vr
f bande *f* magnétique standard
e cinta *f* magnética patrón
i nastro *m* magnetico campione
n normale magneetband
d Normalmagnetband *n*

2409 IDEAL RADIATOR ct
A radiator capable of emitting all wavelengths in the visible, ultraviolet and infrared parts of the spectrum.
f radiateur *m* idéal
e radiador *m* ideal
i radiatore *m* ideale
n ideale straler
d idealer Strahler *m*

IDENT SIGNAL
see: FIELD IDENTIFICATION SIGNAL

2410 IDENTIFICATION CIRCUIT svs
f circuit *m* d'identification
e circuito *m* de identificación
i circuito *m* d'identificazione
n herkenningscircuit *n*
d Kennungskreis *m*

2411 IDENTIFICATION GENERATOR, tv
IDENTIFICATION SOURCE
f générateur *m* d'identification
e generador *m* de identificación
i generatore *m* d'identificazione
n herkenningsgenerator
d Kennungsgeber *m*

2412 IDENTIFICATION LEADER fi/tv
f bande *f* d'identification
e tira *f* de identificación
i coda *f* d'identificazione
n herkenningsstrook
d Allonge *f*,
Kennungsband *n*

2413 IDENTIFICATION SIGNAL tv
Signal placed on all important sound and picture links on major broadcasts to minimize errors and failures during the lining-up period.
f signal *m* d'identification
e señal *f* de identificación
i segnale *m* d'identificazione
n herkenningssignaal *n*
d Kennungssignal *n*

2414 IDLE ROLLER, fi/vr
IDLER
f diabolo *m*, galet *m* libre
e rodillo *m*, roldana *f*
i ingranaggio *m* di rinvio
n loopwiel *n*
d freilaufende Rolle *f*, Laufrolle *f*

2415 I.F., cpl
INTERMEDIATE FREQUENCY
The frequency to which the signal carrier frequency is changed in superheterodyne circuits.
f fréquence *f* intermédiaire,
moyenne fréquence *f*
e frecuencia *f* intermedia
i frequenza *f* intermedia,
media frequenza *f*
n middenfrequentie,
tussenfrequentie
d ZF *f*,
Zwischenfrequenz *f*

2416 I.F. AMPLIFIER, cpl
INTERMEDIATE FREQUENCY AMPLIFIER
Amplifier stage of a superheterodyne circuit which amplifies at a constant frequency to which the signal frequency has been changed by the mixer.
f amplificateur *m* de moyenne fréquence
e amplificador *m* de frecuencia intermedia
i amplificatore *m* di media frequenza
n middenfrequentversterker
d ZF-Verstärker *m*,
Zwischenfrequenzverstärker *m*

2417 I.F. REJECTION, aea/cpl
INTERMEDIATE FREQUENCY REJECTION
f pénétration *f* de la moyenne fréquence,
suppression *f* de la moyenne fréquence
e penetración *f* de la frecuencia intermedia,
supresión *f* de la frecuencia intermedia
i penetrazione *f* della media frequenza,
soppressione *f* della media frequenza
n doorstraling van de middenfrequentie,
verwerping van de middenfrequentie
d ZF-Durchschlagsicherheit *f*,
Zwischenfrequenzsicherheit *f*

2418 I.F. REJECTION FACTOR, aea/cpl
INTERMEDIATE FREQUENCY REJECTION FACTOR
The factor indicating how many times greater a signal of intermediate frequency must be across the aerial (antenna) terminals to give the same output value as a signal of the desired frequency.
f facteur *m* de pénétration de la moyenne fréquence
e factor *m* de penetración de la frecuencia intermedia
i fattore *m* di penetrazione della media frequenza
n middenfrequentiedoorstralingsfactor
d ZF-Durchschlagsicherheitsfaktor *m*,
Zwischenfrequenzsicherheitsfaktor *m*

2419 I.F. SIGNAL, cpl/tv
INTERMEDIATE FREQUENCY SIGNAL
A modulated signal whose carrier frequency is the intermediate frequency value of the receiver.
f signal *m* de moyenne fréquence
e señal *f* de frecuencia intermedia
i segnale *m* di media frequenza
n middenfrequentiesignaal *n*
d ZF-Signal *n*,
Zwischenfrequenzsignal *n*

2420 IGNITION INTERFERENCE, dis
IGNITION NOISE
Interference experienced in TV caused by radiation from the leads of the ignition system of motor vehicles.
f interférence *f* d'allumage
e ruido *m* de encendido
i disturbo *m* dovuto ad accensione
n ontstekingsstoring
d Funkenstörung *f*,
Zündstörung *f*

2421 ILLUMINANCE, ge/stu
ILLUMINATION
The quotient of the luminous flux incident on an infinitesimal element of surface containing the point under consideration by the area of that element.
f éclairement *m*
e iluminación *f*
i illuminamento *m*
n verlichtingssterkte
d Beleuchtungsstärke *f*

2422 ILLUMINANT ct
Radiant energy with a relative spectral power distribution defined over the wavelength range that influences object colo(u)r perception.
f illuminant *m*
e iluminante *m*
i illuminante *m*
n illuminant
d Illuminant *m*

2423 ILLUMINANT C ctv
The reference white of colo(u)r TV, closely matching average daylight.
f illuminant *m* C
e iluminante *m* C
i illuminante *m* C
n C-illuminant
d Weisspunkt *m*

ILLUMINANT E
see: EQUAL-ENERGY WHITE

2424 ILLUMINANT METAMERISM ct
Phenomenon occurring when surface colo(u)rs which match under one illuminant do not match under a different illuminant.
f erreur *f* chromatique due à illuminant erroné
e metamerismo *m* de iluminante
i errore *m* cromatico dovuto ad illuminante errato

n chromatische fout door verkeerde illuminant
d Farbartfehler *m* durch Fehlilluminant

2425 ILLUMINATION CONTROL stu
The control of the lighting condition in e.g. a studio.
f réglage *m* de l'éclairement
e regulación *f* de la iluminación
i regolazione *f* dell'illuminamento
n verlichtingssterkteregeling
d Beleuchtungsstärkeregelung *f*

2426 ILLUMINATION DESK, LIGHTING CONSOLE stu
f pupitre *m* de l'éclairage, pupitre *m* de l'éclairement
e pupitre *m* de la iluminación, pupitre *m* del alumbrado
i tavolo *m* dell'illuminamento, tavolo *m* dell'illuminazione
n verlichtingssterkteregeltafel, verlichtingstafel
d Lichtorgel *f*

2427 ILLUMINATION INTENSITY, INTENSITY OF ILLUMINATION, LUMINOUS INTENSITY ct/stu
The flux density incident upon a surface.
f intensité *f* lumineuse
e intensidad *f* luminosa
i intensità *f* luminosa
n lichtsterkte
d Lichtstärke *f*

2428 ILLUMINATION LEVEL stu
f niveau *m* d'éclairage
e nivel *m* de alumbrado
i livello *m* d'illuminamento
n verlichtingsniveau *n*
d Beleuchtungspegel *m*

2429 ILLUMINATION SENSITIVITY ct/tv
The signal output current divided by the incident illumination on a camera tube or a phototube.
f sensibilité *f* lumineuse
e sensibilidad *f* luminosa
i sensibilità *f* d'un tubo di ripresa
n stralingsgevoeligheid
d Strahlungsempfindlichkeit *f*

2430 IMAGE opt
An optical counterpart of an object, as a real image or a virtual image.
f image *f*
e imagen *f*
i immagine *f*
n afbeelding, beeld
d Abbildung *f*, Bild *n*

2431 IMAGE, PICTURE crt/rep/tv
Electron image or charge pattern formed in a TV camera tube by distributed charges built up on a photo-emissive mosaic when the light impinges on the surface.
f image *f*
e imagen *f*
i immagine *f*
n beeld *n*
d Bild *n*, Schirmbild *n*

2432 IMAGE ANALYZER, PICTURE ANALYZER rec/tv
f analyseur *m* de l'image
e analizador *m* de la imagen
i analizzatore *m* dell'immagine
n beeldontleder
d Bildzerleger *m*

2433 IMAGE AND WAVEFORM MONITOR, PICTURE AND WAVEFORM MONITOR rec/tv
A combination of two cathode-ray tubes, one for controlling the picture and the other for displaying the waveform.
f appareil *m* de contrôle d'image, moniteur *m* d'image et forme d'onde
e monitor *m* de imagen y forma de onda
i monitore *m* d'immagine e forma d'onda
n beeld- en golfvormmonitor
d Bildkontrollgerät *n*

IMAGE ANGLE
see: ANGLE OF IMAGE

2434 IMAGE AREA, PICTURE AREA, PICTURE PLANE opt/tv
The surface filled by the image.
f cadrage *m*, champ *m* de l'image
e campo *m* de la imagen
i campo *m* dell'immagine
n beeldveld *n*, beeldvlak *n*
d Bildausschnitt *m*, Bildfeld *n*

2435 IMAGE BACKGROUND, PICTURE BACKGROUND rec/tv
f fond *m* d'image
e fondo *m* de imagen
i fondo *m* d'immagine
n beeldachtergrond
d Bildhintergrund *m*

2436 IMAGE BLACK, PICTURE BLACK rec/rep/tv
The signal level corresponding to the darkest or blackest part of the picture to be transmitted.
f noir *m* d'une image
e negro *m* de una imagen
i nero *m* di un'immagine
n beeldzwart *n*, niveau *n* van de donkerste beeldpartij
d Bildschwarz *n*

2437 IMAGE BORDER, PICTURE BORDER rep/tv
f bord *m* de l'image
e borde *m* de la imagen
i bordo *m* dell'immagine
n beeldrand
d Bildkante *f*

2438 IMAGE BRIGHTNESS, PICTURE BRIGHTNESS rep/tv

f luminosité *f* de l'image
e luminosidad *f* de la imagen
i luminosità *f* dell'immagine
n beeldhelderheid
d Bildhelligkeit *f*

2439 IMAGE CAMERA TUBE crt
A photo-emissive camera tube having its photo-sensitive electrode separate from the target and in which the photo-electrons emitted are focused to generate a corresponding electron image on the target.
f tube *m* analyseur à transfert d'image
e tubo *m* de cámara de imagen
i tubo *m* da ripresa a trasferimento d'immagine
n tussenbeeldcamerabuis
d Kameraröhre *f* mit Vorabbildung, Zwischenbildkameraröhre *f*

2440 IMAGE CARRIER, rec/rep/tv
PICTURE CARRIER
Carrier wave modulated by the video signal.
f porteuse *f* vidéo
e portadora *f* video
i portante *f* video
n beelddraaggolf
d Bildträger *m*

2441 IMAGE CHANNEL, rec/tv
PICTURE CHANNEL
A one-way path for the vision component of a TV program(me) or of a contribution to a TV program(me).
f canal *m* image
e canal *m* imagen
i canale *m* immagine
n beeldkanaal *n*
d Bildkanal *m*

2442 IMAGE CONTRAST, rep/tv
PICTURE CONTRAST
f contraste *m* d'image
e contraste *m* de imagen
i contrasto *m* d'immagine
n beeldcontrast *n*
d Bildkontrast *m*

2443 IMAGE CONTROL COIL, rep/tv
PICTURE CONTROL COIL
A device in a TV receiver which center(re)s the picture on the face of the picture tube.
f bobine *f* de cadrage
e bobina *f* de encuadre de la imagen
i bobina *f* di regolazione dell'immagine
n beeldinstelspoel
d Bildeinstellungsspule *f*

2444 IMAGE CONVERTER, crt
IMAGE CONVERTER TUBE,
IMAGE VIEWING TUBE
An electronic tube in which an optical image projected on to a photo-emissive surface produces a corresponding image on a luminescent screen.
f convertisseur *m* d'image, tube *m* transformateur d'image
e tubo *m* transformador de imagen
i tubo *m* convertitore d'immagine
n beeldtransformatiebuis, beeldvormbuis
d Bildwandler *m*, Bildwandlerröhre *f*

2445 IMAGE CURRENT, rec/rep/tv
PICTURE CURRENT
f courant *m* d'image
e corriente *f* de imagen
i corrente *f* d'immagine
n beeldstroom
d Bildstrom *m*

2446 IMAGE DEFECTS opt
f défauts *pl* d'image
e defectos *pl* de imagen
i difetti *pl* d'immagine
n afbeeldingsfouten *pl*
d Abbildungsfehler *pl*

2447 IMAGE DEFINITION (GB), rec/tv
IMAGE RESOLUTION (US),
PICTURE DEFINITION (GB),
PICTURE RESOLUTION (US)
f définition *f* d'image
e definición *f* de imagen
i definizione *f* d'immagine
n beelddefinitie
d Bildauflösung *f*, Bildschärfe *f*

2448 IMAGE DETAIL, rec/rep/tv
PICTURE DETAIL
f détail *m* d'image
e detalle *m* de imagen
i dettaglio *m* d'immagine
n beelddetail *n*
d Bilddetail *n*, Bildeinzelheit *f*

2449 IMAGE DETAIL FACTOR, rep/tv
PICTURE DETAIL FACTOR
f facteur *m* de détail d'image
e factor *m* de detalle de imagen
i fattore *m* di dettaglio d'immagine
n beelddetailfactor
d Bilddetailfaktor *m*

2450 IMAGE DETAIL REGION, rec/rep/tv
PICTURE DETAIL REGION
f aire *f* de détails d'image
e área *f* de detalles de imagen
i area *f* di dettagli d'immagine
n beelddetailgebied *n*
d Bilddetailbereich *m*

IMAGE DETERIORATION
see: DEGRADATION

2451 IMAGE DIAGONAL, rec/rep/tv
PICTURE DIAGONAL
Measurement taken across opposite corners of the optical image (picture) formed on the faceplate of a camera or receiver tube.
f diagonale *f* d'écran
e diagonal *f* de pantalla
i diagonale *f* di schermo

n schermdiagonaal
d Schirmdiagonale *f*

2452 IMAGE DIMENSIONS, PICTURE DIMENSIONS rep/tv
f dimensions *pl* de l'image
e dimensiones *pl* de la imagen
i dimensioni *pl* dell'immagine
n beeldafmetingen *pl*
d Bildabmessungen *pl*, Bildgrösse *f*

IMAGE DISSECTOR
see: DISSECTOR

2453 IMAGE DISTANCE opt
The distance of the image behind the focus.
f distance *f* image
e distancia *f* de imagen
i distanza *f* d'immagine
n beeldafstand
d Bildweite *f*

2454 IMAGE DISTORTION, PICTURE DISTORTION dis
f distorsion *f* d'image
e distorsión *f* de imagen
i distorsione *f* d'immagine
n beeldvertekening, beeldvervorming
d Bildverzerrung *f*

2455 IMAGE DRIFT, PICTURE DRIFT rec/rep/tv
An irregularity due to imperfect synchronization.
f oscillation *f* de l'image
e oscilación *f* de la imagen
i oscillazione *f* dell'immagine
n beeldschommeling
d Bildwanderung *f*

2456 IMAGE DROP OUT, PICTURE DROP OUT crt/tv
f interruption *f* image, manque *m* d'image
e interrupción *f* imagen
i interruzione *f* immagine
n uitvallen *n* van het beeld
d Bildausfall *m*

2457 IMAGE DURATION, PICTURE DURATION rep/tv
f durée *f* de l'image
e duración *f* de la imagen
i durata *f* dell'immagine
n beeldduur
d Bilddauer *f*

2458 IMAGE ELEMENT, IMAGE POINT, PICTURE ELEMENT, PICTURE POINT ctv/rec/rep/tv
The smallest portion of the picture area which is capable of resolution by the system in use.
f élément *m* d'image, point *m* d'image
e elemento *m* de imagen
i elemento *m* d'immagine
n beeldelement *n*, beeldpunt *n*
d Bildelement *n*, Bildpunkt *m*, Rasterpunkt *m*

2459 IMAGE FIELD opt
That part of the field of view of a lens where the image has a high definition.
f champ *m* d'image
e campo *m* de imagen
i campo *m* d'immagine
n beeldveld *n*
d Bildfeld *n*

IMAGE FLICKER
see: BRIGHTNESS FLICKER

2460 IMAGE FLYBACK rec/rep/tv
f retour *m* du spot analyseur
e retorno *m* del punto explorador
i ritorno *m* del punto analizzatore
n beeldterugslag
d Bildrücklauf *m*

2461 IMAGE GEOMETRY, PICTURE GEOMETRY rep/tv
f géométrie *f* de l'image
e geometría *f* de la imagen
i geometria *f* dell'immagine
n beeldgeometrie
d Bildgeometrie *f*

2462 IMAGE HEIGHT, PICTURE HEIGHT rep/tv
f hauteur *f* de l'image
e altura *f* de la imagen
i altezza *f* dell'immagine
n beeldhoogte
d Bildhöhe *f*

2463 IMAGE ICONOSCOPE crt
TV camera tube which combines the action of an iconoscope and an image dissector.
f image-iconoscope *m*, supericonoscope *m*
e iconoscopio *m* de imagen
i supericonoscopio *m*
n beeldiconoscoop
d Superikonoskop *n*, Zwischenbildikonoskop *n*

2464 IMAGE I.F. AMPLIFIER, PICTURE I.F. AMPLIFIER rep/tv
f amplificateur *m* de moyenne fréquence image
e amplificador *m* de frecuencia intermedia imagen
i amplificatore *m* di media frequenza immagine
n beeldmiddenfrequentversterker
d Bild-ZF-Verstärker *m*

2465 IMAGE INFORMATION, PICTURE INFORMATION, VIDEO INFORMATION rec/rep/tv

f information *f* d'image
e información *f* de imagen
i informazione *f* d'immagine
n beeldinformatie
d Bildinformation *f*

2466 IMAGE INTENSIFIER TUBE crt
Electron device that converts an image illuminated by visible or other radiation into an electron image, that is in turn accelerated and focused to strike a fluorescent screen.
f tube *m* intensificateur d'image
e tubo *m* intensificador de imagen
i tubo *m* intensificatore d'immagine
n beeldversterkerbuis
d Bildverstärkerröhre *f*

2467 IMAGE INTERFERENCE, rep/tv
PICTURE INTERFERENCE
Interference caused by the video frequency.
f interférence *f* d'image par la fréquence vidéo
e interferencia *f* de imagen por la frecuencia video
i interferenza *f* d'immagine per la frequenza video
n beeldstoring door de videofrequentie
d Störung *f* durch die Bildfrequenz

2468 IMAGE INTERPOLATION, ctv
PICTURE INTERPOLATION
In colo(u)r converters, the interpolation between lines in a complete picture.
f interpolation *f* de l'image
e interpolación *f* de la imagen
i interpolazione *f* dell'immagine
n beeldinterpolatie
d Bildinterpolation *f*

2469 IMAGE INVERSION, opt
INVERSION OF IMAGE
f inversion *f* d'image
e inversión *f* de imagen
i inversione *f* d'immagine
n beeldomkering
d Bildaufrichtung *f*, Bildumkehr *f*

2470 IMAGE LINE, rec/rep/tv
PICTURE LINE
f ligne *f* d'image
e línea *f* de imagen
i linea *f* d'immagine
n beeldleiding
d Bildleitung *f*

2471 IMAGE LINE-AMPLIFIER OUTPUT, tv
PICTURE LINE-AMPLIFIER OUTPUT
The junction between the TV studio facility and the line feeding either a relay transmitter, a visual transmitter or a network.
f connexion *f* ligne image et sortie amplificateur
e conexión *f* línea imagen y salida amplificador
i connessione *f* linea immagine ed uscita amplificatore
n verbinding beeldleiding en versterkeruitgang
d Verbindung *f* Bildleitung und Verstärkerausgang

2472 IMAGE LOCK, rep/tv
PICTURE LOCK
The means whereby the image in a TV receiver is kept in place.
f calage *m* de l'image
e fijación *f* de la imagen
i agganciamento *m* dell'immagine
n vasthouden *n* van het beeld
d Bildhalt *m*

2473 IMAGE MODULATION, rec/rep/tv
PICTURE MODULATION
f modulation *f* d'image
e modulación *f* de imagen
i modulazione *f* d'immagine
n beeldmodulatie
d Bildmodulation *f*

2474 IMAGE MONITOR, crt
PICTURE MONITOR
f écran *m* de contrôle, moniteur *m* d'image
e monitor *m* de imagen
i monitore *m* d'immagine
n beeldmonitor
d Bildkontrollempfänger *m*, Bildmonitor *m*

2475 IMAGE ORBITING FACILITY, rec/rep/tv
PICTURE ORBITING FACILITY
A device which continually shifts the scan impingement spot at a slow rate to prevent ion burn.
f dispositif *m* de translation continue du spot
e dispositivo *m* de translación continua del punto de haz
i dispositivo *m* di traslazione continua del punto di fascio
n apparaat *n* voor continue verplaatsing van de lichtvlek
d Apparat *n* zur kontinuierlichen fortschreitenden Bewegung des Lichtflecks

2476 IMAGE ORIENTATION, rec/rep/tv
PICTURE ORIENTATION
The rotation of the TV image (picture) about its center(re) with the object of producing special effects, by using rotating prisms or by rotating the deflection coils.
f rotation *f* de l'image
e rotación *f* de la imagen
i rotazione *f* dell'immagine
n beeldrotatie
d Bilddrehung *f*

2477 IMAGE ORTHICON crt
TV camera tube in which an electron image is first developed on a photoemissive surface and then focused on a storage target which allows the charges representing the optical image to

accumulate until they are scanned.
f image-orthicon *m*,
superorthicon *m*
e ortinoscopio *m* de imagen
i superorticonoscopio *m*
n beeldorthiconoscoop
d Superorthikon *n*,
Zwischenbildorthikon *n*

2478 IMAGE OUTPUT, rep/tv
PICTURE OUTPUT,
VIDEO TIMEBASE
The timebase for the horizontal scanning of the field.
f base *f* de temps vidéo
e base *f* de tiempo video
i base *f* di tempi video
n beeldtijdbasis
d Bildzeitbasis *f*

2479 IMAGE OUTPUT TRANSFORMER tv
The timebase transformer provided for the scanning of the field.
f transformateur *m* de la base de temps vidéo
e transformador *m* de la base de tiempo de video
i trasformatore *m* della base di tempi video
n beeldtijdbasistransformator
d Bildsperrschwinger *m*,
Bildsperrschwinger-Transformator *m*

IMAGE PATTERN (US)
see: CHARGE IMAGE

2480 IMAGE PLANE opt
f plan *m* image
e plano *m* imagen
i piano *m* immagine
n afbeeldingsvlak *n*
d Abbildungsebene *f*,
Bildebene *f*

2481 IMAGE QUALITY, rep/tv
PICTURE QUALITY
f qualité *f* d'image
e cualidad *f* de imagen
i qualità *f* d'immagine
n beeldkwaliteit
d Bildgüte *f*

2482 IMAGE RECONSTRUCTOR TUBE, crt
IMAGE REPRODUCER TUBE,
KINESCOPE TUBE
Picture tube in a TV receiver.
f cinescope *m*
e cinescopio *m*
i cinescopio *m*
n kinescoop
d Kineskop *n*

2483 IMAGE REGISTRATION (US), rep/tv
PICTURE SUPERIMPOSITION (GB)
f superposition *f* de l'image
e superposición *f* de la imagen
i sovrapposizione *f* dell'immagine
n rastersuperponering
d Rasterdeckung *f*

2484 IMAGE REJECTION rep/tv
The relative strength of a reproduced video or audio signal.
f intensité *f* relative du signal reproduit
e intensidad *f* relativa de la señal reproducida
i intensità *f* relativa del segnale riprodotto
n relatieve intensiteit van het gereproduceerde signaal
d relative Intensität *f* des reproduzierten Signals

2485 IMAGE REPRODUCTION rep/tv
f reproduction *f* de l'image
e reproducción *f* de la imagen
i riproduzione *f* dell'immagine
n beeldreproduktie
d Bildwiederaufbau *m*,
Bildwiedergabe *f*

IMAGE RETENTION
see: BURN

2486 IMAGE SCANNER, rec/rep/tv
PICTURE SCANNER
f analyseur *m* de l'image
e explorador *m* de la imagen
i analizzatore *m* dell'immagine
n beeldaftaster
d Bildfeldzerleger *m*,
Bildzerleger *m*

2487 IMAGE SCANNING, rec/rep/tv
PICTURE SCANNING
f analyse *f* de l'image,
exploration *f* de l'image
e exploración *f* de la imagen
i analisi *f* dell'immagine,
scansione *f* dell'immagine
n beeldaftasting
d Bildabtastung *f*

2488 IMAGE SECTION, rep/tv
PICTURE SECTION
f section *f* image
e sección *f* imagen
i sezione *f* immagine
n beeldgedeelte *n*
d Bildteil *m*

2489 IMAGE SHIFT, rep/tv
PICTURE SHIFT
f déplacement *m* de l'image
e desplazamiento *m* de la imagen
i spostamento *m* dell'immagine
n beeldverschuiving
d Bildverschiebung *f*

2490 IMAGE SIGNAL, rec/rep/tv
PICTURE SIGNAL
The signal which conveys the picture information, as generated by the scanning device.
f signal *m* d'image
e señal *f* de imagen
i segnale *m* d'immagine
n beeldsignaal *n*
d Bildsignal *n*

2491 IMAGE SIGNAL AMPLITUDE, rep/tv
PICTURE SIGNAL AMPLITUDE
The difference between the white peak and the blanking level of a TV signal.
f amplitude *f* du signal d'image
e amplitud *f* de la señal de imagen
i ampiezza *f* del segnale d'immagine
n beeldsignaalamplitude
d Bildsignalamplitude *f*

2492 IMAGE SIGNAL DISTRIBUTION AMPLIFIER, cpl
PICTURE SIGNAL DISTRIBUTION AMPLIFIER
f distributeur *m* vidéo
e distribuidor *m* video
i distributore *m* video
n verdelerversterker van het beeldsignaal
d Bildsignalverteilerverstärker *m*

2493 IMAGE SIGNAL GENERATOR, cpl
PICTURE SIGNAL GENERATOR
f générateur *m* de signal d'image, générateur *m* vidéo
e generador *m* de señal de imagen, generador *m* video
i generatore *m* di segnale d'immagine, generatore *m* video
n beeldsignaalgenerator
d Bildsignalgenerator *m*

2494 IMAGE SIGNAL POLARITY, rep/tv
PICTURE SIGNAL POLARITY
The polarity of the signal voltage representing a dark area of a scene, with respect to the signal voltage representing a light area.
f polarité *f* du signal d'image
e polaridad *f* de la señal de imagen
i polarità *f* del segnale d'immagine
n beeldsignaalpolariteit
d Bildsignalpolarität *f*

2495 IMAGE SOURCES, rec/tv
PICTURE SOURCES
The sources supplying the illustrations for the news items about events occurring all over the world.
f sources *pl* des images
e fuentes *pl* de las imágenes
i sorgenti *pl* delle immagini
n beeldbronnen *pl*
d Bildquellen *pl*

2496 IMAGE SPACE opt
The space in which the image from the object in the object space is reproduced.
f espace *m* d'image
e espacio *m* de imagen
i spazio *m* d'immagine
n beeldruimte
d Bildraum *m*

2497 IMAGE SPOT SIZE rec/tv
f dimension *f* du spot
e dimensión *f* del punto
i dimensione *f* del punto
n beeldpuntgrootte
d Bildpunktgrösse *f*

2498 IMAGE STABILIZATION rec/tv
f stabilisation *f* de l'image
e estabilización *f* de la imagen
i stabilizzazione *f* dell'immagine
n beeldstabilisatie
d Bildstabilisierung *f*

2499 IMAGE STORAGE TUBE, rec/rep/tv
PICTURE STORAGE TUBE,
STORAGE CAMERA TUBE
A form of electron tube containing a mosaic electrode upon which an optical or electric image of the scene is focused and then scanned by the electron beam.
f tube *m* analyseur à accumulation, tube *m* image à mémoire
e tubo *m* de cámara de almacenaje, tubo *m* imagen de memoria
i tubo *m* di ripresa ad accumulazione, tubo *m* immagine a memoria
n geheugenbeeldbuis, katodestraalbuis met geheugenscherm
d Bildspeicherröhre *f*

2500 IMAGE STRIP, rec/tv
PICTURE STRIP,
SCANNING LINE
f ligne *f* d'analyse
e línea *f* de exploración
i linea *f* d'analisi
n aftastlijn
d Abtastzeile *f*

2501 IMAGE SWEEP FREQUENCY rec/tv
f fréquence *f* d'analyse vidéo
e frecuencia *f* de exploración video
i frequenza *f* d'analisi video
n beeldaftastfrequentie
d Bildabtastfrequenz *f*

2502 IMAGE SWITCHING, tv
PICTURE SWITCHING
f changement *m* d'images, escamotage *m* d'images
e conmutación *f* de imágenes, substitución *f* de imágenes
i commutazione *f* d'immagini, sostituzione *f* d'immagini
n beeldverwisseling
d Bildwechsel *m*

2503 IMAGE/SYNCHRONIZING RATIO, rep/tv
PICTURE/SYNCHRONIZING RATIO
The ratio of the maximum value of the picture signal to that of the synchronizing signal.
f rapport *m* d'amplitude de synchronisation, taux *m* de synchronisation
e relación *f* de sincronización
i rapporto *m* d'ampiezza fra sincronismo ed immagine
n beeld/synchronisatiesignaalverhouding
d Bild-Synchronsignalverhältnis *n*

IMAGE SYNTHESIS
see: BUILDING-UP OF IMAGE

2504 IMAGE TRANSFER CONVERTER, rec/rep/tv
PICTURE TRANSFER CONVERTER
Standards converter which displays an image (picture) at one standard on a suitable monitor and rephotographs it at another standard with a TV camera.
f convertisseur *m* d'image par transfert
e convertidor *m* de imagen por transferencia
i convertitore *m* d'immagine per trasferimento
n beeldomzetter met overdracht
d Bildwandler *m* mit Übertragung

2505 IMAGE TRANSMISSION, tv
PICTURE TRANSMISSION
The electric transmission of a picture having a gradation of shade values.
f transmission *f* d'image
e transmisión *f* de imagen
i trasmissione *f* d'immagine
n beeldoverdracht
d Bildübertragung *f*

2506 IMAGE TRANSMITTER, rec/tv
PICTURE TRANSMITTER
f transmetteur *m* d'image
e transmisor *m* de imagen
i trasmettitore *m* d'immagine
n beeldzender
d Bildsender *m*

2507 IMAGE TRANSMITTER POWER, rec/tv
PICTURE TRANSMITTER POWER
f puissance *f* du transmetteur d'image
e potencia *f* en el transmisor de imagen
i potenza *f* del trasmettitore d'immagine
n beeldzendervermogen *n*
d Bildsendeleistung *f*

IMAGE TUBE (GB)
see: CAMERA TUBE

2508 IMAGE WHITE, rep/tv
PICTURE WHITE
The level of the picture signal corresponding to white.
f blanc *m* d'image
e blanco *m* de imagen
i bianco *m* d'immagine
n witwaarde
d Weisswert *m*

2509 IMAGE WIDTH, rep/tv
PICTURE WIDTH
f largeur *f* d'image
e ancho *m* de imagen
i larghezza *f* d'immagine
n beeldbreedte
d Bildbreite *f*

2510 IMAGINARY PART OF THE COMPLEX CHROMATICITY SIGNAL ctv
f partie *f* imaginaire du signal complexe de chrominance
e parte *f* imaginaria de la señal compleja de crominancia
i parte *f* immaginaria del segnale complesso di crominanza
n imaginair gedeelte *n* van het complexe kleursoortsignaal
d Imaginärteil *m* des komplexen Farbartsignals

IMAGINARY PRIMARY COLO(U)RS
see: FICTITIOUS PRIMARIES

2511 IMF, crt
INTERNAL MAGNETIC FOCUS TUBE
f tube *m* cathodique à focalisation magnétique interne
e tubo *m* catódico de enfoque magnético interno
i tubo *m* catodico a focalizzazione magnetica interna
n katodestraalbuis met inwendige magnetische focussering
d Katodenstrahlröhre *f* mit innerer magnetischer Fokussierung

2512 IMMERSION ELECTRON LENS opt
An electrostatical lens usually consisting of two electrodes whereby the electrons emerging from the lens have a different energy as when leaving.
f lentille *f* électronique à immersion
e lente *f* electrónica de immersión
i lente *f* elettronica ad immersione
n immersie-elektronenlens
d Immersionselektronenlinse *f*

2513 IMPACT, crt
IMPINGEMENT
f impact *m*
e choque *m*, impacto *m*
i rimbalzo *m*, urto *m*
n botsing, optreffen *n*
d Aufschlag *m*, Auftreffen *n*

2514 IMPACT FLUORESCENCE, crt
IMPACT RADIATION
The fluorescence created by particle impingement.
f fluorescence *f* par impact
e fluorescencia *f* por choque, fluorescencia *f* por impacto
i fluorescenza *f* per rimbalzo, fluorescenza *f* per urto
n botsingsfluorescentie
d sensibilisierte Fluoreszenz *f*

2515 IMPACT SOUND dis/stu
Sound caused e.g. by footsteps on an uncovered floor, machinery vibrations and slamming doors.
f bruit *m* de pas
e ruido *m* de paso
i rumore *m* di passo
n loopgeluid *n*
d Trittschall *m*

2516 IMPERFECT TAPE vr
f bande *f* défectueuse
e cinta *f* defectuosa
i nastro *m* difettoso
n foutenband
d Fehlband *n*

2517 IMPLODE (TO) crt
To burst inward.
f imploser v
e hacer v implosión
i implodere v
n imploderen v
d implodieren v

2518 IMPLOSION crt
Mechanical inward collapse of a cathode-ray tube.
f implosion *f*
e implosión *f*
i implosione *f*
n implosie
d Implosion *f*

2519 IMPLOSION GUARD crt/rep/tv
f glace *f* de protection,
glace *f* de sécurité,
pareimplosion *m*
e guarda *f* de implosión,
placa *f* de vidrio de protección
i vetro *m* di protezione
n glasplaat,
veiligheidsruit
d Schutzscheibe *f*,
Sicherheitsscheibe *f*

2520 IMPLOSION PROOF crt
f inimplosible adj
e no implosivo
i non implosivo
n implosievrij adj
d implosionssicher adj

IMPREGNATED TAPE
see: DISPERSED-POWDER MAGNETIC TAPE

2521 IMPULSIVE NOISE SIGNAL cpl/dis
Undesired signal on a video system which consists of a series of pulses.
f parasites *pl* impulsives
e parásitos *pl* impulsivos
i perturbazioni *pl* impulsive
n impulsieve storingen *pl*
d Impulsstörungen *pl*

2522 IMPULSIVE RESPONSE cpl
f réponse *f* impulsive
e respuesta *f* impulsiva
i risposta *f* impulsiva
n impulsieve responsie
d aperiodischer Frequenzgang *m*

2523 IN-EDIT, vr
IN-GOING SPLICE,
IN-POINT
f point *m* d'entrée
e entrada *f* de la secuencia
i inizio *m* dell'inserto,
punto *m* d'ingresso
n beginpunt *n*
d eingehender Schnitt *m*

2524 IN-LINE HEADS, vr
STACKED HEADS
Two magnetic-tape heads so mounted that their gaps are in exact vertical alignment.
f têtes *pl* magnétiques alignées verticalement
e cabezas *pl* magnéticas alineadas verticalmente
i testine *pl* magnetiche allineate verticalmente
n verticaal uitgelijnde magneetkoppen *pl*
d vertikal ausgerichtete Magnetköpfe *pl*

2525 IN-LINE STEREOPHONIC TAPE vr
Magnetic stereophonic tape on which a record is made by using in-line heads.
f bande *f* magnétique stéréophonique parcourue entre têtes magnétiques à alignement vertical
e cinta *f* magnética estereofónica pasada entre cabezas magnéticas de alineación vertical
i nastro *m* magnetico stereofonico passato tra testine magnetiche ad allineamento verticale
n met verticaal uitgelijnde magneetkoppen vervaardigde stereofonische magneetband
d mit vertikal ausgerichteten Magnetköpfen hergestelltes stereophonisches Magnetband *n*

2526 IN-PHASE ge
f en phase
e de fases iguales, en fase
i in fase
n in faze
d gleichphasig adj, in Phase

IN-PHASE COMPONENT
see: ACTIVE COMPONENT

INCIDENCE ANGLE
see: ANGLE OF INCIDENCE

2527 INCIDENT BEAM aea/crt/ge
Any wave or particle beam the path of which intercepts a surface of discontinuity.
f faisceau *m* incident
e haz *m* incidente
i fascio *m* incidente
n invallende bundel
d einfallender Strahl *m*,
Einfallsstrahl *m*

2528 INCIDENT LIGHT opt
f lumière *f* incidente
e luz *f* incidente
i luce *f* incidente
n invallend licht *n*
d einfallendes Licht *n*

2529 INCIDENT RAY opt
f rayon *m* incident
e rayo *m* incidente
i raggio *m* incidente
n invallende straal
d einfallender Strahl *m*

2530 INCIDENTAL MUSIC stu/tv
Music occurring incidentally in the spoken play.

f musique *f* d'accompagnement
e música *f* acompañante
i musica *f* d'accompagnamento
n begeleidende muziek
d Begleitmusik *f*,
untermalende Musik *f*

2531 INCIDENTAL SOUNDS aud
f sons *pl* illustratifs
e sonidos *pl* incidentales
i suoni *pl* illustrativi
n illustrerende bijgeluiden *pl*
d illustrierende Nebenläute *pl*

2532 INCOMING LINE tv
f ligne *f* d'arrivée
e línea *f* de entrada
i linea *f* d'arrivo
n inkomende leiding
d ankommende Leitung *f*

2533 INCOMING SIGNAL rep/tv
f signal *m* d'entrée
e señal *f* recibida a la entrada
i segnale *m* di ricezione
n inkomend signaal *n*
d Empfangssignal *n*

2534 INCREMENTAL INDUCTANCE TUNER tv
f sélecteur *m* de canaux à syntonisation inductive
e selector *m* de canales con sintonización inductiva
i selettore *m* di canali a sincronizzazione indottiva
n kanalenkiezer met inductieve afstemming
d Kanalwähler *m* mit induktiver Abstimmung

2535 INDENTED ROLLER vr
Device used in magnetic tape coating.
f rouleau *m* dentelé
e rodillo *m* dentado
i rotolo *m* dentato
n getande rol
d gezahnte Rolle *f*

2536 INDEPENDENT SIDEBAND TRANSMISSION cpl
A method of operation in which the two sidebands correspond to two independent signals.
f émission *f* à bandes latérales indépendantes
e emisión *f* de bandas laterales independientes
i emissione *f* a bande laterali indipendenti
n transmissie met onafhankelijke zijbanden
d Übertragung *f* mit unabhängigen Seitenbändern

2537 INDEPENDENT SIDEBAND TRANSMITTER cpl
A transmitter, generally with a low-level carrier whose upper and lower sidebands can each provide one or more independent channels.
f émetteur *m* à bandes latérales indépendantes
e emisor *m* de bandas laterales independientes
i emettitore *m* a bande laterali indipendenti
n zender met onafhankelijke zijbanden
d Sender *m* mit unabhängigen Seitenbändern

2538 INDEPENDENT SYNCHRONIZATION rec/rep/tv
f synchronisation *f* indépendante
e sincronización *f* independiente
i sincronizzazione *f* indipendente
n onafhankelijke synchronisatie
d unabhängige Synchronisierung *f*

2539 INDEX STRIPS ctv
Strips arranged vertically behind the red phosphor strips of a picture display tube of the apple and zebra type serving to modulate the frequency of the pilot beam.
f filets *pl* colorés,
rubans *pl* d'indice
e cintas *pl* de índice,
tiras *pl* coloradas
i striscie *pl* colorate,
striscie *pl* d'indice
n indexstroken *pl*
d Indexstreifen *pl*

2540 INDEX TUBE ctv
A class of colo(u)r display tube distinguished by having a single picture-writing electron gun and a referencing or index device designed to correct errors and keep the beam on the appropriate phosphor as the receiver signal demands from moment to moment.
f tube *m* à rubans fluorescents,
tube *m* index
e tubo *m* de cintas fluorescentes,
tubo *m* índice
i tubo *m* a striscie fluorescenti,
tubo *m* indice
n indexbuis,
kleurbeeldbuis met fluorescerende stroken
d Farbbildröhre *f* mit fluoreszierenden Streifen,
Indexröhre *f*

2541 INDEXING ELECTRON BEAM, PILOT BEAM ctv
The electron beam which sweeps over the colo(u)r strips with red, green and blue phosphors in colo(u)r display tubes of the apple type.
f faisceau-pilote *m*
e haz-piloto *m*
i fascio-pilota *m*
n stuurbundel
d Steuerstrahl *m*

2542 INDEXING SIGNAL ctv
Signal generated by a beam-indexing colo(u)r cathode-ray tube.
f signal *m* de tube à index
e señal *f* de tubo de índice
i segnale *m* di tubo d'indice
n indexsignaal *n*
d Indexsignal *n*

2543 INDICATOR, vr
MODULATION INDICATOR
f indicateur *m* de la modulation
e indicador *m* de la modulación
i indicatore *m* della modulazione
n indicator,
modulatiemeter
d Anzeigeinstrument *n*,
Modulationsanzeiger *m*

2544 INDICATOR SWITCH vr
Used to check the video signal on the indicator.
f contrôle *m* modulation image
e control *m* de modulación imagen
i controllo *m* di modulazione immagine
n druktoets voor geluid/video-modulatie,
indicatortoets
d Anzeigetaste *f*

2545 INDIRECT COLORIMETRY ct
Indirect measurement of colo(u)rs.
f colorimétrie *f* indirecte
e colorimetría *f* indirecta
i colorimetria *f* indiretta
n indirecte kleurmeting
d mittelbare Farbmessung *f*

2546 INDIRECT LIGHTING ct
Lighting by means of fittings with a light distribution such that no more than 10% of the emitted luminous flux reaches the illuminated plane directly, assuming that the plane is unbounded.
f éclairage *m* indirect
e alumbrado *m* indirecto
i illuminazione *f* indiretta
n indirecte verlichting
d indirekte Beleuchtung *f*

2547 INDIRECT RADIATION aea/tv
Radiation which follows an indirect path as e.g. the radiation of TV signals reflected from near-by hills or other elevated objects.
f rayonnement *m* indirect
e radiación *f* indirecta
i radiazione *f* indiretta
n indirecte straling
d mittelbare Strahlung *f*

2548 INDIRECT SCANNING tv
A scanning method whereby a narrow beam is moved over the dark scanning field.
f analyse *f* indirecte
e exploración *f* indirecta
i analisi *f* indiretta
n indirecte aftasting
d mittelbare Abtastung *f*

2549 INDIRECT SYNCHRONIZATION tv
f synchronisation *f* indirecte
e sincronización *f* indirecta
i sincronizzazione *f* indiretta
n indirecte synchronisatie
d mittelbare Synchronisierung *f*

2550 INDIRECT TRANSMISSION tv
f émission *f* indirecte
e emisión *f* indirecta
i emissione *f* indiretta
n indirecte uitzending
d indirekte Übertragung *f*,
mittelbare Sendung *f*

2551 INDOOR AERIAL, aea/tv
INDOOR ANTENNA
f antenne *f* intérieure
e antena *f* interior
i antenna *f* interiore
n binnenantenne
d Innenantenne *f*,
Zimmerantenne *f*

2552 INDOOR RECORDING aud
f enregistrement *m* en intérieur
e registro *m* al interior
i registrazione *f* all'interiore
n binnenopname
d Innenaufnahme *f*

2553 INDUCED NOISE dis
Caused by some form of coupling between the noise source and the equipment.
f bruit *m* induit
e ruido *m* inducido
i disturbo *m* dovuto a correnti forti,
rumore *m* indotto
n geïnduceerde ruis
d induziertes Geräusch *n*,
Starkstromgeräusch *n*

2554 INDUCTION FIELD aea
That part of the field of an aerial (antenna) which is associated with a pulsation of energy to and fro between the aerial (antenna) and the medium.
f champ *m* inducteur
e campo *m* inductor
i campo *m* d'induzione
n inductieveld *n*
d Induktionsfeld *n*

2555 INDUCTION ZONE, aea
NEAR REGION,
NEAR ZONE
The region immediately surrounding a transmitting aerial (antenna), extending out to a distance of wavelength.
f zone *f* d'induction
e zona *f* de inducción
i zona *f* d'induzione
n gebied *n* van het nabije veld,
quasistationnair gebied *n*
d Nahbereich *m*,
Nahgebiet *n*

2556 INDUCTIVE SAWTOOTH cpl/tv
GENERATOR
f générateur *m* inductif de dents de scie
e generador *m* inductivo de dientes de sierra
i generatore *m* induttivo di denti di sega
n inductieve zaagtandgenerator
d induktiver Sägezahngenerator *m*

2557 INDUSTRIAL TELEVISION, tv
ITV
Closed-circuit TV used in industry for the inspection of processes or examination of materials where direct examination is unsafe or difficult.
f télévision *f* industrielle
e televisión *f* industrial
i televisione *f* industriale
n industriële televisie
d Betriebsfernsehen *n*,
industrielles Fernsehen *n*

2558 INFORMATION (US), rec/rep/tv
SIGNAL COMPLEX (GB)
The sound, picture or controlling pulses together, with their details, which modulate the carrier wave in a TV system.
f ensemble *m* des signaux
e conjunto *m* de las señales
i complesso *m* dei segnali
n signaalcomplex *n*,
signaalpak *n*
d Signalfahrplan *m*,
Signalhaushalt *m*

2559 INFORMATION BANDWIDTH, aud
INTELLIGENCE BANDWIDTH
The sum of the audio or video frequency bandwidths of the one or more channels.
f largeur *f* de bande de l'information
e ancho *f* de banda de la información
i larghezza *f* di banda dell'informazione
n informatiebandbreedte
d Informationsbandbreite *f*

2560 INFORMATION CAPACITY tv
With a frame size of 18 x 24mm and a practical limit to resolution of about 60 lines per mm the capacity is about 15 x 10^5 elements.
f capacité *f* de l'information
e capacidad *f* de la información
i capacità *f* dell'informazione
n informatiecapaciteit
d Informationskapazität *f*

2561 INFORMATION STORAGE tv
Taking place in conversion systems and carried out e.g. in the afterglow of a cathode-ray tube phosphor or in the target of a camera either as a semiconductor lag or as capacity storage.
f emmagasinage *m* de l'information
e almacenamiento *m* de la información
i immagazzinamento *m* dell'informazione
n informatieopslag
d Informationsspeicherung *f*

2562 INFORMATION SYSTEMS tv
A term covering the various fields of application of closed circuit systems.
f systèmes *pl* d'information
e sistemas *pl* de información
i sistemi *pl* d'informazione
n informatiesystemen *pl*
d Informationssysteme *pl*

2563 INFORMATION TRANSMISSION tv
f transmission *f* de l'information
e transmisión *f* de la información
i trasmissione *f* dell'informazione
n informatietransmissie
d Informationsübertragung *f*

INFRABLACK
see: BLACKER-THAN-BLACK

INFRABLACK LEVEL
see: BLACKER-THAN-BLACK LEVEL

INFRABLACK REGION
see: BLACKER-THAN-BLACK REGION

2564 INFRARED CAMERA, rec/tv
INFRARED SENSITIVE TELEVISION CAMERA
f caméra *f* de télévision sensible aux rayons infrarouges
e cámara *f* de televisión sensible para rayos infrarrojos
i camera *f* di televisione sensibile ai raggi infrarossi
n voor infrarood gevoelige televisiecamera
d infrarotempfindliche Fernsehkamera *f*

2565 INFRARED IMAGE CONVERTER crt
An electron tube which converts an invisible scene illuminated with infrared rays into a visible image on a fluorescent screen.
f tube *m* convertisseur d'image à éclairement par rayons infrarouges
e tubo *m* convertidor de imagen por iluminación de rayos infrarrojos
i tubo *m* convertitore d'immagine per illuminazione a raggi infrarossi
n infrarood-beeldomvormbuis
d Infrarotbildwandlerröhre *f*

2566 INFRARED LINK ctv
Used in colo(u)r TV outside broadcasts.
f liaison *f* par rayons infrarouges
e enlace *m* por rayos infrarrojos
i ponte *m* a raggi infrarossi
n infraroodverbinding
d Infrarotverbindung *f*

2567 INFRARED MASER, tv
IRASER,
LASER
An optical maser using a small crystal of ruby generating intermittent but very pure and concentrated beams of infrared light. Used in holography.
f laser *m*
e laser *m*
i laser *m*
n laser
d Laser *m*

INFRARED RADIATION
see: HEAT RADIATION

2568 INHERENT NOISE, SET NOISE — dis
f bruit *m* de fond propre, bruit *m* inhérent
e ruido *m* de fondo propio, ruido *m* inherente
i rumore *m* di fondo proprio, rumore *m* inerente
n eigen ruis
d Eigengeräusch *n*

INJECT (TO)
see: FEED (TO)

2569 INLAID CAPTION — rec/tv
f titre *m* inséré
e título *m* de inserción
i titolo *m* d'intarsio
n inlastitel
d Einblendtitel *m*

2570 INLAY, KEYED INSERTION, STATIC MATTE — tv
In TV, an electronic method of combining in one picture selected areas of two pictures obtained from separate sources, the required areas of each picture being determined by the shape and position of an opaque mask being placed in a silhouette generator.
f procédé *m* des caches électroniques, système *m* électronique d'insertion
e inserción *f* electrónica
i intarsio *m* elettronica, mediante maschera ottica
n insnijden *n*, inzet
d Inlay-Trickmischung *f*

2571 INLAY CAMERA — rec/tv
A special camera containing a vidicon tube or a flying-spot scanner used for making inlays.
f caméra *f* à caches
e cámara *f* para tomar inserciones
i camera *f* per prendere intarsi
n inzetcamera
d Inlay-Kamera *f*

2572 INLAY MASK — tv
A mechanical cut-off for determining the shape and timing of the keying signal electronically generated by means of a flying spot optical system.
f masque *m* de caches
e máscara *f* de inserciones
i maschera *f* d'intarsi
n inzetmasker *n*
d Inlay-Maske *f*

2573 INPUT — ge
f entrée *f*
e entrada *f*
i entrata *f*, ingresso *m*
n ingang
d Eingang *m*

2574 INPUT AMPLIFIER — cpl
f amplificateur *m* d'entrée, préamplificateur *m*
e amplificador *m* de entrada, preamplificador *m*
i amplificatore *m* d'entrata, preamplificatore *m*
n ingangsversterker, voorversterker
d Eingangsverstärker *m*, Vorverstärker *m*

2575 INPUT IMPEDANCE — cpl
f impédance *f* d'entrée
e impedancia *f* de entrada
i impedenza *f* d'entrata
n ingangsimpedantie
d Eingangsimpedanz *f*

2576 INPUT SIGNAL — ge
f signal *m* d'entrée
e señal *f* de entrada
i segnale *m* d'entrata
n ingangssignaal *n*
d Eingangssignal *n*

2577 INPUT STAGE — cpl
f étage *m* d'entrée
e etapa *f* de entrada
i stadio *m* d'entrata
n ingangstrap
d Eingangsstufe *f*

2578 INPUT UHF AERIAL, INPUT UHF ANTENNA — aea
f prise *f* pour antenne UHF
e terminal *m* para antena UHF
i morsetto *m* per antenna UHF
n ingang UHF antenne
d UHF-Antennenanschluss *m*

2579 INPUT VHF-AERIAL, INPUT VHF-ANTENNA — aea
f prise *f* pour antenne VHF
e terminal *m* para antena VHF
i morsetto *m* per antenna VHF
n ingang VHF antenne
d VHF-Antennenanschluss *m*

2580 INSERT — tv
Intercut shot of a static object, e.g. books, letters, clock on mantelpiece, view from window, etc. produced separately from the main sequence of action.
f insertion *f* statique
e inserción *f* estática
i inserzione *f* statica
n statische inlas
d statische Einlage *f*, statischer Einsatz *m*

2581 INSERT CAMERA — tv
f caméra *f* d'insertions
e cámara *f* de inserciones
i camera *f* d'inserzioni
n inlascamera
d Einlagenkamera *f*, Einsatzkamera *f*

2582 INSERT EARPHONE, aud
INSERT EARPIECE
f écouteur *m* auriculaire
e receptor *m* auricular
i microtelefono *m* auriculare
n oortelefoon
d Ohrhörer *m*

2583 INSERTION tv
In TV special effects, the substitution of one picture for another over a part of the camera field (inlays and overlap).
f insertion *f*
e inserción *f*
i inserzione *f*
n inlas
d Einlage *f*

2584 INSERTION GAIN cpt
Gain resulting from the insertion of a transducer in a transmission system.
f gain *m* par insertion
e ganancia *f* de inserción
i guadagno *m* d'inserzione
n inschakelversterking
d Einfügungsgewinn *m*

2585 INSERTION LOSS cpl
Loss resulting from the insertion of a transducer in a transmission system.
f perte *f* par insertion
e atenuación *f* de inserción
i attenuazione *f* d'inserzione
n inschakelverlies *n*
d Einfügungsdämpfung *f*,
Einfügungsverlust *m*

2586 INSTANT TALK-IN SYSTEM stu
A system combining the omnibus talkback system with facilities for immediate reply from each remote location to the studio director.
f système *m* de communication à réponse instantanée
e sistema *m* de comunicación de respuesta instantánea
i sistema *m* di comunicazione a risposta istantanea
n ruggespraaksysteem *n* met directe beantwoording
d Rücksprachesystem *n* mit direkter Beantwortung

2587 INSTANTANEOUS cpl
CHARACTERISTIC
The response of an amplifier to an occasional transient.
f caractéristique *f* instantanée
e característica *f* instantánea
i caratteristica *f* istantanea
n momentkarakteristiek
d Momentancharakteristik *f*

2588 INSTRUCTIVE TELEVISION tv
f télévision *f* instructive
e televisión *f* instructiva
i televisione *f* istruttiva
n instructieve televisie
d belehrendes Fernsehen *n*,
instruktives Fernsehen *n*

2589 INTEGRATING DIVIDER, tv
INTEGRATING FREQUENCY DIVIDER
f diviseur *m* de fréquence à intégration
e divisor *m* de frecuencia integrador
i divisore *m* di frequenza integratore
n integrerende frequentiedeler
d integrierender Frequenzteiler *m*

2590 INTEGRATOR, cpl
INTEGRATOR CIRCUIT
A circuit whose output is substantially proportional to the time integral of the input.
f circuit *m* intégrateur,
intégrateur *m*
e circuito *m* integrador,
integrador *m*
i circuito *m* integratore,
integratore *m*
n integrator,
integratorschakeling
d Integrationsschaltung *f*,
Integrator *m*

2591 INTELLIGENCE ge
The information contained in a signal.
f information *f* du signal,
message *m*
e contenido *m* de la señal,
información *f* de la señal
i informazione *f* nel segnale
n signaalinhoud
d Nachrichtengehalt *m*

INTELLIGENCE BANDWIDTH
see: INFORMATION BANDWIDTH

2592 INTELLIGIBILITY aud
Of a system used for transmitting or reproducing speech, the percentage number of simple ideas correctly received over the system.
f intelligibilité *f*
e inteligibilidad *f*
i intelligibilità *f*
n verstaanbaarheid
d Sinnverständlichkeit *f*

2593 INTENSIFIER, crt
INTENSIFIER ELECTRODE,
POST-DEFLECTION ACCELERATOR
The electrode of a cathode-ray tube which increases the acceleration of the electrons after the beam has been deflected.
f électrode *f* postaccélératrice
e electrodo *m* postacelerador
i elettrodo *m* d'accelerazione addizionale,
elettrodo *m* postacceleratore
n naversnellingselektrode
d Nachbeschleunigungselektrode *f*

2594 INTENSIFIER RING crt
A metallic ring-shaped coating on the inside of the glass envelope of a cathode-ray tube near the fluorescent screen.
f anneau *m* postaccélérateur
e anillo *m* postacelerador
i anello *m* postacceleratore

n naversnellingsring
d Nachbeschleunigungsring *m*

2595 INTENSIFY (TO) crt
To increase the brilliance of an image on the screen of a cathode-ray tube.
f intensifier v, renforcer v
e intensificar v, reforzar v
i rinforzare v
n versterken v
d verstärken v

INTENSITY CONTROL
see: BRIGHTNESS CONTROL

2596 INTENSITY MODULATION crt
Modulation of the luminosity of the spot on the fluorescent screen of a kinescope, by variation of the current in the beam.
f modulation *f* de luminosité
e modulación *f* en luminosidad
i comando *m* di luminosità, modulazione *f* d'intensità
n helderheidsmodulatie
d Helligkeitssteuerung *f*, Intensitätskontrolle *f*

INTENSITY OF ILLUMINATION
see: ILLUMINATION INTENSITY

2597 INTERACTION OF SATURATION AND CONTRAST svs
f interaction *f* saturation /contraste
e interacción *f* saturación /contraste
i interazione *f* saturazione /contrasto
n meeloop verzadiging met contrast
d Mitnahme *f* Sättigung mit Kontrast

2598 INTERCARRIER BEAT dis
An interference pattern that appears on TV pictures when the 4.5mc beat frequency of an intercarrier sound system gets through the video amplifier to the video input of the picture tube.
f interférence *f* de l'interporteuse
e interferencia *f* de la interportadora
i interferenza *f* della portante differenziale
n interdraaggolfstoring
d Differenzträgerstörung *f*

2599 INTERCARRIER NOISE SUPPRESSOR, dis
INTERSTATION NOISE SUPPRESSOR,
NOISE SUPPRESSOR,
SQUELCH CIRCUIT
A circuit that blocks the audiofrequency amplifier of a receiver automatically when no carrier is being received.
f réglage *m* silencieux
e regulación *f* silenciosa
i regolazione *f* silenziosa
n stille regeling
d automatische Geräuschsperre *f*, Rauschunterdrückung *f*

INTERCARRIER SOUND SYSTEM
see: CARRIER DIFFERENCE SYSTEM

2600 INTERCHANGEABLE LENS opt/tv
f objectif *m* interchangeable
e objetivo *m* intercambiable
i obiettivo *m* intercambiabile
n verwisselbaar objectief *n*
d Wechselobjektiv *n*

2601 INTERCITY RELAY SYSTEM tv
A system for the transmission of TV relay signals by fixed stations from one service area to another.
f système *m* interurbain à stations-relais
e retransmisión *f* televisiva interurbana
i ritrasmissione *f* televisiva interurbana
n interlokaal relayeersysteem *n*
d Fernballsendung *f*

2602 INTERCOM MASTER SET cpl
f poste *m* principal
e aparato *m* principal
i apparecchio *m* principale
n hoofdapparatuur
d Hauptapparatur *f*

2603 INTERCOMMUNICATING SYSTEM cpl/stu
Loudspeaking telephones for use in communication between two rooms.
f système *m* d'intercommunication
e sistema *m* de intercomunicación
i sistema *m* d'intercomunicazione
n interfooninstallatie, ruggespraaksysteem *n*
d Gegensprechanlage *f*, Interfonanlage *f*

INTERCUT
see: CROSS-CUT

2604 INTERDOT FLICKER ctv
f papillotement *m* multiple entre les points
e parpadeo *m* múltiple entre los puntos
i sfarfallio *m* multiplo tra i punti
n interpuntflikkering
d Zwischenpunktsflimmern *n*

2605 INTERFERENCE dis
The disturbance of signals due to extraneous energy which tends to interfere with the reception of desired signals.
f brouillage *m*, interférence *f*, parasites *pl*
e interferencia *f*, perturbación *f*, ruido *m* parásito
i disturbo *m*, interferenza *f*, perturbazione *f*
n storing
d Störung *f*

2606 INTERFERENCE ELIMINATOR, dis
INTERFERENCE-SUPPRESSION FILTER,
INTERFERENCE TRAP
f filtre *m* antiparasite
e eliminador *m* de interferencias
i dispositivo *m* antiparassita, soppressore *m* di disturbi

n storingsonderdrukkingsfilter *n*
d Störungsunterdrücker *m*,
Störungsunterdrückungsfilter *n*

INTERFERENCE FIGURE,
INTERFERENCE PATTERN
see: BEAT PATTERN

2607 INTERFERENCE FILTER opt
f filtre *m* interférentiel
e filtro *m* interferencial
i filtro *m* interferenziale
n interferentiefilter *n*
d Interferenzfilter *n*

2608 INTERFERENCE GUARD BANDS cpl/dis
The two bands of frequencies additional to, and on either side of, the communication band and frequency tolerance, which may be provided in order to minimize the possibility of interference.
f bandes *pl* de fréquence de protection, espace *m* libre entre deux canaux
e bandas *pl* de protección
i bande *pl* di protezione
n afschermfrequentiebanden *pl*
d Schutzfrequenzbänder *pl*

INTERFERENCE INVERSION
see: BLACK SPOTTING

INTERFERENCE INVERTER
see: BLACK SPOTTER

2609 INTERFERENCE LEVEL dis
f niveau *m* de perturbation
e nivel *m* de perturbación
i livello *m* di perturbazione
n storingsniveau *n*
d Störpegel *m*,
Störspiegel *m*

2610 INTERFERENCE LIMITER dis
f limiteur *m* de parasites
e limitador *m* de parásitos
i limitatore *m* di parassiti
n storingsbegrenzer
d Störbegrenzer *m*

2611 INTERFERENCE PATTERN, INTERFERENCE STRIPES dis
f barres *pl* d'interférences
e barras *pl* de interferencia
i barre *pl* d'interferenza
n stoorstrepen *pl*
d Störstreifen *pl*

2612 INTERFERENCE PULSE dis
f impulsion *f* parasite
e impulso *m* parásito
i impulso *m* parassita
n stoorimpuls
d Störimpuls *m*

2613 INTERFERENCE SUPPRESSION FOR PICTURE AND SOUND dis
f montage *m* suppresseur de parasites pour l'image et le son
e circuito *m* de supresión de parásitos para la imagen y el sonido
i circuito *m* soppressore di parassiti per l'immagine ed il suono
n storingsonderdrukkingsschakeling voor beeld en geluid
d Störungsunterdrückungsschaltung *f* für Bild und Ton

2614 INTERFERENCE THRESHOLD cpl/dis
A measure of the signal-to-noise requirement for virtually free transmission and reception.
f seuil *m* de perturbation
e umbral *m* de perturbación
i soglia *f* di perturbazione
n storingsdrempel
d Störungsschwelle *f*

2615 INTERFERENCE VOLTAGE dis
In TV, any extraneous voltage present in the signal channel.
f tension *f* de perturbation
e tensión *f* de perturbación
i tensione *f* di perturbazione
n stoorspanning
d Störspannung *f*

INTERFIELD CUT
see: CUT IN BLANKING

2616 INTERLACE CONTROL rec/rep/tv
f ajustage *m* de l'analyse entrelacée
e ajuste *m* de la exploración entrelazada
i aggiustaggio *m* dell'analisi interlacciata
n instelling van de interliniëring
d Einstellung *f* der Zwischenzeilenabtastung

INTERLACE SEQUENCE,
INTERLACING ORDER
see: FIELD SEQUENCE

2617 INTERLACED SCANNING, INTERLACING tv
In TV, a scanning process in which the distance from center(re) to center(re) of successively scanned lines is two or more times the nominal line width, and in which the adjacent lines belong to successive fields.
f analyse *f* à intercalage, analyse *f* entrelacée
e exploración *f* entrecalada, exploración *f* entrelazada
i analisi *f* interlacciata, esplorazione *f* a linee intercalate, esplorazione *f* interlineata, scansione *f* interlacciata
n interliniëring
d Abtastung *f* im Zeilensprungverfahren, Zwischenzeilenabtastung *f*

2618 INTERLACED SPOT SCANNING rec/tv
f système *m* à entrelacement de points
e sistema *m* entrelazado de puntos
i sistema *m* interlacciato di punti
n interpunctering
d Punktsprungverfahren *n*

2619 INTERLEAVE vr
Alternation of parts from different and unrelated messages on a tape.
f altération *f* de signaux d'information
e inserciones *pl* sin relación
i alternanza *f* di segnali d'informazione
n ongewenste afwisseling van informatiesignalen
d Verschachtelung *f* von Informationssignalen

2620 INTERLEAVED TRANSMISSION SIGNAL tv
Type of colo(u)r TV signal in which luminance and chrominance signals are compressed into the same video band of frequencies.
f signal *m* de transmission cocanalisé
e señal *f* de transmisión cocanalizada
i segnale *m* di trasmissione cocanalizzato
n signaalvervlechting
d Signalverschachtelung *f*

2621 INTERLEAVING ctv
A technique of colo(u)r TV transmission in which luminance and chrominance signals occupy the same channel.
f cocanalisation *f*
e cocanalización *f*
i cocanalizzazione *f*
n vervlechting
d Verschachtelung *f*

2622 INTERLINE FLICKER ctv
f papillotement *m* interligne
e parpadeo *m* interlineal
i sfarfallio *m* interlineare
n interlijnflikkering
d Zeilenflimmern *n*,
Zwischenzeilenflimmern *n*

2623 INTERLOCK cpl/tv
A device or circuit designed to make the operation of one piece of apparatus dependent on the fulfilment of predetermined conditions by another piece of apparatus.
f enclenchement *m*,
verrouillage *m*
e enclavamiento *m*,
enganche *m*
i bloccaggio *m*,
blocco *m*
n vergrendeling
d Verriegelung *f*

2624 INTERMEDIATE FILM METHOD tv/vr
Early TV transmission system in which a film camera was used to record a studio or outside scene. After development the wet film was scanned, electronically reversed to a positive and then transmitted.
f télévision *f* par film intermédiaire
e sistema *m* de película intermedia
i sistema *m* a pellicola intermedia
n tussenfilmsysteem *n*
d Zwischenfilmverfahren *n*

INTERMEDIATE FREQUENCY
see: I.F.

INTERMEDIATE FREQUENCY AMPLIFIER
see: I.F. AMPLIFIER

INTERMEDIATE FREQUENCY REJECTION
see: I.F. REJECTION

INTERMEDIATE FREQUENCY REJECTION FACTOR
see: I.F. REJECTION FACTOR

INTERMEDIATE FREQUENCY SIGNAL
see: I.F. SIGNAL

2625 INTERMEDIATE MULTIPLE, RINGING (GB), SPLIT IMAGE (US) rep/tv
In a video circuit, a damped oscillatory response to a picture pulse, resulting in a series of closely spaced black and white images of decreasing density.
f franges *pl*,
suroscillation *f*
e sobreoscilación *f*
i sovraoscillazione *f*
n uitslingereffect *n*
d Überschwingen *n*

2626 INTERMITTENT SCANNING rec/rep/tv
f analyse *f* intermittente
e exploración *f* intermitente
i analisi *f* intermittente
n niet-chronologische aftasting
d zeitlich unregelmässige Abtastung *f*

2627 INTERMODULATION cpl
The modulation of the components of a complex wave by each other as a result of which are produced waves which have frequencies, among others, equal to the sums and differences of those of the components of the original complex wave.
f intermodulation *f*
e intermodulación *f*
i intermodulazione *f*
n intermodulatie
d Intermodulation *f*,
Zwischenmodulation *f*

2628 INTERMODULATION DISTORTION cpl
A constituent of nonlinearity distortion consisting of the occurrence, in the response to coexistent sinusoidal excitations, of sinusoidal components whose frequencies are sums of differences of the excitation frequencies or of integral multiples of these frequencies.
f distorsion *f* par intermodulation
e distorsión *f* por intermodulación
i distorsione *f* per intermodulazione
n vervorming door intermodulatie
d Intermodulationsverzerrung *f*,
Zwischenmodulationsverzerrung *f*

2629 INTERMODULATION INTERFERENCE, INTERMODULATION NOISE cpl/dis
f bruit *m* par intermodulation
e ruido *m* por intermodulación
i rumore *m* per intermodulazione
n intermodulatieruis
d Klirrgeräusch *n*

2630 INTERNAL INTERFERENCE dis
f perturbation *f* interne
e perturbación *f* interna
i perturbazione *f* interna
n inwendige storing
d Innenstörung *f*

INTERNAL MAGNETIC FOCUS TUBE
see: IMF

2631 INTERNAL NOISE dis
f bruit *m* propre
e ruido *m* propio
i rumore *m* proprio
n eigen ruis
d Eigenrauschen *n*

2632 INTERNAL-NOISE BANDWIDTH dis
f largeur *f* de bande du bruit propre
e ancho *m* de banda del ruido propio
i larghezza *f* di banda del rumore proprio
n eigen-ruisbandbreedte
d Bandbreite *f* des Eigenrauschens

2633 INTERNATIONAL NETWORKS tv
Comprise the Eurovision coaxial network and the Intertel network of Eastern Europe.
f réseaux *pl* internationaux
e redes *pl* internacionales
i reti *pl* internazionali
n internationale netwerken *pl*
d internationale Netzwerke *pl*

2634 INTERPHONE aud/cpl/stu
An intercommunication circuit using headphones and microphones for communication between adjoining or nearby studios.
f interphone *m*
e interfono *m*
i interfono *m*
n intercommunicatieketen
d Interfonanlage *f*, Sprechanlage *f*

2635 INTERPOLATION tv
When a new TV standard is going to be used, its raster will lie in the spaces between the old raster and will coincide only rarely at the crossover points of input and output raster and thus must be interpolated.
f interpolation *f*
e interpolación *f*
i interpolazione *f*
n interpolatie
d Interpolation *f*

2636 INTERSATELLITE SERVICE tv
f service *m* de radiocommunication entre satellites
e servicio *m* de radiocomunicación entre satélites
i servizio *m* di radiocomunicazione tra satelliti
n radiocommunicatiedienst tussen satellieten
d Intersatellitenfunkdienst *m*

INTERSTATION NOISE SUPPRESSOR
see: INTERCARRIER NOISE SUPPRESSOR

2637 INTERVAL tv
Time between the end of a program(me) and the start of the next program(me).
f interlude *m*
e intervalo *m*, pausa *f*
i intervallo *m*, pausa *f*
n pauze
d Pause *f*

2638 INTOLERABLE NOISE dis
f parasites *pl* intolérables
e perturbación *f* inaguantable
i disturbo *m* intollerabile
n onverdragelijke storing
d unerträgliche Störung *f*

2639 INTRACITY RELAY SYSTEM tv
A system for the transmission of TV relay signals by fixed stations within a given city or service area for purposes other than those of studio-to-transmitter relay system.
f système *m* local à stations-relais
e retransmisión *f* televisiva local
i ritrasmissione *f* televisiva locale
n lokaal relayeersysteem *n*
d Nahballsendung *f*

INVERSION OF IMAGE
see: IMAGE INVERSION

2640 INVERTED FIELD PULSES cpl
f impulsions *pl* renversées de trame
e impulsos *pl* invertidos de cuadro
i impulsi *pl* invertiti di trama
n omgekeerde rasterimpulsen *pl*
d umgekehrte Teilbildimpulse *pl*

2641 INVERTING AMPLIFIER cpl
Used in TV to produce a positive picture from a negative and vice versa.
f amplificateur *m* pour inverser la polarité du signal
e amplificador *m* inversor de la polaridad de la señal
i amplificatore *m* per invertire la polarità del segnale
n versterker voor omkering van de signaalpolariteit
d Verstärker *m* zur Umwandlung der Signalpolarität

2642 ION BURNS crt/tv
In cathode-ray tubes, areas of reduced luminosity produced by the destruction

of some of the active material of the screen by the impact of negative ions.
f brûlures *pl* ioniques
e quemaduras *pl* iónicas
i bruciature *pl* ioniche
n ionenbrandvlekken *pl*
d Ionenbrennflecke *pl*

2643 ION SPOT, NEGATIVE ION BLEMISH crt
The deformation of target or cathode by ion bombardment, leading in camera tubes to a spurious signal.
f tache *f* ionique
e mancha *f* iónica
i macchia *f* ionica
n ionenvlek
d Ionenfleck *m*

2644 ION TRAP crt
Device employed to prevent ion burn of the fluorescent coating of a cathode-ray tube.
f grille *f* d'arrêt, piège *m* à ions
e trampa *f* de iones
i trappola *f* d'ioni
n ionenval
d Ionenfalle *f*

ION TRAP MAGNET
see: BEAM BENDER

2645 IONOSPHERE ge
That part of the earth's outer atmosphere where ions and free electrons are normally present in quantities sufficient to effect the propagation of radio waves..
f ionosphère *f*
e ionosfera *f*
i ionosfera *f*
n ionosfeer
d Ionosphäre *f*

IRASER
see: INFRARED MASER

2646 IRIS, IRIS DIAPHRAGM opt
Adjustable opening formed by thin overlapping plates usually placed between the elements of the camera lens.
f diaphragme *m* à iris
e diafragma *m* de iris
i diaframma *m* ad iride
n irisdiafragma *n*
d Irisblende *f*

2647 IRIS WIPE ani/opt/tv
Effect providing a transition from one scene to another at the boundary of an enlarging or diminishing circle.
f changement *m* roulant de séquence à cercle
e cambio *m* giratorio con círculo
i dissolvenza *f* giratoria a circolo
n rolovergang met cirkel
d rollender Schnitt *m* mit Kreis

IRON MAN
see: ALL-PURPOSE CAMERA

2648 ISOCHROMATIC STIMULI ctv
Colo(u)r stimuli which, when acting simultaneously in adjacent fields, give rise to identical colo(u)r sensations.
f stimuli *pl* isochromatiques
e estímulos *pl* isocromáticos
i stimoli *pl* isocromatici
n isochromatische stimuli *pl*
d isochromatische Reize *pl*

2649 ISOLATING AMPLIFIER cpl/rep/tv
An amplifier of which the function is to isolate the mixer matrix from other external apparatus.
f amplificateur *m* séparateur
e amplificador *m* separador
i amplificatore *m* separatore
n scheidingsversterker
d Trennverstärker *m*

2650 ISOLATING TRANSFORMER svs
Transformer used in servicing TV receivers.
f transformateur *m* séparateur
e transformador *m* separador
i trasformatore *m* separatore
n scheidingstransformator
d Trenntransformator *m*

ISOLATION AMPLIFIER
see: BUFFER AMPLIFIER

2651 ISOTROPIC AERIAL, ISOTROPIC ANTENNA, ISOTROPIC RADIATOR aea
An aerial (antenna) which radiates uniformly in all directions.
f antenne *f* sphérique
e antena *f* isótropa
i antenna *f* isotropa
n bolvormige antenne
d Kugelantenne *f*

ITV
see: INDUSTRIAL TELEVISION

J

2652 JACKETED LAMP stu
Tungsten halogen studio lamp in which the filament tube is enclosed in an outer envelope of glass, the space between being filled with inert gas.
f lampe *f* à chemise
e lámpara *f* enchaquetada
i lampada *f* a camicia
n ommantelde lamp
d ummantelte Lampe *f*

2653 JITTER cpl/dis
Small, rapid aberrations in the size or position of a repetitive display.
f instabilité *f* d'image
e inestabilidad *f* de la imagen
i tremolio *m*
n beeldbibber
d Bildstandschwankungen *pl*

2654 JITTER vr
f instabilité *f* de phase
e inestabilidad *f* de fase
i pendolamento *m*
n faze-instabiliteit
d Jitter *m*,
Zitterbewegung *f*

JITTER (US)
see: HORIZONTAL HUNTING

2655 JOCKEY ROLLER vr
f galet *m* tendeur
e rodillo *m* tensor
i rullo *m* tenditore
n spanrol
d Spannrolle *f*

2656 JOINING MAGNETIC TAPE, vr
MAGNETIC TAPE SPLICING
Carried out by applying adhesive tape to the back of the film.
f collage *m* de bande magnétique
e empalme *m* de cinta magnética
i incollatura *f* di nastro magnetico
n magneetbandlas
d Magnetbandkleben *n*

2657 JOINT PROGRAM(ME) tv
f programme *m* relayé
e programa *m* colectivo
i programma *m* collettivo
n gemeenschappelijk programma *n*
d Gemeinschaftsprogramm *n*

2658 JOINT TRANSMITTERS tv
f émetteurs *pl* relayeurs
e emisores *pl* afiliados
i emettitori *pl* affiliati
n aangesloten zenders *pl*
d angeschlossene Sender *pl*

2659 JUMP CUT tv
Cut interrupting normal sequence of action and realized e.g. by removing a section of the film to be televised and joining the ends together.
f décadrage *m* image
e salto *m* de imagen
i salto *m* d'immagine
n beeldsprong
d Bildsprung *m*

2660 JUMP IN BRIGHTNESS rep/tv
f contraste *m* brusque
e salto *m* de negro a blanco
i contrasto *m* massimo,
scatto *m* di luminosità
n helderheidssprong
d Helligkeitssprung *m*

JUMP SCANNER
see: HOPPING FILM SCANNER

2661 JUMP SCANNING rec/rep/tv
Method of film scanning which uses a displaced flying spot raster to scan a frame of film twice.
f analyse *f* double de cadre
e exploración *f* doble de fotograma
i analisi *f* doppia di fotogramma
n dubbele beeldaftasting
d doppelte Einzelbildabtastung *f*

JUMPING (US)
see: BOUNCING

2662 JUST PERCEPTIBLE DIFFERENCE, ct
MINIMUM PERCEPTIBLE DIFFERENCE
The smallest perceptible change of a colo(u)r stimulus.
f seuil *m* différentiel de chromaticité
e umbral *m* diferencial de cromaticidad
i soglia *f* differenziale di cromaticità
n kleinst waarneembaar kleursoortverschil *n*
d kleinstwahrnehmbarer Farbartunterschied *m*

2663 JUST PERCEPTIBLE NOISE, dis
MINIMUM PERCEPTIBLE NOISE
f parasites *pl* à peine perceptibles
e perturbación *f* apenas perceptible
i disturbo *m* appena percettibile
n nauwelijks waarneembare ruis
d kaum wahrnehmbares Rauschen *n*

2664 JUST TOLERABLE NOISE, dis
MAXIMUM TOLERABLE NOISE
f parasites *pl* à peine tolérables
e perturbación *f* apenas tolerable
i disturbo *m* appena tollerabile
n nauwelijks toelaatbare ruis
d kaum zulässiges Rauschen *n*

K

2665 K AERIAL, aea
K ANTENNA
Modified form of H aerial (antenna) in which the two arms constitute a dipole and the vertical element a reflector.
f antenne *f* en K
e antena *f* en K
i antenna *f* in K
n K-antenne
d K-Antenne *f*

KELVIN DEGREE
see: DEGREE K

KEY
see: GRADATION

2666 KEY ANIMATION ani
f intermédiaire *m*
e dibujo *m* principal
i disegno *m* principale
n kerntekening
d Hauptphase *f*

2667 KEY ANIMATOR ani
f séquencier *m*
e jefe-animador *m*
i capo-animatore *m*
n hoofdtekenaar
d Hauptphasenzeichner *m*

KEY LIGHTING
see: HOT LIGHTING

KEY LIGHTS
see: HOT LIGHTS

2668 KEYED AUTOMATIC GAIN CONTROL, cpl
PULSED AUTOMATIC GAIN CONTROL
f commande *f* automatique de gain par impulsions
e mando *m* automático de ganancia por impulsos
i comando *m* automatico di guadagno per impulsi
n automatische door impulsen bestuurde versterkingsregeling
d getastete Regelung *f*

2669 KEYED D.C. RESTORER cpl
f restaurateur *m* manipulé de la composante continue
e restaurador *m* manipulado de la componente continua
i restauratore *m* manipolato della componente continua
n gesleutelde hersteller van de werkzame nulcomponent
d getastete Schwarzwerthaltung *f*

KEYED INSERTION
see: INLAY

2670 KEYING PULSE ctv
A pulse used to separate the burst from all other signal components and derived from the line-deflection circuit of the receiver.
f impulsion *f* de découpage,
impulsion *f* de séparation
e impulso *m* de separación
i impulso *m* di separazione
n scheidingsimpuls
d Tastimpuls *m*,
Trennimpuls *m*

2671 KEYING SIGNAL tv
1. A signal used to actuate an electronic switch in the production of special effects.
2. A signal that enables or disables a network during selected time intervals.
f signal *m* commutateur,
signal *m* déclencheur
e señal *f* conmutadora
i segnale *m* commutatore
n schakelsignaal *n*
d Schaltsignal *n*,
Tastsignal *n*

2672 KEYSTONE DISTORTION, rec/rep/tv
TRAPEZIUM DISTORTION
Geometrical distortion of the rectangular image in a TV camera tube which occurs when the electron gun is inclined at an angle to the target or the screen.
f distorsion *f* trapézoïdale
e distorsión *f* trapezoidal
i distorsione *f* trapezoidale
n trapeziumvervorming
d Trapezverzeichnung *f*,
Trapezverzerrung *f*

KICKBACK
see: FLYBACK

KICKBACK POWER SUPPLY
see: FLYBACK EHT SUPPLY

KICKER LIGHT
see: FILL-IN LIGHT

2673 KILLER, crt
POISON
A constituent of a luminescent material which impairs luminous efficiency.
f poison *m*,
substance *f* affaiblisseuse
e impureza *f* debilitadora
i impurità *f* riduttrice di luminescenza
n luminescentieverzwakker
d Lumineszenzgift *n*

2674 KILLING OF FLUORESCENCE crt
f affaiblissement *m* de la luminescence
e debilitamiento *m* de la fluorescencia
i riduzione *f* della fluorescenza

n fluorescentieonderdrukking
d Fluoreszenzunterdrückung *f*

2675 KINESCOPE RECORDING rec/tv/vr
(US),
TELERECORDING (GB),
TELEVISION FILM RECORDING (GB)
The making of a motion picture film by photographing images on the face of the picture tube in a TV monitor or receiving to permit repeating the same TV program(me) later and at different TV stations.
f enregistrement *m* d'images de télévision d'un cinescope sur film,
enregistrement *m* sur vidigraphe,
téléenregistrement *m*
e filmación *f* de imágenes televisivas de un cinescopio,
telerregistro *m*
i il filmare *m* d'immagini televisive d'un cinescopio,
teleregistrazione *f*
n op film opnemen *n* van een televisie-programma,
verfilmen *n* van kinescoopbeelden
d FAZ *f*,
Fernsehaufzeichnung *f* auf Film,
Verfilmung *f* von Kineskopbildern

KINESCOPE TUBE
see: IMAGE RECONSTRUCTOR TUBE

2676 KNEE ge
A point or region of inflection on the characteristic curve, where the slope representing the rate of change alters.
f coude *m*
e codo *m*
i gomito *m*
n knik
d Knick *m*

2677 KNIFE COATING vr
Application of the dope on the base film by means of a knife-shaped device.
f revêtement *m* à lame
e rivestimiento *m* a cuchilla
i ricoprimento *m* a coltello
n aanbrengen *n* van een laag met een strijkmes
d Beschichtung *f* mittels Messer

2678 KNOB-A-CHANNEL MIXER vr
A mixer which utilizes a fader bank with a fader knob on each input channel.
f pupitre *m* de mélange à réglage indépendant des canaux
e pupitre *m* de mezclado con regulación independiente de los canales
i tavolo *m* di mescolanza con regolazione indipendente dei canali
n mengtafel met afzonderlijke regelknoppen
d Mischpult *n* mit separaten Kanalregel-knöpfen

L

2679 LAG crt
A persistence of the electrical-charge image for a small number of fields.
f effet *m* de rémanence
e efecto *m* de remanencia
i effetto *m* di rimanenza
n remanentie-effect *n*
d Nachhinken *n*

2680 LAG COMPENSATION cpl
f compensation *f* de rémanence
e compensación *f* de remancencia
i compensazione *f* di rimanenza
n remanentiecompensatie
d Kompensation *f* des Nachhinkens

2681 LAG ERROR cpl
f erreur *f* de rémanence
e error *m* de remanencia
i errore *m* di rimanenza
n remanentiefout
d Nachhinkfehler *m*

2682 LANYARD MICROPHONE, LAVALLIER MICROPHONE, PERSONAL MICROPHONE aud
A microphone placed on the person, usually secured on the chest by a neck halter.
f microphone *m* lavallière
e micrófono *m* acollador
i microfono *m* portato al collo
n omhangmicrofoon
d Lavalliermikrophon *n*

2683 LAP DISSOLVE rep/tv
Change-over from one TV scene to another in such a way that the new picture appears gradually at the same rate at which the previous picture disappears.
f fondu *m* enchaîné
e fundido *m* encadenado
i dissolvenza *f* incrociata
n mengovergang, overvloeiing
d weiche Überblendung *f*

2684 LAP DISSOLVE SHUTTER stu
f obturateur *m* à rideau pour le fondu enchaîné
e obturador *m* de cortinilla para el fundido encadenado
i otturatore *m* a tendina per la dissolvenza incrociata
n gordijnsluiter voor de mengovergang
d Rollverschluss *m* für weiche Überblendung, Überblendeinrichtung *f*

LAP SWITCHING
see: BREAK-BEFORE-MAKE SWITCHING

2685 LAPEL MICROPHONE aud
f microphone-boutonnière *m*
e micrófono *m* de solapa
i microfono *m* per il risvolto
n knoopsgatmicrofoon
d Knopflochmikrophon *n*

LAPPING SWITCH
see: BREAK-BEFORE-MAKE SWITCH

2686 LARGE AREA CONTRAST rep/tv
f contraste *m* gros
e contraste *m* grueso
i contrasto *m* grosso
n grofcontrast *n*
d Grobkontrast *m*

2687 LARGE-SCREEN TELEVISION PROJECTION tv
f projection *f* de télévision à grand écran
e proyección *f* de televisión de gran imagen
i proiezione *f* di televisione a grande schermo
n groot-beeld-televisieprojectie
d Fernsehgrossbildprojektion *f*

2688 LARGE-SCREEN TELEVISION PROJECTOR tv
f projecteur *m* de télévision à grand écran
e proyector *m* televisivo de gran imagen
i ricevitore *m* proiettore a grande schermo
n groot-beeld-televisieprojector
d Fernsehgrossbildprojektor *m*

LASER
see: INFRARED MASER

2689 LATENT ELECTRONIC IMAGE crt
f image *f* électronique latente
e imagen *f* electrónica latente
i immagine *f* elettronica latente
n latent ladingsbeeld *n*
d gespeichertes Bild *n*

2690 LATERAL CHROMATISM opt
f chromatisme *m* latéral
e cromatismo *m* lateral
i cromatismo *m* laterale
n lateraal chromatisme *n*
d lateraler Chromatismus *m*

2691 LATERAL-CORRECTION MAGNET crt
An auxiliary component of a three-gun picture tube which employs the magnetic convergence principle.
f aimant *m* de correction latérale
e imán *m* de corrección lateral
i magnete *m* di correzione laterale
n instelmagneet
d Schiebemagnet *m*

2692 LATERAL DOLLY SHOT tv
f travelling *m* latéral
e toma *f* de vistas en movimiento lateral

i ripresa *f* in movimento laterale
n opname in zijwaartse beweging
d Parallelfahrt *f*

2693 LATERAL INVERSION rep/tv
Defect in a reproduced TV image, the picture being reversed, the right-hand side appearing on the left, due to a reversal in the connections from the line scanning generator.
f inversion *f* latérale
e inversión *f* lateral
i inversione *f* laterale
n zijdelingse omkering
d Seitenumkehr *f*

2694 LATERAL MAGNETIC FIELD, TRANSVERSAL MAGNETIC FIELD crt
f champ *m* magnétique latéral
e campo *m* magnético lateral
i campo *m* magnetico laterale
n magnetisch dwarsveld *n*
d Magnetquerfeld *n*,
transversales Magnetfeld *n*

2695 LATERAL MAGNIFYING POWER opt
The ratio of the distance between image point and principal axis to that of the congegrate object and the axis.
f agrandissement *m* latéral
e ampliación *f* lateral
i ingrandimento *m* laterale
n dwarsvergroting
d Quervergrösserung *f*

LAWRENCE TUBE
see: CHROMATRON

2696 LAY-ON ROLLER vr
f galet *m* presseur
e rodillo *m* de presión
i rullo *m* di pressione
n aandrukrol
d Andruckrolle *f*

2697 LAZY ARM stu
Small microphone boom.
f girafe *f* petite
e jirafa *f* pequeña
i giraffa *f* piccola
n korte hengel
d kleiner Galgen *m*

LEAD IN
see: DOWN LEAD

2698 LEAD-OXIDE CAMERA TUBE crt
A special tube of the photoconductive type.
f tube *m* cathodique à oxyde de plomb
e tubo *m* catódico a óxido de plomo
i tubo *m* catodico ad ossido di piombo
n katodestraalbuis met loodoxyde
d Bleioxyd-Katodenstrahlröhre *f*

2699 LEAD-OXIDE SCREEN cct/rec/tv
f écran *m* à oxyde de plomb
e pantalla *f* a óxido de plomo
i schermo *m* ad ossido di piombo
n loodoxydescherm *n*
d Bleioxydschirm *m*

2700 LEADER vr
Length of blank magnetic tape at the beginning of a reel.
f amorce *f*
e cabecera *f* protectora,
cinta *f* de arranque
i coda *f* iniziale,
nastro *m* d'inizio
n beginstrook,
startband
d Startband *n*

2701 LEADER FOR TELEVISION FILM, TELEVISION FILM LEADER tv
f amorce *f* de film de télévision
e tira *f* protectora para película televisiva
i coda *f* iniziale di pellicola televisiva
n beginstrook
d Fernsehfilmstartband *n*

2702 LEADING BLACK rec/rep/tv
f précurseur *m* du noir
e negro *m* delantero
i nero *m* d'anticipo
n voorloper van het zwart
d Schwarzvorläufer *m*

2703 LEADING EDGE cpl/ge
Rising slope of a pulse, such as a TV sync pulse, as displayed graphically.
f flanc *m* avant,
front *m* d'impulsion
e flanco *m* frontal
i fronte *f* anteriore
n voorflank
d Vorderflanke *f*

2704 LEADING-EDGE PULSE TIME cpl
The time at which the instantaneous amplitude of a pulse first reaches a stated fraction of the peak pulse amplitude.
f temps *m* de montée du front d'impulsion
e tiempo *m* de crecimiento del flanco frontal de impulso
i tempo *m* di salita della fronte anteriore d'impulso
n stijgtijd van de impulsvoorflank
d Anstiegzeit *f* der Impulsvorderflanke

2705 LEADING GHOST rep/tv
A ghost displaced to the left of the image on a TV screen.
f image *f* fantôme à gauche
e imagen *f* fantasma a la izquierda
i immagine *f* fantasma alla sinistra
n linkse geest
d linker Geist *m*

2706 LENS crt/opt
1. Device for focusing radiation.
2. An arrangement of electrodes used to produce an electric field that serves to focus electrons into a beam.
f lentille *f*
e lente *f*
i lente *f*

n lens
d Linse *f*

LENS APERTURE
see: APERTURE

2707 LENS DISK tv
Early TV scanning device.
f disque *m* à lentilles
e disco *m* de lentes
i disco *m* a lenti
n lenzenschijf
d Linsenscheibe *f*

2708 LENS DISTORTION opt
Occurs when the magnification of the system varies at different parts of the image.
f distorsion *f* de lentille
e distorsión *f* de lente
i distorsione *f* di lente
n vertekening van een lens
d Verzeichnung *f* einer Linse

2709 LENS DRUM tv
f tambour *m* à lentilles
e tambor *m* de lentes
i tamburo *m* a lenti
n lenzentrommel
d Linsentrommel *f*

LENS HOOD (GB)
see: CAMERA HOOD

LENS SCREEN (GB)
see: FLAG

LENS STOP
see: APERTURE

2710 LENS TURRET opt
f tourelle *f* d'objectifs
e torrecilla *f* para objetivos
i torretta *f* portaobiettivo
n objectiefrevolver
d Objektivrevolver *m*

2711 LEVEL aud/ge
Of a quantity related to power, the ratio, expressed in decibels, of the magnitude of the quantity to a specified reference magnitude.
f niveau *m*
e nivel *m*
i livello *m*
n niveau *n*
d Niveau *n*, Pegel *m*

2712 LEVEL BREAKDOWN cpl
f perte *f* de niveau
e pérdida *f* de nivel
i perdita *f* di livello
n niveauverlies *n*
d Pegeleinbruch *m*,
Pegelverlust *m*

2713 LEVEL COMPENSATION, cpl/ge
LEVEL EQUALIZATION
A circuit to compensate automatically the effects caused by amplitude variations of a received signal.
f égalisation *f* du niveau
e igualación *f* del nivel
i uguagliamento *m* del livello
n niveaucompensatie,
niveauvereffening
d Pegelausgleich *m*

2714 LEVEL CONTROL, rec/tv
LEVEL MONITORING
The operation of keeping a watch on the level of the audio and/or video signals, using direct reading equipment.
f contrôle *m* du niveau
e control *m* del nivel
i verifica *f* continua del livello
n niveaucontrole
d Höhenschrittkontrolle *f*,
Pegelaussteuerung *f*,
Pegelüberwachung *f*

2715 LEVEL-DEPENDENT GAIN cpl
f gain *m* dépendant du niveau
e ganancia *f* dependiente del nivel
i guadagno *m* dipendente del livello
n van het niveau afhankelijke versterking
d pegelabhängige Verstärkung *f*

2716 LEVEL-DEPENDENT PHASE cpl
f phase *f* dépendant du niveau
e fase *f* dependiente del nivel
i fase *f* dipendente del livello
n van het niveau afhankelijke faze
d pegelabhängige Phase *f*

2717 LEVEL DEVIATION cpl
f variation *f* de niveau
e variación *f* de nivel
i variazione *f* di livello
n niveauafwijking
d Pegelabweichung *f*

2718 LEVEL DIFFERENCE cpl
f différence *f* de niveau
e diferencia *f* de nivel
i differenza *f* di livello
n niveauverschil *n*
d Pegeldifferenz *f*

2719 LEVEL DISTORTION cpl
f distorsion *f* de niveau
e distorsión *f* de nivel
i distorsione *f* di livello
n niveauvervorming
d Pegelverzerrung *f*

2720 LEVEL INDICATOR, cpl
PROGRAMMETER,
VOLUME INDICATOR
f décibelmètre *m*,
indicateur *m* de la profondeur de modulation
e indicador *m* de la profundidad de modulación
i indicatore *m* della profondità di modulazione
n modulatiedieptemeter
d Aussteuerungsmesser *m*

2721 LEVEL INSTABILITY cpl
f instabilité *f* du niveau
e inestabilidad *f* del nivel
i instabilità *f* del livello
n niveau-instabiliteit
d Pegelinstabilität *f*

2722 LEVEL SETTING rep/tv
Provision for adjusting the base voltage for an irregular waveform, e.g. in TV scanning circuit voltages and signals.
f alignement *m* du niveau
e alineación *f* del nivel
i allineamento *m* del livello
n niveau-instelling
d Einpegeln *n*,
Pegeleinstellung *f*

2723 LEVEL(L)ING, cpl
SMOOTHING
Use of an RC filter to level out fluctuations of a bias voltage.
f filtrage *m*
e aplanamiento *m*
i spianamento *m*
n afvlakken *n*
d Abflachung *f*

LIFT
see: BLACK LIFT

LIFT CONTROL
see: BLACK LIFT CONTROL

LIFT LEVEL
see: BLACK LIFT LEVEL

2724 LIGHT ct
Radiant power capable of stimulating the eye to produce visual sensation.
f lumière *f*
e luz *f*
i luce *f*
n licht *n*
d Licht *n*

2725 LIGHT ABSORPTION ctv
f absorption *f* de lumière
e absorción *f* de luz
i assorbimento *m* di luce
n lichtabsorptie
d Lichtabsorption *f*

2726 LIGHT APPLICATION BAR dis/rec/tv
A horizontal black bar floating up or down the TV picture occurring when the film frame is not being illuminated during part of the active TV field scanning intervals.
f barre *f* noire flottante
e barra *f* negra flotante
i barra *f* nera flottante
n drijvende zwarte balk
d schwebender schwarzer Balken *m*

2727 LIGHT APPLICATION RATIO, rec/tv
LIGHT APPLICATION TIME
The ratio of illuminated area to illuminated time of a film frame.
f rapport *m* de temps d'exposition
e relación *f* de tiempo de exposición
i rapporto *m* di tempo d'esposizione
n belichtingstijdverhouding
d Belichtungszeitverhältnis *n*

2728 LIGHT BARRIER stu
f barrière *f* lumineuse
e barrera *f* luminosa
i barriera *f* luminosa
n lichtbarrière
d Lichtschranke *f*

2729 LIGHT BOX stu
f boîte *f* à lumière
e casa *f* de luz
i cassa *f* a luce
n lichtkast
d Lichtfeld *n*, Lichtkasten *m*

2730 LIGHT CONTROL TAPE stu
f bande *f* d'étalonnage de lumière
e cinta *f* de mando de luz
i nastro *m* di comando di luce
n lichtbesturingsband
d Lichtsteuerband *n*

2731 LIGHT CORRECTION FILTER stu
f filtre *m* correcteur de lumière
e filtro *m* corrector de luz
i filtro *m* correttore di luce
n lichtcorrectiefilter *n*
d Lichtausgleichfilter *n*

2732 LIGHT CURRENT, crt
PHOTOCURRENT,
PHOTOELECTRIC CURRENT
The total current in a photoemissive cell, which is equal to the sum of the current produced by excitation and the dark current.
f courant *m* photoélectrique
e corriente *f* fotoeléctrica
i corrente *f* fotoelettrica
n foto-elektrische stroom,
fotostroom,
lichtelektrische stroom
d lichtelektrischer Strom *m*,
photoelektrischer Strom *m*,
Photostrom *m*

2733 LIGHT DISTRIBUTION stu
f répartition *f* de la lumière
e repartición *f* de la luz
i ripartizione *f* della luce
n lichtverdeling
d Lichtverteilung *f*

LIGHT EMISSION,
see: EMISSION OF LIGHT

2734 LIGHT ENERGY ct
DISTRIBUTION CURVE
f caractéristique *f* de la fonction de distribution pour l'énergie lumineuse
e característica *f* de la función de distribución para la energía luminosa
i caratteristica *f* della funzione di distribuzione per la energia luminosa
n kromme voor de verdelingsfunctie voor lichtenergie

d Kurve *f* der Lichtenergieverteilungsfunktion

2735 LIGHT MODULATION crt
Used for influencing the electron beam current of a cathode-ray tube of which the light spot serves as a light source.
f modulation *f* de la lumière
e modulación *f* de la luz
i modulazione *f* della luce
n lichtmodulatie
d Lichtmodulation *f*

2736 LIGHT MODULATOR tv
f modulateur *m* de la lumière
e modulador *m* de la luz
i modulatore *m* della luce
n lichtmodulator
d Lichtmodulator *m*,
Lichtsteuergerät *n*

2737 LIGHT-SENSITIVE CELL crt
f cellule *f* photosensible
e célula *f* fotosensible
i cellula *f* fotosensibile
n lichtgevoelige cel
d lichtempfindliche Zelle *f*

2738 LIGHT-SENSITIVE MOSAIC crt
Sensitive surface on which light falls in a TV camera tube such as a vidicon or iconoscope.
f mosaïque *f* photosensible
e mosaico *m* fotosensible
i mosaico *m* fotosensibile
n lichtgevoelig mozaïek *n*
d lichtempfindliches Mosaik *n*

2739 LIGHT SPOT crt
f spot *m* lumineux
e punto *m* luminoso
i punto *m* luminoso
n lichtpunt, lichtstip
d Lichtpunkt *m*, Lichtfleck *m*

LIGHT-SPOT SCANNING
see: FLYING-SPOT SCANNING

2740 LIGHT STIMULUS ct
Physically defined radiation entering the eye and producing a sensation of light.
f stimulus *m* de lumière
e estímulo *m* de luz
i stimolo *m* di luce
n lichtprikkel, lichtstimulus
d Lichtreiz *m*

2741 LIGHT TRANSFER CHARACTERISTIC ct
Relation between light input and voltage output.
f caractéristique *f* de transfert pour la lumière
e característica *f* de transferencia para la luz
i caratteristica *f* di trasferimento per la luce
n overgangskarakteristiek voor licht
d Übergangskennlinie *f* für Licht

2742 LIGHT TRANSMISSION, TRANSMISSION OF LIGHT opt
f transmission *f* de lumière
e transmisión *f* de luz
i trasmissione *f* di luce
n lichtdoorlating
d Lichtdurchlass *m*

2743 LIGHT VALUE rec/rep/tv
f valeur *f* de la lumière
e valor *m* de la luz
i valore *m* della luce
n lichtwaarde
d Lichtwert *m*

2744 LIGHTING ctu
f éclairage *m*
e alumbrado *m*
i illuminazione *f*
n verlichting
d Beleuchtung *f*

2745 LIGHTING BRIDGE stu
f passerelle *f* pour les éclairagistes
e puente *m* de alumbrado
i ponte *m* d'illuminazione
n verlichtingsbrug
d Beleuchtungsbrücke *f*

LIGHTING CAMERAMAN
see: DIRECTOR OF PHOTOGRAPHY

LIGHTING CONSOLE
see: ILLUMINATION DESK

2746 LIGHTING CONTRAST stu
f contraste *m* d'éclairement
e contraste *m* de iluminación
i contrasto *m* d'illuminazione
n verlichtingscontrast *n*
d Beleuchtungskontrast *m*

2747 LIGHTING CONTRAST RATIO stu
The ratio between the light value of the key light plus filler and the filler alone.
f rapport *m* de l'éclairement
e relación *f* de iluminación
i rapporto *m* d'illuminamento
n verlichtingsverhouding
d Beleuchtungsverhältnis *n*

2748 LIGHTING CONTROL stu
f commande *f* de l'éclairage
e mando *m* del alumbrado
i comando *m* dell'illuminazione
n verlichtingsregeling
d Beleuchtungssteuerung *f*,
Lichtregie *f*,
Lichtsteuerung *f*

2749 LIGHTING CONTROL CONSOLE stu
f pupitre *m* de commande de l'éclairage
e pupitre *m* de mando del alumbrado
i tavolo *m* di comando dell'illuminazione
n regellessenaar voor de verlichting
d Lichtregiepult *n*

2750 LIGHTING CONTROL ROOM rec/tv
That part of the studio where adjustments

to lighting levels and switching of lighting set-ups are made.
f salle *f* de commande de l'éclairage
e sala *f* de mando del alumbrado
i sala *f* di comando dell'illuminazione
n regelkamer voor de verlichting
d Lichtregieraum *m*

2751 LIGHTING EQUIPMENT stu
f équipement *m* d'éclairage, installation *f* d'éclairage
e equipo *m* de alumbrado
i equipo *m* d'illuminazione
n verlichtingsapparatuur, verlichtingsinstallatie
d Beleuchtungsanlage *f*, Beleuchtungsgeräte *pl*

2752 LIGHTING GRID, LIGHTING SUSPENSION GRID fi/th/tv
f gril *m* d'éclairage
e reja *f* de alumbrado
i inferriata *f* d'illuminazione
n verlichtingsrooster *n*
d Gitterdecke *f*, Gitterrostdecke *f*

2753 LIGHTING LAYOUT, LIGHTING SET-UP stu
f projet *m* de l'éclairage
e proyecto *m* del alumbrado
i progetto *m* dell'illuminazione
n verlichtingsontwerp *n*
d Beleuchtungsplan *m*

2754 LIGHTING LEVEL stu
f niveau *m* d'éclairage
e nivel *m* de alumbrado
i livello *m* d'illuminazione
n verlichtingsniveau *n*
d Beleuchtungspegel *m*

2755 LIGHTING MAN stu
f éclairagiste *m*
e electricista *m* de alumbrado
i addetto *m* alle luci, elettricista *m* d'illuminazione
n verlichtingsmonteur
d Beleuchter *m*

2756 LIGHTNESS ct
The attribute of visual perception in accordance with which a body seems to transmit or reflect diffusely a greater or smaller fraction of the incident light.
f brillance *f*, clarté *f*
e brillo *m*, claridad *f*
i brillantezza *f*, chiarezza *f*
n helderheid
d Helligkeit *f*

2757 LIGHTNESS OF A COLO(U)R ct
The proportion of total light which appears to be reflected by the area of colo(u)r seen.
f clarté *f* d'une couleur
e claridad *f* de un color
i chiarezza *f* d'un colore
n kleurhelderheid
d Farbhelligkeit *f*

2758 LIGHTWEIGHT CAMERA med/tv
A camera used in medical TV.
f caméra *f* de poids léger
e cámara *f* de peso ligero
i camera *f* leggera
n lichtgewichtcamera
d Leichtkamera *f*

2759 LIMEN aud/ct
Smallest difference in pitch (frequency) or intensity of a sound or colo(u)r which can be perceived by the senses.
f valeur *f* de seuil
e valor *m* de umbral
i valore *m* di soglia
n drempelwaarde
d Schwellenwert *m*

2760 LIMIT OF PERCEPTIBILITY ct
f limite *f* de perceptibilité
e límite *m* de perceptibilidad
i limite *m* di percettibilità
n waarneembaarheidsgrens
d Wahrnehmbarkeitsgrenze *f*

2761 LIMITATION OF INTERFERENCE aud/tv
f limitation *f* d'interférence
e limitación *f* de interferencia
i limitazione *f* d'interferenza
n storingsbegrenzing
d Störbegrenzung *f*

2762 LIMITED ACCESS SATELLITE tv
f satellite *m* à accès limité
e satélite *m* de acceso limitado
i satellite *m* ad accesso limitato
n satelliet met beperkte toegang
d Satellit *m* für begrenzten Zugang

2763 LIMITER cpl/ge
Any transducer which sets or tends to set some boundary value upon a signal.
f limiteur *m*
e limitador *m*
i limitatore *m*
n begrenzer
d Begrenzer *m*

2764 LIMITER DIODE cpl
f diode *f* limitatrice
e diodo *m* limitador
i diodo *m* limitatore
n begrenzerdiode
d Begrenzerdiode *f*

2765 LIMITING AMPLIFIER, PEAK LIMITER, PEAK LIMITING AMPLIFIER cpl
A device that automatically limits the peak output to a predetermined maximum value.
f amplificateur *m* limitateur de crêtes
e amplificador *m* con limitación de crestas, amplificador *m* limitador
i amplificatore *m* limitatore di creste
n begrenzingsversterker
d Begrenzungsverstärker *m*

2766 LIMITING DEFINITION (GB), LIMITING RESOLUTION (US) — tv
Highest definition (resolution) which a system or device can produce under the limitations imposed by physical characteristics of lenses, etc.
f limite *f* de définition
e límite *m* de definición
i limite *m* di definizione
n definitiegrens
d Auflösungsgrenze *f*

2767 LIMITING OF FREQUENCY BAND — tv
f limitation *f* de la bande de fréquences
e limitación *f* de la banda de frecuencias
i limitazione *f* della banda di frequenze
n frequentiebandbegrenzing
d Frequenzbandbegrenzung *f*

2768 LIMITING STAGE — cpl
f étage *m* limitateur
e etapa *f* limitadora
i stadio *m* limitatore
n begrenzertrap
d Begrenzerstufe *f*

2769 LINE — cpl/ge
Transmission line, coaxial, balanced, pair or earth return for electric power, signals, or modulation currents.
f ligne *f*
e línea *f*
i linea *f*
n leiding, lijn
d Leitung *f*

2770 LINE, TRACE — rec/rep/tv
Single scan in a TV picture transmission system.
f ligne *f*
e línea *f*
i linea *f*, riga *f*
n lijn
d Zeile *f*

2771 LINE AMPLIFIER, PROGRAM(ME) AMPLIFIER — cpl
An audio amplifier that feeds a program(me) to a transmission line at a specified signal level.
f amplificateur *m* de ligne
e amplificador *m* de línea
i amplificatore *m* di linea
n lijnversterker
d Leitungsverstärker *m*

LINE AMPLITUDE
see: HORIZONTAL AMPLITUDE

LINE AMPLITUDE ADJUSTMENT
see: HORIZONTAL AMPLITUDE ADJUSTMENT

LINE AMPLITUDE CONTROL (GB)
see: HORIZONTAL SIZE CONTROL

LINE BLANKING
see: HORIZONTAL BLANKING

LINE BLANKING INTERVALS
see: HORIZONTAL BLANKING INTERVALS

LINE BLANKING LEVEL
see: HORIZONTAL BLANKING LEVEL

2772 LINE BEND, LINE BEND CORRECTION — dis/rep
An approximately parabolic waveform, recurring at line frequency, which constitutes a component of the video signal introduced to compensate for inherent distortion in the camera output.
f signal *m* parabolique de correction de ligne
e señal *f* parabólica de corrección de línea
i correzione *f* parabolica di luminosità di linea
n lijnparaboolcorrectie
d parabolische Zeilensignale *pl*

2773 LINE BROADCASTING (GB), WIRE BROADCASTING (US) — cpl/ge
The transmission by means of line (wire) communication of a program(me) of sound, vision or facsimile for general reception.
f télédiffusion *f*
e teledifusión *f*
i filodiffusione *f*, telediffusione *f*
n draadomroep
d Drahtfunk *m*, Drahtrundfunk *m*

2774 LINE BROADENING — rec/rep/tv
f élargissement *m* des lignes
e ensanchamiento *m* de las líneas
i allargamento *m* delle linee
n lijnverbreding
d Zeilenverbreiterung *f*

2775 LINE-BY-LINE SCANNING — rec/rep/tv
Scanning in which the sweep is effected in straight, substantially horizontal strips over the entire width of the picture.
f analyse *f* ligne par ligne
e exploración *f* línea a línea
i analisi *f* a linea a linea
n lijn-voor-lijnaftasting
d zeilenweise Abtastung *f*

2776 LINE CLAMP AMPLIFIER, SIGNAL STABILIZATION AMPLIFIER — cpl
f amplificateur *m* stabilisateur du signal
e amplificador *m* estabilizador de la señal
i amplificatore *m* stabilizzatore del segnale
n stabiliserende signaalversterker
d stabilisierender Signalverstärker *m*

2777 LINE CRAWL — ctv
A peculiar and undesirable effect found in line-sequential colo(u)r TV systems whereby a vertical colo(u)r effect is seen for each line of the picture.
f déformation *f* de la structure des lignes

e deformación *f* de la estructura de las líneas
i deformazione *f* della struttura delle linee
n karneffect *n*
d Zeilenstrukturstörung *f*

LINE DEFINITION (GB),
LINE RESOLUTION (US)
see: HORIZONTAL DEFINITION

2778 LINE DIFFUSER rep/tv
An oscillator used in a TV monitor or receiver to produce small vertical oscillations of the spot on the screen so as to make the line structure of the image less noticeable at short viewing distances.
f circuit *m* d'estompage de ligne
e circuito *m* de esfumación de línea
i circuito *m* di sfumatura di linea
n lijndoezelaar
d Zeilenverwischer *m*

2779 LINE DIFFUSION tv
f estompage *m* de ligne
e esfumación *f* de línea
i sfumatura *f* di linea
n lijnverdoezeling
d Zeilenverwischung *f*

2780 LINE DISPLACEMENT, dis/rep/tv
LINE OFFSET
f décalage *m* de lignes
e deslizamiento *m* de líneas
i spostamento *m* di linee
n lijnverschuiving,
lijnverzet *n*
d Zeilenoffset *m*,
Zeilenversatz *m*,
Zeilenverschiebung *f*

LINE DISPLAY MONITOR
see: HORIZONTAL DISPLAY MONITOR

LINE DIVIDER,
LINE FREQUENCY DIVIDER
see: HORIZONTAL FREQUENCY DIVIDER

2781 LINE DRIVE, cpl
LINE TRIGGERING CIRCUIT
A horizontal timing circuit used to trigger circuits in the picture generating equipment.
f circuit *m* de déclenchement de ligne
e circuito *m* de desenganche de línea
i circuito *m* di sblocco di linea
n lijndeblokkeerkring
d Zeilenanstosskreis *m*

2782 LINE DRIVE PULSE, cpl
LINE TRIGGERING PULSE
Signal at line frequency distributed from the station pulse generator to allow triggering of the line scanning circuits at a time suitable to allow insertion of blanking without loss of picture information.
f impulsion *f* de déclenchement de ligne
e impulso *m* de desenganche de línea
i impulso *m* di sblocco di linea
n lijndeblokkeerimpuls
d Zeilenanstossimpuls *m*

2783 LINE DRIVE SIGNAL, cpl
LINE TRIGGERING SIGNAL
Signal used to establish line sync in studio systems.
f signal *m* de déclenchement de ligne
e señal *f* de desenganche de línea
i segnale *m* di sblocco di linea
n lijndeblokkeersignaal *n*
d Zeilenanstosssignal *n*

2784 LINE DURATION, rec/rep/tv
LINE PERIOD
f durée *f* de ligne
e duración *f* de línea
i durata *f* di linea
n lijnduur
d Zeilendauer *f*

2785 LINE-FEEDING AMPLIFIER stu
f amplificateur *m* de sortie
e amplificador *m* de salida
i amplificatore *m* d'uscita
n uitgangsversterker
d Ausgangsverstärker *m*

2786 LINE FLICKER rep/tv
The alternating of the brightness of the lines on the screen of a picture tube.
f papillotement *m* de lignes
e parpadeo *m* de líneas
i sfarfallio *m* di linee
n lijnflikkeren *n*
d Zeilenflimmern *n*

LINE FLYBACK (GB),
LINE RETRACE (US)
see: HORIZONTAL FLYBACK

2787 LINE FOCUS crt
A defect in cathode-ray tubes in which the electron beam meets screen along a line and not at a point.
f foyer *m* linéaire
e foco *m* lineal
i fuoco *m* lineare
n streepfocus
d Strichfokus *m*

LINE FREQUENCY,
LINE SCANNING FREQUENCY
see: HORIZONTAL FREQUENCY

LINE-FREQUENCY BLANKING PULSE
see: HORIZONTAL BLANKING PULSE

LINE FREQUENCY DEVIATION
see: HORIZONTAL FREQUENCY DEVIATION

2788 LINE FREQUENCY TO rec/rep
FIELD FREQUENCY RATIO (GB),
NUMBER OF SCANNING LINES (US)
In TV, the ratio of line frequency to field frequency.
f rapport *m* fréquence lignes/fréquence trame
e relación *f* frecuencia de líneas/frecuencia de cuadro
i rapporto *m* frequenza di linee/frequenza di trama
n verhouding lijnfrequentie/rasterfrequentie

d Zeilen/Teilbildfrequenzverhältnis *n*

2789 LINE GATE rec/rep/tv
f porte *f* de lignes
e compuerta *f* de líneas
i porta *f* di linee
n lijnpoort
d Zeilentor *n*

2790 LINE HEIGHT rec/rep/tv
f hauteur *f* de ligne
e altitud *f* de línea
i altezza *f* di linea
n lijnhoogte
d Zeilenhöhe *f*

2791 LINE INCREMENT vr
f accroissement *m* de ligne
e incremento *m* de línea
i incremento *m* di linea
n lijnvergroting
d Zeilenanwachs *m*

2792 LINE INTERLACED SCANNING, rec/tv
LINE JUMP SCANNING
f analyse *f* interlacée de lignes
e exploración *f* entrelazada de líneas
i analisi *f* a linee intercalate
n interliniëringsaftasting
d Zeilensprungabtastung *f*

LINE KEYSTONE CORRECTION,
LINE KEYSTONE WAVEFORM
see: HORIZONTAL KEYSTONE CORRECTION

LINE LINEARITY CONTROL
see: HORIZONTAL LINEARITY CONTROL

2793 LINE LOCK (US), cpl
MAINS LOCKING (GB)
f verrouillage *m* de fréquences trame et réseau
e enganche *m* de frecuencias cuadro y red
i blocco *m* di frequenze trama e rete
n raster- en netfrequentiekoppeling
d Teilbild- und Netzfrequenzverkopplung *f*

2794 LINE MONITOR tv
f moniteur *m* de câblage
e monitor *m* de cableado
i monitore *m* di cablaggio
n kabelmonitor
d Kabelmonitor *m*

LINE NON-LINEARITY
see: HORIZONTAL NON-LINEARITY

2795 LINE OSCILLATOR rec/tv
f oscillateur *m* de lignes
e oscilador *m* de líneas
i oscillatore *m* di linee
n lijnoscillator
d Zeilenoszillator *m*

2796 LINE OSCILLATOR COIL rec/tv
f bobine *f* d'oscillateur de lignes
e bobina *f* de oscilador de líneas
i bobina *f* d'oscillatore di linee
n lijnoscillatorspoel
d Zeilenoszillatorspule *f*

LINE OUTPUT
see: HORIZONTAL OUTPUT

LINE OUTPUT PENTODE
see: HORIZONTAL OUTPUT PENTODE

LINE OUTPUT STAGE
see: HORIZONTAL OUTPUT STAGE

LINE OUTPUT TRANSFORMER
see: FLYBACK TRANSFORMER

LINE OUTPUT VOLTAGE
see: HORIZONTAL OUTPUT VOLTAGE

LINE PULSE
see: HORIZONTAL PULSE

2797 LINE RATE CONVERTER rec/tv
A device used in standards conversion operating in a picture basis, i.e. it interpolates between adjacent lines on the complete picture and is not confined to adjacent lines in one field.
f convertisseur *m* de normes de télévision à fréquences différentes de lignes et à fréquences identiques de trames
e convertidor *m* de normas televisivas de frecuencias diferentes de líneas y de frecuencias idénticas de cuadros
i convertitore *m* di norme televisive a frequenze differenti di linee ed a frequenze identiche di trame
n omzetter voor televisienormen met verschillende lijnfrequenties en gelijke rasterfrequenties
d Fernsehnormenwandler *m* bei ungleichen Zeilenfrequenzen und bei gleichen Teilbildfrequenzen

2798 LINE RINGING dis/rep/tv
Damped oscillations set up in the line output circuit of a TV receiver by the inductance and stray capacitance of the scanning coils or line output transformers, resulting in the appearance of vertical bars on the left-hand side of the picture.
f barres *pl* verticales à gauche
e barras *pl* verticales a la izquierda
i barre *pl* verticali a sinistra
n linkse verticale balken *pl*
d an der linken Bildseite auftretende vertikale Balken *pl*

2799 LINE SAWTOOTH cpl
Waveform which rises at a constant rate and then falls rapidly, at the frequency of line deflection.
f forme *f* d'onde en dents de scie à fréquence de déviation de ligne
e forma *f* de onda en dientes de sierra de frecuencia de desviación de línea
i forma *f* d'onda in denti di sega a frequenza di deviazione di linea
n zaagtandgolfvorm met lijnafbuigfrequentie
d Sägezahnwellenform *f* mit Zeilenablenkfrequenz

2800 LINE SCAN CIRCUIT tv
f circuit *m* de déviation de ligne
e circuito *m* de desviación de línea
i circuito *m* di deflessione di linea, circuito *m* di deviazione di linea
n lijnafbuigschakeling
d Zeilenkippschaltung *f*

LINE SCANNING
see: HORIZONTAL SCANNING

2801 LINE SEQUENCE rec/rep/tv
f séquence *f* de lignes
e secuencia *f* de líneas
i sequenza *f* di linee
n lijnvolgorde
d Zeilenfolge *f*

2802 LINE SEQUENTIAL ctv/tv
f à séquence de lignes
e de secuencia de líneas
i a sequenza di linee
n met volglijnen
d zeilensequentiell

2803 LINE SEQUENTIAL COLO(U)R TELEVISION ctv
Sequential colo(u)r TV in which the signals derived from the primary colo(u)rs are transmitted in rotation, line by line.
f système *m* de séquence à lignes pour télévision couleur
e sistema *m* de secuencia de líneas para televisión en colores
i sistema *m* a sequenza di linee per televisione a colori
n volglijnkleurensysteem *n*
d Zeilenfolgesystem *n* für Farbfernsehen

2804 LINE SEQUENTIAL SWITCHING ctv/rec
Used in a Japanese-made portable receiver with the grid potential switched during the horizontal blanking interval to reduce radiation and power requirements.
f système *m* séquentiel à variation de la tension de grille
e sistema *m* secuencial de variación de la tensión de rejilla
i sistema *m* sequenziale a variazione della tensione di griglia
n volglijnsysteem *n* met variatie van de roosterspanning
d Zeilenfolgesystem *n* mit Variation der Gitterspannung

2805 LINE SLIP rep/tv
An apparent horizontal displacement or movement of all or part of the reproduced picture resulting from loss of synchronism between the line frequency of a scanning system and the line frequency of the signal.
f défilement *m* horizontal, glissement *m* horizontal
e deslizamiento *m* horizontal
i slittamento *m* orizzontale
n lijnslip, lijnwiebel
d Zeilenschlupf *m*

2806 LINE SPACING crt/rec/rep
f espacement *m* de lignes
e paso *m* de exploración
i passo *m* d'analisi, spaziatura *f* tra le linee d'analisi
n lijnafstand
d Zeilenabstand *m*

2807 LINE STORE CONVERTER cpl/tv
Consists of a means of writing the incoming video signal into a suitable store and then reading out the stored information at a rate appropriate to the output.
f convertisseur *m* de normes de télévision basé sur la mémorisation du signal de ligne
e convertidor *m* de la frecuencia de línea mediante almacenamiento de la señal de línea
i convertitore *m* di norme televisive basato sulla memorizzazione del segnale di linea
n lijntijdgeheugen *n*, omzetter voor televisienormen
d Fernsehnormenwandler *m* mittels Zeilensignalspeicherung

2808 LINE STROBE rep/tv
Special type of waveform monitor for examining TV signals.
f monitor *m* de forme d'onde
e monitor *m* de forma de onda
i monitore *m* di forma d'onda
n golfvormmonitor
d Wellenformmonitor *m*

2809 LINE STRUCTURE rep/tv
Appearance of the picture or raster on a cathode-ray tube.
f structure *f* du canevas
e estructura *f* de la cuadrícula
i struttura *f* del quadro rigato
n rasterstructuur
d Rasterstruktur *f*

LINE SYNC CONTROL
see: HORIZONTAL SYNC CONTROL

LINE SYNC GENERATOR
see: HORIZONTAL SYNC GENERATOR

LINE SYNC PULSE
see: HORIZONTAL SYNC PULSE

2810 LINE TILT, LINE TILT CORRECTION rep/tv
A sawtooth waveform, recurring at line frequency, introduced into the picture signal to compensate for inherent distortion in the camera output.
f signal *m* linéaire de correction de ligne
e señal *f* lineal de corrección de línea
i correzione *f* a denti di sega di luminosità della linea
n lijnhellingcorrectie
d Sägezahnzeilensignale *pl*

LINE TIMEBASE, LINEAR TIMEBASE
see: HORIZONTAL TIMEBASE

LINE TIMEBASE GENERATOR
see: HORIZONTAL TIMEBASE GENERATOR

2811 LINE TRANSFORMER, LINE TRANSLATOR tv
f transformateur *m* du nombre de lignes
e transformador *m* del número de líneas
i trasformatore *m* del numero di linee
n lijnenvertaler, relineator
d Zeilentransformator *m*, Zeilenübertrager *m*

2812 LINE TRANSLATION tv
f transformation *f* du nombre de lignes
e transformación *f* del número de líneas
i trasformazione *f* del numero di linee
n relineatie
d Zeilentransformation *f*, Zeilenumsetzung *f*

2813 LINE TRAVERSING dis/tv
f décrochage *m* de lignes
e caminamiento *m* de líneas
i sorpassata *f* di linee
n doorlopen *n* van de lijnen
d Zeilendurchlauf *m*

LINE TRIGGERING CIRCUIT
see: LINE DRIVE

LINE TRIGGERING PULSE
see: LINE DRIVE PULSE

LINE TRIGGERING SIGNAL
see: LINE DRIVE SIGNAL

2814 LINE-UP stu
f mise *f* au point du studio, préparation *f*
e preparación *f*, puesta *f* a punto del estudio
i messa *f* a punto dello studio, preparazione *f*
n voorbereiding
d Studioanordnung *f*, Vorbereitung *f*

2815 LINE-UP TIME, LINING-UP PERIOD stu
The time required for warming up and adjustment in a TV studio.
f temps *m* de mise au point du studio
e tiempo *m* para puesta a punto del estudio
i tempo *m* da messa a punto dello studio
n tijd voor het gereedmaken van de studio
d Studioanordnungszeit *f*

2816 LINE UTILIZATION FACTOR rec/rep/tv
f coefficient *m* d'utilisation de ligne
e coeficiente *m* de utilidad de línea
i coefficiente *m* d'utilizzazione di linea
n gebruiksfactor voor lijnen
d Zeilenausnützungskoeffizient *m*

2817 LINE VISIBILITY rep/tv
A picture fault in which the line sweep structure is conspicuous to a normally positioned viewer.
f visibilité *f* des lignes
e visibilidad *f* de las líneas
i visibilità *f* delle linee
n zichtbare lijnstructuur
d Zeilensichtbarkeit *f*

2818 LINE WIDTH rec/rep/tv
The reciprocal of the number of lines per unit length in the direction of line progression.
f largeur *f* de ligne
e anchura *f* de línea
i larghezza *f* di linea
n lijnbreedte
d Zeilenbreite *f*

2819 LINEAR ELECTRONIC MATRIXING ctv
f matrixation *f* électronique linéaire
e representación *f* electrónica lineal por matrices
i matrizzazione *f* elettronica lineare
n elektronisch lineair matriceren *n*
d elektronische lineare Matrizierung *f*

2820 LINEAR MATRIX ctv
A linear network forming part of colo(u)r TV coders and decoders whereby the three primary signals are correctly apportioned.
f matrice *f* linéaire
e matriz *f* lineal
i matrice *f* lineare
n lineaire matrix
d lineare Matrize *f*

2821 LINEAR PHASE DISTORTION dis
f distorsion *f* linéaire de phase
e distorsión *f* lineal de fase
i distorsione *f* lineare di fase
n lineaire fazevervorming
d lineare Phasenverzerrung *f*

2822 LINEARITY ge
Relationship between two quantities which can be graphically represented as a straight line.
f linéarité *f*
e linealidad *f*
i linearità *f*
n lineariteit
d Linearität *f*

2823 LINEARITY CONTROL dis
Control to correct for non-linearity of the timebases.
f dispositif *m* de réglage de la linéarité
e regulador *m* de la linealidad
i regolatore *m* della linearità
n lineariteitsregelaar
d Linearitätsregler *m*

2824 LINEARITY CONTROL dis/tv
A control system to correct for non-linearity of the timebases.
f réglage *m* de la linéarité
e regulación *f* de la linealidad
i regolazione *f* della linearità

n linealiteitsregeling
d Linearitätsregelung *f*

2825 LINEARITY CORRECTION crt
The adjustment of one or more constants in a timebase generator circuit to ensure linearity in the timebase waveform.
f ajustage *m* de la linéarité,
correction *f* de la linéarité
e ajuste *m* de la linealidad,
corrección *f* de la linealidad
i aggiustaggio *m* della linearità,
correzione *f* della linearità
n lineariteitscorrectie,
lineariteitsinstelling
d Linearitätseinstellung *f*,
Linearitätskorrektur *f*

2826 LINEARITY DISTORTION, dis
LINEARITY ERROR
f distorsion *f* de linéarité
e distorsión *f* de linealidad
i distorsione *f* di linearità
n lineariteitsvervorming
d Linearitätsfehler *m*

2827 LINEARITY SLEEVE rep/tv
Foil loop mounted inside a former and placed around the neck of a TV receiver tube to provide horizontal linearity adjustment.
f bande *f* d'ajustage de la linéarité
e banda *f* de ajuste de linealidad
i banda *f* d'aggiustaggio della linearità
n instelband voor de lineariteit
d Einstellschleife *f* für die Linearität

LINEARITY TEST CARD
see: GEOMETRY TEST CARD

2828 LINEARITY TEST SIGNAL dis
f signal *m* pour essai de linéarité
e señal *f* para prueba de linealidad
i segnale *m* per prova di linearità
n lineariteitstoetssignaal *n*
d Linearitätsprobesignal *n*

LINING-UP THE CAMERA (US)
see: CAMERA LINE-UP

2829 LINK ge
A communication path or specified characteristics between two points.
f liaison *f*,
voie *f* de transmission
e canal *m*,
vía *f* de transmisión
i canale *m*,
via *f* di trasmissione
n kanaal *n*,
transmissieweg
d Kanal *m*,
Übertragungsweg *m*

2830 LINK TRANSMITTER cpl
Transmitter used to bring a TV signal from a remote location to a studio center(re) or a point in a network.
f émetteur *m* à couplage link,
émetteur *m* à relais
e retransmisor *m* de haz concentrado
i emettitore *m* a radiocollegamento
n koppelzender,
zender met gerichte bundel
d Linksender *m*

LIP MICROPHONE
see: CLOSE-TALKING MICROPHONE

2831 LIP-SYNCHRONIZED SPOT rec/tv
Film or TV shot taken so close to an actor that perfect synchronization is essential.
f prise *f* à haute qualité de synchronisation
e toma *f* de alta calidad de sincronización
i ripresa *f* ad alta qualità di sincronizzazione
n opname met maximale synchronisatie
d Aufnahme *f* mit maximaler Synchronisation

2832 LISTENING CHAIN aud
f chaîne *f* d'écoute
e cadena *f* de escucha
i catena *f* d'ascolto
n meeluisterkring
d Abhörkette *f*

2833 LISTENING ZONE aud/stu
f zone *f* "auditeur",
zone *f* d'écoute
e zona *f* de audibilidad
i zona *f* d'udibilità
n hoorbaarheidsgebied *n*
d Hörbarkeitsbereich *m*

2834 LIVE, rec/rep/tv
LIVE BROADCAST
The broadcast of a TV program(me) at the instant of its production.
f émission *f* de télévision en direct
e transmisión *f* en directa
i ripresa *f* dal vivo,
ripresa *f* in diretta
n directe uitzending,
live-uitzending
d Direktübertragung *f*,
Live-Übertragung *f*

2835 LIVE END stu
The end of a studio that gives almost complete reflection of sound waves.
f paroi *f* réfléchissante
e pared *f* reflectante
i parete *f* riflettente
n reflecterende wand
d Prallwand *f*,
Reflexionswand *f*

2836 LIVE INJECT, rec/tv
LIVE INSERT
That part of a broadcast, generally from records, which is live.
f insertion *f* actuelle,
séquence *f* en direct
e inserción *f* en directa
i inserto *m* in diretta

n actuele inlas,
live-inlas
d aktuelle Einlage *f*,
Live-Beitrag *m*

2837 LIVE ROOM aud
A room having a minimum of sound-absorbing material.
f salle *f* réverbérante
e sala *f* reverberante
i sala *f* riverberante
n ruimte met lange nagalm
d Hallraum *m*

2838 LIVE STUDIO stu
A studio having a comparatively long reverberation time.
f studio *m* réverbérant
e estudio *m* reverberante
i studio *m* riverberante
n studio met lange nagalm
d Studio *n* mit Nachhall

2839 LOBE aea
A portion of the directional pattern bounded by one or two cones of nulls.
f lobe *m*,
oreille *f*,
pétale *f*
e lóbulo *m*
i lobo *m*
n lob,
lus
d Keule *f*,
Strahlungslappen *m*,
Zipfel *m*

2840 LOCAL-DISTANT CONTROL cpl/tv
Pre-set control provided in an automatic gain control line to permit a TV receiver to be adjusted to suit service or fringe area reception conditions.
f réglage *m* local-à distance
e regulación *f* local-de distancia
i regolazione *f* locale-a distanza
n regeling van dichtbij en veraf
d Nahe-Abstand-Regelung *f*

LOCAL OSCILLATOR
see: BEATING OSCILLATOR

2841 LOCAL TELEVISION LINE tv
f ligne *f* locale pour télévision
e línea *f* local para televisión
i linea *f* locale per televisione
n lokale leiding voor televisie
d Fernsehortsleitung *f*

2842 LOCATION rec/tv
Place other than the studio or studio lot of a film or TV production organization, where one of its units is shooting pictures.
f aire *f* de tournage à l'extérieur
e terreno *m* para exteriores
i area *f* per le riprese esterne
n buitenopnameterrein *n*
d Aufnahmegelände *n*

2843 LOCATION SHOOTING rec/tv
f prise *f* de vues à l'extérieur,
tournage *m* à l'extérieur
e toma *f* de vistas al exterior
i girare *m* in esterno
n buitenopnamen *pl*
d Aussenaufnahmen *pl*

2844 LOCATION SOUND RECORDING rec/tv
f enregistrement *m* de son à l'extérieur
e registro *m* de sonido al exterior
i registrazione *f* di suono all'esterno
n buitenopname van geluid
d Tonaussenaufnahme *f*

2845 LOCK-IN ge
Generally, to synchronize one oscillator with another.
f enclenchement *m*,
verrouillage *m*
e enclavamiento *m*,
enganche *m*,
fijación *f* en circuito
i blocco *m*
n vergrendeling
d Verkopplung *f*,
Verriegelung *f*

LOCK-IN RANGE
see: CAPTURE RANGE

2846 LOCK-OUT cpl
f perte *f* de synchronisation
e pérdida *f* de sincronización
i perdita *f* di sincronizzazione
n uit synchronisatie vallen *n*
d Synchronisationsverlust *m*

2847 LOCKED OSCILLATOR cpl
An oscillator held to a specific frequency by an external source.
f oscillateur *m* bloqué
e oscilador *m* clavado
i oscillatore *m* bloccato,
oscillatore *m* pilotato
n gesynchroniseerde oscillator
d Mitlaufgenerator *m*,
Mitnahmegenerator *m*

2848 LOCKED SOUND HEAD rec/tv
f tête *f* magnétique à verrouillage mécanique
e cabeza *f* magnética de enclavamiento mecánico
i testina *f* magnetica a blocco meccanico
n mechanisch vergrendelde magneetkop
d mechanisch verriegelter Magnetkopf *m*

2849 LOCKING cpl/tv
The control of frequency of an oscillating circuit by means of an applied signal of constant frequency.
f réglage *m* de fréquence par un signal à fréquence constante
e regulación *f* de frecuencia por una señal de frecuencia constante
i regolazione *f* di frequenza per un segnale a frequenza costante
n frequentieregeling door een signaal met constante frequentie

d Frequenzregelung *f* durch ein Signal mit konstanter Frequenz

2850 LOCKING, rec/rep/tv
f synchronisation *f* du réseau
e sincronización *f* de la red
i sincronizzazione *f* della rete
n netsynchronisatie
d Netzsynchronisierung *f*

2851 LOFT AERIAL, aea
LOFT ANTENNA
An aerial (antenna) which may be mounted in the loft of a house.
f antenne *f* de grenier
e antena *f* de desván
i antenna *f* da solaio
n zolderantenne
d Dachbodenantenne *f*

2852 LOG-BOOK fi/tv
f relevé *m* des numéros de bord
e esquema *m* de montaje
i elenco *m* di montaggio
n montageschema *n*
d Schnittliste *f*

2853 LOG-SHEET, fi/tv
REPORT SHEET
f rapport *m* script
e reporte *m* de tomas
i rapporto *m* di riprese
n opnamebericht *n*
d Drehbericht *m*

2854 LONG-COIL ELECTROMAGNETIC crt
FOCUSING,
LONG-COIL MAGNETIC FOCUSING
f focalisation *f* magnétique à bobine longue
e enfoque *m* magnético de bobina larga
i focalizzazione *f* magnetica a bobina lunga
n magnetische focussering met lange spoel
d magnetische Fokussierung *f* mit langer Magnetspule

2855 LONG-DISTANCE TELEVISION tv
CIRCUIT,
TELEVISION LINE
f ligne *f* de télévision à grande distance
e línea *f* de televisión de gran distancia
i linea *f* di televisione a gran distanza
n televisieleiding op grote afstand
d Fernsehfernleitung *f*,
Fernsehleitungskette *f*

2856 LONG-FOCUS LENS, tv
TELEPHOTO LENS
f téléobjectif *m*
e objetivo *m* de foco largo
i teleobiettivo *m*
n telelens
d Teleobjektiv *n*

2857 LONG-PERSISTENCE crt
CATHODE-RAY TUBE
f tube *m* cathodique à écran à longue persistance
e tubo *m* catódico de pantalla de persistencia larga
i tubo *m* catodico a schermo luminescente a lunga persistenza
n katodestraalbuis met lang nalichtend scherm
d Katodenstrahlröhre *f* mit Schirm langer Nachleuchtung

2858 LONG-PERSISTENCE SCREEN crt
A screen specially prepared to retain its luminance for an appreciable time after the stimulus has been reduced or removed.
f écran *m* à longue persistance
e pantalla *f* de larga persistencia
i schermo *m* a luminescenza a lunga persistenza
n lang nalichtend scherm *n*
d Schirm *m* langer Nachleuchtung

2859 LONG-PLAYING TAPE vr
f bande *f* de longue durée
e cinta *f* de larga duración
i nastro *m* a lunga durata
n langspeelband
d Langspielband *n*

2860 LONG SHOT, opt/tv
VISTA SHOT
f plan *m* d'ensemble,
prise *f* de vues à distance
e plano *m* distante,
toma *f* de vistas a distancia
i campo *m* lungo,
ripresa *f* di scene lontane
n opname op afstand
d Fernaufnahme *f*,
Gesamtaufnahme *f*,
Totale *f*

2861 LONG STREAKING dis/tv
In a horizontal line-by-line scanning system, a fault which is characterized by long horizontal bands which appear to be attached to strongly contrasted objects.
f traînage *m* long
e arrastre *m* largo
i striscionamento *m* lungo
n lange vegen *pl*
d lange Fahnen *pl*

2862 LONG-TERM TAPE aud/vr
SPEED VARIATION
f variation *f* lente de la vitesse de bande
e variación *f* lenta de la velocidad de cinta
i variazione *f* lenta della velocità di nastro
n langzame bandsnelheidsvariatie
d langsame Bandgeschwindigkeitsschwankung *f*

2863 LONG WHITE SMEAR dis
f traînage *m* long blanc
e arrastre *m* largo blanco
i striscionamento *m* lungo bianco
n lange witte veeg
d lange weisse Fahne *f*

2864 LONGITUDINAL MAGNETIZATION vr
Magnetization of a magnetic recording medium in a direction essentially

parallel to the line of travel.
f magnétisation *f* longitudinale
e magnetización *f* longitudinal
i magnetizzazione *f* longitudinale
n magnetisatie in de lengterichting
d Längsmagnetisierung *f*

2865 LONGITUDINAL MAGNIFICATION opt/tv
The limiting ratio of the distance between two image points on the principal axis to that between the corresponding object points when the latter approach zero.
f amplification *f* longitudinale
e ampliación *f* longitudinal
i ingrandimento *m* longitudinale
n asvergroting
d Achsenvergrösserung *f*

2866 LONGITUDINAL REGISTRATION aud
f régistration *f* longitudinale
e grabación *f* longitudinal, registro *m* longitudinal
i registrazione *f* longitudinale
n longitudinale opname, longitudinale optekening
d Längsspuraufzeichnung *f*, Längsspurverfahren *n*

LOOP
see: ANTINODAL POINT

2867 LOOSE FRAMING rep/tv
f décadrage *m*
e encuadrado *m* holgado
i inquadratura *f* larga
n wijde omlijsting
d weite Umrahmung *f*

LOSS MODULATION
see: ABSORPTION MODULATION

2868 LOSS OF PICTURE LOCK dis
f décrochage *m* vertical de l'image
e pérdida *f* de sincronización vertical
i perdita *f* di sincronizzazione verticale
n verlies *n* van de verticale synchronisatie
d Bildkippen *n*

2869 LOT rec/tv
f terrain *m* de studio
e plató *m* de estudio
i stabilimento *m* di posa
n studioterrein *n*
d Ateliergelände *n*

2870 LOUD TRAILER aud
A high-power directional loudspeaker for dessiminating verbal instructions.
f hautparleur *m* directionnel de grande puissance
e altavoz *m* direccional de alta potencia
i altoparlante *m* direzionale di grande potenza
n gerichte krachtluidspreker
d Hochleistungsrichtlautsprecher *m*

2871 LOUDNESS aud
The intensive attribute of an auditory sensation in terms of which sound may be ordered on a scale extending from soft to loud.
f intensité *f* sonore, puissance *f*
e intensidad *f* sonora
i intensità *f* della sensazione sonora, sonorità *f*
n luidheid
d Lautheit *f*

2872 LOUDNESS CONTROL, VOLUME CONTROL aud
f régleur *m* d'intensité sonore, régleur *m* de puissance
e regulador *m* de volumen
i regolatore *m* di volume
n volumeregelaar
d Lautstärkeregler *m*

LOUDNESS LEVEL
see: EQUIVALENT LOUDNESS

2873 LOUDSPEAKER aud
f hautparleur *m*
e altavoz *m*
i altoparlante *m*
n luidspreker
d Lautsprecher *m*

2874 LOUDSPEAKER-MICROPHONE aud/stu
A loudspeaker which can be used as a microphone.
f hautparleur-microphone *m*
e altavoz-micrófono *m*
i altoparlante-micrófono *m*
n luidspreker-microfoon
d Lautsprecher-Mikrophon *n*

2875 LOUDSPEAKER MONITOR aud
f hautparleur *m* moniteur
e altavoz *m* monitor
i altoparlante *m* monitore
n monitorluidspreker
d Monitorlautsprecher *m*

2876 LOUDSPEAKER PLACEMENT, LOUDSPEAKER POSITIONING aud/stu
f répartition *f* des hautparleurs
e repartición *f* de las altavoces
i ripartizione *f* degli altoparlanti
n luidsprekeropstelling
d Lautsprecheranordnung *f*

2877 LOW-ANGLE SHOT opt
f contreplongée *f*
e toma *f* angular desde abajo
i inquadratura *f* dal basso
n laag hoekshot
d Froschperspektive *f*

2878 LOW-CONTRAST PICTURE tv
f image *f* trop peu contrastée
e imagen *f* de poco contraste
i immagine *f* senza contrasto
n kontrastarm beeld *n*
d kontrastarmes Bild *n*

2879 LOW-DEFINITION TELEVISION (GB), LOW-RESOLUTION TELEVISION (US) tv

TV system using less than 200 scanning lines per picture.

f télévision *f* à basse définition
e televisión *f* de baja definición
i televisione *f* a bassa definizione
n televisie met klein aantal lijnen
d niedrigzeiliges Fernsehen *n*

LOW-ELECTRON-VELOCITY CAMERA TUBE
see: CATHODE-POTENTIAL STABILIZED CAMERA TUBE

LOW-FREQUENCY AMPLIFIER
see: AUDIO AMPLIFIER

2880 LOW-KEY IMAGE, LOW-KEY PICTURE stu

Style of tonal rendering of a scene marked by predominance of dark tones with rich shadow detail.

f image *f* à gradations de timbre obscures
e imagen *f* con gradaciones de tono obscuras
i immagine *f* con gradazioni di tono oscure
n donker beeld *n* met sterke contrasten
d dunkles Bild *n* mit starken Kontrasten

2881 LOW-LEVEL MODULATION, LOW-POWER MODULATION cpl

Modulation of the carrier of a transmitter effected in a stage prior to the final radiofrequency amplifier.

f modulation *f* dans l'anode d'un étage intermédiaire
e modulación *f* en bajo nivel
i modulazione *f* in stadio intermedio
n tussentrapmodulatie
d Modulation *f* auf niedrigem Pegel, Vorstufenmodulation *f*

2882 LOW-LUMINOSITY PICTURE rep/tv

f image *f* de basse luminosité
e imagen *f* de baja luminosidad
i immagine *f* a scarsa luminosità
n niet-helder beeld *n*
d lichtschwaches Bild *n*

2883 LOW-MODULATION TRACK vr

f piste *f* sonore à basse modulation
e pista *f* sonora de baja modulación
i pista *f* sonora a bassa modulazione
n zwak gemoduleerd geluidsspoor *n*
d schwach ausgesteuerte Tonspur *f*

2884 LOW-PASS FILTER cpl

A filter that transmits alternating currents below a given cutoff frequency and substantially attenuates all other currents.

f filtre *m* passe-bas
e filtro *m* paso bajo
i filtro *m* passo-basso
n laagdoorlatend filter *n*
d Tiefpass *m*, Tiefpassfilter *n*

2885 LOW-POWER REPEATER STATION tv

f station *f* amplificatrice à basse puissance
e estación *f* amplificadora de baja potencia
i stazione *f* amplificatrice a bassa potenza
n versterkerstation *n* met laag vermogen
d Verstärkerstation *f* niedriger Leistung

2886 LOW-POWER VIDEO TRANSMITTER cpl

f émetteur *m* vidéo à basse puissance
e emisor *m* video de baja potencia
i emettitore *m* video a bassa potenza
n beeldzender met laag vermogen
d Bildsender *m* niedriger Leistung

2887 LOW-VELOCITY SCANNING crt

The scanning of a target with electrons of velocity less than the minimum velocity needed to give a secondary-emission ratio of unity.

f analyse *f* avec électrons de basse vitesse
e exploración *f* con electrones de baja velocidad
i analisi *f* con elettroni di bassa velocità
n aftasting met langzame elektronen
d Abtastung *f* mit langsamen Elektronen

2888 LUMINANCE, PHOTOMETRIC BRIGHTNESS ct

The quotient of the luminous intensity in the given direction of an infinitesimal element of the surface containing the point under consideration, by the orthogonally projected area of the element on a plane perpendicular to the given direction.

f luminance *f*
e luminancia *f*
i brillanza *f* luminosa, luminanza *f*
n luminantie
d Helligkeit *f*, Leuchtdichte *f*, Luminanz *f*

2889 LUMINANCE AXIS ctv

The axis in colo(u)r space, containing the reference white, along which the luminance component is plotted.

f axe *m* de luminance
e eje *m* de luminancia
i asse *m* di luminanza
n luminantieas
d Leuchtdichteachse *f*

2890 LUMINANCE BAND ct

The band occupied by the luminance signal.

f bande *f* du signal de luminance
e banda *f* de la señal de luminancia
i banda *f* del segnale di luminanza
n luminantiesignaalband
d Helligkeitssignalband *n*

2891 LUMINANCE CHANNEL ctv

In a colo(u)r TV system, any path intended to carry the luminance signal.

f canal *m* de luminance
e canal *m* de luminancia
i canale *m* di luminanza
n luminantiekanaal *n*
d Helligkeitskanal *m*

2892 LUMINANCE COEFFICIENTS ctv
In any trichromatic system, the three coefficients expressing the luminance of unit amounts of the three tristimulus values.
f coefficients *pl* de luminance
e coeficientes *pl* de luminancia
i coefficienti *pl* di luminanza
n luminantiecoëfficiënten *pl*
d Helligkeitskoeffizienten *pl*

2893 LUMINANCE DELAY ctv
f retard *m* du signal de luminance
e retardo *m* de la señal de luminancia
i ritardo *m* del segnale di luminanza
n vertraging van het luminantiesignaal
d Verzögerung *f* des Helligkeitssignals

2894 LUMINANCE DELAY LINE, LUMINANCE LINE cpl
f ligne *f* à retard pour le signal de luminance
e línea *f* de retardo para la señal de luminancia
i linea *f* a ritardo per il segnale di luminanza
n vertragingslijn voor het luminantiesignaal
d Verzögerungsleitung *f* des Helligkeitssignals

2895 LUMINANCE DIFFERENCE ct
f différence *f* de luminance
e diferencia *f* de luminancia
i differenza *f* di luminanza
n luminantieverschil *n*
d Leuchtdichtedifferenz *f*

2896 LUMINANCE DIFFERENCE SIGNAL ctv
The signal which must be subtracted from the luminance signal to give the monochrome signal.
f signal *m* de différence de luminance
e señal *f* de diferencia de luminancia
i segnale *m* di differenza di luminanza
n luminantieverschilsignaal *n*
d Leuchtdichtedifferenzsignal *n*

LUMINANCE DIFFERENCE THRESHOLD
see: BOUNDARY CONTRAST

2897 LUMINANCE DISTORTION dis
A distortion characterized by the fact that an object of uniform luminosity is reproduced with a varying luminosity in the vertical or horizontal sense.
f distorsion *f* de luminance
e distorsión *f* de luminancia
i variazione *f* di luminanza
n helderheidsverloop *n*
d Bildabschattung *f*

2898 LUMINANCE ENHANCEMENT tv
f accentuation *f* de la luminance
e acentuación *f* de la luminancia
i accentuazione *f* della luminanza
n verhoging van de helderheid
d Helligkeitsanhebung *f*

2899 LUMINANCE FACTOR ct
Of a non-luminance body the ratio of the luminance of the body considered when illuminated and observed under specified conditions of illumination and observation to the luminance of a non-absorbing perfect-diffuser receiving the same illumination.
f facteur *m* de luminance
e factor *m* de luminancia
i fattore *m* di luminanza
n luminantiefactor, remissiefactor
d Remissionsgrad *m*

2900 LUMINANCE FLICKER ctv
Flicker which results from fluctuation of luminance.
f papillotement *m* de luminance
e parpadeo *m* de luminancia
i sfarfallio *m* di luminanza
n luminantieflikkering
d Helligkeitsflimmern *n*

2901 LUMINANCE LEVEL ct
f niveau *m* de luminance
e nivel *m* de luminancia
i livello *m* di luminanza
n luminantieniveau *n*
d Helligkeitspegel *m*

2902 LUMINANCE NOTCH ctv
A depression introduced in the amplitude/frequency response of the luminance channel and used either to reduce the visibility of the subcarrier or to reduce cross-colo(u)r.
f encoche *f* de luminance
e hendidura *f* de luminancia
i intaglio *m* di luminanza
n luminantiekerf
d Leuchtdichtekerbe *f*

2903 LUMINANCE PLANE ct
f plan *m* de luminance
e plano *m* de luminancia
i piano *m* di luminanza
n luminantievlak *n*
d Helligkeitsebene *f*

2904 LUMINANCE PRIMARY ctv
The colo(u)r stimulus, usually achromatic, defined by the direction of the luminous axis.
f primaire *f* prépondérante
e primario *m* de luminancia
i primario *m* di luminanza
n primaire kleursoort voor luminantie
d Leuchtdichtefarbart *f*

2905 LUMINANCE RANGE ctv
f plage *f* de luminance
e campo *m* de luminancia

i campo *m* di luminanza
n luminantiegebied *n*
d Helligkeitsbereich *m*

2906 LUMINANCE SIGNAL, Y SIGNAL — ctv
1. A signal intended to have exclusive control of the luminance of a picture.
2. A signal expressing in electrical form the luminance of the elements of the object, in accordance with a well-defined law.
f signal *m* de luminance
e señal *f* de luminancia
i segnale *m* di luminanza
n luminantiesignaal *n*
d Helligkeitssignal *n*, Leuchtdichtesignal *n*

2907 LUMINANCE SIGNAL TRANSITION, LUMINANCE TRANSITION — ctv
f saut *m* de luminance
e señal *f* de luminancia
i segnale *m* di luminanza
n luminantiesprong
d Helligkeitssprung *m*

2908 LUMINANCE TRANSMISSION CONSTANT — ct
That type of transmission in which the transmission primaries are a luminance primary and two chrominance primaries.
f constante *f* de transmission de luminance
e constante *f* de transmisión de luminancia
i costante *f* di trasmissione di luminanza
n luminantietransmissieconstante
d Helligkeitsübertragungskonstante *f*

2909 LUMINANCE VALUE — ct
f valeur *f* de luminance
e valor *m* de luminancia
i valore *m* di luminanza
n luminantiewaarde
d Helligkeitswert *m*

2910 LUMINESCENCE — ct
The process whereby matter emits radiation which, for certain wavelengths or restricted regions of the spectrum, is in excess of that attributable to the thermal state of the material and the emissivity of its surface.
f luminescence *f*
e luminiscencia *f*
i luminescenza *f*
n luminescentie
d Lumineszenz *f*

2911 LUMINESCENCE THRESHOLD, THRESHOLD OF LUMINESCENCE — ct
The lowest frequency of radiation that is capable of exciting a luminescent material.
f seuil *m* de luminescence
e umbral *m* de luminiscencia
i soglia *f* di luminescenza
n luminescentiedrempel
d Lumineszenzschwelle *f*

2912 LUMINESCENT — crt
f luminescent adj
e luminiscente adj
i luminescente adj
n luminescerend adj
d lumineszierend adj

LUMINESCENT MATERIAL
see: FLUORESCENT MATERIAL

LUMINESCENT SCREEN
see: CATHODE-RAY SCREEN

2913 LUMINESCENT SCREEN SPECTRAL CHARACTERISTIC — crt
The relation between wavelength and emitted radiant power per unit wavelength interval.
f caractéristique *f* spectrale d'un écran luminescent
e característica *f* espectral de una pantalla luminiscente
i caratteristica *f* spettrale d'uno schermo luminescente
n spectrale karakteristiek van een luminescerend scherm
d Spektralcharakteristik *f* eines Leuchtschirmes

2914 LUMINESCENT SCREEN TUBE — crt
f tube *m* cathodique à écran luminescent
e tubo *m* catódico de pantalla luminiscente
i tubo *m* catodico a schermo luminescente
n buis met luminescerend scherm
d Leuchtschirmröhre *f*

LUMINOSITY
see: APPARENT BRIGHTNESS

2915 LUMINOSITY COEFFICIENTS — ct
The constant multipliers for the respective tristimulus values of any colo(u)r, such that the sum of the three products is the luminance of the colo(u)r.
f facteurs *pl* d'efficacité lumineuse, facteurs *pl* de luminosité
e coeficientes *pl* de luminosidad
i coefficienti *pl* di luminosità
n helderheidscoëfficiënten *pl*
d Helligkeitskoeffizienten *pl*, Leuchtdichtebeiwerte *pl*

2916 LUMINOSITY CURVE — ct
A graphical relationship between wavelength of light in the visible range, and reciprocal of the radiance required to produce visual sensations of equal brightness.
f courbe *f* de luminosité
e curva *f* de luminosidad
i curva *f* di luminosità
n helderheidskromme
d Helligkeitskennlinie *f*

2917 LUMINOSITY FUNCTION — ct/tv
f fonction *f* de la sensation de luminosité
e función *f* de la sensación de luminosidad
i funzione *f* della sensazione di luminosità
n functie van de helderheidswaarneming
d Helligkeitsempfindungsfunktion *f*

2918 LUMINOSITY INFORMATION ctv
f information *f* de la luminosité
e información *f* de la luminosidad
i informazione *f* della luminosità
n helderheidsinformatie
d Helligkeitsinformation *f*

2919 LUMINOUS EDGE rep/tv
Reflection reducing light filter around the picture surface of a TV set.
f filtre *m* optique encadrant
e recuadro *m* filtrante de luz
i quadro *m* filtrante di luce
n randlichtfilter *n*
d magischer Rahmen *m*

2920 LUMINOUS EFFICIENCY ct
The ratio of the luminous flux to the radiant flux.
f efficacité *f* lumineuse
e eficacia *f* luminosa
i coefficiente *m* di visibilità
n visueel rendement *n*
d photometrisches Strahlungsäquivalent *n*

2921 LUMINOUS FLUX ct
The total visible light energy per second by a light source.
f flux *m* lumineux
e flujo *m* luminoso
i flusso *m* luminoso
n lichtstroom
d Lichtstrom *m*

LUMINOUS INTENSITY
see: ILLUMINATION INTENSITY

2922 LUMINOUS SENSITIVITY crt
The output current of a phototube or camera tube divided by the incident luminous flux.
f photosensibilité *f*, sensibilité *f* lumineuse
e fotosensibilidad *f*, sensibilidad *f* luminosa
i fotosensibilità *f*, sensibilità *f* luminosa
n lichtgevoeligheid
d Lichtempfindlichkeit *f*

M

2923 MACHINE CONTROL, TELECINE MACHINE CONTROL tv
The control of telecine machines carrying prerecorded program(me)s and used during station breaks for filling the gap.
f contrôle *m* de projecteurs de télécinéma
e control *m* de proyectores telecinematográficos
i controllo *m* di proiettori di riprese filmate
n controle van de projectoren voor tussenfilmtelevisiesysteem
d Kontrolle *f* von Projektionsgeräte für Zwischenfilmfernsehbetrieb

2924 MACROSHOT, PACK-SHOT opt
f macrophotographie *f*
e macrofotografía *f*
i macrofotografia *f*
n macro-opname
d Makroaufnahme *f*

2925 MACROSHOT LENS, PACK-SHOT LENS opt
f objectif *m* à double tirage
e objetivo *m* para macrofotografía
i obiettivo *m* per macrofotografia
n macrokilaar *n*
d Makrokilar *m*

2926 MACULA LUTEA, YELLOW SPOT ct
f macula lutea, tache *f* jaune
e macula lútea
i macula lutea
n gele vlek
d gelber Fleck *m*

2927 MAGENTA, MINUS GREEN ct
A colo(u)r in which the spectral components are confined mainly to the 400-500 and 600-700 nm bands of the spectrum, i.e. the 500-600 nm (green) band is missing.
f magenta, moins vert
e magenta, menos verde
i magenta, meno verde
n magenta, minus groen
d magenta, minus grün

2928 MAGIC BALANCE, MAGNETIC BALANCE aud/vr
An instrument for measuring the strength of a magnetic field by balancing the force exerted on a conductor carrying current in the field.
f balance *f* magnétique
e balanza *f* magnética
i bilancia *f* magnetica
n magnetische balans
d magische Waage *f*

2929 MAGIC LINE crt
f ligne *f* magique
e línea *f* mágica
i linea *f* magica
n magische lijn
d magische Linie *f*

2930 MAGNETIC BALANCE TRACK aud/vr
f piste *f* magnétique de compensation
e pista *f* magnética de compensación
i pista *f* magnetica di compensazione
n magnetisch compensatiespoor *n*
d Magnetausgleichsspur *f*

2931 MAGNETIC BIAS aud/vr
Steady magnetic field added to signal field in magnetic recording.
f polarisation *f* magnétique
e polarización *f* magnética
i polarizzazione *f* magnetica
n voormagnetisatie
d Vormagnetisierung *f*

2932 MAGNETIC CENTER(RE) TRACK aud/vr
f piste *f* magnétique centrale
e pista *f* magnética central
i pista *f* magnetica centrale
n centraal magnetisch spoor *n*
d Magnetmittenspur *f*

2933 MAGNETIC COATING vr
f enduit *m* magnétique
e capa *f* magnética
i strato *m* magnetico
n magnetische laag
d magnetische Schicht *f*

2934 MAGNETIC CONVERGENCE crt
f convergence *f* magnétique
e convergencia *f* magnética
i convergenza *f* magnetica
n magnetische convergentie
d magnetische Konvergenz *f*

MAGNETIC DEFLECTION
see: ELECTROMAGNETIC DEFLECTION

MAGNETIC DEFLECTION SENSITIVITY
see: ELECTROMAGNETIC DEFLECTION SENSITIVITY

2935 MAGNETIC FIELD ge
Any space or region in which a magnetic force is exerted on moving electric charges.
f champ *m* magnétique
e campo *m* magnético

i campo *m* magnetico
n magneetveld *n*,
magnetisch veld *n*
d Magnetfeld *n*,
magnetisches Feld *n*

2936 MAGNETIC FILM vr
f bande *f* magnétique perforée
e cinta *f* magnética perforada
i nastro *m* magnetico perforato
n geperforeerde magneetband
d Cordband *n*,
Magnetperfoband *n*,
Magnetsplitfilm *m*

MAGNETIC FILM REPRODUCER
see: FILM REPRODUCER

2937 MAGNETIC FLUX vr
f flux *m* magnétique
e flujo *m* magnético
i flusso *m* magnetico
n magnetische flux
d Magnetfluss *m*,
magnetischer Fluss *m*

MAGNETIC FOCUSING
see: ELECTROMAGNETIC FOCUSING

MAGNETIC HEAD
see: HEAD

2938 MAGNETIC IRON OXIDE vr
A material used in magnetic tape manufacture.
f oxyde *m* de fer magnétique
e óxido *m* de hierro magnético
i ossido *m* di ferro magnetico
n magnetisch ijzeroxyde *n*
d magnetisches Eisenoxyd *n*

2939 MAGNETIC LAMINATING TAPE vr
f bande *f* pour piste marginale
e cinta *f* para pista marginal
i nastro *m* per pista marginale
n randspoorband
d Kaschierband *n*

2940 MAGNETIC LAMINATION, vr
MAGNETIC STRIPING
f coucher *m* de piste magnétique
e aplicación *f* de pista marginal
i applicazione *f* di pista marginale
n aanbrengen *n* van een randspoor
d Magnetrandbeschichten *n*,
Magnetrandbespuren *n*

MAGNETIC LENS
see: ELECTROMAGNETIC LENS

2941 MAGNETIC MASTER, aud/vr
MASTER COPY
The original recording on magnetic tape from which copies can be made.
f bande *f* magnétique originale,
enregistrement *m* magnétique original
e cinta *f* magnética madre,
registro *m* magnético original
i nastro *m* magnetico campione,
registrazione *f* magnetica originale
n originele magnetische opname
d Hauptband *n*,
Stammband *n*

2942 MAGNETIC PARTICLE aud/rec/vr
Needle-shaped crystals approximately 1 micron in length and 0.2 micron in width, used for producing magnetic tape.
f particule *f* magnétique
e partícula *f* magnética
i particella *f* magnetica
n magnetisch deeltje *n*
d magnetisches Teilchen *n*

2943 MAGNETIC PARTICLE aud/vr
ORIENTATION,
ORIENTATION OF MAGNETIC
PARTICLES
f orientation *f* des particules magnétiques
e orientación *f* de las partículas magnéticas
i orientazione *f* delle particelle magnetiche
n richten *n* van de magnetische deeltjes
d Ausrichtung *f* der magnetischen Teilchen

2944 MAGNETIC PICKUP COIL vr
f bobine *f* magnétique enregistreuse
e bobina *f* magnética registradora
i bobina *f* magnetica di registrazione
n magnetische opneemspoel
d magnetische Aufnahmespule *f*

2945 MAGNETIC PICTURE vr
f image *f* magnétique
e imagen *f* magnética
i immagine *f* magnetica
n magnetisch beeld *n*
d Magnetbild *n*

2946 MAGNETIC PICTURE rec/tv/vr
TRACING,
VIDEORECORDING,
VIDEOTAPE RECORDING,
VTR
f enregistrement *m* vidéo,
magnétoscopie *f*
e magnetoscopía *f*,
registro *m* video,
videorregistro *m*
i magnetoscopia *f*,
registrazione *f* video,
videoregistrazione *f*
n beeldopname,
video-opname,
VTR
d Magnetbildverfahren *n*,
MAZ-Aufzeichnung *f*,
Video-Aufzeichnung *f*

2947 MAGNETIC POLE, crt
POLE
That part of the surface of a magnet from which the magnetic flux emanates or to which it returns.
f pôle *m*
e polo *m*
i polo *m*
n pool
d Pol *m*

2948 MAGNETIC POLE FACE, crt
POLE FACE
Surface at the pole of a magnet or core of an electromagnet through which the flux passes.
f face *f* polaire
e cara *f* polar
i faccia *f* polare
n poolschoenvlak *n*
d Polfläche *f*,
Polschuhfläche *f*

2949 MAGNETIC POLE PIECE, crt
POLE PIECE,
POLE SHOE
That part of the core of an electromagnet which terminates at the air gap.
f masse *f* polaire,
pièce *f* polaire
e expansión *f* polar,
pieza *f* polar
i espansione *f* polare,
scarpa *f* polare
n poolschoen
d Polschuh *m*

2950 MAGNETIC POLE TIPS, vr
POLE TIPS
f pièces *pl* polaires
e piezas *pl*
i espansioni *pl* polari
n pooltips *pl*
d Polschuhspitzen *pl*

MAGNETIC-POWDER COATED TAPE
see: COATED MAGNETIC TAPE

MAGNETIC-POWDER IMPREGNATED TAPE
see: DISPERSED-POWDER MAGNETIC TAPE

2951 MAGNETIC PRINT-THROUGH, vr
MAGNETIC TAPE CROSSTALK
The permanent transfer of a recorded signal from a section of a magnetic recording medium to another section of the same or another medium when these sections are brought in proximity.
f écho *m* magnétique,
effet *m* d'écho
e eco *m* magnético,
efecto *m* de eco
i eco *m* magnetico,
effetto *m* d'eco
n echo-effect *n*,
magnetische echo
d Kopiereffekt *m*

2952 MAGNETIC READING HEAD vr
A magnetic head used to transform magnetic variations in magnetic tape into corresponding voltage or current variations.
f tête *f* magnétique de lecture
e cabeza *f* magnética de lectura
i testina *f* magnetica di lettura
n magnetische leeskop
d magnetischer Lesekopf *m*

2953 MAGNETIC RECORDER vr
Equipment incorporating an electromagnetic transducer and means for moving a magnetic recording medium relative to the transducer for recording electric signals as magnetic variations in the medium.
f enregistreur *m* magnétique,
magnétophone *m*
e grabador *m* sobre cinta,
magnetófono *m*
i magnetofono *m*,
registratore *m* magnetico
n magnetische opneemapparatuur
d Magnettongerät *n*,
MTG *n*

2954 MAGNETIC RECORDING aud/vr
Recording of sound or video signals by longitudinal magnetization of a uniformly moving plastic tape impregnated with iron oxide or other substance.
f enregistrement *m* magnétique
e grabación *f* magnética,
registro *m* magnético
i registrazione *f* magnetica
n magnetische opname
d magnetische Tonaufzeichnung *f*

2955 MAGNETIC RECORDING HEAD vr
A magnetic head used to convert magnetic variations on magnetic media into electric variations.
f tête *f* d'enregistrement magnétique
e cabeza *f* de grabación magnética
i testina *f* di registrazione magnetica
n magnetische opneemkop,
magnetische schrijfkop
d magnetischer Aufnahmekopf *m*,
magnetischer Schreibkopf *m*

2956 MAGNETIC RECORDING vr
INSTALLATION
f installation *f* d'enregistrement sur bande magnétique
e instalación *f* de grabación sobre cinta magnética
i stabilimento *m* di registrazione su strato magnetico
n magnetische opneeminstallatie
d Magnettonanlage *f*

2957 MAGNETIC RECORDING aud/vr
MATERIAL
f matériel *m* d'enregistrement magnétique
e material *m* de grabación magnética
i materiale *m* di registrazione magnetica
n magnetisch opneemmateriaal *n*
d magnetisches Aufnahmematerial *n*

2958 MAGNETIC RECORDING vr
MEDIUM
A magnetizable material used in a magnetic recorder.
f support *m* magnétique d'enregistrement
e soporte *m* magnético de grabación
i sopporto *m* magnetico di registrazione
n magnetisch opneemmedium *n*
d Magnettonträger *m*

2959 MAGNETIC REPRODUCER aud/vr
f reproducteur *m* magnétique
e reproductor *m* magnético
i riproduttore *m* magnetico
n magnetische weergever
d magnetischer Geber *m*

2960 MAGNETIC REPRODUCING HEAD aud/vr
f tête *f* magnétique de reproduction
e cabeza *f* magnética de reproducción
i testina *f* magnetica di riproduzione
n magnetische weergeefkop
d magnetischer Hörkopf *m*,
magnetischer Wiedergabekopf *m*

2961 MAGNETIC SHEET vr
f feuille *f* magnétique
e hoja *f* magnética
i foglio *m* magnetico
n magneetblad *n*,
magneetfoelie
d Magnettonfolie *f*,
Magnettonplatte *f*

2962 MAGNETIC SOUND aud
f son *m* magnétique
e sonido *m* magnético
i suono *m* magnetico
n magnetisch geluid *n*
d Magnetton *m*

2963 MAGNETIC SOUND METHOD vr
f système *m* de magnétophone
e sistema *m* de magnetófono
i sistema *m* di magnetofono
n magnetische opneemmethode
d Magnettonverfahren *n*

2964 MAGNETIC SOUND TRACK, MAGNETIC TRACK aud/vr
f piste *f* magnétique de son
e pista *f* magnética de sonido
i pista *f* magnetica di suono
n magnetisch geluidsspoor *n*
d magnetische Tonspur *f*

MAGNETIC SOUND TRACK DIMENSIONS
see: DIMENSIONS OF MAGNETIC SOUND TRACKS

2965 MAGNETIC TACHOMETER aud/vr
f tachymètre *m* magnétique
e taquímetro *m* magnético
i tachimetro *m* magnetico
n magnetische tachymeter
d magnetisches Tachymeter *n*

2966 MAGNETIC TAPE vr
Flexible plastic tape coated on one side with a dispersed magnetic material.
f bande *f* magnétique,
ruban *m* magnétique
e cinta *f* magnética
i nastro *m* magnetico
n magneetband,
magnetische band
d Magnetband *n*,
Magnettonband *n*,
Tonband *n*

2967 MAGNETIC TAPE ARCHIVE vr
f archives *pl* de bandes magnétiques
e archivo *m* de cintas magnéticas
i archivio *m* di nastri magnetici
n magneetbandarchief *n*
d Bandarchiv *n*,
Magnetbandarchiv *n*

2968 MAGNETIC TAPE CONVERTER, MAGNETIC TAPE STANDARDS CONVERTER aud/vr
The conversion of one standard to another one by means of a magnetic tape, i.e. a video tape.
f convertisseur *m* de système à bande magnétique
e convertidor *m* di sistema de cinta magnética
i convertitore *m* di sistema a nastro magnetico
n systeemomzetter met behulp van magneetband
d Systemumsetzer *m* mit Hilfe von Magnetband

2969 MAGNETIC TAPE EDITING vr
f montage *m* de bande magnétique
e montaje *m* de cinta magnética
i montaggio *m* di nastro magnetico
n magneetbandmontage
d Bandschnitt *m*,
Magnetbandschnitt *m*

2970 MAGNETIC TAPE LEADER vr
f amorce *f* de bande magnétique
e cinta *f* de guía,
cinta *f* de prolongación delantera
i coda *f* iniziale di nastro magnetico
n magneetbandbeginstrook
d Vorband *n*,
Vorspannband *n*

2971 MAGNETIC TAPE NOISE dis/vr
f bruit *m* de bande
e ruido *m* de cinta magnética
i rumore *m* di nastro magnetico
n magneetbandruis
d Bandrauschen *n*

2972 MAGNETIC TAPE RECORDER vr
A magnetic recorder that uses magnetic tape as the recording medium.
f appareil *m* à bande magnétique
e aparato *m* de cinta magnética
i apparecchio *m* a nastro magnetico
n geluidsbandapparaat *n*
d Tonbandgerät *n*

2973 MAGNETIC TAPE SLITTING, SLITTING OF MAGNETIC TAPE aud/vr
f découpage *m* de bandes magnétiques
e corte *m* de cintas magnéticas
i taglio *m* di nastri magnetici
n op maat snijden *n* van magnetische banden
d Schneiden *n* von Magnetbändern

MAGNETIC TAPE SPLICING
see: JOINING MAGNETIC TAPE

2974 MAGNETIC TAPE TENSION vr
f tension *f* de bande magnétique
e tensión *f* de tracción de cinta magnética
i tensione *f* di trazione di nastro magnetico
n magneetbandtrekspanning
d Bandzug *m*,
Magnetbandzugspannung *f*

2975 MAGNETIC TAPE VELOCITY vr
f vitesse *f* de bande magnétique
e velocidad *f* de cinta magnética
i velocità *f* di nastro magnetico
n magneetbandsnelheid
d Bandgeschwindigkeit *f*

2976 MAGNETIC TRANSFER au/fi/tv
f report *m* optique
e conversión *f* de registro óptico en registro magnético
i conversione *f* di registrazione in registrazione magnetica
n omzetten *n* van optische in magnetische registratie
d Überspielung *f* von Licht- auf Magnetton

2977 MAGNETIC TRANSFER CHARACTERISTIC aud
f caractéristique *f* de transfert magnétique
e característica *f* de transferencia magnética
i caratteristica *f* di trasferimento magnetico
n magnetische overdrachtskarakteristiek
d magnetische Übergangskennlinie *f*

2978 MAGNETICALLY CONFINED ELECTRON BEAM crt
f faisceau *m* électronique à convergence magnétique
e haz *m* electrónico de convergencia magnética
i fascio *m* elettronico a convergenza magnetica
n magnetisch geconvergeerde bundel
d magnetisch konvergierter Strahl *m*

2979 MAGNETIZATION vr
f magnétisation *f*
e magnetización *f*
i magnetizzazione *f*
n magnetisatie,
magnetisering
d Magnetisierung *f*

2980 MAGNETOMETRY aud
f magnétométrie *f*
e magnetometría *f*
i magnetometria *f*
n magnetometrie
d Magnetometrie *f*

2981 MAGNETOPHONE CONNECTION aud/vr
f prise *f* magnétophone
e conexión *f* para magnetófono
i presa *f* per magnetofono
n bandrecorderaansluiting
d Magnettongerätanschluss *m*

2982 MAIN DISPATCHING CENTER(RE) tv
f centre *m* de modulation
e centro *m* de modulación
i centro *m* di modulazione
n schakelcentrum *n* voor beeld en geluid
d Bild- und Tonschaltraum *m*

2983 MAIN DISTRIBUTION FRAME stu
f centre *m* de commutation
e centro *m* de conmutación
i centro *m* di commutazione
n schakelcentrum *n*
d Hauptschaltraum *m*,
Schaltzentrale *f*

2984 MAIN STUDIO CENTER(RE), MASTER STUDIO stu
f studio *m* principal
e estudio *m* principal
i studio *m* principale
n hoofdstudio
d Hauptstudio *n*

2985 MAIN TITLE, TITLE rep/tv
Title which gives the name of a TV program(me).
f titre *m*
e título *m*
i titolo *m*
n titel
d Titel *m*

2986 MAINS FLUCTUATIONS, MAINS ONDULATIONS dis
f variations *pl* de la tension de réseau
e variaciones *pl* de la tensión de red
i fluttuazioni *pl* della tensione di rete
n netschommelingen *pl*
d Netzschwankungen *pl*

MAINS HOLD
see: LOCKING

MAINS LOCKING (GB)
see: LINE LOCK

2987 MAINS SWITCH rep/tv
f interrupteur *m* secteur
e interruptor *m* de línea
i interruttore *m* di rete
n netschakelaar
d Netzschalter *m*

2988 MAINS VOLTAGE cpl
f tension *f* secteur
e tensión *f* de red
i tensione *f* di rete
n netspanning
d Netzspannung *f*

2989 MAJOR BROADCASTS tv
Broadcasts of the main items of a TV program(me).
f émissions *pl* grand écran,
programme *m* principal
e espacios *pl* principales del programa,
partes *pl* principales del programa
i centri *pl* principali
n hoofdprogramma *n*
d Hauptprogramm *n*

MAKE-BEFORE-BREAK SWITCH
see: GAPPING SWITCH

MAKE-BEFORE-BREAK SWITCHING
see: GAP SWITCHING

2990 MAKE-UP fi/th/tv
f maquillage *m*
e maquillaje *m*
i truccatura *f* del viso,
trucco *m*
n make-up,
schmink
d Maske *f*,
Schminke *f*

2991 MAKE-UP ARTIST,
MAKE-UP SUPERVISOR fi/stu
f maquilleur *m*
e maquillador *m*
i truccatore *m*
n grimeur
d Maskenbildner *m*,
Schminkmeister *m*

2992 MAN-MADE NOISE dis
Interference produced by various man-made sources.
f parasites *pl* industriels
e parásitos *pl* industriales
i parassiti *pl* industriali
n vonkstoringen *pl*
d industrielle Störgeräusche *pl*,
Industriestörungen *pl*

2993 MANUAL EDITING vr
f montage *m* manuel
e montaje *m* manual
i montaggio *m* meccanico
n mechanisch monteren *n*
d mechanischer Bandschnitt *m*

2994 MARGINAL MAGNETIC TRACK vr
f piste *f* couchée en marge
e pista *f* magnética marginal
i pista *f* magnetica marginale
n magnetisch randspoor *n*
d Magnetrandspur *f*

MARRIED OPERATION
see: AUDIO-FOLLOW-VIDEO OPERATION

2995 MARRIED SOUND rec/rep/tv
f mélange *m* image-son
e mezclado *m* imagen-sonido
i mescolanza *f* immagine-suono
n beeldgeluidsmenging
d Bild-Tonmischung *f*

2996 MASER cpl
An extremely stable low-noise amplifier or oscillator, operating by the interaction between radiation and atomic particles.
f maser *m*
e maser *m*
i maser *m*
n maser
d Maser *m*

2997 MASK opt
A piece of opaque paper used to cover any part of a negative, lantern-slide or print which it is desired to obscure.
f cache *f*
e desvanecedor *m*
i mascherino *m*
n masker *n*,
vignet *n*
d Abdeckung *f*,
Kasch *m*,
Vignette *f*

2998 MASK rep/tv
A frame used in front of a TV picture tube to conceal the rounded edges of the screen.
f masque *f*
e máscara *f*
i maschera *f*
n masker *n*
d Maske *f*

2999 MASK IMAGE SIGNAL ctv
f signal *m* de masquage d'image
e señal *f* de enmascaramiento de imagen
i segnale *m* di mascheramento d'immagine
n beeldmaskeringssignaal *n*
d Bildmaskierungssignal *n*

MASK LINE
see: FRAME BAR

3000 MASKING rep/tv
A special effect in which part of the picture signal is suppressed electronically with a view of its substitution by another picture signal.
f découpage *m* électronique
e enmascaramiento *m*
i intarsio *m* elettronico
n partiële beeldmaskering,
schablooneffect *n*
d Einmischung *f* eines Silhouettensignals

MASKING
see: AUDIO MASKING

MASKING
see: ELECTRICAL MASKING

3001 MASKING AUDIOGRAM dis
A graphical representation of the masking due to a stated noise.
f audiogramme *m* de masquage
e audiograma *m* de enmascaramiento
i audiogramma *m* di mascheramento
n maskeringsaudiogram *n*
d Verdeckungsaudiogramm *n*

3002 MASKING PLATE tv
f plaque *f* de masquage
e placa *f* de enmascaramiento
i placca *f* di mascheramento
n maskeerplaat
d Blendenfolie *f*,
Maskierplatte *f*,
Rasterblende *f*

3003 MASKING SOUND aud
f son *m* de masquage
e sonido *m* de enmascaramiento
i suono *m* di mascheramento
n maskeringsgeluid *n*
d Verdeckungsschall *m*

3004 MASKING TENSION dis
f tension *f* de masquage
e tensión *f* de enmascaramiento
i tensione *f* di mascheramento
n maskeringsspanning
d Verdeckungsspannung *f*

3005 MASS-TONE,
SELF-TONE ct
The colo(u)r by reflected light of a bulk of undiluted pigment.
f timbre *m* propre
e matiz *f* propia
i tinta *f* propria
n eigen tint
d Eigenfarbton *m*

3006 MASTER,
MASTER COPY vr
The first tape ready to be used for transmission.
f programme *m* prêt à diffuser
e montaje *m* original
i copia *f* di trasmissione
n programmaband
d Sendeband *n*

3007 MASTER BLACK CONTROL ctv
A single control for adjusting the picture black.
f dispositif *m* de réglage du noir
e regulador *m* del negro
i regolatore *m* del nero
n zwartregelaar
d Schwarzregler *m*

3008 MASTER BRIGHTNESS CONTROL ctv
A variable resistor that adjusts the grid bias on all guns of a three-gun colo(u)r picture tube simultaneously.
f dispositif *m* de réglage commun de la luminosité
e regulador *m* común de la luminosidad
i regolatore *m* collettivo della luminosità
n gemeenschappelijke helderheidsregelaar
d gemeinsamer Helligkeitsregler *m*

MASTER CONTROL
see: CENTRAL CONTROL

MASTER CONTROL AUTOMATION
see: AUTOMATED MASTER SWITCHING

MASTER CONTROL DESK
see: CENTRAL CONTROL DESK

3009 MASTER CONTROL OPERATOR rep/tv
Operator of the controls which switch inputs from studios, videotape recorders, telecine, etc., to one or more outputs.
f technicien *m* de la salle de réglage principal
e operador *m* del control principal
i controllore *m* principale per la selezione in uscita dei vari segnali d'ingresso
n hoofdregelkamertechnicus
d Endkontrolltechniker *m*

MASTER COPY
see: MAGNETIC MASTER

3010 MASTER FADER rep/tv
A device located on the vision mixer control panel and operating on its own fading amplifier which forms part of a following line clamp or stabilizing amplifier.
f atténuateur *m* principal
e atenuador *m* principal,
debilitador *m* principal
i attenuatore *m* principale
n hoofdverzwakker
d Hauptabschwächer *m*

3011 MASTER GAIN CONTROL cpl/ctv
A combined control for adjusting the gain.
f réglage *m* principal du gain
e control *m* principal de ganancia
i controllo *m* principale di guadagno
n hoofdversterkingsregeling
d Hauptverstärkungsregelung *f*

3012 MASTER LIFT CONTROL ctv
Control for adjusting the overall black level of a colo(u)r display.
f régleur *m* principal du niveau du noir
e regulador *m* principal del nivel del negro
i regolatore *m* principale del livello del nero
n hoofdzwartregelaar
d Hauptschwarzwertsteller *m*

3013 MASTER MONITOR,
ON-THE-AIR MONITOR (US),
TRANSMISSION MONITOR (GB) rep/tv
Final monitor controlling the image which is being broadcast.
f moniteur *m* d'émission,
moniteur *m* final
e monitor *m* de emisión,
monitor *m* principal
i monitore *m* di trasmissione
n eindcontroleapparaat *n*,
zendermonitor
d Endkontrollgerät *n*,
Sendermonitor *m*

3014 MASTER OSCILLATOR tv
f maître-oscillateur *m*,
oscillateur-pilote *m*
e oscilador *m* maestro,
oscilador *m* principal
i oscillatore *m* principale
n hoofdzender
d Hauptsender *m*

3015 MASTER PICTURE MONITOR rep/tv
A precision monitor placed at a keypoint in the control system and providing the operator with his main source of information.

f moniteur *m* principal d'image
e monitor *m* principal de imagen
i monitore *m* principale d'immagine
n hoofdbeeldmonitor
d Hauptbildmonitor *m*

MASTER PROGRAM(ME) CONTROL ROOM
see: CENTRAL PROGRAM(ME) CONTROL ROOM

3016 MASTER SOUND TRACK aud/vr
f piste *f* originale de son
e pista *f* original de sonido
i pista *f* originale di suono
n origineel geluidsspoor *n*
d Mustertonspur *f*

MASTER STUDIO
see: MAIN STUDIO CENTER(RE)

3017 MASTER SWITCHING CENTER(RE) rec/tv
f salle *f* centrale de commutation
e sala *f* central de conmutación
i sala *f* centrale di commutazione
n hoofdschakelcentrum *n*
d Hauptschaltraum *m*

3018 MASTER TRANSMITTER, PARENT STATION tv
f émetteur *m* pilote
e emisor *m* piloto
i emettitore *m* pilota
n stuurzender
d Muttersender *m*, Steuersender *m*

3019 MATCHING OF A TELEVISION AERIAL CABLE, MATCHING OF A TELEVISION ANTENNA CABLE aec
f impédance *f* terminale d'un câble de télévision
e impedancia *f* terminal de un cable de televisión
i impedenza *f* terminale d'un cavo di televisione
n eindimpedantie van een televisiekabel
d Abschluss *m* eines Fernsehkabels

3020 MATCHING STIMULI ct
The defined stimuli of an additive colorimeter.
f stimuli *pl* d'adaptation
e estímulos *pl* de adaptación
i stimoli *pl* d'adattamento
n aanpassingsstimuli *pl*
d Anpassungsreize *pl*

3021 MATRIX ctv
An array of coefficients symbolic of a colo(u)r co-ordinate transformation.
f matrice *f*
e matriz *f*
i matrice *f*
n matrix
d Matrize *f*

3022 MATRIX (TO) ctv
To perform a colo(u)r co-ordinate transformation by electrical optical or other means.
f matricer v
e representar v por matrices
i rappresentare v per matrici
n matriceren v
d matrizieren v

3023 MATRIX CIRCUIT, MATRIX UNIT, MATRIXER ctv
In colo(u)r TV, a device which performs a colo(u)r co-ordinate transformation by electrical, or optical, or other means.
f dispositif *m* de matriçage
e unidad *f* matricial
i unità *f* di matrici
n matricator
d Matrizenschaltung *f*

3024 MATRIX SIZE tv
f dimensions *pl* de matrice
e dimensiones *pl* de matriz
i dimensioni *pl* di matrice
n matrixafmetingen *pl*
d Matrizenabmessungen *pl*

3025 MATRIXING ctv
In colo(u)r TV, the NTSC process of combining and decombining the three primary signals representing one luminance and two chrominance components and vice versa.
f matrixation *f*
e representación *f* por matrices
i matrizzazione *f*
n matriceren *n*
d Matrizierung *f*

3026 MAVAR, PARAMETRIC AMPLIFIER cpl
An amplifier used in amplifying weak satellite signals in which energy is fed from a pumping oscillator into signal through a varying reactance, made to act as a negative resistance by the pump signal.
f amplificateur *m* paramétrique
e amplificador *m* paramétrico
i amplificatore *m* parametrico
n parametrische versterker
d parametrischer Verstärker *m*

3027 MAXIMUM AVERAGE POWER OUTPUT tv
The maximum radio-frequency output power that can occur under any combination of signals transmitted, averaged over the longest repetitive modulation cycle.
f puissance *f* maximale moyenne de sortie
e potencia *f* máxima media de salida
i potenza *f* massima media d'uscita
n maximaal gemiddeld uitgangsvermogen *n*
d maximale mittlere Ausgangsleistung *f*

3028 MAXIMUM CARRIER LEVEL cpl
f niveau *m* maximal de porteuse

e nivel *m* máximo de portadora
i livello *m* massimo di portante
n maximaal draaggolfniveau *n*
d maximaler Trägerpegel *m*

3029 MAXIMUM CONTRAST svs
f contraste *m* maximal
e contraste *m* máximo
i contrasto *m* massimo
n maximaal contrast *n*
d grösster Kontrast *m*

3030 MAXIMUM DEFLECTION ANGLE crt
f angle *m* maximal de déviation
e ángulo *m* máximo de desviación
i angolo *m* massimo di deflessione, angolo *m* massimo di deviazione
n maximale afbuighoek
d maximaler Ablenkwinkel *m*

3031 MAXIMUM DEFLECTION SPEED OF LIGHT SPOT crt
f vitesse *f* maximale de déviation du spot lumineux
e velocidad *f* máxima de desviación del punto luminoso
i velocità *f* massima di deviazione del punto luminoso
n maximale afbuigsnelheid van de lichtstip
d höchste Ablenkgeschwindigkeit *f* des Leuchtflecks

3032 MAXIMUM OUTPUT LEVEL cpl
f niveau *m* maximal de sortie
e nivel *m* máximo de salida
i livello *m* massimo d'uscita
n maximaal uitgangsniveau *n*
d höchstzulässiger Ausgangspegel *m*

3033 MAXIMUM PERMISSIBLE NOISE LEVEL dis
f niveau *m* de bruit admissible maximal
e nivel *m* de ruido permisible máximo
i livello *m* di rumore permissibile massimo
n maximaal toelaatbaar ruisniveau *n*
d höchstzulässiger Störpegel *m*

3034 MAXIMUM SATURATION svs
f saturation *f* maximale
e saturación *f* máxima
i saturazione *f* massima
n maximale verzadiging
d grösste Sättigung *f*

MAXIMUM TOLERABLE NOISE
see: JUST TOLERABLE NOISE

MAXWELL TRIANGLE
see: COLO(U)R TRIANGLE

3035 MEAN PICTURE LEVEL tv
The mean d.c. level of the video signal.
f niveau *m* moyen du signal vidéo
e nivel *m* medio de la señal video
i livello *m* medio del segnale video
n gemiddeld niveau *n* van het beeldsignaal
d mittlerer Bildsignalpegel *m*

3036 MEASURING POINTS svs
f points *pl* de mesure
e puntos *pl* de medida
i punti *pl* di misura
n meetpunten *pl*
d Messpunkte *pl*

3037 MEASURING SIGNAL, TEST SIGNAL ge
f signal *m* de mesure
e señal *f* de medida
i segnale *m* di misura
n meetsignaal *n*
d Messignal *n*

3038 MEASURING SIGNAL GENERATOR, TEST SIGNAL GENERATOR ge
f générateur *m* de signal de mesure
e generador *m* de señal de medida
i generatore *m* di segnale di misura
n meetsignaalgenerator
d Messignalgenerator *m*

3039 MECHANICAL CENT(E)RING crt/rep
f centrage *m* mécanique
e centrado *m* mecánico
i centratura *m* meccanica
n mechanische centrering
d mechanische Zentrierung *f*

3040 MECHANICAL ERRORS vr
In colo(u)r videotape recording occurring errors e.g. by the positioning of the playback head with respect to the tape resulting in a timing error.
f erreurs *pl* dues au mécanisme
e errores *pl* debidos al mecanismo
i errori *pl* dovuti al meccanismo
n door het mechanisme veroorzaakte fouten *pl*
d mechanisch verursachte Fehler *pl*

3041 MECHANICAL FLYING SPOT SCANNER rec/rep/tv
f analyseur *m* mécanique à spot lumineux
e explorador *m* mecánico de punto luminoso
i analizzatore *m* meccanico a punto luminoso
n mechanische lichtstipaftaster
d mechanischer Lichtpunktabtaster *m*

3042 MECHANICAL SCANNING tv
A system using a beam of light controlled e.g. by a rotating mirror or a rotating scanning disk.
f analyse *f* mécanique
e exploración *f* mecánica
i analisi *f* meccanica
n mechanische aftasting
d mechanische Abtastung *f*

3043 MEDICAL TELEVISION med/tv
f télévision *f* médicale
e televisión *f* médica
i televisione *f* medica
n medische televisie, televisie voor medische doeleinden
d Fernsehen *n* für medizinische Zwecke

MEDIUM CLOSE SHOT,
MEDIUM SHOT
see: CLOSE MEDIUM SHOT

3044 MEDIUM CLOSE-UP opt
f plan *m* rapproché
e medio primer plano *m*
i mezzo primo piano *m*
n half-grootopname
d Halbnah *f*

3045 MEDIUM LONG SHOT rec/tv
f plan *m* demigénéral
e campo *m* semigeneral, plano *m* medio
i campo *m* medio
n half-totaalopname
d Halbtotale *f*

3046 MEDIUM-POWER VIDEO TRANSMITTER tv
f émetteur *m* vidéo à puissance moyenne
e emisor *m* video de potencia media
i emettitore *m* video a potenza media
n beeldzender van gemiddeld vermogen
d Bildsender *m* mittlerer Leistung

3047 MEDIUM SHOT rec/tv
f plan *m* moyen
e plano *m* medio
i mezza figura *f*
n opname van gemiddelde afstand
d Mittelaufnahme *f*

3048 MERIDIONAL FOCAL LINE opt
The focal line at right angles to the plane passing through the principal axis of the lens and the axis of the beam of light.
f ligne *f* focale tangentielle
e línea *f* focal meridional
i linea *f* focale meridionale
n meridionale brandlijn
d meridionale Brennlinie *f*

3049 MESH EFFECT tv
A type of blemish on the screen of the image orthicon and vidicon tubes which have a target mesh electrode and produce a visible mesh pattern under specified circumstances.
f effet *m* de maille
e efecto *m* de malla
i effetto *m* di maglia
n roostereffect *n*
d Gittereffekt *m*

3050 MESOPIC VISION ct
Vision intermediate between photopic and scotopic vision.
f vision *f* mésopique
e visión *f* mesópica
i visione *f* mesopica
n mesopisch zien *n*
d Dämmerungssehen *n*, mesopisches Sehen *n*, Übergangssehen *n*

3051 METAL BACKING, METAL COATING crt
f revêtement *m* métallique
e recubrimiento *m* metálico
i ricoprimento *m* metallico
n metaalbekleding
d Metallhinterlegung *f*, Metallschicht *f*

3052 METAL-BACKING SCREEN, METAL(L)IZED SCREEN crt
f écran *m* métallisé
e pantalla *f* metalizada
i schermo *m* metallizzato
n gemetalliseerd scherm *n*
d metallisierter Schirm *m*

3053 METAL CONE crt
f cône *m* métallique
e cono *m* metálico
i cono *m* metallico
n metaalconus
d Metallkonus *m*

3054 METAL CONE TUBE crt
A TV picture tube having a metal cone between the glass faceplate and the glass neck of the tube.
f tube *m* image à cône métallique
e tubo *m* imagen de cono metálico
i tubo *m* immagine a cono metallico
n beeldbuis met metaalconus
d Bildröhre *f* mit Konusteil aus Metall

3055 METAMERIC COLO(U)R STIMULI ct
Spectrally different radiations that produce the same colo(u)r under the same viewing conditions.
f stimuli *pl* métamères de couleurs
e estímulos *pl* metámeros de colores
i stimoli *pl* metameri di colori
n metamere kleurstimuli *pl*
d metamere Farbreize *pl*

3056 METAMERIC MATCH ct
Visual equivalence of physically different colo(u)rs.
f équivalence *f* métamère
e equivalencia *f* metámera
i equivalenza *f* metamera
n metamere kleurgelijkheid
d metamere Farbgleichheit *f*

3057 METAMERIC OBJECTS ct
Objects having different spectrophotometric curves that match if illuminated with some particular quality of light.
f objets *pl* métamères
e objetos *pl* metámeros
i oggetti *pl* metameri
n metamere voorwerpen *pl*
d metamere Gegenstände *pl*, metamere Objekte *pl*

3058 METAMERS ct
Radiations producing the same visual effect but having different spectral compositions.
f métamères *pl*
e metámeros *pl*
i metameri *pl*
n metameren *pl*
d Metamere *pl*

3059 MICROPHONE aud
f microphone *m*
e micrófono *m*
i microfono *m*
n microfoon
d Mikrophon *n*

3060 MICROPHONE DIRECTIVITY aud
f directivité *f* de microphone
e directividad *f* de micrófono
i direttività *f* di microfono
n richtvermogen *n* van een microfoon
d Mikrophonrichtvermögen *n*

3061 MICROPHONE NOISE aud/dis
f bruit *m* de microphone
e ruido *m* de micrófono
i rumore *m* di microfono
n microfoonruis
d Mikrofonrauschen *n*

3062 MICROPHONE OUTPUT aud
f puissance *f* de sortie d'un microphone
e potencia *f* de salida de un micrófono
i potenza *f* d'uscita d'un microfono
n uitgangsvermogen *n* van een microfoon
d Mikrophonrauschen *n*

3063 MICROPHONE OUTPUT TERMINALS aud
f bornes *pl* de sortie de microphone
e bornes *pl* de salida de micrófono
i morsetti *pl* d'uscita di microfono
n uitvoerpennen *pl* van een microfoon
d Ausfuhrklemmen *pl* eines Mikrophons

3064 MICROPHONE PLACEMENT, MICROPHONE POSITIONING aud/stu
f répartition *f* des microphones
e repartición *f* de los micrófonos
i ripartizione *f* dei microfoni
n microfoonopstelling
d Mikrophonanordnung *f*

3065 MICROWAVE AERIAL ASSEMBLY, MICROWAVE ANTENNA ASSEMBLY aea
f réseau *m* d'antennes microondes
e sistema *m* de antenas microondas
i sistema *m* d'antenne microonde
n microgolfantennestelsel *n*
d Mikrowellenantennenanordnung *f*

3066 MICROWAVE LINK tv
A radio system using the 3-30 GHz band, and used for point-to-point TV transmission.
f faisceau *m* hertzien,
liaison *f* microondes,
voie *f* microondes
e canal *m* microondas,
vía *f* microondas
i canale *m* microonde
n microgolfkanaal *n*
d Dezistrecke *f*,
Mikrowellenkanal *m*

3067 MICROWAVE TELEVISION RECEIVER rep/tv
f téléviseur *m* à microondes
e televisor *m* de microondas
i televisore *m* a microonde
n microgolftelevisieontvanger
d Mikrowellenfernsehgerät *n*

3068 MID-RANGE LIGHT VALUE ct
f valeur *f* de lumière à demiportée
e valor *m* de luz de medio alcance
i valore *m* di luce a media portata
n lichtwaarde in gebiedsmidden
d Lichtwert *m* in Bereichsmitte

3069 MIDDLE-KEY PICTURE rep/tv
f image *f* normale
e imagen *f* normal
i immagine *f* normale
n normaal beeld *n*
d normales Bild *n*

3070 MILLER INTEGRATOR tv
Used in TV in combination with a transitron to obtain a sawtooth-oscillator having very special properties.
f intégrateur *m* de Miller
e integrador *m* de Miller
i integratore *m* di Miller
n millerintegrator
d Miller-Integrator *m*

3071 MINIMUM BRIGHTNESS svs
f luminosité *f* minimale
e luminosidad *f* mínima
i luminosità *f* minima
n minimale helderheid
d Mindesthelligkeit *f*

3072 MINIMUM CARRIER LEVEL cp
f niveau *m* minimal de porteuse
e nivel *m* mínimo de portadora
i livello *m* minimo di portante
n minimumniveau *n* van de draaggolf
d niedrigster Trägerpegel *m*

MINIMUM PERCEPTIBLE DIFFERENCE
see: JUST PERCEPTIBLE DIFFERENCE

MINIMUM PERCEPTIBLE NOISE
see: JUST PERCEPTIBLE NOISE

3073 MINOR BROADCASTS tv
Broadcasts of minor importance in the TV program(me).
f émissions *pl* petit écran,
programme *m* de complément
e complementos *pl* del programa,
espacios *pl* secundarios del programa
i centri *pl* di radiodiffusione secondari
n bijprogramma *n*
d Nebenprogramm *n*

3074 MINUS BLUE, YELLOW ct
A colo(u)r in which the spectral components are confined mainly to the 500-700 nm band of the spectrum, i.e. the 400-500 nm (blue) band is missing.
f jaune,
moins bleu
e amarillo,
menos azul

i giallo,
meno blu
n geel,
minus blauw
d gelb,
minus blau

3075 MINUS COLO(U)RS ct
Colo(u)rs in which only the spectral components associated with the colo(u)r named are not present to any substantial extent, e.g. a minus red.
f couleurs *pl* moins
e colores *pl* menos
i colori *pl* meno
n minuskleuren *pl*
d Minusfarben *pl*

MINUS GREEN
see: MAGENTA

MINUS RED
see: CYAN

3076 MIRROR aea
A reflecting surface used as a component in an aerial (antenna).
f miroir *m*,
réflecteur *m*
e espejo *m*,
reflector *m*
i riflettore *m*,
specchio *m*
n reflector,
spiegel
d Reflektor *m*,
Spiegel *m*

3077 MIRROR DRUM tv
f roue *f* à miroirs,
roue *f* de Weiler,
tambour *m* à miroirs
e rueda *f* de espejos,
rueda *f* de Weiler,
tambor *m* de espejos
i ruota *f* di specchi,
ruota *f* di Weiler,
tamburo *m* a specchi
n rad *n* van Weiler,
spiegeltrommel
d Spiegeltrommel *f*,
Weiler-Rad *n*

MIRROR GRID
see: GRID OF MIRRORS

MIRROR REFLECTION
see: DIRECT REFLECTION

3078 MIRROR SCREW tv
f hélice *f* à miroirs
e tornillo *m* de espejos
i specchio *m* elicoidale,
spirale *f* di specchi
n spiegelschroef
d Spiegelschraube *f*

3079 MIRROR SHUTTER, opt
REFLEX SHUTTER
f obturateur *m* à miroir
e obturador *m* de espejo
i otturatore *m* a specchio
n reflexsluiter
d Spiegelblende *f*,
Spiegelreflexblende *f*

3080 MISCELLANEOUS INTERFERENCES, cpl/rep/tv
SIGNAL CHANNEL MISCELLANEOUS INTERFERENCES
Any extraneous voltage present in a signal channel, found in both audio signal and picture signal TV channels.
f interférences *pl* miscellanées,
interférences *pl* variées
e interferencias *pl* diversas,
interferencias *pl* varias
i interferenze *pl* diverse
n verschillende storingen *pl*
d verschiedene Störungen *pl*

3081 MISREGISTRATION, ctv
REGISTRATION FAULT
The presence of wrong colo(u)r or colo(u)rs in a picture tube due to some faults in the tube, in components directly associated with the tube, or in some of the adjustments.
f défaut *m* de calage,
fausse superposition *f*
e registro *m* falso,
superposición *f* errónea
i registrazione *f* errata,
sovrapposizione *f* erronea
n foute superpositie,
foutieve registratie
d Fehlüberdeckung *f*

3082 MIX (TO) aud/tv
In sound broadcasting, to combine the outputs of several sound sources at the desired relative volumes.
In TV broadcasting, to superimpose a picture fade-in over fade-out of equal duration.
f mélanger v
e mezclar v
i mescolare v
n mengen v
d mischen v

3083 MIXED BLANKING SIGNALS rec/tv
f signaux *pl* de suppression mélangés,
suppression *f* mélangée
e señales *pl* de supresión mezcladas
i segnali *pl* di cancellazione mescolati
n gemengde onderdrukkingssignalen *pl*
d Austastgemisch *n*,
gemischte Austastsignale *pl*

3084 MIXED HIGHS ctv
Those high-frequency components of the picture signal which are intended to be reproduced achromatically in a colo(u)r picture.
f hautes fréquences *pl* mixtes
e altas frecuencias *pl* mezcladas
i mescolanza *f* delle componenti d'alta frequenza

n gemengde hoge frequenties *pl*
d gemischte Höhen *pl*

3085 MIXED SIGNAL rec/tv
f mélange *m* de signaux,
signal *m* mixte
e mezclado *m* de señales,
señal *f* mezclada
i mescolanza *f* di segnali,
segnale *m* mescolato
n gemengd signaal *n*,
signaalmengsel *n*
d gemischtes Signal *n*,
Signalgemisch *n*

3086 MIXED SYNCS tv
A synchronizing signal consisting of line sync pulses and field sync pulses but containing no other information.
f signaux *pl* de synchronisation mixtes
e señales *pl* de sincronización mezcladas
i segnali *pl* di sincronizzazione mescolati
n gemengde synchronisatiesignalen *pl*
d gemischte Synchronisierungssignale *pl*

3087 MIXER rep/tv
An apparatus by means of which the outputs of several channels can be faded up and down independently, selected individually, or combined at any desired relative volumes.
f mélangeur *m* image/son
e mezclador *m* imagen/sonido
i mescolatore *m* di segnali
n beeld/geluidmenger
d Bild/Schallmischer *m*

3088 MIXER, rep/tv
MIXING CONSOLE
A desk grouping within the reach of a single operator all the controls required for mixing.
f pupitre *m* de mélange
e pupitre *m* de control,
pupitre *m* mezclador
i banco *m* di controllo centrale,
tavolo *m* di mescolanza
n menglessenaar,
mengtafel
d Kontrollpult *n*,
Mischpult *n*

3089 MIXER AMPLIFIER aud/cpl
f amplificateur *m* de mélange
e amplificador *m* de mezclado
i amplificatore *m* di mescolatura
n mengversterker
d Mischverstärker *m*

MIXES
see: DISSOLVE

3090 MIXING cpl/tv
The process of superimposing the outputs of two or more sound or video program(me) sources to produce desired effects.
f mélange *m*
e mezcla *f*
i mescolanza *f*
n menging
d Mischung *f*

3091 MIXING EQUIPMENT rec/rep/tv
f installation *f* de mélange
e instalación *f* de mezcla
i stabilimento *m* di mescolanza
n menginstallatie
d Mischanlage *f*

3092 MIXING POINT, cpl/tv
MIXING STAGE
A point in a control system where an output is obtained from two independent inputs.
f étage *m* de mélange
e etapa *f* de mezcla
i stadio *m* di mescolanza
n mengtrap
d Mischstufe *f*

3093 MOBILE BACKGROUND ani
f fond *m* mobile
e fondo *m* móvil
i fondo *m* mobile
n verplaatsbare achtergrond
d fahrbarer Hintergrund *m*

MOBILE CAMERA
see: CREEPIE PEEPIE

3094 MOBILE CONTROL ROOM, rec/tv
OB VEHICLE
f car *m* régie
e carro *m* de control
i carro *m* di regia
n regiewagen
d Regiewagen *m*

3095 MOBILE EQUIPMENT ctv/tv/vr
Equipment that is readily transportable and operable in transit.
f équipement *m* mobile
e equipo *m* móvil
i apparecchiatura *f* mobile
n verplaatsbare uitrusting
d fahrbare Ausrüstung *f*

3096 MOBILE GENERATOR rec/tv
f générateur *m* mobile
e generador *m* móvil
i generatore *m* mobile
n verplaatsbare generator
d fahrbarer Generator *m*

3097 MOBILE LINKS CONTROL rec/tv
ROOM
f salle *f* de contrôle pour les connexions mobiles
e sala *f* de control para las conexiones móviles
i sala *f* di controllo per le connessioni mobili
n controlekamer voor de verplaatsbare verbindingen
d Kontrollraum *m* für die fahrbaren Verbindungen

3098 MOBILE MOUNTING UNIT tv
f plateforme *f* à roulettes

e plataforma *f* móvil
i piattaforma *f* mobile
n verplaatsbaar platform *n*
d fahrbarer Untersatz *m*

3099 MOBILE RECORDING UNIT fi/tv
f car *m* de reportage
e carro *m* de televisión
i carro *m* per riprese
n reportagewagen
d Aufnahmewagen *m*,
Fernsehaufnahmewagen *m*

3100 MOBILE SOUND RECORDING UNIT rec/tv
f enregistreur *m* de son mobile
e registrador *m* de sonido móvil
i registratore *m* di suono mobile
n verplaatsbaar geluidsopnameapparaat *n*
d fahrbares Magnettongerät *n*

3101 MOBILE TRANSMISSION EQUIPMENT rec/tv
f équipement *m* de transmission mobile
e equipo *m* de transmisión móvil
i apparecchiatura *f* di trasmissione mobile
n verplaatsbare zendinstallatie
d fahrbare Sendeanlage *f*

3102 MOBILE VIDEOTAPE RECORDER vr
f magnétoscope *m* mobile
e magnetoscopio *m* móvil
i magnetoscopio *m* mobile
n verplaatsbare beeldbandopnemer
d fahrbares Videogerät *n*,
MAZ-Wagen *m*

3103 MODEL,
MODEL SHOT opt
f maquette *f*
e maqueta *f*
i abbozzo *m*, modello *m*
n maquette
d Modell *n*

3104 MODEL(L)ING LIGHT stu
f projecteur *m* de l'ambiance
e proyector *m* del ambiente
i proiettore *m* dell'ambiente
n omgevingsprojector
d Umgebungsprojektor *m*

3105 MODULATE (TO) ge
1. To vary the amplitude, frequency, or phase of a radio or electric wave by impressing one wave (signal) on another wave of constant properties (carrier).
2. To vary the velocity of the electrons in an electron beam.
f moduler v
e modular v
i modulare v
n moduleren v
d modeln v, modulieren v

3106 MODULATED CARRIER ge
A radio-frequency carrier, some characteristics have been varied in accordance with the intelligence that is to be transmitted.
f porteuse *f* modulée
e portadora *f* modulada
i portante *f* modulata
n gemoduleerde draaggolf
d modulierter Träger *m*

3107 MODULATED STAGE cpl/ge
f étage *m* modulé
e etapa *f* modulada
i stadio *m* modulato
n gemoduleerde trap
d modulierte Stufe *f*

3108 MODULATING WAVE cpl
f onde *f* modulante
e onda *f* moduladora
i onda *f* modulante
n modulerende golf
d modulierende Welle *f*

3109 MODULATION ge
f modulation *f*
e modulación *f*
i modulazione *f*
n modulatie
d Modelung *f*, Modulation *f*

3110 MODULATION BRIDGE cpl/tv
f modulateur *m* en pont,
pont *m* de modulation
e modulador *m* en puente,
puente *m* de modulación
i modulatore *m* a ponte,
ponte *m* di modulazione
n modulatiebrug
d Modulationsbrücke *f*

3111 MODULATION DEPTH,
MODULATION FACTOR cpl
The fractional ratio between the difference and sum of the numerical values of the largest and smallest amplitude encountered in one cycle of modulation.
f facteur *m* de modulation,
profondeur *f* de modulation
e grado *m* de modulación,
profundidad *f* de modulación
i grado *m* di modulazione,
profondità *f* di modulazione
n modulatiediepte,
modulatiefactor
d Modulationsfaktor *m*,
Modulationsgrad *m*,
Modulationstiefe *f*

3112 MODULATION DISTORTION cpl
f distorsion *f* de modulation
e distorsión *f* de modulación
i distorsione *f* di modulazione
n modulatievervorming
d Modulationsklirrfaktor *m*

3113 MODULATION ELECTRODE,
MODULATION GRID crt
Cylindrical grid for controlling the intensity of the electron beam in a cathode-ray tube.

f électrode *f* de commande,
grille *f* de modulation
e electrodo *m* de mando,
rejilla *f* de modulación
i elettrodo *m* di comando,
griglia *f* di modulazione
n modulatie-elektrode,
stuurrooster *n*
d Steuerelektrode *f*,
Steuerzylinder *m*

3114 MODULATION ENVELOPE cpl
f enveloppe *f* de la modulation
e envolvente *m* de la modulación
i inviluppo *m* della modulazione
n omhullende van de modulatie
d Modulationshüllkurve *f*,
Modulationsumhüllende *f*,
Modulationsverlauf *m*

3115 MODULATION HUM cpl/dis
f ronflement *m* de modulation
e zumbido *m* de modulación
i ronzio *m* di modulazione
n modulatiebrom
d Modulationsbrumm *m*

3116 MODULATION INDEX cpl
The ratio of the frequency deviation to the rated system deviation of a frequency modulated wave.
f indice *m* de modulation
e índice *m* de modulación
i indice *m* di modulazione
n modulatie-index
d Modulationsindex *m*

MODULATION INDICATOR
see: INDICATOR

3117 MODULATION LEVEL cpl
f niveau *m* de modulation
e nivel *m* de modulación
i livello *m* di modulazione
n modulatieniveau *n*
d Modulationspegel *m*

3118 MODULATION LEVEL METER cpl
f appareil *m* de mesure du niveau de modulation
e medidor *m* del nivel de modulación
i dispositivo *m* di misura del livello di modulazione
n modulatieniveaumeter
d Modulationsaussteuerungsmesser *m*

3119 MODULATION NOISE, cpl/dis
NOISE BEHIND THE SIGNAL
Noise in a modulated carrier in excess of the signal but varying with it.
f bruit *m* de modulation
e ruido *m* de modulación
i rumore *m* di modulazione
n modulatieruis
d Modulationsrauschen *n*

3120 MODULATION PERCENTAGE, cpl
PERCENTAGE MODULATION
The modulation factor expressed as a percentage.
f pourcentage *m* de modulation
e porcentaje *m* de modulación
i percentuale *f* di modulazione
n modulatiepercentage *n*
d Modulationsgrad *m* in Prozent

3121 MODULATION SIGNAL, tv
PROGRAM(ME) SIGNAL
One of the electrical signals representing a program(me), used to modulate the transmitter.
f signal *m* de modulation,
signal *m* modulant
e señal *f* de programa,
señal *f* moduladora
i segnale *m* di modulazione,
segnale *m* modulante
n modulatiesignaal *n*,
programmasignaal *n*
d Modulationssignal *n*

3122 MODULATION STANDARD cpl
f norme *f* de modulation
e norma *f* de modulación
i norma *f* di modulazione
n modulatienorm
d Modulationsnorm *f*

3123 MODULATION TRANSFER cpl
FUNCTION
f fonction *f* de transfert de modulation
e función *f* de transferencia de modulación
i funzione *f* di trasferimento di modulazione
n overdrachtsfunctie van de modulatie
d Modulationsübertragungsfunktion *f*

3124 MODULATION TRIANGLE stu
f triangle *m* de modulation
e triángulo *m* de modulación
i triangolo *m* di modulazione
n modulatiedriehoek
d Modulationsdreieck *n*

3125 MODULATION VOLTAGE, crt
MODULATOR VOLTAGE
Of a cathode-ray tube, the value of steady voltage applied to the modulator to produce a specified intensity of light or other radiation from the screen.
f tension *f* de modulation
e tensión *f* de modulación
i tensione *f* di modulazione
n modulatiespanning
d Modulationsspannung *f*

3126 MOIRÉ, dis/rep/tv
MOIRÉ PATTERN,
WATERED-SILKS EFFECT
The spurious pattern in the reproduced picture resulting from interference beats between two sets of periodic structures in the image.
f moirage *m*, moirure *f*
e moiré *m*
i marezzatura *f*
n moiré *n*
d Moiré *n*, Riffelmuster *n*

3127 MONITOR ge
f dispositif *m* de contrôle automatique, moniteur *m*
e monitor *m*
i dispositivo *m* automatico per la verifica della qualità, monitore *m*
n monitor
d Kontrollempfänger *m*, Monitor *m*, Normempfänger *m*, Warngerät *n*

3128 MONITOR FILMING med/tv
The making of cineradiographic films from a monitor in medical TV.
f filmage *m* d'images du monitor
e filmación *f* de imágenes del monitor
i il filmare *m* d'immagini del monitore
n filmen *n* van monitorbeelden
d Filmen *n* von Monitorbildern

3129 MONITOR HEAD vr
An additional playback head provided on tape recorders to permit playing back the recorded sounds off the tape while the recording is being made.
f tête *f* de reproduction de contrôle
e cabeza *f* de reproducción de control
i testina *f* di riproduzione di controllo
n terugspeelkop
d Rückspielkopf *m*

3130 MONITOR SIZE stu
f dimensions *pl* du moniteur
e dimensiones *pl* del monitor
i dimensioni *pl* del monitore
n afmetingen *pl* van de monitor
d Monitorabmessungen *pl*

3131 MONITOR TUBE crt
f tube *m* controleur d'images, tube *m* moniteur
e tubo *m* de control de imágenes, tubo *m* monitor
i tubo *m* di controllo d'immagini, tubo *m* monitore
n beeldmonitor
d Bildkontrollröhre *f*

3132 MONITORING ge
Observation of the characteristics of transmitted signals.
f contrôle *m* automatique
e control *m* automático
i verifica *f* automatica
n controleren *n*
d Kontrollieren *n*

3133 MONITORING AERIAL, MONITORING ANTENNA aea
f antenne *f* du moniteur de réception
e antena *f* del monitor de recepción
i antenna *f* del monitore ricevente
n antenne van de controleontvanger
d Antenne *f* des Kontrollempfängers

3134 MONITORING OF TRANSMISSION LEVEL cpl/ge
f contrôle *m* de modulation
e control *m* de profundidad de modulación
i controllo *m* del livello di trasmissione
n modulatiediepteregeling
d Aussteuerungskontrolle *f*

3135 MONITORING PICTURE rep/tv
f image *f* de contrôle
e imagen *f* de control
i immagine *f* di controllo
n controlebeeld *n*
d Kontrollbild *n*

3136 MONITORING ROOM aud/rec/rep/tv
f salle *f* des moniteurs
e sala *f* de los monitores
i sala *f* dei monitori
n monitorkamer
d Monitorraum *m*

MONKEY CHATTER
see: ADJACENT CHANNEL INTERFERENCE

3137 MONO-ACCELERATION CATHODE-RAY TUBE crt
An electrostatic cathode-ray tube in which all acceleration of the electron beam is produced before the beam passes through the deflection electrodes.
f tube *m* cathodique à préaccélération
e tubo *m* catódico de preaceleración
i tubo *m* catodico a preaccelerazione
n katodestraalbuis met voorversnelling
d Katodenstrahlröhre *f* mit Vorbeschleunigung

3138 MONOCHROMATISM ct
Form of colo(u)r blindness.
f monochromatisme *m*
e monocromatismo *m*
i monocromatismo *m*
n monochromatisme *n*
d Monochromatismus *m*

3139 MONOCHROME BANDWIDTH, MONOCHROME CHANNEL BANDWIDTH tv
The bandwidth of the path intended to carry the monochrome signal.
f largeur *f* de bande du signal monochrome
e anchura *f* de banda de la señal monocroma
i larghezza *f* di banda del segnale monocromo
n bandbreedte van het monochrome signaal
d Bandbreite *f* des Monochromsignals

MONOCHROME CHANNEL
see: BLACK-AND-WHITE CHANNEL

3140 MONOCHROME INFORMATION tv
f information *f* monochrome
e información *f* monocroma
i informazione *f* monocroma
n monochrome informatie
d Monochrominformation *f*

MONOCHROME PICTURE
see: BLACK-AND-WHITE PICTURE

3141 MONOCHROME RECEIVER rep/tv
f récepteur *m* monochrome
e receptor *m* monocromo
i ricevitore *m* monocromo
n monochroomontvanger
d Monochromempfänger *m*

3142 MONOCHROME REPRODUCTION rep/tv
f reproduction *f* monochrome
e reproducción *f* monocroma
i riproduzione *f* monocroma
n monochrome weergave
d monochrome Wiedergabe *f*

3143 MONOCHROME SIGNAL tv
A signal wave for controlling the luminance values in the picture.
f signal *m* monochrome
e señal *f* monocroma
i segnale *m* monocromo
n monochroomsignaal *n*, zwart-witsignaal *n*
d Schwarz-Weiss-Signal *n*

3144 MONOCHROME SIGNAL BANDWIDTH tv
In TV, the video bandwidth required to transmit a monochrome signal.
f largeur *f* de bande du signal monochrome
e anchura *f* de banda de la señal monocroma
i larghezza *f* di banda del segnale monocromo
n bandbreedte van het monochroomsignaal
d Bandbreite *f* des Schwarz-Weiss-Signals

MONOCHROME TELEVISION
see: BLACK-AND-WHITE TELEVISION

MONOCHROME TRANSMISSION
see: BLACK-AND-WHITE TRANSMISSION

3145 MONOSCOPE, MONOTRON, PHASMAJECTOR, VIDEOTRON crt
An electron beam tube capable of providing a picture signal derived from a given image on an electrode inside the tube.
f monoscope *m*
e monoscopio *m*
i monoscopio *m*
n monoscoop
d Monoskop *n*

3146 MONOSTABLE CIRCUIT cpl
A circuit with only one stable state of operation.
f circuit *m* monostable
e circuito *m* monoestable
i circuito *m* monostabile
n monostabiele kring
d monostabiler Kreis *m*

3147 MONTAGE (US), MOUNTING (GB) tv
In TV, the juxtaposition or partial superposition of a number of individual scenes to form the complete picture.
f montage *m*
e montaje *m*
i montaggio *m*
n montage
d Montage *f*

3148 MOP-UP WAVEFORM CORRECTOR dis
Amplitude and phase response corrector installed in a system to cater for slight inaccuracy of correction of individual items of equipment, the error of each of which may be negligible.
f postcorrection *f* de la forme d'onde
e postcorrección *f* de la forma de onda
i postcorrezione *f* della forma d'onda
n nacorrectie van de golfvorm
d Nachkorrektur *f* der Wellenform

MOSAIC
see: ICONOSCOPE MOSAIC

3149 MOSAIC COAT crt
f couche *f* en mosaïque
e capa *f* en mosaica
i strato *m* in mosaico
n rasterlaag
d gerasterte Schicht *f*

3150 MOSAIC ELECTRODE crt
f électrode *f* à mosaïque
e electrodo *m* de mosaico
i elettrodo *m* a mosaico
n mozaïekelektrode
d Mosaikelektrode *f*

3151 MOSAIC SCREEN crt
f écran *m* en mosaïque
e pantalla *f* en mosaico
i schermo *m* in mosaico
n rasterscherm *n*
d Rasterschirm *m*

MOTION PICTURE PICKUP
see: FILM SCANNER

MOTION PICTURE PICKUP
see: FILM SCANNING

3152 MOTORIZED PAN AND TILT HEAD rep/tv
f tourelle *f* universelle motorisée
e cabeza *f* motorizada para movimiento panorámico y de cabeceo
i piattaforma *f* doppia motorizzata
n door motor aangedreven panoramakop
d Schwenkkopf *m* mit Motorantrieb

3153 MOUNTING PLAN cpl
Instructions accompanying component parts of TV sets and aerials (antennae).
f instructions *pl* de montage
e instrucciones *pl* de montaje
i istruzioni *pl* di montaggio
n montage-instructies *pl*
d Montageanweisungen *pl*

MOVABLE SCENE (GB)
see: GETAWAY

3154 MOVING COIL, aud
SPEECH COIL (GB),
VOICE COIL (US)
f bobine *f* mobile
e bobina *f* móvil
i bobina *f* mobile
n spreekspoel
d Schwingspule *f*

3155 MOVING MATTE, tv
OVERLAY,
SELF-KEYED INSERTION
An electronic method of combining in one picture a foreground taken from one source and a background from another source, the unwanted parts of each portion of the picture being suppressed by masking.
f procédé *m* de transparence électronique
e superposición *f* electrónica
i sovrapposizione *f* elettronica
n overlappen *n*
d Overlay-Trickmischung *f*

MUG SHOT (US)
see: CLOSE SHOT

3156 MULTIBAND AERIAL, aea
MULTIBAND ANTENNA
f antenne *f* multibande
e antena *f* multibanda
i antenna *f* multibanda
n antenne voor meer banden,
brede-bandantenne
d Mehrbandantenne *f*,
Mehrbereichantenne *f*

3157 MULTIBURST ctv
Video test signal having each line divided into a number of trains of specified frequency, the mean level being set in the region of mid-grey.
f salve *f* multiple
e señal *f* múltiple de sincronización de color
i salva *f* multipla,
segnale *m* di prova televisivo composto da pacchi di sinusoidi di differenti frequenze
n meervoudig salvo *n*
d Mehrfachburst *m*

3158 MULTICHANNEL MIXER ctv
f mélangeur *m* de plusieurs canaux
e mezclador *m* de múltiples canales
i mescolatore *m* di più canali
n meerkanalenmenger
d Mehrkanalenmischer *m*

MULTICHANNEL SELECTOR
see: CHANNEL SELECTOR

3159 MULTICHANNEL TELEVISION tv
f télévision *f* à plusieurs canaux
e televisión *f* de múltiples canales
i televisione *f* a più canali
n meerkanaaltelevisie
d Mehrkanalfernsehen *n*

3160 MULTIGUN CATHODE-RAY crt
TUBE,
MULTIPLE GUN CATHODE-RAY
TUBE
A cathode-ray tube containing two or more separate electron beam systems.
f tube *m* cathodique à plusieurs canons
e tubo *m* catódico de múltiples cañones
i tubo *m* catodico a più cannoni
n elektronenstraalbuis met meer kanonnen
d Mehrstrahlelektronenröhre *f*

3161 MULTILATERAL PROGRAM(ME)S tv
Program(me)s originating from many sources and distributed to many destinations both national and foreign.
f programmes *pl* multilatéraux
e programas *pl* multilaterales
i programmi *pl* televisivi formati
con segnali d'immagine provenienti da differenti sorgenti
n multilaterale schakelprogramma's *pl*
d multilaterale Programme *pl*

3162 MULTILAYER PHOSPHOR crt
SCREEN
f écran *m* luminescent à plusieurs couches
e pantalla *f* luminiscente de múltiples capas
i schermo *m* luminescente a multipli strati
n meerlagig beeldscherm *n*
d Mehrschichtenbildschirm *m*

MULTIPATH EFFECT
see: DOUBLE IMAGE

3163 MULTIPATH SIGNALS tv
Signals reflected from the ionosphere back to earth by various combinations of paths.
f signaux *pl* à plusieurs voies
e señales *pl* de múltiples vías
i segnali *pl* a più canali
n signalen *pl* met verschillende terugweg
d mehrwegige Signale *pl*

3164 MULTIPLE ACCESS SATELLITE tv
f satellite *m* à accès multiple
e satélite *m* de acceso múltiple
i satellite *m* ad accesso multiplo
n satelliet met meervoudige toegang
d Satellit *m* für vielfachen Zugang

3165 MULTIPLE AERIAL, aea
MULTIPLE ANTENNA
f antenne *f* multiple
e antena *f* múltiple
i antenna *f* multipla
n meervoudige antenne
d Mehrfachantenne *f*,
Strahlergruppe *f*

3166 MULTIPLE AVERAGING ctv
f détermination *f* multiple de valeurs moyennes
e determinación *f* múltiple de valores medios
i determinazione *f* multipla di valori medi

n meervoudig middelen *n*
d mehrfache Mittelwertbestimmung *f*

3167 MULTIPLE INTERLACED SCANNING rec/tv
f analyse *f* à intercalage multiple
e exploración *f* entrecalada múltiple
i analisi *f* interlacciata multipla
n meervoudige interliniëring
d Mehrfachzeilensprungverfahren *n*

3168 MULTIPLE SCANNING tv
f analyse *f* multiple
e exploración *f* múltiple
i analisi *f* multipla
n meervoudige aftasting
d Mehrfachabtastung *f*

3169 MULTIPLE SOUND TRACK aud/vr
f enregistrement *m* sonore à plusieurs pistes
e registro *m* sonoro de múltiples pistas
i registrazione *f* sonora a più piste
n meersporige geluidsregistratie
d Mehrfachtonspur *f*

3170 MULTIPLE TECHNIQUES ani
A combination of animation with live-action photography by means of superimposition, split screen arrangement and back projection.
f techniques *pl* d'animation combinées
e técnicas *pl* combinadas
i tecniche *pl* combinate
n gecombineerde technieken *pl*
d kombinierte Techniken *pl*

3171 MULTIPLE TRANSMISSION rep/tv
f transmission *f* en multiplex
e transmisión *f* de multidifusión, transmisión *f* múltiplex
i trasmissione *f* multiplex
n multiplexoverdracht
d Multiplexübertragung *f*

3172 MULTIPLEXER opt
An optical device for combining pictures from several sources into a common generating channel.
f multiplexeur *m*
e multiplexor *m*
i apparato *m* ottico per la combinazione di differenti immagini provenienti da differenti sorgenti
n multiplexer
d Multiplexer *m*

3173 MULTIPLEXING SOUND ON VISION rec/tv
Achieved by frequency modulating the sound signal on to a subcarrier whose frequency is above the upper limit of the video signal but within the pass-band of the link equipment.
f multiplexage *m* du son dans le canal vidéo
e transmisión *f* en múltiplex de sonido-imagen,
multiplaje *m* de sonido-imagen
i inserzione *f* del suono nel canale video
n onderbrengen *n* van het geluid in het videokanaal
d Toneinführung *f* ins Videokanal

3174 MULTIPLICATION POINT, MULTIPLICATION STAGE cpl/tv
A mixing point (stage) where the two independent inputs are multiplied.
f point *m* de mixage à multiplication
e punto *m* de mezclado de multiplicación
i punto *m* di mescolanza a moltiplicazione
n vermenigvuldigingsmengpunt *n*
d Vervielfachungsmischpunkt *m*

3175 MULTIPLICATIVE MIXING rep/tv
f mixage *m* à multiplication
e mezclado *m* de multiplicación
i mescolanza *f* a moltiplicazione
n vermenigvuldigingsmenging
d Vervielfachungsmischung *f*

3176 MULTIPLIER PHOTOTUBE, PHOTOMULTIPLIER crt
A sensitive detector of light in which the initial electric current derived from photoelectric emission is amplified by successive stages secondary electron emission.
f tube *m* photomultiplicateur
e fototubo *m* multiplicador
i tubo *m* moltiplicatore fotoelettronico
n multiplicatorfotobuis
d Photoelektronenvervielfacher *m*

3177 MULTIPURPOSE STUDIO stu
f studio *m* à plusieurs buts
e estudio *m* de varios usos
i studio *m* ad usi multipli
n studio voor verschillende doeleinden
d Mehrzweckstudio *n*

3178 MULTISPIRAL tv
f hélice *f* multiple
e espiral *f* múltiple
i elice *f* multipla
n meervoudige spiraal
d Mehrfachspirale *f*

3179 MULTISPIRAL SCANNING DISK rec/tv
f disque *m* d'analyse à hélice multiple
e disco *m* de exploración de espiral múltiple
i disco *m* analizzatore ad elice multipla
n meervoudige-spiraalschijf
d Mehrfachspiralscheibe *f*

3180 MULTISTANDARD SYSTEM tv
A piece of equipment capable of working on more than one TV system of line and field scanning rates.
f système *m* de télévision à plusieurs nombres de lignes
e sistema *m* televisivo de múltiples números de líneas
i sistema *m* televisivo a più numeri di linee
n televisiesysteem *n* voor verschillende lijngetallen
d Fernsehsystem *n* für verschiedene Zeilenzahlen

3181 MULTITRACK TAPE aud
f bande *f* multipiste,
bande *f* sonore à plusieurs pistes
e cinta *f* sonora de múltiples pistas
i nastro *m* sonoro a più piste
n meersporenband
d Mehrspurband *n*

3182 MULTITRACK TAPE RECORDER aud/vr
f enregistreur *m* à plusieurs pistes
e registrador *m* de múltiples pistas
i registratore *m* a più piste
n meersporenopneemapparaat *n*
d Mehrspurgerät *n*

3183 MULTIVIBRATOR cpl
A form of relaxation oscillator which comprises two stages, so coupled that the input of each one is derived from the output of the other.
f basculateur *m*,
multivibrateur *m*
e multivibrador *m*
i basculatore *m*,
multivibratore *m*
n multivibrator
d Multivibrator *m*

MUNSELL CHROMA
see: CHROMA

3184 MUNSELL SYSTEM ct
f système *m* de Munsell
e sistema *m* de Munsell
i sistema *m* di Munsell
n munsellsysteem *n*
d Munsell-System *n*

3185 MUNSELL VALUE ct
f valeur *f* de Munsell
e valor *m* de Munsell
i valore *m* di Munsell
n munsellwaarde
d Munsell-Wert *m*

3186 MUSH dis/rep/tv
Fading and distortion in reception in an area through the interaction of waves from two or more synchronized transmitters.
f brouillage *m*, distorsion *f*
e distorsión *f*,
perturbación *f*
i confusione *f*,
disturbo *m*
n interferentie,
verwarring
d Interferenz *f*,
Verwirrung *f*

3187 MUSH AREA rep/tv
A region in which fading and distortion occur owing to interaction between two or more synchronized transmitters.
f zone *f* de brouillage,
zone *f* de distorsion
e zona *f* de distorsión,
zona *f* de perturbación
i zona *f* di confusione,
zona *f* di disturbo
n interferentiegebied *n*,
verwarringsgebied *n*
d Interferenzgebiet *n*,
Verwirrungsgebiet *n*

3188 MUSIC DESK aud/stu
The desk of the music producer in a concert hall charged with the sound pickup of the orchestra, i.e. the correct placing of the microphones, the mixing of the microphone signals and the control of level.
f pupitre *m* musicien
e pupitre *m* de control músico
i tavolo *m* di controllo musico
n muziekcontrolelessenaar
d Musikkontrollpult *n*

3189 MUSIC LINE, aud
PROGRAM(ME) LINE
In sound broadcasting, a circuit for carrying the modulating signals.
f câble *m* de modulation,
ligne *f* de modulation
e línea *f* de programa
i linea *f* musicale
n muzieklijn,
programmalijn
d Tonleitung *f*

3190 MUSIC STUDIO stu
f studio *m* de musique
e estudio *m* de música
i studio *m* di musica
n muziekstudio
d Musikstudio *n*

3191 MUTUAL CONDUCTANCE (GB), cpl
SLOPE,
TRANSCONDUCTANCE (US)
Gradient of a device characteristic, e.g. the incident light versus signal current characteristic of a camera pickup tube.
f pente *f*
e conductancia *f* mutua,
pendiente *f*
i conduttanza *f* mutua,
pendenza *f*
n steilheid
d Steilheit *f*

N

3192 NARROW-ANGLE LENS opt
f objectif *m* à angle étroit
e objetivo *m* de ángulo estrecho
i obiettivo *m* ad angolo stretto
n objectief *n* met een kleine beeldhoek
d Objektiv *n* mit kleinem Bildwinkel

3193 NARROW BAND cpl
Contrasting term to broad band, used in TV transmission channels and multi-channel communication systems.
f bande *f* étroite
e banda *f* angosta,
banda *f* estrecha
i banda *f* stretta
n smalle band
d Schmalband *n*

3194 NARROW-BAND AXIS,
Q AXIS ctv
In a system of the NTSC type, the chrominance axis along which the chrominance component corresponding to the coarse chrominance primary is plotted.
f axe *m* de la primaire transmise à bande étroite,
axe *m* Q
e eje *m* de banda angosta
i asse *m* del primario trasmesso a banda stretta,
asse *m* di banda stretta
n Q-as,
smalle-bandas
d Q-Achse *f*,
Schmalbandachse *f*

3195 NARROW-BANDPASS FILTER cpl
f filtre *m* à bande étroite
e filtro *m* de banda angosta
i filtro *m* a banda stretta
n smalle-bandfilter *n*
d Filter *n* geringer Durchlassbreite,
Schmalbandfilter *n*

3196 NARROW-BANDWIDTH
INTERPOLATION cpl
The separation of wanted and unwanted information by a simple low-pass filter in standards conversion.
f interpolation *f* d'information par filtre passe-bas
e interpolación *f* de información por filtro paso bajo
i interpolazione *f* d'informazione per filtro passo-basso
n informatie-interpolatie door laagdoorlatend filter
d Informationsinterpolation *f*. durch Tiefpassfilter

3197 NARROW BEAM crt
f faisceau *m* filiforme
e haz *m* filiforme
i fascio *m* filiforme
n smalle bundel
d schmaler Strahl *m*

3198 NARROW-CUT FILTER ct/ctv
A colo(u)r filter which transmits a limited group of wavelengths only, absorbing all others, so that the result is a pure saturated colo(u)r.
f filtre *m* chromatique à bande étroite
e filtro *m* cromático de banda angosta
i filtro *m* cromatico a banda stretta
n een klein aantal golflengten doorlatend kleurenfilter *n*
d eine geringe Zahl von Wellenlängen durchlassendes Farbfilter *n*

3199 NATURAL FREQUENCY aea
Of an aerial (antenna) the lowest frequency of free oscillation.
f fréquence *f* propre
e frecuencia *f* propia
i frequenza *f* propria
n eigenfrequentie
d Eigenfrequenz *f*

3200 NATURAL PERIOD aea
The reciprocal of the natural frequency.
f période *f* propre
e período *m* propio
i periodo *m* proprio
n eigenperiode
d Eigenperiode *f*

3201 NATURAL WAVELENGTH aea
The fundamental wavelength at which an open aerial (antenna) will oscillate by virtue of its distributed inductance and capacitance and sustain a standing wave of current.
f longueur *f* d'onde propre
e longitud *f* de onda propia
i lunghezza *f* d'onda propria
n eigengolflengte
d Eigenwellenlänge *f*

NEAR FIELD
see: FRESNEL REGION

NEAR REGION,
NEAR ZONE
see: INDUCTION ZONE

3202 NECK,
NECK OF THE TUBE crt
Of a cathode-ray tube the tubular part of the envelope which contains the electron gun.
f col *m* du tube
e cuello *m* del tubo
i collo *m* del tubo
n buissteel, steel
d Hals *m*, Röhrenhals *m*

3203 NECK REGION crt
f zone *f* du col
e zona *f* del cuello
i zona *f* del collo
n steelgebied *n*
d Halsgebiet *n*

3204 NECK SHADOW crt
f ombre *f* du col
e sombra *f* del cuello
i ombra *f* del collo
n steelschaduw
d Halsschatten *m*

NECKLACE MICROPHONE
see: CHEST MICROPHONE

3205 NEEDLE-SHAPED PARTICLE vr
f particule *f* aciculaire
e partícula *f* acicular
i particella *f* acicolare
n naaldvormig deeltje *n*
d nadelförmiges Teilchen *n*

3206 NEGATIVE AMPLITUDE MODULATION, cpl/ge/tv
NEGATIVE LIGHT MODULATION,
NEGATIVE PICTURE MODULATION
A form of modulation in which the amplitude of the vision signal decreases with increasing luminosity of the picture elements.
f modulation *f* d'amplitude négative
e modulación *f* de amplitud negativa
i modulazione *f* d'ampiezza negativa
n negatieve amplitudemodulatie
d negative Amplitudenmodulation *f*

3207 NEGATIVE COLO(U)RS ct
f couleurs *pl* négatives
e colores *pl* negativos
i colori *pl* negativi
n negatieve kleuren *pl*
d negative Farben *pl*

3208 NEGATIVE ECHO, dis
NEGATIVE GHOST IMAGE
A ghost image in which white is black and vice versa, due to particular phase and amplitude relations between the direct path and reflection path signals.
f image *f* fantôme négative
e imagen *f* fantasma negativa
i immagine *f* fantasma negativa
n negatief echobeeld *n*
d negatives Echobild *n*

3209 NEGATIVE FREQUENCY MODULATION cpl
A form of modulation in which the frequency of the video signal decreases with increasing brightness of the picture elements.
f modulation *f* de fréquence négative
e modulación *f* de frecuencia negativa
i modulazione *f* di frequenza negativa
n negatieve frequentiemodulatie
d negative Frequenzmodulation *f*

3210 NEGATIVE GOING SYNC PULSES tv
f impulsions *pl* de synchronisation à potentiel négatif
e impulsos *pl* de sincronización de potencial negativo
i impulsi *pl* di sincronizzazione a potenziale negativo
n synchronisatie-impulsen *pl* met negatieve potentiaal
d Synchronisationsimpulse *pl* mit negativem Potential

3211 NEGATIVE IMAGE, dis
NEGATIVE PICTURE,
REVERSED IMAGE
Negative reproduction of a TV picture with blacks and whites interchanged.
f image *f* négative
e imagen *f* negativa
i immagine *f* negativa
n negatief beeld *n*
d negatives Bild *n*

NEGATIVE ION BLEMISH,
see: ION SPOT

3212 NEGATIVE MODULATION cpl
f modulation *f* négative
e modulación *f* negativa
i modulazione *f* negativa
n negatieve modulatie
d Negativmodulation *f*

3213 NEGATIVE PICTURE PHASE, cpl/tv
NEGATIVE PICTURE POLARITY
For a TV signal, the condition in which increases in brightness make the picture signal voltage swing below the zero level in a negative direction.
f polarité *f* négative du signal image
e polaridad *f* negativa de la señal imagen
i polarità *f* negativa del segnale immagine
n negatieve polariteit van het beeldsignaal
d negative Polarität *f* des Bildsignals

3214 NEGATIVE PICTURE SIGNAL, cpl tv
NEGATIVE VIDEO SIGNAL
A signal in which only the picture excursions during the active part of each line must be inverted.
f signal *m* image à potentiel négatif
e señal *f* imagen de potencial negativo
i segnale *m* immagine a potenziale negativo
n negatief beeldsignaal *n*
d negatives Bildsignal *n*

3215 NEGATIVE SCANNING rec/tv
Scanning a photographic negative, with reversal in the circuits, so that the reproduced image is the normal positive.
f analyse *f* d'images négatives
e exploración *f* negativa
i analisi *f* negativa
n negatieve aftasting
d Negativabtastung *f*

3216 NEGATIVE TRANSMISSION tv
f transmission *f* à modulation négative
e transmisión *f* de modulación negativa

i trasmissione *f* a modulazione negativa
n transmissie met negatieve modulatie
d Übertragung *f* mit negativer Modulation

3217 NEGATIVE TRANSMISSION POLARITY tv
f polarité *f* négative de la transmission
e polaridad *f* negativa de la transmisión
i polarità *f* negativa della trasmissione
n negatieve transmissiepolariteit
d negative Übertragungspolarität *f*

NEMO
see: FIELD PICKUP

3218 NET TRANSMISSION EQUIVALENT (US), OVERALL ATTENUATION (GB) cpl/tv
The total attenuation in a transmission equipment due to leakage absorption or radiation.
f équivalent *m* de transmission
e atenuación *f* total, equivalente *m* de transmisión
i attenuazione *f* residua, equivalente *m* di trasmissione
n restdemping
d Restdämpfung *f*

3219 NETWORK ge
1. A plurality of interrelated circuits.
2. A circuit, or part of a circuit, containing a number of branches, which is considered as a unit.
3. A combination of elements.
f réseau *m*
e red *f*
i rete *f*
n netwerk *n*
d Netzwerk *n*

3220 NETWORK CONTROL ROOM tv
f salle *f* de contrôle du réseau
e sala *f* de control de la red
i sala *f* di controllo della rete
n netwerkcontrolekamer
d Netzwerkkontrollraum *m*

3221 NETWORK SWITCHING CENTER(RE) tv
A center(re) which provides an interconnection point between the local channels and studios, program(me) switching center(re)s and transmitters and the main channels to other cities and countries.
f centre *m* de commutation du réseau
e centro *m* de conmutación de la red
i centro *m* di commutazione della rete
n netwerkschakelcentrum *n*
d Netzwerkschaltzentrum *n*

3222 NEUTRAL-DENSITY FACEPLATE crt
A TV picture-tube faceplate in which a neutral density filter has been incorporated to increase picture contrast by attenuating external light reflected from the screen.
f fond *m* à filtre neutre
e fondo *m* de filtro neutro
i piastra *f* frontale a filtro grigio
n frontplaat met neutraal filter
d Schirmträger *m* mit Graufilter

NEUTRAL-DENSITY FILTER
see: GREY FILTER

3223 NEUTRAL-DENSITY GLASS crt
f verre *m* neutre
e vidrio *m* neutro
i vetro *m* neutro
n neutraal glas *n*
d Neutralglas *n*

NEUTRAL WEDGE
see: GREY WEDGE

3224 NEWS CAMERA rec/tv
f caméra *f* d'actualités
e cámara *f* de actualidades
i camera *f* per il telegiornale
n camera voor de nieuwsreportage
d Kamera *f* für den Nachrichtendienst

3225 NEWS COMMUNICATIONS rep
The supplying of the news studio and/or operator with fresh communications as soon as they come in.
f transmission *f* d'actualités
e transmisión *f* de mensajes
i trasmissione *f* d'informazioni, trasmissione *f* di notizie
n nieuwsinformatie
d Nachrichteninformation *f*, Nachrichtenübermittlung *f*

3226 NEWS EDITING, NEWS FILM EDITING rec/tv
f montage *m* du journal filmé
e montaje *m* del diario filmado
i montaggio *m* del telegiornale
n montage van de nieuwsfilm
d Nachrichtenfilmschnitt *m*

3227 NEWS MATERIAL PREPARATION, PREPARATION OF NEWS MATERIAL tv
f préparation *f* du journal filmé
e preparación *f* del diario filmado actualidades
i preparazione *f* del telegiornale
n gereedmaken *n* van de nieuwsfilm
d Anfertigung *f* des Nachrichtenfilms

3228 NEWS NETWORK tv
f réseau *m* de télécommunication d'informations
e red *f* de telecomunicación de informaciones
i rete *f* di telecomunicazione d'informazioni
n nieuwsberichtennetwerk *n*
d Nachrichtennetzwerk *n*

NEWS READER, NEWSCASTER
see: ANNOUNCER

3229 NEWS STUDIO tv
f studio *m* d'information
e estudio *m* del servicio de actualidades

i studio *m* del telegiornale
n nieuwsstudio
d Nachrichtenstudio *n*

3230 NEXT-CHANNEL MIXER rep/tv
A vision mixer in which only two banks A and B are used, each feeding a fading amplifier and thence to the contacts of a common cut relay.
f mélangeur *m* de canaux adjacents
e mezclador *m* de canales adyacentes
i apparato *m* per la mescolanza di due segnali televisivi assegnati su canali differenti
n tweekanaalsmenger
d Zweikanalenmischer *m*

3231 NEXT-CHANNEL SWITCHER, PRESET TYPE SWITCHER rep/tv
A switcher consisting of two identical channels which are used alternately as transmission and preview prelisten outputs.
f commutateur *m* de canaux adjacents
e conmutador *m* de canales adyacentes
i commutatore *m* inserito su due canali d'ingresso per la scelta del segnale di trasmissione
n kanaalomschakelaar
d Kanalumschalter *m*

NIGHT-BLINDNESS
see: HEMERALOPIA

3232 NIPKOW DISK tv
The early scanning disk of mechanical TV invented by P. Nipkow.
f disque *m* de Nipkow
e disco *m* de Nipkow
i disco *m* di Nipkow
n nipkowschijf
d Lochscheibe *f*,
Nipkow-Scheibe *f*,
Spirallochscheibe *f*

3233 NIR COLO(U)R TELEVISION SYSTEM tv
Variant of the SECAM system introduced by the Russians.
f système *m* de télévision couleur NIR
e sistema *m* de televisión en colores NIR
i sistema *m* di televisione a colori NIR
n NIR-kleurentelevisiesysteem *n*
d NIR-Farbfernsehsystem *n*

3234 NOCTOVISION tv
A TV system in which the image to be transmitted is illuminated with infrared instead of visible light.
f télévision *f* nocturne par rayons infrarouges
e televisión *f* nocturna
i televisione *f* notturna
n voor infrarood gevoeligtelevisiesysteem *n*
d infrarotempfindliches Fernsehsystem *n*,
Nachtsehverfahren *n*

3235 NOCTOVISOR tv
f appareil *m* de télévision nocturne par rayons infrarouges
e televisor *m* nocturno
i televisore *m* notturno
n voor infrarood gevoelig televisieapparaat *n*
d Nachtsehgerät *n*

3236 NOCTOVISOR SCAN tv
f analyse *f* avec l'appareil de télévision nocturne
e exploración *f* con el televisor nocturno
i analisi *f* col televisore notturno
n infraroodaftasting
d Infrarotabtastung *f*

3237 NODAL POINTS opt
Two points on the principal axes of a lens such that an incident ray of light directed towards one of them emerges from the lens as if from the other in a direction parallel to that of the incident ray.
f points *pl* nodaux
e puntos *pl* nodales
i punti *pl* nodali
n knooppunten *pl*
d Knotenpunkte *pl*

3238 NODE ge
Of a standing wave, any point, line or surface in a distribution field at which some specified variable attains a zero or minimum magnitude.
f noeud *m*
e nodo *m*
i nodo *m*
n knoop
d Knoten *m*,
Knotenpunkt *m*

3239 NOISE dis/tv
Interference in a communication channel.
f bruit *m*,
parasites *pl*,
perturbation *f*
e parásitos *pl*,
perturbación *f*,
ruido *m*
i disturbo *m*,
parassiti *pl*,
perturbazione *f*,
rumore *m*
n ruis,
storing
d Geräusch *n*,
Rauschen *n*,
Störung *f*

NOISE BEHIND THE SIGNAL
see: MODULATION NOISE

3240 NOISE BLANKING dis
f suppression *f* des parasites
e supresión *f* de los parásitos
i cancellazione *f* dei parassiti
n storingsonderdrukking
d Störaustastung *f*

NOISE-CANCEL(L)ING MICROPHONE
see: CLOSE-TALKING MICROPHONE

3241 NOISE CURRENT dis
That part of a signal current conveying noise power.
f courant *m* du bruit
e corriente *f* del ruido
i corrente *f* del rumore
n ruissignaal *n*
d Rauschstrom *m*

3242 NOISE FACTOR, NOISE FIGURE dis
The ratio of the total noise power per unit bandwidth delivered by a system into an output termination to the portion thereof engendered at the input frequency by the input terminals whose noise temperature is standard.
f facteur *m* de bruit
e factor *m* de ruido
i fattore *m* di rumore
n ruisfactor
d Rauschfaktor *m*, Rauschzahl *f*

3243 NOISE FIELD dis
The electromagnetic radiation field due to all sources which can interfere with reception of the required signal.
f champ *m* perturbateur
e campo *m* perturbador
i campo *m* perturbatore
n storend veld *n*
d Störfeld *n*

3244 NOISE FILTER dis
f filtre *m* antiparasite
e filtro *m* antiparásito
i filtro *m* antiparassita
n antistoringsfilter *n*
d Störschutzfilter *n*

3245 NOISE-FREE EQUIVALENT AMPLIFIER cpl/dis/ge
An ideal amplifier having no internally generated noise that has the same gain and input output characteristics as the actual amplifier.
f amplificateur *m* équivalent libre de parasites
e amplificador *m* equivalente libre de ruidos
i amplificatore *m* equivalente senza disturbi
n storingsvrije equivalente versterker
d rauschfreier äquivalenter Verstärker *m*

3246 NOISE GENERATOR dis
Device for producing a controlled noise for test purposes.
f générateur *m* de bruit
e generador *m* de ruido
i generatore *m* di rumore
n ruisgenerator
d Rauschgenerator *m*

3247 NOISE INTENSITY dis
The intensity of the noise field in a specific frequency band.
f intensité *f* du bruit
e intensidad *f* del ruido
i intensità *f* del rumore
n ruisintensiteit
d Geräuschstärke *f*

3248 NOISE INVERSER, NOISE LIMITER, NOISE SUPPRESSOR dis
f inverseur *m* des parasites, limiteur *m* des parasites
e invertidor *m* de los parásitos, limitador *m* de los parásitos, supresor *m* del ruido
i inversore *m* dei parassiti, limitatore *m* dei disturbi impulsivi
n ruisonderdrukker, storingsonderdrukker
d Störaustaster *m*, Störbegrenzer *m*

3249 NOISE INVERSION, NOISE LIMITING, NOISE SUPPRESSION dis
f limitation *f* des parasites
e limitación *f* de parásitos, supresión *f* de ruidos
i limitazione *f* dei disturbi impulsivi
n ruisonderdrukking
d Störbegrenzung *f*

3250 NOISE LEVEL dis
The noise power density spectrum in the frequency range of interest.
f niveau *m* du bruit
e nivel *m* del ruido
i livello *m* del rumore
n ruisniveau *n*
d Rauschpegel *m*, Störpegel *m*

3251 NOISE MEASUREMENT dis
f mesure *f* du bruit
e medición *f* del ruido
i misura *f* del rumore
n ruismeting
d Rauschmessung *f*

3252 NOISE METER, PSOPHOMETER dis
f mesureur *m* de la tension de bruit, psophomètre *m*
e medidor *m* de la tensión de ruido, sofómetro *m*
i misuratore *m* della tensione di rumore, psofometro *m*
n psofometer, ruisspanningsmeter
d Psophometer *n*, Rauschspannungsmesser *m*

3253 NOISE MODULATION dis
The modulation of a carrier wave with a noise signal.
f modulation *f* par signal de bruit
e modulación *f* por señal de ruido
i modulazione *f* per segnale di rumore
n modulatie door het storend signaal
d Modulation *f* durch das Störsignal

3254 NOISE OUTPUT (GB), NOISE POWER (US) dis

The power in a system by all signals present.
f puissance *f* du bruit
e potencia *f* del ruido
i potenza *f* del rumore
n ruisvermogen *n*
d Rauschleistung *f*

3255 NOISE PEAKS dis
f crêtes *pl* de bruit
e crestas *pl* de ruido
i creste *pl* di rumore
n ruistoppen *pl*
d Rauschspitzen *pl*

3256 NOISE PREVENTION dis
f prévention *f* de bruit
e prevención *f* de ruido
i prevenzione *f* di rumore
n voorkomen *n* van ruis
d Vorbeugen *n* von Rausch

3257 NOISE REDUCTION dis
f réduction *f* du bruit
e reducción *f* del ruido
i riduzione *f* del rumore
n ruisreductie
d Rauschreduktion *f*

3258 NOISE RESISTANCE dis
f résistance *f* de souffle
e resistencia *f* de ruido
i resistenza *f* di rumore
n ruisweerstand
d Rauschwiderstand *m*

3259 NOISE SIGNAL dis
f signal *m* de bruit
e señal *f* de ruido
i segnale *m* di ruido
n stoorsignaal *n*, storend signaal *n*
d Störsignal *n*

3260 NOISE SOURCE dis
The object or system from which the noise originates.
f source *f* de bruit
e fuente *f* de ruido
i sorgente *f* di rumore
n storingsbron
d Störquelle *f*

NOISE SUPPRESSOR
see: INTERCARRIER NOISE SUPPRESSOR

3261 NOISE TEMPERATURE dis
The temperature of a passive system having an available noise power per unit bandwidth equal to that of the actual terminals.
f température *f* de bruit
e temperatura *f* de ruido
i temperatura *f* di rumore
n ruistemperatuur
d Rauschtemperatur *f*

3262 NOISE TRACK dis
f piste *f* bruitée
e pista *f* disturbada
i pista *f* disturbata
n ruisspoor *n*
d Rauschspur *f*

3263 NOISY BLACKS dis
Term used for non-uniformity of the dark areas of a picture due to fluctuations of signal level caused by noise.
f noir *m* perturbé
e negro *m* perturbado
i nero *m* perturbato
n gestoord zwart *n*,
ruis in de donkere delen van het beeld
d gestörtes Schwarz *n*

NOMINAL BLACK SIGNAL
see: ARTIFICIAL BLACK SIGNAL

3264 NOMINAL LINE HEIGHT rec/tv
The average separation between center(re)s of adjacent lines forming a raster.
f hauteur *f* nominale des lignes
e altitud *f* nominal de las líneas
i altezza *f* nominale delle linee
n nominale lijnhoogte
d Nennzeilenhöhe *f*,
Zeilennennhöhe *f*

3265 NOMINAL LINE WIDTH tv
The average separation between center(re)s of adjacent scanning or recording lines.
f largeur *f* nominale de ligne
e anchura *f* nominal de línea
i larghezza *f* nominale di linea
n nominale lijnbreedte
d Nennzeilenbreite *f*,
Zeilennennbreite *f*

NOMINAL WHITE SIGNAL
see: ARTIFICIAL WHITE SIGNAL

3266 NON-ANIMATED PICTURE, STILL rep/tv
f image *f* fixe
e imagen *f* fija
i immagine *f* fissa
n stilstaand fotobeeld *n*
d Standbild *n*,
Stehbild *n*

3267 NON-COMPOSITE COLO(U)R-PICTURE SIGNAL ctv
The electric signal that represents complete colo(u)r picture information including set-up and the colo(u)r burst, but excluding all other synchronizing signals.
f signal *m* incomplet de l'image couleur
e señal *f* de video en colores no compuesta,
señal *f* no completa de la imagen en colores
i segnale *m* non composito dell'immagine a colori
n kleurenvideosignaal *n* zonder synchronisatieimpulsen
d Farbbildsignal *n* ohne Synchronisierungsimpulse

3268 NON-COMPOSITE SIGNAL tv
A picture signal lacking its accompanying (composite) sync signals, but having blanking information.
f signal *m* incomplet de l'image
e señal *f* de video no compuesta, señal *f* no completa de la imagen
i segnale *m* non composito dell'immagine
n beeldsignaal *n*
d unvollständiges Bildsignal *n*

3269 NON-COMPOSITE VIDEO MIXING rec/rep/tv
Mixing process whereby the sync pulses are added to the video component at the output from master control.
f mélange *m* signal vidéo
e mezcla *f* de video no compuesta, mezclado *m* de video no completo
i mescolanza *f* di segnali video non compositi
n beeldsignaalmenging
d Bildsignalmischung *f*

3270 NON-DIRECTIONAL AERIAL (GB), OMNIDIRECTIONAL ANTENNA (US) aea
f antenne *f* omnidirectionnelle
e antena *f* omnidireccional
i antenna *f* omnidirezionale
n rondstralende antenne
d Rundstrahlantenne *f*

3271 NON-DIRECTIONAL MICROPHONE (GB), OMNIDIRECTIONAL MICROPHONE (US) aud
f microphone *m* omnidirectionnel
e micrófono *m* omnidireccional
i microfono *m* omnidirezionale
n ongerichte microfoon
d Allrichtungsmikrophon *n*

3272 NON-LINEAR AMPLIFIER cpl/tv
Amplifier in which the output is not proportional to the input signal.
f amplificateur *m* nonlinéaire
e amplificador *m* no lineal
i amplificatore *m* non lineare
n niet-lineaire versterker
d nichtlinearer Verstärker *m*

3273 NON-LINEAR DISTORTION, NON-LINEARITY DISTORTION dis
That part of the distortion arising in a non-linear system which is due to the non-linearity of the system.
f distorsion *f* nonlinéaire
e distorsión *f* no lineal
i distorsione *f* non lineare
n niet-lineaire vervorming
d nichtlineare Verzerrung *f*

3274 NON-LINEAR DISTORTION FACTOR dis
f facteur *m* de distorsion nonlinéaire
e factor *m* de distorsión no lineal
i fattore *m* di distorsione non lineare
n niet-lineaire vervormingsfactor
d nichtlinearer Verzerrungsfaktor *m*

3275 NON-LINEARITY dis
Lack of proportionality between output and input currents and voltages of a network, amplifier, or transmission line, resulting in distortion of a passing signal.
f nonlinéarité *f*
e falta *f* de linealidad, no linealidad *f*
i nonlinearità *f*
n niet-leariteit
d Nichtlinearität *f*

NON-OBJECT PERCEIVED COLO(U)R
see: APERTURE COLO(U)R

NON-PHYSICAL PRIMARIES
see: FICTITIOUS PRIMARIES

NON-SELECTIVE ABSORBER
see: GREY FILTER

NON-SELECTIVE RADIATOR
see: GREY BODY

3276 NON-SELF-LUMINOUS COLO(U)R, SURFACE COLO(U)R ct
Colo(u)r perceived to belong to a non-self-luminous object.
f couleur *f* superficielle
e color *m* de reflexión
i superficie *f* colorata ma non autoluminosa
n kleur van een oppervlak, oppervlaktekleur
d Körperfarbe *f*, nichtselbstleuchtende Farbe *f*, Oberflächenfarbe *f*

3277 NON-SELF-LUMINOUS OBJECT ct
f objet *m* sans propre luminosité
e objeto *m* no autoluminoso
i oggetto *m* non autoluminoso
n niet-zelflichtend voorwerp *n*
d nichtselbstleuchtendes Ding *n*

3278 NON-STORAGE CAMERA TUBE crt
Camera tube in which output signal depends on incident light intensity, and not on its integration over a defined period of time between scannings.
f tube *m* de prise de vues sans mémoire
e tubo *m* de toma de vistas sin memoria
i tubo *m* di ripresa senza memoria
n geheugenloze opneembuis
d Aufnahmeröhre *f* ohne Speicherung

3279 NON-SYNC SIGNALS rec/tv
Those signals which contain no synchronizing information.
f signaux *pl* sans information de synchronisation
e señales *pl* sin información de sincronización
i segnali *pl* senza informazione di sincronizzazione
n signalen *pl* zonder synchronisatie-informatie
d Signale *pl* ohne Synchronisationsinformation

3280 NON-SYNC SOUND TRACK, WILD TRACK aud
Sound track recorded otherwise than with a synchronized picture.
f piste *f* sonore nonsynchronisée
e pista *f* sonora no sincronizada
i pista *f* sonora non sincronizzata
n niet-gesynchroniseerd geluidsspoor *n*
d nichtsynchronisierte Tonspur *f*

3281 NORMAL COLO(U)R SIGHT ct
f perceptivité *f* normale des couleurs
e perceptibilidad *f* normal de los colores
i percettività *f* normale dei colori
n normaal kleurwaarnemingsvermogen *n*
d Farbnormalsichtigkeit *f*

3282 NORMAL D.C. RESTORATION cpl
f restauration *f* normale de la composante continue
e restauración *f* normal de la componente continua
i restaurazione *f* normale della componente continua
n normaal herstellen *n* van de werkzame nulcomponent
d einfache Schwarzwerthaltung *f*

3283 NORMAL-REVERSE SWITCH rec/rep/tv
A switch which reverses the horizontal and vertical scanning currents through the deflection coils in a TV camera, resulting in a lateral or vertical inversion of the image.
f commutateur *m* inverseur d'image
e conmutador *m* inversor de imagen
i commutatore *m* invertitore d'immagine
n beeldomkeerschakelaar
d Bildumkehrschalter *m*

3284 NORMAL TRICHROMATIC VISION, NORMAL TRICHROMATISM ct
Vision in which the additive mixture of three stimuli is necessary and sufficient for colo(u)r matching and in which ability to discriminate colo(u)rs is normal.
f vision *f* trichromatique normale
e visión *f* tricromática normal
i visione *f* tricromatica normale
n normaal trichromatisch zien *n*
d normales trichromatisches Sehen *n*

3285 NORMALIZING WHITE, WHITE REFERENCE ctv
The chromaticity of the picture produced by a correctly aligned receiver when the amplitude of the chrominance signal is zero.
f blanc *m* de référence
e blanco *m* normalizador
i bianco *m* di riferimento, bianco *m* normalizzato
n referentiewit *n*
d Vergleichsweiss *n*

3286 NOTCH FILTER cpl/tv
A band-rejection filter that produces a sharp notch in the frequency response curve of a system.
f filtre *m* de réjection à flancs raides
e filtro *m* supresor de banda de flancos empinados
i filtro *m* di blocco di banda a ripidità di fronte
n bandstopfilter *n* met steile flanken
d Fallenfilter *n*, Lochfilter *n*

3287 NTSC COLO(U)R TELEVISION SYSTEM ctv
A compatible system of 252-line colo(u)r TV in which a luminance signal is transmitted as amplitude modulation of the vision carrier and two chrominance signal components are transmitted simultaneously as quadrature modulation of a subcarrier.
f système *m* NTSC
e sistema *m* NTSC
i sistema *m* NTSC
n NTSC-systeem *n*
d NTSC-System *n*

3288 NTSC TRIANGLE ct/ctv
On a chromaticity diagram, a triangle which defines the gamut of colo(u)r obtainable through the use of phosphors.
f triangle *m* NTSC
e triángulo *m* NTSC
i triangolo *m* NTSC
n NTSC-driehoek
d NTSC-Dreieck *m*

3289 NUMBER OF ACTIVE LINES rec/tv
f nombre *m* de lignes utiles
e número *m* de líneas útiles
i numero *m* di linee attive
n aantal *n* aftastlijnen
d Abtastzeilenzahl *f*

3290 NUMBER OF FIELDS rec/rep/tv
The number of fields per picture.
f ordre *m* d'entrelacement
e orden *m* del entrelazado
i numero *m* di trame per quadro
n aantal *n* rasters per beeld
d Teilbildzahl *f*

3291 NUMBER OF IMAGE ELEMENTS, NUMBER OF PICTURE ELEMENTS crt
f nombre *m* d'éléments composant l'image
e número *m* de elementos de imagen
i numero *m* d'elementi d'immagine
n aantal *n* beeldelementen
d Bildpunktzahl *f*

3292 NUMBER OF LOOPS crt
The number of maxima in the beam diameter between the electron gun and the target or between a point on the photocathode and the target.
f nombre *m* d'antinodes
e número *m* de antinodos
i numero *m* d'antinodi
n buikgetal *n*
d Bauchzahl *f*

NUMBER OF SCANNING LINES (US)
see: LINE FREQUENCY TO FIELD FREQUENCY RATIO

3293 NUMERICAL APERTURE crt
The square root of the difference of the square of the index of refraction of the fiber(re) core, and the square of the index of refraction of the cladding material.
f ouverture *f* numérique
e abertura *f* numérica
i apertura *f* numerica
n numerieke apertuur
d numerische Apertur *f*

3294 NYQUIST DEMODULATOR cpl
f démodulateur *m* à talon
e desmodulador *m* de talón
i demodulatore *m* di Nyquist
n nyquistdemodulator
d Nyquist-Demodulator *m*

3295 NYQUIST INTERVAL BANDWIDTH cpl
A parameter representing the width of the frequency band in which, in a TV receiver, a progressive attenuation is produced.
f talon *m*
e talón *m*
i larghezza *f* di banda dell'intervallo di Nyquist
n bandbreedte van de nyquistflank
d Bandbreite *f* der Nyquist-Flanke

O

OB
see: FIELD PICKUP

OB APPARATUS
see: FIELD APPARATUS

3296 OB PRODUCER, OUTSIDE BROADCAST PRODUCER rep/tv
f producteur *m* d'extérieurs
e productor *m* de transmisiones de exteriores
i produttore *m* di riprese esterne
n produktieleider van buitenopnamen
d Aussenaufnahmenproduzent *m*

3297 OB PROGRAM(ME) CIRCUITS, OUTSIDE BROADCAST PROGRAM(ME) CIRCUITS rec/tv
f circuits *pl* pour programmes de transmissions à l'extérieur
e circuitos *pl* para programas de transmisiones exteriores
i circuiti *pl* per programmi di riprese esterne
n schakelingen *pl* voor buitenopname-programma's
d Schaltungen *pl* für Aussenaufnahme-programme

3298 OB SITE CONNECTIONS, OUTSIDE BROADCAST SITE CONNECTIONS rec/tv
f connexions *pl* de l'aire de tournage à l'extérieur
e conexiones *pl* del terreno para exteriores
i connessioni *pl* dell'area per riprese esterne
n terreinverbindingen *pl* voor de buiten-opnamen
d Verbindungen *pl* für das Aufnahmegelände

3299 OB VAN, OUTSIDE BROADCAST VAN ctv
f car *m* de reportage pour prise de vues extérieure
e carro *m* para transmisión de exteriores
i carro *m* per riprese esterne
n reportagewagen
d Übertragungswagen *m*

OB VEHICLE
see: MOBILE CONTROL ROOM

3300 OBJECT METAMERISM ct
Phenomenon occurring when surface colo(u)rs match under one illuminant but do not match under a different illuminant.
f métamérisme *m* d'un objet
e metamerismo *m* de un objeto
i metamerismo *m* d'un oggetto
n metamerisme *n* van een voorwerp
d Metamerismus *m* eines Dings

3301 OBJECT PERCEIVED COLO(U)R ct
Colo(u)r perceived to belong to an object either self-luminous or non-self-luminous.
f couleur *f* perçue d'un objet
e color *m* percibido de un objeto
i colore *m* percepito d'un oggetto
n waargenomen kleur van een voorwerp
d wahrgenommene Farbe *f* eines Dings

3302 OBJECT SPACE opt
A mathematical conception covering both the region lying in front of the system and behind the system.
f espace *m* de l'objet
e espacio *m* del objeto
i spazio *m* dell'oggetto
n voorwerpruimte
d Dingraum *m*

3303 OBJECTIVE opt/tv
In an optical or electronic system, the first lens through which the light rays pass.
f objectif *m*
e objetivo *m*
i obiettivo *m*
n objectief *n*
d Objektiv *n*

3304 OBLIQUE-INCIDENCE TRANSMISSION cpl
The transmission of a radio wave at a slant to the ionosphere and back to earth, as in long-distance radiocommunication.
f émission *f* sous incidence oblique
e emisión *f* sobre incidencia oblicua
i emissione *f* su incidenza obliqua
n transmissie bij schuine inval
d Übertragung *f* bei schrägem Einfall

3305 OBLIQUE MAGNETIZATION vr
f magnétisation *f* oblique
e magnetización *f* oblicua
i magnetizzazione *f* obliqua
n schuine magnetisatie
d Schrägmagnetisierung *f*

3306 OCTAGONAL COIL crt
A coil used for demagnetizing the metal parts of a mask tube.
f bobine *f* octagonale
e bobina *f* octagonal
i bobina *f* ottagonale
n achthoekige spoel
d Achterspule *f*

OCULAR ACCOMODATION
see: ACCOMODATION

3307 OCW, ORANGE-CYAN AXIS, ORANGE-CYAN SIDEBAND ct
f axe *m* de large bande de l'orange vers le cyan

e eje *m* de banda ancha del naranjado hacia el ciano
i asse *m* di banda larga dall'arancio verso il ciano
n brede-bandas van oranje naar cyaan
d Breitbandachse *f* von Orange nach Zyan

3308 ODD-LINE INTERLACE rec/tv
Interlace in which each field contains an extra half line; in the standard 525-line TV picture, each field contains 262.5 lines.
f analyse *f* entrelacée à nombre impair de lignes
e exploración *f* entrecalada de número impar de líneas
i analisi *f* interlacciata a numero dispari di linee
n aftasting met oneven aantal lijnen
d ungeradzeilige Abtastung *f*

3309 ODD NUMBER OF LINES rec/tv
f nombre *m* impair de lignes
e número *m* impar de líneas
i numero *m* dispari di linee
n oneven aantal *n* lijnen
d ungerade Zeilenzahl *f*

3310 OF HIGH DEFINITION rep/tv
f à haute définition
e de alta definición
i ad alta definizione
n gedetailleerd adj
d detailreich adj

3311 OFF-SWITCH vr
Used in winding the tape back into the cassette and switching off the recorder.
f commande *f* de retrait de bande
e botón *m* de desconexión
i pulsante *m* di disinnesto
n uitschakeltoets
d Ausschalttaste *f*

OFF-WHITE
see: BROKEN WHITE

3312 OFFSET CARRIER SYSTEM tv
System in which the frequency of the image carrier waves of a number of transmitters has such a value that the mutual interference is reduced to a minimum.
f système *m* à ondes porteuses décalées
e sistema *m* de portadoras desplazadas
i sistema *m* a portanti sfalsate
n systeem *n* met ten opzichte van elkaar verschoven draaggolven,
systeem *n* met verzette draaggolven
d Offsetbetrieb *m*,
versetztes Trägerwellensystem *n*

3313 OFFSET CARRIERS dis
f porteuses *pl* décalées
e portadoras *pl* desplazadas
i portanti *pl* sfalsate
n ten opzichte van elkaar verschoven draaggolven *pl*,
verzette draaggolven *pl*
d gegeneinander versetzte Träger *pl*

3314 OFFSET HEADS, aud/vr
STAGGERED HEADS
Magnetic-tape heads that are staggered or displaced 3.09 cm apart.
f têtes *pl* magnétiques décalées
e cabezas *pl* magnéticas desplazadas
i testine *pl* magnetiche sfalsate
n ten opzichte van elkaar verschoven magneetkoppen *pl*
d gegeneinander versetzte Magnetköpfe *pl*

3315 OFFSET SIGNAL cpl
f signal *m* décalé
e señal *f* desplazada
i segnale *m* sfalsato
n verzet signaal *n*
d Offsetsignal *n*,
versetztes Signal *n*

3316 OFFSET SIGNAL METHOD rep/tv
f système *m* à signal décalé
e sistema *m* de señal desplazada
i sistema *m* a segnale sfalsato
n systeem *n* met signaalverschuiving,
systeem *n* met verzetsignaal
d Signalversetzungsverfahren *n*

3317 OFFSET STEREOPHONIC TAPE, aud/vr
STAGGERED STEREOPHONIC TAPE
A stereophonic tape recorded with the head gaps spaced 3.09 cm along the length of the tape.
f bande *f* stéréophonique à enregistrements décalés
e cinta *f* estereofónica de registro desplazado
i nastro *m* stereofonico a registrazione sfalsata
n stereofonische magneetband met ten opzichte van elkaar verschoven geluidssporen
d stereophonisches Magnetband *n* mit gegeneinander versetzten Tonspuren

3318 OFFSET SUBCARRIER SYSTEM ctv
f système *m* à porteuse de chrominance décalée
e sistema *m* de portadora de crominancia desplazada
i sistema *m* a portante di crominanza sfalsata
n systeem *n* met verzette chrominantiedraaggolf
d System *n* mit versetztem Chrominanzträger

3319 OIL FILM tv
Used in eidophor projection apparatus.
f film *m* d'huile,
pellicule *f* d'huile
e nata *f* de aceite,
película *f* de aceite
i pellicola *f* d'olio,
velo *m* d'olio
n oliefilm,
olievlies *n*
d Ölfilm *m*
Ölhäutchen *n*

3320 OMEGA LOOP, vr
OMEGA WRAP
f guidage *m* alpha
e bucle *m* omega
i doppino *m* omega
n omegalus
d Omega-Bandführung *f*

3321 OMNIBUS CUE CIRCUIT, stu
OMNIBUS TALKBACK
An arrangement of talkback circuits used in televising special events under supervision of the control studio director.
f circuits *pl* d'interphone interconnectés,
circuit *m* en anneau d'interphones
e circuito *m* cerrado de interfonos,
circuitos *pl* de interfono interconectados
i circuiti *pl* d'interfono omnibus,
linea *f* circolare d'interfoni
n onderling verbonden intercommunicatieketens *pl*,
verzamelruggespraakleiding
d Kommandoringleitung *f*,
Rückspracheverbundnetz *n*,
Sammelrückspracheleitung *f*

OMNIDIRECTIONAL ANTENNA (US)
see: NON-DIRECTIONAL AERIAL

OMNIDIRECTIONAL MICROPHONE (US)
see: NON-DIRECTIONAL MICROPHONE

3322 ON-OFF SWITCH rep/tv
f commutateur *m* marche-arrêt
e conmutador *m* conectado-desconectado
i interruttore *m* spento-acceso
n aan/uitschakelaar
d Ein/Aus-Schalter *m*

3323 ON-SWITCH vr
Used in switching on the recorder and inserting the tape.
f commande *f* d'insertion de bande
e botón *m* de conexión
i pulsante *m* d'inserzione
n inschakeltoets
d Einschalttaste *f*

3324 ON-THE-AIR cpl/tv
Transmitting a radio signal.
f en émission
e en el aire,
en emisión
i in emissione,
in esercizio
n in bedrijf,
in de ether
d in Betrieb

ON-THE-AIR MONITOR (US)
see: MASTER MONITOR

3325 ONE-CYCLE MULTIVIBRATOR, rec/tv
SINGLE-SHOT MULTIVIBRATOR,
START-STOP MULTIVIBRATOR
f multivibrateur *m* monostable
e multivibrador *m* monoestable
i multivibratore *m* monostabile
n monostabiele multivibrator
d monostabiler Multivibrator *m*

OPENING CREDITS
see: FRONT CREDITS

OPERATING INSTRUCTIONS
see: DIRECTIONS FOR USE

3326 OPERATING VOLTAGE ge
f tension *f* de fonctionnement,
tension *f* de régime
e tensión *f* de régimen,
tensión *f* de trabajo
i tensione *f* d'esercizio,
tensione *f* di regime
n bedrijfsspanning
d Arbeitsspannung *f*,
Betriebsspannung *f*

3327 OPERATIONS CENTER(RE) tv
f centrale *f* de commandes
e central *f* de mandos
i centrale *f* di comandi
n commandocentrale
d Kommandozentrale *f*

OPERATIONS ROOM
see: APPARATUS ROOM

3328 OPPOSITE FIELD rep/tv
f trame *f* jointive
e cuadro *m* ceñido
i trama *f* allacciante
n aansluitend raster *n*
d anschliessendes Halbbild *n*

3329 OPPOSITE PHASE, cpl
OPPOSITION
f opposition *f* de phase
e contrafase *f*,
oposición *f* de fase
i controfase *f*,
opposizione *f* di fase
n tegenfaze
d Gegenphase *f*

OPTIC LIGHT FILTER (US)
see: AMBIENT LIGHT FILTER

3330 OPTICAL AXIS, opt
PRINCIPAL AXIS
The line which passes through the center(re) of curvature of a lens surface.
f axe *m* optique
e eje *m* óptico
i asse *m* ottico
n optische as
d optische Achse *f*

3331 OPTICAL DISTORTION dis/opt
f distorsion *f* optique
e distorsión *f* óptica
i distorsione *f* ottica
n optische vertekening,
optische vervorming
d optische Verzeichnung *f*,
optische Verzerrung *f*

3332 OPTICAL FILTER ctv/rec
A filter ensuring that the brightness of the light reaching each camera tube corresponds to the relative brightness

required for the appropriate primary in the display system.
f filtre *m* optique
e filtro *m* óptico
i filtro *m* ottico
n optisch filter *n*
d optisches Filter *n*

3333 OPTICAL IMAGE rec/rep/tv
The image produced by an optical device, on the photosensitive electrode of a TV camera tube.
f image *f* optique
e imagen *f* óptica
i immagine *f* ottica
n optisch beeld *n*
d optisches Bild *n*

3334 OPTICAL INTERMITTENT rep/tv
A device which optically arrests the image on a continually moving length of film, so that it can be seen by projection or direct viewing as if the film has been brought physically to rest by a true intermittent movement.
f appareil *m* d'intermittence optique
e aparato *m* de intermitencia óptica
i apparecchio *m* d'intermettenza ottica
n optisch stilstandapparaat *n*
d optischer Stillstandapparat *m*

3335 OPTICAL MASER cpl
A maser in which the stimulating frequency is visible or infrared radiation.
f maser *m* optique
e maser *m* óptico
i maser *m* ottico
n optische maser
d optischer Maser *m*

3336 OPTICAL SOUND aud
f son *m* optique
e sonido *m* óptico
i suono *m* ottico
n optisch geluid *n*
d Lichtton *m*

3337 OPTICAL SOUND RUSHES aud
f épreuves *pl* de son optique
e rodaje *m* diario de sonido óptico
i giornalieri *pl* di suono ottico
n werkkopie van optisch geregistreerd geluid
d optische Schallmuster *pl*

OPTICAL SPLITTER
see: BEAM SPLITTER

OPTICAL SPLITTING SYSTEM
see: BEAM SPLITTING SYSTEM

3338 OPTICAL TRANSFER CHARACTERISTIC, OTF opt
A characteristic in optics equivalent to frequency and phase distortion curves in electronics.
f caractéristique *f* de transfert optique
e característica *f* de transferencia óptica
i caratteristica *f* di trasferimento ottico
n optische overdrachtskarakteristiek
d optische Übertragungskennlinie *f*

3339 OPTICAL VIEWFINDER opt
f viseur *m* direct, viseur *m* optique
e visor *m* directo, visor *m* óptico
i mirino *m* diretto, mirino *m* ottico
n optische zoeker
d Direktsucher *m*, optischer Sucher *m*

3340 OPTIMIZATION vr
f optimisation *f*
e optimización *f*
i ottimizzazione *f* delle testine, taratura *f* della corrente in registrazione
n optimale afregeling, optimatisering
d Kopfstromoptimierung *f*

3341 OPTIMUM FOCUSING FOR 625 AND 819 LINE SYSTEMS ctv
f focalisation *f* optimale pour systèmes de 625 et 819 lignes
e enfoque *m* óptimo para sistemas de 625 y 819 líneas
i focalizzazione *f* ottima per sistemi di 625 e 819 linee
n optimale focussering voor 625 en 819 lijnensystemen
d optimale Fokussierung *f* für 625- und 819-Zeilensysteme

3342 OPTIMUM PERFORMANCE, PEAK PERFORMANCE rec/tv
f représentation *f* parfaite
e representación *f* perfecta
i rappresentazione *f* perfetta
n uitvoering van hoge kwaliteit
d erstklassige Darbietung *f*

ORANGE-CYAN AXIS, ORANGE-CYAN SIDEBAND
see: OCW

3343 ORANGE-CYAN AXIS TRANSFORMATION ctv
The colo(u)r co-ordinate transformation to a system with I and Q axes.
f transformation *f* de l'axe orange-cyan
e transformación *f* del eje naranjado-ciano
i trasformazione *f* dell'asse arancio-ciano
n transformatie van de oranje-cyaanas
d Transformation *f* der Orange-Zyan-Achse

3344 ORANGE-CYAN LINE ctv
f ligne *f* orange-cyan
e línea *f* naranjado-ciano
i linea *f* arancio-ciano
n oranje-cyaanlijn
d Orange-Zyan-Linie *f*

3345 ORBIT tv
f orbite *f*
e órbita *f*
i orbita *f*

n omloopbaan,
satellietenbaan
d Satellitenbahn *f*,
Umlaufbahn *f*

3346 ORBIT AROUND THE EARTH tv
f orbite *f* autour de la terre
e órbita *f* alrededor de la tierra
i orbita *f* intorno alla terra
n baan rond de aarde
d Bahn *f* um die Erde

3347 ORBITING SATELLITES tv
f mise *f* en orbite de satellites
e colocación *f* de satélites en órbita,
orbitación *f* de satélites
i messa *f* in orbita di satelliti
n satellieten in een baan brengen
d in eine Bahn bringen von Satelliten

3348 ORCHESTRA aud/th
f orchestre *m*
e orquesta *f*
i orchestra *f*
n orkest *n*
d Orchester *n*

3349 ORCHESTRA LAY-OUT, rec/tv
ORCHESTRAL ARRANGEMENT
The placement of the various instruments and the microphones for each instrument or group of instruments.
f arrangement *m* de l'orchestra
e disposición *f* de la orquesta
i disposizione *f* dell'orchestra
n opstelling van het orkest
d Aufstellung *f* des Orchesters

3350 ORCHESTRA RECORDING aud
f enregistrement *m* orchestral
e registro *m* orquestal
i registrazione *f* orchestrale
n orkestopname
d Orchesteraufnahme *f*

3351 ORCHESTRAL BROADCASTING STUDIO rec
f studio *m* pour diffusion orchestrale
e estudio *m* para difusión orquestal
i studio *m* per emissione orchestrale
n orkeststudio
d Orchesterregieraum *m*,
Orchesterstudio *n*

3352 ORDER SIGNAL stu
A pre-arranged signal sent to a studio or any other broadcast program(me) source to bring about an intervention in the normal course of the program(me).
f signal *m* de commande
e señal *f* de mando
i segnale *m* di comando
n interventiesignaal *n*
d Befehlssignal *n*,
Kommandozeichen *n*

ORIENTATION OF MAGNETIC PARTICLES
see: MAGNETIC PARTICLE ORIENTATION

3353 ORIGIN DISTORTION crt/dis
Distortion in a gas-focused cathode-ray tube due to reduced deflection sensitivity when the beam is in the region of the axis or undeflected position.
f anomalie *f* du point zéro,
distorsion *f* de l'origine
e anomalía *f* de punto cero,
distorsión *f* de origen
i anomalia *f* di punto zero,
distorsione *f* d'origine
n nulpuntanomalie,
nulpuntvervorming
d Nullpunktverzerrung *f*

ORIGINAL
see: FIRST GENERATION TAPE

3354 ORIGINAL BROADCAST tv
f émission *f* originale,
inédit *m* de télédiffusion
e emisión *f* original
i emissione *f* originale
n oorspronkelijke uitzending
d Originalsendung *f*

3355 ORIGINAL LOCATION, tv
OUTDOOR LOCATION
f décor *m* naturel
e decoración *f* natural
i allestimento *m* naturale
n natuurlijk decor *n*
d realer Aussendekor *m*

3356 ORTHICON, crt
ORTHICONOSCOPE
A camera tube in which a beam of low-velocity electrons scans a photo-emissive mosaic capable of storing an electric-charge system.
f orthiconoscope *m*
e orticonoscopio *m*
i orticonoscopio *m*
n orthicon *n*
d Orthikon *n*

3357 ORTHOGONAL SCANNING rec rep tv
System of magnetic scanning used in low-velocity camera tubes in which the electron beam is subjected to both axial and lateral magnetic fields.
f analyse *f* orthogonale
e exploración *f* ortogonal
i analisi *f* ortogonale
n orthogonale aftasting
d orthogonale Abtastung *f*

3358 OSCILLATORY SCANNING rec/rep/tv
A scanning method in which the scanning spot moves repeatedly to and fro across the image so that successive lines are scanned in opposite directions.
f analyse *f* oscillante
e exploración *f* oscilante
i analisi *f* oscillante
n oscillerende aftasting
d oszillierende Abtastung *f*

3359 OSCILLOGRAM svs

f oscillogramme *m*
e oscilograma *m*
i oscillogramma *m*
n oscillogram *n*
d Oszillogramm *n*

3360 OSCILLOGRAPH cpl
f oscillographe *m*
e oscilógrafo *m*
i oscillografo *m*
i oscillograaf
d Oszillograph *m*

3361 OSCILLOSCOPE svs
f oscilloscope *m*
e osciloscopio *m*
i oscilloscopio *m*
n oscilloscoop
d Oszilloskop *n*

3362 OUT-EDIT, OUT-GOING SPLICE, OUT-POINT vr
f point *m* de sortie
e salida *f* de la secuencia
i fine *f* dell'inserto, punto *m* d'uscita
n eindpunt *n*
d eingehender Schnitt *m*

3363 OUT OF FRAME rep/tv
f décadré adj
e desencuadrado adj
i a spostamento del cuadro
n met rasterverschuiving
d mit Zeilenzugverschiebung

3364 OUT OF PHASE cpl/ge
Said of voltages, currents or other quantities having the same frequency and waveform which vary rhythmically but are out of step with each other, so that they do not reach corresponding values at the same time.
f déphasé adj
e defasado adj
i fuori fase, sfasato adj
n in faze verschoven
d ausser Phase, phasenverschoben adj

3365 OUTDOOR AERIAL, OUTDOOR ANTENNA aea
f antenne *f* extérieure
e antena *f* exterior
i antenna *f* esteriore
n buitenantenne
d Aussenantenne *f*

OUTDOOR LOCATION
see: ORIGINAL LOCATION

3366 OUTGOING LINE cpl
f ligne *f* de départ
e línea *f* de salida
i linea *f* di partenza
n uitvoerleiding
d Ausgangsleitung *f*

3367 OUTGOING SIGNAL rec/tv
f signal *m* de sortie
e señal *f* de salida
i segnale *m* d'uscita
n uitgaand signaal *n*
d Sendesignal *n*

3368 OUTPUT ge
The current, voltage, power or driving force delivered by a circuit or device.
f sortie *f*
e salida *f*
i uscita *f*
n uitgang
d Ausgang *m*

3369 OUTPUT AMPLIFIER cpl
f amplificateur *m* de sortie, amplificateur *m* final
e amplificador *m* de salida, amplificador *m* final
i amplificatore *m* d'uscita, amplificatore *m* finale
n eindversterker, uitgangsversterker
d Ausgangsverstärker *m*, Endverstärker *m*

3370 OUTPUT FILTER rec/tv
A lowpass filter in the output leads to eliminate overshoots.
f filtre *m* de sortie
e filtro *m* de salida
i filtro *m* d'uscita
n uitgangsfilter *n*
d Ausgangsfilter *n*

3371 OUTPUT IMPEDANCE cpl
The impedance presented by the electron tube (valve) under its operating conditions, to the external circuit connected between the terminals of the tube (valve) giving access respectively to an output electrode under consideration and to a reference point which is the cathode, earth, or some other reference point.
f impédance *f* de sortie
e impedancia *f* de salida
i impedenza *f* d'uscita
n uitgangsimpedantie
d Ausgangsimpedanz *f*

OUTPUT MONITOR
see: ACTUAL MONITOR

3372 OUTPUT SIGNAL ge
f signal *m* de sortie
e señal *f* de salida
i segnale *m* d'uscita
n uitgangssignaal *n*
d Ausgangssignal *n*

3373 OUTPUT STAGE cpl
f étage *m* final
e etapa *f* final
i stadio *m* finale
n eindtrap
d Endstufe *f*

3374 OUTPUT VARIATION tv
In TV, the change in peak amplitude during a period not exceeding one field in length.
f variation *f* de sortie
e variación *f* de salida
i variazione *f* d'uscita
n uitgangsvariatie
d Ausgangsvariation *f*

OUTSIDE BROADCAST
see: FIELD PICKUP

OUTSIDE BROADCAST APPARATUS
see: FIELD APPARATUS

OUTSIDE BROADCAST PRODUCER
see OB PRODUCER

OUTSIDE BROADCAST PROGRAM(ME) CIRCUITS
see: OB PROGRAM(ME) CIRCUITS

OUTSIDE BROADCAST SITE CONNECTIONS
see: OB SITE CONNECTIONS

OUTSIDE BROADCAST VAN
see: OB VAN

3375 OVAL-SHAPED TRACK vr
f ovalisation *f* de la piste
e ovalización *f* de la pista
i ovalizzazione *f* della pista
n onrondheid van het spoor
d Unrundheit *f* der Spur

OVERALL ATTENUATION (GB)
see: NET TRANSMISSION EQUIVALENT

3376 OVERALL BRIGHTNESS TRANSFER CHARACTERISTIC, OVERALL BRILLIANCE TRANSFER CHARACTERISTIC tv
Ratio between the brightness of a screen and the brightness of the reproduced image.
f contraste *m* des luminosités
e característica *f* total de transmisión de luminosidad
i caratteristica *f* globale di trasferimento della luminosità
n helderheidsreproduktiekarakteristiek
d Helligkeitsverhältnis *n*

3377 OVERALL CONTRAST RATIO rec/tv
Ratio of the light to the dark areas of a TV screen on which a raster is reproduced having one half adjusted to maximum brightness and the other half to black level.
f rapport *m* global de contraste
e contraste *m* nominal
i rapporto *m* globale di contrasto
n totale contrastverhouding
d Gesamtkontrastverhältnis *n*, Gradationsverhältnis *n*

OVERALL GAMMA
see: GAMMA

3378 OVERALL HARMONIC DISTORTION dis
f distorsion *f* harmonique globale, distorsion *f* nonlinéaire globale
e distorsión *f* harmónica total, distorsión *f* no lineal total
i distorsione *f* armonica globale, distorsione *f* non lineare globale
n totale niet-lineaire vervorming
d totale nichtlineare Verzerrung *f*

3379 OVERALL TOLERANCE IN GAIN VARIATION cpl
f tolérance *f* globale de variation de gain
e tolerancia *f* total de variación de ganancia
i tolleranza *f* globale di variazione di guadagno
n totale tolerantie van de versterkingsvariatie
d Gesamttoleranz *f* der Verstärkungsänderung *f*

3380 OVERFLOW OF WHITES dis
f surintensité *f* des blancs
e sobreintensidad *f* de los blancos
i lampeggiamento *m* dei bianchi
n oplichten *n* van het wit
d Aufleuchten *n* von Weiss

OVERHEAD LIGHTING BATTENS
see: BANK OF LAMPS

3381 OVERHEAD PROJECTOR aud/vr
An audio-visual means for projecting instructive texts on to a board or screen.
f rétroprojecteur *m*
e retroproyector *m*
i retroproiettore *m*
n schrijfprojector
d Tageslichtprojektor *m*

3382 OVERLAPPING CHANNELS cpl/tv
f canaux *pl* partiellement superposés
e canales *pl* parcialmente superposicionados
i canali *pl* parzialmente sovrapposti
n gedeeltelijk overlappende kanalen *pl*
d teilweise überlappende Kanäle *pl*

3383 OVERLAPPING SPLICE fi/tv
f collure *f* à recouvrement
e empalme *m* superpuesto
i incollatura *f* a sovrapposizione
n overlappende las
d überlappender Stoss *m*, Überlappungsklebestelle *f*

OVERLAY
see: MOVING MATTE

3384 OVERLOADING, rep/tv
1. The condition in which the volume of a program(me) output exceeds, during program(me) peaks, the limit which the equipment is designed to carry.
2. Non-linear distortion resulting from this condition.
f surcharge *f*, surmodulation *f*
e sobremodulación *f*
i sovramodulazione *f*
n overbelasting,

overmodulatie,
oversturing
d Übersteuerung *f*

3385 OVERRUN tv
f dépassement *m* du temps d'émission
e sobrepasado *m* del tiempo de emisión
i oltrepassato *m* del tempo d'emissione
n overschrijding van de zendtijd
d Sendezeitüberschreitung *f*

3386 OVERSCANNING crt/rec/tv
Deflection of the electron beam of a cathode-ray tube outside the angle subtending the screen area.
f analyse *f* dépassante
e sobreexploración *f*
i analisi *f* oltrepassante
n overschrijdende aftasting
d überschreitende Abtastung *f*

3387 OVERSHOOT, dis
OVERSWING
The initial transient response to a unidirectional change in input that exceeds the steady-state response.
f dépassement *m* balistique,
suroscillation *f*
e sobreoscilación *f*
i sovraelongazione *f*
n doorschieten *n*,
doorschot *n*
d Überschwingen *n*

3388 OVERSHOOT DISTORTION, dis
OVERSWING DISTORTION
Distortion that occurs when the maximum amplitude of a signal wavefront exceeds the steady-state amplitude of the signal wave.
f distorsion *f* de suroscillation
e distorsión *f* de sobreoscilación
i distorsione *f* di sovraelongazione
n doorschietvervorming
d Überschwingungsverzerrung *f*

3389 OVERSHOOT RATIO, dis
OVERSWING RATIO
Is represented by the formula

$$\frac{\text{maximum value-final value}}{\text{final value}}.$$

f rapport *m* de suroscillation
e relación *f* de sobreoscilación
i rapporto *m* di sovraelongazione
n doorschietverhouding
d Überschwingungsverhältnis *n*

OXIDE PAINT
see: DOPE

3390 OXIDE SHEDDING vr
f perte *f* d'oxyde,
poudrage *m*
e desprendimientos *pl* de la emulsión
i perdita *f* d'ossido
n poederen *n*
d Schichtablösung *f*,
Schichtabrieb *m*

P

3391 PACK-CARRIER TELEVISION STATION — tv
f émetteur *m* de télévision portatif
e emisor *m* de televisión portátil
i emettitore *m* di televisione portatile
n draagbare televisiezender
d Rücksacksender *m*

PACK-SHOT
see: MACROSHOT

PACK-SHOT LENS
see: MACROSHOT LENS

3392 PACKING (US), PICTURE COMPRESSION — rep/tv
The tendency for portions of the TV picture to pack or compress in either the horizontal or vertical plane, the result of a non-linear sawtoothed scanning wave.
f compression *f* de l'image
e compresión *f* de la imagen
i compressione *f* dell'immagine
n beeldcompressie
d Bildkompression *f*, Bildverdichtung *f*

3393 PAIRING, TWINNING — rec/tv
Imperfection of interlace caused by an incorrect displacement of the set of lines of one field relative to those of the preceding field in a direction normal to the line scan.
f pairage *m*
e apareamiento *m*, solapado *m* de líneas
i appaiamento *m* delle linee
n lijnenparing, paren *n* van de lijnen
d Zeilenpaarung *f*

3394 PAIRING OFF — rep/tv
f disparition *f* des trames
e desenganche *m* de los cuadros
i sparizione *f* delle trame
n verdwijnen *n* van de rasters
d Aussertrittfallen *n* der Teilbilder

3395 PAL-COLO(U)R SYSTEM, PAL-SYSTEM, PHASE-ALTERNATION LINE SYSTEM — ctv
A compatible system of colo(u)r TV similar to the NTSC system, except that one of the two chrominance signal components is reversed in phase, for the duration of alternate scanning lines.
f système *m* de télévision couleur PAL
e sistema *m* de televisión en colores PAL
i sistema *m* di televisione a colori PAL
n PAL-kleurentelevisiesysteem *n*
d PAL-Farbfernsehsystem *n*

PALING OUT
see: DESATURATION

3396 PAM, PULSE AMPLITUDE MODULATION — cpl
f modulation *f* d'impulsion en amplitude
e modulación *f* de impulsos en amplitud
i modulazione *f* d'ampiezza d'impulso
n impulsamplitudemodulatie
d IAM *f*, Impuls-Amplitudenmodulation *f*

3397 PAN, PANNING — rec/rep/tv
To tilt or otherwise move a TV camera vertically and horizontally to keep it trained on a moving object or secure a panoramic effect.
f panoramique *m*
e movimiento *m* panorámico de la cámara, panoramización *f* de la cámara
i movimento *m* panoramico della camera
n panoramicabeweging van de camera
d Kameraschwenkung *f*

3398 PAN AND TILT HEAD, PANNING HEAD — rec/tv
Component part of the panning mechanism.
f tête *f* panoramique, tourelle *f* universelle
e cabeza *f* para movimiento panorámico y de cabeceo
i piattaforma *f* doppia
n panoramakop, zwenk- en neigkop
d Schwenkkopf *m*

3399 PAN DOWN (GB), PAN UP, TILTING (US) — tv
f panoramique *m* vertical
e panorámica *f* vertical
i panoramica *f* verticale
n verticale panoramica
d Senkrechtschwenk *m*

3400 PAN SHOT, PANNING SHOT — rep/tv
f panoramique *m*
e toma *f* durante movimiento panorámico de la cámara
i panoramica *f*
n panoramaopname
d Schwenkaufnahme *f*

3401 PARABOLIC AERIAL, PARABOLIC ANTENNA — aea
A directional microwave aerial (antenna) using some form of parabolic reflector to give directional characteristics.
f antenne *f* parabolique
e antena *f* parabólica
i antenna *f* parabolica

n paraboolantenne
d Parabolantenne *f*

3402 PARABOLIC MICROPHONE, PARABOLIC REFLECTOR MICROPHONE aud
A microphone used at the focal point of a parabolic sound reflector to give improved sensitivity and high directivity.
f microphone *m* en réflecteur
e micrófono *m* posicionado en el punto focal de un reflector parabólico
i microfono *m* in riflettore parabolico
n microfoon in paraboolreflector
d Mikrophon *n* in Parabolreflektor

3403 PARABOLIC MIRROR opt
f miroir *m* parabolique
e espejo *m* parabólico
i specchio *m* parabolico
n paraboolspiegel
d Parabolspiegel *m*

3404 PARABOLIC REFLECTOR aea
f réflecteur *m* parabolique
e reflector *m* parabólico
i riflettore *m* parabolico
n parabolische reflector
d Parabolreflektor *m*

3405 PARABOLIC SHADING dis
Non-uniformity of brightness across the raster between the center(re) and the corner when a fixed input signal is applied.
f irrégularité *f* parabolique dans le noir
e irregularidad *f* parabólica en el negro
i irregolarità *f* parabolica nel nero
n parabolische onregelmatigheid in het zwart
d parabolische Bildabschattung *f*, parabolische Ungleichmässigkeit *f* im Bildschwarz

3406 PARABOLIC WAVEFORM SWITCHING SIGNAL rec/tv
Used in electronic generation of an iris wipe.
f signal *m* commutateur à forme d'onde parabolique
e señal *f* conmutadora de forma de onda parabólica
i segnale *m* commutatore a forma d'onda parabolica
n schakelsignaal *n* met parabolische golfvorm
d Schaltsignal *n* mit parabolischer Wellenform

3407 PARALLEL (US), ROSTRUM (GB) stu/th
f praticable *m*
e plataforma *f* elevada móvil
i praticabile *m*
n praktikabel *n*
d Praktikabel *n*

3408 PARALLEL DISTORTION, SKEW DISTORTION dis
f distorsion *f* en parallélogramme
e distorsión *f* en paralelograma
i distorsione *f* in parallelogramma
n parallellogramvervorming
d Parallelogramm-Verzeichnung *f*

PARALLELS
see: GUIDE RAILS

3409 PARAMETER ge
A quantity which is a constant under a set of circumstances but which may have a different value under other circumstances.
f paramètre *m*
e parámetro *m*
i parametro *m*
n parameter
d Parameter *m*

PARAMETRIC AMPLIFIER
see: MAVAR

3410 PARAPHASE AMPLIFIER cpl
Amplifier in which a single input is divided into two inputs, coupled in opposite phase relationship.
f amplificateur *m* déphaseur, amplificateur *m* en contrephase
e amplificador *m* desfasador, amplificador *m* en contrafase
i amplificatore *m* in controfase, amplificatore *m* sfasatore
n versterker in tegenfaze
d Paraphasenverstärker *m*, Verstärker *m* mit Phasenumkehrung

3411 PARASITIC AERIAL, PARASITIC ANTENNA aea
f antenne *f* passive
e antena *f* pasiva
i antenna *f* passiva
n passieve antenne
d passiver Strahler *m*, strahlungsgekoppelte Antenne *f*

3412 PARASITIC ELEMENT aea
f élément *m* secondaire
e elemento *m* secundario
i elemento *m* secondario
n secundair element *n*
d Sekundärelement *n*, strahlungsgekoppeltes Element *n*

3413 PARASITIC OSCILLATIONS, SPURIOUS OSCILLATIONS dis
Undesired oscillations in a network.
f oscillations *pl* parasites
e oscilaciones *pl* parásitas
i oscillazioni *pl* parassite
n parasitaire trillingen *pl*
d Streuschwingungen *pl*, wilde Schwingungen *pl*

3414 PARASITIC RADIATION dis/tv
f rayonnement *m* parasite
e radiación *f* parásita
i irradiazioni *pl* parassite
n parasitaire straling
d parasitäre Strahlung *f*

3415 PARC NOISE ctv
A noise analogous to the cross-colo(u)r effects produced by high frequency luminance signals.
f bruit *m* de chrominance dû à la diaphonie de bruit de luminance,
bruit *m* de parc
e ruido *m* de crominancia debido a la diafonía de ruido de luminancia
i effetto *m* disturbante causato dalle componenti in alta frequenza del segnale di luminanza, demodulate nel canale di crominanza
n door overspreken van luminantieruis ontstane chrominantieruis
d durch Übersprechen des Helligkeitsrauschens verursachtes Chrominanzrauschen

PARENT STATION
see: MASTER TRANSMITTER

3416 PARTIAL COMPATIBILITY rep/tv
System of TV transmission for which both monochrome and colo(u)r reception are possible, but in which the chrominance information is outside the black-and-white video band of frequencies.
f compatibilité *f* partielle
e compatibilidad *f* parcial
i compatibilità *f* parziale
n gedeeltelijke compatibiliteit
d Teilkompatibilität *f*

PARTIAL DEUTERANOPIA
see: DEUTERANOMALOUS VISION

3417 PARTIAL PROTANOPIA, ct
PROTANOMALOUS VISION
A form of anomalous trichromatic vision in which more red is required in a mixture of red and green to match a spectral yellow than is the case for the normal trichromat.
f protanopie *f* partielle
e visión *f* protanómala
i protanopia *f* parziale
n gedeeltelijke protanopie
d Teilprotanopie *f*,
Rotblindheit *f*

3418 PARTIAL STORED FIELD SYSTEM vr
A telerecording system in which the pull-down time is reduced to about 5 msec corresponding to 45° and phosphor afterglow is used to record the part of the picture occurring during that period.
f système *m* de mise en mémoire partielle de l'image
e sistema *m* con almacenamiento parcial de la imagen,
sistema *m* de imagen parcialmente almacenada
i sistema *m* d'immagazzinamento parziale dell'informazione di quadro
n systeem *n* met gedeeltelijke opslag van het beeld
d System *n* mit teilweiser Speicherung des Bildes

3419 PARTIAL TRITANOPIA, ct
TRITANOMALOUS VISION
A form of anomalous trichromatic vision intermediate between normal trichromatism and tritanopia, but about which detailed information is lacking.
f tritanopie *f* partielle
e visión *f* tritanómala
i tritanopia *f* parziale
n gedeeltelijke tritanopie
d Teiltritanopie *f*

PARTITION NOISE
see: DISTRIBUTION NOISE

3420 PASS-BAND cpl
The band of a selective network which freely transmits frequencies in sound recording and reproduction.
f bande *f* passante
e banda *f* de paso,
banda *f* de transmisión libre
i banda *f* passante
n doorlaatgebied *n*
d Durchlassbereich *m*

3421 PASSIVE RETRANSMISSION tv
f réémission *f* passive
e retransmisión *f* pasiva
i riemissione *f* passiva
n passieve heruitzending
d passive Weitersendung *f*

3422 PASSIVE SATELLITE ge/tv
A satellite used for sky wave communication, which is not fitted with receiver or transmitter equipment.
f satellite *m* passif
e satélite *m* pasivo
i satellite *m* passivo
n passieve satelliet
d passiver Satellit *m*

3423 PASSIVE SATELLITE REPEATER tv
A repeater operating on the principle that a reflecting surface returns a portion of the transmitted energy to the receiver.
f répéteur *m* de satellite passif
e repetidor *m* de satélite pasivo
i ripetitore *m* di satellite passivo
n versterker voor passieve satelliet
d Verstärker *m* für passiven Satellit

3424 PASSIVE TUNING UNIT rep/tv
f unité *f* de syntonie passive
e unidad *f* de sintonía pasiva
i unità *f* di sintonia passiva
n passieve afstemeenheid
d passives Abstimmglied *n*

3425 PATCHING aud
In sound recording equipment, the bringing of equalizer, echo chamber and similar devices into circuit.
f insertion *f* dans un circuit
e inserción *f* en un circuito
i inserzione *f* in un circuito
n in een keten inschakelen
d in einen Kreis hineinschalten

3426 PATH ATTENUATION aea/rep/tv
Loss in decibels between the power leaving the aerial (antenna) and that reaching the receiving aerial (antenna) per unit area.
f affaiblissement *m* de propagation
e atenuación *f* de propagación
i attenuazione *f* di percorso, attenuazione *f* di propagazione
n wegdemping
d Streckendämpfung *f*

3427 PATTERN rep/tv
The test pattern produced on the screen of a cathode-ray tube by a pattern generator.
f mire *f*
e imagen *f* de prueba
i immagine *f* di prova
n testbeeld *n*, toetsbeeld *n*
d Testbild *n*

PATTERN DISTORTION
see: DEFLECTION DISTORTION

3428 PATTERN GENERATOR mea/tv
A device which provides an artificial video signal for a simple well-defined picture, without making use of the normal transmitter scanning equipment.
f générateur *m* de mire
e generador *m* de imagen de prueba
i generatore *m* d'immagine di prova
n toetsbeeldgenerator
d Bildmustergenerator *m*, Testbildgenerator *m*

3429 PATTERN NOISE dis
f tensions *pl* de bruit périodiques
e tensiones *pl* perturbadoras periódicas
i tensioni *pl* perturbatrici periodiche
n periodieke stoorspanningen *pl*
d periodische Störspannungen *pl*

3430 PATTERN SELECTOR rep/tv
f sélecteur *m* de mire
e selector *m* de imagen de prueba
i selettore *m* d'immagine di prova
n toetsbeeldkiezer
d Musterwähler *m*

3431 PATTERNING dis/tv
Background pattern superimposed on a TV picture as a result from interference of an unwanted transmission or radiation from a neighbo(u)ring receiver.
f surimposition *f* de fond faux
e sobreposición *f* de fondo falso
i sovrapposizione *f* di fondo falso
n valse-achtergrondsuperpositie
d falsche Hintergrundüberlagerung *f*

PAY-AS-YOU-SEE TELEVISION
see: COIN-FREED TELEVISION

PEAK BLACK (GB)
see: BLACK PEAK

3432 PEAK BRIGHTNESS dis
Brightness of a small portion of a cathode-ray tube phosphor when excited by a peak white signal.
f hyperluminosité *f* d'aire réduite
e hiperluminosidad *f* de superficie reducida
i iperluminosità *f* di superficie ridotta
n kleine intensieve lichtvlek
d kleiner intensiver Lichtfleck *m*

3433 PEAK DISTORTION dis/tv
A form of distortion resulting from amplitude non-linearity affecting the higher amplitude portions of the video signal.
f distorsion *f* de la crête de l'amplitude
e distorsión *f* de la cresta de la amplitud
i distorsione *f* della cresta dell'ampiezza
n amplitudetopvervorming
d Amplitudenspitzenverzerrung *f*

3434 PEAK ENVELOPE POWER, PEAK POWER cpl
The average power output of a transmitter measured over one radio-frequency cycle at the crest of the modulation envelope.
f puissance *f* de crête, puissance *f* en crête de modulation
e potencia *f* de cresta
i potenza *f* sulla cresta dell'inviluppo di modulazione
n PEP, piek-omhullend vermogen *n*
d Spitzenleistung *f*

3435 PEAK EXCURSION rec/rep/tv
The full excursion above or below the sync level, according to whether modulation is positive or negative.
f excursion *f* maximale
e excursión *f* máxima
i sovraelongazione *f* massima
n maximale uitschieting
d maximale Überschwingung *f*

PEAK LIMITER, PEAK LIMITING AMPLIFIER
see: LIMITING AMPLIFIER

3436 PEAK LIMITING dis
The reduction of the dynamic sound range when it exceeds the capacity of the recording medium.
f limitation *f* de la dynamique
e limitación *f* del margen dinámico
i limitazione *f* della dinamica
n dynamiekbegrenzing
d Begrenzung *f* des Lautstärkeumfangs

PEAK PERFORMANCE
see: OPTIMUM PERFORMANCE

3437 PEAK POWER cpl
f puissance *f* de crête
e potencia *f* de cresta
i potenza *f* di cresta
n topvermogen *n*
d Spitzenleistung *f*

3438 PEAK-POWER OUTPUT cpl
In a modulated carrier system, the output power, averaged over a carrier cycle, at the maximum amplitude which can occur with any combination of signals to be transmitted.
f puissance *f* de sortie de crête
e potencia *f* de salida de cresta
i potenza *f* d'uscita di cresta
n topuitgangsvermogen *n*
d Spitzenausgangsleistung *f*

3439 PEAK-POWER OUTPUT ADJUSTMENT rep/tv
In TV, the control, either manual or automatic, to maintain the transmitter output power within definite limits over long-time intervals.
f réglage *m* de la puissance de sortie de crête
e regulación *f* de la potencia de salida de cresta
i regolazione *f* della potenza d'uscita di cresta
n regeling van het topuitgangsvermogen
d Regelung *f* der Spitzenausgangsleistung

3440 PEAK PROGRAM(ME) LEVEL rec/tv
The sound level laid down with particular values which should not be exceeded.
f niveau *m* de la dynamique de crête
e nivel *m* de la dinámica de cresta
i livello *m* della dinamica di cresta
n maximaal toelaatbare dynamiek
d maximaler Lautstärkepegel *m*

3441 PEAK PROGRAM(ME) METER mea
An instrument designed to measure the volume of program(me) in a sound channel in terms of the peaks averaged over a specified period.
f voltmètre *m* de crête
e voltímetro *m* de cresta
i voltmetro *m* di cresta
n niveaupiekmeter
d Aussteuerungsmesser *m*, Spitzenzeiger *m*

3442 PEAK PULSE AMPLITUDE tv
The maximum absolute peak value of the pulse excluding those portions considered to be unwanted, such as spikes.
f amplitude *f* maximale d'une impulsion
e amplitud *f* máxima de un impulso
i ampiezza *f* massima d'un impulso
n maximale impulssterkte
d Spitzenimpulsstärke *f*

3443 PEAK SIDEBAND POWER, PSP cpl
The average sideband power of a transmitter measured over one radio-frequency cycle at the highest crest of the modulation envelope.
f puissance *f* de crête d'une bande latérale
e potencia *f* de cresta de una banda lateral
i potenza *f* di cresta d'una banda laterale
n topvermogen *n* van een zijband
d Seitenbandspitzenleistung *f*

3444 PEAK SIGNAL, PEAK SIGNAL LEVEL cpl
f niveau *m* de crête du signal
e nivel *m* de cresta de la señal
i livello *m* di cresta del segnale
n maximaal signaalniveau *n*
d maximaler Signalpegel *m*

3445 PEAK-TO-PEAK SIGNAL AMPLITUDE cpl
f amplitude *f* crête-à-crête du signal
e amplitud *f* de cresta a cresta de la señal
i ampiezza *f* da cresta a cresta del segnale
n seinspan
d Signalamplitude *f* Spitze-zu-Spitze

3446 PEAK WHITE, WHITE PEAK rep/tv
The level in the vision signal corresponding to white level.
f crête *f* du blanc
e cresta *f* de blanco
i bianco *m* limite
n witniveau *n* van het hoogfrequent beeldzendersignaal
d Maximum *n* an Weiss, Spitzenweisspegel *n*, Weissspitze *f*

3447 PEAK-WHITE RASTER, WHITE LEVEL RASTER, WHITE RASTER rep/tv
f canevas *m* à niveau du blanc
e cuadrícula *f* de nivel del blanco
i quadro *m* rigato a livello del bianco
n raster *n* op witniveau
d Raster *n* auf Weisspegel

3448 PEAKING CIRCUIT cpl
Circuit employed in video amplifiers which introduces a resonance peak in the interstage coupling to extend the frequency response at the higher frequencies.
f circuit *m* d'augmentation de la pente
e circuito *m* de aumento de la pendiente
i circuito *m* d'aumento della ripidità
n schakeling voor steilheidsverhoging
d Versteilerungsschaltung *f*

PEAKING COIL
see: CORRECTING COIL

3449 PEAKING NETWORK rep/tv
An interstage coupling network used in a video amplifier to enhance the output at some portion of the frequency range.
f circuit *m* de différentiation
e circuito *m* diferenciador
i circuito *m* differenziatore
n differentieerschakeling
d Differenzierschaltung *f*

3450 PEAKS aud
Momentary high volume levels during a broadcast program(me) making the volume indicator in the studio or transmitter swing upward.

f crêtes *pl* du niveau
e crestas *pl* del nivel
i creste *pl* del livello
n niveaupieken *pl*,
uitschieters *pl*
d Spitzen *pl*

3451 PEDESTAL rep/tv
In the picture signal, the separation in level between the black level and the blanking level.
f décollement *m* du niveau de noir
e pedestal *m*
i piedistallo *m*
n pedestal
d Schwarzabhebung *f*

3452 PEDESTAL ADJUSTMENT rep/tv
f ajustage *m* du décollement de noir
e ajuste *m* del pedestal
i aggiustaggio *m* del piedistallo
n pedestalinstelling
d Einstellung *f* der Schwarzabhebung

PEDESTAL LEVEL
see: BLACK-OUT LEVEL

PEDESTAL SIGNAL
see: BLACK-OUT SIGNAL

3453 PEG BAR ani
f barre *f* à chevilles
e barra *f* de espigas
i barra *f* a carichi
n pennenstang
d Zapfenstange *f*

3454 PEG TRACK ani
f voie *f* de chevilles
e ranura *f* de espigas
i gola *f* di carichi
n pennengoot
d Zapfenrille *f*

3455 PELLICLE rec/vr
Extremely thin semi-reflecting film for telecine multiplexer systems which reduces almost completely the double reflections encountered when a thicker support is used.
f film *m*,
membrane *f*,
pellicule *f*
e membrana *f*,
nata *f*,
película *f*
i membrana *f*,
pellicola *f*,
velo *m*
n film,
membraan *n*,
vlies *n*
d Film *m*,
Häutchen *n*,
Membran *f*

3456 PENCIL crt
Narrow cylindrical beam of moving electrons in an electron optical system.
f faisceau *m* étroit,
pinceau *m*
e haz *m* estrecho
i pennello *m*
n smalle bundel
d Büschel *m*

3457 PENTODE ELECTRON GUN crt
A cathode-ray tube electron gun which has a cathode, control grid and three successive anodes.
f canon *m* electronique à trois anodes
e cañón *m* electrónico de tres ánodos
i cannone *m* elettronico a tre anodi
n elektronenkanon *n* met drie anoden
d Dreianodenelektronenkanone *f*

3458 PERCEIVED ACHROMATIC COLO(U)R ct
Perceived colo(u)r devoid of hue.
f plage *f* des gris perçue
e estímulo *m* acromático percibido
i stimolo *m* acromatico percepito
n waargenomen achromatische stimulus
d wahrgenommenes Unbunt *n*

3459 PERCEIVED COLO(U)R ct
Aspect of visual perception by which an observer may distinguish between two fields of the same size, shape and structure, such as may be caused by differences in the spectral composition of the radiation concerned in the observation.
f couleur *f* perçue
e color *m* percibido
i colore *m* percepito
n waargenomen kleur
d wahrgenommene Farbe *f*

3460 PERCEIVED OBJECT COLO(U)R ct
The colo(u)r perceived to belong to an object.
f couleur *f* d'un objet perçue
e color *m* de un objeto percibido
i colore *m* d'un oggetto percepito
n waargenomen kleur van een voorwerp
d wahrgenommene Dingfarbe *f*

PERCENTAGE BEAM MODULATION
see: BEAM MODULATION PERCENTAGE

PERCENTAGE MODULATION
see: MODULATION PERCENTAGE

3461 PERCENTAGE SYNCHRONIZATION tv
The ratio of the difference in amplitude between sync peaks and blanking level to the difference in amplitude between sync peaks and reference white level.
f pourcentage *m* de synchronisation
e porcentaje *m* de sincronizacion
i percentuale *f* di sincronizzazione
n synchronisatiepercentage *n*
d Synchronisierungsnutzeffekt *m*

3462 PERCENTAGE TILT dis
f pourcentage *m* de déclivité,
taux *m* d inclinaison
e caída *f* porcentual

i avvallamento *m* percentuale
n procentueel sprongverval *n*
d prozentuale Dachschräge *pl*

3463 PERCEPTION ct
Complex effect in the field of consciousness derived from sense impressions supplemented by the memory.
f perception *f*
e percepción *f*
i percezione *f*
n perceptie, waarneming
d Empfindung *f*, Wahrnehmung *f*

PERCEPTION OF DEPTH
see: DEPTH PERCEPTION

3464 PERCEPTUAL CONSTANCIES ct
The constancies which play a role in vision, viz. brightness, size and shape constancies.
f constances *pl* de perception
e constancias *pl* de percepción
i costanze *pl* di percezione
n perceptieconstantheden *pl*, waarnemingsconstantheden *pl*
d Wahrnehmungskonstanzen *pl*

3465 PERCEPTUAL MECHANISMS ct
f mécanismes *pl* de la perception
e mecanismos *pl* de la percepción
i meccanismi *pl* della percezione
n perceptiemechanismen *pl*, waarnemingsmechanismen *pl*
d Wahrnehmungsmechanismen *pl*

3466 PERFORATING vr
The making of holes in magnetic tape used to accompany photographic films after synchronization.
f perforation *f*
e perforación *f*
i perforazione *f*
n perforatie
d Lochung *f*

3467 PERFORATION, SPROCKET HOLE fi/tv/vr
f perforation *f*
e perforación *f*
i perforazione *f*
n perforatie
d Perforationsloch *n*

3468 PERFORATION PITCH fi/tv
f pas *m* de perforation
e paso *m* de perforación
i passo *m* di perforazione
n perforatiespoed
d Lochschritt *m*

3469 PERFORMANCE ge/tv
f représentation *f*
e representación *f*
i rappresentazione *f*
n opvoering
d Aufführung *f*, Vorstellung *f*

3470 PERIOD ge
A variable quantity whose characteristics are reproduced at equal intervals of an independent variable is said to be periodic and the minimum interval after which the same characteristics recur is known as the period.
f période *f*
e período *m*
i periodo *m*
n periode
d Periode *f*

3471 PERIODIC cpl
Said of a quantity that repeats at regular intervals of time.
f périodique adj
e periódico adj
i periodico adj
n periodiek adj
d periodisch adj

3472 PERIODIC FOCUSING FIELDS crt
A method of electron beam focusing which uses a number of short focusing fields that vary periodically along the electron stream, and constitute a series of thick or thin electron lenses.
f champs *pl* de focalisation à variations périodiques
e campos *pl* de enfoque de variaciones periódicas
i campi *pl* di focalizzazione a variazioni periodiche
n periodiek veranderlijke focusseringsvelden *pl*
d periodisch veränderliche Fokussierungsfelder *pl*

3473 PERIODIC NOISE dis
Repetitive undesired signal on an audio or video network.
f bruit *m* périodique
e ruido *m* periódico
i rumore *m* periodico
n periodieke ruis *n*
d periodisches Rauschen *n*

3474 PERIODIC POLARITY REVERSAL ctv
A polarity reversal of one colo(u)r difference signal in the PAL-system used to minimize the effects of distortion.
f inversion *f* périodique de polarité
e inversión *f* periódica de polaridad
i inversione *f* periodica di polarità
n periodieke ompoling
d periodische Umpolung *f*

3475 PERIODIC TIME ge
The period when the independent variable is time.
f période *f* de temps
e período *m* de tiempo
i periodo *m* di tempo
n tijdsperiode
d Zeitperiode *f*

3476 PERIPHERAL VISION ct
The seeing of objects displaced from the primary line of sight and outside the central visual field.

f vision *f* périphérique
e visión *f* periférica
i visione *f* periferica
n zien *n* van de periferie
d Peripheriesehen *n*

3477 PERIPHERAL VISUAL FIELD ct
That portion of the visual field that falls outside the region corresponding to the foveal portion of the retina.
f champ *m* de vue périphérique
e campo *m* visual periférico
i campo *m* visuale periferico
n periferiegezichtsveld *n*
d Peripheriegesichtsfeld *n*

3478 PERITRON crt
A cathode-ray tube in which the fluorescent screen is displaced harmonically along the z-axis using a drive motor and crank assembly in order to produce a three-dimensional representation.
f peritron *m*,
tube *m* cathodique pour images en trois dimensions
e peritrón *m*,
tubo *m* catódico para imágenes en tres dimensiones
i peritrone *m*,
tubo *m* catodico per immagini in tre dimensioni
n katodestraalbuis voor driedimensionele beelden,
peritron *n*
d Katodenstrahlröhre *f* für dreidimensionale Bilder,
Peritron *n*

3479 PERMANENT MAGNET CENT(E)RING crt
Cent(e)ring of the image on the screen of a TV picture tube by means of magnetic fields produced by permanent magnets mounted around the neck of the tube.
f centrage *m* de l'image par aimants permanents
e centraje *m* de la imagen por imanes permanentes
i centratura *f* dell'immagine per magneti permanenti
n beeldcentrering door permanente magneten
d Bildzentrierung *f* durch Permanentmagnete

3480 PERMISSIBLE NOISE LEVEL dis
f niveau *m* de bruit admissible
e nivel *m* sonoro admisible
i livello *m* sonoro ammissibile
n toelaatbaar ruisniveau *n*
d zulässiger Rauschpegel *m*

3481 PERPENDICULAR MAGNETIZATION vr
Magnetizing of the recording medium in a direction perpendicular to the line of travel and parallel to the smallest cross-sectional dimensions of the medium.
f magnétisation *f* perpendiculaire
e magnetización *f* perpendicular
i magnetizzazione *f* perpendicolare
n dwarsmagnetisatie,
verticale magnetisatie
d Quermagnetisierung *f*

PERSISTENCE
see: AFTERGLOW

3482 PERSISTENCE CHARACTERISTIC crt
Of a cathode-ray tube, the relation between the time elapsing after removal of the stimulus and the intensity of light or other radiation obtained from the screen.
f caractéristique *f* de persistance
e característica *f* de persistencia
i caratteristica *f* di persistenza
n nalichtkarakteristiek
d Nachleuchtcharakteristik *f*

3483 PERSISTENCE OF VISION ct
The ability of the eye to retain images for a definite period after removal of the stimulus so that it does not respond to changes occurring at a higher rate than 25 per second.
f persistance *f* de vision,
persistance *f* rétinienne
e persistencia *f* de visión
i persistenza *f* di visione,
persistenza *f* retinica
n nawerking van de ontvangen lichtindruk,
traagheid van het oog
d Augenträgheit *f*,
Nachwirkung *f* des empfangenen Lichteindrucks,
Visionspersistenz *f*

3484 PERSISTENT CHARACTERISTIC OF CAMERA TUBES crt
The temporal step response of a camera tube to illumination.
f caractéristique *f* d'illumination,
durée *f* du point lumineux
e característica *f* de iluminación,
duración *f* del punto luminoso
i caratteristica *f* d'illuminazione,
durata *f* del punto luminoso
n beeldpuntduur,
oplichtkarakteristiek
d Aufleuchtcharakteristik *f*,
Bildpunktdauer *f*

3485 PERSISTENT PHOSPHOR crt
Phosphor in which the image persists after excitation and whose decay law is such that a usable or viewable image remains for TV purposes over the intervals commonly encountered.
f substance *f* luminescente à persistance
e substancia *f* luminiscente de persistencia
i sostanza *f* luminescente a persistenza
n nalichtende luminescerende stof
d nachleuchtender Leuchtstoff *m*

3486 PERSISTENT SCREEN tv
f écran *m* à persistance
e pantalla *f* de persistencia
i schermo *m* a persistenza
n nalichtscherm *n*
d Speicherschirm *m*

PERSONAL MICROPHONE
see: LANYARD MICROPHONE

3487 PERSUADER crt
Electrode in the image orthicon camera tube which deflects electrons returned from the target into the electron multiplier.
f déflecteur *m* d'électrons
e deflector *m* de electrones
i diflettore *m* d'elettroni
n elektronendeflector
d Elektronendeflektor *m*

3488 PFM, cpl
PULSE FREQUENCY MODULATION
f modulation *f* de fréquence de l'impulsion
e modulación *f* de frecuencia del impulso
i modulazione *f* di frequenza dell'impulso
n impulsfrequentiemodulatie
d IFM *f*,
Impulsfrequenzmodulation *f*,
PFM *f*,
Pulsfrequenzmodulation *f*

3489 PG, tv
PULSE GENERATOR
f génerateur *m* d'impulsions
e generador *m* de impulsos
i generatore *m* d'impulsi
n impulsgenerator
d Impulsgeber *m*,
Taktgeber *m*

3490 PHANTOM CIRCUIT cpl
A superposed circuit derived from two suitably arranged pairs of wires, called side circuits, the two wires of each pair being effectively in parallel.
f circuit *m* fantôme
e circuito *m* fantasma
i circuito *m* virtuale
n fantoomkring,
vierdraadslus
d Phantomleitung *f*,
Viererkreis *m*

3491 PHASE ge
Relationship between current and voltage in a reactive circuit.
f phase *f*
e fase *f*
i fase *f*
n faze
d Phase *f*

3492 PHASE ADJUSTMENT, rec/rep/tv
PHASING
The process by which the formation of an image is brought point by point into the same time and position relationship as the scanned elements of the original scene.
f mise *f* en phase
e puesta *f* en fase
i messa *f* in fase
n in faze brengen *n*
d Phasenabgleich *m*,
Phaseneinstellung *f*

PHASE-ALTERNATION LINE SYSTEM
see: PAL-COLO(U)R SYSTEM

3493 PHASE AMPLITUDE dis
DISTORTION
In a linear system, the lack of constancy of the phase difference between the output and the input of a system at different amplitudes of the input.
f distorsion *f* de phase/amplitude
e distorsión *f* de fase/amplitud
i distorsione *f* di fase/ampiezza
n faze/amplitudevervorming
d Phasen/Amplitudenverzerrung *f*

3494 PHASE ANGLE ge
The phase difference between two voltages or currents having the same frequency which are out of step with each other or the angular difference between the two vectors representing the quantities, expressed in degrees or radians.
f angle *m* de phase
e ángulo *m* de fase
i angolo *m* di fase
n fazehoek
d Phasenwinkel *m*

3495 PHASE BANDWIDTH cpl
Of an amplifier, the extent of continuous range of frequencies over which the phase/frequency characteristic of the amplifier does not depart from linearity by more than a specified amount.
f largeur *f* de bande de phase
e anchura *f* de banda de fase
i larghezza *f* di banda di fase
n fazebandbreedte,
fazedoorlaatband
d Phasenbandbreite *f*

3496 PHASE COMPARATOR cpl
A circuit in which the relative phase between the sync pulses and the sawtooth voltage of the timebase generator are compared.
f comparateur *m* de phase
e comparador *m* de fase
i comparatore *m* a regolazione di fase
n fazevergelijker
d Phasenvergleicher *m*

3497 PHASE CONSTANT cpl
Coefficient of phase change.
f constante *f* de phase
e constante *f* de fase
i costante *f* di fase
n fazeconstante
d Phasenkonstante *f*

PHASE CONTROL
see: HUE CONTROL

3498 PHASE DELAY cpl
In transmission between two points, the time equal to the phase shift between those two points divided by the angular frequency of the sinusoidal wave current.

f déphasage *m* fréquence,
temps *m* de propagation de phase
e tiempo *m* de propagación de fase
i ritardo *m* di fase-frequenza,
tempo *m* di propagazione di fase
n fazelooptijd
d Phasenlaufzeit *f*

3499 PHASE DETECTOR, cpl
PHASE DISCRIMINATOR
In a phase-modulation receiver, a device the output of which is substantially proportional to the deviation of the phase of an alternating input from some predetermined value.
f discriminateur *m* de phase
e detector *m* de fase
i discriminatore *m* di fase
n fazediscriminator
d Phasenbrücke *f*,
Phasendiskriminator *m*

3500 PHASE DEVIATION cpl
The peak difference between the instantaneous phase angle of the modulated wave and that of the sine-wave carrier.
f déviation *f* de phase
e desviación *f* de fase
i deviazione *f* di fase
n fazeafwijking
d Phasenhub *m*

3501 PHASE DIFFERENCE cpl
The time in electrical degrees by which one wave leads or lags another.
f déphasage *m*
e desfasaje *m*,
diferencia *f* de fase
i differenza *f* di fase,
sfasamento *m*
n fazeverschil *n*
d Gangunterschied *m*,
Phasendifferenz *f*,
Phasenunterschied *m*

3502 PHASE DISPLACEMENT, cpl
PHASE SHIFT
f décalage *m* de phase
e desplazamiento *m* de fase
i spostamento *m* di fase
n fazeverschuiving
d Phasendrehung *f*,
Phasenverschiebung *f*

3503 PHASE DISTORTION, dis
PHASE FREQUENCY DISTORTION
An undesired variation with frequency of the phase difference between a sinusoidal excitation and the fundamental component of the response.
f distorsion *f* de phase
e distorsión *f* de fase
i distorsione *f* di fase
n fazevervorming
d Phasenverzerrung *f*

3504 PHASE DISTORTION CORRECTION, ctv
PHASE FREQUENCY DISTORTION
CORRECTION
f correction *f* de distorsion de phase
e corrección *f* de distorsión de fase
i correzione *f* di distorsione di fase
n fazevervormingscorrectie
d Phasenentzerrung *f*

3505 PHASE DISTORTION CORRECTOR, tv
PHASE FREQUENCY DISTORTION
CORRECTOR
Apparatus for correcting faults due to phase distortion.
f correcteur *m* de phase
e corrector *m* de fase
i correttore *m* di fase
n fazecorrector
d Phasenentzerrer *m*

3506 PHASE EQUALIZATION tv
f compensation *f* du temps de propagation de phase
e compensación *f* del tiempo de propagación de fase
i compensazione *f* del tempo di propagazione di fase
n fazelooptijdcompensatie
d Phasengangentzerrung *f*,
Phasenlaufzeitausgleich *m*

3507 PHASE EQUALIZER dis
A network designed to compensate for phase-frequency distortion within a specified frequency band.
f compensateur *m* du temps de propagation de phase
e compensador *m* del tiempo de propagación de fase
i compensatore *m* del tempo di propagazione di fase
n fazelooptijdcompensator
d Phasengangentzerrer *m*

3508 PHASE ERROR dis
f erreur *f* de phase
e error *m* de fase
i errore *m* di fase
n fazefout
d Phasenfehler *m*

3509 PHASE ERROR CORRECTION ctv
f correction *f* d'erreur de phase
e corrección *f* de error de fase
i correzione *f* d'errore di fase
n fazefoutcorrectie
d Phasenfehlerkorrektur *f*

3510 PHASE FAILURE dis
f manque *m* de phase
e interrupción *f* de fase
i mancanza *f* di fase
n faze-uitval
d Phasenausfall *m*

3511 PHASE FREQUENCY cpl
f fréquence *f* de phase
e frecuencia *f* de fase
i frequenza *f* di fase
n fazefrequentie
d Phasenfrequenz *f*

3512 PHASE-FREQUENCY RESPONSE CHARACTERISTIC, PHASE RESPONSE CHARACTERISTIC cpl
A curve showing phase displacement versus frequency for a network or system.
f rapport *m* phase/fréquence vidéo, réponse *f* en phase
e relación *f* fase/frecuencia video, respuesta *f* de fase
i rapporto *m* fase/frequenza video, risposta *f* in fase
n verhouding faze/videofrequentie
d Phasendispersion *f*, Phasengang *m*

3513 PHASE JITTER dis
f bruit *m* de phase
e ruido *m* de fase
i rumore *m* di fase
n fazeruis
d Phasengeräusch *n*, Zittern *n* in der Phasenlage

3514 PHASE-LOCKED LOOP tv/vr
Feedback system in which the phase of the signal is the most important factor, employed in picture locking and video tape recorder.
f boucle *f* de réaction à accrochage de phase
e anillo *m* de reacción con encadenación de fase
i anello *m* di reazione a concatenazione di fase
n terugkoppellus met fazevergrendeling
d Rückkopplungsschleife *f* mit Phasenverkettung

3515 PHASE LOCKING cpl
Two devices are phase-locked when they are synchronized not only in speed but also in time.
f accrochage *m* de phase
e encadenacion *f* de fase
i concatenazione *f* di fase
n fazevergrendeling
d Phasenverkettung *f*

3516 PHASE MODULATION, PM cpl
Angle modulation in which the phase angle of a sine wave carrier is caused to depart from the carrier angle by an amount proportional to the instantaneous magnitude of the modulating wave.
f modulation *f* de phase
e modulación *f* de fase
i modulazione *f* di fase
n fazemodulatie
d Phasenmodulation *f*, PM *f*

3517 PHASE PRECOMPENSATION, PHASE PRECORRECTION tv
f précorrection *f* de phase
e precorrección *f* de fase
i precorrezione *f* di fase
n fazevoorcorrectie
d Phasenvorentzerrung *f*

3518 PHASE QUADRATURE, QUADRATURE cpl/tv
A phase-shift of 90°.
f quadrature *f* de phase
e cuadratura *f* de fase
i quadratura *f* di fase
n fazekwadratuur
d 90°-Verschiebung *f*, 90°-Vor-oder Nacheilung

3519 PHASE REFERENCE ctv
A reference in the encoded colo(u)r TV signal for the chrominance detector in the receiver.
f référence *f* de phase
e referencia *f* de fase
i riferimento *m* di fase
n fazereferentie
d Phasenbezug *m*

3520 PHASE REVERSAL tv
f inversion *f* de phase
e inversión *f* de fase
i inversione *f* di fase
n fazeomkering
d Phasensprung *m* am 180°, Phasenumkehr *f*

PHASE SHIFT
see: PHASE DISPLACEMENT

3521 PHASE STABILITY cpl
f stabilité *f* de phase
e estabilidad *f* de fase
i stabilità *f* di fase
n fazestabiliteit
d Phasenstabilität *f*

3522 PHASE VELOCITY cpl
Of a progressive wave consisting of a single sinusoidal component, the product of the wavelength and the frequency.
f vitesse *f* de phase
e velocidad *f* de fase
i velocità *f* di fase
n fazesnelheid
d Phasengeschwindigkeit *f*

3523 PHASELESS BOOST rec/tv
Complex circuit free from phase distortions in the signal.
f récupération *f* d'énergie sans distorsion
e recuperación *f* de energía sin distorsión
i ricupero *m* d'energia senza distorsione
n vervormingsloze energieterugwinning
d entzerrungsfreie Energierückgewinnung *f*

3524 PHASER rep/tv
f cadreur *m*
e sincronizador *m* de encuadramiento
i sincronizzatore *m* di centratura
n inraaminrichting
d Einstellvorrichtung *f* für die Phasenlage des Bildes

PHASING
see: PHASE ADJUSTMENT

3525 PHASING LINE rep/tv
In TV, that portion of the length of

scanning line set aside for the phasing signal.
f ligne *f* de cadrage
e línea *f* de encuadramiento
i linea *f* di centratura
n inramingslijn
d Phasenlagezeile *f*

3526 PHASING SIGNAL rep/tv
In TV, a signal used for adjustment of the picture position along the scanning line.
f signal *m* de cadrage
e señal *f* de encuadramiento
i segnale *m* di centratura
n inramingssignaal *n*
d Phasenlagesignal *n*

PHASMAJECTOR
see: MONOSCOPE

3527 PHASOR ge
A complex number used in connection with steady state a.c. phenomenon.
f vecteur *m* de phase
e vector *m* de fase
i vettore *m* di fase
n fazevector
d Phasenvektor *m*

3528 φ PHENOMENON ct
One of the two basic characteristics of the visual system, the other one being persistence of vision.
f phénomène *m* φ
e fenómeno *m* φ
i fenomeno *m* φ
n φ-verschijnsel *n*
d φ-Erscheinung *f*

PHOSPHOR
see: FLUORESCENT MATERIAL

3529 PHOSPHOR CLUSTER, crt
PHOSPHOR DOT TRIO,
TRIAD
f triade *f* à substance luminescente
e tríada *f* de substancia luminiscente
i triada *f* a sostanza luminescente
n luminescerende driekleurenstip
d Leuchtstoffdreier *m*,
Trüffeldreier *m*

3530 PHOSPHOR COLO(U)R RESPONSE crt
f réponse *f* chromatique d'une substance luminescente
e respuesta *f* cromática de una substancia luminiscente
i risposta *f* cromatica d'una sostanza luminescente
n kleurresponsie van een luminescerende stof
d Farbempfindlichkeit *f* eines Leuchtstoffs

3531 PHOSPHOR DOT crt
One of the tiny dots of phosphor material that are used in groups of three, one for each primary colo(u)r, on the screen of a colo(u)r TV picture tube.
f point *m* de substance luminescente
e punto *m* de substancia luminiscente
i punto *m* di sostanza luminescente
n luminescerende punt,
luminescerende stip
d Leuchtstoffpunkt *m*

3532 PHOSPHOR-DOT FACEPLATE crt
f écran *m* luminescent à points
e pantalla *f* luminiscente de puntos
i schermo *m* luminescente a punti
n stippenscherm *n*
d Punktschirm *m*

3533 PHOSPHOR EFFICIENCY crt
f rendement *m* de la substance luminescente
e rendimiento *m* de la substancia luminiscente
i rendimento *m* della sostanza luminescente
n rendement *n* van de luminescerende stof
d Leuchtstoffwirkungsgrad *m*,
Phosphor-Koeffizient *m*

3534 PHOSPHOR GRAIN crt
f grain *m* de substance luminescente
e grano *m* de substancia luminiscente
i grano *m* di sostanza luminescente
n korrel van een luminescerende stof
d Leuchtstoffkorn *n*

3535 PHOSPHOR GRANULARITY crt
An important cause of image degradation if the size of the particles is not small enough.
f granularité *f* d'une substance luminescente
e granularidad *f* de una substancia luminiscente
i granularità *f* d'una sostanza luminescente
n korreligheid van een luminescerende stof
d Leuchtstoffkörnigkeit *f*

3536 PHOSPHOR SATURATION crt
The state where further excitation of a phosphor produces no increase of light output.
f émission *f* maximale d'une substance luminescente
e emisión *f* máxima de una substancia luminiscente
i emissione *f* massima d'una sostanza luminescente
n maximale lichtemissie van een luminescerende stof
d maximale Lichtemission *f* eines Leuchtstoffs

3537 PHOSPHOR SCREEN crt
f écran *m* luminescente
e pantalla *f* luminiscente
i schermo *m* luminescente
n luminescerend scherm *n*
d Leuchtschirm *m*

3538 PHOSPHOR SCREEN BRIGHTNESS crt
f luminosité *f* d'un écran luminescent
e luminosidad *f* de una pantalla luminiscente
i luminosità *f* d'uno schermo luminescente
n schermhelderheid
d Schirmhelligkeit *f*

3539 PHOSPHOR STRIP crt
Component part of colo(u)r display tubes.
f ruban *m* luminescent
e tira *f* luminiscente
i striscia *f* luminescente
n luminescerende strook
d Leuchtstoffstreifen *m*

3540 PHOTICON crt
Miniature form of image iconoscope camera tube in which electrons emitted from a photosensitive electrode are focused onto a separate target to generate a corresponding electron image.
f iconoscope-image *m* à dimensions réduites, photicon *m*
e fotición *m*, iconoscopio-imagen *m* de dimensiones reducidas
i foticone *m*, iconoscopio-immagine *m* a dimensioni ridotte
n foticon *n*, miniatuurbeeldiconoscoop
d Miniaturbildikonoskop *n*, Photikon *n*

3541 PHOTOCATHODE, PHOTOELECTRIC CATHODE crt
In a camera tube, a semitransparent plate on which the optical image is focused and which emits electrons under the influence of light.
f photocathode *f*
e fotocátodo *m*
i fotocatodo *m*
n fotokatode
d Photokatode *f*

3542 PHOTOCELL, PHOTOELECTRIC CELL, PHOTOTUBE crt
An electronic tube in which one of the electrodes emits electrons under the influence of light or other electromagnetic radiation.
f cellule *f* photoélectrique
e célula *f* fotoeléctrica, fotocélula *f*
i cellula *f* fotoelettrica, fotocella *f*
n fotocel, lichtelektrische cel
d lichtelektrische Zelle *f*, Photozelle *f*

3543 PHOTOCONDUCTIVE CAMERA TUBE crt
A camera tube in which the photosensitive electrode is photoconductive.
f tube *m* de caméra photoconductif
e tubo *m* de cámara fotoconductor
i tubo *m* da ripresa fotoconduttivo
n camerabuis met fotogeleiding
d Aufnahmeröhre *f* mit innerem Photoeffekt

PHOTOCURRENT, PHOTOELECTRIC CURRENT
see: LIGHT CURRENT

3544 PHOTODIODE cpl
A semiconductor with a light-sensitive junction.
f photodiode *f*
e fotodiodo *m*
i fotodiodo *m*
n fotodiode
d Photodiode *f*

3545 PHOTOELECTRIC EMISSION, PHOTOEMISSION crt
Electron emission from a surface due directly to the incidence of radiant energy.
f émission *f* photoélectrique
e emisión *f* fotoeléctrica
i emissione *f* fotoelettrica
n foto-elektrische emissie
d äusserer Photoeffekt *m*

3546 PHOTOELECTRIC SCANNER rec/tv
f analyseur *m* photoélectrique
e explorador *m* fotoeléctrico
i analizzatore *m* fotoelettrico
n foto-elektrische aftaster
d photoelektrischer Abtaster *m*

3547 PHOTOELECTRIC SCANNING rec/tv
f analyse *f* photoélectrique
e exploración *f* fotoeléctrica
i analisi *f* fotoelettrica
n foto-elektrische aftasting
d photoelektrische Abtastung *f*, Photozellenabtastung *f*

3548 PHOTOEMISSIVE CAMERA TUBE crt
A camera tube in which the photosensitive electrode is photo-emissive.
f tube *m* analyseur photoémissif
e tubo *m* de cámara fotoemisor
i tubo *m* da ripresa fotoemettitore
n camerabuis met foto-emissie
d Aufnahmeröhre *f* mit äusserem Photoeffekt

3549 PHOTOLUMINESCENCE crt
Light emitted by visible, infrared or ultraviolet rays after irradiation.
f photoluminescence *f*
e fotoluminiscencia *f*
i fotoluminescenza *f*
n fotoluminescentie
d Photolumineszenz *f*

PHOTOMETRIC BRIGHTNESS
see: LUMINANCE

3550 PHOTOMOSAIC crt
Sheet of material, usually mica, which is covered by a large number of minute photocells.
f mosaïque *m* photoélectrique
e mosaico *m* fotoeléctrico
i mosaico *m* fotoelettrico
n foto-elektrisch mozaïek *n*
d photoelektrisches Mosaik *n*

PHOTOMULTIPLIER
see: MULTIPLIER PHOTOTUBE

3551 PHOTOPIC RESPONSE ct
f réponse *f* photopique
e respuesta *f* fotópica
i risposta *f* fotopica
n fotopische responsie
d Reaktion *f* des Tagessehens, Tagessehenempfindlichkeit *f*

3552 PHOTOPIC RESPONSE CURVE ct
Graphical representation of the sensitivity of the average human eye to light of different wavelengths or colo(u)rs.
f caractéristique *f* de réponse photopique
e característica *f* de respuesta fotópica
i caratteristica *f* di risposta fotopica
n fotopische responsiekromme
d Empfindlichkeitskennlinie *f* für Tagessehen

3553 PHOTOPIC VISION ct
Vision experienced by the normal eye when adapted to the higher levels of luminance.
f vision *f* photopique
e visión *f* fotópica
i fotopsia *f*
n fotopsie, kegeltjeszien *n*
d Tagessehen *n*

3554 PHOTO-PULSE stu
A disturbance resulting in a heavy black and white horizontal bar in the camera output.
f photo-impulsion *f*
e foto-impulso *m*
i foto-impulso *m*
n foto-impuls
d Photoimpuls *m*

3555 PHYSICAL COLO(U)RS ct
f couleurs *pl* physiques, couleurs *pl* réelles
e colores *pl* físicos, colores *pl* reales
i colori *pl* fisici, colori *pl* reali
n fysische kleuren *pl*, reële kleuren *pl*
d physikalische Farben *pl*

3556 PHYSICAL EDITING vr
Cutting and splicing audio or video tape mechanically, as distinct from electronic editing.
f montage *m* mécanique
e montaje *m* mecánico
i montaggio *m* meccanico
n mechanische montage
d mechanischer Schnitt *m*

3557 PHYSICAL TAPE TESTS aud/vr
f essais *pl* du matériel de bande
e pruebas *pl* del material de cinta
i prove *pl* del materiale di nastro
n bandmateriaalonderzoek *n*
d Bandmaterialprüfung *f*

3558 PHYSIOLOGICAL COLO(U)RS ct
f couleurs *pl* physiologiques
e colores *pl* fisiológicos
i colori *pl* fisiologici
n fysiologische kleuren *pl*
d physiologische Farben *pl*

3559 PIANO KEYBOARD rec/tv
A device used in a news studio to select a desired registered effect in a video cassette.
f clavier *m*
e teclado *m*
i tastiera *f*
n toetsenbord *n*
d Tastenbrett *n*

3560 PICKUP EQUIPMENT rec/tv
f appareil *m* de prise de vues
e aparato *m* de toma de vistas
i apparecchio *m* di ripresa
n beeldopneemapparaat *n*
d Bildaufnahmegerät *n*

3561 PICKUP LINK TRANSMITTER rec/tv
f émetteur *m* lié à la caméra
e emisor *m* de enlace con la cámara
i emettitore *m* con connessione alla telecamera
n zender met cameraverbinding
d Kameraverbindungssender *m*

3562 PICKUP SENSITIVITY rec/tv
f sensibilité *f* de prise de vues
e sensibilidad *f* de toma de vistas
i sensibilità *f* di ripresa
n opneemgevoeligheid
d Aufnahmeempfindlichkeit *f*

3563 PICKUP SPECTRAL CHARACTERISTIC rec/tv
f caractéristique *f* spectrale de prise de vues
e característica *f* espectral de captación
i caratteristica *f* spettrale di ripresa
n spectrale opneemkarakteristiek
d spektrale Aufnahmecharakteristik *f*

PICKUP TUBE (US)
see: CAMERA TUBE

3564 PICKUP VELOCITY (US), SCANNING SPEED (GB) rec/tv
f vitesse *f* d'analyse
e velocidad *f* de exploración
i velocità *f* d'analisi
n aftastsnelheid
d Abtastgeschwindigkeit *f*

PICTURE (GB)
see: FRAME

PICTURE
see: IMAGE

PICTURE ALTITUDE (GB), PICTURE HEIGHT (GB)
see: FRAME AMPLITUDE

PICTURE ALTITUDE CONTROL (GB)
see: FRAME-AMPLITUDE CONTROL

PICTURE ANALYZER
see: IMAGE ANALYZER

PICTURE AND WAVEFORM MONITOR,
see: IMAGE AND WAVEFORM MONITOR

PICTURE AREA,
PICTURE PLANE
see: IMAGE AREA

PICTURE BACKGROUND
see: IMAGE BACKGROUND

PICTURE BLACK
see: IMAGE BLACK

PICTURE BORDER
see: IMAGE BORDER

PICTURE BRIGHTNESS
see: IMAGE BRIGHTNESS

PICTURE CARRIER
see: IMAGE CARRIER

PICTURE CARRIER TRAP
see: ADJACENT PICTURE CARRIER TRAP

PICTURE CHANNEL
see: IMAGE CHANNEL

PICTURE COIL (GB)
see: FRAME COIL

PICTURE COMPRESSION
see: PACKING

PICTURE CONTRAST
see: IMAGE CONTRAST

PICTURE CONTROL COIL
see: IMAGE CONTROL COIL

PICTURE CURRENT
see: IMAGE CURRENT

PICTURE DEFINITION (GB),
PICTURE RESOLUTION (US)
see: IMAGE DEFINITION

PICTURE DEFLECTION (GB)
see: FRAME DEFLECTION

PICTURE DETAIL
see: IMAGE DETAIL

PICTURE DETAIL FACTOR
see: IMAGE DETAIL FACTOR

PICTURE DETAIL REGION
see: IMAGE DETAIL REGION

PICTURE DIAGONAL
see: IMAGE DIAGONAL

PICTURE DIMENSIONS
see: IMAGE DIMENSIONS

PICTURE DISTORTION (GB)
see: FRAME DISTORTION

PICTURE DISTORTION
see: IMAGE DISTORTION

PICTURE DRIFT
see: IMAGE DRIFT

PICTURE DROP OUT
see: IMAGE DROP OUT

PICTURE DURATION (US)
see: FRAME RATE

PICTURE DURATION
see: IMAGE DURATION

PICTURE EDGE (GB)
see: FRAME BORDER

PICTURE ELEMENT,
PICTURE POINT
see: IMAGE ELEMENT

PICTURE FLYBACK TIME (GB)
see: FRAME RETRACE TIME

PICTURE FREQUENCY (GB),
PICTURE REPETITION RATE (GB)
see: FRAME FREQUENCY

PICTURE GATE
see: CAMERA APERTURE

3365 PICTURE GENERATOR tv
f générateur *m* d'image artificielle
e generador *m* de imagen artificial
i generatore *m* d'immagine artificiale
n kunstbeeldgenerator
d Kunstbildgeber *m*

PICTURE GEOMETRY
see: IMAGE GEOMETRY

PICTURE GEOMETRY FAULT
see: GEOMETRIC DISTORTION

PICTURE HEIGHT
see: IMAGE HEIGHT

PICTURE IF AMPLIFIER
see: IMAGE IF AMPLIFIER

PICTURE INFORMATION
see: IMAGE INFORMATION

PICTURE INPUT TRANSFORMER (GB)
see: FRAME INPUT TRANSFORMER

PICTURE INTERFERENCE
see: IMAGE INTERFERENCE

PICTURE INTERPOLATION
see: IMAGE INTERPOLATION

3566 PICTURE JOIN vr
A method to overcome the difficulties in telerecording a film of which the frame rate is 24 per second by a TV camera whose picture rate is 30 per second.
f égalisation *f* des vitesses d'images

e compensación *f* de las velocidades de cuadros y imágenes
i compensazione *f* delle velocità di fotogrammi e d'immagini
n vereffening van het verschil in snelheid van film- en televisiebeelden
d Ausgleich *m* der Geschwindigkeitsdifferenz von Film- und Fernsehbilder

PICTURE LINE
see: IMAGE LINE

PICTURE LINE-AMPLIFIER OUTPUT
see: IMAGE LINE-AMPLIFIER OUTPUT

3567 PICTURE LINE-UP GENERATING EQUIPMENT, PLUGE rep/tv
Equipment which generates a waveform specially designed for rapid and accurate adjustment of the operational controls of a picture monitor.
f générateur *m* du signal du moniteur
e generador *m* de la señal del monitor
i generatore *m* del segnale del monitore
n monitorsignaalgenerator
d Monitoreichpegelgeber *m*

PICTURE LINEARITY CONTROL (GB)
see: FRAME LINEARITY CONTROL

PICTURE LOCK
see: IMAGE LOCK

3568 PICTURE LOCKING TECHNIQUES rec/tv
Techniques serving to synchronize different and often remote TV picture sources so that transitions between them by way of mixes and special effects may be made without disturbance to the received picture.
f techniques *pl* de verrouillage de l'image
e técnicas *pl* de enganche de la imagen
i tecniche *pl* di blocco dell'immagine
n beeldvergrendelingstechnieken *pl*
d Bildverkopplungstechniken *pl*

PICTURE MODULATION
see: IMAGE MODULATION

PICTURE MONITOR
see: IMAGE MONITOR

PICTURE ORBITING FACILITY
see: IMAGE ORBITING FACILITY

PICTURE ORIENTATION
see: IMAGE ORIENTATION

PICTURE OUTPUT
see: IMAGE OUTPUT

PICTURE OUTPUT TRANSFORMER (GB)
see: FRAME OUTPUT TRANSFORMER

PICTURE PERIOD (GB)
see: FRAME PERIOD

3569 PICTURE PICKUP SYSTEM, TELEVISION PICKUP SYSTEM tv
The operation and process by which the luminous characteristics of the object to be televised are translated into electrical signals.
f prise *f* de vues en télévision
e toma *f* de vistas en televisión
i presa *f* televisiva, ripresa *f* televisiva
n beeldopname voor televisie, optisch-elektrische beeldopname
d Fernsehbildaufnahme *f*

PICTURE QUALITY
see: IMAGE QUALITY

PICTURE RATIO
see: APERTURE RATIO

PICTURE SCANNER
see: IMAGE SCANNER

PICTURE SCANNING
see: IMAGE SCANNING

PICTURE SECTION
see: IMAGE SECTION

PICTURE SEQUENTIAL SYSTEM (GB)
see: FRAME SEQUENTIAL SYSTEM

PICTURE SHIFT
see: IMAGE SHIFT

PICTURE SIGNAL
see: IMAGE SIGNAL

PICTURE SIGNAL AMPLITUDE
see: IMAGE SIGNAL AMPLITUDE

PICTURE SIGNAL DISTRIBUTION AMPLIFIER
see: IMAGE SIGNAL DISTRIBUTION AMPLIFIER

PICTURE SIGNAL GENERATOR
see: IMAGE SIGNAL GENERATOR

PICTURE SIGNAL POLARITY
see: IMAGE SIGNAL POLARITY

PICTURE SIMULTANEOUS SYSTEM (GB)
see: FRAME SIMULTANEOUS SYSTEM

PICTURE SIZE (GB)
see: FRAME SIZE

PICTURE SLIP (GB)
see: FRAME SLIP

PICTURE SOURCES
see: IMAGE SOURCES

PICTURE STORAGE TUBE
see: IMAGE STORAGE TUBE

PICTURE STRIP
see: IMAGE STRIP

PICTURE SUPERIMPOSITION (GB)
see: IMAGE REGISTRATION

PICTURE SWEEP UNIT (GB)
see: FRAME SWEEP UNIT

PICTURE SWEEP VOLTAGE (GB)
see: FRAME SWEEP VOLTAGE

PICTURE SWITCHING
see: IMAGE SWITCHING

PICTURE SYNC PULSE (GB)
see: FRAME SYNC PULSE

PICTURE SYNCHRONIZATION (GB)
see: FRAME SYNCHRONIZATION

PICTURE SYNCHRONIZATION CONTROL (GB)
see: FRAME SYNCHRONIZATION CONTROL

PICTURE SYNCHRONIZING PULSE (GB)
see: FRAME SYNCHRONIZING PULSE

PICTURE SYNCHRONIZING RATIO
see: IMAGE SYNCHRONIZING RATIO

PICTURE TIMEBASE (GB)
see: FRAME TIMEBASE

PICTURE TIMEBASE OSCILLATION (GB)
see: FRAME TIMEBASE OSCILLATION

3570 PICTURE-TO-SYNCHRONIZING RATIO
f rapport *m* vidéo-synchronisation
e relación *f* video-sincronización
i rapporto *m* immagine-sincronizzazione
n video-synchronisatieverhouding
d Bild-Synchronverhältnis *n*

PICTURE TRANSFER CONVERTER
see: IMAGE TRANSFER CONVERTER

PICTURE TRANSMISSION
see: IMAGE TRANSMISSION

PICTURE TRANSMITTER
see: IMAGE TRANSMITTER

PICTURE TRANSMITTER POWER
see: IMAGE TRANSMITTER POWER

3571 PICTURE TUBE crt
f tube *m* image
e tubo *m* imagen
i tubo *m* immagine
n beeldbuis
d Bildröhre *f*, Bildwiedergaberöhre *f*

PICTURE TUBE GAMMA
see: GAMMA OF PICTURE TUBE

3572 PICTURE TUBE TRANSFER CHARACTERISTIC crt/rec/tv
f caractéristique *f* de transfert du tube image
e característica *f* de transferencia del tubo imagen
i caratteristica *f* di trasferimento del tubo immagine
n overdrachtskarakteristiek van de beeldbuis
d Übertragungscharakteristik *f* der Bildröhre

PICTURE WHITE
see: CARRIER REFERENCE WHITE LEVEL

PICTURE WHITE
see: IMAGE WHITE

PICTURE WIDTH
see: IMAGE WIDTH

3573 PILLOW DISTORTION, PINCUSHION DISTORTION crt/dis
Distortion of a rectangular display into one in which the sides cave inwards, due to distortion of the deflecting field and/or screen curvature.
f distorsion *f* en coussinet
e distorsión *f* en cojín
i distorsione *f* a cuscinetto
n kussenvervorming, kussenvormige vertekening
d Kissenverzerrung *f*

3574 PILLOW DISTORTION EQUALIZING, PINCUSHION DISTORTION EQUALIZING dis
f correction *f* de la distorsion en coussinet
e corrección *f* de la distorsión en cojín
i correzione *f* della distorsione a cuscinetto
n opheffen *n* van de kussenvervorming
d Kissenentzerrung *f*

3575 PILOT, PILOT CARRIER, PILOT SIGNAL cpl/tv
A low-level signal at carrier frequency transmitted for receiver-control purposes, e.g. tuning control and automatic gain control simultaneously with the required signal.
f signal *m* de commande
e señal *f* de mando
i segnale *m* di comando
n stuursignaal *n*
d Steuersignal *n*

PILOT BEAM
see: INDEXING ELECTRON BEAM

3576 PILOT LAMP vr
f lampe-témoin *f*
e lámpara *f* testigo
i lampadina *f* spia
n controlelamp, indicatielamp
d Kontrollampe *f*

3577 PILOT TONE aud
A method used in field sound recording or recording a control track in addition to the normal sound on a magnetic recorder.

f ton *m* de commande
e tono *m* de mando
i tono *m* di comando
n stuurtoon
d Steuerton *m*

3578 PIPED PROGRAM(ME) tv
A radio or TV program(me) sent over commercial transmission lines.
f programme *m* diffuse par téléphone
e programa *m* transmitido por teléfono
i programma *m* trasmesso per telefono
n over het telefoonnet uitgezonden programma *n*
d Sendung *f* über das öffentliche Telephonieleitungsnetz

3579 PIRATE VIEWER tv
f téléspectateur *m* clandestin
e telespectador *m* clandestino
i telespettatore *m* clandestino
n zwartkijker
d Schwarzseher *m*

3580 PIRATE VIEWER DETECTION tv
f dépistage *m* de téléspectateurs clandestins
e descubrimiento *m* de telespectadores clandestinos
i rivitracciamento *m* di telespettatori clandestini
n opsporing van zwartkijkers
d Schwarzseherfahndung *f*

3581 PIT PATTERN vr
f configuration *f* de piqûres
e configuración *f* de hoyos
i configurazione *f* di rilievi
n putjespatroon *n*
d längliche Vertiefungen *pl* der Bildspur

3582 PITCH CONTROL tv
f réglage *m* de la distance interligne
e regulación *f* de la distancia entre líneas
i regolazione *f* della distanza tra linee
n regeling van de onderlinge lijnafstand
d Regelung *f* der Zeilendichte

PLANAR MASK
see: APERTURE MASK

3583 PLANCKIAN LOCUS ct
The line in a chromaticity diagram representing full radiators of different temperatures.
f courbe *f* de Planck,
lieu *m* des corps noirs
e lugar *m* geométrico de Planck
i luogo *m* geometrico di Planck
n kleursoortenkromme van de zwarte stralers
d Ortskurve *f* der schwarzen Strahler

3584 PLANE REFLECTOR AERIAL, aea
PLANE REFLECTOR ANTENNA
f antenne *f* à réflecteur plan
e antena *f* de reflector plano
i antenna *f* a riflettore piatto
n antenne met vlakke reflector
d Flachreflektorantenne *f*

PLANNING FOR CONTINUITY
see: CONTINUITY PLANNING

3585 PLASTIC EFFECT, rep/tv
RELIEF EFFECT
An appearance of relief in the reproduced picture due to accidental or intentional modification of the picture signal.
f plastique *f*
e efecto *m* plástico
i effetto *m* di rilievo
n reliëfeffect *n*
d Bildplastik *f*,
Plastikeffekt *m*,
Raumeffekt *m*

PLATE-VOLTAGE-STABILIZED CAMERA TUBE (US)
see: ANODE-VOLTAGE-STABILIZED CAMERA TUBE

3586 PLATEN ani
Component part of animation equipment.
f plaque *f* en verre
e placa *f* en vidrio
i piastra *f* en vetro
n glasplaat
d Glasplatte *f*

3587 PLAYBACK, vr
REPLAY,
REPRODUCTION
The process of reproducing recorded information.
f lecture *f*,
reproduction *f*
e reproducción *f*
i lettura *f*,
riproduzione *f*
n afspelen *n*,
reproduktie,
weergave
d Wiedergabe *f*

3588 PLAYBACK AMPLIFIER aud/cpl
f amplificateur *m* de reproduction
e amplificador *m* de reproducción
i amplificatore *m* di riproduzione
n terugspeelversterker
d Rückspielverstärker *m*,
Wiedergabeverstärker *m*

3589 PLAYBACK BUTTON vr
f touche *f* de reproduction
e tecla *f* de reproducción
i tasto *m* di riproduzione
n terugspeeltoets
d Rückspieltaste *f*,
Wiedergabetaste *f*

3590 PLAYBACK CHARACTERISTIC, vr
REPLAY CHARACTERISTIC,
REPRODUCTION CHARACTERISTIC
f caractéristique *f* de reproduction
e característica *f* de reproducción
i caratteristica *f* di riproduzione
n weergeefkarakteristiek
d Wiedergabecharakteristik *f*

3591 PLAYBACK DURATION vr
f durée *f* d'audition, durée *f* de reproduction
e duración *f* de reproducción
i durata *f* di riproduzione
n terugspeeltijd
d Abspieldauer *f*

3592 PLAYBACK EQUALIZING vr
f égalisation *f* de la reproduction
e igualación *f* de la reproducción
i uguagliamento *m* della riproduzione
n weergavecorrectie
d Wiedergabeentzerrung *f*

3593 PLAYBACK HEAD, REPRODUCER HEAD vr
A head that converts a changing magnetic field on magnetic tape into corresponding electric signals.
f tête *f* de reproduction
e cabeza *f* de reproducción
i testina *f* di riproduzione
n weergeefkop
d Wiedergabekopf *m*

3594 PLAYBACK LOSS vr
f pertes *pl* de reproduction
e pérdidas *pl* de reproducción
i perdite *pl* di riproduzione
n weergeefverliezen *pl*
d Abspielfehler *pl*, Wiedergabeverluste *pl*

PLAYBACK LOUDSPEAKER (US)
see: BACKGROUND LOUDSPEAKER

3595 PLAYBACK ON MONO vr
f reproduction *f* d'enregistrements monophoniques
e reproducción *f* de registros monofónicos
i riproduzione *f* di registrazioni monofonici
n monoweergave
d Wiedergabe *f* von Monoaufnahmen

3596 PLAYBACK ON STEREO vr
f reproduction *f* d'enregistrements stéréophoniques
e reproducción *f* de registros estereofónicos
i riproduzione *f* di registrazioni stereofonici
n stereoweergave
d Wiedergabe *f* von Stereoaufnahmen

3597 PLUG-IN CIRCUITRY cpl
f carte *f* enfichable
e circuitos *pl* enchufables
i pannello *m* ad innesto
n insteekbedrading
d Einschubbedrahtung *f*, Einsteckleiterplatte *f*

3598 PLUG-IN SWITCHING BLOCKS, PLUG-IN UNITS cpl
f blocs *pl* de commutation enfichables
e bloques *pl* de conmutación enchufables
i blocchi *pl* di commutazione ad innesto
n insteekbare schakelblokken *pl*
d Einschubschalteinheiten *pl*

3599 PLUMBICON (REG. T.M.) crt/ctv
Photoconductive type of TV camera tube similar to the vidicon in principle of operation but giving improved results by the use of a semiconductor lead monoxide target which is doped three sub-layers forming a p-i-n diode which is reverse biased.
f ---
e ---
i ---
n ---
d ---

PM
see: PHASE MODULATION

3600 PNEUMATIC FAST PULLDOWN MECHANISM vr
f mécanisme *m* à entraînement rapide
e mecanismo *m* de transporte rápido
i meccanismo *m* di trasporto rapido
n snel transportmechanisme *n*
d Schnellschaltwerk *n*

3601 PNEUMATIC PULLDOWN vr
Film feeding device in telecine machine using compressed air as moving agent.
f transporteur *m* pneumatique
e transportador *m* neumático
i trasportatore *m* pneumatico
n pneumatische transporteur
d pneumatisches Transportglied *n*

POINT GAMMA
see: CONTRAST GRADIENT

3602 POINT-TO-POINT TRANSMISSION rep/tv
f transmission *f* entre deux points fixes
e transmisión *f* fija de punto a punto
i trasmissione *f* fissa da punto a punto
n transmissie tussen twee vaste punten
d Punkt-zu-Punkt-Übertragung *f*

POISON
see: KILLER

3603 POISONING crt
f empoisonnement *m*
e envenenamiento *m*
i avvelenamento *m*
n inbranden *n*, vergiftiging
d Einbrennen *n*

POLAR DIAGRAM
see: DIRECTIONAL PATTERN

3604 POLARITY rec/rep/tv
f polarité *f*
e polaridad *f*
i polarità *f*
n polariteit
d Polarität *f*

3605 POLARITY INVERSION ctv
f inversion *f* de polarité
e inversión *f* de polaridad
i inversione *f* di polarità
n ompoling
d Umpolung *f*

POLE
see: MAGNETIC POLE

POLE FACE
see: MAGNETIC POLE FACE

POLE PIECE,
POLE SHOE
see: MAGNETIC POLE PIECE

POLE TIPS
see: MAGNETIC POLE TIPS

3606 POLISHING OF THE TAPE, vr
TAPE POLISHING
One of the operations needed to realize a perfect surface of the magnetic tape.
f polissage *m* de la bande magnétique
e pulimento *m* de la cinta magnética
i pulitura *f* del nastro magnetico
n polijsten *n* van de magneetband
d Polieren *n* des Magnetbandes

3607 POLYROD AERIAL, aea
POLYROD ANTENNA
f antenne *f* diélectrique en barre
e antena *f* dieléctrica en barra
i antenna *f* dielettrica in barra
n staafvormige diëlektrische antenne
d Polystyrolstabantenne *f*

3608 POOR LANDING, crt
PORTHOLE
In camera tubes, a defect in a properly aligned camera tube emplying low-velocity scanning, resulting in an increase in target-cutoff voltage and a decrease in sensitivity toward the corners of the picture.
f oblitération *f* des coins
e obliteración *f* de las esquinas
i obliterazione *f* angolare
n patrijspoort
d Eckenverwaschnung *f*

3609 POP STRANDING dis/vr
f décalage *m* vertical de spires
e decalaje *m* vertical de las espiras
i spire *pl* disallineate
n omhooglopen *n* van enige windingen
d Hochlaufen *n* einzelner Windungen im Bandwickel

3610 PORCH rec/rep/tv
f palier *m*
e rellano *m*
i pianerottolo *m*
n stoep
d Schwarzschulter *f*,
Schwarztreppe *f*

3611 POROUS ABSORBERS aud
Materials such as felt, glass wool and acoustic plaster, used to absorb sound energy.
f absorbants *pl* poreux
e absorbentes *pl* porosos
i assorbenti *pl* porosi
n poreuse absorberende stoffen *pl*
d poröses Schluckmaterial *n*

3612 POROUS STRUCTURE aud
Sound absorbing materials in form of blankets, perforated or slotted surfaces, etc.
f structure *f* poreuse
e estructura *f* porosa
i struttura *f* porosa
n poreuse structuur
d poröse Struktur *f*

PORTABLE CAMERA
see: CREEPIE PEEPIE

3613 PORTABLE EYE-WITNESS rec/tv
REPORT TRANSMITTER,
PORTABLE TELEVISION CAMERA TRANSMITTER
f émetteur *m* de reportage de télévision portatif
e emisor *m* de reportaje televisivo portátil
i emettitore *m* di cronaca televisiva portatile,
emettitore *m* di reportage televisivo portatile
n draagbare televisiereportagezender
d Tornistersender *m* für Fernsehreportagen

3614 PORTABLE RECORDER vr
f appareil *m* enregistreur portatif
e aparato *m* registrador portátil
i apparecchio *m* registratore portatile
n draagbaar opneemapparaat *n*,
draagbare recorder
d tragbares Aufnahmegerät *n*

3615 PORTABLE TELEVISION rep/tv
RECEIVER,
WALKIE-LOOKIE
f téléviseur *m* portatif
e televisor *m* portátil
i televisore *m* portatile
n draagbaar televisietoestel *n*
d tragbares Fernsehgerät *n*

3616 PORTABLE TELEVISION rec/tv
STATION
f appareil *m* de prise de vues portatif
e aparato *m* de toma de vistas portátil
i apparecchio *m* di riprese portatile
n draagbaar televisieopneemapparaat *n*
d tragbares Fernsehaufnahmegerät *n*

3617 PORTABLE TELEVISION rec/tv
TRANSMITTER,
WALKIE-TALKIE
f émetteur *m* de télévision portatif
e emisor *m* de televisión portátil
i emettitore *m* di televisione portatile
n draagbare televisiezender
d tragbarer Fernsehsender *m*

3618 POSITIONAL CONTROL SERVO opt
Means for remotely controlling a mechanism, such as the focus ring of a lens.
f dispositif *m* de téléréglage
e dispositivo *m* de telerregulación
i dispositivo *m* di teleregolazione
n afstandsregelaar
d Fernregler *m*

3619 POSITIONAL CROSSTALK rep/tv
The variation in the path followed by any one electron beam as a result of a change impressed on any other beam in a multi-beam cathode-ray tube.
f interpénétration *f* du parcours
e influencia *f* mutua de haces
i influenza *f* mutua di fasci
n onderlinge baanbeïnvloeding
d gegenseitige Bahnbeeinflussung *f*

POSITIONAL HUM
see: GEOMETRY HUM

3620 POSITIVE AMPLITUDE MODULATION, cpl/ge/tv
POSITIVE LIGHT MODULATION,
POSITIVE PICTURE MODULATION
A form of modulation in which the amplitude of the vision signal increases with increasing luminosity of the picture elements.
f modulation *f* d'amplitude positive
e modulación *f* de amplitud positiva
i modulazione *f* d'ampiezza positiva
n positieve amplitudemodulatie
d positive Amplitudenmodulation *f*, Positivmodulation *f*, Steuerung *f* auf Hell

3621 POSITIVE ECHO, dis
POSITIVE GHOST
A ghost image having the same tonal variations as the primary TV image.
f image *f* fantôme positive
e imagen *f* fantasma positiva
i immagine *f* fantasma positiva
n positief echobeeld *n*
d positives Echobild *n*

3622 POSITIVE FEEDBACK, cpl
REGENERATIVE FEEDBACK
The process by which a part of the power in the output circuit of an amplifying device reacts upon the input circuit in such a manner as to reinforce the initial power, thereby increasing the amplification.
f réaction *f* positive
e reacción *f* positiva
i reazione *f* positiva
n meekoppeling, positieve terugkoppeling
d Mitkopplung *f*

3623 POSITIVE FREQUENCY MODULATION cpl/ge/tv
A form of modulation in which the frequency of the vision signal increases with increasing the brightness of the picture elements.
f modulation *f* de fréquence positive
e modulación *f* de frecuencia positiva
i modulazione *f* di frequenza positiva
n positieve frequentiemodulatie
d positive Frequenzmodulation *f*

3624 POSITIVE GOING SYNC PULSES tv
f impulsions *pl* de synchronisation à potentiel positif
e impulsos *pl* de sincronización de potencial positivo
i impulsi *pl* di sincronizzazione a potenziale positivo
n synchronisatie-impulsen *pl* met positieve potentiaal
d Synchronisationsimpulse *pl* mit positivem Potential

3625 POSITIVE IMAGE, rep/tv
POSITIVE PICTURE
A picture as normally seen on a TV picture tube, having the same rendition of light and shade as in the original scene being televised.
f image *f* positive
e imagen *f* positiva
i immagine *f* positiva
n positief beeld *n*
d Positivbild *n*

3626 POSITIVE-NEGATIVE SWITCH rec/rep/tv
Switch on a camera to change the picture from positive to negative or on a telecine machine to enable it to produce a positive picture from a negative film.
f commutateur *m* positif-négatif
e conmutador *m* positivo-negativo
i commutatore *m* positivo-negativo
n positief-negatief-schakelaar
d Positiv-Negativ-Schalter *m*

3627 POSITIVE PICTURE PHASE, rep/tv
POSITIVE PICTURE POLARITY
A condition in which increase in brilliance of the object being televised causes the picture signal voltage to increase above the zero level in a positive direction.
f polarité *f* positive du signal image
e polaridad *f* positiva de la señal imagen
i polarità *f* positiva del segnale immagine
n positieve polariteit van het beeldsignaal
d positive Polarität *f* des Bildsignals

3628 POSITIVE TRANSMISSION tv
A TV system in which maximum radiated power from the transmitter corresponds to maximum white area in the picture.
f transmission *f* de télévision en modulation positive
e transmisión *f* televisiva mandada
i trasmissione *f* televisiva comandata
n televisieoverdracht met positieve modulatie
d Fernsehübertragung *f* mit positiver Modulation

3629 POSITIVE VIDEO SIGNAL rec/rep/tv
A video signal in which increase in amplitude corresponds to increasing light value in the transmitted image.
f signal *m* vidéo positif
e señal *f* video positiva
i segnale *m* video positivo
n positief videosignaal *n*
d positives Videosignal *n*

POST-ACCELERATION
see: AFTER-ACCELERATION

POST-ACCELERATION CATHODE-RAY TUBE,
POST-DEFLECTION CATHODE-RAY TUBE
see: AFTER-ACCELERATION CATHODE-RAY TUBE

POST-DEFLECTION ACCELERATOR
see: INTENSIFIER

POST-DEFLECTION FOCUS TUBE
see: CHROMATRON

POST-DEFLECTION FOCUSING
see: AFTER-DEFLECTION FOCUSING

POST-EMPHASIS
see: DE-ACCENTUATION

POST-EMPHASIS CIRCUIT
see: DE-ACCENTUATOR

POST-EMPHASIS NETWORK
see: DE-ACCENTUATION NETWORK

3630 POST-EQUALIZATION aud/dis
A method of overcoming distortion due to gramophone pickups, loudspeakers and microphones by inserting the equal and opposite gain correction after the distortion has occurred.
f postégalisation *f*
e postigualación *f*
i postuguagliamento *m*
n nacorrectie
d Nachentzerrung *f*

3631 POST-SYNC FIELD-BLANKING INTERVAL rec/rep/tv
The interval of time immediately following the field sync pulse and during which blanking level and line sync pulses are transmitted.
f intervalle *m* de suppression de trame après synchronisation
e intervalo *m* de supresión de campo después de sincronización
i intervallo *m* di cancellazione di trama dopo sincronizzazione
n rastersynchronisatieachterstoep
d Austastzeit *f* nach dem Vertikal-synchronsignal

3632 POST-SYNCHRONIZATION fi/tv
f postsynchronisation *f*
e postsincronización *f*
i postsincronizzazione *f*
n nasynchronisatie
d Doubeln *n*,
Nachsynchronisation *f*

3633 POWER AMPLIFIER cpl
Amplifier designed to deliver a relatively large output current into a low impedance in order to obtain a power gain.
f amplificateur *m* de puissance
e amplificateur *m* de potencia
i amplificatore *m* di potenza
n krachtversterker
d Leistungsverstärker *m*

3634 POWER CONSUMPTION cpl
f consommation *f* électrique,
puissance *f* absorbée
e consumo *m*
i consumo *m*
n netbelasting,
opgenomen vermogen *n*
d Leistungsaufnahme *f*

POWER EQUALIZER (US)
see: ECHO EQUALIZER

POWER FEEDBACK
see: BOOSTER CIRCUIT

3635 POWER LEVEL cpl
f niveau *m* de puissance
e nivel *m* de potencia
i livello *m* di potenza
n vermogensniveau *n*
d Leistungspegel *m*

3636 POWER RATING OF A TRANSMITTER cpl
The radio-frequency power applied to the aerial (antenna), feeder, or equivalent artificial load under specified conditions.
f régime *m* nominal de la puissance de l'émetteur
e régimen *m* nominal de la potencia del emisor
i regime *m* nominale della potenza dell'emettitore
n nominale vermogensgegevens *pl* van de zender
d Leistungsziffer *f* des Senders

3637 POWER SEPARATION FILTERS cpl
Filters which separate power supplies from signals enable both to be carried on the same conductors.
f filtres *pl* séparateurs
e filtros *pl* separadores
i filtri *pl* separatori
n scheidingsfilters *pl*
d Trennfilter *pl*

3638 POWER SUPPLY cpl
f alimentation *f*
e alimentación *f*
i alimentazione *f*
n voeding
d Speisung *f*

3639 POWER UNIT aea/tv
f bloc *m* d'alimentation, groupe *m* électrogène
e unidad *f* de alimentación
i unità *f* d'alimentazione
n voedingseenheid
d Speiseaggregat *n*, Speisungseinheit *f*

3640 POWER VAN rec/tv
f voiture *f* d'alimentation
e máquina *f* de alimentación
i macchina *f* d'alimentazione
n voedingswagen
d Speisungskraftwagen *m*

3641 PREAMP, PREAMPLIFIER cpl
An amplifier used ahead of a main amplifier.
f préamplificateur *m*
e preamplificador *m*
i preamplificatore *m*
n voorversterker
d Vorverstärker *m*

3642 PRECISION BLANKING AND SYNC SIGNAL MIXER cpl
f mélangeur *m* de mesure
e mezclador *m* de medida
i mescolatore *m* di misura
n meetmenger
d Messmischer *m*

3643 PRECISION MONITOR rep/tv
A monitor, e.g. the master picture monitor which is placed at a keypoint in the control system.
f moniteur *m* de précision
e monitor *m* de precisión
i monitore *m* di precisione
n precisiemonitor
d Präzisionsmonitor *m*

3644 PRECISION OFFSET cpl
f décalage *m* de précision
e desplazamiento *m* de precisión
i spostamento *m* di precisione
n precisieverzet *n*
d Präzisionsoffset *m*

3645 PREDETERMINED SEQUENCE rec/rep/tv
f séquence *f* prédéterminée
e secuencia *f* predeterminada
i sequenza *f* predeterminata
n voorafbepaalde volgorde
d vorgewählte Reihenfolge *f*

3646 PREDUBBING fi/tv
f présynchronisation *f*
e presincronización *f*
i sincronizzazione *f* preventiva
n voorsynchronisatie
d Vorsynchronisation *f*

PRE-EMPHASIS
see: ACCENTUATION

PRE-EMPHASIS CIRCUIT
see: ACCENTUATOR

PRE-EMPHASIS FILTER
see: ACCENTUATION FILTER

3647 PRE-EQUALIZATION aud/dis
A method of overcoming distortion due to gramophone pickups, loudspeakers and microphones by anticipating the distortion which occurs further down the chain and provides equal and opposite gain variation before the distortion occurs.
f préégalisation *f*
e preigualación *f*
i preuguagliamento *m*
n voorcorrectie
d Vorentzerrung *f*

3648 PREFADE LISTENING rep/tv
Listening to a program(me) output for control purposes before it is faded up for transmission.
f écoute *f* de test
e escucha *f* de prueba
i ascolto *m* di prova
n meeluisteren *n* vóór de uitzending
d Vorhören *n*

3649 PRELISTEN aud
Facility which enables a sound operator to listen to a sound source before that source is connected to the main output of a sound mixer.
f dispositif *m* d'écoute de test
e dispositivo *m* de escucha de prueba
i dispositivo *m* d'ascolto di prova
n meeluisterapparaat *n*
d Vorhörapparat *m*

3650 PREMIXED ELEMENTS aud/stu
f éléments *pl* prémixés
e elementos *pl* premezclados
i elementi *pl* premescolati
n vooraf gemengde elementen *pl*
d vorgemischte Elemente *pl*

3651 PREMIXING aud
A technique for selecting groups of an inconveniently large number of sound tracks to be combined and mixing them; the smaller number of pre-mixed tracks is combined into the final sound track.
f prémixage *m*
e premezclado *m*
i premescolanza *f*
n voormenging
d Vormischung *f*

PREPARATION OF NEWS MATERIAL
see: NEWS MATERIAL PREPARATION

3652 PRERECORD (TO) tv
To record program(me) material before it is required for transmission.
f préenregistrer v
e prerregistrar v.
i preregistrare v
n vooraf opnemen v
d vorher aufnehmen v

3653 PRERECORDED ELEMENTS stu
f éléments *pl* préenregistrés
e elementos *pl* prerregistrados
i elementi *pl* preregistrati
n vooraf opgenomen elementen *pl*
d vorheraufgenommene Elemente *pl*

3654 PRERECORDED TAPE, RECORDED TAPE vr
A recording that is commercially available on magnetic tape.
f bande *f* à préenregistrement
e cinta *f* de prerregistro
i nastro *m* a preregistrazione
n vooraf bedrukte band
d vorbespieltes Band *n*

3655 PRERECORDED TELEVISION PROGRAM(ME) tv
f programme *m* de télévision préenregistré
e programa *m* televisivo prerregistrado
i programma *m* televisivo preregistrato
n vooraf opgenomen televisieprogramma *n*
d vorher aufgenommenes Fernsehprogramm *n*

3656 PREROLL TIME, RUN-UP TIME vr
The time necessary to engage a telecine machine for filling the gap due to a breakdown of the program(me).
f intervalle *m* de garde,
temps *m* de mise en jeu,
temps *m* de mise en service
e tiempo *m* de mesa en funcionamiento
i tempo *m* d'assestamento
n opsteltijd
d Startvorlaufzeit *f*

3657 PRESELECTION rec/tv
f présélection *f*
e preselección *f*
i preselezione *f*
n voorkeuze
d Vorwahl *f*

3658 PRESELECTION KEYBOARD cpl
f clavier *m* de présélection
e teclado *m* de preselección
i tastiera *f* di preselezione
n voorkeuzetoetsenbord *n*
d Vorwahltastenbrett *n*

3659 PRESELECTOR rec/tv
f présélecteur *m*
e preselector *m*
i preselettore *m*
n voorkiezer
d Vorwähler *m*

3660 PRESENTATION rep/tv
Technique of presenting a program(me) in continuity.
f présentation *f*,
régie *f* principale
e presentación *f*
i presentazione *f*
n presentatie
d Programmregie *f*,
Vorführung *f*

3661 PRESENTATION CONTROL rep/tv
f réglage *m* de la présentation
e regulación *f* de la presentación
i regolazione *f* della presentazione
n presentatieregeling
d Vorführungsregelung *f*

3662 PRESENTATION CONTROLLER, PROGRAM(ME) CONTROLLER rep/tv
f dispositif *m* de réglage de la présentation
e dispositivo *m* de regulación de la presentación
i dispositivo *m* di regolazione della presentazione
n presentatieregelaar
d Vorführungsregler *m*

3663 PRESENTATION MIXER rep/tv
f Sound and vision equipment used to present program(me)s in continuity.
f mélangeur *m* pour la présentation
e mezclador *m* para la presentación
i mescolatore *m* per la presentazione
n presentatiemenger
d Vorführungsmischer *m*

3664 PRESENTATION STUDIO stu
f studio *m* d'annonces,
studio *m* de complément
e estudio *m* de presentación
i studio *m* di presentazione
n presentatiestudio
d Ansagestudio *n*

PRESENTATION SUITE
see: CENTRAL CONTROL

3665 PRESET FADERS stu
f résistances *pl* chutrices préréglées
e reductores *pl* de luz de ajuste previo
i oscuratori *pl* graduali a predisposizione
n vooraf ingestelde lichtintensiteitsregelaars *pl*
d voreingestellte Abblendwiderstände *pl*

PRESET TYPE SWITCHER
see: NEXT-CHANNEL SWITCHER

3666 PRESETTING THE CHANNELS rep/tv
f présélection *f* des canaux
e preselección *f* de los canales
i preselezione *f* dei canali
n eerste afstemming
d Vorabstimmung *f*

3667 PRESSED GLASS BASE crt
A base in which heated powdered glass is pressed around the electrode leads and supports.
f base *f* en verre pressé
e base *f* de vidrio prensado
i base *f* in vetro stampato
n persglasvoet
d Pressglassockel *m*

3668 PRETUNING CONTROLS vr
Used for tuning to six different TV channels.
f recherche *f* de programme
e reguladores *pl* de presintonización

i ricerca *f* di programma
n voorkeuzeafstemregelaars *pl*
d Programmspeicher *m*

3669 PREVIEW rep/tv
Advance view of a TV picture produced by a camera on a monitor at a control console for observation before connecting the camera to the transmission network.
f première vision *f*, preview *m*
e comprobación *f* previa, control *m* previo, primera visión *f*
i antiprima *f*
n voorbezichtiging
d Vorschau *f*

PREVIEW MONITOR (US)
see: CAMERA MONITOR

3670 PREVIEW MONITORING rep/tv
f première vision *f* par moniteur, preview *m* par moniteur
e comprobación *f* previa en el monitor
i antiprima *f* nel monitore
n voorbezichtiging in de monitor
d Monitorvorschau *f*

3671 PREVIEW SELECTOR cpl/tv
f sélecteur *m* de première vision, sélecteur *m* de preview
e selector *m* de comprobación previa
i selettore *m* d'antiprima
n voorbezichtigingskiezer
d Vorschauwähler *m*

3672 PREVIEW SWITCHER rec/rep/tv
f commutateur *m* de première vision, commutateur *m* de preview
e conmutador *m* de comprobación previa
i commutatore *m* d'antiprima
n voorbezichtigingsschakelaar
d Vorschauschalter *m*

PRIMARY ADDITIVE COLO(U)RS
see: ADDITIVE PRIMARIES

PRIMARY COLO(U)R FIELD
see: COLO(U)R FIELD

3673 PRIMARY COLO(U)R SIGNAL, PRIMARY SIGNAL ctv
f signal *m* primaire de couleur
e señal *f* primaria de color
i segnale *m* primario di colore
n primair kleursignaal *n*
d Farbauszugssignal *n*

3674 PRIMARY COLO(U)R UNIT ctv
The area within a colo(u)r cell in a colo(u)r picture tube that is occupied by one primary colo(u)r.
f élément *m* d'une couleur primaire
e elemento *m* de un color primario
i elemento *m* di uno colore primario
n element *n* van één primaire kleur
d Element *n* einer Primärfarbe

PRIMARY COLO(U)RS
see: COLO(U)R PRIMARIES

3675 PRIMARY LIGHT SOURCE ct
A body or object emitting light by virtue of a transformation of energy into radiant energy within itself.
f source *f* primaire
e fuente *f* luminosa primaria
i sorgente *f* luminosa primaria
n primaire lichtbron
d Primärlichtquelle *f*, Selbstleuchter *m*

PRIMARY RADIATOR
see: ACTIVE AERIAL

3676 PRIMARY SERVICE AREA
The area within which the field strength of the ground wave from a broadcasting transmitter is great enough, in comparison with that of interference and with that of indirect rays from the transmitter, for reception to be consistently good by day and by night.
f zone *f* de service primaire, zone *f* primaire
e área *f* de servicio primaria
i area *f* di servizio primaria
n primair bereik *n*
d Primärversorgungsbereich *m*

3677 PRIMARY SIGNAL MONITOR, TRISTIMULUS VALUES MONITOR ctv
f appareil *m* de contrôle des signaux primaires, oscilloscope *m* triple
e aparato *m* de control de las señales primarias, osciloscopio *m* triple
i apparecchio *m* di controllo dei segnali primari, oscilloscopio *m* triplo
n controleapparaat *n* voor de primaire signalen, kleursoortoscilloscoop
d Farbwertkontrollgerät *n*, Farbwertoszilloskop *n*

3678 PRIMARY SUBTRACTIVE COLO(U)RS, SUBTRACTIVE PRIMARIES ct
Minimum number of spectral colo(u)rs (cyan, magenta, yellow) which, when subtracted in the right intensity from a given white, result in a match with a given colo(u)r.
f couleurs *pl* primaires soustractives, primaires *pl* soustractives
e colores *pl* primarios subtractivos
i colori *pl* primari sottrattivi
n subtractieve primaire kleuren *pl*
d Primärfarbensubtrahenden *pl*, subtraktive Primärfarben *pl*

PRINCIPAL AXIS
see: OPTICAL AXIS

3679 PRINCIPAL DESK aud/stu

A desk for mixing the international program(me) with the other modulation sources, viz. reproducing machines, external lines and announcer's microphones.
f pupitre *m* principal
e pupitre *m* principal
i tavolo *m* principale
n hoofdlessenaar
d Hauptpult *n*

3680 PRINCIPAL FOCUS opt
f foyer *m* principal
e foco *m* principal
i fuoco *m* principale
n hoofdbrandpunt *n*
d Hauptbrennpunkt *m*

3681 PRINCIPAL IMAGE COMPONENTS, PRINCIPAL PICTURE COMPONENTS rep/tv
Line period, picture frequency, field suppression, active lines, line suppression period, active line time, aspect ratio and line frequency.
f composantes *pl* principales de l'image
e componentes *pl* principales de la imagen
i componenti *pl* principali dell'immagine
n hoofdkenmerken *pl* van het beeld
d Hauptkennzeichen *pl* des Bildes

3682 PRINCIPAL PLANE opt
f plan *m* principal
e plano *m* principal
i piano *m* principale
n hoofdvlak *n*
d Hauptebene *f*

3683 PRINCIPAL POINTS opt
f points *pl* principaux
e puntos *pl* principales
i punti *pl* principali
n hoofdpunten *pl*
d Hauptpunkte *pl*

3684 PRINTED CIRCUIT ge/tv
Circuit of an equipment, formed into a unit which can be realized by copper conductors laminated on a phenol base.
f circuit *m* imprimé
e circuito *m* impreso
i circuito *m* stampato
n gedrukte schakeling
d gedruckte Schaltung *f*

3685 PRINTED PANEL cpl
f panneau *m* imprimé
e tablero *m* impreso
i pannello *m* stampato
n print
d Platine *f*

3686 PRINTED WIRING CHASSIS cpl
f châssis *m* à câblage imprimé
e chasis *m* de cableado impreso
i telaio *m* di montaggio a cablaggio stampato
n chassis *n* met gedrukte bedrading
d Chassis *n* mit gedruckter Verdrahtung

3687 PROBE MICROPHONE, TEST MICROPHONE aud/stu
f microphone *m* sonde
e micrófono *m* sonda
i microfono *m* sonda
n sondemicrofoon
d Sondenmikrophon *n*

3688 PRODUCER, PRODUCTION MANAGER tv
f producteur *m*
e director *m* de producción, productor *m*
i capo *m* di produzione, produttore *m*
n produktieleider
d Produzent *m*

3689 PRODUCTION CONTROL ROOM rec/tv
That part of the studio where the program(me) director views the output of the cameras on preview monitors and listens to the sound on a loudspeaker.
f salle *f* de contrôle de la production, salle *f* régie
e sala *f* de control de la producción
i sala *f* di controllo della produzione, sala *f* di regia
n produktieregelingsruimte, regiekamer
d Produktionsregelungsraum *m*, Regieraum *m*

3690 PRODUCTION PLANNING rec/tv
f service *m* de programmation
e servicio *m* de programación
i servizio *m* di programmazione
n produktieplanning, programmaplan *n*
d Produktionsplanung *f*, Programmdisposition *f*

3691 PROGRAM(ME) tv
A sequence of audio and video signals, transmitted for entertainment or information.
f programme *m*
e programa *m*
i programma *m*
n programma *n*
d Programm *n*

PROGRAM(ME) AMPLIFIER
see: LINE AMPLIFIER

3692 PROGRAM(ME) ASSEMBLY SWITCHER rec/tv
f mélangeur *m* du programme
e mezclador *m* del programa
i mescolatore *m* del programma
n programmamenger
d Programmischer *m*

3693 PROGRAM(ME) CHANNEL, PROGRAM(ME) TRANSMISSION CHANNEL rec/tv
A one-way path for the electric currents corresponding to a program(me) or to a contribution to a program(me).

f canal *m* du programme
e canal *m* del programa
i canale *m* del programma
n programmakanaal *n*
d Programmkanal *m*

3694 PROGRAM(ME) CONSOLE cpl/stu
f console *m* de programme
e pupitre *m* de programa
i tavolo *m* di programma
n programmalessenaar
d Programmpult *n*

3695 PROGRAM(ME) CONTINUITY stu
f cabine *f* de programme
e cabina *f* de programa
i sala *f* di regia
n regiezaal
d Senderegie *f*

3696 PROGRAM(ME) CONTINUITY tv
f continuité *f* du programme
e continuidad *f* del programa
i continuità *f* del programma
n programmavoortgang
d Programmkontinuität *f*

3697 PROGRAM(ME) CONTROL rec/stu/tv
UNITS
f dispositifs *pl* de réglage du programme
e dispositivos *pl* de regulación del programa
i dispositivi *pl* di regolazione del programma
n programmaregelaars *pl*
d Programmregler *pl*

PROGRAM(ME) CONTROLLER
see: PRESENTATION CONTROLLER

3698 PROGRAM(ME) EXCHANGE tv
f échange *m* des programmes
e cambio *m* de los programas
i scambio *m* dei programmi
n programma-uitwisseling
d Programmaustausch *m*

3699 PROGRAM(ME) FOR LOCAL tv
OR REGIONAL TRANSMISSION
f programme *m* à diffusion nationale et régionale
e programa *m* de difusión local y regional
i programma *m* di diffusione locale e regionale
n programma *n* voor lokale en regionale transmissie
d Programm *n* für lokale und regionale Übertragung

3700 PROGRAM(ME) INDICATOR WITH vr
ZERO RESET BUTTON
A device for rapid location of recordings.
f compteur *m* à bouton pressoir
e contador *m* con botón de ajuste
i contatore *m* con botone di regolazione
n telwerk *n* met instelknop
d Zählwerk *n* mit Einstellknopf

3701 PROGRAM(ME) LEVEL rep/tv
The level of programming signals as indicated by a programmeter in volume units.
f niveau *m* des signaux du programme
e nivel *m* de las señales del programa
i livello *m* dei segnali del programma
n programmasignaalniveau *n*
d Programmsignalpegel *m*

PROGRAM(ME) LINE
see: MUSIC LINE

3702 PROGRAM(ME) LOG, rec/tv
SWITCHING SCHEDULE
f horaire *m* du programme
e horario *m* del programa
i orario *m* del programma
n programmatijdschema *n*
d Programmfahrplan *m*

3703 PROGRAM(ME) MONITOR rep/tv
f moniteur *m* du programme
e monitor *m* del programa
i monitore *m* del programma
n programmamonitor
d Programmonitor *m*

3704 PROGRAM(ME) REPEATER tv
Amplifier which is of sufficiently high-grade performance for insertion into transmission lines for relaying broadcast program(me)s.
f amplificateur -relais *m* de programmes
e amplificador -repetidor *m* de programas
i amplificatore *m* di rimbalzo
n relayeerversterker
d Ballverstärker *m*

3705 PROGRAM(ME) ROUTING rec/tv
f acheminement *m* du programme
e enrutado *m* del programa
i via *f* del programma
n programmaroute
d Programmleitweg *m*

PROGRAM(ME) SIGNAL
see: MODULATION SIGNAL

3706 PROGRAM(ME) SWITCHER rec/tv
f commutateur *m* de programme
e conmutador *m* de programa
i commutatore *m* di programma
n programmaschakelaar
d Programmschalter *m*

3707 PROGRAM(ME) SWITCHING rec/tv
CENTER(RE),
PSC
A center(re) in which program(me) items are switched in order to provide a complete program(me).
f salle *f* de commutation du programme
e sala *f* de conmutación del programa
i sala *f* di commutazione del programma
n programmaschakelkamer
d Programmschaltraum *m*

3708 PROGRAM(ME) TRANSMISSION tv
f émission *f* du programme
e emisión *f* del programa
i emissione *f* del programma
n programma-uitzending
d Programmsendung *f*,
Programmübertragung *f*

3709 PROGRAMMED CAMERA INSTRUCTIONS ani/rec/tv
f instructions *pl* programmées de la caméra
e instrucciones *pl* programadas de la cámara
i istruzioni *pl* programmate della camera
n geprogrammeerde camerainstructies *pl*
d programmierte Kameraanweisungen *pl*

PROGRAMMETER
see: LEVEL INDICATOR

3710 PROGRESSIVE INTERLACE rec/tv
Normal scanning of a TV image, whereby all the odd lines are scanned and then the even lines and so on.
f analyse *f* entrelacée progressive
e exploración *f* entrelazada progresiva
i analisi *f* interlacciata progressiva
n progressieve interliniëring
d fortlaufendes Zeilensprungverfahren *n*

3711 PROGRESSIVE SCANNING, SEQUENTIAL SCANNING rec/tv
A method of scanning in which the lines forming the picture are traced successively and contiguously.
f analyse *f* ligne par ligne non entrelacée
e exploración *f* secuencial
i analisi *f* sequenziale
n niet-geïnterlinieerde aftasting, sequentiële aftasting
d zeilenweise Abtastung *f* ohne Zwischensprung

PROJECTION CATHODE-RAY TUBE
see: CATHODE-RAY TUBE PROJECTOR

3712 PROJECTION OPTICS (GB), REFLECTIVE OPTICS (US) tv
A system of mirrors and lenses used to project the image onto a screen in projection TV.
f optique *f* de projection
e óptica *f* de proyección
i ottica *f* di proiezione
n projectie-optiek
d Projektionsoptik *f*

3713 PROJECTION RECEIVER, PROJECTION TELEVISION RECEIVER rep/tv
A receiver in which an image, usually enlarged, of the TV picture is projected optically on the viewing screen.
f récepteur *m* de télévision à projection
e receptor-proyector *m* de televisión
i ricevitore *m* televisivo a proiezione
n projectietelevisieontvanger
d Projektionsempfänger *m*

3714 PROJECTION TELEVISION tv
System of TV reproduction in which an enlarged image of the picture is optically projected, usually on the back of a flat translucent screen.
f télévision *f* à projection
e televisión *f* de proyección
i televisione *f* a proiezione
n projectietelevisie
d Projektionsfernsehen *n*

3715 PROJECTIONIST fi/tv
f projectionniste *m*
e proyectionista *m*
i proiezionista *m*
n operateur
d Vorführer *m*

3716 PROMPTER rep/tv
Device which displays on a cathode-ray tube or on a paper roll the words of a newsreader or actor in rough synchronization with the action.
f autocue *m*, projecteur *m* du texte
e proyector *m* del texto
i proiettore *m* del testo
n tekstprojector
d Teleprompter *m*, Textprojektor *m*

3717 PROP PLOT (US), PROPERTIES PLOT (US), SCENERY PLOT (GB) stu
f liste *f* des accessoires
e lista *f* de los accesorios
i elenco *m* dell'attrezzeria
n rekwisietenlijst
d Requisitenliste *f*

3718 PROPAGATION ge
The travel of electromagnetic waves or sound waves through a medium.
f propagation *f*
e propagación *f*
i propagazione *f*
n voortplanting
d Ausbreitung *f*, Fortpflanzung *f*

3719 PROPERTIES (US), PROPS (US), SCENERY (GB) stu
f accessoires *pl*
e accesorios *pl*
i attrezzeria *f* della scena
n rekwisieten *pl*
d Requisiten *pl*

3720 PROPERTY MAN fi/th/tv
f accessoiriste *m*
e encargado *m* de los accesorios
i trovaroba *m*
n rekwisiteur
d Innenrequisiteur *m*, Requisiteur *m*

3721 PROPERTY MANAGER fi/th/tv
f décorateur-ensemblier *m*, ensemblier *m*
e encargado *m* de los accesorios
i attrezzista *m* di scena
n toneelmeester
d Requisitenmeister *m*

PROTANOMALOUS VISION
see: PARTIAL PROTANOPIA

3722 PROTANOPE ct
One who possesses protanopia.
f protanope *m*

e protanope *m*
i protanope *m*
n protanoop
d Protanop *m*

3723 PROTANOPIA ct
A form of dichromatic vision in which colo(u)rs can be matched by a mixture of yellow and blue stimuli, but in which the relative luminous efficiency of the red and orange regions of the spectrum is much less than normal.
f protanopie *f*
e protanopía *f*
i protanopia *f*
n protanopie
d Protanopie *f*

3724 PROTECTIVE LEADER fi/tv/vr
f amorce *f* de protection
e cola *f* de protección
i coda *f* di protezione
n beschermband
d Schutzband *n*

3725 PSEUDO-EQUALIZING PULSES tv
Pulses used to derive the leading edges of line-sync pulses, equalizing pulses and frame-sync pulses from a common source.
f pseudosignaux *pl* d'égalisation
e seudoimpulsos *pl* de igualación
i pseudoimpulsi *pl* d'uguagliamento
n pseudo-egalisatie-impulsen *pl*
d Pseudoausgleichsimpulse *pl*

3726 PSEUDO-FIELD-SYNC PULSES tv
f pseudoimpulsions *pl* de synchronisation de trame
e seudoimpulsos *pl* de sincronización de cuadro
i pseudoimpulsi *pl* di sincronizzazione di trama
n pseudorastersynchronisatie-impulsen *pl*
d Pseudohalbbildsynchronisationsimpulse *pl*

3727 PSEUDO-LINE-SYNC PULSES tv
f pseudoimpulsions *pl* de synchronisation de ligne
e seudoimpulsos *pl* de sincronización de línea
i pseudoimpulsi *pl* di sincronizzazione di linea
n pseudolijnsynchronisatie-impulsen *pl*
d Pseudozeilensynchronisationsimpulse *pl*

3728 PSEUDO-LOCK ctv
A method for making an incoming unlocked source yield signals locked to the station sync pulses which do not depend on controlling the generators involved in any way.
f pseudoenclenchement *m*
e seudoenganche *m*
i pseudoblocco *m*
n pseudovergrendeling
d Pseudoverkopplung *f*,
Pseudoverriegelung *f*

3729 PSEUDO-RANDOM SCANNING tv
f analyse *f* pseudostatique
e exploración *f* seudoestática
i analisi *f* pseudostatica
n pseudostatische aftasting
d pseudostatische Abtastung *f*

PSOPHOMETER
see: NOISE METER

PSP
see: PEAK SIDEBAND POWER

3730 PSYCHOPHYSICAL ACHROMATIC COLO(U)R ct
Psychophysical colo(u)r of zero purity.
f couleur *f* achromatique psychophysique
e color *m* acromático psicofísico
i colore *m* acromatico psicofisico
n psychofysieke achromatische kleur
d psychophysikalische achromatische Farbe *f*

3731 PSYCHOPHYSICAL CHROMATIC COLO(U)R ct
Psychophysical colo(u)r of greater than zero excitation purity and hence possessing a dominant or complementary wavelength.
f couleur *f* chromatique psychophysique
e color *m* cromático psicofísico
i colore *m* cromatico psicofisico
n psychofysieke chromatische kleur
d psychophysikalische chromatische Farbe *f*

3732 PSYCHOPHYSICAL COLO(U)R ct
Characteristic of a visible radiation by which an observer may distinguish differences between two fields of view of the same size, shape and structure, such as may be caused by differences in the spectral composition of the radiation concerned in the observation.
f couleur *f* psychophysique
e color *m* psicofísico
i colore *m* psicofisico
n psychofysieke kleur
d psychophysikalische Farbe *f*

3733 PULLDOWN vr
The action of moving the film from one frame to the next in a camera or projector.
f entraînement *m* du film,
escamotage *m*
e arrastre *m* de la película
i avanzamento *m* della pellicola
n stapsgewijze voortbeweging van de film
d Filmfortschaltung *f*,
ruckweises Weiterbewegen *n* des Films

3734 PULLDOWN CLAW fi/tv/vr
f griffe *f* d'entraînement,
griffe *f* d'escamotage
e garra *f* de arrastre
i griffa *f* d'avanzamento
n transportgrijper
d Transportgreifer *m*

PULL-IN RANGE
see: CAPTURE RANGE

3735 PULLING dis/tv
f décalage *m* de ligne
e desplazamiento *m* de lînea
i scatto *m* d'ampiezza
n lijnverschuiving
d Zeilenversatz *m*

3736 PULLING ON WHITES dis/tv
Faulty TV reproduction in which vertical lines are momentarily displaced to the right following a white area at the right hand side of the picture.
f filage *m* horizontal
e prolongación *f* irregular de las lîneas horizontales
i trascinamento *m*
n horizontaal naslepen *n*
d Nachziehen *n*

3737 PULSE ge
An electrical disturbance whose duration is short in relation to the time scale of interest, and whose initial and final values are the same.
f impulsion *f*
e impulso *m*
i impulso *m*
n impuls
d Impuls *m*

3738 PULSE AMPLIFIER cpl
Amplifier with very wide frequency response which can amplify pulses without distortion of the very short rise time of the leading edge.
f amplificateur *m* d'impulsions
e amplificador *m* de impulsos
i amplificatore *m* d'impulsi
n impulsversterker
d Impulsverstärker *m*

3739 PULSE AMPLITUDE ge
The peak value of a pulse.
f amplitude *f* d'impulsion
e amplitud *f* de impulso
i ampiezza *f* d'impulso
n impulsamplitude
d Impulsamplitude *f*,
Impulsstärke *f*

PULSE AMPLITUDE MODULATION
see: PAM

3740 PULSE ANALYZER cpl/ge
An equipment used for analyzing pulses in order to determine their amplitude, duration, shape, and other characteristics.
f analyseur *m* d'impulsion
e analizador *m* de impulso
i analizzatore *m* d'impulso
n impulsanalysator
d Impulsanalysator *m*

3741 PULSE CARRIER ge
A pulse train used as a carrier.
f train *m* porteur
e portadora *f* de impulsos,
tren *m* de impulsos portador
i treno *m* d'impulsi portante
n impulstrein als draaggolf
d Impulsreihe *f* als Träger

3742 PULSE CLIPPER rec/rep/tv
f écrêteur *m* d'impulsion
e recortador *m* de impulso
i limitatore *m* d'impulso
n impulsbegrenzer
d Impulsabtrennstufe *f*

3743 PULSE CLIPPING cpl
f écrêtage *m* d'impulsion
e recorte *m* de impulso
i limitazione *f* d'impulso
n impulsbegrenzing
d Impulsabtrennung *f*

3744 PULSE CORRECTOR, cpl
PULSE STRETCHER
A non-storage device for correcting the shape of pulses.
f correcteur *m* d'impulsion
e corrector *m* de costados de los impulsos,
corrector *m* de impulso
i correttore *m* d'impulso
n impulscorrector
d Impulskorrektor *m*

PULSE DECAY TIME,
PULSE FALL TIME
see: DECAY TIME

3745 PULSE DISTRIBUTION cpl
AMPLIFIER
f distributeur *m* d'impulsions
e distribuidor *m* de impulsos
i distributore *m* d'impulsi
n impulsverdelerversterker
d Impulsverteilerverstärker *m*

3746 PULSE DURATION, ge
PULSE WIDTH
The interval between the first and last instant at which the instantaneous value of a pulse or its envelope reaches a specified fraction of the peak.
f durée *f* d'impulsion
e anchura *f* de impulso,
duración *f* de impulso
i durata *f* d'impulso
n impulsbreedte,
impulsduur
d Impulsbreite *f*,
Impulsdauer *f*,
Impulslänge *f*

3747 PULSE DUTY FACTOR (US), ge
PULSE RATE FACTOR (GB)
The ratio of the average pulse duration to the average pulse spacing.
f taux *m* d'impulsions
e factor *m* de trabajo del impulso
i fattore *m* d'utilizzazione dell'impulso
n impulswerktijdverhouding
d Impulstastverhältnis *n*,
Wirkungsperiode *f*

PULSE EDGE
see: EDGE OF A PULSE

3748 PULSE EHT GENERATOR cpl
The means of generating the EHT supply for a TV tube by the rectification of the voltage pulses occurring in the anode circuit of an electron tube (valve) when the anode current is suddenly cut off.
f générateur *m* d'impulsions à très haute tension
e generador *m* de impulsos de tensión extraalta
i generatore *m* d'impulsi a tensione altissima
n hoogspanningsimpulsgenerator
d Hochspannungsimpulsgenerator *m*, Impulshochspannungsgenerator *m*

PULSE FREQUENCY MODULATION
see: PFM

PULSE GENERATOR
see: PG

3749 PULSE INTERLACING, PULSE INTERLEAVING cpl/tv
A process in which pulses from two or more sources are combined in time-division multiplex for transmission over a common path.
f transmission *f* simultanée d'impulsions différentes sur une seule voie
e entrelazado *m* de impulsos
i interlacciamento *m* d'impulsi
n impulsvervlechting
d Impulsverflechtung *f*, Impulsverschachtelung *f*

3750 PULSE JITTER dis
A relatively small variation of the pulse spacing in a pulse train.
f instabilité *f* de l'impulsion
e inestabilidad *f* del impulso
i instabilità *f* dell'impulso
n impulsonstabiliteit
d Impulsinstabilität *f*

3751 PULSE MIXING rec/tv
f mixage *m* d'impulsions
e mezclado *m* de impulsos
i mescolanza *f* d'impulsi
n impulsmenging
d Impulsmischung *f*

3752 PULSE MODULATION cpl
f modulation *f* d'impulsions
e modulación *f* de impulsos
i modulazione *f* d'impulsi
n impulsmodulatie
d Impulsmodulation *f*

3753 PULSE PHASING cpl
A measure taken preparatory to locking two or more picture sources so that they may be intercut without relative movement.
f mise *f* en phase d'impulsions
e puesta *f* en fase de impulsos
i messa *f* in fase d'impulsi
n in faze brengen van impulsen
d Phasenabgleich *m* von Impulsen

3754 PULSE REGENERATION ge
The process of correcting a series of pulses to a desired timing, shape and magnitude.
f remise *f* en forme d'impulsions, rétablissement *m* de la forme d'impulsion
e restablecimiento *m* de la forma de impulso
i ricostituzione *f* d'impulsi
n impulsherstel *n*, impulsregeneratie
d Impulsregenerierung *f*

3755 PULSE REGENERATION UNIT tv
f dispositif *m* de rétablissement de la forme d'impulsion
e dispositivo *m* de restablecimiento de la forma del impulso
i dispositivo *m* di ricostituzione d'impulsi
n impulshersteller, impulsregenerator
d Auffrischer *m*, Regeneriergerät *n*

3756 PULSE REPETITION FREQUENCY ge
The pulse repetition rate when this is independent of the interval of time over which it is measured.
f fréquence *f* de répétition d'impulsions
e frecuencia *f* de repetición de impulsos
i frequenza *f* di ricorrenza d'impulsi
n impulsherhalingsfrequentie
d Impulsfolgefrequenz *f*

3757 PULSE RESPONSE cpl/ctv
f réponse *f* d'impulsion
e respuesta *f* de impulso
i risposta *f* d'impulso
n impulsresponsie
d Impulsverhalten *n*

3758 PULSE RESTORATION cpl
f régénération *f* d'impulsions
e restauración *f* de impulsos
i restaurazione *f* d'impulsi
n impulsherstel *n*
d Impulserneuerung *f*

3759 PULSE RISE TIME cpl
Time required for amplitude to rise from 0.10 to 0.90 of its maximum value.
f temps *m* de montée de l'impulsion
e duración *f* de establecimiento del impulso
i tempo *m* di formazione dell'impulso
n impulsstijgtijd
d Impulsanstiegzeit *f*

3760 PULSE SEPARATOR rep/tv
In a TV receiver the circuit that separates the horizontal from the vertical sync pulses in the received signal.
f séparateur *m* d'impulsions
e separador *m* de impulsos
i separatore *m* d'impulsi
n impulsscheider
d Impulstrenner *m*

3761 PULSE SHAPER, PULSE SHAPING CIRCUIT cpl
f conformateur *m* d'impulsions

e conformador *m* de impulsos,
variador *m* de la forma del impulso
i conformatore *m* d'impulsi
n impulsvormer
d Impulsformer *m*

3762 PULSE SOUND cpl
Alternative to separate sound carrier used in satellite communication to conserve bandwidth.
f impulsion *f* pour son
e impulso *m* para sonido
i impulso *m* per suono
n impuls voor geluid
d Impuls *m* für Ton

3763 PULSE SPACING ge
The interval between the times of corresponding features of two consecutive pulses.
f intervalle *m* d'impulsions
e intervalo *m* de impulsos
i intervallo *m* d'impulsi
n impulsafstand,
impulsinterval *n*
d Impulsabstand *m*,
Impulsintervall *m*

3764 PULSE SPECTRUM ge
The distribution as a function of frequency, of the magnitudes of the Fourier components of a pulse.
f spectre *m* d'impulsions
e espectro *m* de impulsos
i spettro *m* d'impulsi
n impulsspectrum *n*
d Impulsspektrum *n*

3765 PULSE SPIKE,
SPIKE dis
An unwanted pulse of relatively short duration superimposed on the main pulse.
f impulsion *f* parasite
e impulso *m* parásito
i impulso *m* parassita
n storende impulspiek
d Störspitze *f*

PULSE STRETCHER
see: PULSE CORRECTOR

3766 PULSE SYNC tv
Method of synchronization between separate picture and sound recording cameras by generating in the picture camera a series of regular pulses recorded on tape to form electronic sprocket holes.
f synchronisation *f* par impulsions
e sincronización *f* por impulsos
i sincronizzazione *f* per impulsi
n synchronisatie met behulp van impulsen
d Synchronisation *f* mittels Impulse

3767 PULSE TAIL,
TAIL OF PULSE cpl
f traîne *f* de l'impulsion
e cola *f* del impulso
i coda *f* dell'impulso
n impulsstaart
d hintere Flanke *f*,
Impulsschwanz *m*

3768 PULSE TRAILING EDGE cpl
The major portion of the decay of a pulse.
f flanc *m* arrière de l'impulsion
e flanco *m* posterior del impulso
i fronte *m* posteriore dell'impulso
n impulsachterflank
d Impulsrückflanke *f*

3769 PULSE TRAIN ge
A sequence of pulses of similar characteristics.
f train *m* d'impulsions
e tren *m* de impulsos
i treno *m* d'impulsi
n impulstrein
d Impulsreihe *f*,
Impulsserie *f*

3770 PULSE TRIGGERING cpl
f déclenchement *m* de circuit par impulsions
e desconexión *f* de circuito por impulsos
i sblocco *m* di circuito per impulsi
n deblokkering van een circuit door impulsen,
impulstrekkerschakeling
d Impulstriggerschaltung *f*

PULSE WIDTH
see: PULSE DURATION

PULSED-AUTOMATIC GAIN CONTROL
see: KEYED AUTOMATIC GAIN CONTROL

3771 PULSED-DISCHARGE TUBE,
PULSED LAMP vr
Type of high intensity, short duration discharge tube.
f tube *m* à décharge modulé par impulsions
e tubo *m* de descarga modulado por impulsos
i tubo *m* a scarica modulato per impulsi
n door impulsen gemoduleerde ontladingsbuis
d durch Impulse gesteuerte Entladungsröhre *f*

3772 PULSED-LIGHT SYSTEM vr
f système *m* d'éclairage modulé par impulsions
e sistema *m* de alumbrado modulado por impulsos
i sistema *m* d'illuminazione modulato per impulsi
n door impulsen gemoduleerd lichtsysteem *n*
d durch Impulse gesteuertes Beleuchtungssystem *n*

3773 PUPIL ct
Variable aperture in the iris.
f pupille *f*
e pupila *f*
i pupilla *f*
n pupil
d Pupille *f*

3774 PURE COLO(U)R ct
Colo(u)r with CIE co-ordinates lying on the spectrum locus or on the purple boundary.
f couleur *f* pure
e color *m* puro
i colore *m* puro
n zuivere kleur
d reine Farbe *f*

3775 PURIFYING MAGNET ctv
f aimant *m* purificateur
e imán *m* purificador
i magnete *m* purificatore
n kleurzuiveringsmagneet
d Farbreinheitsmagnet *m*

PURITY
see: COLO(U)R PURITY

PURITY COIL
see: COLO(U)R PURITY COIL

PURITY CONTROL
see: COLO(U)R PURITY CONTROL

PURITY MAGNET
see: COLO(U)R PURITY MAGNET

PURITY RING
see: COLO(U)R PURITY RING

3776 PURKINJE PHENOMENON ct
Reduction in the luminosity of a red light relative to that of a blue light when the luminances are reduced in the same proportions without changing the respective spectral distributions.
f effet *m* Purkinje
e efecto *m* Purkinje
i effetto *m* Purkinje
n purkinje-effect *n*
d Purkinje-Phänomen *n*

3777 PURPLE BOUNDARY ct
Straight line joining the ends of the spectrum locus on a chromaticity diagram.
f ligne *f* des pourpres
e línea *f* de las púrpuras
i linea *f* delle porpore
n purpurgrens
d Purpurgrenze *f*

3778 PUSH-ON BLACK tv
f poussée *f* vers le noir
e golpe *m* hacia el negro
i spinta *f* verso il nero
n opduw naar zwart
d Schub *m* nach schwarz

3779 PUSH-ON WHITE tv
f poussée *f* vers le blanc
e golpe *m* hacia el blanco
i spinta *f* verso il bianco
n opduw naar wit
d Schub *m* nach weiss

3780 PUSH-OVER WIPE tv
f fondu *m* effacé
e fundido *m* lateral
i dissolvenza *f* laterale
n schuifovergang
d Schiebeblende *f*,
seitliche Überblendung *f*

3781 PUSHBUTTON SELECTION rep/tv
f sélection *f* de canal par boutons-pressoir
e selección *f* de canal con botones
i selezione *f* di canale con pulsanti
n kanaalkeuze met drukknoppen
d Kanalenwahl *f* mit Druckknöpfen

3782 PUSHBUTTON UNIT rep/tv
f bloc *m* touches
e unidad *f* botones
i unità *f* pulsanti
n druktoetseneenheid
d Drucktasteneinheit *f*

3783 PUSHBUTTONS vr
Used for reception from stations pretuned by means of the pretuning controls.
f touches *pl* de sélection du programme
e botones *pl* de preselección,
teclas *pl* de preselección
i pulsanti *pl* di preselezione
n drukknoppen *pl* voor voorkeurzenders
d Stationsdrucktasten *pl*

Q

3784 Q AERIAL, aea
Q ANTENNA,
STUB-MATCHED AERIAL,
STUB-MATCHED ANTENNA
Combination of a dipole with a quarter-wavelength of thin-wire transmission line for the purpose of matching the impedance of the feeder with that of the dipole.
f antenne *f* Q,
antenne *f* à adaptateur d'impédance
e antena *f* de adaptador de impedancia,
antena *f* dipolo Q
i antenna *f* Q,
dipolo *m* con adattatore
n dipool met aanpasleiding,
Q-antenne
d mit Stichleitung angepasster Dipol *m*,
Q-Antenne *f*

Q AXIS
see: NARROW-BAND AXIS

3785 Q COMPONENT ctv
That component of the chrominance signal formed by modulating the chrominance subcarrier with the Q signal.
f composante *f* Q
e componente *f* Q
i componente *f* Q
n Q-component
d Q-Komponente *f*

3786 Q DEMODULATOR ctv
The demodulator in which the chrominance signal and voltage from the colo(u)r-burst oscillator signal are combined to recover the Q signal in a colo(u)r TV receiver.
f démodulateur *m* du signal Q
e demodulador *m* de la señal Q
i demodulatore *m* del segnale Q
n Q-signaaldemodulator
d Q-Signaldemodulator *m*

Q FACTOR
see: FACTOR OF MERIT

3787 Q METER cpl
An instrument for measuring the Q factor of an oscillatory circuit or of a circuit element.
f Q-mètre *m*
e medidor *m* del factor de calidad
i apparecchio *m* di misura del fattore di qualità
n Q-meter
d Q-Meter *n*,
Gütefaktormessgerät *n*

3788 Q MULTIPLIER cpl
A filter that gives a sharp response peak or a deep rejection notch at a particular frequency, equivalent to boosting the Q of a tuned circuit at that frequency.
f multiplicateur *m* Q
e multiplicador *m* Q
i moltiplicatore *m* Q
n Q-vermenigvuldiger
d Q-Vervielfacher *m*

3789 Q SIGNAL ctv
The signal representing the component of chrominance information along the green magenta axis of the colo(u)r diagram (NTSC system).
f signal *m* Q
e señal *f* Q
i segnale *m* Q
n Q-signaal *n*
d Q-Signal *n*

3790 QAM, cpl
QUADRATURE AMPLITUDE MODULATION
f modulation *f* d'amplitude en quadrature de phase
e modulación *f* de amplitud en cuadratura de fase
i modulazione *f* d'ampiezza in quadratura di fase
n amplitudemodulatie met 90° fazeverschuiving
d Quadraturamplitudenmodulation *f*,
Quadraturmodulation *f*

3791 QUAD cpl
Cable consisting of four insulated conductors twisted together in one lay in an overall single envelope.
f câble *m* à quartes
e cable *m* en cuadretes
i cavo *m* a quattro fili,
cavo *m* bicoppia
n quad
d Adernvierer *m*,
Vierer *m*

3792 QUADRANT AERIAL, aea
QUADRANT ANTENNA
Form of transmitting aerial (antenna) which is made up of two wings forming a quadrant.
f antenne *f* quadrant
e antena *f* de cuadrante
i antenna *f* quadrante
n kwadrantantenne
d Quadrantantenne *f*

QUADRATURE (GB)
see: OPPOSITION

3793 QUADRATURE AMPLIFIER cpl
A stage used to supply two signals having the same frequency and having phase angles which differ by 90 electrical degrees.
f amplificateur *m* en quadrature de phase

e amplificador *m* en cuadratura de fase
i amplificatore *m* in quadratura di fase
n versterker met 90° fazeverschuiving
d Verstärker *m* mit 90° Phasenverschiebung

3794 QUADRATURE DETECTOR cpl
Form of f.m. detector in which two resonant circuits tuned to the signal, one in the control grid and the other in the suppressor grid in the case of a pentode, are 90° degrees out of phase.
f détecteur *m* en quadrature de phase
e detector *m* en cuadratura de fase
i rivelatore *m* in quadratura di fase
n detector met 90° fazeverschuiving
d Detektor *m* mit 90° Phasenverschiebung

3795 QUADRATURE ERROR, QUADRATURE PORTION ctv
That portion of the chrominance signal having the same phase or the opposite phase of the subcarrier modulated by the Q signal.
f défaut *m* de quadrature
e error *m* de cuadratura
i errore *m* di quadratura
n fout bij 90° fazeverschuiving
d Quadraturfehler *m*

3796 QUADRATURE INFORMATION CORRELATOR, QUADRICORRELATOR ctv
Contraction of quadrature and information correlator; a circuit sometimes added to the automatic phase control loop in a colo(u)r TV receiver to obtain improved performance under severe interference conditions.
f quadricorrélateur *m*
e cuadricorrelador *m*
i quadricorrelatore *m*
n quadricorrelator
d Quadrikorrelator *m*

3797 QUADRATURE PHASE tv
f phase *f* rectangulaire
e fase *f* rectangular
i fase *f* rettangolare
n rechthoekfaze
d Rechtwinkelphase *f*

3798 QUADRATURE PHASE SUBCARRIER SIGNAL ctv
The portion of the chrominance signal that leads or lags the in-phase portion by 90°.
f signal *m* de sousporteuse en phase rectangulaire
e señal *f* de subportadora en fase rectangular
i segnale *m* di sottoportante in fase rettangolare
n hulpdragersignaal *n* in rechthoekfaze
d Hilfsträgersignal *n* in Rechtwinkelphase

3799 QUADROPHONY vr
f quadrophonie *f*
e cuadrofonía *f*
i quadrofonia *f*
n quadrofonie
d Quadrophonie *f*

3800 QUADROSONIC RECORDER aud
f magnétophone *m* autonome tétraphonique
e magnetófono *m* tetrafónico
i magnetofono *m* tetrafonico
n tetrafone magnetische opneemapparatuur
d tetraphonisches Magnettongerät *n*

3801 QUADRUPLE SCANNING, QUADRUPLED SCANNING INTERLACE rec/tv
A TV scanning system which requires four frames to be scanned to give a complete picture.
f entrelacement *m* quadruple
e exploración *f* entrelazada cuádruple
i analisi *f* interlacciata quadrupla
n viervoudige interliniëring
d vierfaches Zwischenzeilenverfahren *n*

3802 QUADRUPLE SPIRAL SCANNING DISK tv
f disque *m* analyseur à quatre hélices
e disco *m* de exploración de cuatro espirales
i disco *m* d'analisi a quattro spirali
n aftastschijf met vier spiralen
d Vierfachspiralenlochscheibe *f*

3803 QUADRUPLEX VIDEOTAPE RECORDER, TRANSVERSE-TRACK TELEVISION TAPE RECORDER vr
f magnétoscope *m* à pistes transversales, magnétoscope *m* à quatre têtes
e magnetoscopio *m* de cabeza cuádruple, magnetoscopio *m* de pista transversal
i registratore *m* video a quattro testine
n vierkops-videorecorder
d Querspuraufzeichnungsanlage *f*, Vierkopfaufzeichnungsgerät *n*

3804 QUALITY CHECK tv
f supervision *f* du programme
e supervisión *f* del programa
i supervisione *f* del programma
n programma-eindcontrole
d Bildendkontrolle *f*, Programmüberwachung *f*

3805 QUANTITATIVE MONITORING aud/stu
f contrôle *m* quantitatif
e control *m* cuantitativo
i controllo *m* quantitativo
n kwantitatieve controle
d quantitative Kontrolle *f*

3806 QUANTITY OF RADIATION ge
The total radiated energy passing through a unit area per unit of time.
f quantité *f* de rayonnement
e cantidad *f* de radiación
i quantità *f* di radiazione
n stralingshoeveelheid
d Strahlungsmenge *f*

3807 QUANTIZATION cpl
A process in which the range of values of a wave is divided into a finite number of smaller subranges, each of which is represented by an assigned or quantized value within the subrange.

f quantification *f*
e cuantificación *f*, cuantización *f*
i quantizzazione *f*
n quantificatie
d Quantelung *f*, Quantisierung *f*

3808 QUANTIZATION DISTORTION, QUANTIZATION NOISE dis
The distortion produced in the process of quantization.
f distorsion *f* par quantification
e distorsión *f* de cuantización
i distorsione *f* per quantizzazione
n quantificatievervorming
d Quantenverzerrung *f*, Quantisierungsverzerrung *f*

3809 QUANTIZATION LEVEL cpl
One of the subrange values obtained by quantization.
f niveau *m* de quantification
e nivel *m* de cuantización
i livello *m* di quantizzazione
n quantificatieniveau *n*
d Quantisierungspegel *m*

3810 QUARTER TRACK vr
f quart *m* de piste
e cuarto *m* de pista
i quarto *m* di pista
n kwartspoor *n*
d Viertelspur *f*

3811 QUARTER-WAVE AERIAL, QUARTER-WAVE ANTENNA aea
An aerial (antenna) electrically a quarter-wavelength long and resonating at wavelength slightly less than four times its physical length. Used as an indoor TV aerial (antenna).
f antenne *f* quart d'onde
e antena *f* en cuarto de onda
i antenna *f* per 1/4 d'onda
n kwartgolflengteantenne
d Viertelwellenantenne *f*

R

3812 R, ct
RED PRIMARY
f couleur *f* primaire rouge,
rouge *m* primaire
e color *m* primario rojo,
rojo *m* primario
i colore *m* fondamentale rosso,
rosso *m* primario
n primaire kleur rood,
rode grondkleur
d rote Primärfarbe *f*

3813 R BLACK LEVEL, ctv
RED BLACK LEVEL
The minimum permissible level of the R signal.
f niveau *m* minimal pour le signal rouge
e nivel *m* mínimo para la señal roja
i livello *m* minimo per il segnale rosso
n minimumniveau *n* voor het signaal rood
d Mindestpegel *m* für das Rotsignal

3814 R PEAK LEVEL, ctv
RED PEAK LEVEL
The maximum permissible level of the R signal.
f niveau *m* maximal pour le signal rouge
e nivel *m* máximo para la señal roja
i livello *m* massimo per il segnale rosso
n maximumniveau *n* voor het signaal rood
d Höchstpegel *m* für das Rotsignal

3815 R-Y AXIS, ctv
RED COLO(U)R DIFFERENCE AXIS
f axe *m* R-Y
e eje *m* R-Y
i asse *m* R-Y
n as van het rode-kleurverschilsignaal,
R-Y-as
d R-Y-Achse *f*

3816 R-Y MATRIX, ctv
RED COLO(U)R DIFFERENCE MATRIX
f matrice *f* R-Y
e matriz *f* R-Y
i matrice *f* R-Y
n matrix voor het rode-kleurverschilsignaal,
R-Y-matrix
d R-Y-Matrize *f*

3817 R-Y MODULATOR, ctv
RED COLO(U)R DIFFERENCE MODULATOR
f modulateur *m* R-Y
e modulador *m* R-Y
i modulatore *m* R-Y
n modulator voor het rode-kleurverschilsignaal,
R-Y-modulator
d R-Y-Modulator *m*

3818 R-Y-SIGNAL, ctv
RED COLO(U)R DIFFERENCE SIGNAL
f signal *m* R-Y
e señal *f* R-Y
i segnale *m* R-Y
n R-Y-signaal *n*,
rode-kleurverschilsignaal *n*
d R-Y-Signal *n*

3819 RADIAL DEFLECTING ELECTRODE crt
Used in certain types of cathode-ray tubes and located in the center(re) of the screen.
f électrode *f* à déviation radiale
e electrodo *m* de desviación radial
i elettrodo *m* a deflessione radiale,
elettrodo *m* a deviazione radiale
n radiaal afbuigende elektrode
d Elektrode *f* für Radialablenkung

3820 RADIANCE ct
The radiant flux per unit solid angle per unit of projected area of the source.
f radiance *f*
e radiancia *f*
i radianza *f*
n stralingsdichtheid
d spezifische Lichtausstrahlung *f*,
Strahldichte *f*

3821 RADIANT ENERGY ge
Energy transmitted in the form of electromagnetic radiation.
f énergie *f* radiante,
énergie *f* rayonnante
e energía *f* radiante
i energia *f* radiante
n stralingsenergie
d Strahlungsenergie *f*

3822 RADIANT ENERGY DENSITY ge
Radiant energy per unit volume, expressed in such units as ergs /cm^3.
f densité *f* de l'énergie radiante,
densité *f* de l'énergie rayonnante
e densidad *f* de la energía radiante
i densità *f* dell'energia radiante
n dichtheid van de stralingsenergie
d Strahlungsenergiedichte *f*

3823 RADIANT FLUX ge
The time rate of flow of radiant energy.
f flux *m* énergétique
e flujo *m* energético
i flusso *m* energetico
n stralingsstroom
d Strahlungsfluss *m*,
Strahlungsstrom *m*

3824 RADIANT FLUX DENSITY aea
A measure of the radiant power per unit area that flows across a surface.
f densité *f* du flux énergétique
e densidad *f* del flujo energético
i densità *f* del flusso energetico
n dichtheid van de stralingsstroom

d Strahlungsflussdichte *f*,
Strahlungsstromdichte *f*

3825 RADIANT INTENSITY, RADIATION INTENSITY ct
The energy emitted per unit time, per unit solid angle about the direction considered.
f intensité *f* radiante
e intensidad *f* de radiación
i intensità *f* d'irradiazione,
intensità *f* di radiazione
n stralingsintensiteit
d Strahlungsintensität *f*,
Strahlungsstärke *f*

3826 RADIANT SENSITIVITY crt
The signal output current of a camera tube or phototube divided by the incident radiant flux at a given wavelength.
f sensibilité *f* du tube de prise de vues
e sensibilidad *f* del tubo de toma de vistas
i sensibilità *f* del tubo di ripresa
n gevoeligheid van een beeldopnamebuis
d Empfindlichkeit *f* einer Bildaufnahmeröhre

3827 RADIATE (TO) ge
To send out energy, such as electromagnetic waves, into space.
f émettre v,
rayonner v
e emitir v,
irradiar v
i emettere v,
irradiare v
n uitstralen v,
uitzenden v,
zenden v
d ausstrahlen v,
senden v,
strahlen v

3828 RADIATED POWER aea
The actual power level of the radio signals transmitted by an aerial (antenna).
f puissance *f* de rayonnement,
puissance *f* émise
e potencia *f* de radiación,
potencia *f* emitida
i potenza *f* di radiazione,
potenza *f* irradiata
n uitgestraald vermogen *n*
d ausgestrahlte Leistung *f*,
Strahlungsleistung *f*

3829 RADIATING CIRCUIT aea
Any circuit capable of sending out power, in the form of electro-magnetic waves, into space, especially the aerial (antenna) circuit of a radio transmitter.
f circuit *m* de rayonnement
e circuito *m* de radiación
i circuito *m* di radiazione
n stralingscircuit *n*
d Strahlungskreis *m*

3830 RADIATING ELEMENT, RADIATION ELEMENT aea
A basic subdivision of an aerial (antenna) that in itself is capable of radiating or receiving radio-frequency energy.
f élément *m* rayonnant primaire
e radiador *m* primario
i radiatore *m* primario
n primaire straler
d Primärstrahler *m*

3831 RADIATION cpl
The emission of energy in the form of electromagnetic waves.
f radiation *f*,
rayonnement *m*
e radiación *f*
i radiazione *f*
n uitstraling
d Ausstrahlung *f*,
Strahlung *f*

3832 RADIATION ABSORBER aea
An insulating material in sheet form and having a conductive backing, used as dielectric and reflecting elements for absorbing unwanted radio-frequency energy.
f absorbant *m* de rayonnement
e absorbente *m* de radiación
i assorbitore *m* di radiazione
n stralingabsorberend middel *n*
d strahlungsabsorbierendes Mittel *n*

3833 RADIATION ANGLE aea
The angle between the horizontal plane and the line of radiation from a directional aerial (antenna).
f angle *m* de rayonnement
e ángulo *m* de radiación
i angolo *m* di radiazione
n uitstralingshoek
d Abstrahlwinkel *m*

RADIATION EFFICIENCY
see: AERIAL EFFICIENCY

3834 RADIATION FIELD aea
That part of the field of an aerial (antenna) which is associated with an outward flow of energy.
f champ *m* de rayonnement
e campo *m* de radiación
i campo *m* di radiazione,
campo *m* irradiato
n stralingsveld *n*
d Strahlungsfeld *n*

3835 RADIATION LOBE aea
The portion of a radiation pattern that is bounded by one or two cones of nulls.
f lobe *m* de rayonnement,
pétale *f* de rayonnement
e lóbulo *m* de radiación
i lobo *m* di radiazione
n stralingslob,
stralingslus
d Strahlungskeule *f*,
Strahlungslappen *m*

3836 RADIATION LOSSES aea
That part of the transmission loss, due to

radiation of radio-frequency power from a transmission system.
f pertes *pl* par rayonnement
e pérdidas *pl* por radiación
i perdite *pl* per irradiazione
n stralingsverliezen *pl*
d Strahlungsverluste *pl*

RADIATION PATTERN
see: DIRECTIONAL PATTERN

3837 RADIATION PRESSURE aud
The extremely small pressure exerted on a surface or interface by a sound wave.
f pression *f* de radiation
e presión *f* de radiación
i pressione *f* d'irradiazione
n stralingsdruk
d Strahlungsdruck *m*

3838 RADIATION RESISTANCE aea
In transmission usage, the resistance equal to the power radiated by an aerial (antenna) divided by the square of the effective aerial (antenna) current referred to a specified point.
f résistance *f* de rayonnement
e resistencia *f* de radiación
i resistenza *f* d'irradiazione
n stralingsweerstand
d Strahlungswiderstand *m*

3839 RADIATOR aea
That part of an aerial (antenna) or transmission line that radiates electromagnetic energy either directly into space or against a reflector for focusing or directing.
f antenne *f* active,
radiateur *m*
e radiador *m*
i radiatore *m*
n straler
d Strahler *m*

RADIO CAMERA
see: ELECTRON CAMERA

3840 RADIOCOMMUNICATION ge
Telecommunication using radio waves, not guided between transmitter and receiver by artificial boundaries, such as wires or waveguides.
f liaison *f* radioélectrique,
radiocommunication *f*
e enlace *m* radioeléctrico,
radiocomunicación *f*
i collegamento *m* radioelettrico,
radiocomunicazione *f*
n radiocommunicatie,
radioverbinding
d drahtlose Verbindung *f*,
Funkverbindung *f*

RADIO CONTROL RECEIVER
see: EAR-PLUG

3841 RADIO FREQUENCY, R.F. cpl
Any frequency at which electromagnetic radiation is used for telecommunication.
f haute fréquence *f*
e alta frecuencia *f*
i alta frequenza *f*
n hoge frequentie,
radiofrequentie
d Hochfrequenz *f*

3842 RADIO-FREQUENCY AMPLIFIER, cpl
R.F. AMPLIFIER
f amplificateur *m* haute fréquence
e amplificador *m* alta frecuencia
i amplificatore *m* alta frequenza
n hoogfrequentversterker
d Hochfrequenzverstärker *m*

3843 RADIO-FREQUENCY cpl
INTERMODULATION DISTORTION,
R.F. INTERMODULATION DISTORTION
A constituent of nonlinearity distortion consisting of the occurrence in the respons of components arising from intermodulation distortion in the radio-frequency stages of the receiver.
f distorsion *f* par intermodulation dans les étages de haute fréquence
e distorsión *f* por intermodulación en las etapas de alta frecuencia
i distorsione *f* per intermodulazione negli stadi d'alta frequenza
n vervorming door intermodulatie in het hoogfrequentgedeelte
d Intermodulationsverzerrung *f* im Hochfrequenzteil

3844 RADIO-FREQUENCY PULSE, cpl
R.F. PULSE
A train of radio-frequency oscillations whose envelope has the form of a pulse.
f impulsion *f* haute fréquence
e impulso *m* de alta frecuencia
i impulso *m* d'alta frequenza
n hoogfrequentimpuls
d Hochfrequenzimpuls *m*

3845 RADIO LINK tv
f liaison *f* hertzienne,
voie *f* de radiocommunication
e vía *f* de radiocomunicación
i collegamento *m* radio,
radiocollegamento *m*
n gerichte radioverbinding,
radiolink
d Funkübertragungsweg *m*,
Richtfunkstrecke *f*,
Richtfunkzubringerlinie *f*

3846 RADIO-OPTICAL LINE OF ge
DISTANCE,
RADIO-OPTICAL RANGE
Maximum distance, with only normal refraction in the atmosphere, at which transmitter and receiver aerials (antennae) of a certain height can work together.
f portée *f* radiooptique
e alcance *m* radioóptico
i portata *f* radioottica
n radio-optische dracht
d radiooptische Reichweite *f*

3847 RADIO-RELAY AERIAL, RADIO-RELAY ANTENNA aea/rep
f antenne *f* pour station-relais
e antena *f* para estación-relé
i antenna *f* per trasmissione a fasci direttivi
n richtstraalantenne
d Richtfunkantenne *f*

3848 RADIO-RELAY SYSTEM tv
Transmission system including relay stations where the signal is received and retransmitted by radio to compensate for losses.
f système *m* station-relais
e sistema *m* de estación-relé
i trasmissione *f* a fasci direttivi
n richtstraalzendersysteem *n*
d Richtfunksystem *n*

RADIOVISION
see: CCTV

3849 RAGGED IMAGE, RAGGED PICTURE rep/tv
Fault due to continual changes in the edge steepness of the sync pulses, by which the sync moment becomes irregular.
f image *f* ondulante
e imagen *f* desgarrada
i immagine *f* ondulata
n rafelig beeld *n*
d welliges Bild *n*

3850 RAINBOW GENERATOR rep/tv
A signal generator that generates a signal which, when fed into a colo(u)r TV receiver, produces the entire colo(u)r spectrum on the screen.
f générateur *m* de mire en arc en ciel
e generador *m* de imagen de prueba en arco iris
i generatore *m* di figura di prova in arcobaleno
n regenboogtoetsbeeldgenerator
d Regenbogentestbildgenerator *m*

3851 RAINBOW TEST PATTERN ctv
f mire *f* en arc en ciel
e imagen *f* de prueba en arco iris
i figura *f* di prova in arcobaleno
n regenboogtoetsbeeld *n*
d Regenbogentestbild *n*

3852 RAMP FUNCTION tv
A function of linear increase of a signal.
f échelon *m* de vitesse, fonction *f* rampe
e escalón *m* de velocidad, función *f* rampa
i funzione *f* rampa
n stijgingsfunctie
d Anstiegfunktion *f*

3853 RANDOM INTERLACE rec/tv
Interlace based on less precise timing of sweep frequencies than is required for TV broadcast.
f analyse *f* entrelacée approximative
e exploración *f* entrelazada aproximada
i analisi *f* interlacciata approssimativa
n grove interliniëring
d grobes Zwischenzeilenverfahren *n*

RANDOM NOISE
see: BACKGROUND NOISE

3854 RANDOM NOISE WEIGHTING NETWORK cpl
f filtre *m* de pondération de bruit, filtre *m* vidéométrique
e filtro *m* de ponderación de ruido
i filtro *m* di ponderazione di rumore
n ruisevaluatiefilter *n*
d Rauschbewertungsfilter *n*

3855 RANDOM PHASE ERRORS cpl/dis
Disturbances caused by the effect of noise on the video signal, e.g. jitter.
f erreurs *pl* de phase aléatoires
e errores *pl* de fase fortuitos
i errori *pl* di fase casuali
n toevallige fazefouten *pl*
d zufällige Phasenfehler *pl*

3856 RANGE tv
The distance from a transmitter over which communication is effective.
f portée *f*
e alcance *m*
i portata *f*
n draagwijdte, dracht
d Reichweite *f*

3857 RASTER rep/tv
A predetermined pattern of scanning lines which provides substantially uniform coverage of an area.
f canevas *m*
e cuadrícula *f*, trama *f*
i quadro *m*, quadro *m* rigato
n raster *n*
d Zeilenraster *n*

3858 RASTER BURN crt/dis
In camera tubes, a change in the characteristics of that area of the target which has been scanned, resulting in a spurious signal.
f endommagement *m* de canevas
e daño *m* de cuadrícula
i danneggiamento *m* di quadro rigato
n rasteraantasting
d Rasterbeschädigung *f*

3859 RASTER GENERATOR tv
f générateur *m* de canevas
e generador *m* de cuadrícula
i generatore *m* di quadro rigato
n rastergenerator
d Rastergenerator *m*

3860 RASTER LINEARITY AND GEOMETRY crt
f linéarité *f* et géométrie *f* du canevas
e linealidad *f* y geometría *f* de la cuadrícula

i linearità *f* e geometria *f* del quadro rigato
n rasterlineariteit en -geometrie
d Rasterlinearität *f* und -geometrie *f*

3861 RASTER PITCH rec/tv
f espacement *m* des lignes
e espaciado *m* de las líneas
i distanziamento *m* delle linee
n rasterlijnafstand
d Rasterlinienabstand *m*

3862 RASTER SHADING dis
Non-uniformity of brightness across the raster when a fixed input signal is applied.
f canevas *m* à luminosité inégale
e cuadrícula *f* de luminosidad dispareja
i quadro *m* rigato a luminosità disuguale
n ongelijke rasterhelderheid
d ungleiche Rasterhelligkeit *f*

3863 RATCHET TIMEBASE tv
Timebase which gives a certain delay to the light spot in a cathode-ray tube.
f base *f* de temps à encliquetage
e base *f* de tiempo de trinquete
i base *f* dei tempi ad arresto
n blokkeertijdbasis
d Sperrzeitbasis *f*

3864 RATED POWER SUPPLY rec/tv
In TV, the rated power supply of the transmitter is described by specifying the voltage, the number of phases, and the frequency of the supply with which the transmitter shall be required to meet all applicable standards of performance.
f puissance *f* nominale
e potencia *f* nominal
i potenza *f* nominale
n nominale zendenergie
d Nennleistung *f*

3865 RATIO DETECTOR cpl
Circuit for demodulation of phase- or frequency-modulated signals.
f détecteur *m* de rapport
e detector *m* de relación
i rivelatore *m* di rapporto
n verhoudingsdetector
d Brückendemodulator *m*,
Ratiodetektor *m*,
Verhältnisdetektor *m*

3866 REACTANCE CIRCUIT svs
f circuit *m* de réactance
e circuito *m* de reactancia
i circuito *m* di reattanza
n reactantiecircuit *n*
d Reaktanzkreis *m*

3867 REACTANCE TUBE (US), cpl
REACTANCE VALVE (GB)
A stage connected in parallel with the tank circuit of an oscillator so that the signal produced by the tube (valve) will either lead or lag the signal produced by the tank circuit.
f tube *m* de glissement,
tube *m* de réactance
e válvula *f* de reactancia
i valvola *f* di reattanza
n reactantiebuis
d Blindröhre *f*,
Hubröhre *f*,
Reaktanzröhre *f*

3868 REACTANCE TUBE OSCILLATOR (US), cpl
REACTANCE VALVE OSCILLATOR (GB)
f oscillateur *m* à tube de réactance
e oscilador *m* de válvula de reactancia
i oscillatore *m* a valvola di reattanza
n reactantiebuisoscillator
d Reaktanzröhrenoszillator *m*

3869 READ THROUGH OF THE SCRIPT rec/tv
f lecture *f* du scénario
e lectura *f* del escenario
i lettura *f* della sceneggiatura
n doorlezen *n* van het draaiboek
d Durchlesen *n* des Drehbuches

READY FOR TRANSMISSION
see: CLEAN START

3870 REAL IMAGE opt
An image formed in such a way that the light which passes through an optical system from a point of the object actually passes through or close to a point of the image.
f image *f* réelle
e imagen *f* real
i immagine *f* reale
n reëel beeld *n*
d reelles Bild *n*

3871 REAL PART OF THE COMPLEX CHROMATICITY SIGNAL ctv
f partie *f* réelle du signal complexe de chrominance
e parte *f* real de la señal compleja de crominancia
i parte *f* reale del segnale complesso di crominanza
n reëel gedeelte *n* van het complexe kleursoortsignaal
d reeller Teil *m* des komplexen Farbartsignals

REAR LIGHT
see: BACK LIGHT

3872 REAR PANEL rep/tv
f panneau *m* arrière
e panel *m* posterior
i piastra *f* posteriore
n achterwand
d Rückwand *f*

REAR PROJECTION
see: BACK PROJECTION

REAR PROJECTOR
see: BACK PROJECTOR

3873 REBROADCAST, tv
RETRANSMISSION
Repetition of a TV program(me) at a later time.
f réémission *f*,
retransmission *f*
e reemisión *f*,
retransmisión *f*
i riemissione *f*,
ritrasmissione *f*
n heruitzending
d wiederholte Übertragung *f*

3874 REBROADCAST RECEIVER, rep/tv
RETRANSMISSION RECEIVER
Receiver developed to connect low and medium-power vision transmitters to a main broadcast network.
f récepteur *m* connecteur,
récepteur *m* de retransmission
e receptor *m* conector,
receptor *m* de retransmisión
i ricevittore *m* connettore,
ricevittore *m* di ritrasmissione
n verbindingsontvanger
d Ballempfänger *m*,
Verbindungsempfänger *m*

3875 REBROADCASTING TRANSMITTER tv
An equipment which receives and retransmits a program(me) broadcast by another broadcasting transmitter.
f émetteur *m* relais,
réémetteur *m*
e reemisor *m*,
transmisor *m* de repetición
i emettitore *m* ripetitore,
ripetitore *m*
n omroeprelaiszender
d Relaissender *m*,
Umsetzer *m*

3876 RECEIVER AERIAL, aea
RECEIVER ANTENNA,
RECEIVING AERIAL,
RECEIVING ANTENNA
f antenne *f* de réception
e antena *f* de recepción
i antenna *f* ricevente
n ontvangantenne
d Empfangsantenne *f*

3877 RECEIVER BANDPASS rep/tv
f filtre *m* passe-bande de récepteur
e filtro *m* pasabanda de receptor
i filtro *m* passabanda di ricevittore
n ontvangerbandfilter *n*
d Empfängerbandfilter *n*

RECEIVER PRIMARIES
see: DISPLAY PRIMARIES

3878 RECEIVER TRANSFER rep/tv
CHARACTERISTIC
f caractéristique *f* de transfert du récepteur
e característica *f* de transferencia del receptor
i caratteristica *f* di trasferimento del ricevittore
n overdrachtskarakteristiek van de ontvanger
d Übertragungscharakteristik *f* des Empfängers

3879 RECEIVING TERMINAL STATION tv
Part of a TV relay system which accepts a TV relay signal and delivers a TV relay output signal.
f terminal *m* de réception
e estación *f* receptora terminal
i stazione *f* ricevente terminale
n eindontvangstation *n*
d Endempfangsstation *f*

3880 RECONNAISSANCE SATELLITE tv
An earth satellite designed to provide strategic information as by TV or photography.
f satellite *m* de reconnaissance
e satélite *m* de reconicimiento
i satellite *m* per ricerca
n verkenningssatelliet
d Aufklärungssatellit *m*

3881 RECORD CURRENT vr
f courant *m* d'enregistrement
e corriente *f* de grabación
i corrente *f* di registrazione
n opneemstroom
d Aufnahmestrom *m*

3882 RECORD KEY, vr
RECORDING BUTTON
Used to record.
f touche *f* enregistrement
e botón *m* de registro
i pulsante *m* di registrazione
n opneemtoets
d Aufnahmetaste *f*

3883 RECORDED BROADCAST tv
The broadcasting of a previously recorded radio or TV program(me).
f radiodiffusion *f* différée
e transmisión *f* diferida
i trasmissione *f* d'un programma registrato
n uitzending van een opgenomen programma
d Wiedergabe *f* einer Aufzeichnung

3884 RECORDED PROGRAM(ME) tv
A program(me) that uses e.g. magnetic tapes or other means of reproduction.
f programme *m* enregistré
e programa *m* registrado
i programma *m* registrato
n opgenomen programma *n*
d aufgezeichnetes Programm *n*

RECORDED TAPE
see: PRERECORDED TAPE

3885 RECORDER aud/vr
Instrument employed in sound recording.
f enregistreur *m*
e grabador *m*,
registrador *m*
i registratore *m*
n opneemapparaat *n*,
recorder
d Aufnahmegerät *n*

RECORDING (GB)
see: CAMERA SHOOTING

3886 RECORDING AMPLIFIER aud/vr
The amplifier preceding the recording heads of any type of recorder.
f amplificateur *m* pour enregistrement sonore
e amplificador *m* para registro sonoro
i amplificatore *m* per registrazione sonora
n versterker voor geluidsopname
d Verstärker *m* für Tonaufzeichnung

3887 RECORDING AND REPRODUCING HEAD, vr
RECORDING-PLAYBACK HEAD
f tête *f* d'enregistrement et de reproduction
e cabeza *f* de grabación y de reproducción,
cabeza *f* de registro y de reproducción
i testina *f* di registrazione e di riproduzione
n opneem-weergeefkop
d Aufnahme-Wiedergabekopf *m*

3888 RECORDING CHAIN vr
f chaîne *f* de lecture
e cadena *f* de reproducción
i catena *f* di riproduzione
n opneemkanaal *n*
d Wiedergabekanal *m*

3889 RECORDING CHANNEL rec/tv
Equipment containing two identical recording units, together with associated apparatus, by means of which a continuous recording of a program(me) can be made.
f canal *m* d'enregistrement
e canal *m* de registro
i canale *m* di registrazione
n opneemkanaal *n*
d Aufnahmekanal *m*

3890 RECORDING CHARACTERISTIC vr
f caractéristique *f* d'enregistrement
e característica *f* de grabación
i caratteristica *f* di registrazione
n opneemkarakteristiek
d Aufzeichnungscharakteristik *f*

3891 RECORDING DENSITY vr
f densité *f* d'enregistrement
e densidad *f* de registro
i densità *f* di registrazione
n opneemdichtheid
d Schreibdichte *f*

3892 RECORDING EQUALIZER vr
f correcteur *m* de distorsion de l'enregistrement sonore
e corrector *m* de distorsión del registro sonoro
i correttore *m* di distorsione della registrazione sonora
n vereffenaar van besprekingsvervorming,
vervormingsvereffenaar
d Aufsprechentzerrer *m*

3893 RECORDING HEAD vr
A magnetic head used only for recording.
f tête *f* d'enregistrement
e cabeza *f* de registro
i testina *f* di registrazione
n opneemkop
d Sprechkopf *m*

3894 RECORDING LEVEL vr
The amplifier output level required to drive a particular recorder.
f niveau *m* de modulation de la sortie de l'amplificateur
e nivel *m* de modulación de la salida del amplificador
i livello *m* di modulazione dell'uscita dell'amplificatore
n modulatieniveau *n* van de versterker-uitgang
d Aussteuerung *f*

3895 RECORDING NOISE vr
Noise that is introduced during a recording process.
f bruit *m* d'enregistrement
e ruido *m* de grabación,
ruido *m* de registro
i rumore *m* di registrazione
n opneemruis
d Aufnahmerauschen *n*

3896 RECORDING ROOM stu
A room containing the equipment required to record audio and/or video signals and to monitor this recording.
f cellule *f* d'enregistrement
e sala *f* de registro
i sala *f* di registrazione
n opneemkamer,
opneemruimte
d Aufnahmeraum *m*,
Tonstudio *n*

3897 RECORDIST aud/vr
Operator of the controls which determine the amplitude of electric currents which control a sound-recording device.
f ingénieur *m* d'enregistrement de son
e operador *m* de registro sonoro,
operador *m* de sonido
i operatore *m* di controllo dei segnali sonori registrati
n geluidsopnametechnicus
d Tonaufnahmetechniker *m*

3898 RECTANGULAR PICTURE TUBE crt
A TV picture tube having an essentially rectangular faceplate and screen.
f tube *m* image à écran rectangulaire
e tubo *m* imagen de pantalla rectangular
i tubo *m* immagine a schermo rettangolare
n beeldbuis met rechthoekig scherm
d Bildröhre *f* mit Rechteckschirm

3899 RECTANGULAR SCANNING rec/tv
Any scanning system producing a rectangular field.
f analyse *f* rectangulaire
e exploración *f* rectangular
i analisi *f* rettangolare
n rechthoekige aftasting
d rechteckige Abtastung *f*

3900 RECTILINEAR PROPAGATION ge
f propagation *f* rectiligne
e propagación *f* rectilínea
i propagazione *f* rettilinea
n rechtlijnige voortplanting
d geradlinige Fortpflanzung *f*

3901 RECTILINEAR SCANNING, ZONE TELEVISION tv
In TV, the process of scanning an area in a predetermined sequence of narrow, straight parallel strips.
f analyse *f* par lignes
e exploración *f* por líneas, exploración *f* rectilínea
i analisi *f* per fasci paralleli
n stripaftasting
d Streifenabtastung *f*

3902 RECURRENCE RATE, REPETITION FREQUENCY, REPETITION RATE ge
1. The number of times a periodic quantity is repeated per second.
2. The rate at which recurrent signals are produced or transmitted.
f fréquence *f* de répétition
e frecuencia *f* de repetición
i frequenza *f* di ripetizione
n herhalingsfrequentie
d Folgefrequenz *f*

3903 RECURRENT NOISE dis
f parasites *pl* de récurrence
e perturbación *f* recurrente
i disturbo *m* ricorrente
n terugkerende ruis
d sich wiederholendes Rauschen *n*

3904 RED ADDER ctv
f circuit *m* mélangeur pour le rouge
e circuito *m* aditivo para el rojo
i circuito *m* combinatore per il rosso
n mengkring voor rood, optelschakeling voor rood
d Rotbeimischer *m*

3905 RED APEX ctv
f point *m* de couleur de la primaire rouge
e punto *m* de color del primario rojo
i punto *m* di colore del primario rosso
n kleurpunt *n* van de rode primaire kleur
d Farbort *n* der roten Primärfarbe

3906 RED BEAM crt /ctv
f faisceau *m* pour le rouge
e haz *m* para el rojo
i fascio *m* per il rosso
n bundel voor rood
d Strahl *m* für Rot

3907 RED BEAM MAGNET crt /ctv
A small permanent magnet used as a convergence adjustment to change the direction of the electron beam for red phosphor dots in a three-gun colo(u)r TV tube.
f aimant *m* du faisceau pour le rouge
e imán *m* del haz para el rojo
i magnete *m* del fascio per il rosso
n rode-bundelmagneet
d Rotstrahlmagnet *m*

RED BLACK LEVEL crt
see: R BLACK LEVEL

3908 RED CATHODE
f cathode *f* rouge
e cátodo *m* rojo
i catodo *m* rosso
n rode katode
d rote Katode *f*

RED COLO(U)R DIFFERENCE AXIS
see: R-Y AXIS

RED COLO(U)R DIFFERENCE MATRIX
see: R-Y MATRIX

RED COLO(U)R DIFFERENCE MODULATOR
see: R-Y MODULATOR

RED COLO(U)R DIFFERENCE SIGNAL
see: R-Y SIGNAL

3909 RED CONSCIOUS rec/tv
Said of an electron camera which is unduly sensitive to light of long wavelengths.
f hypersensible pour le rouge
e hipersensible para el rojo
i ipersensibile per il rosso
n overgevoelig voor rood
d überempfindlich für Rot

3910 RED CONVERGENCE CIRCUIT crt /ctv
f circuit *m* de convergence pour le rouge
e circuito *m* de convergencia para el rojo
i circuito *m* di convergenza per il rosso
n rode-convergentieschakeling
d Rotkonvergenzschaltung *f*

3911 RED CONVERGENCE POLE PIECE, RED GUN POLE PIECE crt
f bloc *m* de convergence pour le rouge
e bloque *m* de convergencia para el rojo
i blocco *m* di convergenza per il rosso
n convergentie-eenheid voor rood
d Konvergenzeinheit *f* für Rot

3912 RED ELECTRON GUN, RED GUN crt /ctv
The electron gun whose beam strikes phosphor dots emitting the red primary colo(u)r in a three-gun colo(u)r TV picture tube.
f canon *m* du rouge
e cañón *m* del rojo
i cannone *m* del rosso
n kanon *n* voor het rood
d Rotstrahlsystem *n*

3913 RED GAIN CONTROL crt /ctv
A variable resistor used in the matrix of a three-gun colo(u)r picture TV receiver to adjust the intensity of the red primary signal.
f régleur *m* d'intensité pour le rouge

e regulador *m* de intensidad para el rojo
i regolatore *m* d'intensità per il rosso
n intensiteitsregelaar voor rood
d Intensitätsregler *m* für Rot

3914 RED GRID crt
f grille *f* rouge
e rejilla *f* roja
i griglia *f* rossa
n rood rooster *n*
d rotes Gitter *n*

3915 RED HIGHS crt/ctv
f hautes fréquences *pl* pour le rouge
e altas frecuencias *pl* para el rojo
i componenti *pl* d'alta frequenza per il rosso
n hoge frequenties *pl* voor rood
d höhe Frequenzen *pl* für Rot,
Rothöhe *f*

3916 RED HORIZONTAL SHIFT MAGNET, crt
RED LATERAL SHIFT MAGNET,
RED POSITIONING MAGNET
f aimant *m* latéral pour le rouge
e imán *m* lateral para el rojo
i magnete *m* laterale per il rosso
n lateraalmagneet voor het rood
d Rotschiebemagnet *m*

3917 RED LOWS crt/ctv
f basses fréquences *pl* pour le rouge
e bajas frecuencias *pl* para el rojo
i componenti *pl* di bassa frequenza per il rosso
n lage frequenties *pl* voor rood
d niedrige Frequenzen *pl* für Rot,
Rottiefe *f*

RED PEAK LEVEL
see: R PEAK LEVEL

RED PRIMARY
see: R

3918 RED PRIMARY INFORMATION ctv
f information *f* du rouge primaire
e información *f* del rojo primario
i informazione *f* del rosso primario
n informatie van de rode primaire kleur
d Information *f* der roten Primärfarbe

3919 RED PRIMARY SIGNAL crt/ctv
f signal *m* du rouge primaire
e señal *f* del rojo primario
i segnale *m* del rosso primario
n signaal *n* van de rode primaire kleur
d Signal *n* der roten Primärfarbe

3920 RED RESTORATION crt
The d.c. restoration for the red channel of a three-gun colo(u)r picture tube circuit.
f restauration *f* de la composante de courant continu du niveau du rouge
e restauración *f* de la componente de corriente continua del nivel del rojo
i reinserzione *f* della componente di corrente continua del livello del rosso
n herstel *n* van het roodniveau
d Wiederherstellung *f* des Rotpegels

3921 RED SATURATION SCALE SIGNAL cpl
f échelle *f* de rouges
e escala *f* de rojos
i scala *f* di rossi
n roodtrap
d Rotskala *f*

3922 RED SCREEN-GRID crt
f grille-écran *m* rouge
e rejilla-pantalla *f* roja
i griglia-schermo *m* rosso
n rood schermrooster *n*
d rotes Schirmgitter *n*

3923 RED SHIFT crt
f décalage *m* du canevas rouge
e desplazamiento *m* de la cuadrícula roja
i spostamento *m* del quadro rigato rosso
n verschuiving van het rode raster
d Verschiebung *f* des roten Rasters

3924 RED VIDEO VOLTAGE cpl/ctv
The signal voltage output from the red section of a colo(u)r TV camera, or the signal voltage between the matrix and the grid of the red gun in a three-gun colo(u)r TV picture tube.
f tension *f* vidéo pour le rouge
e tensión *f* video para el rojo
i tensione *f* video per il rosso
n beeldspanning voor rood
d Bildspannung *f* für Rot

3925 REDUCTION OF DETECTION EFFICIENCY cpl/tv
Result of single sideband reception in the detection circuit for the lower and the higher modulation frequencies.
f affaiblissement *m* du rendement détecteur
e pérdida *f* de nivel de la modulación
i attenuazione *f* del rendimento rivelatore
n modulatieverondieping
d Herabsetzung *f* der Aussteuerung

3926 REEL vr
A container consisting of a core and flanged ends, used for magnetic tape.
f bobine *f*,
dévidoir *m*
e bobina *f*,
devanador *m*
i bobina *f*,
rocchetto *m*
n rol,
spoel
d Rolle *f*,
Spule *f*

3927 REEL-TO-REEL RECORDER vr
f magnétophone *m* à deux bobines
e magnetófono *m* de dos bobinas
i magnetofono *m* a due bobine
n magneetbandrecorder met twee spoelen
d Magnetbandgerät *n* mit zwei Rollen

3928 REELING ON TO SPOOLS vr
f bobinage *m*
e devanado *m*
i avvolgimento *m*

n opspoelen *n*
d Aufspulen *n*

3929 REFERENCE AUDIO LEVEL, REFERENCE SOUND LEVEL, REFERENCE SURFACE INDUCTION FOR THE SOUND TRACK vr
f niveau *m* de référence audio
e nivel *m* de referencia audio
i livello *m* di riferimento audio
n audioreferentieniveau *n*, geluidsreferentieniveau *n*
d Tonbezugspegel *m*

3930 REFERENCE BEAM tv
One of the light beams obtained by splitting the light emitted by a laser.
f faisceau *m* de référence
e haz *m* de referencia
i fascio *m* di riferimento
n referentiebundel
d Bezugsbündel *n*

3931 REFERENCE BLACK LEVEL rep/tv
The level corresponding to specified maximum excursion of the luminance signal in the black direction.
f niveau *m* du noir de référence
e nivel *m* de negro de referencia
i livello *m* di riferimento per il nero
n referentieniveau *n* voor het zwart
d Vergleichsschwarzpegel *m*

3932 REFERENCE CIRCUIT cpl
Circuit along which a signal is sent which is used as a carrier to be reinserted in suppression-carrier signals.
f circuit *m* de référence
e circuito *m* de referencia
i circuito *m* di riferimento
n referentiecircuit *n*
d Bezugskreis *m*

3933 REFERENCE COLO(U)RS ct
f couleurs *pl* de référence
e colores *pl* de referencia
i colori *pl* di riferimento
n referentiekleuren *pl*
d Bezugsfarben *pl*

3934 REFERENCE EDGE OF TAPE vr
f bord *m* de référence de la bande
e borde *m* de referencia de la cinta
i bordo *m* di riferimento del nastro
n referentiezijde van de band
d Bandbezugskante *f*

3935 REFERENCE FREQUENCY cpl
A frequency available from the main city generating station which ensures that in case of supply breakdown the sync pulse generators stay locked to the frequency of the national electricity supply grid.
f fréquence *f* de référence, fréquence *f* étalon
e frecuencia *f* de referencia, frecuencia *f* patrón
i frequenza *f* di paragone, frequenza *f* di riferimento
n referentiefrequentie, vergelijkingsfrequentie
d Bezugsfrequenz *f*, normierte Frequenz *f*, Vergleichsfrequenz *f*

3936 REFERENCE-GENERATOR PERFORMANCE ctv
f conduite *f* du générateur de la porteuse de chrominance
e comportamiento *m* del generador de la portadora de crominancia
i comportamento *m* del generatore della portante di crominanza
n gedrag *n* van de chrominantiedraaggolf-generator
d Benehmen *n* des Chrominanzträger-generators

3937 REFERENCE-GENERATOR RADIATION ctv
f rayonnement *m* du générateur de la porteuse de chrominance
e radiación *f* del generador de la portadora de crominancia
i radiazione *f* del generatore della portante di crominanza
n straling van de chrominantiedraaggolf-generator
d Strahlung *f* des Chrominanzträger-generators

3938 REFERENCE INPUT, REFERENCE INPUT SIGNAL ctv
f signal *m* d'entrée de référence
e señal *f* de entrada de referencia
i segnale *m* d'ingresso di riferimento
n referentieingangssignaal *n*
d Bezugseingangssignal *n*

3939 REFERENCE LEVEL ge
f niveau *m* de référence
e nivel *m* de referencia
i livello *m* di riferimento
n referentieniveau *n*
d Bezugspegel *m*

3940 REFERENCE MONITOR cpl
f moniteur *m* de référence
e monitor *m* de referencia
i monitore *m* di riferimento
n referentiemonitor
d Bezugsmonitor *m*

3941 REFERENCE NOTE, REFERENCE TONE aud
A pure tone of 1.000Hz and the reference intensity is 10^{-16} watts per sq.cm., which corresponds to a pressure of 0.0002 dynes per sq.cm.
f note *f* de référence, ton *m* de référence
e nota *f* de referencia, tono *m* de referencia
i nota *f* di riferimento, tono *m* di riferimento
n referentienoot, referentietoon
d Bezugsnote *f*, Vergleichston *m*

3942 REFERENCE OSCILLATOR cpl
f oscillateur *m* de référence
e oscilador *m* de referencia
i oscillatore *m* di riferimento
n referentieoscillator
d Bezugsoszillator *m*

3943 REFERENCE PHASE cpl
f phase *f* de référence
e fase *f* de referencia
i fase *f* di riferimento
n referentiefaze
d Bezugsphase *f*

3944 REFERENCE RECORDING rep/tv
Recording a program(me) for future reference.
f enregistrement *m* pour consultation ultérieure
e registro *m* para consultación ulterior
i registrazione *f* per consultazione ulteriore
n opname voor latere raadpleging
d Aufnahme *f* für zukünftige Bezugnahme

3945 REFERENCE RECORDING HEAD rec tv vr
f tête *f* d'enregistrement de référence
e cabeza *f* de registro de referencia
i testina *f* di registrazione di riferimento
n referentie-opneemkop
d Bezugssprechkopf *m*

3946 REFERENCE SIGNAL ctv
f signal *m* de référence
e señal *f* de referencia
i segnale *m* di riferimento
n referentiesignaal *n*
d Bezugssignal *n*

3947 REFERENCE STIMULI ct
The three standard stimuli of a trichromatic system of colo(u)r specification.
f stimuli *pl* de référence, stimuli *pl* primaires de couleur
e estímulos *pl* de referencia, estímulos *pl* primarios de color
i stimoli *pl* di riferimento, stimoli *pl* primari di colore
n primaire kleurstimuli *pl*, referentiestimuli *pl*
d Bezugsreize *pl*, Primärvalenzen *pl*

REFERENCE SUBCARRIER
see: CHROMINANCE SUBCARRIER REFERENCE

3948 REFERENCE TAPE, STANDARD TAPE vr
f bande *f* de référence, bande-étalon *f*
e cinta *f* de referencia, cinta *f* patrón
i nastro *m* campione, nastro *m* di riferimento
n referentieband, standaardband
d Bezugsband *n*

3949 REFERENCE VOLUME aud
The audio volume level that gives a reading of O on a standard volume indicator.
f niveau *m* de volume sonore
e nivel *m* de volumen sonoro
i livello *m* di volume sonoro
n geluidsvolumeniveau *n*
d Schallvolumenpegel *m*

3950 REFERENCE WHITE, REFERENCE WHITE STANDARD stu
The chromaticity of the white card, as used in a studio for the alignment of cameras and coder.
f blanc *m* de référence
e blanco *m* de referencia
i bianco *m* di riferimento
n referentiewit *n*
d Vergleichsweiss *n*

3951 REFERENCE WHITE LEVEL ct
The level corresponding to the specified maximum excursion of the luminance signal in the white direction.
f niveau *m* du blanc de référence
e nivel *m* del blanco de referencia
i livello *m* del bianco di riferimento
n niveau *n* van het referentiewit
d Vergleichsweisspegel *m*

3952 REFLECTANCE, REFLECTION FACTOR ct
Ratio of reflected to incident flux.
f facteur *m* de réflexion, réflectance *f*
e factor *m* de reflexión
i fattore *m* di riflessione
n reflectiefactor, remissiefactor
d Reflexionsgrad *m*

3953 REFLECTED BEAM KINESCOPE, REFLECTION TUBE crt
A cathode-ray tube in which the electron beam is reflected back from the transparent glass front to the inside rear curved surface to provide a deflection angle of 180°.
f tube *m* cathodique à réflexion du faisceau
e tubo *m* catódico de reflexión del haz
i tubo *m* catodico a riflessione del fascio
n katodestraalbuis met bundelreflectie
d Katodenstrahlröhre *f* mit Strahlreflexion

3954 REFLECTED LIGHT opt
f lumière *f* réfléchie
e luz *f* reflejada
i luce *f* riflessa
n gereflecteerd licht *n*
d reflektiertes Licht *n*

3955 REFLECTED SIGNAL dis
The cause of ghost images.
f signal *m* réfléchi
e señal *f* reflejada
i segnale *m* riflesso
n gereflecteerd signaal *n*
d reflektiertes Signal *n*

3956 REFLECTION ge
The return or change in direction of light, sound, radar or radio waves striking a surface or travel(l)ing from one medium into another.
f réflexion *f*
e reflexión *f*
i riflessione *f*
n reflectie, weerkaatsing
d Reflexion *f*, Rückstrahlung *f*

REFLECTION ANGLE
see: ANGLE OF REFLECTION

3957 REFLECTION COLO(U)R TUBE crt/ctv
A colo(u)r picture which produces an image by the technique of electron reflection in the screen region.
f tube *m* image couleur à réflexion
e tubo *m* imagen en colores de reflexión
i tubo *m* immagine a colori a riflessione
n reflectiekleurenbeeldbuis
d Reflexionsfarbbildröhre *f*

3958 REFLECTION TELEVISION RECEIVER rep/tv
f téléviseur *m* à réflexion
e televisor *m* de reflexión
i televisore *m* a riflessione
n reflectietelevisieontvanger
d Reflexionsfernsehempfänger *m*

3959 REFLECTIONS rep/tv
Electromagnetic waves that have been reflected from an obstacle during their travel to a TV receiving aerial (antenna), causing ghost images to appear on the screen.
f ondes *pl* réfléchies
e ondas *pl* reflejadas
i onde *pl* riflesse
n gereflecteerde golven *pl*
d reflektierte Wellen *pl*

3960 REFLECTIONS FROM PHOSPHOR PARTICLES crt
f réflexions *pl* de particules luminescentes
e reflexiones *pl* de partículas luminiscentes
i riflessioni *pl* da particelle luminescenti
n reflecties *pl* van luminescerende deeltjes
d Reflexionen *pl* von lumineszierenden Teilchen

3961 REFLECTIVE NETWORK cpl
A network which provides an incorrect termination of a line and thereby causes a reflection.
f réseau *m* réflectant
e red *f* reflectante
i rete *f* riflettente
n reflecterend netwerk *n*
d reflektierendes Netzwerk *n*

REFLECTIVE OPTICS (US)
see: PROJECTION OPTICS

3962 REFLECTIVE TELEVISION rep/tv
f système *m* de télévision à réflexion
e televisión *f* de reflexión
i telericezione *f* a riflessione
n reflectietelevisie
d Reflexionsfernsehen *n*

3963 REFLECTOR aea
A secondary radiator, or an array of secondary radiators, or a reflecting surface, placed behind a primary radiator or an array of primary radiators, or a feed, in order to increase forward and reduce backward radiation from the aerial (antenna).
f réflecteur *m*
e reflector *m*
i riflettore *m*
n reflector
d Reflektor *m*

3964 REFLECTOR (GB), REFLECTOR ELECTRODE (GB), REPELLER (US) crt
An electrode whose primary function is to reverse the direction of an electron beam.
f électrode *f* de réflexion
e electrodo *m* de reflexión
i elettrodo *m* di riflessione
n reflectie-elektrode
d Reflektorelektrode *f*

3965 REFLECTOR ELEMENT aea
A single rod or other parasitic element serving as a reflector in an aerial (antenna) array.
f élément *m* réflecteur
e elemento *m* reflectante
i elemento *m* riflettente
n reflectorelement *n*
d Reflektorelement *n*

3966 REFLECTOR LAMP stu
Type of incandescent projection lamp which contains a built-in reflector.
f lampe *f* à réflecteur
e lámpara *f* de reflector
i lampada *f* a riflettore
n reflectorlamp
d Reflektorlampe *f*

3967 REFLEX AMPLIFICATION cpl
Amplification in a circuit wherein the same tube (valve) acts as an amplifier of given signals both before and after detection and or frequency changing.
f amplification *f* réflex
e amplificación *f* reflex
i amplificazione *f* reflex
n reflexversterking
d Reflexverstärkung *f*

REFLEX SHUTTER
see: MIRROR SHUTTER

REGENERATIVE FEEDBACK
see: POSITIVE FEEDBACK

3968 REGISTER, tv
REGISTRATION
The condition of accurate superposition of one or more partial pictures on another.
f calage *m*,
registre *m*
e coincidencia *f*,
registro *m*,
superposición *f*
i registro *m*
n dekking,
het in register zijn *n*,
superpositie
d Deckung *f*,
Registerhalten *n*

3969 REGISTRATION ACCURACY ctv
f précision *f* de calage
e precisión *f* de registro
i precisione *f* di registro
n nauwkeurige dekking
d Deckungsgenauigkeit *f*

3970 REGISTRATION CONTROL ctv
f contrôle *m* de calage
e control *m* de registro
i controllo *m* di registro
n dekkingscontrole
d Deckungskontrolle *f*

3971 REGISTRATION DRIFT ctv/rec/tv
f dérive *f* de calage
e desviación *f* de registro
i deviazione *f* di registro
n dekkingsafwijking
d Deckungsabwanderung *f*

REGISTRATION FAULT
see: MISREGISTRATION

REGULAR REFLECTION
see: DIRECT REFLECTION

3972 REHEARSALS ge
f répétitions *pl*
e ensayos *pl*
i prove *pl*
n repetities *pl*
d Proben *pl*

3973 REJECTION OF THE aud
ACCOMPANYING SOUND,
SOUND REJECTION (GB),
TAKE-OFF (US)
Suppression of the sound-carrier wave in the TV receiver without impairing its pass characteristic on the side where the sound-carrier wave is situated.
f réjection *f* de la porteuse son
e supresión *f* de la portadora sonido
i soppressione *f* della portante suono
n geluidsdraaggolfonderdrukking
d Tonträgerunterdrückung *f*

3974 REJECTOR CIRCUIT cpl
A closed resonant circuit consisting of inductance and capacitance in parallel.
f circuit *m* bouchon,
circuit *m* éliminateur
e circuito *m* de supresión
i circuito *m* soppressore
n parallel-resonantiekring
d Sperrkreis *m*

3975 RELATED PERCEIVED COLO(U)R ct
Colo(u)r perceived to belong to an area or object in relation to other perceived colo(u)rs in the visual field.
f couleur *f* perçue en ambiance colorée
e color *m* percibido en ambiente colorado
i colore *m* percipito in ambiente colorato
n waargenomen kleur in gekleurde omgeving
d in farbiger Umgebung wahrgenommene Farbe *f*

3976 RELATIVE LUMINOSITY ct
The ratio of the value of the luminosity at particular wavelength to the value at the wavelength of maximum luminosity.
f luminosité *f* relative
e luminosidad *f* relativa
i luminosità *f* relativa
n relatieve helderheid
d relative Hellempfindlichkeit *f*

3977 RELATIVE TIME DELAY tv
The difference in time delay encountered by the audio signal and the composite picture signal or between the components of the picture signal travel(l)ing over a TV relay system.
f retard *m* relatif,
temps *m* de transit relatif
e retraso *m* relativo,
tiempo *m* de tránsito relativo
i ritardo *m* relativo,
tempo *m* di transito relativo
n relatieve looptijd,
relatieve vertraging
d relative Laufzeit *f*,
relative Verzögerung *f*

3978 RELAXATION GENERATOR, cpl
RELAXATION OSCILLATOR
A generator of oscillations involving an asymptotic variation of the electrical state.
f générateur *m* à relaxation,
oscillateur *m* à relaxation
e generador *m* de relajación,
oscilador *m* de relajación
i generatore *m* a rilassamento,
oscillatore *m* a rilassamento
n relaxatiegenerator,
relaxatieoscillator
d Kippgenerator *m*,
Kippgerät *n*

3979 RELAY BROADCAST STATION tv
A station licensed to transmit, from points where wire facilities are not available, program(me)s for broadcast by one or more broadcast stations.
f station-relais *m* de radiodiffusion
e estación *f* de retransmisión
i stazione *f* di rimbalzo
n relaiszender
d Ballsendungsstation *f*

3980 RELAY CHANNEL tv
The band of frequencies used in transmitting a single TV relay signal, including the guard bands.
f canal *m* du signal relais
e canal *m* de la señal relé
i canale *m* del segnale relè
n relaissignaalkanaal *n*
d Ballsignalkanal *m*

3981 RELAY FROM ABROAD rec/tv
f transmission *f* de l'étranger
e transmisión *f* del estranjero
i trasmissione *f* dall'estero
n uitzending uit het buitenland
d Übertragung *f* aus dem Ausland

3982 RELAY LENS ctv/opt
f lentille *f* auxiliaire
e lente *f* auxiliar
i lente *f* ausiliaria
n relaislens
d Relaisoptik *f*,
Zwischenabbildungsobjektiv *n*

3983 RELAY OPTICS opt/tv
f système *m* optique à relais
e sistema *m* óptico de relé
i sistema *m* ottico a relè
n optisch relaissysteem *n*
d optisches Relaissystem *n*

RELAY RECEIVER
see: BALL RECEIVER

RELAY TELEVISION
see: BALL RECEPTION

3984 RELAY TRANSMITTER cpl/tv
f transmetteur *m* à relais
e transmisor *m* de relé
i trasmettitore *m* a relè
n relaiszender
d Ballsender *m*,
Relaissender *m*,
Zwischensender *m*

3985 RELEASE KEY vr
Used to open the cassette holder.
f touche *f* éjection de la cassette
e botón *m* de desenclavamiento de la caseta
i pulsante *m* di liberazione della cassetta
n ontgrendeltoets,
toets voor cassettelift
d Entriegeltaste *f*

RELIEF EFFECT
see: PLASTIC EFFECT

3986 REMAKE,
RETAKE fi/tv
f nouvelle mise *f* en scène
e toma *f* nueva
i rifacimento *m*
n heropname,
nieuwe opname
d Neuproduktion *f*

3987 REMANENCE,
RESIDUAL MAGNETISM ge
The magnetism remaining in a substance after the magnetizing force has been removed.
f rémanence *f*
e remanencia *f*
i rimanenza *f*
n remanentie
d Remanenz *f*

REMOTE
see: FIELD PICKUP

3988 REMOTE CONTROL ge
Control, usually by electric or radio signals, carried out from a distance in response to information provided by monitoring instruments.
f commande *f* à distance,
télécommande *f*
e mando *m* a distancia,
telemando *m*,
telemanejo *m*
i comando *m* a distanza,
telecomando *m*
n afstandsbediening,
afstandsregeling
d Fernbedienung *f*,
Fernsteuerung *f*

3989 REMOTE GAIN CONTROL cpl/tv
f téléréglage *m* de l'amplification
e telerregulación *f* de la amplificación
i teleregolazione *f* dell'amplificazione
n afstandsregeling van de versterking
d Fernverstärkungsregelung *f*

3990 REMOTE LINE stu/tv
A program(me) transmission line running between a remote pickup point and a broadcast studio or transmitter site.
f ligne *f* de connexion entre le studio et l'extérieur
e línea *f* de conexión entre el estudio y el exterior
i linea *f* di connessione tra lo studio e la ripresa esterna
n verbindingslijn studio-buitenopname
d Verbindungslinie *f* Studio-Aussenaufnahme

3991 REMOTE STUDIO stu
f studio *m* à l'extérieur
e estudio *m* al exterior
i studio *m* all'esterno
n buitenstudio,
veldstudio
d Aussenstudio *n*

3992 REMOTELY CONTROLLED CAMERA,
REMOTELY OPERATED CAMERA rec/tv
f caméra *f* à téléopération
e cámara *f* de teleoperación
i camera *f* a teleoperazione
n op afstand bediende camera
d fernbediente Kamera *f*,
ferngesteuerte Kamera *f*

3993 REMOTELY CONTROLLED MICROPHONES aud/stu
f microphones *pl* télécommandés

e microfonos *pl* telemandados
i microfoni *pl* telecomandati
n op afstand bediende microfonen *pl*
d fernbediente Mikrophone *pl*

3994 REMOTELY CONTROLLED STATION tv
A radio station controlled from a distant point.
f station *f* télécommandée
e estación *f* telemandada
i stazione *f* telecomandata
n op afstand bediend station *n*
d ferngesteuerte Station *f*

3995 REPEAT KEY vr
f touche *f* répétition
e tecla *f* de repetición
i pulsante *f* a ripetizione
n herhalingstoets
d Wiederholtaste *f*

3996 REPEATER rep/tv
Remote amplifier in a communications satellite, microwave chain, landline, etc., which receives signals, amplifies them, if necessary modifies them, and then retransmits them.
f amplificateur *m*, répéteur *m*
e amplificador *m*, repetidor *m*
i amplificatore *m*, ripetitore *m*
n versterker
d Verstärker *m*

REPELLER (US)
see: REFLECTOR

REPETITION FREQUENCY, REPETITION RATE
see: RECURRENCE RATE

3997 REPETITIVE SIGNAL cpl/dis
The cause of periodic noise in an audio or video network.
f signal *m* de récurrence
e señal *f* de recurrencia
i segnale *m* ricorrente
n zich herhalend signaal *n*
d wiederkehrendes Signal *n*

REPLAY
see: PLAYBACK

REPLAY CHARACTERISTIC
see: PLAYBACK CHARACTERISTIC

3998 REPLAY HEAD, REPRODUCING HEAD vr
The reproducing head of a magnetic-tape recorder.
f tête *f* de reproduction
e cabeza *f* de reproducción
i testina *f* di riproduzione
n weergeefkop
d Wiedergabekopf *m*

REPORT SHEET
see: LOG-SHEET

REPRODUCER HEAD
see: PLAYBACK HEAD

3999 REPRODUCING CHAIN vr
f chaîne *f* de reproduction
e cadena *f* de reproducción
i catena *f* di riproduzione
n weergeefkanaal *n*
d Wiedergabekanal *m*

REPRODUCTION
see: PLAYBACK

REPRODUCTION CHARACTERISTIC
see: PLAYBACK CHARACTERISTIC

4000 REPRODUCTION FIDELITY tv
f fidélité *f* de reproduction
e fidelidad *f* de reproducción
i fedeltà *f* di riproduzione
n reproduktiegetrouwheid, weergavegetrouwheid
d Wiedergabegüte *f*

4001 REQUIRED SIGNAL tv
f signal *m* désiré
e señal *f* deseada
i segnale *m* desiderato
n gewenst signaal *n*
d verlangtes Signal *n*

4002 RERECORDED PROGRAM(ME) tv/vr
A program(me) that begins as a videotape recording and is then transferred to film by a process called telerecording.
f programme *m* réenregistré
e programa *m* de rerregistro
i programma *m* reregistrato
n overgespeeld programma *n*
d überspieltes Programm *n*, umspieltes Programm *n*

4003 RERECORDING aud/vr
Recording acoustic waveforms immediately upon reproduction from the same, or any other type of recording medium as that in use.
f réenregistrement *m*
e rerregistro *m*
i reregistrazione *f*
n overspelen *n*
d Überspielen *n*, Umspielung *f*

4004 RERECORDING ROOM aud
f salle *f* de mixage
e sala *f* de mezcla
i sala *f* di mescolanza
n mengkamer
d Mischraum *m*

4005 RERECORDING SYSTEM aud/vr
A system of reproducers, mixers, amplifiers, and recorders used to combine or modify various sound recordings to provide a final sound record.
f appareillage *m* de mixage sonore
e equipo *m* de mezcla sonora

i apparecchiatura *f* di mescolanza sonora
n geluidsmengingsapparatuur
d Tonmischapparatur *f*

4006 RESIDUAL CHARGE crt/svs
A charge in a cathode-ray tube which must be shortcircuited when servicing the TV receiver.
f charge *f* résiduelle
e carga *f* residua
i carica *f* residua
n restlading
d Restladung *f*

4007 RESIDUAL FLUX DENSITY crt/vr
Flux density remaining after exciting magnetic field has been removed.
f densité *f* de flux résiduel
e densidad *f* de flujo residuo
i densità *f* di flusso residuo
n restfluxdichtheid
d Restflussdichte *f*

4008 RESIDUAL HUM dis
f ronflement *m* résiduel
e zumbido *m* residual
i ronzio *m* residuo
n restbrom
d Restbrumm *m*

RESIDUAL MAGNETISM
see: REMANENCE

4009 RESIDUAL SIGNAL tv
f signal *m* résiduel
e señal *f* residual
i segnale *m* residuo
n restsignaal *n*
d Restsignal *n*

RESOLUTION (US)
see: DEFINITION

RESOLUTION PATTERN (US)
see: DEFINITION TEST CARD

4010 RESOLUTION RESPONSE crt/tv
The ratio of the peak-to-peak signal amplitude given by a test pattern consisting of alternate black and white vertical bars of equal widths corresponding to a specified line number to the peak-to-peak signal amplitude given by large-area blacks and large-area whites having the same luminosity, as the black and white bars in the test pattern.
f réponse *f* de la résolution
e respuesta *f* de la resolución
i risposta *f* della risoluzione
n scheidingsresponsie
d Auflösungsverhältnis *n*

4011 RESOLUTION WEDGE tv
A group of gradually converging lines on a test pattern, used to measure resolution in TV.
f coin *m* de définition
e cuña *f* de resolución
i cuneo *m* di definizione,
cuneo *m* di risoluzione
n scheidingswig
d Auflösungskeil *m*

4012 RESOLVING POWER TEST TARGET opt/tv
Consists of groups of black stripes on a white background or of white stripes on a black background.
f mire *f* de résolution pour lentilles
e mira *f* de resolución para lentes
i figura *f* di prova per lenti
n lenzentoetsbeeld *n*
d Linsentestbild *n*

4013 RESONANCE aud
The condition in which a vibrating system responds with maximum amplitude to a periodically applied force, or an alternating circuit oscillates at maximum voltage.
f résonance *f*
e resonancia *f*
i risonanza *f*
n resonantie
d Resonanz *f*

4014 RESONANCE CHARACTERISTIC, RESONANCE CURVE aud
An amplitude-frequency response curve showing the current or voltage response of a tuned circuit to frequencies at and near the resonant frequency.
f courbe *f* de résonance
e curva *f* de resonancia
i curva *f* di risonanza
n doorlaatkromme,
resonantiekromme
d Durchlasskurve *f*,
Resonanzkurve *f*,
Resonanzverlauf *m*

4015 RESONANCE FREQUENCY rec
The frequency of the absorber adjusted by a suitable choice of constructional details of that absorber.
f fréquence *f* de résonance
e frecuencia *f* de resonancia
i frequenza *f* di risonanza
n resonantiefrequentie
d Resonanzfrequenz *f*

4016 RESONANCE PEAK cpl
f crête *f* de résonance
e cresta *f* de resonancia
i cresta *f* di risonanza
n resonantiepiek
d Resonanzspitze *f*

4017 RESONANT ABSORBERS rec
f absorbants *pl* résonnants
e absorbentes *pl* resonantes
i assorbitori *pl* risonanti
n resonerende absorptiemiddelen *pl*
d resonierende Schluckmittel *pl*

4018 RESONANT AERIAL, RESONANT ANTENNA aea
An aerial (antenna) tuned to the frequency it is designed to transmit or receive.
f antenne *f* résonnante

e antena *f* resonante
i antenna *f* risonante
n resonerende antenne
d resonierende Antenne *f*

4019 RESONATOR MODE cpl
Of a microwave oscillator, a condition of operation corresponding to a particular radio-frequency field configuration for which the electron stream introduces a negative conductance into the circuit to which it is coupled.
f mode *m* de résonateur
e modo *m* de resonador
i modo *m* di risonatore
n resonatormodus, trilholtemodus
d Hohlraumresonatormodus *m*, Resonatorschwingungsart *f*

4020 RESPONSE ge
Of a device or system, a quantitative expression of the output as a function of the input under conditions which must be explicitly stated.
f réponse *f*
e respuesta *f*
i risposta *f*
n responsie
d Übertragungsfaktor *m*

RESPONSE CHARACTERISTIC
see: AMPLITUDE CHARACTERISTIC

RESTING FREQUENCY
see: CENTER(RE) FREQUENCY

RETAINED IMAGE
see: BURNED-IN IMAGE

RETAINING ZONE,
RETENTION RANGE (US)
see: HOLD RANGE

RETARDING ELECTRODE
see: DECELERATING ELECTRODE

4021 RETINA ct
The complicated surface of the eye.
f rétine *f*
e retina *f*
i retina *f*
n netvlies *n*, retina
d Netzhaut *f*, Retina *f*

RETINAL CONES
see: CONES

4022 RETINAL ILLUMINANCE ct
A psycho-physiological quantity, particularly correlated with the brightness attribute of visual sensation.
f éclairement *m* rétinien
e iluminación *f* de la retina
i illuminamento *m* della retina
n verlichtingssterkte op het netvlies
d Netzhautbeleuchtungsstärke *f*

4023 RETINAL PERSISTANCE ct
f persistance *f* rétinienne
e persistencia *f* de la retina
i persistenza *f* della retina
n netvliestraagheid
d Netzhautträgheit *f*, Netzträgheit *f*

4024 RETINAL RODS, ct
RODS
Non-colo(u)r sensitive light-perceptive elements on periphery of human retina.
f bâtonnets *pl*
e bastoncillos *pl*
i bastoncini *pl*
n staafjes *pl*
d Stäbchen *pl*

RETRACE (GB),
RETURN TRACE (US)
see: FLYBACK

RETRACE BLANKING
see: FLYBACK BLANKING

RETRANSMISSION
see: REBROADCAST

RETRANSMISSION RECEIVER
see: REBROADCAST RECEIVER

4025 RETUNE (TO) rep/tv
f réajuster v, retoucher v le réglage
e resintonizar v, volver v a sintonizar
i resintonizzare v
n bijstemmen v
d nachstimmen v

4026 RETURN GHOST rep/tv
A ghost image produced in a TV receiver, screen during retrace periods caused by insufficient blanking of the camera tube at the transmitter.
f image *f* fantôme par retour du spot
e imagen *f* fantasma por retorno del punto
i immagine *f* fantasma per ritorno del punto
n terugslagechobeeld *n*
d Rücklaufgeisterbild *n*

4027 RETURN INTERVAL, crt
RETURN PERIOD,
RETURN TUNE
In cathode-ray tubes, the interval corresponding to the direction of sweep not used for delineation.
f durée *f* de retour du spot
e duración *f* de retorno del punto
i durata *f* di ritorno del punto
n terugslagtijd
d Rücklaufzeit *f*

4028 RETURN LIGHT stu
A pre-arranged light signal in answer to a cue from a studio or other program(me) source.
f signal *m* de réponse
e señal *f* de respuesta

i segnale *m* di risposta
n antwoordsignaallicht *n*
d Bereitschaftssignal *n*

4029 RETURN LINE, RETURN TRACE crt/rec/rep/tv
The line traced by the electron beam in a cathode-ray tube in going from the end of one line or field to the start of the next line or field.
f trace *f* de retour du spot
e traza *f* de retorno del punto
i traccia *f* di ritorno del punto
n terugslagspoor *n*
terugslagstreep
d Rücklaufspur *f*,
Rücklaufstrich *m*

4030 RETURN SCANNING BEAM crt/rec/tv
The scanning beam in an image orthicon which after discharging the target is reversed and returned to the cathode.
f faisceau *m* analyseur reconduit à la cathode
e haz *m* explorador restituido al cátodo
i fascio *m* analizzatore ricondotto al catodo
n naar de katode teruggevoerde aftastbundel
d zur Katode rückgeleiteter Abtaststrahl *m*

4031 REVERBERANT CAVITY aud/stu
f cavité *f* réverbérante
e cavidad *f* reverberante
i cavità *f* riverberante
n resonerende holle ruimte
d Hohlraumresonator *m*

4032 REVERBERATION aud
The persistence of a sound in an enclosed or partly enclosed space after interruption of the acoustic source.
f réverbération *f*
e reverberación *f*
i riverberazione *f*
n nagalm
d Nachhall *m*,
Widerhall *m*

4033 REVERBERATION ABSORPTION COEFFICIENT aud
The absorption coefficient of a large plane uniform surface when the incident sound wave is of random intensity and direction, as is the reverberant field in an enclosure.
f coefficient *m* d'absorption de la réverbération
e coeficiente *m* de absorción de la reverberación
i coefficiente *m* d'assorbimento della riverberazione
n absorptiecoëfficiënt bij nagalm
d Schluckgrad *m* bei Nachhall

4034 REVERBERATION BRIDGE aud
Method of measuring the reverberation time in an enclosure.
f pont *m* de mesure de la réverbération
e puente *m* de medida de la reverberación
i ponte *m* di misura della riverberazione
n nagalmmeetbrug
d Nachhallmessbrücke *f*

4035 REVERBERATION CHAMBER, REVERBERATION ROOM aud
A room with the minimum acoustic absorption.
f chambre *f* réverbérante
e cámara *f* reverberante
i camera *f* riverberante
n nagalmruimte
d Hallraum *m*,
Nachhallraum *m*

4036 REVERBERATION CONTROLLED GAIN CIRCUIT cpl
f montage *m* d'expansion du contraste
e expandor *m* del contraste
i circuito *m* d'espansione del contrasto
n contrastexpansieschakeling
d Kontrastdehner *m*

4037 REVERBERATION KEY aud/stu
f touche *f* de réverbération
e botón *m* de reverberación
i pulsante *m* di riverberazione
n nagalmtoets
d Nachhalltaste *f*

4038 REVERBERATION REFLECTION COEFFICIENT aud
The sound reflection coefficient when the distribution of incident sound is completely random.
f coefficient *m* de réflexion de la réverbération
e coeficiente *m* de reflexión de la reverberación
i coefficiente *m* di riflessione della riverberazione
n nagalmreflectiecoëfficiënt
d Reflexionskoeffizient *m* bei Nachhall

4039 REVERBERATION RESPONSE aud
The response of a microphone for reverberant sound.
f sensibilité *f* du microphone pour la réverbération
e sensibilidad *f* del micrófono para la reverberación
i sensibilità *f* del microfono per la riverberazione
n nagalmgevoeligheid van de microfoon
d Nachhallempfindlichkeit *f* des Mikrophons

4040 REVERBERATION RESPONSE CURVE aud
f courbe *f* de sensibilité du microphone pour la réverbération
e curva *f* de sensibilidad del micrófono para la reverberación
i curva *f* di sensibilità del microfono per la riverberazione
n nagalmgevoeligheidskromme van de microfoon
d Frequenzkurve *f* eines Mikrophons unter Berücksichtigung der Reflexions- und Interferenzeinflüsse

4041 REVERBERATION TIME rec
The time taken for the sound intensity to fall by 60 dB from the equilibrium intensity.
f durée *f* de la réverbération
e duración *f* de la reverberación
i durata *f* della riverberazione
n nagalmtijd
d Nachhallzeit *f*

4042 REVERBERATION TIME METER aud
f réverbéromètre *m*
e reverberómetro *m*
i riverberometro *m*
n nagalmtijdmeter
d Nachhallzeitmessgerät *n*

4043 REVERBERATION TRANSMISSION COEFFICIENT aud
The sound transmission coefficient when the distribution of incident sound is completely random.
f coefficient *m* de transmission de la réverbération
e coeficiente *m* de transmisión de la reverberación
i coefficiente *m* di trasmissione della riverberazione
n nagalmtransmissiecoëfficiënt
d Durchlässigkeitskoeffizient *m* bei Nachhall

4044 REVERBERATION UNIT aud
f bloc *m* de réverbération
e unidad *f* de reverberación
i complesso *m* di riverberazione
n nagalmeenheid
d Nachhalleinheit *f*, Nachhallgerät *n*

4045 REVERSE ANGLE, REVERSE SHOT opt
f contrechamp *m*, inversion *f* d'axe
e contracampo *m*
i controcampo *m*
n tegeninstelling, tegenopname
d Achsensprung *m*, Gegeneinstellung *f*, Gegenschuss *m*

4046 REVERSE COMPATIBILITY ctv
Of a colo(u)r TV system, the attribute which permits a colo(u)r TV receiver to reproduce a monochrome picture from a transmitted monochrome signal.
f compatibilité *f* inverse, rétrocompatibilité *f*
e compatibilidad *f* inversa, retrocompatibilidad *f*
i compatibilità *f* inversa, retrocompatibilità *f*
n omgekeerde compatibiliteit
d Rekompatibilität *f*

4047 REVERSE PROGRAM(ME) SOUND aud/rec/tv
In broadcasting of special events an additional sound distribution network is set up carrying complete program(me) sound from the studio to all the participating outside broadcast units.
f son *m* de programme en direction inverse
e sonido *m* de programa en dirección inversa
i suono *m* di programma in direzione inversa
n programmageluid *n* in tegengestelde richting
d Sendeton *m* in entgegengesetzter Richtung

REVERSED IMAGE
see: NEGATIVE IMAGE

4048 REVIEW ROOM, VIEWING ROOM stu
A room for screening films for TV use.
f salle *f* de projection-contrôle, salle *f* de visionnage
e sala *f* de proyección preliminar
i sala *f* di proiezione preliminare
n filmbeoordelingsruimte
d Vorschauraum *m*

4049 REWIND, REWINDER vr
The components on a magnetic tape recorder that serve to return the tape to the supply reel at high speed.
f embobineuse *f*
e rebobinador *m*
i riavvolgitore *m*
n terugspoelapparaat *n*
d Rückspulgerät *n*

4050 REWIND (TO) vr
To return a magnetic tape to its starting position.
f embobiner v
e rebobinar v
i riavvolgere v
n terugspoelen v
d rückspulen v

4051 REWIND KEY vr
Used to rewind at speed to the start of a recording or the start of the tape.
f touche *f* réembobinage rapide
e botón *m* de rebobinado rápido
i pulsante *m* di riavvolgimento rapido
n snelterugspoeltoets
d Rückspultaste *f*

4052 REWINDING vr
f réembobinage *m*
e rebobinado *m*
i riavvolgimento *m*
n terugspoelen *n*
d Rückspulen *n*

R.F.
see: RADIOFREQUENCY

R.F. AMPLIFIER
see: RADIO-FREQUENCY AMPLIFIER

R.F. INTERMODULATION DISTORTION
see: RADIO-FREQUENCY INTERMODULATION DISTORTION

R.F. PULSE
see: RADIO-FREQUENCY PULSE

4053 RGB SIGNALS, TRISTIMULUS SIGNALS ctv
A set of three signals each of which represents one of the red, green or blue transmission primaries.
f signaux *pl* RVB
e señales *pl* RVB
i segnali *pl* RVB
n RGB-signalen *pl*
d RGB-Signale *pl*

4054 RIBBON FEEDER aea
f ruban *m* plat
e cable *m* cinta
i conduttore *m* a piattina
n lintkabel
d Bandleitung *f*, Flachkabel *n*

RIFLE MIKE
see: GUN MIKE

4055 RIGGER stu/th
f machiniste *m*
e obrero *m* escenógrafo
i macchinista *m*
n toneelknecht
d Bühnenarbeiter *m*

4056 RIGGING stu
Placing studio lights in their preliminary position before the accurate adjustment.
f montage *m* préliminaire des lampes
e montaje *m* preliminario de las lámparas
i montaggio *m* preliminario delle lampade
n voorlopige lampenmontage
d vorläufige Lampenmontage *f*

4057 RIGGING TENDER rec/tv
Vehicle carrying material for rigging a TV show.
f voiture *f* de montage
e carro *m* de montaje
i carro *m* di montaggio
n montagewagen
d Montagewagen *m*

RIM MAGNET
see: FIELD-NEUTRALIZING MAGNET

4058 RING HEAD, RING HEAD PICKUP vr
A magnetic head in which the magnetic material forms an enclosure having one or more air gaps.
f tête *f* magnétique annulaire
e cabeza *f* magnética anular
i testina *f* magnetica anulare
n ringvormige magneetkop
d Ringkopf *m*, Ringmagnetkopf *m*

RING MAGNET
see: ANNULAR MAGNET

4059 RINGING dis
In TV, a damped oscillatory response to a picture pulse, resulting in a series of closely-spaced alternate black and white images of decreasing density.
f franges *pl*, suroscillation *f*
e sobreoscilación *f*
i sovraoscillazione *f*
n uitslingereffect *n*
d Überschwingen *n*

RINGING (GB)
see: INTERMEDIATE MULTIPLE

4060 RIPPLE EFFECT rec/tv
A special effect in which the picture dissolves into ripples as though reflected in disturbed water.
f effet *m* d'ondulation
e efecto *m* de ondulación
i effetto *m* d'ondulazione
n golfeffect *n*
d Welleneffekt *m*

RISE TIME
see: BUILD-UP TIME

4061 RISE TIME CORRECTION cpl
f correction *f* du temps de montée
e corrección *f* del tiempo de establecimiento
i correzione *f* del tempo di formazione
n stijgtijdcorrectie
d Anstiegzeitentzerrung *f*

4062 RISE TIME DISTORTION cpl
f distorsion *f* du temps de montée
e distorsión *f* del tiempo de establecimiento
i distorsione *f* del tempo di formazione
n stijgtijdvervorming
d Anstiegzeitverzerrung *f*

RODS
see: RETINAL RODS

4063 ROLL dis/tv
Slow upward or downward movement of the entire image on the screen of a TV receiver, due to a lack of vertical synchronization.
f mouvement *m* lent en bas et en haut
e cabeceo *m*
i lento scorrimento *m* verticale
n langzame op- en neerbeweging
d langsame Bewegung *f* nach oben und nach unten

4064 ROLL TITLES, ROLLER CAPTION, ROLLER TITLES tv
f générique *m* déroulant
e didascalia *f* ascendente, título *m* ascendente
i titolo *m* ascendente, titolo *m* evolvente
n titelrol
d Rolltitel *m*

4065 ROLLER COATING vr
Application of the dope on the base film by means of rollers.

f revêtement *m* appliqué par rouleaux
e revestimiento *m* aplicado con rodillos
i rivestimento *m* applicato con rulli
n opwalsen *n*
d Aufwalzen *n*

4066 ROLLING rec/tv
Tilting the camera on the horizontal axis sideways.
f décentrement *m* horizontal
e inclinación *f* transversal
i inclinazione *f* trasversale
n transversale cameradraaiing
d Transversaldrehung *f*

ROSTRUM (GB)
see: PARALLEL

4067 ROTATING DISK ANALYZER, ROTATING DISK SCANNER rec/tv
f analyseur *m* à disque tournant
e explorador *m* de disco giratorio
i analizzatore *m* a disco girevole
n draaischijfaftaster
d Drehscheibenabtaster *m*

4068 ROTATING DISK FIELD STORE ctv
f mémoire *f* de trame à disque tournant
e memoria *f* de cuadro de disco giratoria
i memoria *f* di trama a disco girevole
n rastergeheugen *n* met draaiende schijf
d Teilbildspeicher *m* mit Drehscheibe

4069 ROTATING-FIELD AERIAL, ROTATING-FIELD ANTENNA aea
f antenne *f* à champ tournant
e antena *f* de campo giratorio
i antenna *f* a campo girevole
n draaiveldantenne
d Drehfeldantenne *f*

4070 ROTATING FLOOR fi/th/tv
f plateau *m* tournant
e escenario *m* giratorio
i palcoscenico *m* girevole
n draaibare toneelvloer, draaitoneel *n*
d Drehbühne *f*

4071 ROTATING MIRROR ANALYZER, ROTATING MIRROR SCANNER rec/tv
f analyseur *m* à miroirs tournants
e explorador *m* de espejos giratorios
i analizzatore *m* a specchi girevoli
n draaispiegelaftaster
d Drehspiegelabtaster *m*

4072 ROUGH CUT fi/tv
f montage *m* bout-à-bout, premier montage *m*, prémontage *m*
e primero montaje *m*
i montaggio *m* preliminare, primo montaggio *m*
n ruwe montage
d Rohschnitt *m*

4073 ROVING PREVIEW rep/stu
Facility of viewing at will and without selecting for transmission, any input or output of a video-switching system.
f preview *m* superficiel
e visión *f* previa superficial
i antiprima *f* superficiale
n oppervlakkige voorbezichtigung
d oberflächige Vorschau *f*

4074 RUN-OUT TRAILER fi/tv/vr
f amorce *f* de fin
e cola *f* final
i coda *f* finale
n eindband
d Endband *n*

RUN-UP TIME
see: PRE-ROLL TIME

4075 RUNNING TIME fi/tv
f durée *f* de projection
e duración *f* de proyección
i durata *f* di proiezione
n looptijd
d Laufzeit *f*

4076 RUSHES fi/tv
f épreuves *pl* de tournage
e producción *f* diaria, rodaje *m* diario
i copia *f* rapida, giornalieri *pl*
n eerste print, werkkopie
d Bildmuster *pl*

S

4077 S CORRECTION dis
A correction to ensure good picture linearity.
f correction *f* S
e corrección *f* S
i correzione *f* S
n S-correctie
d S-Korrektur *f*

4078 S DISTORTION, SPIRAL DISTORTION crt
Of camera or image tubes using magnetic focusing, a distortion in which image rotation varies with distance from the axis of symmetry of the electron optical system.
f distorsion *f* en S
e distorsión *f* en S
i distorsione *f* in S
n S-vervorming, spiraalvormige vertekening
d S-Verzeichnung *f*, S-Verzerrung *f*

4079 S METER, SIGNAL STRENGTH METER cpl
A meter used in communication receivers for measuring relative signal strength.
f S-mètre *m*
e S-metro *m*
i S-metro *m*
n S-meter, veldsterktemeter
d Feldstärkeanzeiger *m*, S-Meter *n*

4080 SAFE AREA fi/tv
f cadre *m*, format *m* d'image vu sur un récepteur
e campo *m* visible en el receptor
i area *f* utile dell'immagine
n kader *n*, zichtbaar deel *n* van het beeld
d Rahmen *m*

4081 SAFE AREA GENERATOR fi/tv
f générateur *m* de cadre, traceur *m* de cadre télévision
e generador *m* del campo visible
i generatore *m* di segnali di locazione sull'area utile
n cadreergenerator
d Rahmengeber *m*

4082 SAFE AREA GENERATOR MIXER fi/tv
f mélangeur *m* du traceur du cadre
e mezclador *m* generador del campo visible
i mescolatore *m* del segnale di locazione sull'area utile con il segnale d'immagine
n cadreerschakeling
d Rahmeneinblender *m*

4083 SAFETY GLASS crt
f verre *m* de sécurité
e vidrio *m* de protección
i vetro *m* di protezione
n glasruit
d Schutzscheibe *f*, Sicherheitsscheibe *f*

4084 SAGITTAL FOCAL LINE opt
f ligne *f* focale sagittale
e línea *f* focal sagital
i linea *f* focale sagittale
n sagittale brandlijn
d sagittale Brennlinie *f*

4085 SAMPLE INTELLIGENCE cpl
A part of a signal used as evidence of the equality of the signal.
f échantillon *m* de signal
e muestra *f* de señal
i campione *m* di segnale
n signaalmonster *n*
d Signalmuster *n*

4086 SAMPLER ctv
An electronic switch which scans the momentary amplitude of the video signal of each colo(u)r 3.6 x 10^6 times per second and passes these pulses to the modulator.
f discriminateur *m* chromatique
e discriminador *m* cromático
i commutatore *m* elettronico di colori
n kleurenschakelaar
d Abfrageschalter *m*, Farbschalter *m*

4087 SAMPLING ctv
In a three-colo(u)r TV system, a method of deriving the colo(u)r intensity or tone during the scanning of the mosaic.
f discrimination *f* chromatique
e discriminación *f* cromática
i commutazione *f* elettronica di colori
n kleurenschakeling
d Abfragemethode *f*, Farbschaltung *f*

4088 SAMPLING FREQUENCY cpl
f fréquence *f* de commutation
e frecuencia *f* de conmutación
i frequenza *f* di commutazione
n schakelfrequentie
d Schaltfrequenz *f*

4089 SATCOM, SATELLITE COMMUNICATION tv
System employing space satellite stations for relaying long-distance telephony and TV.
f communication *f* par satellite
e comunicación *f* por satélite
i comunicazione *f* per satellite
n communicatiesysteem *n* met een satelliet
d Satellitennachrichtensystem *n*

4090 SATELLITE ge
Circulating vehicle at a weight above the earth to reflect or transmit back radio waves as a means of communication.
f satellite *m*
e satélite *m*
i satellite *m*
n satelliet
d Satellit *m*

4091 SATELLITE AERIAL, SATELLITE ANTENNA aea
f antenne *f* de satellite
e antena *f* de satélite
i antenna *f* di satellite
n satellietantenne
d Satellitenantenne *f*

4092 SATELLITE CIRCUIT, SATELLITE LINK tv
f circuit *m* spatial, liaison *f* spatiale
e circuito *m* espacial
i circuito *m* spaziale
n satellietverbinding
d Satellitenstrecke *f*, Satellitenverbindung *f*

4093 SATELLITE ORBITS tv
f orbites *pl* de satellites
e órbitas *pl* de satélites
i orbite *pl* di satelliti
n satellietbanen *pl*
d Satellitenbahnen *pl*

4094 SATELLITE POWER SUPPLY cpl/tv
f alimentation *f* de satellite
e alimentación *f* de satélite
i alimentazione *f* di satellite
n satellietvoeding
d Satellitenspeisung *f*

4095 SATELLITE STATION tv
A subsidiary transmitting TV station that takes the vision signals by direct radio reception from a main station and rebroadcasts them together with the associated sound signals.
f station *f* relais
e estación *f* retransmisora
i stazione *f* relè
n satellietstation *n*
d Relaisstelle *f*

4096 SATELLITE TRANSMISSION tv
f transmission *f* par satellite
e transmisión *f* por satélite
i trasmissione *f* per satellite
n overbrenging door satelliet
d Übertragung *f* durch Satellit

4097 SATELLITE TRANSMITTER cpl
f émetteur *m* satellite
e emisor *m* satélite
i emettitore *m* satellite
n satellietzender *m*
d Satellitensender *m*

4098 SATELLITE TRANSMITTER, SLAVE TRANSMITTER tv
f transpondeur *m*
e transmisor *m* ausiliar, transmisor *m* satélite
i trasmettitore *m* ausiliario, trasmettitore *m* satellite
n hulpzender, satellietzender
d Hilfssender *m*

4099 SATURABLE CORE REACTOR, SATURABLE REACTOR ge/stu
Variable inductance in which the reactance of the load circuit is varied by altering the magnetic flux in the core by means of d.c. flowing in an auxiliary winding. Used in studio lighting circuits.
f bobinage *m* à saturation, transducteur *m*
e reactor *m* de núcleo saturable, transductor *m*
i trasduttore *m*
n magnetische versterker, transductor
d magnetischer Verstärker *m*

4100 SATURATED COLO(U)R ct
A pure colo(u)r, not contaminated by white.
f couleur *f* saturée
e color *m* saturado
i colore *m* saturo
n verzadigde kleur
d satte Farbe *f*

4101 SATURATION ct
The attribute of a visual sensation which permits a judgment to be made of the proportion of pure chromatic colo(u)r in the total sensation.
f saturation *f*
e saturación *f*
i saturazione *f*
n verzadiging
d Sättigung *f*

SATURATION
see: CHROMA

4102 SATURATION BANDING dis
f saturation *f* striée
e saturación *f* rayada
i saturazione *f* striata
n gestreepte verzadiging
d Sättigungsstreifigkeit *f*

4103 SATURATION CONTROL ctv
f régleur *m* de la saturation
e regulador *m* de la saturación
i regolatore *m* della saturazione
n verzadigingsregelaar
d Sättigungseinsteller *m*

SATURATION CONTROL
see: CHROMA CONTROL

4104 SATURATION LIGHTING stu
The provision of a surplus number of luminaires regularly spaced above a studio to avoid delay in rigging them to the required positions.

f éclairage *m* à surplus de luminaires
e alumbrado *m* a exceso de lámparas
i illuminazione *f* ad eccesso di lampade
n verlichting met lampenoverschot
d Beleuchtung *f* mit Lampenüberschuss

4105 SATURATION SCALE, SATURATION SCALE SIGNAL ct
Minimum visual steps of saturation, varying with wavelength.
f échelle *f* de saturation
e escala *f* de saturación
i scala *f* di saturazione
n verzadigingstrap
d Sättigungstreppe *f*

4106 SAWTOOTH COUPLING cpl
f couplage *m* en dents de scie
e acoplado *m* en dientes de sierra
i accoppiamento *m* in denti di sega
n zaagtandkoppeling
d Sägezahnkopplung *f*

4107 SAWTOOTH CURRENT cpl
A current having a sawtooth waveform.
f courant *m* à dents de scie
e corriente *f* de dientes de sierra
i corrente *f* a denti di sega
n zaagtandstroom
d sägezahnförmiger Kippstrom *m*, Sägezahnstrom *m*

4108 SAWTOOTH GENERATOR, SAWTOOTH OSCILLATOR cpl
A generator of oscillations each cycle of which consists of a variation of voltage or current substantially proportional to time, followed by a relatively rapid variation of opposite sense.
f générateur *m* de dents de scie
e generador *m* de dientes de sierra
i generatore *m* di denti di sega
n zaagtandgenerator
d Sägezahngenerator *m*, Sägezahnoszillator *m*

4109 SAWTOOTH OSCILLATION, SAWTOOTH WAVE cpl
Periodic wave of sawtooth shape whose amplitude increases in a roughly linear manner with time to a maximum value and returns more rapidly to a minimum.
f onde *f* en dents de scie, oscillation *f* en dents de scie
e onda *f* en dientes de sierra, oscilación *f* en dientes de sierra
i onda *f* in denti di sega, oscillazione *f* in denti di sega
n zaagtandgolf, zaagtandtrilling
d Kippschwingung *f*, Sägezahnschwingung *f*, Sägezahnwelle *f*

4110 SAWTOOTH PULSE cpl
f impulsion *f* en dents de scie
e impulso *m* en dientes de sierra
i impulso *m* in denti di sega
n zaagtandimpuls
d Sägezahnimpuls *m*

4111 SAWTOOTH VOLTAGE cpl
A voltage having a sawtooth waveform.
f tension *f* à dents de scie
e tensión *f* de dientes de sierra
i tensione *f* a denti di sega
n zaagtandspanning
d Kippspannung *f*, Sägezahnspannung *f*

4112 SAWTOOTH WAVEFORM cpl
A waveform characterized by a slow rise time and a sharp fall, resembling the tooth of a saw.
f forme *f* d'onde en dents de scie
e forma *f* de onda en dientes de sierra
i forma *f* d'onda in denti di sega
n zaagtandgolfvorm
d Sägezahnwellenform *f*

4113 SCALLOP, SCALLOPING dis/vr
f feston *m*
e efecto *m* festón
i distorsione *f* degli elementi verticali dell'immagine
n vervorming van de verticale beeld-elementen
d Verzerrung *f* der vertikalen Bild-elementen

4114 SCAN (TO) tv
To examine an area or a region point by point in an ordered sequence as when converting a scene or image to an electrical signal.
f analyser v, balayer v, explorer v
e analizar v, barrer v, explorar v
i analizzare v, esplorare v, scandire v
n aftasten v
d abtasten v

4115 SCAN BURNS tv
Blemishes caused by overloading phosphors or other scanned substances with excess current, or by repeated use.
f brûlures *pl* d'analyse
e quemaduras *pl* de exploración
i bruciature *pl* d'analisi
n ingebrand raster *n*
d eingebrannter Raster *m*

4116 SCAN CONVERSION DEVICE tv
Used e.g. in closed circuit systems in passenger ships using vidicon cameras for 405-line and 819-line reception.
f convertisseur *m* du nombre de lignes
e convertidor *m* del número de líneas
i convertitore *m* del numero di linee
n lijnvertaler
d Zeilenzahlumsetzer *m*

4117 SCAN GENERATOR rec/rep/tv
Generator of line or field frequency sawtooth voltage or current waveform,

used to deflect a cathode ray beam in both cameras and receivers.
f générateur *m* de dents de scie pour l'analyse
e generador *m* de dientes de sierra para la exploración
i generatore *m* di denti di sega per l'analisi
n zaagtandgenerator voor aftasting
d Sägezahngenerator *m* für Abtastung

4118 SCAN LINEARITY, SCANNING LINEARITY — rep/tv
Accuracy of position both horizontally and vertically, of any point in a reproduced image.
f linéarité *f* de balayage
e linealidad *f* de barrido
i linearità *f* di deflessione, linearità *f* di deviazione
n lineariteit van de afbuiging
d Linearität *f* der Ablenkung

4119 SCAN PROTECTION — crt/tv
Action of reducing or turning off the current in the scanning beam of a TV receiver or camera tube in the event of failure in the scanning fields.
f interruption *f* du courant du faisceau, suppression *f* du courant du faisceau
e interrupción *f* de la corriente del haz, supresión *f* de la corriente del haz
i interruzione *f* della corrente del fascio, soppressione *f* della corrente del fascio
n bundelstroomonderbreking, bundelstroomonderdrukking
d Strahlstromunterbrechung *f*, Strahlstromunterdrückung *f*

4120 SCAN REGISTRATION — ctv
Registration of the blue, red and Y signals with the green signal.
f régistration *f* des signaux chromatiques
e registro *m* de las señales cromáticas
i registrazione *f* dei segnali cromatici
n het in register zijn van de kleursignalen
d Farbsignalüberdeckung *f*

4121 SCAN RINGS — dis
f suroscillation *f* en début de ligne
e sobreoscilación *f* en principio de línea
i sovraoscillazione *f* in principio di linea
n doorschot *n* aan het begin van een lijn
d Zeileneinschwinger *m*

4122 SCANNED AREA — rep/tv
f surface *f* analysée, zone *f* analysée
e zona *f* explorada
i zona *f* analizzata
n afgetast oppervlak *n*
d abgetastete Oberfläche *f*

4123 SCANNED PICTURE — tv
f image *f* analysée
e imagen *f* explorada
i immagine *f* analizzata
n afgetast beeld *n*
d abgetastetes Bild *n*

4124 SCANNER, SCANNING DEVICE — tv
f analyseur *m*, dispositif *m* d'analyse
e dispositivo *m* de exploración, explorador *m*
i analizzatore *m*, dispositivo *m* d'analisi
n aftastapparaat *n*, aftaster
d Abtastvorrichtung *f*, Bildabtaster *m*

4125 SCANNER TUBE — tv
Cathode-ray tube used for producing the flying spot raster with optically flat screens and operating at high voltages (30k V), high beam currents (300 mA) to provide a very bright scan, whereas the tube face is cooled by air blast.
f tube *m* analyseur
e tubo *m* explorador
i tubo *m* analizzatore
n aftastbuis
d Abtaströhre *f*

4126 SCANNING (GB), SWEEP (US) — tv
In TV, scanning is the process of analyzing or synthesizing successively, according to a predetermined method, the light values or equivalent characteristics of elements constituting a picture area.
f analyse *f*, balayage *m*, exploration *f*
e analización *f*, barrido *m*, exploración *f*
i analisi *f*, scansione *f*
n aftasten *n*, aftasting
d Abtastung *f*

4127 SCANNING AMPLIFIER — rec/rep/tv
f amplificateur *m* d'analyse
e amplificador *m* de exploración
i amplificatore *m* d'analisi
n aftastversterker
d Abtastverstärker *m*

4128 SCANNING ANGLE — crt/rec/tv
f angle *m* d'analyse
e ángulo *m* de exploración
i angolo *m* d'analisi
n aftasthoek
d Abtastwinkel *m*

4129 SCANNING APERTURE, SCANNING DIAPHRAGM — tv
f diaphragme *m* d'analyse
e diafragma *m* de exploración
i diaframma *m* d'analisi
n aftastdiafragma *n*
d Abtastblende *f*

4130 SCANNING BEAM — tv
f faisceau *m* analyseur

e haz *m* explorador
i fascio *m* analizzatore
n aftastbundel,
aftaststraal
d abtastender Elektronenstrahl *m*,
Abtaststrahl *m*

4131 SCANNING BEAM CURRENT crt
Electron current in the beam of a cathode-ray tube reaching the target.
f courant *m* du faisceau analyseur
e corriente *f* del haz explorador
i corrente *f* del fascio analizzatore
n aftastbundelstroom,
bundelstroom,
straalstroom
d Abtaststrahlstrom *m*,
Strahlstrom *m*

SCANNING COIL
see: DEFLECTION COIL

4132 SCANNING CYCLE rec/rep/tv
f cycle *m* d'analyse
e ciclo *m* de exploración
i ciclo *m* d'analisi
n aftastcyclus
d Abtastzyklus *m*

4133 SCANNING DIRECTION tv
One of the items tested by standard test cards.
f direction *f* d'analyse
e dirección *f* de exploración
i direzione *f* d'analisi
n aftastrichting
d Abtastrichtung *f*

4134 SCANNING DISK, ctv/tv
SPIRAL DISK
In mechanical TV systems, e.g. a Nipkow disk.
f disque *m* analyseur
e disco *m* de exploración
i disco *m* analizzatore
n aftastschijf
d Abtastscheibe *f*

4135 SCANNING ELEMENT, rec/rep/tv
SCANNING POINT
f élément *m* d'image
e elemento *m* de imagen
i elemento *m* d'immagine
n beeldelement *n*, beeldpunt *n*
d Bildpunkt *m*

4136 SCANNING EQUIPMENT rec/tv
f dispositif *m* analyseur
e equipo *m* explorador
i dispositivo *m* d'analisi
n aftastapparatuur
d Abtastgeräte *pl*

4137 SCANNING FIELD tv
The area of the mosaic of a TV camera or the screen of a cathode-ray tube explored by scanning.
f champ *m* d'analyse
e campo *m* de exploración
i campo *m* d'analisi
n afgetast veld *n*
d abgetastetes Gebiet *n*

4138 SCANNING FREQUENCY crt
Number of times an image is scanned each second.
f fréquence *f* d'analyse
e frecuencia *f* de exploración
i frequenza *f* d'analisi
n aftastfrequentie
d Abtastfrequenz *f*

4139 SCANNING GENERATOR (GB), crt
SWEEP GENERATOR (US)
Timebase used for controlling the scanning process of a picture tube.
f générateur *m* de balayage,
générateur *m* de déviation
e generador *m* de barrido,
generador *m* de desviación
i generatore *m* di deflessione,
generatore *m* di deviazione
n afbuigingsgenerator
d Kippgenerator *m*

4140 SCANNING HOLES rec/tv
f trous *pl* d'analyse
e agujeros *pl* de exploración
i fori *pl* d'analisi
n aftastgaten *pl*
d Abtastlöcher *pl*

4141 SCANNING IN RECEPTION rep/tv
The process of building up the image from elements derived in succession from the received signal.
f restitution *f* de l'image,
synthèse *f* de l'image
e exploración *f* en recepción
i scansione *f* dell'immagine in ricezione,
sintesi *f* dell'immagine in ricezione
n beeldopbouw
d Bildzusammensetzung *f*

4142 SCANNING IN TRANSMISSION rec/tv
The process of analyzing the scene or object into picture elements which are then represented in succession by the magnitude of an electric signal.
f analyse *f* de l'image,
analyse *f* de l'objet
e exploración *f* en transmisión
i analisi *f* dell'immagine in trasmissione,
scansione *f* dell'immagine in trasmissione
n beeldontleding
d Bildzerlegung *f*

4143 SCANNING INTERFERENCE dis
f interférence *f* dans le canevas
e interferencia *f* en la cuadrícula
i interferenza *f* nel quadro rigato
n rasterstoring
d Rasterstörung *f*

4144 SCANNING LINE rec/tv
A sequence of picture elements extending throughout one dimension of the picture and represented by successive signal values.

f ligne *f* d'analyse
e línea *f* de exploración
i linea *f* d'analisi,
riga *f* d'analisi
n aftastlijn,
beeldlijn
d Abtastzeile *f*,
Bildzeile *f*

SCANNING LINE
see: IMAGE STRIP

4145 SCANNING-LINE LENGTH tv
The length of the path traced by the scanning or recording spot in moving from a point on one line to a corresponding point on the next following line.
f longueur *f* de la ligne d'analyse
e longitud *f* de la línea de exploración
i lunghezza *f* della linea d'analisi
n beeldlijnlengte
d Bildzeilenlänge *f*

4146 SCANNING METHODS rec/rep/tv
f méthodes *pl* d'analyse
e métodos *pl* de exploración
i metodi *pl* d'analisi
n aftastsystemen *pl*
d Abtastsysteme *pl*

4147 SCANNING PATTERN tv
f diagramme *m* d'analyse
e forma *f* de exploración
i disegno *m* d'analisi
n aftastschema *n*
d Abtastschema *n*

4148 SCANNING PITCH tv
f espacement *m* des lignes d'analyse
e paso *m* de exploración
i passo *m* d'analisi
n onderlinge afstand van de beeldlijnen
d Zeilenabstand *m*,
Zeilendichte *f*

SCANNING POINT
see: SCANNING ELEMENT

4149 SCANNING PROCESS rec/rep/tv
f procédé *m* d'analyse
e procedimiento *m* de exploración
i procedimento *m* d'analisi
n aftastproces *n*
d Abtastverfahren *n*

4150 SCANNING RASTER rec/rep/tv
f canevas *m* d'analyse
e cuadrícula *f* de exploración
i quadro *m* rigato d'analisi
n aftastraster *n*
d Abtastraster *m*

4151 SCANNING REVERSAL rec/rep/tv
The reversal of the direction of the currents in the deflection coils in a TV system by means of a normal reverse switch.
f inversion *f* du courant d'analyse
e inversión *f* de la corriente de exploración
i inversione *f* della corrente d'analisi
n aftaststroomompoling
d Abtaststromumpolung *f*

4152 SCANNING SEQUENCE rec/rep/tv
f séquence *f* d'analyse
e secuencia *f* de exploración
i sequenza *f* d'analisi
n aftastvolgorde
d Reihenfolge *f* der Abtastung

4153 SCANNING SIGNAL, rec/tv
SCANNING WAVEFORM
f signal *m* d'analyse
e señal *f* de exploración
i segnale *m* d'analisi
n aftastsignaal *n*
d Abtastsignal *n*

SCANNING SPEED (GB)
see: PICKUP VELOCITY

4154 SCANNING SPOT tv
1. The area that is viewed instantaneously by the pickup system of a TV camera.
2. The concentrated spot of light produced by the impact of electrons on the phosphor coating of a cathode-ray tube.
f spot *m* analyseur
e punto *m* explorador
i punto *m* d'analisi
n aftastpunt,
lichtstip
d Abtastfleck *m*

SCANNING SPOT CONTROL
see: FOCUS CONTROL

4155 SCANNING STAGE rec/tv
f étage *m* d'analyse
e etapa *f* de exploración
i stadio *m* d'analisi
n aftasttrap
d Bildabtaststufe *f*

4156 SCANNING STANDARDS rep/tv
f normes *pl* d'analyse
e patrones *pl* de exploración
i norme *pl* d'analisi
n aftastnormen *pl*
d Abtastnormen *pl*

4157 SCANNING TIME tv
f temps *m* d'analyse
e tiempo *m* de exploración
i tempo *m* d'analisi
n aftasttijd
d Abtastzeit *f*

4158 SCANNING VOLTAGE (GB), rec/tv
SWEEPS (US)
f tension *f* d'analyse
e tensión *f* de exploración
i tensione *f* d'analisi
n aftastspanning
d Abtastspannung *f*

SCANNING YOKE
see: DEFLECTION SYSTEM

4159 SCATTERED LIGHT stu
f lumière *f* dispersée
e luz *f* dispersada
i luce *f* dispersa
n strooilicht *n*
d Streulicht *n*

4160 SCENARIO WRITER, SCENARIST, SCRIPT WRITER fi/tv
f auteur *m* de scénario, scénariste *m*
e autor *m* de escenario, escenógrafo *m*
i sceneggiatore *m*
n draaiboekschrijver
d Drehbuchautor *m*

4161 SCENE ge
A term sometimes used to denote the setting for a series of spots and sometimes for an individual camera set-up.
f scène *f*
e escena *f*
i scena *f*
n scène, toneel *n*
d Bühne *f*, Szene *f*

4162 SCENE CHANGING, SCENE SHIFTING stu
f change *m* de scène
e cambio *m* de escena
i cambio *m* di scena
n decorwisseling
d Kulissenverschiebung *f*, Szenenwechsel *m*

4163 SCENE ILLUMINATION stu
f éclairage *m* de la scène
e alumbrado *m* de la escena
i illuminazione *f* della scena
n toneelverlichting
d Szenenbeleuchtung *f*

SCENE SHIFTER (GB)
see: GRIP

SCENERY (GB)
see: PROPERTIES

SCENERY PLOT (GB)
see: PROP PLOT

4164 SCHLIEREN PROJECTION SYSTEM tv
System used in the Eidophor system and in the reproduction of thermoplastic recordings.
f système *m* de projection de Schlieren
e sistema *m* de proyección de Schlieren
i sistema *m* di proiezione di Schlieren
n projectiesysteem *n* van Schlieren
d Schlieren-Projektionssystem *n*

4165 SCHMIDT OPTICAL SYSTEM tv
Optical system used for projection TV in which light from a high-intensity picture tube situated in the focal plane of a concave mirror is reflected back through a correcting plate or lens, which eliminates spherical distortion, and projected on to a screen.
f optique *f* de Schmidt
e óptica *f* de Schmidt
i ottica *f* di Schmidt
n optisch stelsel *n* van Schmidt
d Schmidt-Optik *f*

SCHOOL AND COLLEGES TELEVISION
see: EDUCATIONAL TELEVISION

4166 SCOOP stu
Floodlight used in film and TV studios and shaped like a grocer's scoop.
f lampe *f* en réflecteur ouvert
e lámpara *f* en reflector abierto
i lampada *f* in riflettore aperto
n lamp in open reflector
d Lampe *f* in offenem Reflektor

4167 SCOPHONY SYSTEM tv
An early mirror-drum projection system recently revived used in TV based on Schlieren optics.
f système *m* Scophony
e sistema *m* Scophony
i sistema *m* Scophony
n scophonysysteem *n*
d Scophony-Verfahren *n*

4168 SCOTOPHOR crt
A material, usually potassium chloride, which darkens under electron bombardment.
f substance *f* noircissante
e substancia *f* de ennegrecimiento
i sostanza *f* d'annerimento
n zwartingsstof
d Schwärzungsstoff *m*

4169 SCOTOPIC VISION ct
Vision experienced by the normal eye when adapted to very low levels of luminance.
f vision *f* scotopique
e visión *f* escótica
i visione *f* scotopica
n scotopisch zien *n*, staafjeszien *n*
d Nachtsehen *n*

4170 SCRAPE FLUTTER dis/vr
Flutter in tape recording created by longitudinal oscillation of the tape arising from scraping against the guides.
f effet *m* de grattement
e efecto *m* de raspamiento
i effetto *m* di grattamento
n schaafeffect *n*
d Schabeeffekt *m*

4171 SCREEN crt
Of a cathode-ray tube the surface of the tube on which the visible pattern is produced.
f écran *m*
e pantalla *f*
i schermo *m*
n scherm *n*
d Schirm *m*

4172 SCREEN ge
Metallic envelope surrounding a component or circuit intended to shield it from external magnetic or electric fields.
f blindage *m*
e blindaje *m*
i schermatura *f*, schermo *m*
n afscherming
d Abschirmung *f*

4173 SCREEN BRIGHTNESS, SCREEN LUMINANCE fi/tv
f luminance *f* de l'écran
e luminancia *f* de la pantalla
i luminanza *f* dello schermo
n schermhelderheid
d Bildwandhelligkeit *f*

4174 SCREEN BURN, SCREEN BURNING crt
Of a cathode-ray tube, an area of reduced luminosity produced by the destruction of some of the active material of the screen by electron or ion impact.
f brûlure *f* de l'écran
e quemadura *f* de la pantalla
i bruciatura *f* dello schermo
n schermverbranding
d Schirmeinbrennung *f*

4175 SCREEN CHARACTERISTICS crt
f caractéristiques *pl* d'écran
e características *pl* de pantalla
i caratteristiche *pl* di schermo
n schermkarakteristieken *pl*
d Schirmkennlinien *pl*

4176 SCREEN EFFICIENCY crt
The ratio of the light or other radiation intensity of an excited area on the screen of a cathode-ray tube to the product of beam current and final accelerated voltage.
f rendement *m* d'écran
e rendimiento *m* de pantalla
i rendimento *m* di schermo
n schermrendement *n*
d Schirmausbeute *f*

4177 SCREEN LUMINANCE PERSISTENCE CHARACTERISTIC
f caractéristique *f* de persistance
e característica *f* de persistencia
i caratteristica *f* di persistenza
n nalichtkarakteristiek
d Nachleuchtcharakteristik *f*

4178 SCREEN MATERIAL crt
f matériel *m* d'écran
e material *m* de pantalla
i materiale *m* di schermo
n schermmateriaal *n*
d Leuchtschirmmaterial *n*

4179 SCREEN SATURATION crt
Limitation of the brightness of a cathode-ray tube fluorescent screen by the rate at which energy from the electron beam can be transformed into light.
f limitation *f* de la luminosité de l'écran
e limitación *f* de la luminosidad de la pantalla
i limitazione *f* della luminosità dello schermo
n begrenzing van de schermhelderheid
d Begrenzung *f* der Schirmhelligkeit

4180 SCREEN SETTLING crt
Manufacture of a cathode-ray tube screen by allowing the phosphor to settle from a suspension in a liquid medium that is then drawn off.
f précipitation *f* d'une substance luminescente
e precipitación *f* de una substancia luminiscente
i precipitazione *f* d'una sostanza luminescente
n inleggen *n* van een luminescerende stof
d Fällung *f* einer lumineszierender Substanz

4181 SCREEN WIDTH crt
f largeur *f* de l'écran
e anchura *f* de la pantalla
i larghezza *f* dello schermo
n schermbreedte
d Schirmbreite *f*

4182 SCREENED AERIAL, SCREENED ANTENNA aea
f antenne *f* blindée
e antena *f* blindada
i antenna *f* schermata
n afgeschermde antenne
d abgeschirmte Antenne *f*

4183 SCREENING crt
The manufacturing process of a luminescent screen.
f manufacture *f* de l'écran
e manufactura *f* de la pantalla
i fabbricazione *f* dello schermo
n schermvervaardiging
d Schirmanfertigung *f*

4184 SCREENING CAN ge
f chemise *f* de blindage
e caja *f* de blindaje
i scatola *f* schermante
n afschermbus, afschermkap
d Abschirmbuchse *f*, Abschirmkappe *f*

SCRIM
see: CHEESE CLOTH

4185 SCRIPT fi/tv
f scénario *m*, scripte *m*/f
e guión *m* de película
i copione *m*, sceneggiatura *f*
n draaiboek *n*
d Drehbuch *n*

SCRIPT GIRL
see: CONTINUITY GIRL

SCRIPT WRITER
see: SCENARIO WRITER

4186 SEC VIDICON crt/ctv
A pickup tube for colo(u)r TV in the research stage of the bombardment-induced conductivity type.
f vidicon *m* SEC
e vidicón *m* SEC
i vidicon *m* SEC
n SEC-vidicon *n*
d SEC-Vidikon *n*

4187 SECAM COLO(U)R SYSTEM, ctv
SECAM SYSTEM
A compatible system of colo(u)r TV in which a luminance signal is transmitted as amplitude modulation of the vision carrier and colo(u)r information is transmitted, line sequentially, using a subcarrier.
f système *m* SECAM
e sistema *m* SECAM
i sistema *m* SECAM
n SECAM-systeem *n*
d SECAM-System *n*

4188 SECOND ANODE, crt
ULTOR
Of a cathode-ray tube employing electrostatic deflection, the electrode used to accelerate the electron beam before it is deflected and after it has been focused by the first anode.
f anode *f* accélératrice
e ánodo *m* acelerador
i anodo *m* acceleratore
n versnellingsanode
d Beschleunigungsanode *f*,
Endanode *f*,
Hochspannungsanode *f*

SECOND CAMERAMAN
see: FOCUS PULLER

SECOND CHANNEL ATTENUATION
see: ADJACENT CHANNEL ATTENUATION SELECTANCE

4189 SECOND SHOWING tv
f deuxième émission *f*
e segunda emisión *f*
i seconda emissione *f*
n tweede uitzending
d Zweitsendung *f*

4190 SECONDARY ELECTRON cpl
An electron emitted as a result of bombardment of a material by an incident electron.
f électron *m* secondair
e electrón *m* secundario
i elettrone *m* secondario
n secundair elektron *n*
d Sekundärelektron *n*

4191 SECONDARY EMISSION cpl
Electron emission as a result of bombardment of a surface by electrons or ions.
f émission *f* secondaire
e emisión *f* secundaria
i emissione *f* secondaria
n secundaire emissie
d Sekundäremission *f*

4192 SECONDARY EMISSION COEFFICIENT, cpl
SECONDARY EMISSION FACTOR,
SECONDARY EMISSION RATIO
The ratio of the number of secondary electrons to the number of incident primary electrons.
f coefficient *m* d'émission secondaire
e coeficiente *m* de emisión secundaria
i coefficiente *m* d'emissione secondaria
n secundaire emissiecoëfficiënt
d Sekundäremissionskoeffizient *m*

SECONDARY EMISSION MULTIPLIER
see: ELECTRON MULTIPLIER

4193 SECONDARY LIGHT SOURCE ct
A body or object transmitting or reflecting light falling on it from another source, whether primary or secondary.
f source *f* secondaire
e fuente *f* luminosa secundaria
i sorgente *f* luminosa secondaria
n secundaire lichtbron
d Fremdleuchter *m*,
Sekundärlichtquelle *f*

4194 SECONDARY RECEIVER tv
f deuxième récepteur *m*
e receptor *m* adicional
i secondo ricevitore *m*
n tweede toestel *n*
d Zweitempfänger *m*

4195 SECONDARY SERVICE AREA tv
The area within which satisfactory reception of a broadcasting transmitter is possible, the ground wave being attenuated to a field strength substantially less than that of the indirect rays.
f zone *f* de service secondaire,
zone *f* secondaire
e área *f* de servicio secundario
i area *f* di servizio secondario
n secundair gebied *n*
d Sekundär-Versorgungsbereich *m*

4196 SELECTED ORDINATE METHOD ct
OF COLORIMETRIC CALCULATION
The numerous multiplications indicated under computational colorimetry can be avoided by summing the spectral distribution data for specially selected, non-uniformly distributed wavelengths.
f méthode *f* des ordonnées sélectionnées
e método *m* de las ordenadas seleccionadas
i metodo *m* delle ordinate selezionate
n colorimetrische berekening volgens de methode der uitgekozen golflengten
d kolorimetrische Berechnung *f* nach dem System der ausgewählten Wellenlängen

4197 SELECTIVE FILTER, SELECTIVE SCREEN ctv
f filtre *m* sélectif
e filtro *m* selectivo
i filtro *m* selettivo
n selectief filter *n*
d Farbauszugfilter *n*, Farbfilter *n*

4198 SELECTIVE RADIATOR ct
A radiator whose spectral sensitivity depends on the wavelength (in the visible part of the spectrum).
f radiateur *m* sélectif
e radiador *m* selectivo
i radiatore *m* selettivo
n selectieve thermische straler
d Selektivstrahler *m*

4199 SELECTOR BUTTON, SELECTOR PUSHBUTTON rep/tv
f bouton-poussoir *m* sélecteur, touche *f* sélectrice
e botón *m* selector, tecla *f* selector
i pulsante *m* selettore, tasto *m* selettore
n programmakeuzedrukknop, programmakeuzetoets
d Wählerdruckknopf *m*, Wählertaste *f*

4200 SELENICON, SELENIUM-LAYER VIDICON crt/ctv
A pick-up tube for colo(u)r TV in the research stage.
f vidicon *m* à couche de sélénium
e vidicón *m* con capa de selenio
i vidicon *m* con strato di selenio
n vidicon *n* met seleniumlaag
d Vidikon *n* mit Selenschicht

4201 SELF-ERASING vr
f à autoeffacement
e de autoborrado
i ad autocancellazione
n zelfwissend adj
d selbstlöschend adj

4202 SELF-FOCUSED PICTURE TUBE crt
A TV picture tube having automatic electrostatic focus incorporated into the design of the electron gun.
f tube *m* image à autofocalisation
e tubo *m* imagen de enfoque automático
i tubo *m* immagine ad autofocalizzazione
n beeldbuis met automatische focussering
d Bildröhre *f* mit Selbstfokussierung

SELF-KEYED INSERTION
see: MOVING MATTE

4203 SELF-LUMINOUS OBJECT ct
f objet *m* à luminosité propre
e objeto *m* autoluminoso
i oggetto *m* autoluminoso
n zelflichtend voorwerp *n*
d selbstleuchtender Gegenstand *m*

SELF-TONE
see: MASS-TONE

4204 SELF-WHISTLES dis
Interference caused by a beat signal between the oscillator signal, or its harmonics, and harmonics of the desired radio-frequency vision or sound signals.
f autosifflements *pl*
e autosilbido *m*
i fischio *m* d'interferenza
n eigen fluittoon
d Eigenpfiffe *pl*, Überlagerungspfiffe *pl*

4205 SEMI-ATTENDED STATION tv
A station which is normally not manned but which is manned when necessary.
f station *f* semisurveillée
e estación *f* semiatendida
i stazione *f* parzialmente sorvegliata
n semi-bemand station *n*, semi-bewaakt station *n*
d zeitweise bemannte Station *f*

4206 SEMI-AUTOMATIC SWITCHER rep/tv
Device used in master control equipment.
f commutateur *m* semiautomatique
e conmutador *m* semiautomático
i commutatore *m* semiautomatico
n halfautomatische schakelaar
d halbautomatischer Schalter *m*

4207 SEMICONDUCTOR cpl
f semiconducteur *m*
e semiconductor *m*
i semiconduttore *m*
n halfgeleider
d Halbleiter *m*

4208 SEMIREFLECTOR, SEMISILVERED MIRROR ctv/opt
A semireflecting surface is one which partly reflects light, the proportions either being equal or having some predetermined ratio.
f miroir *m* semiargenté
e reflector *m* semiplateado
i riflettore *m* semiargentato
n halfverzilverde reflector
d halbverspiegelter Reflektor *m*

4209 SEMITRANSPARENT CATHODE, SEMITRANSPARENT PHOTOCATHODE crt
In camera tubes or phototubes, a photocathode in which radiant flux incident on one side produces photoelectric emission from both sides.
f photocathode *f* semitransparente
e fotocátodo *m* semitransparente
i fotocatodo *m* semitrasparente
n halfdoorlatende fotokatode
d halbdurchlässige Katode *f*

4210 SENDING-RECEIVING TERMINAL STATION tv
Part of a TV relay system where the functions of both a sending terminal

station and receiving terminal station are, or may be, performed.
f station *f* terminale transmission-réception
e estación *f* terminal transmisión-recepción
i stazione *f* terminale trasmissione-ricezione
n zend-ontvang-eindstation *n*
d Sende-Empfang-Endstation *f*

4211 SENDING TERMINAL STATION tv
Part of a TV relay system which accepts a TV relay input signal and radiates a TV relay signal.
f station *f* terminale transmission
e estación *f* terminal transmisión
i stazione *f* terminale trasmissione
n zendeindstation *n*
d Sendeendstation *f*

4212 SENSATION ct
Any operation of one of the senses.
f sensation *f*
e sensación *f*
i sensazione *f*
n sensatie, zintuiglijke gewaarwording
d Sinnesempfindung *f*, Sinneswahrnehmung *f*

4213 SENSATION FROM A CONSTANT STIMULUS ct
f sensation *f* par stimulus constant
e sensación *f* por estímulo constante
i sensazione *f* da stimolo costante
n zintuigelijke waarneming door een constante stimulus
d Sinneswahrnehmung *f* durch konstanten Reiz

4214 SENSITIVITY crt
Of a camera tube, the signal current developed per unit incident radiation density.
f sensibilité *f*
e sensibilidad *f*
i sensibilità *f*
n gevoeligheid
d Empfindlichkeit *f*

SENSITIVITY OF DEFLECTION
see: DEFLECTION SENSITIVITY

SENSITIZATION (GB)
see: ACTIVATION

SENSITIZER (GB)
see: ACTIVATOR

4215 SENSORY CHARACTERISTICS, VISUAL SENSORY CHARACTERISTICS ct
f caractéristiques *pl* sensorielles
e características *pl* sensorias
i caratteristiche *pl* sensorie
n waarnemingskarakteristieken *pl*
d Empfindungscharakteristiken *pl*

4216 SEPARATE LUMINANCE CAMERA ctv
A colo(u)r TV camera in which the luminance signal is derived independently of the tristimulus signals.
f caméra *f* pour télévision couleur à dérivation indépendante du signal de luminance
e cámara *f* de televisión en colores con derivación independiente de la señal de luminancia
i camera *f* di televisione a colori con derivazione indipendente del segnale di luminanza
n luminantiescheidingscamera
d Farbfernsehkamera *f* mit unabhängig abgeleitetem Leuchtdichtesignal

4217 SEPARATE MESH VIDICON crt
Improved form of vidicon which contains an extra electrode in the form of a wire mesh to improve the resolution of the tube.
f vidicon *m* à électrode maillée additionnelle
e vidicón *m* de electrodo de mallas adicional
i vidicon *m* ad elettrodo a maglie addizionale
n vidicon *n* met extra gaaselektrode
d Vidikon *n* mit zusätzlicher Maschenelektrode

4218 SEPARATE OPTICAL SOUND TRACK, SEPOPT vr
f piste *f* sonore optique séparée, sepopt *m*
e pista *f* sonora óptica separada, sepopt *m*
i pista *f* sonora ottica separata, sepopt *m*
n gescheiden optisch geluidsspoor *n*, sepopt
d getrennte optische Schallspur *f*, Sepopt *m*

4219 SEPARATE SOUND AND PICTURE RECORD, SEPMAG vr
f piste *f* sonore magnétique séparée, sepmag *m*
e pista *f* sonora magnética separada, sepmag *m*
i pista *f* sonora magnetica separata, sepmag *m*
n gescheiden magnetisch geluidsspoor *n*, sepmag
d getrennte magnetische Schallspur *f*, Sepmag *m*

4220 SEPARATE SYNC MONITORS, rec/tv
f moniteurs *pl* séparés pour la synchronisation
e monitores *pl* separados para la sincronización
i monitori *pl* separati per la sincronizzazione
n afzonderlijke synchronisatiemonitors *pl*
d getrennte Synchronisationsmonitoren *pl*

4221 SEPARATION, SYNC PULSE SEPARATION cpl
Separation of the composite synchronization signal from the video signal.
f séparation *f* d'impulsions de synchronisation
e separación *f* de impulsos de sincronización
i separazione *f* d'impulsi di sincronizzazione
n impulsscheiding
d Impulstrennung *f*

4222 SEPARATION CIRCUIT cpl
A circuit that sorts signals according to amplitude, frequency, or some other characteristic.
f circuit *m* de triage de signaux
e circuito *m* seleccionador de señales
i circuito *m* selettore di segnali
n signaalscheidingskring
d Signaltrennungskreis *m*

4223 SEPARATION FILTER cpl
A filter used to separate one band of frequencies from another, as in carrier system.
f coupleur *m* sélectif, filtre *m* séparateur
e filtro *m* separador
i filtro *m* separatore
n scheidingsfilter *n*
d Weiche *f*

SEPARATION OF COLO(U)R IMAGES
see: COLO(U)R IMAGE SEPARATION

4224 SEPARATION SIGNAL ctv
f signal *m* de séparation
e señal *f* de separación
i segnale *m* di separazione
n scheidingssignaal *n*
d Trennsignal *n*

4225 SEPDUMAG vr
f sepdumag *m*, son *m* magnétique double piste
e sepdumag *m*, sonido *m* magnético doble pista
i sepdumag *m*, suono *m* magnetico doppia pista
n apart magnetisch geluidsspoor *n* op dubbelspoor, sepdumag
d separater Magnetton *m* auf Zweispur, Sepdumag *m*

4226 SEQUENTIAL COLO(U)R DIFFERENCE SIGNALS ctv
f signaux *pl* séquentiels de différence de couleur
e señales *pl* secuenciales de diferencia de color
i segnali *pl* sequenziali di differenza di colore
n kleurverschilsignalen *pl* in volgorde, op elkaar volgende kleurverschilsignalen *pl*
d Farbdifferenzsignale *pl* in Reihenfolge

4227 SEQUENTIAL COLO(U)R TELEVISION ctv
Colo(u)r TV in which the components of the signal derived from the primary colo(u)rs are transmitted sequentially.
f système *m* de télévision couleur à séquence de trames
e sistema *m* televisivo por sucesión de colores
i sistema *m* di televisione a colori sequenziali
n informatiewisselsysteem *n*, kleurwisselsysteem *n*
d Farbwechselverfahren *n*, Zeitfolgeverfahren *n* beim Farbfernsehen

4228 SEQUENTIAL INTERLACE rec/rep/tv
Interlace in which the lines of one field fall directly under the corresponding lines of the preceding field.
f analyse *f* entrelacée séquentielle
e exploración *f* entrelazada secuencial
i analisi *f* interlacciata sequenziale
n ononderbroken interliniëring
d sequentielles Zwischenzeilenverfahren *n*

4229 SEQUENTIAL MONITORING rec/tv
Monitoring of short samples from a number of program(me)s one after another in a continuous cycle.
f contrôle *m* cyclique de programme
e control *m* cíclico de programa
i controllo *m* ciclico di programma
n continue programmacontrole
d zyklische Programmkontrolle *f*

4230 SEQUENTIAL RASTER tv
f canevas *m* séquentiel
e cuadrícula *f* secuencial
i quadro *m* rigato sequenziale
n volgraster *n*
d sequentieller Raster *m*

SEQUENTIAL SCANNING
see: PROGRESSIVE SCANNING

4231 SEQUENTIAL SIGNAL tv
f signal *m* séquentiel
e señal *f* secuencial
i segnale *m* sequenziale
n volgsignaal *n*
d Folgesignal *n*

4232 SEQUENTIAL TRANSMISSION rec
A method of transmitting pictures in which the picture elements are selected in regular time intervals and delivered in correct sequence to the communication channel.
f transmission *f* en séquence normale
e transmisión *f* de secuencia normal
i trasmissione *f* a sequenza normale
n transmissie in de juiste volgorde
d Übertragung *f* in der richtigen Reihenfolge

4233 SEQUENTIAL WORKING st/tv
TV transmission in which not all information representing any given detail or element of the picture image is transmitted at the same time.

f opération *f* non-simultanée
e operación *f* no simultánea
i operazione *f* non simultanea
n ongelijktijdige werkwijze
d nichtgleichzeitiges Verfahren *n*

4234 SERIES BROADCASTS tv
f série *f* d'émissions
e serie *f* de emisiones
i serie *f* d'emissioni
n vervolguitzendingen *pl*
d Fortsetzungsreihe *f*,
Sendereihe *f*

4235 SERIES CAPACITOR CLAMP cpl
A clamp in which the input video waveform is fed to the grid circuit of an electron tube (valve) via a series capacitor with the usual grid leak omitted.
f circuit *m* de verrouillage à condensateur en série
e circuito *m* de bloqueo con condensador en serie
i circuito *m* di bloccaggio a condensatore in serie
n klemschakeling met seriecondensator
d Klemmschaltung *f* mit Reihenkondensator

SERIES-EFFICIENCY DIODE
see: BOOSTER DIODE

4236 SERIES-RESONANT CIRCUIT cpl
f circuit *m* à résonance série
e circuito *m* de resonancia en serie
i circuito *m* a risonanza in serie
n serieresonantieketen
d Reihenresonanzkreis *m*

4237 SERRATED PULSE cpl
A pulse having notches or sawtooth indentations in its waveform.
f impulsion *f* à crête fractionnée
e impulso *m* fraccionado
i impulso *m* dentellato,
impulso *m* segato
n ingezaagde impuls
d gezahnter Impuls *m*

4238 SERRATED VERTICAL PULSE cpl
A vertical sync pulse that is broken up by five notches extending down to the black level of a TV signal, to give six component pulses, each lasting about 0.4 line and serving to keep the horizontal sweep circuits in step during the vertical sync pulse interval.
f impulsion *f* verticale à crête fractionnée
e impulso *m* vertical fraccionado
i impulso *m* verticale dentellato,
impulso *m* verticale segato
n ingezaagde verticale impuls
d gezahnter Vertikalimpuls *m*

4239 SERVICE ADJUSTMENT svs
Preset adjustment intended to be used during repair or maintenance operations.
f prépositionnement *m* de service
e ajuste *m* previo de servicio
i preregolazione *f* di servizio
n service-instelling
d Voreinstellung *f* für Kundendienstarbeiten

4240 SERVICE AREA tv
The area within which satisfactory reception of a broadcasting transmitter is normally possible.
f zone *f* de service
e área *f* de servicio
i area *f* di servizio
n reikwijdte
d Versorgungsbereich *m*

4241 SERVICE AREA DIAGRAM tv
A diagram indicating the area within which satisfactory reception of a broadcasting transmitter is normally possible.
f carte *f* de la zone de service
e diagrama *m* del área de servicio
i mappa *f* dell'area di servizio
n reikwijdtekaart
d Versorgungsbereichdiagramm *n*

4242 SERVICE BAND cpl
The band allocated in the frequency spectrum and specified for a definite class of radio service, for which there may be a number of channels.
f bande *f* accordée à un service déterminé
e banda *f* de frecuencia para servicios especiales
i banda *f* di frequenza per servizi speciali
n frequentieband voor speciale doeleinden
d Frequenzband *n* für spezielle Dienste

4243 SERVICE CONTRACT svs/tv
f contrat *m* de service
e contrato *m* de servicio
i contratto *m* di servizio
n servicecontract *n*
d Kundendienstvertrag *m*

4244 SERVICING svs
Preventive maintenance and repair of equipment.
f dépannage *m*,
entretien *m*
e servicio *m*
i servizio *m*
n onderhoud *n* en reparatie
d Pflege *f*,
Wartung *f*

4245 SERVO-CONTROLLED TAPE MECHANISM vr
f mécanisme *m* de bande asservi
e mecanismo *m* de cinta servorregulado
i meccanismo *m* di nastro a servoregolazione
n bandmechanisme *n* met servoregeling
d Bandmechanismus *m* mit Servosteuerung

4246 SERVO SYSTEM cpl/vr
Auxiliary equipment used i.a. during the recording of a TV signal on magnetic tape.
f système *m* asservi
e servosistema *m*
i servosistema *m*
n servosysteem *n*
d Servosystem *n*

4247 SET ACOUSTICS stu
f acoustique *f* de plateau
e acústica *f* del plató
i acustica *f* dello studio
n akoestiek van de opnameruimte
d Aufnahmeraumakustik *f*

4248 SET DESIGN fi/stu/th
f ébauche *f* du décor
e proyecto *m* de decoración
i progetto *m* dello scenario
n decorontwerp *n*
d Bühnenbildentwurf *m*, Dekorationsentwurf *m*, Szenenbild *n*

SET DESIGNER
see: DESIGNER

4249 SET ERECTION fi/stu/th
f montage *m* du décor
e montaje *m* de la construcción escénica
i montaggio *m* dello scenario
n decorbouw
d Dekorationsaufbau *m*

4250 SET LIGHTING stu
f éclairage *m* du plateau
e alumbrado *m* del plató
i illuminazione *f* dello studio
n verlichting van de opnameruimte
d Aufnahmeraumbeleuchtung *f*

4251 SET LIGHTS, STUDIO LIGHTS stu
f lampes *pl* de prise de vues, lampes *pl* du plateau
e lámparas *pl* de toma de vistas, lámparas *pl* del plató
i lampade *pl* dello studio, lampade *pl* di ripresa
n opnamelampen *pl*
d Aufnahmelampen *pl*

SET NOISE
see: INHERENT NOISE

4252 SET OF COLOUR IMAGE TUBES (GB), TRICONOSCOPE (US) crt
f bloc *m* de trois tubes de projection
e juego *m* de tres tubos de proyección
i blocco *m* di tre tubi di proiezione
n driebuizenstel *n*
d Dreiröhrenaggregat *n*

4253 SET STRIKING fi/stu/th
f démontage *m* du décor
e desmontaje *m* de la construcción escénica
i smontaggio *m* dello scenario
n decorafbraak
d Dekorationsabbau *m*

SET-UP
see: BLACK LIFT

4254 SET-UP A SHOT stu
f mise *f* en scène
e disposición *f* del escenario
i messinscena *f*
n mise-en-scène
d eine Szene einrichten

SET-UP CONTROL
see: BLACK LIFT CONTROL

SET-UP LEVEL
see: BLACK LIFT LEVEL

4255 SETTING UP ani
In animation the positioning of the background and the placing of the appropriate set of celluloids over the background.
f montage *m*
e montaje *m*
i montaggio *m*
n montage
d Montage *f*

4256 SHADE ct
A colo(u)r of the same hue and saturation but lower luminosity.
f teinte *f*
e tinte *f*
i tinta *f*
n tint, nuance
d Farbton *m*, Färbung *f*, Nuance *f*, Schattierung *f*

4257 SHADE OF GREY rep/tv
f marche *f* de gris
e escalón *m* de gris
i scalino *m* di grigio
n grijstrede
d Graustufe *f*

4258 SHADING, SHADING ERROR dis
Variations in brightness in a televised image because of local defects in the signal plate of a camera tube, arising from inadequate discharge.
f effet *m* d'ombrage
e error *m* de sombra
i errore *m* d'ombra
n schaduwverschijnsel *n*
d Bildabschattung *f*

4259 SHADING COMPENSATION SIGNAL, SHADING CORRECTION SIGNAL, SHADING CORRECTOR tv
f signal *m* compensateur d'ombrage
e señal *f* de compensación de sombra
i segnale *m* di compensazione d'ombra
n schaduwcompensatiesignaal *n*, schaduwcorrectiesignaal *n*
d Schattenkompensationssignal *n*

4260 SHADING CONTROL rep/tv
f réglage *m* de l'ombrage
e regulación *f* de la sombra
i regolazione *f* dell'ombra
n schaduwregeling, zwartpartijregeling
d Rauschpegelregelung *f*, Störsignalregelung *f*

4261 SHADING CORRECTION dis
The introduction into the picture signal of externally generated waveforms in order to compensate for amplitude distortion of that signal.
f correction *f* d'ombrage
e corrección *f* de sombra
i correzione *f* della distorsione di luminosità, correzione *f* d'ombra
n schaduwcorrectie, zwartpartijcorrectie,
d Störsignalkompensation *f*

4262 SHADING GENERATOR dis
One of the signal generators used at a TV transmitter to generate waveforms that are 180° out of phase with the undesired shading signals produced by a TV camera.
f générateur *m* de signaux correcteurs d'ombrage
e generador *m* de señales correctoras de sombra
i generatore *m* di segnali correttori d'ombra
n schaduwgenerator
d Schattengenerator *m*

4263 SHADING SIGNAL dis
An undesired signal produced by the TV camera.
f signal *m* d'ombrage
e señal *f* de sombra
i segnale *m* d'ombra
n schaduwsignaal *n*
d Schattensignal *n*

4264 SHADING SYSTEM dis
In a TV transmitter, a system of six waveform generators, which develop in two groups vertical and horizontal scanning signals to balance out unwanted signals arising in the camera tube.
f système *m* générateur de signaux de compensation d'ombrage
e sistema *m* generador de señales de compensación de sombra
i sistema *m* generatore di segnali di compensazione d'ombra
n genereersysteem *n* van schaduwcorrectiesignalen
d Generierungssystem *n* von Schattenkompensationssignalen

4265 SHADING VALUE crt
f valeur *f* de la luminosité
e valor *m* de la luminosidad
i valore *m* della luminosità
n helderheidswaarde
d Helligkeitsstufe *f*, Helligkeitswert *m*

4266 SHADOW BOARD ani
A device used in animation technique to counteract the undesirable reflecting properties of the glass plates.
f écran *m* antiréflexion
e pantalla *f* antirreflexión
i schermo *m* antiriflessione
n antireflectiescherm *n*
d Antireflexionsschirm *m*

SHADOW MASK
see: APERTURE MASK

4267 SHADOW MASK HOLE crt
f trou *m* de masque d'ombre
e foro *m* de máscara de sombra
i orificio *m* di maschera d'ombra
n schaduwmaskergat *n*
d Maskenloch *n*

4268 SHADOW MASK TUBE crt
A three-gun colo(u)r picture tube in which an electrically conductive sheet, containing many holes, effects colo(u)r selection by masking.
f tube *m* à masque
e tubo *m* de máscara
i tubo *m* di ripresa a maschera d'ombra
n schaduwmaskerbuis
d Maskenröhre *f*

SHADOW REGION
see: BLIND AREA

4269 SHAPE CONSTANCY opt
f constance *f* de forme
e constancia *f* de forma
i costanza *f* di forma
n vormconstantheid
d Formbeständigkeit *f*

4270 SHAPE DISTORTION ani
One of the features used in animation techniques.
f distorsion *f* de forme
e distorsión *f* de forma
i distorsione *f* di forma
n vervorming
d Verzerrung *f*

4271 SHAPING NETWORK cpl
Linear network used, in circuitry such as synchronized generators, to form pulses and edges of square waves to the shape required.
f réseau *m* conformateur
e red *f* formadora
i rete *f* conformatrice
n impulsvormend netwerk *n*
d impulsformendes Netzwerk *n*

SHARPNESS
see: ACUITY

4272 SHIELD crt
Of a cathode-ray tube, a cylinder having a small central aperture at one end which surrounds the cathode and forms the electron gun.
f cylindre *m* de la cathode
e cilindro *m* del cátodo
i cilindro *m* del catodo
n katodecilinder
d Katodenzylinder *m*

4273 SHIELD ge
Any screen, mesh, housing, or other structure designed to reduce the effects of electric or magnetic fields on objects placed behind or inside the shield.

f blindage *m*,
gaine *f*
e apantallamiento *m*,
blindaje *m*
i schermatura *f*,
schermo *m*
n afscherming
d Abschirmung *f*

4274 SHIFT crt
Movement of a pattern on a cathode-ray tube phosphor, by imposition of steady voltages, e.g. X-shift, Y-shift.
f déplacement *m*
e desplazamiento *m*
i spostamento *m*
n verschuiving
d Verschiebung *f*

4275 SHIFT OF LINE POSITION dis
f déplacement *m* de ligne
e desplazamiento *m* de línea
i spostamento *m* di linea
n lijnverplaatsing
d Zeilenverlagerung *f*

4276 SHOOTING fi/tv
f prise *f* de vues
e toma *f* de vistas
i presa *f* cinematografica,
ripresa *f* cinematografica
n filmopname
d Filmaufnahme *f*

4277 SHOOTING BRAKE,
TELEVISION CAR rec/tv
f car *m* de télévision
e carro *m* de televisión
i carro *m* per riprese
n opnamewagen,
televisiewagen
d Aufnahmewagen *m*,
Fernsehwagen *m*

4278 SHOOTING PATTERN,
SHOOTING PLAN,
SHOOTING SCHEDULE tv
f plan *m* de prise de vues,
plan *m* de tournage
e plano *m* de toma de vistas
i piano *m* di lavorazione
n opneemplan *n*
d Drehplan *m*,
Drehübersicht *f*

4279 SHOOTING SCRIPT fi/tv
f scénario *m* définitif
e libro *m*
i sceneggiatura *f* definitiva
n definitief draaiboek *n*
d Kamerascript *n*

4280 SHORTCIRCUIT FLUX vr
f flux *m* de court-circuit
e flujo *m* de cortocircuito
i flusso *m* di cortocircuito
n kortsluitflux
d Kurzschlussflux *m*

4281 SHORT-COIL ELECTRO-MAGNETIC FOCUSING,
SHORT-COIL MAGNETIC FOCUSING crt
f focalisation *f* magnétique à bobine courte
e enfoque *m* magnético de bobina corta
i focalizzazione *f* magnetica a bobina corta
n magnetische focussering met korte spoel
d magnetische Fokussierung *f* mit kurzer Magnetspule

4282 SHORT-PERSISTENCE
CATHODE-RAY TUBE crt
f tube *m* cathodique à courte durée de persistance
e tubo *m* catódico de corta duración de persistencia
i tubo *m* catodico a corta durata di persistenza
n kort nalichtende katodestraalbuis
d Katodenstrahlröhre *f* mit kurzer Nachleuchtdauer

4283 SHORT-PERSISTENCE SCREEN crt
A screen whose luminescence decays rapidly after the stimulus has been reduced or removed.
f écran *m* à courte durée de persistance
e pantalla *f* de corta duración de persistencia
i schermo *m* a corta durata di persistenza
n kort nalichtend scherm *n*
d Schirm *m* mit kurzer Nachleuchtdauer

4284 SHORT STREAKING dis
A fault, which in the case of horizontal line by line scanning, gives rise to horizontal lines apparently attached to the edges of strongly contrasted objects.
f traînage *m* court
e arrastre *m* corto
i striscionamento *m* corto
n korte vegen *pl*
d kurze Fahnen *pl*

4285 SHORTWAVE STUDIO stu
f studio *m* ondes courtes
e estudio *m* ondas cortas
i studio *m* onde corte
n studio voor korte golf
d Kurzwellenstudio *n*

SHOT BOX
see: FOCUS UNIT

4286 SHOT EFFECT,
SHOT NOISE dis
Noise due to random emission from a cathode.
f effet *m* de grenaille
e efecto *m* de granalla
i effetto *m* granulare
n hageleffect *n*,
schrooteffect *n*
d Schroteffekt *m*,
Schrotrauschen *n*

SHUNTED MONOCHROME SIGNAL
see: BYPASS MONOCHROME SIGNAL

4287 SIBILANCE aud
f sifflement *m*
e sonido *m* sibilante
i suono *m* sibilante
n sissen *n*,
sissend geluid *n*
d Zischen *n*,
Zischlaut *m*

4288 SIDEBAND cpl
A range of frequencies occupied by the spectral components resulting from modulation of a carrier wave by a signal.
f bande *f* latérale
e banda *f* lateral
i banda *f* laterale
n zijband
d Seitenband *n*

4289 SIDEBAND FILTER cpl
Filter used in a vision transmitter to attenuate one of the sidebands of the transmission.
f filtre *m* de bande latérale
e filtro *m* de banda lateral
i filtro *m* di banda laterale
n zijbandfilter *n*
d Seitenbandfilter *n*

4290 SIDEBAND FREQUENCY,
SIDE FREQUENCY cpl
A single frequency component of a sideband.
f fréquence *f* latérale
e frecuencia *f* lateral
i frequenza *f* di banda laterale
n zijbandfrequentie
d Seitenbandfrequenz *f*,
Seitenfrequenz *f*

SIDEBAND INTERFERENCE,
SIDEBAND SPLASH
see: ADJACENT CHANNEL INTERFERENCE

4291 SIDE-LOCK ctv/dis
In a colo(u)r receiver, a fault condition arising when the subcarrier local oscillator locks to a side frequency of the subcarrier instead of to the subcarrier frequency.
f verrouillage *m* sur fréquence latérale
e enclavamiento *m* sobre banda lateral
i agganciamento *m* sopra banda laterale
n nevenvergrendeling,
vergrendeling op een zijband
d Verriegelung *f* auf einem Seitenband

4292 SIGN-OFF stu
f fin *f* des émissions
e cierre *m* de la emisión
i fine *f* dell'emissione
n einde *n* der uitzending
d Sendeschluss *m*

4293 SIGN-ON stu
f commencement *m* des émissions
e abertura *f* de la emisión
i apertura *f* dell'emissione
n begin *n* der uitzending
d Sendebeginn *m*

4294 SIGNAL cpl/ge
The physical embodiment of a message.
f signal *m*
e señal *f*
i segnale *m*
n signaal *n*
d Signal *n*

4295 SIGNAL ABOVE BLACK LEVEL dis
f défoncement *m* du niveau du noir
e desbordamiento *m* del nivel del negro
i sfondato *m* del livello del nero
n doorstoten *n* van het zwartniveau
d Durchstossen *n* des Schwarzwertes

4296 SIGNAL CHANNEL,
SIGNAL TRANSMISSION PATH cpl
f canal *m* de signal,
voie *f* de signal
e canal *m* de señal
i canale *m* di segnale
n signaalkanaal *n*
d Signalkanal *m*

SIGNAL CHANNEL MISCELLANEOUS INTERFERENCES
see: MISCELLANEOUS INTERFERENCES

4297 SIGNAL CIRCUIT cpl
f circuit *m* de signal
e circuito *m* de señal
i circuito *m* di segnale
n signaalketen
d Signalkreis *m*

SIGNAL COMPLEX (GB)
see: INFORMATION

4298 SIGNAL COMPONENTS cpl
f composantes *pl* du signal
e componentes *pl* de la señal
i componenti *pl* del segnale
n signaalcomponenten *pl*
d Signalkomponenten *pl*

4299 SIGNAL CURRENT cpl
f courant *m* de signal
e corriente *f* de señal
i corrente *f* di segnale
n signaalstroom
d Signalstrom *m*

4300 SIGNAL DISTORTION cpl/dis
f distorsion *f* du signal
e distorsión *f* de la señal
i distorsione *f* del segnale
n signaalvervorming
d Signalverzerrung *f*

4301 SIGNAL ELECTRODE,
SIGNAL PLATE crt
In a storage camera tube, the metal backplate insulated from the mosaic and serving as a common output electrode for the signals from active particles.
f anode *f* collectrice,
plaque *f* collectrice
e ánodo *m* colector,
placa *f* de señal
i placca *f* di segnale

n signaalplaat,
trefplaat
d Speicherplatte *f*

4302 SIGNAL ERRORS dis
f erreurs *pl* du signal
e errores *pl* de la señal
i errori *pl* del segnale
n signaalfouten *pl*
d Signalfehler *pl*

4303 SIGNAL FIELD vr
f champ *m* du signal
e campo *m* de la señal
i campo *m* del segnale
n signaalveld *n*
d Nutzfeld *n*

SIGNAL FIELD PERIOD
see: ACTIVE FIELD PERIOD

4304 SIGNAL FREQUENCY cpl
f fréquence *f* du signal
e frecuencia *f* de la señal
i frequenza *f* del segnale
n signaalfrequentie
d Signalfrequenz *f*

4305 SIGNAL FREQUENCY RANGE cpl
f gamme *f* de fréquences du signal
e gama *f* de frecuencias de la señal
i gamma *f* di frequenze del segnale
n signaalfrequentiegebied *n*
d Signalfrequenzbereich *m*

4306 SIGNAL GENERATOR cpl
A generator of electrical signals, having calibrated controls, and used for testing.
f générateur *m* de mesure
e generador *m* de señales de medida
i generatore *m* di misura
n meetzender,
signaalgenerator
d Messsender *m*

4307 SIGNAL HANDLING EQUIPMENT cpl
f installation *f* de manipulation de signaux
e equipo *m* de manejo de señales
i apparecchiatura *f* di trattamento di segnali
n signaalverwerkingsapparatuur
d Signalverarbeitungsapparatur *f*

4308 SIGNAL INVERSION cpl
Takes place when the polarity of the signal is reversed with respect to itself.
f inversion *f* de signal
e inversión *f* de señal
i inversione *f* di segnale
n signaalompoling
d Signalumpolung *f*

4309 SIGNAL LEVEL cpl
At any point in a transmission system, the difference between the measure of the signal at that point and the measure of an arbitrary signal chosen as a reference.
f niveau *m* de signal
e nivel *m* de señal
i livello *m* di segnale
n signaalniveau *n*
d Signalpegel *m*

4310 SIGNAL MAKE-UP,
SIGNAL PROCESSING ctv
f établissement *m* du signal
e constitución *f* de la señal
i costituzione *f* del segnale
n signaalopbouw
d Signalaufbau *m*

4311 SIGNAL MIXER stu
f mélangeur *m* de signaux
e mezclador *m* de señales
i mescolatore *m* di segnali
n signaalmenger
d Signalmischer *m*

4312 SIGNAL MIXING tv
f mixage *m* de signaux
e mezcla *f* de señales
i mescolanza *f* di segnali
n signaalmenging
d Signalmischung *f*

4313 SIGNAL MIXTURE cpl
f mélange *m* de signaux
e mezclado *m* de señales
i miscela *f* di segnali
n signaalmengsel *n*
d Signalgemisch *n*

4314 SIGNAL NOISE ctv
f bruit *m* de signal
e ruido *m* de señal
i rumore *m* di segnale
n signaalruis
d Signalrauschen *n*

4315 SIGNAL OUTPUT CURRENT crt
Of camera tubes or phototubes, the absolute value of the difference between output current and dark current.
f valeur *f* absolue du courant de sortie du signal
e valor *m* absoluto de la corriente de salida de la señal
i valore *m* assoluto della corrente d'uscita del segnale
n absolute waarde van de signaaluitgangsstroom
d absoluter Wert *m* des Signalausgangsstromes

4316 SIGNAL PROBLEMS vr
The problems which arise when a TV signal in video tape recording in colo(u)r has to be changed into a form suitable for recording on to tape, and reconstituting this signal on replay without distortion.
f problèmes *pl* de change de signaux
e problemas *pl* de cambio de señales
i problemi *pl* di cambio di segnali
n signaalomzettingsproblemen *pl*
d Signalumwandlungsprobleme *pl*

4317 SIGNAL ROUTING cpl
f acheminement *m* du signal

e encaminamiento *m* de la señal
i istradamento *m* del segnale
n signaalroutebepaling
d Signalleitwegbestimmung *f*

4318 SIGNAL SEPARATOR cpl
f séparateur *m* de signaux
e separador *m* de señales
i separatore *m* di segnali
n signaalscheider
d Signaltrennstufe *f*

4319 SIGNAL SPLITTER aea/tv
f répartiteur *m* de signaux
e distribuidor *m* de señales, divisor *m* de señales
i divisore *m* di segnali, separatore *m* di segnali
n signaalverdeler
d Signalverteiler *m*

SIGNAL STABILIZATION AMPLIFIER
see: LINE CLAMP AMPLIFIER

SIGNAL STRENGTH METER
see: S-METER

4320 SIGNAL TAIL, TAIL OF SIGNAL cpl
f queue *f* de signal
e cola *f* de señal
i coda *f* di segnale
n signaalstaart
d Signalschwanz *m*

4321 SIGNAL-TO-NOISE RATIO dis
For given conditions of adjustment and signal input, the ratio of the magnitude of some specified feature of the wanted response to the magnitude of the appropriate feature of the co-existent noise.
f rapport *m* signal-bruit
e relación *f* señal-ruido
i rapporto *m* segnale-rumore
n signaal-ruisverhouding
d Geräuschabstand *m*, Rauschabstand *m*

4322 SIGNAL TRANSITION tv
f transition *f* de signal
e transición *f* de señal
i transizione *f* di segnale
n signaalovergang
d Signalübergang *m*

4323 SIGNAL VALUE cpl
f valeur *f* du signal
e valor *m* de la señal
i valore *m* del segnale
n signaalwaarde
d Signalwert *m*

4324 SIGNAL VOLTAGE cpl
f tension *f* efficace de signal
e tensión *f* eficaz de señal
i tensione *f* utile di segnale
n nuttige signaalspanning
d Nutzspannung *f*, Signalspannung *f*

4325 SIGNAL VOLTAGE GENERATOR cpl
f générateur *m* de la tension efficace de signal
e generador *m* de la tensión eficaz de señal
i generatore *m* della tensione utile di segnale
n generator van de nuttige signaalspanning
d Nutzspannungsgenerator *m*

4326 SIGNAL WAVE cpl
A wave whose characteristic permits some intelligence or message to be conveyed.
f onde *f* de travail
e onda *f* de trabajo
i onda *f* di lavoro
n werkgolf
d Betriebswelle *f*, Signalwelle *f*

SIGNAL WAVEFORM ELEMENTS
see: ELEMENTS OF LINE SIGNAL WAVEFORM

4327 SIMULTANEOUS BROADCAST tv
Broadcast by a number of stations of the same program(me) at the same time.
f radiodiffusion *f* par réseau d'émetteurs, transmission *f* simultanée
e transmisión *f* simultánea
i trasmissione *f* in rete
n gemeenschappelijke programmauitzending
d Gemeinschaftssendung *f*, Simultanübertragung *f*

4328 SIMULTANEOUS COLO(U)R TELEVISION ctv
Colo(u)r TV in which the components of the signal derived from the primary colo(u)rs are transmitted simultaneously.
f système *m* additif de télévision couleur
e sistema *m* aditivo de televisión en colores
i sistema *m* additivo di televisione a colori
n simultaan kleurentelevisiesysteem *n*
d Simultanfarbfernsehen *n*

4329 SIMULTANEOUS DISPLAY ctv
Display in which all picture points are illuminated simultaneously.
f présentation *f* simultanée
e presentación *f* simultánea
i presentazione *f* simultanea
n simultane weergeving
d Simultandarstellung *f*

SIMULTANEOUS FREQUENCY AND AMPLITUDE MODULATION
see: FAM

4330 SINE-SQUARED PULSE cpl
A pulse which has the shape defined by the equations:
$V = O$ for $t < O$
$V = \sin^2 t$ for $O < t$
$V = O$ for $t > \pi$
f impulsion *f* à sinus carré
e impulso *m* de seno cuadrado
i impulso *m* a seno quadrato

n sinus²-impuls
d Sin²-Impuls *m*

4331 SINE WAVE CONVERGENCE crt
Use of a sine wave of current to effect dynamic convergence of the beam of a shadow-mask tube.
f convergence *f* par courant sinusoïdal
e convergencia *f* por corriente sinusoidal
i convergenza *f* per corrente sinusoidale
n convergentie d.m.v. een sinusvormige stroom
d Konvergenz *f* mittels Sinuswelle

4332 SINGLE ACCESS SATELLITE tv
f satellite *m* à simple accès
e satélite *m* de acceso simple
i satellite *m* ad accesso semplice
n satelliet met enkelvoudige toegang
d Satellit *m* für einfachen Zugang

SINGLE BROADSIDE
see: BABY CAN

4333 SINGLE ECHO dis
A signal which is a sample of the original signal and arriving later than the original does, owing to a reflection.
f écho *m* simple
e eco *m* simple
i eco *m* semplice
n eenvoudige echo
d Einfachecho *n*

4334 SINGLE-GUN COLO(U)R TUBE crt
f tube *m* image couleur à canon unique
e cromoscopio *m* de cañón electrónico único
i tubo *m* di ripresa a colori ad cannone unico
n kleurenbeeldbuis met één kanon
d Fernsehbildröhre *f* mit einem Elektronenstrahl

4335 SINGLE-KNOB CONTROL tv
f monocommande *f*
e mando *m* único, monocontrol *m*
i monocomando *m*
n eenknopsbediening
d Einknopfbedienung *f*

4336 SINGLE POLE-PIECE MAGNETIC HEAD vr
f tête *f* magnétique à pièce polaire unique
e cabeza *f* magnética de pieza polar única
i testina *f* magnetica a scarpa polare unica
n magneetkop met één poolschoen
d Magnetkopf *m* mit einem Polstück

4337 SINGLE-SHOT BLOCKING OSCILLATOR cpl/rec/tv
f oscillateur *m* monostable de blocage
e oscilador *m* de bloqueo de ciclo simple
i oscillatore *m* monostabile di rilassamento a bloccaggio
n monostabiele blokkeeroscillator
d monostabiler Sperrschwinger *m*

SINGLE-SHOT MULTIVIBRATOR
see: ONE-CYCLE MULTIVIBRATOR

4338 SINGLE-SHOT TRIGGER CIRCUIT rec/tv
f circuit *m* de déclenchement à cycle simple
e circuito *m* de disparo a período simple
i circuito *m* di sgancio a singolo periodo
n eenperiodetrekkerimpulsketen
d monostabile Triggerschaltung *f*

4339 SINGLE SIDEBAND cpl
Used in the technique of reducing the signal bandwidth by entirely suppressing the frequencies produced by modulation on one side of the carrier, and leaving information to reconstitute the signal.
f bande *f* latérale unique
e banda *f* lateral única
i banda *f* laterale unica
n eenzijband
d Einseitenband *n*

4340 SINGLE-SIDEBAND MODULATION cpl
Modulation whereby the spectrum of the modulating wave is translated in frequency by a specified amount.
f modulation *f* à bande latérale unique
e modulación *f* de banda lateral única
i modulazione *f* a banda laterale unica
n eenzijbandmodulatie
d Einseitenbandmodulation *f*

4341 SINGLE-SIDEBAND MODULATOR cpl
f modulateur *m* à bande latérale unique
e modulador *m* de banda lateral única
i modulatore *m* a banda laterale unica
n eenzijbandmodulator
d Einseitenbandmodulator *m*

4342 SINGLE-SIDEBAND SUPPRESSED-CARRIER SYSTEM, SSSC cpl
f système *m* d'émission sur bande latérale unique et suppression de l'onde porteuse
e sistema *m* de emisión sobre banda lateral única con portadora suprimida
i sistema *m* d'emissione su banda laterale unica con portante soppressa
n eenzijbandsysteem *n* met onderdrukte draaggolf
d Einseitenbandsystem *n* mit unterdrücktem Träger

4343 SINGLE-SIDEBAND TRANSMISSION cpl
That method of operation in which either the upper or lower sideband, as produced by the process of modulation, is transmitted.
f transmission *f* sur bande latérale unique
e transmisión *f* sobre banda lateral única
i trasmissione *f* su banda laterale unica
n eenzijbandtransmissie
d Einseitenbandübertragung *f*

4344 SINGLE-SIDEBAND TRANSMITTER cpl
A transmitter using a method of operation in which either the upper or lower sideband, as produced by the process of modulation, is suppressed.

f émetteur *m* sur bande latérale
e emisor *m* sobre banda lateral
i emettitore *m* su banda laterale
n eenzijbandzender
d Einseitenbandsender *m*

4345 SINGLE-SYSTEM SOUND RECORDING aud
Method of sound recording in which the sound is originally recorded at the same time and on the same strip of film as the picture image.
f enregistrement *m* simultané de son et image
e registro *m* simultáneo de sonido y imagen
i registrazione *f* simultanea di suono ed immagine
n gelijktijdige beeld- en geluidsopname
d gleichzeitige Bild- und Schallaufnahme *f*

4346 SIZE CONSTANCY opt
f constance *f* de dimensions
e constancia *f* de dimensiones
i costanza *f* di dimensioni
n constantheid van afmetingen
d Abmessungenkonstanz *f*

4347 SIZE CONTROL rep/tv
A control provided on a TV receiver for changing the size of a picture either horizontally or vertically.
f réglage *m* du format
e regulación *f* del formato
i regolazione *f* del formato
n regeling van het formaat
d Formatregelung *f*

4348 SIZE OF PICTURE POINT crt
f largeur *f* du spot
e tamaño *m* del punto de imagen
i larghezza *f* del punto d'immagine
n beeldpuntgrootte
d Bildpunktgrösse *f*

4349 SKEW vr
f enroulement *m* en biais de la bande
e sesgadura *f* de la cinta
i il sbiecare *m* del nastro
n scheeflopen *n* van de band
d Bandschräglauf *m*

SKEW DISTORTION
see: PARALLEL DISTORTION

SKIATRON
see: DARK-TRACE TUBE

4350 SKIN TONE ctv
f coloration *f* de la peau
e color *m* de la piel
i colore *m* della pelle
n huidkleur
d Hautfarbe *f*

4351 SLAVE cpl
That unit of two coupled ones that is controlled by the other.
f appareil *m* asservi
e aparato *m* secundario
i apparecchio *m* secondario
n nevenapparaat *n*
d Nebenapparat *m*

SLAVE TRANSMITTER
see: SATELLITE TRANSMITTER

4352 SLAVELOCK cpl
A term used to denote a number of remote TV stations running synchronously with a main station by means of radiated signals from the main station.
f synchronisation *f* d'émetteurs secondaires
e sincronización *f* de emisores secundarios
i sincronizzazione *f* d'emettitori secondari
n synchronisatie van hulpzenders
d Synchronisierung *f* von Hilfssendern

4353 SLAVING crt
The operation of making a local synchronizing pulse generator operate in lock with an incoming video signal in both line and field.
f synchronisation *f* externe
e sincronización *f* externa
i sincronizzazione *f* esterna
n synchronisatie door inkomend signaal
d Fremdsynchronisation *f*

4354 SLEWING tv
f panoramique *m* latéral
e panorámica *f* lateral
i panoramica *f* laterale
n zijwaartse panoramica
d seitlicher Schwenk *m*

4355 SLIDE ADVERTISING, SLIDE PUBLICITY tv
f publicité *f* par diapositives
e publicidad *f* por diapositivas
i pubblicità *f* per diapositive
n televisiereclame met dia's
d Diawerbung *f*

4356 SLIDE PROJECTOR tv
f projecteur *m* de diapositives
e proyector *m* de diapositivas
i proiettore *m* di diapositive
n diaprojector
d Diaprojektor *m*

4357 SLIDE SCANNER rep/tv
f analyseur *m* de diapositives
e explorador *m* de diapositivas
i analizzatore *m* di diapositive
n dia-aftaster
d Diaabtaster *m*

4358 SLIP-ON FILTER cpl
f filtre *m* interchangeable
e filtro *m* intercambiable
i filtro *m* intercambiabile
n opsteekfilter *n*
d Aufsteckfilter *n*

4359 SLIPPAGE OF SOUND TO PICTURE, SOUND TRACK ADVANCE dis
f décalage *m* image-son
e deslizamiento *m* imagen-sonido
i scorrimento *m* immagine-suono
n beeld-geluidverzet *n*
d Bild-Tonversatz *m*

SLITTING OF MAGNETIC TAPE
see: MAGNETIC TAPE SLITTING

SLOPE
see: MUTUAL CONDUCTANCE

4360 SLOT AERIAL, SLOT ANTENNA aea
f antenne *f* à fente, antenne *f* fendue
e antena *f* ranurada
i antenna *f* a fessura
n spleetantenne
d Schlitzantenne *f*

4361 SLOW-FREQUENCY DRIFT dis
f glissement *m* graduel de la fréquence
e corrimiento *m* gradual de la frecuencia
i deriva *f* tardiva della frequenza
n langzame frequentiedrift
d Frequenzverwerfung *f*, langsame Frequenzwanderung *f*

SLOW GENLOCK
see: CONTROLLED RATE GENLOCK

4362 SLOW-MOTION SHOOTING fi/tv
f tournage *m* d'un ralenti
e sistema *m* de toma de vistas con retardador
i sistema *m* di ripresa al rallentatore
n vertraagde-opnamesysteem *n*
d Zeitlupenverfahren *n*

4363 SLOW-SCAN TELEVISION SYSTEM tv
f système *m* de télévision à analyse lente
e sistema *m* de televisión de exploración lenta
i sistema *m* di televisione ad analisi lenta
n televisiesysteem *n* met langzame aftasting
d Fernsehsystem *n* mit langsamer Abtastung

4364 SMEAR dis
Loss of TV image definition due to lack of sufficiently high video-frequency response, or due to smear ghosts.
f maculage *m*
e borrosidad *f*
i sbavatura *f*
n smeren *pl*
d Fahnen *pl*

4365 SMEAR GHOST dis
Loss of definition of a TV picture due to a ghost image following the primary image so closely as to partially merge the outlines.
f image *f* fantôme à maculage
e imagen *f* fantasma de borrosidad
i immagine *f* fantasma a sbavatura
n geestbeeld *n* met smeren
d Fahnengeisterbild *n*

4366 SMEARER cpl
A circuit used to cancel the overshoot of a pulse.
f circuit *m* neutraliseur du dépassement balistique
e circuito *m* neutralizador de la sobreoscilación
i circuito *m* neutralizzatore della sovraelongazione
n doorschotopheffer
d Überschwungbeseitiger *m*

4367 SMEARING dis
f effet *m* de traînage horizontal
e efecto *m* de prolongación irregular de las líneas horizontales
i effetto *m* di trascinamento orizzontale
n versmering
d Nachzieheffekt *m*, Verschmierung *f*

SMOOTHING
see: LEVEL(L)ING

4368 SNOW crt/dis
Interference resembling falling snow which may appear on a TV screen, in the absence of a signal or in the presence of a weak signal, owning to noise in the equipment.
f neige *f*
e nieve *f*
i neve *f*
n sneeuw
d Schnee *m*

4369 SOFT IMAGE, SOFT PICTURE dis
f image *f* à faible contraste
e imagen *f* débil
i immagine *f* debole
n contrastarm beeld *n*
d weiches Bild *n*

4370 SOLAR BATTERY, SOLAR CELL cpl
A battery principally used as a source of power supply for communication satellites, e.g. a silicon junction photovoltaic cell.
f cellule *f* solaire
e batería *f* solar
i batteria *f* solare
n zonnecel
d Sonnenbatterie *f*

4371 SOLID ANGLE ge
f angle *m* solide
e ángulo *m* sólido
i angolo *m* solido
n ruimtehoek
d räumlicher Winkel *m*, Raumwinkel *m*

4372 SOLID STATE CIRCUIT cpl
Electrical circuit using solid state components.
f circuit *m* d'état solide
e circuito *m* de estado sólido
i circuito *m* di corpo solido

n vaste-stofcircuit *n*
d Festkörperkreis *m*

4373 SOLID STATE COMPONENTS cpl
Transistors, semiconductor diodes and similar components.
f éléments *pl* d'état solide
e elementos *pl* de estado sólido
i elementi *pl* di corpo solido
n vaste-stofelementen *pl*
d Festkörperelemente *pl*

4374 SOLOIST aud/th
f soliste *m*, *f*
e solista *m*, *f*
i solista *m*, *f*
n solist(e)
d Solist *m*,
Solistin *f*

4375 SOOT AND WHITE WASH opt
f image *f* trop contrastée
e imagen *f* demasiado contrastante
i immagine *f* troppo contrastante
n beeld *n* met te hoog contrast
d knochiges Bild *n*

4376 SOUND aud
Mechanically transmitted stimuli in the range of $10H_Z$ to $20000H_Z$ received by the ear.
f son *m*
e sonido *m*
i suono *m*
n geluid *n*
d Schall *m*, Ton *m*

4377 SOUND ABSORPTION aud
f absorption *f* acoustique
e absorción *f* acústica
i assorbimento *m* acustico
n geluidsabsorptie
d Schallschluck *m*

4378 SOUND ARRANGEMENTS aud/vr
f arrangements *pl* sonores
e disposiciones *pl* acústicas
i disposizioni *pl* acustiche
n geluidsverzorging
d akustische Anordnung *f*

4379 SOUND BACKGROUND aud/stu
f illustrations *pl* sonores
e ilustraciones *pl* sonoras
i illustrazioni *pl* sonore
n begeleidend geluid *n*
d Begleitton *m*

4380 SOUND BANDWIDTH aud
f largeur *f* bande de son
e anchura *f* de la banda de sonido
i larghezza *f* della banda audio
n bandbreedte audio,
geluidsbandbreedte
d NF-Bandbreite *f*,
Tonbandbreite *f*

4381 SOUND BAR dis
One of the two or more alternate dark and bright horizontal bars that appear in a TV picture when audio-frequency voltage reaches the video input circuit of the picture tube.
f barre *f* de son
e barra *f* de sonido
i barra *f* di suono
n geluidsbalk
d Tonbalken *m*

4382 SOUND BRANCH AMPLIFIER aud/cpl
f amplificateur *m* son à deux sorties
e amplificador *m* de sonido con dos salidas
i amplificatore *m* di suono a due uscite
n geluidsversterker met twee uitgangen
d Tonabzweigverstärker *m*

4383 SOUND BREAKDOWN aud/tv
f interruption *f* son
e interrupción *f* sonido
i interruzione *f* suono
n uitvallen *n* van het geluid
d Tonausfall *m*

SOUND CARRIER
see: AUDIO CARRIER

4384 SOUND CARRIER ATTENUATOR aud/cpl
f palier *m* son
e atenuador *m* de la portadora sonido
i attenuatore *m* della portante suono
n geluidsdragerverzwakker
d Tontreppe *f*

SOUND CARRIER FREQUENCY,
SOUND CENTER(RE) FREQUENCY
see: AUDIO CENTER(RE) FREQUENCY

4385 SOUND CARRIER SWEEP tv
f excursion *f* de porteuse son
e excursión *f* de portadora sonido
i escursione *f* di portante suono
n geluidsdraaggolfzwaai
d Tonträgerhub *m*

4386 SOUND CARRIER TRAP aud/cpl
f piège *m* son
e trampa *f* sonido
i trappola *f* suono
n geluidsdragerval
d Tonträgersperre *f*

4387 SOUND CENTER(RE) I.F. cpl
f fréquence *f* intermédiaire nominale son
e frecuencia *f* intermedia nominal sonido
i frequenza *f* intermedia nominale suono
n centrale geluidstussenfrequentie
d mittlere Tonzwischenfrequenz *f*

4388 SOUND CHANNEL tv
The carrier frequency with its associated sidebands, involved in the transmission of sound in TV.
f canal *m* son
e canal *m* sonido
i canale *m* suono
n geluidskanaal *n*
d Tonkanal *m*

4389 SOUND-CHROMINANCE BEAT ctv/dis
Interference between the chrominance channel and the co-channel sound signal.
f interférence *f* entre porteuse son et porteuse chrominance
e interferencia *f* entre portadora sonido y portadora crominancia
i interferenza *f* tra portante suono e portante crominanza
n interferentie tussen geluids- en chrominantiedraaggolf
d Interferenz *f* zwischen Ton- und Chrominanzträger

4390 SOUND CLARITY, aud
SOUND DEFINITION
f clarté *f* du son
e claridad *f* del sonido
i chiarezza *f* del suono
n helderheid van het geluid
d Schallklarheit *f*

SOUND COLORATION
see: COLORATION

SOUND COMPRESSION
see: COMPRESSION

SOUND COMPRESSOR
see: COMPRESSOR

4391 SOUND CONSOLE, aud
SOUND MIXING CONSOLE
Sound controlling and adjusting apparatus housing all the controls required by a sound mixer to make up a complete program(me).
f pupitre *m* de contrôle son
e pupitre *m* de control del sonido
i tavolo *m* di regia audio
n geluidslessenaar
d Tonregiepult *n*

SOUND CONTROL ENGINEER,
SOUND ENGINEER
see: AUDIO CONTROL ENGINEER

4392 SOUND CONTROL ROOM rec/tv
That part of the studio where the output of the various microphones is mixed to form a homogeneous whole.
f salle *f* de mixage du son,
salle *f* de régie son
e sala *f* de control de sonido,
sala *f* de mezclado de sonido
i sala *f* di mescolanza di suono,
sala *f* di regia di suono
n geluidsmengkamer,
geluidsregiekamer
d Tonmischungsraum *m*,
Tonregieraum *m*

4393 SOUND CUTTER aud
Operator whose task is to build up a set of dubbing tracks which complement the effect of the picture.
f opérateur *m* de montage des pistes sonores
e operador *m* de montaje de las pistas sonoras
i operatore *m* di montaggio delle piste sonore
n geluidsmonteur
d Tonschnittmeister *m*

4394 SOUND CUTTING vr
f montage *m* du son
e montaje *m* del sonido
i montaggio *m* del suono
n geluidsmontage
d Tonschnitt *m*

4395 SOUND DETECTOR aud
f détecteur *m* acoustique
e detector *m* acústico
i rivelatore *m* acustico
n geluidsdetector
d Schalldetektor *m*

4396 SOUND DISCRIMINATOR aud
f redresseur *m* de la fréquence sonore
e discriminador *m* de la frecuencia sonora
i raddrizzatore *m* della frequenza sonora
n geluidsfrequentiegelijkrichter
d Tonfrequenzgleichrichter *m*

4397 SOUND DISTORTION aud/dis
f distorsion *f* acoustique
e distorsión *f* acústica
i distorsione *f* acustica
n geluidsvervorming
d Klangverzerrung *f*

4398 SOUND DISTURBANCE aud/dis
EFFECTS
f effets *pl* de perturbations acoustiques
e efectos *pl* de perturbaciones acústicas
i effetti *pl* di disturbi acustici
n akoestische storingseffecten *pl*
d akustische Störungseffekte *pl*

4399 SOUND EFFECTS aud
f bruitage *m*,
effets *pl* sonores
e efectos *pl* sonoros
i effetti *pl* sonori
n geluidseffecten *pl*
d Geräusche *pl*,
Schalleffekte *pl*

4400 SOUND ENERGY aud
f énergie *f* acoustique,
énergie *f* sonore
e energía *f* acústica,
energía *f* del sonido
i energia *f* acustica,
energia *f* sonora
n geluidsenergie
d Schallenergie *f*

4401 SOUND FREQUENCY RANGE aud
f gamme *f* des fréquences acoustiques
e gama *f* de las frecuencias acústicas
i gamma *f* delle frequenze acustiche
n gebied *n* van de geluidsfrequenties
d Tonfrequenzbereich *m*

4402 SOUND GATE aud
f fente *f* de lecture

e ventanilla *f* de lectura
i fenditura *f* di lettura,
finestra *f* sonora
n geluidsspleet
d Tonabtastung *f*,
Tonschleuse *f*

4403 SOUND HEAD aud/vr
That part of a cinematograph projector or telecine equipment that contains apparatus for reproducing the sound from the sound track of the film.
f tête *f* sonore
e cabeza *f* sonora
i testina *f* sonora
n geluidskop
d Tonkopf *m*

4404 SOUND/IMAGE TAPE-RECORDER vr
f appareil *m* d'enregistrement sur bande image/son
e aparato *m* de registro sobre cinta imagen/sonido
i apparecchio *m* di registrazione su nastro immagine/suono
n geluids/beeldbandapparaat *n*
d Ton/Bildbandgerät *n*

4405 SOUND INSULATION, stu
SOUND ISOLATION
A measure taken in the studio to ascertain that only the sound from the artist or speaker plus the reverberation reaches the microphone.
f insonorisation *f*,
isolation *f* phonique
e aislamiento *m* fónico,
insonorización *f*
i isolamento *m* acustico
n geluidsisolatie
d Schallabdichtung *f*,
Schallisolation *f*

4406 SOUND INTENSITY aud
The average sound power passing through unit area normal to the direction of propagation at a given point.
f intensité *f* sonore
e intensidad *f* acústica
i intensità *f* acustica
n geluidssterkte
d Schallintensität *f*,
Schallstärke *f*

4407 SOUND INTENSITY DECAY aud
f diminution *f* de l'intensité sonore
e disminución *f* de la intensidad acústica
i diminuzione *f* della intensità acustica
n achteruitgang van de geluidssterkte
d Abklingen *n*,
Ausklingen *n*

4408 SOUND INTENSITY LEVEL, aud
SOUND LEVEL
The ratio, expressed in decibels, of sound power to a zero reference level.
f niveau *m* du son
e nivel *m* del sonido
i livello *m* del suono
n geluidsniveau *n*
d Schallpegel *m*

4409 SOUND INTERCARRIER cpl
DETECTOR
f détecteur *m* à interporteuse
e desmodulador *m* de interportadora
i rivelatore *m* a frapportante,
rivelatore *m* a portante differenziale
n interdraaggolfdemodulator
d Differenzträgerdemodulator *m*

4410 SOUND INTERFERENCE aud/dis
BANDS,
SOUND INTERFERENCE BARS
f barres *pl* d'interférence par le son
e barras *pl* de interferencia por el sonido
i barre *pl* d'interferenza dovute al suono
n geluidsstrepen *pl*
d Tonstreifen *pl*

4411 SOUND LIMITER aud
f limiteur *m* du son
e limitador *m* del sonido
i limitatore *m* del suono
n geluidsbegrenzer
d Schallbegrenzer *m*

4412 SOUND LOCK stu
Sound-proof doors enclosing an area.
f sas *m* son
e labirinto *m* sonoro
i labirinto *m* sonoro
n geluidssluis
d Schallschleuse *f*

SOUND MIXER
see: AUDIO MIXER

4413 SOUND MIXING aud
f mixage *m* son
e mezcla *f* de sonido
i mescolatura *f* di suono
n geluidsmenging
d Tonmischung *f*

4414 SOUND-OFF SWITCH tv
f interrupteur *m* son
e interruptor *m* sonido
i interruttore *m* suono
n pauzeschakelaar
d Pausenschalter *m*,
Tonausschalter *m*

4415 SOUND ON VISION, dis
SOUND-PICTURE SPOILATION
Descriptive term for interference between sound and vision signals in a TV receiver, manifested as intermittent horizontal black and white streaks in the picture.
f son *m* dans l'image
e sonido *m* en la imagen
i suono *m* nell'immagine
n geluid *n* in het beeld
d Ton *m* im Bild,
Tonstreifen *pl*

4416 SOUND PERCEPTION aud
f sensation *f* du son

e sensación *f* del sonido
i sensazione *f* del suono
n geluidswaarneming
d Schallwahrnehmung *f*

4417 SOUND PRELISTEN aud
Arrangement which enables an operator to check the sound produced before broadcasting or recording.
f préécoute *f*
e comprobación *f* auditiva previa
i prova *f* auditiva previa
n vooraf horen *n*
d Vorabhören *n*,
Vorhören *n*

4418 SOUND QUALITY aud
f qualité *f* acoustique
e calidad *f* acústica
i qualità *f* acustica
n kwaliteit van de akoestiek
d Qualität *f* der Akustik

SOUND RECORDER
see: AUDIO TAPE RECORDER

4419 SOUND RECORDING aud
f enregistrement *m* de son
e registro *m* de sonido
i registrazione *f* di suono
n geluidsopname
d Tonaufnahme *f*

4420 SOUND RECORDING SYSTEM aud
The system comprising transducers and associated equipment needed for storing sound in a form suitable for subsequent reproduction.
f système *m* d'enregistrement sonore
e sistema *m* de grabación sonora,
sistema *m* de registro sonoro
i sistema *m* di registrazione sonora
n geluidsopnamesysteem *n*
d Tonaufnahmesystem *n*

4421 SOUND REINFORCEMENT aud/stu
f amplification *f* du son
e amplificación *f* del sonido
i amplificazione *f* del suono
n geluidsversterking
d Schallverstärkung *f*

SOUND REJECTION (GB)
see: REJECTION OF THE ACCOMPANYING SOUND

4422 SOUND REPRODUCING SYSTEM aud
The system comprising transducers and equipment for sound reproduction.
f système *m* de reproduction sonore
e sistema *m* de reproducción sonora
i sistema *m* di riproduzione sonora
n geluidsweergavesysteem *n*
d Tonwiedergabesystem *n*

4423 SOUND SCREEN stu
f écran *m* sonore
e pantalla *f* acústica
i schermo *m* acustico
n geluidsscherm *n*
d Schallschirm *m*

SOUND SIGNAL
see: AUDIO SIGNAL

4424 SOUND STABILITY ac
f stabilité *f* sonore
e estabilidad *f* sonora
i stabilità *f* sonora
n geluidsstabiliteit
d Tonstabilität *f*

4425 SOUND TAKE-OFF aud
Point in a receiver circuit at which the sound signal is separated from the composite received signal.
f point *m* de dérivation du signal son
e punto *m* de ramificación de la señal sonido
i punto *m* di derivazione del segnale suono
n aftappunt *n* van het geluidssignaal
d Abzweigstelle *f* des Tonsignals

4426 SOUND TALK-BACK aud
A two-way communication between sound control and the microphone boom operators on the studio floor.
f interphone *m* du studio
e interfono *m* del estudio
i interfono *m* dello studio
n studioruggespraak
d Studiorücksprache *f*

4427 SOUND TRACK aud/rec/tv
The narrow band of magnetized surface of magnetic tape which carrier impressions corresponding to the characteristics of recorded sound.
f piste *f* sonore
e pista *f* sonora
i pista *f* sonora
n geluidsspoor *n*
d Schallspur *f*

SOUND TRACK ADVANCE
see: SLIPPAGE OF SOUND TO PICTURE

4428 SOUND TRANSFER aud
Duplicated copy of a sound track.
f copie *f* de piste sonore
e copia *f* de pista sonora
i copia *f* di pista sonora
n kopie van een geluidsspoor
d Kopie *f* einer Tonspur

SOUND TRANSMITTER
see: AUDIO TRANSMITTER

4429 SOUND TRAP aud/tv
A filter used in TV receivers to reduce interference of the vision signal by the sound signal.
f circuit *m* atténuateur de la porteuse son,
piège *m* du son
e trampa *f* del sonido
i trappola *f* del suono
n geluidssperkring
d Tonfalle *f*,
Tonsperrkreis *f*

4430 SOUND TRAVEL aud
f voie *f* du son
e trayecto *m* del sonido
i percorso *m* del suono
n geluidsweg
d Schallweg *m*, Tonweg *m*

4431 SOUND TREATMENT stu
f traitement *m* acoustique
e tratamiento *m* acústico
i trattamento *m* acustico
n akoestische behandeling
d akustische Massnahmen *pl*

4432 SOUND VOLUME aud
Magnitude of a complex audio-frequency wave in an electrical circuit measured with a specified instrument, i.e. a volume indicator.
f volume *m* acoustique
e volumen *m* acústico
i volume *m* acustico
n geluidsvolume *n*
d Klangfülle *f*,
Schallvolumen *n*

4433 SPACE PATTERN dis
A geometric pattern in a test chart, used to measure geometric distortion in a TV equipment.
f mire *f* de linéarité
e imagen *f* de prueba de linealidad
i figura *f* di prova di linearità
n lineariteitstoetsbeeld *n*
d Linearitätstestbild *n*

4434 SPACE STATION tv
A communication station located on an object that is beyond, or intended to go beyond the major portion of the earth's atmosphere.
f station *f* orbitale,
station *f* spatiale
e estación *f* espacial
i stazione *f* spaziale
n ruimtestation *n*
d Raumstation *f*

4435 SPARKING crt
Internal disturbance in cathode-ray tubes from the anode to the low voltage electrode nearest to it.
f décharge *f* disruptive
e chisporroteo *m*,
descarga *f* disruptiva
i formazione *f* di scintille
n vonken *n*
d Funken *n*,
Funkenstörung *f*

4436 SPATIAL FREQUENCY cpl
A frequency which indicates the number of times a pattern is repeated in a given distance.
f fréquence *f* spatiale
e frecuencia *f* espacial
i frequenza *f* spaziale
n ruimtelijke frequentie
d raümliche Frequenz *f*

4437 SPEAKING CLOCK stu
Auxiliary sound source of the studio.
f horloge *f* parlante
e reloj *m* parlante
i orologio *m* parlante
n akoestische tijdmeldingsklok
d Zeitansage-Uhr *f*

4438 SPECIAL EFFECTS rec/rep/tv
Those effects which are based on the technique of electronic picture insertion.
f effets *pl* de trucage,
effets *pl* spéciaux
e efectos *pl* especiales
i effetti *pl* speciali
n speciale effecten *pl*
d Spezialeffekte *pl*,
Trickeffekte *pl*

4439 SPECIAL EFFECTS AMPLIFIER tv
An amplifier which provides a suitable form of signal for keying or switching purposes in the production of electronic special effects.
f amplificateur *m* d'effets spéciaux
e amplificador *m* de efectos especiales
i amplificatore *m* d'effetti speciali
n versterker van speciale effecten
d Trickmischer *m*

4440 SPECIAL EVENTS tv
Events to be broadcast of special importance, e.g. elections.
f événements *pl* spéciaux
e eventos *pl* especiales
i avvenimenti *pl* speciali
n belangrijke gebeurtenissen *pl*,
buitengewone gebeurtenissen *pl*
d wichtige Ereignisse *pl*

4441 SPECTRAL CHARACTERISTICS ctv
In signal formation equipment, the set of spectral responses of the colo(u)r separation channels with respect to wavelength.
f caractéristiques *pl* spectrales
e características *pl* espectrales
i caratteristiche *pl* spettrali
n spectrale karakteristieken *pl*
d spektrale Charakteristiken *pl*

4442 SPECTRAL COLO(U)R, ct
SPECTRUM COLO(U)R
A colo(u)r that appears in the spectrum of white light.
f couleur *f* spectrale
e color *m* espectral
i colore *m* spettrale
n spectrale kleursoort
d Spektralfarbe *f*

4443 SPECTRAL DISTRIBUTION, ct
SPECTRAL ENERGY DISTRIBUTION,
SPECTRAL POWER DISTRIBUTION
f répartition *f* relative du rayonnement
e repartición *f* relativa de la radiación
i ripartizione *f* relativa della radiazione
n spectrale energieverdeling
d relative spektrale Strahlungsverteilung *f*

4444 SPECTRAL DISTRIBUTION GRAPH cpl
A plot of relative energy versus wavelength.
f courbe *f* de répartition spectrale
e curva *f* de repartición espectral
i curva *f* di ripartizione spettrale
n spectrale verdelingskarakteristiek
d spektrale Verteilungscharakteristik *f*

4445 SPECTRAL HUE ct
A hue that can be defined by a point on the spectral locus of a chromaticity diagram.
f teinte *f* dominante
e matiz *m* dominante
i tinta *f* dominante
n dominerende kleurtoon
d dominierende Farbart *f*

4446 SPECTRAL PURITY OF A COLO(U)R ct
f pureté *f* spectrale de couleur
e pureza *f* espectral de color
i purezza *f* spettrale di colore
n spectrale kleurzuiverheid
d spektrale Farbreinheit *f*

4447 SPECTRAL REFLECTANCE (US), SPECTRAL REFLECTION FACTOR (GB) ct
The ratio of the spectral concentration of the reflected luminous flux to that of the incident flux for a given wavelength.
f facteur *m* de réflexion spectrale
e factor *m* de reflexión espectral
i fattore *m* di riflessione spettrale
n spectrale reflectiefactor
d spektraler Reflexionsgrad *m*

4448 SPECTRAL RESPONSE, SPECTRAL SENSITIVITY CHARACTERISTIC tv
The relation between the radiant sensitivity and the wavelength of the incident radiation of a camera tube under specified conditions of irradiation.
f courbe *f* de sensibilité spectrale
e curva *f* de sensibilidad espectral
i curva *f* di sensibilità spettrale
n spectrale gevoeligheidskromme
d Farbempfindlichkeitskurve *f*

4449 SPECTRAL SENSITIVITY ct
f sensibilité *f* spectrale
e sensibilidad *f* espectral
i sensibilità *f* al colore
n spectrale gevoeligheid
d Farbempfindlichkeit *f*

4450 SPECTRUM LOCUS ct
The locus of points representing the chromaticities of spectrally pure stimuli in a chromaticity diagram.
f lieu *m* des couleurs spectrales
e lugar *m* de los colores espectrales
i luogo *m* dei colori spettrali
n kromme van de spectrale kleursoorten
d Spektralkurve *f*

SPECULAR REFLECTION
see: DIRECT REFLECTION

4451 SPEECH aud
f parole *f*
e palabra *f*, voz *f*
i parola *f*, voce *f*
n spraak
d Sprache *f*

4452 SPEECH CHANNEL, TELEPHONY CHANNEL aud
Channel suitable for telephony, with a frequency range of 250 to 3.400H_z maximum.
f bande *f* de conversation, liaison *f* téléphonique
e banda *f* de conversación, canal *m* telefónico, comunicación *f* telefónica
i banda *f* di conversazione, canale *m* telefonico, collegamento *m* telefonico
n spreekverbinding
d Sprechweg *m*

SPEECH COIL (GB)
see: MOVING COIL

4453 SPEECH/MUSIC BUTTON rep/tv
f touche *f* parole/musique
e tecla *f* palabra/música
i tasto *m* parola/musica
n spraak/muziektoets
d Sprache/Musikknopf *m*

4454 SPEECH/MUSIC SWITCH aud
f commutateur *m* parole/musique
e conmutador *m* palabra/música
i commutatore *m* parola/musica
n spraak/muziekomschakelaar
d Sprache/Musikumschalter *m*

4455 SPEED SELECTOR SWITCH vr
f commutateur *m* de vitesse
e conmutador *m* de velocidad
i commutatore *m* di velocità
n snelheidsomschakelaar
d Geschwindigkeitsumschalter *m*

4456 SPEED TUNING CIRCUIT rep/tv
f montage *m* de syntonisation accélérée
e circuito *m* de sintonía acelerada
i circuito *m* di sintonizzazione accelerata
n snelafstemschakeling
d Schnellabstimmschaltung *f*

4457 SPHERICAL ABERRATION opt
f aberration *f* sphérique
e aberración *f* esférica
i aberrazione *f* sferica
n sferische aberratie
d sphärische Aberration *f*

4458 SPHERICAL FACEPLATE crt
A TV picture tube faceplate that is a portion of a spherical surface.
f fenêtre *f* courbée
e ventana *f* curvada
i finestra *f* curvata
n gewelfd venster *n*
d gewölbter Schirmträger *m*

4459 SPIDER BOX stu
A junction box in which two, four or more lamps may be quickly plugged for studio or location use.
f douille *f* multiple
e portalámpara *m* múltiple
i portalampada *m* multiplo
n meervoudige lamphouder
d Mehrfachfassung *f*

SPIKE
see: PULSE SPIKE

SPIRAL DISK
see: SCANNING DISK

SPIRAL DISTORTION
see: S-DISTORTION

SPIRAL SCANNING
see: HELICAL RECORDING

4460 SPLICE fi/tv/vr
f collage *m*
e empalme *m*
i incollatura *f*
n las
d Klebestelle *f*

4461 SPLICE (TO) vr
f coller v
e empalmar v
i incollare v
n lassen v, lijmen v, plakken v
d kleben v

SPLICE BUMP
see: BLOOP

4462 SPLICELESS TAPE vr
f bande *f* vidéo sans collage
e cinta *f* video sin empalme
i nastro *m* video senza incollatura
n lasloze beeldband
d Bildband *n* ohne Klebstelle

4463 SPLICER fi/tv
f colleuse *f*
e empalmadora *f*
i incollatrice *f*
n lasapparaat *n*
d Klebepresse *f*,
Schneidelehre *f*

4464 SPLIT-BEAM CATHODE-RAY TUBE crt
A tube containing only one electron gun, but with the beam subdivided so that two traces are obtained on the screen.
f tube *m* cathodique à faisceau divisé
e tubo *m* catódico con haz dividido
i tubo *m* catodico con fascio diviso
n katodestraalbuis met gesplitste bundel
d Katodenstrahlröhre *f* mit Strahlspaltung

4465 SPLIT FIELD PICTURE tv
Result of electrically dividing a picture field into two parts, each part being supplied with a different picture from separate sources.
f trame *f* à deux demi-images différentes
e cuadro *m* de dos semiimágenes diferentes,
imagen *f* rajada por la mitad y desplazada
i trama *f* a due semiimmagini differenti
n raster *n* met twee verschillende halve beelden
d geteiltes Halbbild *n*,
Vollbild *n* mit zwei verschiedenen Teilbildern

4466 SPLIT FOCUS opt/rec/tv
Focus on a point between two objects, widely separated in their distance from the camera, so that both are included within the depth of field with approximately equal definition.
f ajustage *m* du foyer à un point entre deux objets
e ajuste *m* del foco a un punto entre dos objetos,
foco *m* dividido
i aggiustaggio *m* del fuoco ad un punto tra due oggetti
n brandpunt *n* tussen twee ver uit elkaar gelegen voorwerpen
d Brennpunkt *m* zwischen zwei weit auseinanderliegenden Objekten

4467 SPLIT IMAGE, dis
SPLIT PICTURE
Fault due to defectuous synchronization.
f image *f* déchirée,
image *f* découpée
e imagen *f* despuesta,
imagen *f* dividida
i immagine *f* divisa
n gedeeld beeld *n*
d geteiltes Bild *n*

SPLIT IMAGE (US)
see: INTERMEDIATE MULTIPLE

4468 SPLIT SCREEN EFFECT tv
Effect shot where two separate image areas are exposed or printed on each frame; in TV this effect is produced by combining the output of two picture sources.
f découpage *m* électronique
e efecto *m* de pantalla dividida
i presentazione *f* contemporanea sullo schermo di due immagini
n deelbeeldtrucage
d Tricküberblendung *f*

4469 SPLIT SCREEN VIEWING tv
f télévision *f* à écran divisé
e televisión *f* de pantalla dividida
i televisione *f* a schermo diviso
n televisie met gedeeld scherm
d Fernsehen *n* mit geteiltem Schirm

4470 SPLITTING ELECTRODE crt
f électrode *f* diviseuse du faisceau
e electrodo *m* divisor del haz
i elettrodo *m* divisore del fascio
n bundelverdelende elektrode,
vertakkingselektrode
d strahlspaltende Elektrode *f*,
strahlteilende Elektrode *f*

4471 SPONSOR tv
The advertiser who pays part or all of the cost of a TV program(me).
f commanditaire *m*
e patrocinador *m*
i patrocinante *m*
n sponsor
d Sponsor *m*

4472 SPONSORED EMISSION tv
f émission *f* commanditée,
émission *f* patronnée
e emisión *f* patrocinada
i emissione *f* patrocinata
n sponsorzending
d Sponsorsendung *f*

4473 SPONSORED PROGRAM(ME) tv
f programme *m* commandité,
programme *m* patronné
e programa *m* patrocinado
i programma *m* patrocinato
n sponsorprogramma *n*
d Sponsorprogramm *n*

SPONSORED TELEVISION (US)
see: COMMERCIAL TELEVISION

4474 SPOT crt
1. The small area of the screen of a cathode-ray tube instantaneously affected by the impact of the electron beam.
2. The light produced by the impact of the slender beam of electrons on the fluorescent screen of a cathode-ray tube.
f point *m* lumineux,
spot *m*
e punto *m* luminoso
i macchia *f* luminosa,
punto *m* luminoso,
spot *m*
n lichtstip,
stip
d Fleck *m*,
Leuchtfleck *m*,
Punkt *m*

4475 SPOT vr
f élément *m* de piste sonore
e elemento *m* de pista sonora
i elemento *m* di pista sonora
n geluidsspoorelement *n*
d Spurelement *n*

4476 SPOT,
SPOTLAMP,
SPOTLIGHT stu
Focusable lamp, the beam of which can be narrowed to a spot to provide a key light.
f lampe *f* à lumière ponctuelle,
projecteur *m* à faisceau concentré,
spot *m*
e lámpara *f* para luz concentrada
i lampada *f* a luce puntiforme
n spotlight *n*,
zoeklamp
d Punktlicht *n*,
Punktstrahllampe *f*

SPOT
see: COMMERCIAL

4477 SPOT ACCELERATION crt
f accélération *f* du spot
e aceleración *f* del punto luminoso
i accelerazione *f* del punto luminoso
n lichtstipversnelling
d Punktbeschleunigung *f*

SPOT BRIGHTNESS
see: BRIGHTNESS OF THE SPOT

4478 SPOT DISPLACEMENT crt
f déplacement *m* du spot
e desplazamiento *m* del punto luminoso
i spostamento *m* del punto luminoso
n lichtstipverschuiving
d Leuchtfleckverschiebung *f*

4479 SPOT DISTORTION crt
f distorsion *f* du spot
e distorsión *f* del punto luminoso
i distorsione *f* del punto luminoso
n lichtstipvervorming
d Fleckverzerrung *f*

4480 SPOT LIMITER dis
f limiteur *m* de la visibilité d'interférence impulsive
e limitador *m* de la visibilidad de interferencia impulsiva
i limitatore *m* di visibilità d'interferenza impulsiva
n storingsbegrenzer,
zichtbaarheidsbegrenzing van de impulsieve storing
d Sichtbarheitsbegrenzung *f* der impulsiven Störung

4481 SPOT SHAPE CORRECTION crt
Carried out by means of an aperture-correcting amplifier.
f correction *f* de la configuration du spot
e corrección *f* de la configuración del punto luminoso
i correzione *f* della configurazione del punto luminoso
n lichtstipvormcorrectie
d Leuchtfleckformkorrektur *f*

4482 SPOT SIZE crt
Of a cathode-ray tube, the effective size of the spot.
f diamètre *m* du spot,
ouverture *f* du faisceau
e abertura *f* del haz, tamaño *m* del punto
i **apertura *f* del fascio, diametro *m* del punto**
n bundelopening, lichtstipafmeting
d Bündelöffnung *f*, Leuchtfleckgrösse *f*

4483 SPOT SPEED crt
In TV, the product of the number of spots in the scanning line by the number of scanning lines per second.
f vitesse *f* d'analyse
e velocidad *f* de exploración
i velocità *f* d'analisi
n aftastsnelheid
d Abtastgeschwindigkeit *f*

4484 SPOT SUPPRESSOR dis
Circuit used with cathode-ray devices to prevent burning effects from an undeflected spot on switchoff.
f suppresseur *m* de tache de brûlage
e supresor *m* de mancha de quemadura
i soppressore *m* di macchia di bruciatura
n brandvlekonderdrukker
d Brennfleckunterdrücker *m*

4485 SPOT WOBBLE crt
An oscillatory movement imparted to the spot in the direction of the field scan in order to broaden the lines and render them less distinct.
f vobulation *f* du spot
e vobulación *f* del punto
i oscillazione *f* del punto
n lichtstipwobbel
d Strahlwobbelung *f*, Zeilenwobbelung *f*

4486 SPOTTINESS dis
The effect on the TV screen caused by the variation of the instantaneous light value of the reproduced image.
f parasites *pl* en forme de tâches
e pantalla *f* moteada, parásitos *pl*
i il essere *m* macchiettato, mancanza *f* d'uniformità
n vlekkigheid
d Fleckeneffekt *m*

SPREAD LIGHT
see: BROAD LIGHT

SPROCKET HOLE
see: PERFORATION

4487 SPROCKET-HOLE MODULATION dis/vr
An irregular motion of the tape due to the sprockets by which the tape is driven.
f entraînement *m* irrégulier par galets
e alimentación *f* irregular por rodillos dentados
i alimentazione *f* irregolare per tamburi dentati
n onregelmatige tandrolaandrijving
d unregelmässiger Zahntrommelantrieb *m*

4488 SPROCKETED TAPE vr
f bande *f* à perforations
e cinta *f* con perforaciones
i nastro *m* a perforazioni
n perforatieband
d Lochband *n*

4489 SPUR BAND cpl
Band of frequencies in a multi-channel system which is separated out to feed a spur of the main network.
f bande *f* de fréquence pour ligne dérivée
e banda *f* de frecuencia para línea derivada
i banda *f* di frequenza per linea di derivazione
n frequentieband voor aftakking
d Frequenzband *n* für Abzweigleitung

4490 SPURIOUS FREQUENCY aud
The frequency of the sprocket-hole modulation.
f fréquence *f* irrégulière
e frecuencia *f* irregular
i frequenza *f* irregolare
n onregelmatige frequentie
d unregelmässige Frequenz *f*

SPURIOUS OSCILLATIONS
see: PARASITIC OSCILLATIONS

4491 SPURIOUS RADIATION dis
Any unintended radiation from a transmitter at any frequencies other than those appropriate to its intended purpose.
f émission *f* sur une bande interdite, rayonnement *m* parasite
e radiación *f* parásita
i irradiazione *f* parassita
n ongewenste uitstraling, wilde straling
d Störstrahlung *f*, unerwünschte Ausstrahlung *f*, wilde Strahlung *f*

4492 SPURIOUS SIGNAL cpl
Unrequired signal obtained fortuitously by some imperfect property of the device employed.
f signal *m* parasite
e señal *f* parásita
i segnale *m* parassita
n stoorsignaal *n*
d Störsignal *n*

4493 SQUARE WAVE cpl
Wave of rectangular form.
f onde *f* carrée, onde *f* rectangulaire
e onda *f* cuadrada, onda *f* rectangular
i onda *f* rettangolare
n kanteelgolf
d Rechteckwelle *f*

4494 SQUARE-WAVE ANALYZER tv
f analyseur *m* d'onde rectangulaire
e analizador *m* de onda rectangular
i analizzatore *m* d'onda rettangolare
n kanteelgolfanalysator
d Rechteckwellenanalysierstelle *f*

4495 SQUARE-WAVE GENERATOR cpl
f générateur *m* d'ondes rectangulaires
e generador *m* de ondas rectangulares
i generatore *m* d'onde rettangolari
n kanteelgolfgenerator
d Rechteckwellengenerator *m*

4496 SQUARE-WAVE RESPONSE crt
Of a camera tube, the ratio of the peak-to-peak signal amplitude given by a test pattern consisting of alternate black and white bars of equal width to the difference in signal between large-area blacks and large-area whites having the same illumination as the black and white bars in the pattern.

f réponse *f* d'onde rectangulaire
e respuesta *f* de onda rectangular
i risposta *f* d'onda rettangolare
n responsie op een kanteelgolf
d Einschwingverhalten *n*

4497 SQUARE-WAVE RESPONSE CHARACTERISTIC crt
f caractéristique *f* de réponse d'onde rectangulaire
e característica *f* de respuesta de onda rectangular
i caratteristica *f* di risposta d'onda rettangolare
n kanteelgolfresponsiekromme
d Einschwingcharakteristik *f*

4498 SQUARE-WAVE VOLTAGE cpl
f tension *f* crénelée,
tension *f* d'onde rectangulaire
e tensión *f* de onda rectangular
i tensione *f* d'onda rettangolare
n kanteelspanning
d Rechteckspannung *f*,
Zinnenspannung *f*

4499 SQUARER,
SQUARING CIRCUIT cpl
Symmetrical limiter or slicer used to convert a sinusoidal wave into a square wave.
f circuit *m* conformateur
e circuito *m* de cuadratura
i circuito *m* squadratore
n kanteelgolfvormer
d Rechteckumformer *m*

4500 SQUARING cpl/tv
f conformation *f* d'onde rectangulaire
e conformación *f* de onda rectangular
i conformazione *f* d'onda rettangolare
n dubbelknippen *n*,
kanteelgolfvorming
d Rechteckformung *f*

4501 SQUEGGING cpl
Mode of oscillation of an oscillator when operated under certain conditions, e.g. with excessive resistance in the grid circuit; used in cathode-ray tube timebase generator.
f mode *m* d'oscillation commandée par oscillations de relaxation
e generación *f* de oscilaciones de saturación
i generazione *f* d'oscillazioni bloccate periodicamente
n hikken *n*
d Pendelung *f*

4502 SQUEGGING OSCILLATOR cpl
f oscillateur *m* commandé par oscillations de relaxation
e oscilador *m* de saturación
i oscillatore *m* bloccato periodicamente
n hikoscillator
d Pendeloszillator *m*,
Sperrschwinger *m*

SQUELCH CIRCUIT
see: INTERCARRIER NOISE SUPPRESSOR

SSSC
see: SINGLE-SIDEBAND SUPPRESSED-CARRIER SYSTEM

4503 STABILIZED GLASS crt
Glass which darkens much less than ordinary glass when it is irradiated.
f verre *m* libre de décoloration
e vidrio *m* no descolorante
i vetro *m* non decolorante
n niet-verkleurend glas *n*
d entfärbungsfreies Glas *n*

4504 STABILIZED SUPPLY VOLTAGE cpl
f tension *f* d'alimentation stabilisée
e tensión *f* de alimentación estabilizada
i tensione *f* d'alimentazione stabilizzata
n gestabiliseerde voedingsspanning
d stabilisierte Betriebsspannung *f*

STACKED HEADS
see: IN-LINE HEADS

4505 STAGE,
STUDIO FLOOR rec/tv
That part of the studio where the action takes place.
f plateau *m*, scène *f*
e escena *f*, plataforma *f*
i piattaforma *f*, scena *f*
n toneel *n*
d Bühne *f*

4506 STAGE DIRECTION stu/th/tv
f équipe *m* de réalisation
e dirección *f* de escena
i squadra *f* di regia
n spelleiding,
toneelregie
d Regieführung *f*,
Regiestab *m*

STAGE HAND (GB)
see: GRIP

4507 STAGE MANAGER,
STUDIO MANAGER fi/th/tv
f chef *m* de plateau
e traspunte *m*
i direttore *m* di scena
n inspiciënt
d Inspizient *m*

4508 STAGGER-TUNED cpl
f à circuits de syntonisation décalés
e sintonizado en escalera
i sintonizzato in scaglione
n met verscherfde afstemming
d gegen Bandmitte versetzt,
gegeneinander verstimmt,
gestaffelt abgestimmt

4509 STAGGERED CIRCUITS cpl
f circuits *pl* décalés
e circuitos *pl* escalonados
i circuiti *pl* scaglionati

n verscherfde kringen *pl*
d verstimmte Kreise *pl*

4510 STAGGERED FREQUENCIES cpl
f fréquences *pl* décalées
e frecuencias *pl* escalonadas
i frequenze *pl* scaglionate
n verscherfde frequenties *pl*
d gegeneinander versetzte Frequenzen *pl*

STAGGERED HEADS
see: OFFSET HEADS

4511 STAGGERED PAIR cpl
f deux circuits *pl* décalés
e dos circuitos *pl* escalonados
i due circuiti *pl* scaglionati
n twee verscherfde kringen *pl*
d zwei verstimmte Kreise *pl*

STAGGERED STEREOPHONIC TAPE
see: OFFSET STEREOPHONIC TAPE

4512 STAGGERING cpl
f décalage *m*
e escalonamiento *m*
i scaglione *m*
n verscherving
d Verstimmung *f*

4513 STAIN crt/dis
A fault sometimes present in camera or picture tubes which shows itself as a blemish on the reproduced picture.
f tache *f*
e mancha *f*
i macchia *f*
n vlek
d Fleck *m*

4514 STAIRCASE GENERATOR cpl
f générateur *m* de signaux dégradés
e generador *m* de señales degradadas
i generatore *m* di segnali degradati
n grijstrapsignaalgenerator
d Graukeilsignalgenerator *m*

4515 STAIRCASE SIGNAL tv
A waveform consisting of a series of discrete steps resembling a staircase.
f signal *m* dégradé
e señal *f* degradada
i segnale *m* degradato
n grijstrapsignaal *n*
d Graukeilsignal *n*

4516 STANDARD ILLUMINANT ct
An agreed light source specified in such a way that its energy distribution is reproducible.
f étalon *m* A,
illuminant *m* étalon
e iluminante *m* patrón,
patrón *m* A
i illuminante *m* campione
n standaardilluminant
d Normalilluminant *m*

4517 STANDARD LIGHT SOURCE ct
f source *f* de lumière étalon
e fuente *f* luminosa patrón
i sorgente *f* luminosa campione
n standaardlichtbron
d Normallichtquelle *f*

4518 STANDARD LUMINANCE RESPONSE ct
f réponse *f* de luminance normale
e respuesta *f* de luminancia patrón
i risposta *f* di luminanza campione
n normale luminantieresponsie
d normale Ansprechcharakteristik *f* für Luminanz

4519 STANDARD MEASURING SIGNAL cpl
A signal of standard form used as a picture signal for measuring the characteristics of a transmission channel.
f signal *m* type
e señal *f* tipo
i segnale *m* di prova
n standaardmeetsignaal *n*
d Prüfsignal *n*

4520 STANDARD OBSERVER ct
A hypothetical observer who requires standard amounts of primaries in a colo(u)r mixture to match every colo(u)r.
f observateur *m* moyen
e observador *m* patrón
i osservatore *m* di riferimento
n standaardwaarnemer
d Normalbeobachter *m*

4521 STANDARD PATTERN tv
f mire *f* étalon
e imagen *f* de prueba patrón
i figura *f* di prova campione
n standaardtoetsbeeld *n*
d Normaltestbild *n*

4522 STANDARD PULSE GENERATOR rec/rep/tv
f générateur *m* de l'impulsion de synchronisation
e generador *m* del impulso de sincronización
i generatore *m* del sincronismo,
generatore *m* dell'impulso di sincronizzazione
n synchronisatie-impulsgenerator
d Normgemischgenerator *m*

4523 STANDARD RECORDING TAPE, STANDARD REFERENCE TAPE vr
A tape containing an original recording of three 200 mil tracks on fully coated film containing a series of spot frequencies from 50 to 12.000 H_z.
f bande *f* référence
e cinta *f* de referencia
i nastro *m* di riferimento
n referentieband
d Bezugsband *n*

4524 STANDARD SELECTOR tv
f sélecteur *m* de système de lignes
e selector *m* de sistema de líneas
i selettore *m* di sistema di linee

n lijnsysteemkiezer
d Zeilensystemwähler *m*

4525 STANDARD STIMULUS ct
f stimulus *m* étalon
e estímulo *m* patrón
i stimolo *m* campione
n standaardstimulus
d Eichreiz *m*

4526 STANDARD SYNC SIGNAL tv
f signal *m* de synchronisation étalon
e señal *f* de sincronización patrón
i segnale *m* di sincronizzazione campione, sincronismo *m* campione
n standaardsynchronisatiesignaal *n*
d Normgemisch *n*

STANDARD TAPE
see: REFERENCE TAPE

4527 STANDARD TELEVISION SIGNAL rep/tv
A signal that conforms to TV transmission standards.
f signal *m* de télévision étalon
e señal *f* televisiva patrón
i segnale *m* televisivo campione
n standaardtelevisiesignaal *n*
d normiertes Fernsehsignal *n*

4528 STANDARDS CONVERSION rep/tv
f conversion *f* du nombre de lignes
e conversión *f* del número de líneas
i conversione *f* del numero di linee
n lijnsysteemomzetting, lijnvertaling, normvertaling
d Normwandlung *f*

4529 STANDARDS CONVERTER rep/tv
f convertisseur *m* du nombre de lignes
e convertidor *m* del número de líneas
i convertitore *m* del numero di linee
n normvertaler
d Normwandler *m*

4530 STANDARDS SWITCHING rep/tv
f commutation *f* des systèmes de lignes
e conmutación *f* de los sistemas de líneas
i commutazione *f* dei sistemi di linee
n systeemschakeling
d Normschaltung *f*

4531 STANDBY EQUIPMENT, STANDBY FACILITIES stu
f appareillage *m* de secours
e equipo *m* de reserva
i apparecchiatura *f* di riserva
n reserveapparatuur
d passive Reserve *f*, Reserveapparatur *f*

4532 STANDBY POWER SOURCES stu
f sources *pl* d'énergie de secours
e fuentes *pl* de energía de reserva
i sorgenti *pl* d'energia di riserva
n reserve-energiebronnen *pl*
d Reserveenergiequellen *pl*

4533 STANDBY PROGRAM(ME) tv
f programme *m* de secours
e programa *m* de emergencia
i programma *m* di soccorso
n reserveprogramma *n*
d Reserveprogramm *n*

4534 STANDBY SIGNAL tv
A cue mark electronically keyed into the picture to provide a warning.
f signal *m* d'avertissement incorporé dans l'image
e señal *f* de aviso incorporada en la imagen
i segnale *m* d'avvertimento incorporato nell'immagine
n in beeld ingebouwd waarschuwingssignaal *n*
d im Bilde eingebautes Achtungssignal *n*

4535 STANDBY TRANSMITTER tv
A transmitter installed and maintained for use during periods when a main transmitter is out of service for maintenance or repair.
f émetteur *m* de secours
e emisor *m* de reserva
i emettitore *m* di riserva
n reservezender
d Ersatzsender *m*, Füllsender *m*

4536 STANDING D.C. COMPONENT cpl
A d.c. or slowly varying component of a picture or video signal which is unrelated to the mean luminance of the televised objects.
f composante *f* continue inutile
e componente *f* continua permanente
i componente *f* continua sovrapposta al segnale immagine
n gelijkspanningsrestcomponent
d überlagerter Gleichstromanteil *m*

4537 START KEY vr
Used to start recording or reproducing.
f touche *f* défilement
e botón *m* de arranque
i pulsante *m* d'avvio, tasto *m* d'inizio
n starttoets
d Starttaste *f*

4538 START LEADER fi/tv
f amorce *f* de début
e cinta *f* de arranque
i nastro *m* d'avvio
n startband
d Startvorspann *m*

4539 START MARK fi/tv
f repère *m* de départ
e marca *f* de arranque
i marca *f* d'avvio
n startmerkteken *n*
d Startmarke *f*

START-STOP MULTIVIBRATOR
see: ONE-CYCLE MULTIVIBRATOR

4540 STARTING HUM tv
f ronflement *m* au démarrage
e zumbido *m* de arranque
i ronzio *m* di messa in circuito
n inschakelbrom
d Einschaltbrumm *m*

4541 STARTING TIME vr
f temps *m* de démarrage
e tiempo *m* de arranque
i tempo *m* d'avviamento
n aanlooptijd
d Anlaufzeit *f*

4542 STATIC BACKGROUND ani
f fond *m* permanent
e fondo *m* permanente
i fondo *m* permanente
n vaste achtergrond
d fester Hintergrund *m*

STATIC CONVERGENCE
see: D.C. CONVERGENCE

4543 STATIC CONVERGENCE CORRECTION crt
f correction *f* de convergence statique
e corrección *f* de convergencia estática
i correzione *f* di convergenza statica
n statische-convergentiecorrectie
d statische Konvergenzkorrektur *f*

4544 STATIC FOCUS crt
The focus of the undeflected electron beam in a cathode-ray tube.
f foyer *m* statique
e foco *m* estático
i fuoco *m* statico
n statisch brandpunt *n*
d statischer Brennpunkt *m*

STATIC MATTE
see: INLAY

4545 STATIC PHASE ERROR dis
f erreur *f* statique de phase
e error *m* estático de fase
i errore *m* statico di fase
n statische fazefout
d statischer Phasenfehler *m*

4546 STATIC PHASE MEASUREMENT cpl
f mesure *f* de l'erreur statique de phase
e medida *f* del error estático de fase
i misura *f* dell'errore statico di fase
n meting van de statische fazefout
d Messung *f* des statischen Phasenfehlers

4547 STATION tv
A location at which TV equipment is installed.
f station *f*
e estación *f*
i stazione *f*
n station *n*
d Station *f*, Stelle *f*

4548 STATION ANNOUNCEMENT tv
f annonce *f* de début d'émission
e anuncio *m* de comienzo de emisión
i annunzio *m* d'inizione d'emissione
n aankondiging van het begin van de uitzending
d Stationsansage *f*

4549 STATION BREAK, STATION BREAKDOWN dis/tv
An interruption in the transmission of a program(me).
f interruption *f* de l'émission
e interrupción *f* de la emisión
i difetto *m* di continuità
n zendonderbreking
d Sendeunterbrechung *f*

4550 STATION CAPTION tv
f indicatif *m* de la station
e señal *f* indicativa de la estación
i segnale *m* caratteristico della stazione
n stationskenteken *n*
d Stationskennzeichen *n*

4551 STATION DIA tv
f diapositive *f* d'identification de la station
e diapositiva *f* de identificación de la estación
i diapositiva *f* d'identificazione della stazione
n stationsdia *n*
d Stationsdia *n*

4552 STATION SYNCHRONIZATION tv
The synchronization of different and often remote TV picture sources so that transitions between them by way of mixes and special effects can be realized.
f synchronisation *f* de stations de prise de vues
e sincronización *f* de estaciones de toma de vistas
i sincronizzazione *f* di stazioni di ripresa
n synchronisatie van opnamestations
d Aufnahmestellensynchronisierung *f*

4553 STATION TIMING tv
A process for providing a signal which is exactly to specification, by delaying those signals which would otherwise arrive too early at the outgoing transmission line.
f minutage *m* des signaux d'information
e regulación *f* del tiempo de las señales de información
i regolazione *f* del tempo dei segnali d'informazione
n tijdsduurregeling van de informatiesignalen
d Zeitdauerregelung *f* der Informationssignale

4554 STEADY-STATE AMPLITUDE cpl
f amplitude *f* en régime permanent
e amplitud *f* en régimen permanente
i ampiezza *f* in regime stazionario
n amplitude in de blijvende toestand
d Amplitude *f* bei eingeschwungenem Zustand

4555 STEADY-STATE CHARACTERISTIC, STEADY-STATE RESPONSE cpl
The response to a defined continuous signal.
f réponse *f* stationnaire
e respuesta *f* estacionaria
i risposta *f* stazionaria
n responsie in de blijvende toestand
d Ansprechen *n* bei eingeschwungenem Zustand

STEM
see: FOOT

4556 STEP-BY-STEP MOTION fi/tv
f défilement *m* saccadé
e movimiento *m* paso a paso
i movimento *m* passo a passo
n stapsgewijze voortbeweging
d Schrittlauf *m*

4557 STEP-BY-STEP SWITCH fi/tv
f mécanisme *m* de transport saccadé
e mecanismo *m* para movimiento paso a paso
i meccanismo *m* per movimento passo a passo
n schakelaar voor stapsgewijze voortbeweging
d Schrittschaltwerk *n*

4558 STEP WEDGE opt
f coin *m* à échelons
e cuña *f* escalonada
i cuneo *m* a gradini
n trapvormige wig
d Belichtungskeil *m*, Stufenkeil *m*

4559 STEREO-EIGHT-CASSETTE vr
A cassette for recordings on an eight-track tape.
f cassette *f* stéréophonique à huit pistes sur la bande
e caseta *f* estereofónica con ocho pistas sobre la cinta
i cassetta *f* stereofonica con otto piste sul nastro
n stereofonische cassette met 8 sporen op de band
d stereophonische Kassette *f* mit 8 Spuren auf dem.Band

4560 STEREO-MIXER vr
f pupitre *m* de mélange stéréophonique
e pupitre *m* de mezclado estereofónico
i tavolo *m* di mescolatura stereofonica
n stereomenglessenaar
d Stereomischpult *n*

4561 STEREOPHONIC REPRODUCTION aud/vr
f reproduction *f* stéréophonique
e reproducción *f* estereofónica
i riproduzione *f* stereofonica
n stereofonische weergave
d stereophonische Wiedergabe *f*

4562 STEREOPHONIC SOUND aud
f son *m* stéréophonique
e sonido *m* estereofónico
i suono *m* stereofonico
n stereofonisch geluid *n*
d Raumton *m*

4563 STEREOPHONIC SOUND TAPE, STEREOPHONIC TAPE aud/vr
f bande *f* sonore stéréophonique
e cinta *f* sonora estereofónica
i nastro *m* sonoro stereofonico
n stereofonische geluidsband
d stereophonisches Tonband *n*

4564 STEREO-RECORDED TAPE aud/vr
Recorded magnetic tape having two separate recordings, one for each channel of a stereo sound system.
f bande *f* à enregistrement stéréophonique
e cinta *f* de registro estereofónico
i nastro *m* a registrazione stereofonica
n band met stereo-opname, stereofonieband
d Stereophonieband *n*

4565 STEREOSCOPIC EFFECT tv
f effet *m* stéréoscopique
e efecto *m* estereoscópico
i effetto *m* stereoscopico
n dieptewerking, ruimte-effect *n*, stereoscopisch effect *n*
d Raumeffekt *m*, stereoskopische Wirkung *f*, Tiefenwirkung *f*

4566 STEREOSCOPIC TELEVISION, STEREOTELEVISION tv
f stéréotélévision *f*
e estereotelevisión *f*
i stereotelevisione *f*
n driedimensionale televisie, stereotelevisie
d dreidimensionales Fernsehen *n*, plastisches Fernsehen *n*, stereoskopisches Fernsehen *n*

4567 STEREO-TAPE RECORDER aud/vr
A magnetic tape recorder having two stacked playback heads, for reproduction of stereo-recorded tape.
f appareil *m* d'enregistrement stéréophonique
e aparato *m* de registro estereofónico
i apparecchio *m* di registrazione stereofonica
n stereo-opneemapparaat *n*
d Stereotonbandgerät *n*

STICKING
see: BURN

STICKING PICTURE
see: BURNED-IN IMAGE

4568 STICKING POTENTIAL, STICKING VOLTAGE crt
Of a cathode-ray tube with a non-aluminized screen, the voltage of the

screen, relative to the cathode, above which the secondary emission coefficient of the screen material would be less than unity, that is, the limiting screen voltage.
f tension *f* de blocage, tension *f* limite
e tensión *f* de bloqueo, tensión *f* límite
i tensione *f* di bloccaggio, tensione *f* limite
n blokkeerspanning, grensspanning
d Grenzspannung *f*, Sperrspannung *f*

STILL
see: NON-ANIMATED PICTURE

4569 STIMULATING FREQUENCY cpl
Used in lasers and masers to stimulate emission of radiation.
f fréquence *f* d'excitation
e frecuencia *f* de excitación
i frequenza *f* d'eccitazione
n aanstootfrequentie
d Anregungsfrequenz *f*

4570 STIMULUS ct
Something that excites an organ or tissue to a specific activity or function.
f stimulus *m*
e estímulo *m*
i stimolo *m*
n stimulus
d Reiz *m*

4571 STOP COIL dis
Large number of turns of coaxial cable wound upon a magnetic core to filter supply-frequency interference from a video signal.
f bobine *f* d'arrêt en câble coaxial
e bobina *f* de freno en cable coaxial
i bobina *f* d'arresto in cavo coassiale
n smoorspoel met coaxiale kabel
d Koaxialkabeldrosselspule *f*

STOP FRAME
see: FREEZE FRAME

4572 STOP KEY vr
Used to stop recording, reproducing, fast winding and fast rewinding.
f touche *f* arrêt
e botón *m* de parada
i pulsante *m* d'arresto
n stoptoets
d Stopptaste *f*

4573 STOPPING TIME vr
f temps *m* d'arrêt
e tiempo *m* de parada
i tempo *m* d'arresto
n stoptijd, uitlooptijd
d Nachlaufzeit *f*, Stoppzeit *f*

STORAGE CAMERA TUBE
see: IMAGE STORAGE TUBE

4574 STORAGE CHARACTERISTIC crt
Manner in which a device such as a light-sensitive mosaic or target holds its charge for a period of time.
f caractéristique *f* d'accumulation
e característica *f* de almacenaje
i caratteristica *f* d'accumulazione
n opslagkarakteristiek
d Speicherungscharakteristik *f*

4575 STORAGE OF CHARGES crt
f accumulation *f* de charges
e almacenaje *m* de cargas
i accumulazione *f* di cariche
n opslaan *n* van ladingen
d Ladungsspeicherung *f*

4576 STORAGE OF MAGNETIC MATERIALS rec/rep/tv
f emmagasinage *m* de matériaux magnétiques
e almacenamiento *m* de materiales magnéticos
i magazzinaggio *m* di materiali magnetici
n bewaren *n* van magnetische materialen
d Lagerung *f* von magnetischen Materialen

4577 STORAGE OF PROGRAM(ME) PARTS tv
f conservation *f* de parts du programme
e almacenamiento *m* de partes del programa
i magazzinaggio *m* di parti del programma
n opslag van programmadelen
d Speicherung *f* von Programmteilen

4578 STORAGE PRINCIPLE crt
f principe *m* d'accumulation, principe *m* d emmagasinage
e principio *m* de almacenaje
i principio *m* d'accumulazione
n opslagprincipe *n*
d Speicherungsprinzip *n*

4579 STORE ELEMENT tv
Element having some storage property, electrical, magnetic, acoustic, etc.
f élément *m* d'emmagasinage, élément *m* d'enregistrement
e elemento *m* de almacenaje, elemento *m* de registro
i elemento *m* d'immagazzinamento, elemento *m* di registrazione
n geheugenelement *n*, registreerelement *n*
d Speicherelement *n*

4580 STORED FIELD SYSTEM rec/rep/tv
A telerecording system in which the shutter is obscured during one field period, the field in question being stored by the phosphor screen with a long afterglow, whereas the next field is recorded directly on the film but the first field thanks to the afterglow also.
f système *m* à emmagasinage de trames alternantes
e sistema *m* con almacenamiento de campos alternativos
i sistema *m* con immagazzinamento di trame alternanti

n opslagsysteem *n* met afwisseling van de rasters
d Speichersystem *n* mit Teilbildabwechslung

4581 STORYBOARD vr
Sequence of still pictures, which outlines the story of a sequence or an entire film by highlighting its key points.
f feuille *f* de service
e plan *m* de puntos esenciales
i piano *m* d'andamento
n hoofdlijnenoverzicht *n*
d Ablaufplan *m*,
Storyboard *m*

4582 STRAIGHT SCANNING rec/rep/tv
f analyse *f* continue
e exploración *f* continua
i analisi *f* continua
n doorlopende beeldaftasting
d fortlaufende Bildabtastung *f*

4583 STRATOVISION tv
A proposed system for increasing the range of TV coverage by transmitting from an airplane circling in the atmosphere.
f stratovision *f*
e estratovisión *f*
i stratovisione *f*
n stratovisie
d Scatterverbindung *f*

4584 STRAYLIGHT dis
Light falling on the viewing screen of a TV receiver which impairs the quality of the projected picture.
f lumière *f* parasite
e luz *f* parásita
i luce *f* parassita
n storend licht *n*
d störendes Licht *n*

4585 STRAYLIGHT stu
In a studio, light used to attenuate sharp contrasts.
f lumière *f* diffuse
e luz *f* difusa
i luce *f* diffusa
n diffuus licht *n*
d diffuses Licht *n*

4586 STREAKING dis
In TV, prolongation of the trailing edge of the picture element caused by relative delay in the components of the picture signal.
f traînage *m*
e arrastre *m*
i trascinamento *m*
n veegeffect *n*, vegen *pl*
d Fahnen *pl*

4587 STREAKING EFFECT TEST CARD tv
A test card for assessing streaking effects in a picture transmitted by TV.
f mire *f* de trainage
e imagen *f* de prueba de arrastre
i immagine *f* di prova per lo striscionamento
n veegeffecttoetsbeeld *n*
d Fahneneffekttestbild *n*

4588 STRENGTH OF A COLO(U)R ct
That colo(u)r quality, an increase in which is associated with an increase in the concentration of the colo(u)ring material present, all other conditions remaining the same.
f intensité *f* de couleur
e intensidad *f* de color
i intensità *f* di colore
n kleurintensiteit,
kleursterkte
d Intensität *f* einer Farbe,
Stärke *f* einer Farbe

4589 STRETCH FACTOR cpl
A factor indicating the number of times by which the rise time has increased and indicating also the number of networks in cascade.
f facteur *m* de proportionalité
e factor *m* de ensanchamiento
i fattore *m* d'espansione
n rekfactor
d Dehnungsfaktor *m*

4590 STRIPE FILTER ct/ctv
A filter used in a two-tube camera consisting of a recurring series of very narrow red, blue, green and black stripes.
f filtre *m* à filets
e filtro *m* de rayas
i filtro *m* a striscie
n strepenfilter *n*
d Streifenfilter *n*

4591 STRIPE SIGNAL tv
f signal *m* de filet marginal
e señal *f* de raya marginal
i segnale *m* di striscia marginale
n randstreepsignaal *n*
d Randstreifensignal *n*

4592 STRIPE SIGNAL GENERATOR tv
f générateur *m* de signaux de filet marginal
e generador *m* de señales de raya marginal
i generatore *m* di segnali di striscia marginale
n randstreepsignaalgenerator
d Randstreifensignalgenerator *m*

4593 STROBE WHEEL vr
f roue *f* phonique
e rueda *f* fónica
i ruota *f* fonica
n fonisch rad *n*
d phonisches Rad *n*

4594 STRONGER ct
A difference apparently due to the presence of more colo(u)r than in the original sample.
f plus intensif
e más intensivo
i più intensivo
n intensiever,
sterker
d intenser,
stärker

4595 STRUCTURAL DEFINITION, STRUCTURAL RESOLUTION ctv
The definition (resolution) as limited by the size and shape of the screen elements.
f définition *f* structurelle
e definición *f* estructural
i definizione *f* strutturale
n door schermstructuur begrensd scheidend vermogen *n*
d durch Schirmstruktur begrenzte Auflösung *f*

STRUCTURALLY DUAL NETWORKS
see: DUAL NETWORKS

4596 STUB aea/cpl
f adaptateur *m*, plongeur *m*
e línea *f* de adaptación no disipativa, tetón *m* adaptador
i tronco *m* adattatore
n reactantiepijp, stomp
d Blindleitung *f*, Stichleitung *f*

STUB-MATCHED AERIAL, STUB-MATCHED ANTENNA
see: Q AERIAL

4597 STUDIO rec/tv
A room or hall acoustically treated and equipped for broadcasting.
f studio *m*
e estudio *m*
i studio *m*
n studio
d Studio *n*

4598 STUDIO ABSORPTION aud
f absorption *f* acoustique dans le studio
e absorción *f* acústica en el estudio
i assorbimento *m* acustico nello studio
n geluidsabsorptie in de studio
d Schallschluck *m* im Studio

4599 STUDIO ACOUSTICS rec
f acoustique *f* du studio
e acústica *f* del estudio
i acustica *f* dello studio
n studioakoestiek
d Akustik *f* des Studios

4600 STUDIO ALLOCATIONS SCHEDULE, STUDIO BOOKINGS stu
f plan *m* de charge des studios, plan *m* d'occupation des studios
e plano *m* de ocupación de los estudios
i piano *m* d'occupazione degli studi
n studiobezettingsplan *n*
d Studiobelegungsplan *m*

4601 STUDIO BACKGROUNDS stu
f fonds *pl* du studio
e fondos *pl* del estudio
i fondi *pl* dello studio
n studioachtergronden *pl*
d Studiohintergründe *pl*

4602 STUDIO BROADCAST (GB), STUDIO PICKUP (US) rec/rep/stu
f émission *f* du studio
e transmisión *f* desde el estudio
i trasmissione *f* dallo studio
n studio-uitzending
d Studiosendung *f*, Studioübertragung *f*

4603 STUDIO CAMERA CHAIN stu
f chaîne *f* de caméras pour le studio
e cadena *f* de cámaras en el estudio
i catena *f* della ripresa dallo studio
n cameraketen in de studio
d Studiokameraaggregat *n*

4604 STUDIO CAMERA REHEARSAL rec/tv
f répétition *f* dans le studio
e ensayo *m* en el estudio
i prova *f* nello studio
n studiorepetitie
d Studioprobeaufführung *f*

4605 STUDIO CAPACITY rec
The relation between number of performers and studio volume.
f capacité *f* du studio
e capacidad *f* del estudio
i capacità *f* dello studio
n studiocapaciteit
d Studiokapazität *f*

4606 STUDIO COMMUNICATIONS rep
Communications sent to the studio before or during broadcast.
f communications *pl* au studio
e comunicaciones *pl* al estudio
i comunicazioni *pl* allo studio
n mededelingen *pl* aan de studio
d Mitteilungen *pl* für das Studio

4607 STUDIO CONTROL stu
f contrôle *m* dans le studio
e control *m* en el estudio
i controllo *m* nello studio
n studiocontrole
d Studiokontrolle *f*

4608 STUDIO DECORATIONS stu
f décors *pl* du studio
e decoraciones *pl* del estudio
i attrezzeria *f* dello studio, decorazioni *pl* sceniche dello studio
n studiodecors *pl*
d Studiodekors *pl*

STUDIO DECORATION (GB)
see: ABSTRACT SET

4609 STUDIO DISPLAY FACILITIES stu
f appareillage *m* de reproduction dans le studio
e facilidades *pl* de reproducción en el estudio
i apparecchiatura *f* di riproduzione nello studio
n reproduktieapparatuur in de studio
d Studiowiedergabeausrüstung *f*

4610 STUDIO EQUIPMENT (GB), STUDIO FACILITIES (US) stu
f équipement *m* du studio
e facilidades *pl* del estudio
i apparecchiatura *f* dello studio
n studioapparatuur
d Studioausrüstung *f*

4611 STUDIO EXTERIORS rec/tv
f simulation *f* d'extérieurs dans le studio
e simulación *f* de exteriores en el estudio
i simulazione *f* di ripresa esterna nello studio
n buitenopnamenabootsing in de studio
d Aussenaufnahmennachahmung *f* im Studio

STUDIO FLOOR
see: STAGE

4612 STUDIO FLOOR MANAGER stu
The controller of crowds, visitors, unwanted technicians and others who may hold up studio work.
f directeur *m* technique de prises de vues
e director *m* técnico de escena
i direttore *m* tecnico di scena
n technische opnameleider
d technischer Aufnahmeleiter *m*

4613 STUDIO FOLDBACK CIRCUIT stu
Circuits used in TV studios to feed sound back to the floor for guide or mood control.
f circuit *m* pour ramener le son vers le studio
e circuito *m* para restituir el sonido hacia el studio
i circuito *m* per ricondurre il suono allo studio
n geluidsterugvoercircuit *n* naar de studio
d Schallrückfuhrkreis *m* zum Studio

4614 STUDIO GALLERY stu
f passerelle *f* de studio
e galería *f* de estudio
i galleria *f* di studio
n studiogalerij
d Studiorundgang *m*

4615 STUDIO LIGHT BOARDS stu
f batterie *f* de lampes dans le studio
e batería *f* de luces en el estudio
i batteria *f* di luci, tabella *f* luminosa
n lampenbatterij in de studio
d Beleuchtungsbühne *f*

STUDIO LIGHTS
see: SET LIGHTS

4616 STUDIO LINING aud/stu
f revêtement *m* acoustique du studio
e revestimiento *m* acústico del estudio
i rivestimento *m* acustico dello studio
n studiobekleding
d Studioauskleidung *f*

STUDIO MANAGER
see: STAGE MANAGER

4617 STUDIO OPERATION, STUDIO SOUND OPERATION aud/stu
The operational set-up for converting the sound program(me) into electrical signals.
f prise *f* de son
e toma *f* de sonido
i ripresa *f* sonora
n geluidsopname
d Tonaufnahme *f*

4618 STUDIO PEDESTAL stu
Camera mounting having a single vertical member surmounted by a friction head to permit tilting and panning.
f colonne *f* de caméra
e columna *f* de cámara
i colonna *f* di camera
n camerazuil
d Kamerasäule *f*

4619 STUDIO PLAN stu
f planification *f* du studio
e planeamiento *m* del estudio
i pianificazione *f* dello studio
n studioplanning
d Studioplanung *f*

4620 STUDIO POWER SUPPLY cpl/stu
f alimentation *f* du studio
e alimentación *f* del estudio
i alimentazione *f* dello studio
n studiovoeding
d Studiospeisung *f*

STUDIO RECORDING
see: FLOOR SHOOTING

4621 STUDIO RECORDING PRACTICES stu
f pratiques *pl* d'enregistrement dans le studio
e prácticas *pl* de registro en el studio
i pratiche *pl* di registrazione nello studio
n opnamepraktijken *pl* in de studio
d Aufnahmepraxis *f* im Studio

4622 STUDIO REVERBERATION aud
f réverbération *f* dans le studio
e reverberación *f* en el estudio
i riverberazione *f* nello studio
n studionagalm
d Studionachhall *m*

4623 STUDIO SITING AND LAY-OUT stu
f situation *f* et plan *m* du studio
e emplazamiento *m* y proyecto *m* del estudio
i situazione *f* e progetto *m* dello studio
n terreinkeuze en opzet van de studio
d Bauplatzwahl *f* und Entwurf *m* des Studios

4624 STUDIO STAFF stu
f équipe *f* du studio
e personal *m* del estudio
i personale *m* dello studio
n studiopersoneel *n*
d Studiobelegschaft *f*

4625 STUDIO-TO-TRANSMITTER LINK rep/tv
f liaison *f* entre le studio et l'émetteur
e conexión *f* entre estudio y emisor, enlace *m* del estudio al transmisor
i collegamento *m* tra studio ed emettitore
n studio-zenderverbinding
d Studio-Senderstrecke *f*, Verbindung *f* zwischen Studio und Sender

4626 STUDIO UNIT stu
f bloc *m* studio
e bloque *m* estudio
i blocco *m* studio
n studio-eenheid
d Studioeinheit *f*

4627 STUDIO WHITE stu
Chromaticity of the illuminant used in the studio.
f blanc *m* de studio
e blanco *m* de estudio
i bianco *m* di studio
n studiowit *n*
d Studioweiss *n*

4628 STYLIZED STUDIO DESIGN stu
f projet *m* stylisé du studio
e proyecto *m* estilizado del estudio
i progetto *m* stilizzato dello studio
n gestyleerd studioontwerp *n*
d stilisierter Studioentwurf *m*

4629 SUBASSEMBLY ge
f sousensemble *m*
e semiensamblado *m*, subconjunto *m*
i sottogruppo *m*
n deelopstelling, subsamenstel *n*
d Teilanordnung *f*, Untergruppe *f*

4630 SUBCARRIER cpl/ge
A carrier which is applied as a modulating wave to modulate another circuit.
f sousporteuse *f*
e subportadora *f*
i sottoportante *f*
n hulpdraaggolf
d Hilfsträger *m*

4631 SUBCARRIER BALANCE cpl
Extent to which the chrominance subcarrier is suppressed.
f équilibre *m* de la sousporteuse de la chrominance
e equilibrio *m* de la subportadora de la crominancia
i equilibrio *m* della sottoportante della crominanza
n balanceringsgraad van de chrominantie-draaggolf
d Gleichgewichtsgrad *m* des Chrominanz-hilfsträgers

4632 SUBCARRIER BEAT dis
Interference generated by a subcarrier and another signal or signal component.
f interférence *f* dans la sousporteuse de la chrominance
e interferencia *f* en la subportadora de la crominancia
i interferenza *f* nella sottoportante della crominanza
n interferentie met de chrominantiedraag-golf
d Störung *f* mit dem Chrominanzhilfsträger

4633 SUBCARRIER BUZZ dis
Sound buzz on high-luminance high-purity colo(u)rs.
f bourdonnement *m* dans le signal de luminance
e zumbido *m* en la señal de luminancia
i ronzio *m* nel segnale di luminanza
n ratel door het luminantiesignaal
d Summton *m* ins Helligkeitssignal

4634 SUBCARRIER COMPONENTS cpl/ctv
f composantes *pl* de la sousporteuse
e componentes *pl* de la subportadora
i componenti *pl* della sottoportante
n componenten *pl* van de hulpdraaggolf
d Komponenten *pl* des Hilfsträgers

4635 SUBCARRIER DOTS dis
Dot pattern produced by the chrominance carrier in the luminance signal.
f moiré *m* par le signal de chrominance
e moiré *m* por la señal de crominancia
i marezzatura *f* per il segnale di cromi-nanza
n moiré *n* door het chrominantiesignaal
d Moiré *n* durch Chrominanzsignal

4636 SUBCARRIER DRIVE cpl/ctv
Sine wave signal at subcarrier reference frequency.
f signal *m* sinusoïdal à la fréquence de la sousporteuse de référence
e señal *f* sinusoidal de la frecuencia de la subportadora de referencia
i segnale *m* sinusoidale alla frequenza della sottoportante di riferimento
n sinusvormig signaal *n* met de frequentie van de referentiehulpdraaggolf
d sinusförmiges Signal *n* mit der Frequenz des Bezugshilfsträgers

4637 SUBCARRIER FREQUENCY cpl/ctv
f fréquence *f* de la sousporteuse
e frecuencia *f* de la subportadora
i frequenza *f* della sottoportante
n hulpdraaggolffrequentie
d Hilfsträgerfrequenz *f*

4638 SUBCARRIER GENERATOR, SUBCARRIER OSCILLATOR ctv
Oscillator which generates the colo(u)r subcarrier frequency which is subsequently modulated by colo(u)r signals.
f générateur *m* de la sousporteuse
e generador *m* de la subportadora
i generatore *m* della sottoportante
n generator van de hulpdraaggolf
d Hilfsträgergenerator *m*

4639 SUBCARRIER OFFSET cpl
f décalage *m* de la sousporteuse de la chrominance
e desplazamiento *m* de la subportadora
i spostamento *m* della sottoportante
n chrominantiedraaggolfverzet *n*
d Chrominanzträgerversetzung *f*

4640 SUBCARRIER RECTIFICATION cpl
f redressement *m* du signal de chrominance
e rectificación *f* de la señal de crominancia
i raddrizzamento *m* del segnale di crominanza
n gelijkrichting van het chrominantiesignaal
d Gleichrichtung *f* des Chrominanzsignals

4641 SUBCARRIER REGENERATOR ctv
f régénérateur *m* de la sousporteuse de chrominance
e regenerador *m* de la subportadora de crominancia
i rigeneratore *m* della sottoportante di crominanza
n regenerator van de chrominantiedraaggolf
d Regenerator *m* des Chrominanzhilfsträgers

4642 SUBCARRIER SIDEBAND cpl /ctv
The sideband(s) of the subcarrier in a colo(u)r transmission.
f bande *f* latérale de la sousporteuse
e banda *f* lateral de la subportadora
i banda *f* laterale della sottoportante
n zijband van de hulpdraaggolf
d Hilfsträgerseitenband *n*

SUBJECTIVE BRIGHTNESS
see: APPARENT BRIGHTNESS

4643 SUBJECTIVE LOUDNESS aud
f intensité *f* sonore subjective,
puissance *f* subjective
e intensidad *f* sonora subjetivo
i intensità *f* soggettiva della sensazione sonora,
sonorità *f* soggettiva
n subjectieve luidheid
d subjektive Lautheit *f*

4644 SUBJECTIVE NOISE dis
f bruit *m* subjectif
e ruido *m* subjetivo
i rumore *m* soggettivo
n subjectieve ruis
d subjektives Rauschen *n*

4645 SUBJECTIVE NOISE AMPLITUDE dis
f amplitude *f* du bruit subjectif
e amplitud *f* del ruido subjetivo
i ampiezza *f* del rumore soggettivo
n amplitude van de subjectieve ruis
d Amplitude *f* des subjektiven Rauschens

4646 SUBJECTIVE SHADOW SERIES ct
A series of colo(u)rs of varying luminosity but constant hue and saturation.
f série *f* de teintes subjectives
e serie *f* de tintes subjectivas
i serie *f* di tinte soggettive
n subjectieve tintenserie
d subjektive Farbtonserie *f*

4647 SUBOSCILLATOR svs
f oscillateur *m* secondaire
e oscilador *m* secundario
i oscillatore *m* secondario
n hulposcillator
d Nebenoszillator *m*

SUBSCRIPTION TELEVISION
see: COIN-FREED TELEVISION

4648 SUBTITLE, tv
SUPERIMPOSED TITLE
Title superimposed on a film or TV image for the purpose of translating foreign dialogue.
f soustitre *m*
e título *m* sobreimpreso,
subtítulo *m*
i sottotitolo *m*
n onderschrift *n*
d Unterschrift *f*,
Untertitel *m*

4649 SUBTRACTIVE COLO(U)R SYSTEM ctv
f système *m* soustractif de couleurs
e sistema *m* substractivo de colores
i sistema *m* sottrattivo di colori
n subtractief kleurensysteem *n*
d subtraktives Farbensystem *n*

4650 SUBTRACTIVE COMPLEMENTARY COLO(U)RS ct
f couleurs *pl* soustractives complémentaires
e colores *pl* substractivos complementarios
i colori *pl* sottrattivi complementari
n subtractieve complementaire kleuren *pl*
d subtraktive Komplementärfarben *pl*

4651 SUBTRACTIVE MIXTURE ct
The mixture of absorbing media or the superposition of filters so that the composition of the light stimulus passing through the combination is determined by the simultaneous or successive absorption of parts of the spectrum by each medium present.
f mélange *m* soustractif
e mezcla *f* substractiva
i mescolanza *f* sottrattiva
n subtractief mengsel *n*
d subtraktives Gemisch *n*

SUBTRACTIVE PRIMARIES
see: PRIMARY SUBTRACTIVE COLO(U)RS

4652 SUBTRACTIVE PROCESS ct
Of colo(u)r reproduction, a process of reproducing the colo(u)rs of objects in a picture by means of the subtractive colo(u)r mixture of two or more subtractive primaries, the amounts of which are determined at any point by colo(u)r separation images.
f procédé *m* soustractif
e proceso *m* substractivo
i procedimento *m* sottrattivo
n subtractief procédé *n*
d subtraktives Verfahren *n*

4653 SUCCESSIVE FIELDS rec/rep/tv
f trames *pl* successives
e campos *pl* sucesivos
i trame *pl* successive
n op elkaar volgende rasters *pl*
d aufeinanderfolgende Halbbilder *pl*

4654 SUMMING POINT, SUMMING STAGE cpl/tv
A mixing point (stage) where two independent inputs are added.
f point *m* d'addition de deux entrées
e punto *m* de adición de dos entradas
i punto *m* d'addizione di due ingressi
n samenvoegingspunt *n* van twee ingangen
d Summierstelle *f*

4655 SUPERGROUP cpl
A group of carrier channels with adjacent frequencies.
f supergroupe *m*
e supergrupo *m*
i supergruppo *m*
n supergroep
d Übergruppe *f*

4656 SUPERICONOSCOPE crt
f supericonoscope *m*
e supericonoscopio *m*
i supericonoscopio *m*
n supericonoscoop
d Bildwandlerikonoskop *n*, Superikonoskop *n*

4657 SUPERIMPOSE (TO) tv
To photograph or print one image on top of another in such a way that they keep a constant relation to one another.
f superposer v, surimprimer v
e superponer v
i sovrapporre v
n boven elkaar plaatsen v, superponeren v
d überlagern v

4658 SUPERIMPOSED IMAGES tv
f images *pl* superposées, images *pl* surimprimées
e imágenes *pl* superpuestas
i immagini *pl* sovrapposte
n beelden *pl* in superpositie, boven elkaar geplaatste beelden *pl*
d überlagerte Bilder *pl*

4659 SUPERIMPOSED INTERFERENCE dis
f transmodulation *f*
e interferencia *f* por superposición
i interferenza *f* per sovrapposizione
n storingssuperponering
d Störüberlagerung *f*

4660 SUPERIMPOSER vr
f appareil *m* de surimpression
e aparato *m* de superposición
i apparecchio *m* di sovrapposizione
n ondertitelapparaat *n*
d Einkopiervorrichtung *f*

4661 SUPERIMPOSING tv
f superposition *f*, surimpression *f*
e superposición *f*
i sovrapposizione *f*
n beeldsuperpositie, boven elkaar plaatsen *n*
d Überlagerung *f*

4662 SUPERSYNC SIGNAL tv
A combination of a horizontal and a vertical sync signal transmitted at the end of a TV scanning line to synchronize the operation of a TV receiver with that of the transmitter.
f signal *m* de top de synchronisation, top *m* de synchronisation
e señal *f* supersincrónica
i picco *m* di sincronizzazione
n gecombineerd horizontaal en verticaal synchronisatiesignaal *n*
d kombiniertes horizontales und vertikales Synchronisierungssignal *n*

SUPERTURNSTILE ANTENNE (US)
see: BATWING AERIAL

4663 SUPPLY VOLTAGE cpl
f tension *f* d'alimentation
e tensión *f* de alimentación
i tensione *f* d'alimentazione
n voedingsspanning
d Speisespannung *f*

4664 SUPPRESSED CARRIER SIGNAL cpl
f signal *m* de système à porteuse supprimée
e señal *f* de sistema de portadora suprimida
i segnale *m* di sistema a portante soppressa
n draaggolfonderdrukkingssignaal *n*, signaal *n* bij onderdrukte draaggolf
d Trägerunterdrückungssignal *n*

4665 SUPPRESSED CARRIER SYSTEM cpl
Technique of transmitting information with one sideband and no carrier.
f système *m* à porteuse supprimée
e sistema *m* de portadora suprimida
i sistema *m* a portante soppressa, sistema *m* senza portante
n systeem *n* met onderdrukte draaggolf
d unterdrücktes Trägersystem *n*

4666 SUPPRESSED CARRIER TRANSMITTER cpl
A transmitter which emits sidebands only.
f émetteur *m* à porteuse supprimée
e emisor *m* de portadora suprimida
i emettitore *m* a portante soppressa
n zender met draaggolfonderdrukking
d Sender *m* mit unterdrücktem Träger

4667 SUPPRESSED FIELD SYSTEM tv
A method of telerecording whereby only alternate fields are recorded. This means that when the number of active lines is 377, the suppressed field recordings contain only 188½ lines
f système *m* à suppression de trames alternantes

e sistema *m* con supresión de campos alternativos
i sistema *m* con soppressione di trame alternanti
n opneemsysteem *n* met onderdrukking van afwisselende rasters
d System *n* mit Unterdrückung von abwechselnden Teilbildern

4668 SUPPRESSED SIDEBAND cpl
f bande *f* latérale supprimée
e banda *f* lateral suprimida
i banda *f* laterale soppressa
n onderdrukte zijband
d unterdrücktes Seitenband *n*

SURFACE COLO(U)R
see: NON-SELF-LUMINOUS COLO(U)R

4669 SURFACE INDUCTION vr
f induction *f* superficielle
e inducción *f* de superficie
i induzione *f* superficiale
n oppervlakte-inductie
d Oberflächeninduktion *f*

4670 SURGE CHARACTERISTIC (US), TRANSIENT RESPONSE, UNIT FUNCTION RESPONSE (GB) dis/rep/tv
f caractéristique *f* du saut, réponse *f* au signal unité
e respuesta *f* de fenómenos transitorios
i risposta *f* di fenomeni transitori
n sprongkarakteristiek
d Einschwingverhalten *n*, Sprungkennlinie *f*

SUSCEPTANCE (US)
see: FEEDBACK ADMITTANCE

4671 SWEEP cpl/crt
The steady movement of the electron beam across the screen of a cathode-ray tube, producing a steady bright line when no signal is present.
f mouvement *m* du faisceau
e movimiento *m* del haz
i movimento *m* del fascio
n bundelloop
d Strahlengang *m*

SWEEP (US)
see: SCANNING

4672 SWEEP AMPLIFIER rec/rep/tv
An amplifier used in a TV receiver to amplify the sawtooth output voltage of the sweep oscillator.
f amplificateur *m* de balayage
e amplificador *m* de barrido
i amplificatore *m* di scansione
n afbuigversterker
d Ablenkverstärker *m*

4673 SWEEP CIRCUIT cpl/crt
In TV, the circuit which produces the required rectangular scan by the electron beam.
f circuit *m* de balayage linéaire
e circuito *m* de barrido lineal
i circuito *m* di scansione lineare
n lineaire-afbuigingskring
d Zeitablenkschaltung *f*

4674 SWEEP FREQUENCY crt
The rate at which the electron beam is swept to and fro across the face of a cathode-ray tube.
f fréquence *f* de balayage
e frecuencia *f* de barrido
i frequenza *f* di scansione
n afbuigfrequentie
d Kippfrequenz *f*, Wobbelfrequenz *f*

4675 SWEEP GENERATOR, SWEEP OSCILLATOR, TIMEBASE GENERATOR cpl/tv
A test instrument that generates a radio-frequency voltage whose frequency varies back and forth through a given frequency range at a rapid constant rate.
f générateur *m* de balayage, oscillateur *m* de base de temps
e generador *m* de barrido, oscilador *m* de base de tiempo
i generatore *m* di deflessione, oscillatore *m* di base di tempi
n afbuiggenerator, tijdbasisoscillator
d Ablenkgenerator *m*, Kippgenerator *m*, Kipposzillator *m*, Zeitablenkgenerator *m*

SWEEP GENERATOR (US)
see: SCANNING GENERATOR

4676 SWEEP-OUT OF TELEVISION SCREEN dis
f dépassement *m* de l'écran
e traspaso *m* de la pantalla
i sconfinamento *m* dello schermo
n overschrijding van het scherm
d Überstreichen *n* des Leuchtschirms

4677 SWEEP UNIT tv
f partie *f* de balayage
e parte *f* de barrido
i parte *f* di scansione
n aftastgedeelte *n*
d Abtastteil *m*

4678 SWEEP VOLTAGE, TIMEBASE VOLTAGE crt
f tension *f* de base de temps
e tensión *f* de base de tiempo
i tensione *f* di base di tempi
n tijdbasisspanning
d Kippspannung *f*, Zeitablenkspannung *f*

SWEEPING COIL (US)
see: DEFLECTION COIL

SWEEPS (US)
see: SCANNING VOLTAGE

SWINGING BURST
see: ALTERNATING BURST

4679 SWISH PAN, fi/rec/tv
WHIP PAN,
ZIP PAN
f arraché *m*
e panorámica *f* rápida
i panoramica *f* rapida
n snelle camerazwenking
d Reissschwenkung *f*

4680 SWITCH MATRIX, tv
SWITCHING MATRIX
f dispositif *m* de commutation automatique, matrice *f* de commutation
e matriz *f* de conmutación
i matrice *f* di commutazione
n omschakelautomatiek, schakelmatrix
d Schaltmatrize *f*, Umschaltautomatik *f*

4681 SWITCHING CENTER(RE) tv
f centre *m* de commutation
e centro *m* de conmutación
i centro *m* di commutazione
n schakelcentrum *n*
d Schaltzentrum *n*

4682 SWITCHING SCHEDULE stu
f schéma *m* de commutation
e esquema *m* de conmutación
i schema *m* di commutazione
n schakelschema *n*
d Schaltplan *m*

SWITCHING SCHEDULE
see: PROGRAM(ME) LOG

4683 SWITCHING SIGNAL cpl
f signal *m* de commutation
e señal *f* de conmutación
i segnale *m* di commutazione
n schakelsignaal *n*
d Schaltsignal *n*

4684 SYMMETRICAL AERIAL INPUT, aea
SYMMETRICAL ANTENNA INPUT
f entrée *f* d'antenne symétrique
e entrada *f* de antena simétrica
i entrata *f* d'antenna simmetrica
n symmetrische antenne-invoer
d symmetrischer Antenneneingang *m*

4685 SYMMETRICAL CLIPPER cpl
f écrêteur *m* symétrique
e recortador *m* simétrico
i tosatore *m* simmetrico
n symmetrische drempelwaardebegrenzer
d symmetrische Abschneidestufe *f*

4686 SYMMETRICAL DEFLECTION crt
f déviation *f* symétrique
e desviación *f* simétrica
i deflessione *f* simmetrica, deviazione *f* simmetrica
n symmetrische afbuiging
d symmetrische Ablenkung *f*

4687 SYNC AMPLITUDE, cpl
SYNC SIGNAL AMPLITUDE
f amplitude *f* du signal de synchronisation
e amplitud *f* de la señal de sincronización
i ampiezza *f* del sincronismo
n amplitude van het synchronisatiesignaal
d Synchronamplitude *f*

4688 SYNC COMPARATOR tv
f appareil *m* de contrôle du synchronisme de parties du programme
e aparato *m* para controlar el sincronismo de partes del programa
i apparecchio *m* di controllo del sincronismo di parti del programma
n apparaat *n* voor het controleren van het gelijklopen van programmadelen
d Gerät *n* zur Gleichlaufskontrolle von Programmteilen

4689 SYNC COMPRESSION, cpl
SYNC SIGNAL COMPRESSION
The reduction in gain applied to the sync signal over any part of its amplitude range with respect to the gain at a specified reference level.
f compression *f* du signal de synchronisation
e compresión *f* de la señal de sincronización
i compressione *f* del sincronismo
n compressie van het synchronisatiesignaal
d Kompression *f* des Synchronisierungssignals, Stauchung *f* des Synchronisierungssignals

SYNC CONTROL (GB)
see: HOLD CONTROL

4690 SYNC GENERATOR, tv
SYNC SIGNAL GENERATOR
An electronic generator that supplies sync pulses to TV studio and transmitter equipment.
f générateur *m* du signal de synchronisation
e generador *m* de la señal de sincronización
i generatore *m* del sincronismo
n synchronisatiesignaalgenerator
d Impulszentrale *f*, Synchronisierungsgenerator *m*, Taktgeber *m*

4691 SYNC-IN PULSE tv
f impulsion *f* d'entrée de la synchronisation
e impulso *m* de entrada de la sincronización
i impulso *m* d'ingresso della sincronizzazione
n synchronisatie-ingangsimpuls
d Synchronisierungseingangsimpuls *m*

4692 SYNC INPUT cpl
f entrée *f* de la synchronisation
e entrada *f* de la sincronización
i ingresso *m* della sincronizzazione
n synchronisatie-ingang
d Synchronisierungseingang *m*

4693 SYNC LEVEL, cpl
SYNC SIGNAL LEVEL
The level reached by the tips of the sync pulses.

f niveau *m* du signal de synchronisation
e nivel *m* de la señal de sincronización
i livello *m* del sincronismo
n niveau *n* van het synchronisatiesignaal
d Pegel *m* des Synchronisierungssignals

4694 SYNC LIMITER, SYNC SIGNAL LIMITER cpl/tv
A limiter circuit used in TV to prevent sync pulses from exceeding a predetermined amplitude.
f limiteur *m* de l'amplitude du signal de synchronisation
e limitador *m* de la amplitud de la señal de sincronización
i limitatore *m* dell'ampiezza del sincronismo
n amplitudebegrenzer van het synchronisatiesignaal
d Synchronamplitudenbegrenzer *m*

4695 SYNC-OUT PULSE cpl
f impulsion *f* de sortie de la synchronisation
e impulso *m* de salida de la sincronización
i impulso *m* d'uscita della sincronizzazione
n synchronisatie-uitgangsimpuls
d Synchronisierungsausgangsimpuls *m*

4696 SYNC-OUTPUT cpl
f sortie *f* de la synchronisation
e salida *f* de la sincronización
i uscita *f* della sincronizzazione
n synchronisatie-uitgang
d Synchronisierungsausgang *m*

4697 SYNC PIP aud
f signe *m* de synchronisation
e signo *m* de sincronización
i segno *m* di sincronizzazione
n synchronisatieteken *n*
d Synchronisierungszeichen *n*

4698 SYNC PULSE, SYNCHRONIZING PULSE cpl
The pulse generated in association with the scanning process to keep the receiving system in synchronism with the transmitter.
f impulsion *f* de synchronisation
e impulso *m* de sincronización
i impulso *m* di sincronizzazione, sincronismo *m*
n synchronisatie-impuls
d Synchronisierungsimpuls *m*

4699 SYNC PULSE GENERATOR, SYNCHRONIZING PULSE GENERATOR cpl
f générateur *m* des impulsions de synchronisation
e generador *m* de los impulsos de sincronización
i generatore *m* dei sincronismi
n synchronisatie-impulsgenerator
d Synchronisierungsimpulsgeber *m*

4700 SYNC PULSE REGENERATION, SYNCHRONIZING PULSE REGENERATION dis
f régénération *f* des impulsions de synchronisation
e regeneración *f* de los impulsos de sincronización
i riformazione *f* dei sincronismi
n regeneratie van de synchronisatie-impulsen
d Synchronisierungsimpulsregeneration *f*

4701 SYNC PULSE REGENERATOR, SYNCHRONIZING PULSE REGENERATOR cpl
f régénérateur *m* des impulsions de synchronisation
e regenerador *m* de los impulsos de sincronización
i riformatore *m* dei sincronismi
n regenerator van de synchronisatie-impulsen
d Synchronisierungsimpulsregenerator *m*

4702 SYNC PULSE SELECTION, SYNCHRONIZING PULSE SELECTION tv
f sélection *f* de l'impulsion de synchronisation
e selección *f* del impulso de sincronización
i selezione *f* del sincronismo
n selectie van de synchronisatie-impuls
d Selektion *f* des Synchronisierungsimpulses

SYNC PULSE SEPARATION
see: SEPARATION

4703 SYNC SEPARATOR cpl
A device for extracting the synchronizing signals from the complete video signal.
f séparateur *m* des signaux de synchronisation
e separador *m* de las señales de sincronización
i separatore *m* dei sincronismi
n synchronisatiesignaalscheider
d Separator *m*

SYNC SEPARATOR
see: AMPLITUDE FILTER

4704 SYNC SIGNAL INSERTION cpl
f insertion *f* du signal de synchronisation
e inserción *f* de la señal de sincronización
i inserzione *f* del sincronismo
n invoeging van het synchronisatiesignaal
d Einfügung *f* des Synchronisierungssignals

4705 SYNC SIGNAL RE-INSERTION cpl tv
f réinsertion *f* du signal de synchronisation
e reinserción *f* de la señal de sincronización
i reinserzione *f* del sincronismo
n wederinvoeging van het synchronisatiesignaal
d Wiedereinfügung *f* des Synchronisierungssignals

4706 SYNC STRETCHING tv
f expansion *f* du signal de synchronisation
e expansión *f* de la señal de sincronización
i espansione *f* del sincronismo

n strekken *n* van het synchronisatiesignaal
d Dehnung *f* des Synchronisierungssignal, Pegelstabilisierung *f*

4707 SYNCHRONISM cpl
In TV, identity in frequency and correspondence in phase between the scanning process at two points in the system.
f synchronisme *m*
e sincronismo *m*
i sincronismo *m*
n synchronisme *n*
d Synchronismus *m*

4708 SYNCHRONIZATION BAY cpl
Equipment forming part of one transmitter of a synchronized chain and comprising a very stable master oscillator together with means for comparing its frequency with that of the other transmitters in the chain.
f baie *f* de synchronisation
e bastidor *m* de sincronización
i sezione *f* di sincronizzazione
n gelijkloopapparatuur
d Frequenzsteueranlage *f*

4709 SYNCHRONIZATION LOSS cpl
f perte *f* de synchronisation
e pérdida *f* de sincronización
i perdita *f* di sincronizzazione
n synchronisatieverlies *n*
d Synchronisationsverlust *m*

4710 SYNCHRONIZATION SENSITIVITY cpl
f sensibilité *f* à la synchronisation
e sensibilidad *f* a la sincronización
i sensibilità *f* alla sincronizzazione
n synchronisatiegevoeligheid
d Synchronisationsempfindlichkeit *f*

4711 SYNCHRONIZED SWEEP crt
f analyse *f* synchronisée
e exploración *f* sincronizada
i analisi *f* sincronizzata
n gesynchroniseerde aftasting
d synchronisierte Abtastung *f*

4712 SYNCHRONIZING rep/tv
Process of maintaing the line and frame scanning in a TV receiver in step with that of the camera.
f synchronisation *f*
e sincronización *f*
i sincronizzazione *f*
n synchronisatie
d Synchronisation *f*, Synchronisierung *f*

4713 SYNCHRONIZING LEVEL cpl
f niveau *m* de synchronisation
e nivel *m* de sincronización
i livello *m* di sincronizzazione
n synchroonniveau *n*
d Synchronwert *m*

4714 SYNCHRONIZING TIME tv
f durée *f* de la synchronisation
e duración *f* de la sincronización
i durata *f* della sincronizzazione
n inregeltijd
d Gleichlaufzeit *f*

4715 SYNCHRONOUS DEMODULATOR, SYNCHRONOUS DETECTOR cpl
A demodulator which utilizes a reference signal having the same frequency as the carrier or subcarrier that is to be demodulated.
f démodulateur *m* synchrone
e desmodulador *m* sincrono
i demodulatore *m* sincrono
n synchrone demodulator
d Synchrondemodulator *m*

4716 SYNCHRONOUS SATELLITE, SYNCOM, SYNCSTAT tv
Communications satellite orbiting so as to remain in a fixed position above the earth.
f satellite *m* synchrone
e satélite *m* sincrono
i satellite *m* sincrono
n synchrone kunstmaan
d Synchronsatellit *m*

4717 SYNCHRONOUS SCANNING tv
f analyse *f* synchrone
e exploración *f* sincrona
i analisi *f* sincrona
n gelijkfazige aftasting, synchrone aftasting
d Synchronabtastung *f*

4718 SYSTEM OF BEAMS ctv
f système *m* de faisceaux
e sistema *m* de haces
i sistema *m* di fasci
n bundelsysteem *n*
d Strahlsystem *n*

4719 SYSTEMATIC CONVERGENCE ERROR crt/dis
f erreur *f* systématique de convergence
e error *m* sistemático de convergencia
i errore *m* sistematico di convergenza
n systematische convergentiefout
d systematischer Konvergenzfehler *m*

T

4720 TABLE OF FREQUENCY ALLOCATIONS cpl/ge
f table *f* de division des fréquences
e cuadro *m* de distribución de las frecuencias
i tabella *f* d'allogazione delle frequenze
n frequentietoewijzingstabel
d Frequenzbereichsplan *m*, Frequenzverteilungsplan *m*

4721 TABLE OF RELATIVE LUMINOSITY COEFFICIENTS ct
f table *f* de facteurs de luminosité
e cuadro *m* de coeficientes de luminosidad relativa
i tabella *f* di coefficienti di luminosità relativa
n tabel van de relatieve helderheidscoëfficiënten
d Tafel *f* der spektralen Hellempfindlichkeit

4722 TABLE OF REPLACEABLE PARTS svs
f liste *f* de pièces détachées
e lista *f* de las partes
i elenco *m* dei pezzi
n stuklijst
d Schaltteilliste *f*, Stückliste *f*

4723 TABLE SET rep/tv
f récepteur *m* de table, téléviseur *m* de table
e televisor *m* de sobremesa
i televisore *m* da tavola
n tafeltoestel *n*
d Tischempfänger *m*, Tischgerät *n*

4724 TAIL LEADER fi/tv/vr
f amorce *f* de fin
e cola *f* final
i coda *f* finale
n eindband, eindstrook
d Endband *n*, Nachlauf *m*

TAIL OF PULSE
see: PULSE TAIL

TAIL OF SIGNAL
see: SIGNAL TAIL

TAILING
see: HANGOVER

TAKE-OFF (US)
see: REJECTION OF THE ACCOMPANYING SOUND

4725 TAKE-OFF SPOOL vr
f bobine *f* débitrice
e bobina *f* desenrolladora
i bobina *f* svolgitrice
n afwikkelspoel
d Abwickelspule *f*

4726 TAKE-UP CASSETTE vr
f cassette *f* enrouleuse, cassette *f* réceptrice
e caseta *f* devanadora
i cassetta *f* d'avvolgimento
n opwikkelcassette
d Aufspulkassette *f*, Aufwickelkassette *f*

4727 TAKE-UP SPOOL vr
f bobine *f* enrouleuse, bobine *f* réceptrice
e bobina *f* colectora, bobina *f* devanadora
i bobina *f* d'avvolgimento, bobina *f* raccoglitrice
n opwikkelspoel
d Aufwickelspule *f*

4728 TALKBACK CIRCUIT stu
A circuit enabling spoken directions to be given from a studio, control cubicle, or TV control room, or from a production-panel to a studio or other program(me) source, for the purpose of directing a performance or rehearsal.
f circuit *m* d'ordres, réseau *m* d'ordres
e circuito *m* de intercomunicación
i circuito *m* per trasmissione d'ordini
n ruggespraakleiding
d Kommandoleitung *f*

4729 TALKBACK MICROPHONE aud
f microphone *m* d'ordre
e micrófono *m* de intervención
i microfono *m* d'intervento
n ruggespraakmicrofoon
d Gegensprechmikrophon *n*, Kommandomikrophon *n*

4730 TALK-LISTEN SWITCH cpl/stu
f commutateur *m* parle-écoute
e conmutador *m* habla-escucha
i commutatore *m* parlo-ascolto
n luister-spreekschakelaar
d Hör-Sprech-Schalter *m*

TALLY LIGHTS (US)
see: CUE LIGHTS

TALLY SIGNAL (US)
see: CUE

4731 TAPE aud/vr
f bande *f*, ruban *m*
e cinta *f*
i nastro *m*
n band
d Band *n*

4732 TAPE ASPERITY rec
f rugosité *f* de bande
e aspereza *f* de cinta
i rugosità *f* di nastro
n bandruwheid
d Bandrauheit *f*

4733 TAPE BACKGROUND NOISE, dis/vr
TAPE NOISE
f bruit *m* de bande
e ruido *m* de cinta
i rumore *m* di nastro
n bandruis
d Bandrauschen *n*

4734 TAPE BACKING vr
f dorsale *f* de bande
e dorso *m* de cinta
i dorso *m* di nastro
n bandachterkant
d Bandrückseite *f*

4735 TAPE BASE vr
f support *m* de bande
e soporte *m* de cinta
i sopporto *m* di nastro
n drager van de magnetische laag
d Magnetschichtträger *m*,
Schichtträger *m*

4736 TAPE BREAK vr
f rupture *f* de bande
e ruptura *f* de cinta
i rottura *f* di nastro
n bandbreuk
d Bandriss *m*

4737 TAPE CARTRIDGE (US), aud/vr
TAPE CASSETTE (GB)
A cartridge that holds a length of magnetic tape in such a way that it can be slipped into a tape recorder, without threading the tape.
f cassette *f* à bande
e caseta *f* de cinta
i cassetta *f* a nastro
n bandcassette
d Bandkassette *f*

4738 TAPE CLATTER aud/vr
Vibration occurring as a result of tape sticking to the guides of a tape recorder, which causes flutter in the reproduction.
f fracas *m*
e matraqueo *m*
i fracasso *m*
n ratelen *n*
d Rattern *n*

4739 TAPE COATING vr
f couchage *m* de bande
e revestimiento *m* de cinta
i rivestimento *m* di nastro
n aanbrengen *n* van de magnetische laag
d Bandbeschichtung *f*

4740 TAPE COPY vr
f copie *f* sur bande,
report *m*
e copia *f* en cinta
i copia *f* su nastro
n bandkopie
d Bandkopie *f*,
Tochterband *n*

4741 TAPE COUNTER, vr
TAPE FOOTAGE COUNTER
f compteur *m* de longueur de bande
e contador *m* de longitud de cinta
i contatore *m* di lunghezza di nastro
n bandlengtemeter,
bandtelwerk *n*
d Bandzählwerk *n*

4742 TAPE CUPPING dis/vr
f courbure *f* transversale de la bande
e curvatura *f* transversal de la cinta
i curvatura *f* trasversale del nastro
n transversale kromming van de band
d transversale Bandkrümmung *f*

4743 TAPE CURLING dis/vr
f curling *f* de bande
e enroscadura *f* de cinta
i serpeggiamento *m* di nastro
n kronkelen *n* van de band
d Bandverformung *f*

4744 TAPE CURVATURE, dis/rec
TAPE WEAVE
A deformation occurring when the tape is transported through the cutters with sideways movement.
f courbure *f* de la bande
e curvatura *f* de la cinta
i curvatura *f* del nastro
n sabelvormigheid van de band
d Bandverdehnung *f*,
Sabelförmigkeit *f* des Bandes

TAPE DAMAGE FROM WINDING
see: DAMAGE FROM WINDING

4745 TAPE DECK aud/vr
Platform incorporating essentials for magnetic recording.
f châssis *m*
e chasis *m*
i telaio *m*
n chassis *n*
d Chassis *n*,
Deck *n*

4746 TAPE DISTORTION aud/dis/vr
f distorsion *f* de bande
e distorsión *f* de cinta
i distorsione *f* di nastro
n bandvervorming
d Bandverzerrung *f*

TAPE EDITING
see: EDITING OF THE TAPE

4747 TAPE ERROR vr
f erreur *f* de bande
e error *m* de cinta
i errore *m* di nastro
n bandfout
d Bandfehler *m*

4748 TAPE GUIDANCE vr
f guidage *m* de piste
e conducción *f* de pista
i guidamento *m* di pista
n spoorgeleiding
d Spurhaltung *f*

4749 TAPE GUIDANCE PIN vr
f goupille *f* de guide-bande
e perno *m* de guía-cinta
i spina *f* di guidanastro
n bandgeleidestift
d Bandführungsstift *m*

4750 TAPE GUIDE vr
f guide-bande *m*
e guía cinta *f*
i guidanastro *m*
n bandgeleider
d Bandführer *m*,
Bandführung *f*

4751 TAPE GUIDE SEGMENT vr
f segment *m* de guide-bande
e segmento *m* de guía-cinta
i segmento *m* di guidanastro
n bandgeleidingssegment *n*
d Bandführungssegment *n*

4752 TAPE GUIDE SERVO SYSTEM aud/cpl/vr
f système *m* asservi pour guide-bande
e servosistema *m* para guía-cinta
i servosistema *m* per guidanastro
n bandgeleiderservosysteem *n*
d Bandführungsservosystem *n*

4753 TAPE GUIDING DRUM vr
f tambour *m* de guidage de bande
e tambor *m* de conducción de cinta
i tamburo *m* di guidamento di nastro
n bandgeleidingstrommel
d Bandführungstrommel *f*

4754 TAPE INTERCHANGEABILITY vr
f interchangeabilité *f* de bande
e conmutabilidad *f* de cinta
i intercambialità *f* di nastro
n bandverwisselbaarheid
d Bandauswechselbarkeit *f*

4755 TAPE LEADER vr
f bout *m* mort
e cinta *f* de arranque
i nastro *m* d'avviamento
n startband
d Startband *n*, Vorlaufband *n*

4756 TAPE LENGTH INDICATOR vr
f indicateur *m* de longueur de bande
e indicador *m* de longitud de cinta
i indicatore *m* di lunghezza di nastro
n bandlengteaanwijzer
d Bandlängenanzeiger *m*

4757 TAPE LIBRARY vr
f magnétothèque *f*
e archivo *m* de cintas
i archivio *m* di nastri
n bandarchief *n*
d Bandarchiv *n*

4758 TAPE LIFE vr
The average life of a tape.
f durée *f* de bande
e duración *f* de cinta,
vida *f* útil de cinta
i durata *f* di nastro
n levensduur van de band
d Bandlebensdauer *f*

4759 TAPE LIFTER vr
f décolleur *m* de bande
e llevador *m* de cinta
i sollevatore *m* di nastro
n bandafnemer
d Bandabheber *m*

4760 TAPE LOOP aud/vr
A length of magnetic tape having the ends spliced together to form an endless loop.
f bande *f* sans fin,
boucle *f* de bande
e bucle *m* de cinta,
cinta *f* sin fin
i doppina *f* di nastro,
nastro *m* senza fine
n bandlus,
eindloze band
d Bandschleife *f*,
endloses Band *n*

4761 TAPE LOOP CARTRIDGE (US), TAPE LOOP CASSETTE (GB) vr
f chargeur *m* à bande sans fin
e caseta *f* de cinta sin fin
i cassetta *f* di nastro senza fine
n eindloze-bandcassette
d Endlosbandkassette *f*

4762 TAPE MARK vr
f marque *f* de bande
e marca *f* de cinta
i marca *f* di nastro
n bandmerkteken *n*
d Bandmarke *f*

4763 TAPE NEUTRAL PLANE vr
f plan *m* neutre de la bande
e plano *m* neutro de la cinta
i piano *m* neutro del nastro
n neutraal bandvlak *n*
d neutrale Ebene *f* des Bandlaufes

4764 TAPE PERFORATING aud
f perforation *f* de bande
e perforación *f* de cinta
i perforazione *f* di nastro
n bandperforatie
d Bandlochung *f*

4765 TAPE PLAYBACK vr
f lecture *f* de bande,
reproduction *f*
e lectura *f* de cinta,
reproducción *f*
i lettura *f* di nastro,
riproduzione *f*

n reproduktie,
terugspelen *n*
d Abspielen *n* vom Band,
Wiedergabe *f*

4766 TAPE PLAYER, aud/vr
TAPE REPRODUCER
f dérouleur *m* son
e reproductor *m* de sonido,
tocacinta *f*
i riproduttore *m* di suono
n bandweergeefapparaat *n*
d Bandspieler *m*

TAPE POLISHING
see: POLISHING OF THE TAPE

4767 TAPE POSITION INDICATOR vr
f indicateur *m* de position
e indicador *m* de posición
i indicatore *m* di posizione
n positieaanwijzer
d Bandstellenanzeiger *m*

4768 TAPE PRESSURE vr
f pression *f* de la bande
e presión *f* de la cinta
i pressione *f* del nastro
n bandaandrukking
d Bandandruck *m*

4769 TAPE PRESSURE ERROR vr
f défaut *m* de pression
e defecto *m* de presión
i difetto *m* di pressione
n bandaandrukfout
d Bandandruckfehler *m*

4770 TAPE RECORD vr
f enregistrement *m* sur bande
e registro *m* en cinta
i registrazione *f* su nastro
n bandopname
d Bandaufnahme *f*

TAPE RECORDER
see: AUDIO TAPE RECORDER

TAPE RECORDING
see: AUDIO TAPE RECORDING

4771 TAPE RUN vr
f défilement *m* de la bande,
marche *f* de la bande
e marcha *f* de la cinta
i marcia *f* del nastro
n loop van de band
d Bandlauf *m*

4772 TAPE SCRATCH dis/vr
f rayure *f* de bande
e rascadura *f* de cinta
i graffio *m* di nastro
n bandkras
d Bandschramme *f*

4773 TAPE SLIPPAGE dis/vr
f glissement *m* de la bande
e deslizamiento *m* de la cinta
i scorrimento *m* del nastro
n bandslip
d Bandschlupf *m*

4774 TAPE SPEED vr
f vitesse *f* de la bande
e velocidad *f* de la cinta
i velocità *f* del nastro
n bandsnelheid
d Bandgeschwindigkeit *f*

4775 TAPE SPLICER aud/vr
A device for splicing magnetic tape by using splicing tape or by fusion with heat.
f colleuse *f* de bande
e empalmador *m* de cinta
i incollatrice *f* di nastro
n bandlasser
d Bandschneidelehre *f*

4776 TAPE SPOOL vr
f bobine *f* de bande
e bobina *f* de cinta
i bobina *f* di nastro
n bandspoel
d Bandspule *f*

4777 TAPE STRETCH vr
An effect produced by head intrusion and which can be used to advantage when controlled.
f allongement *m* de la bande
e alargamiento *m* de la cinta
i allungamento *m* del nastro
n banduitzetting
d Banddehnung *f*

4778 TAPE TENSION vr
f tension *f* de la bande
e tensión *f* de la cinta
i tensione *f* del nastro
n bandspanning
d Bandspannung *f*

4779 TAPE TENSION CONTROL vr
f réglage *m* de la tension de la bande
e regulación *f* de la tensión de la cinta
i regolazione *f* della tensione del nastro
n regeling van de bandspanning
d Regelung *f* der Bandspannung

4780 TAPE TEST vr
f essai *m* de la bande
e prueba *f* de la cinta
i prova *f* del nastro
n bandproef
d Bandprüfung *f*

4781 TAPE TRANSPORT aud/vr
f transport *m* de la bande
e transporte *m* de la cinta
i trasporto *m* del nastro
n bandtransport *n*
d Bandtransport *m*

4782 TAPE TRANSPORT LOCKING vr
CATCH
f verrouillage *m* du transport de la bande
e palanca *f* de enclavamiento del transporte
de la cinta

i dispositivo *m* di blocco del trasporto del nastro
n bandtransportgrendel
d Bandtransportriegel *m*

4783 TAPE WIDTH vr
f largeur *f* de bande
e anchura *f* de cinta
i larghezza *f* di nastro
n bandbreedte
d Bandbreite *f*

4784 TARGET,
TARGET ELECTRODE crt
The electrode in a camera tube which is subject to bombardment by the electron beam.
f cible *f*,
plaque *f* à accumulation
e blanco *m*
i bersaglio *m*
n trefplaat
d Fangelektrode *f*,
Speicherplatte *f*

4785 TARGET CAPACITANCE crt
The capacitance between the scanned area of a camera-tube target and the backplate.
f capacité *f* de la cible
e capacitancia *f* del blanco
i reattanza *f* capacitiva del bersaglio
n trefplaatcapacitantie
d Speicherplattenkapazitanz *f*

4786 TARGET CUT-OFF VOLTAGE crt
f tension *f* de coupure de la cible
e tensión *f* de corte del blanco
i tensione *f* di taglio del bersaglio
n afsnijspanning van de trefplaat
d Speicherplattengrenzspannung *f*

4787 TARGET LAYER crt
Layer of material deposited on the glass target of a TV camera tube.
f couche *f* de la cible
e capa *f* del blanco
i strato *m* del bersaglio
n trefplaatlaag
d Speicherplattenschicht *f*

4788 TARGET MESH crt
Fine wire mesh screen, with about 1.000.000 hole per square inch, placed in front of the target of an image orthicon camera tube.
f gaze *f* de cible
e gasa *f* de blanco
i garza *f* di bersaglio
n trefplaatgaas *n*
d Treffplattengaze *f*

4789 TARGET SPOT crt
f tache *f* sur la cible
e mancha *f* sobre el blanco
i macchia *f* sul bersaglio
n vlek op de trefplaat
d Fleck *m* auf der Speicherplatte

4790 TARGET VOLTAGE crt
f tension *f* de la cible
e tensión *f* del blanco
i tensione *f* del bersaglio
n trefplaatspanning
d Bildschirmspannung *f*

TEAM
see: CREW

4791 TEAR OUT,
TEARING dis
Break-up of a section of the TV picture by intermittent failure of the synchronizing circuits.
f déchirage *m* de l'image,
distorsion *f* en drapeau,
drapeau *m*
e efecto *m* bandera
i strappo *m*
n scheuren *n* van het beeld
d Zeilenausreissen *n*

4792 TECHNICAL DIRECTOR,
TECHNICAL OPERATIONS MANAGER stu/tv
f chef *m* de groupe,
directeur *m* des services techniques
e director *m* técnico,
jefe *m* de grupo técnico
i capo *m* di squadra tecnica
n bedrijfsleider
d Betriebsleiter *m*,
technischer Direktor *m*

4793 TECHNICAL OPERATOR stu/tv
An operator responsible for the set-up for converting the sound program(me) into electrical signals.
f musicien *m* metteur en ondes,
preneur *m* de son
e operador *m* de sonido
i tecnico *m* audio,
tecnico *m* per la ripresa sonora
n geluidstechnicus
d Tonmeister *m*

4794 TECHNICAL SUPERVISION fi/tv
f direction *f* technique
e supervisión *f* técnica
i direzione *f* tecnica
n technische leiding
d technische Direktion *f*

TELECAMERA,
TELEVISION CAMERA
see: CAMERA

4795 TELECAST (TO) tv
f diffuser v par la télévision
e difundir v por televisión
i diffondere v per televisione
n via de televisie uitzenden v
d über Fernsehen übertragen v,
über Fernsehen verbreiten v

4796 TELECASTING (US),
TELEVISION BROADCAST (GB) tv

f émission *f* de télévision,
émission *f* télévisée
e transmisión *f* por televisión
i trasmissione *f* per televisione
n televisie-uitzending
d Fernsehsendung *f*

4797 TELECHROME crt/ctv
A special paint for TV decorations.
f téléchrome *m*
e telecromo *m*
i telecromo *m*
n televisieverf
d Fernsehanstrichfarbe *f*

TELECINE,
TELECINE MACHINE,
TELECINE PROJECTOR
see: FILM SCANNER

4798 TELECINE ALIGNMENT ctv
f ajustage *m* du télécinéma
e ajuste *m* del telecinema
i aggiustaggio *m* del telecinema
n instelling van de filmaftaster
d Einstellung *f* des Filmabtasters

TELECINE MACHINE CONTROL
see: MACHINE CONTROL

4799 TELECINE OPERATOR tv
Operator who adjusts and tends the machines which produce the electronic picture for the TV program(me)s carried on film.
f opérateur *m* de télécinéma
e operador *m* de telecinema
i operatore *m* di telecinema
n filmaftasteroperateur
d Filmabtasteroperator *m*

4800 TELECINE SCAN tv
f analyse *f* par télécinéma
e exploración *f* por telecinema
i analisi *f* per telecinema
n filmaftasting
d Filmabtastung *f*

4801 TELECINE TRANSMISSION tv
f transmission *f* par télécinéma
e transmisión *f* por telecinema
i trasmissione *f* per telecinema
n filmoverdracht
d Filmübertragung *f*

4802 TELECONTROL rep/tv
f télécommande *f*
e telemando *m*
i telecomando *m*
n verrebediening
d Fernbedienung *f*

4803 TELEFILM tv
Film specially made for reproduction by a TV system.
f film *m* de télévision,
téléfilm *m*
e telefilm *m*
i telefilm *m*
n film voor televisiedoeleinden
d Fernsehfilm *m*

4804 TELEGENIC tv
f télégénique adj
e telegénico adj
i telegenico adj
n telegeniek adj
d telegen adj

4805 TELEMETRY cpl
Passing of information relating to measurements by radio or wire from a remote site, e.g. satellite to earth.
f télémétrie *f*
e telemetría *f*
i telemetria *f*
n verremeting
d Fernmessung *f*

4806 TELEMICROSCOPE med/tv
An apparatus used in TV microscopy, enabling to obtain magnified views of an object at any distance from 5 cm (2 in.) upwards.
f télémicroscope *m*
e telemicroscopio *m*
i telemicroscopio *m*
n telemicroscoop
d Telemikroskop *n*

TELEPHONY CHANNEL
see: SPEECH CHANNEL

TELEPHOTO LENS
see: LONG-FOCUS LENS

4807 TELEPLAYER, vr
TELEVISION CASSETTE PLAYER
f projecteur *m* de télécassettes
téléplayer *m*
e proyector *m* para casetas de telefilm
i proiettore *m* per cassette di telefilm
n projectieapparaat *n* voor televisiefilm-cassettes
d Fernsehfilmkassettenwiedergabegerät *n*

4808 TELERAN, tv
TELEVISION RADAR AIR NAVIGATION
Blind flying system with course indication by TV from the ground.
f téléran *m*
e telerán *m*
i teleran *m*
n teleran
d Teleran *n*

TELERECORDING (GB)
see: KINESCOPE RECORDING

4809 TELERECORDING EQUIPMENT tv
Equipment for the purpose of recording TV program(me)s either on film or magnetic tape.
f vidigraphe *m*
e telerregistrador *m*
i equipaggiamento *m* di teleregistrazione
n beeldbandopnameapparatuur

d FAZ-Anlage *f*,
Fernsehaufzeichnungsanlage *f*

4810 TELERECORDING EQUIPMENT FOR MAGNETIC TAPE vr
f magnétoscope *m*
e magnetoscopio *m*
i registratore *m* videomagnetico
n beeldbandopnameapparaat *n*, beeldbandrecorder
d Magnetband-FAZ-Anlage *f*

4811 TELESCOPE AND QUARTZ ROD ENDOSCOPE med
f endoscope *m* à télescope et tige en quartz
e endoscopio *m* con telescopio y varilla de cuarzo
i endoscopio *m* con telescopio e baretta di quarzo
n endoscoop met telescoop en kwartsstaaf
d Endoskop *n* mit Teleskop und Quarzstab

4812 TELESCOPE TUBE crt
A cathode-ray tube in which an infrared object is focused on a photoemitting surface.
f télescope *m* électronique à rayons infra-rouges
e telescopio *m* electrónico infrarrojo
i telescopio *m* elettronico a raggi infrarossi
n infrarood-elektronentelescoop
d Infrarot-Elektronenteleskop *n*

4813 TELESCOPIC AERIAL, TELESCOPIC ANTENNA aea
f antenne *f* télescopique
e antena *f* telescópica
i antenna *f* telescopica
n telescoopantenne
d Teleskopantenne *f*

4814 TELETRANSCRIPTION, TELEVISION RECORDING tv
f enregistrement *m* d'émissions télévisées
e registro *m* de emisiones de televisión
i registrazione *f* d'emissione di televisione
n optekenen *n* van televisieuitzendingen
d Aufzeichnung *f* von Fernsehsendungen

4815 TELEVISE (TO) tv
f téléviser v
e televisar v
i teletrasmettere v
n uitzenden v van televisie
d durch Fernsehen übertragen v

4816 TELEVISION tv
f télévision *f*
e televisión *f*
i televisione *f*
n televisie
d Fernsehen *n*

4817 TELEVISION ACOUSTICS rec
f acoustique *f* de télévision
e acústica *f* de televisión
i acustica *f* di televisione
n televisieakoestiek
d Fernsehakustik *f*

4818 TELEVISION ADAPTATION fi/tv
f adaptation *f* pour la télévision
e adaptación *f* para la televisión
i adattamento *m* per la televisione
n televisiebewerking
d Fernsehbearbeitung *f*

4819 TELEVISION ADVERTISING tv
f publicité *f* télévisée
e publicidad *f* por televisión, publicidad *f* televisada
i pubblicità *f* per televisione
n reclame op de televisie, televisiereclame
d Fernsehwerbung *f*

4820 TELEVISION AERIAL, TELEVISION ANTENNE aea/tv
f antenne *f* de télévision
e antena *f* de televisión
i antenna *f* di televisione
n televisieantenne
d Fernsehantenne *f*

4821 TELEVISION ANNOUNCER tv
f speaker *m* de télévision
e anunciador *m* de televisión
i annunziatore *m* di televisione
n televisieomroeper
d Fernsehansager *m*

4822 TELEVISION ARCHIVES tv
f archives *pl* centrales de télévision, téléthèque *f*
e archivo *m* central de televisión
i archivio *m* centrale di televisione
n televisiearchief *n*
d Fernseharchiv *n*

4823 TELEVISION AUDIENCE tv
f téléspectateurs *pl*
e espectadores *pl* de televisión, telespectadores *pl*
i telespettatori *pl*
n televisiekijkers *pl*
d Fernsehzuschauer *pl*

4824 TELEVISION BAND, TELEVISION BROADCAST BAND tv
f bande *f* de télévision
e banda *f* de televisión
i banda *f* di televisione
n televisieband
d Fernsehband *n*

4825 TELEVISION BANDWIDTH LIMITATIONS tv
f limitations *pl* de la largeur de bande en télévision
e limitaciones *pl* de la anchura de banda en televisión
i limitazioni *pl* della larghezza di banda in televisione
n bandbreedtegrenzen *pl* bij televisie
d Bandbreitegrenzen *pl* für Fernsehen

TELEVISION BROADCAST (GB)
see: TELECASTING

4826 TELEVISION BROADCAST STATION tv

f station *f* émettrice de télévision
e estación *f* emisora de televisión
i stazione *f* trasmettitrice di televisione
n televisiezendstation *n*
d Fernsehsendeanlage *f*, Fernsehstation *f*

4827 TELEVISION BROADCASTING tv
f radiodiffusion *f* visuelle
e videodifusión *f*
i radiodiffusione *f* di televisione
n televisieomroep
d Fernsehrundfunk *m*

4828 TELEVISION CABLE cpl/tv
f câble *m* de télévision
e cable *m* de televisión
i cavo *m* di televisione
n televisiekabel
d Fernsehkabel *n*

TELEVISION CAMERA
see: CAMERA

TELEVISION CAMERA FOR INDIRECT LARYNGOSCOPY
see: CAMERA FOR INDIRECT LARYNGOSCOPY

TELEVISION CAMERA LENS
see: CAMERA LENS

TELEVISION CAMERA OPTICS,
see: CAMERA OPTICS

TELEVISION CAR
see: SHOOTING BRAKE

4829 TELEVISION CENTER(RE) tv
f centre *m* de télévision
e centro *m* de televisión
i centro *m* di televisione
n televisiecentrum *n*, televisiecomplex *n*
d Fernsehkomplex *n*, Fernsehzentrum *n*

4830 TELEVISION CHANNEL tv
f canal *m* de télévision
e canal *m* de televisión
i canale *m* di televisione
n televisiekanaal *n*
d Fernsehkanal *m*

4831 TELEVISION CHANNEL USING LOWER SIDEBAND cpl/tv
f canal *m* inversé
e canal *m* inverso
i canale *m* invertito
n televisiekanaal *n* met gebruik van de onderzijbanden
d Fernsehkanal *m* mit spiegelbildlicher Lage der Träger

4832 TELEVISION CHANNEL USING UPPER SIDEBAND tv
f canal *m* direct
e canal *m* directo
i canale *m* diretto
n televisiekanaal *n* met gebruik van bovenzijbanden
d Fernsehkanal *m* mit normaler Lage der Träger

4833 TELEVISION CHART, TELEVISION TEST CARD tv
f mire *f* de télévision
e imagen *f* de prueba de televisión
i immagine *f* di prova televisiva
n televisietoetsbeeld *n*
d Fernsehtestbild *n*

4834 TELEVISION COMMENTATOR tv
f commentateur *m* de télévision
e comentador *m* de televisión
i commentatore *m* di televisione
n televisiecommentator
d Fernsehkommentator *m*

4835 TELEVISION CONTROLLED MISSILE tv
f projectile *m* guidé par télévision
e proyectil *m* mandado por televisión
i proiettile *m* comandato per televisione
n door televisie gestuurd projectiel *n*
d fernsehtechnisch gesteuerter Flugkörper *m*

4836 TELEVISION CURRENT AFFAIRS PROGRAM(ME)S tv
f actualités *pl* télévisées
e actualidades *pl* televisadas
i programmi *pl* d'attualità
n actualiteitenprogramma's
d Zeitgeschehen *n*

4837 TELEVISION DESIGNER tv
The operator concerned with the whole complex operation of bringing TV pictures to the screen.
f programmateur *m*
e planificador *m* del programa
i organizzatore *m* del programma
n programmaontwerper, televisieontwerper
d Fernsehprogrammacher *m*, Fernsehprogrammgestalter *m*

4838 TELEVISION DIRECT TRANSMISSION, TELEVISION LIVE TRANSMISSION tv
A TV system in which the object to be televised is focused directly on to the scanning surface.
f émission *f* de télévision en direct
e emisión *f* televisiva en directo, emisión *f* viva de televisión
i emissione *f* televisiva in diretto, telecronaca *f* diretta
n live-televisie-uitzending, live-uitzending
d Fernseh-Livesendung *f*

4839 TELEVISION DIRECTOR tv
The creative head of an artistic and technical team.
f directeur *m* de la télévision
e director *m* de televisión

i direttore *m* di televisione
n televisiedirecteur
d Fernsehdirektor *m*

4840 TELEVISION DISTRIBUTION SATELLITE tv
f satellite *m* de répartition par télévision
e satélite *m* de repartición por televisión
i satellite *m* di ripartizione per televisione
n televisieverdelingssatelliet
d Fernsehverteilungssatellit *m*

4841 TELEVISION DRAMA tv
f dramatique *f* de télévision,
télédrame *m*
e telecomedia *f*
i teledrama *m*,
teleromanzo *m*
n televisiespel *n*
d Fernsehdrama *n*,
Fernsehspiel *n*

4842 TELEVISION ENGINEER tv
f ingénieur *m* de télévision
e ingeniero *m* de televisión
i ingegnere *m* di televisione
n televisie-ingenieur
d Fernsehingenieur *m*

4843 TELEVISION ENGINEERING tv
f technique *f* de la télévision
e técnica *f* de la televisión
i tecnica *f* della televisione
n televisietechniek
d Fernsehtechnik *f*

4844 TELEVISION EPISCOPE opt/tv
f épiscope *m* de télévision
e episcopio *m* de televisión
i episcopio *m* di televisione
n televisie-episcoop
d Fernsehepiskop *n*

4845 TELEVISION EYE tv
f oeuil *m* de télévision
e ojo *m* de televisión
i occhio *m* di televisione
n televisie-oog *n*
d Fernsehauge *n*

TELEVISION FILM LEADER
see: LEADER FOR TELEVISION FILM

4846 TELEVISION FILM LIBRARY rec
f filmothèque *f* pour la télévision
e cinemateca *f* para la televisión
i filmoteca *f* per la televisione
n televisiefilmotheek
d Fernsehfilmothek *f*,
Fernsehfilmtrésor *m*

TELEVISION FILM RECORDING (GB)
see: KINESCOPE RECORDING

TELEVISION FILM SCANNER
see: FILM SCANNER

4847 TELEVISION FOLDED PICTURE dis
An overlapping, in the vertical or horizontal direction, of portions of the television image.
f recouvrement *m* d'image
e recubrimiento *m* de imagen
i ricoprimento *m* d'immagine
n beeldoverlapping
d Bildüberlappung *f*

4848 TELEVISION FREQUENCY CONVERTER UNIT tv
f convertisseur *m* de canal de télévision
e convertidor *m* de canal de televisión
i convertitore *m* di canale di televisione
n televisiekanaalomzetter
d Fernsehkanalumsetzer *m*

4849 TELEVISION IMAGE, TELEVISION PICTURE tv
f image *f* de télévision
e imagen *f* de televisión
i immagine *f* di televisione
n televisiebeeld *n*
d Fernsehbild *n*

4850 TELEVISION INTERCOMMUNICATION NETWORK tv
f réseau *m* d'intercommunication en télévision
e red *f* de intercomunicación en televisión
i rete *f* d'intercomunicazione in televisione
n ruggespraaknetwerk *n* in televisiesysteem
d Fernsehsprechnetz *n*

4851 TELEVISION INTERFERENCE, TVI dis
Interference produced in TV receivers by amateur radio and other transmitters.
f interférence *f* en télévision
e interferencia *f* en televisión
i interferenza *f* in televisione
n televisiestoring
d Fernsehstörung *f*

4852 TELEVISION LEVEL, TELEVISION SIGNAL LEVEL tv
f niveau *m* du signal de télévision
e nivel *m* de la señal de televisión
i livello *m* del segnale di televisione
n niveau *n* van het televisiesignaal
d Fernsehsignalpegel *m*

TELEVISION LINE
see: LONG-DISTANCE TELEVISION CIRCUIT

4853 TELEVISION LINE NUMBER tv
The ratio of the raster height to the half period of a periodic test pattern.
f lignage *m* en télévision
e número *m* de líneas en televisión
i numero *m* di linee in televisione
n aantal *n* lijnen in televisie
d Fernsehzeilenzahl *f*

4854 TELEVISION LINK cpl/tv
f liaison *f* par lignes de télévision
e enlace *m* por líneas de televisión

i connessione *f* per linee di televisione
n televisieleidingverbinding
d Fernsehleitungsverbindung *f*

4855 TELEVISION MECHANIC tv
f technicien *m* de télévision
e técnico *m* de televisión
i tecnico *m* di televisione
n televisietechnicus
d Fernsehmechanikus *m*, Fernsehtechniker *m*

4856 TELEVISION NETWORK cpl
f réseau *m* de télévision
e red *f* de televisión
i rete *f* di televisione
n televisienet *n*
d Fernsehnetz *n*

4857 TELEVISION NETWORK SWITCHING CENTER(RE) cpl/tv
f centre *m* nodal de télévision
e centro *m* nodal de televisión
i centro *m* nodale di televisione
n knooppunt *n* van televisieleidingen
d Fernsehleitungsschaltstelle *f*

4858 TELEVISION NETWORKS MAINTENANCE tv
f entretien *m* de réseaux de télévision
e entretenimiento *m* de redes de televisión
i manutenzione *f* di reti di televisione
n onderhoud *n* van televisienetwerken
d Wartung *f* von Fernsehnetzen

4859 TELEVISION OF SURGICAL OPERATIONS med
f télévision *f* d'opérations chirurgicales
e televisión *f* de operaciones quirúrgicas
i televisione *f* d'operazioni chirurgiche
n televisieweergave van chirurgische operaties
d Fernsehwiedergabe *f* von chirurgischen Operationen

4860 TELEVISION PERCENTAGE MODULATION rep/tv
For the aural transmitter of TV broadcast stations, a frequency swing of ± 25 KC is defined as 100 per cent modulation.
f pourcentage *m* de modulation en télévision
e porcentaje *m* de modulación en televisión
i percentuale *f* di modulazione in televisione
n modulatiepercentage *n* bij televisie
d Modulationsgrad *m* in Prozent beim Fernsehen

TELEVISION PICKUP SYSTEM
see: PICTURE PICKUP SYSTEM

4861 TELEVISION PRINCIPLES tv
f principes *pl* de la télévision
e principios *pl* de la televisión
i principi *pl* della televisione
n grondslagen *pl* van de televisie
d Grundlagen *pl* des Fernsehens

4862 TELEVISION PROGRAM(ME) tv
f programme *m* de télévision
e programa *m* de televisión
i programma *m* di televisione
n televisieprogramma *n*
d Fernsehprogramm *n*

TELEVISION RADAR AIR NAVIGATION
see: TELERAN

4863 TELEVISION RECEIVER tv
f récepteur *m* de télévision, téléviseur *m*
e receptor *m* de televisión, televisor *m*
i ricevitore *m* di televisione, ricevitore *m* televisivo, televisore *m*
n televisieontvanger, televisietoestel *n*
d Fernsehempfänger *m*

4864 TELEVISION RECEPTION tv
f réception *f* de télévision
e recepción *f* de televisión
i ricezione *f* di televisione
n televisieontvangst
d Fernsehempfang *m*

4865 TELEVISION RECONNAISSANCE tv
Reconnaissance in which TV is used to transmit a scene from the reconnoitering point to another location on the surface or in the air.
f reconnaissance *f* par télévision
e reconocimiento *m* por televisión
i ricognizione *f* per televisione
n televisieverkenning
d Fernsehaufklärung *f*

TELEVISION RECORDING
see: TELETRANSCRIPTION

4866 TELEVISION RELAY SIGNALS tv
Input signals to, or output signals from, a TV relay station, consisting of a standard composite signal and, optionally, the accompanying audio, auxiliary, and control signals available on separate circuits.
f signaux *pl* de station de télévision de retransmission
e señales *pl* de estación de televisión retransmisora
i segnali *pl* di stazione di televisione di ritrasmissione
n relaisstationsignalen *pl*
d Ballsendersignale *pl*

4867 TELEVISION RELAY SYSTEM, TELEVISION REPEATER tv
A repeater that transmits TV signals from point to point by using radio waves in free space as a medium.
f système *m* de télévision à relais
e sistema *m* de televisión de relé
i sistema *m* di televisione a relè
n relaiszendsysteem *n*
d Relaissendesystem *n*

4868 TELEVISION SERVICE svs
f service *m* de télévision

e servicio *m* de televisión
i servizio *m* di televisione
n televisieservice
d Fernsehkundendienst *m*

4869 TELEVISION SHOOTING SCRIPT tv
f découpage *m* pour télévision, scénario *m* pour télévision
e escenario *m* para televisión
i copione *m* per televisione, sceneggiatura *f* per televisione
n televisiedraaiboek *n*
d Fernsehdrehbuch *n*

4870 TELEVISION SIGNAL (US), VISION SIGNAL (GB) tv
f signal *m* de télévision
e señal *f* de televisión
i segnale *m* di televisione
n televisiesignaal *n*
d Fernsehsignal *n*

TELEVISION SIGNAL LEVEL
see: TELEVISION LEVEL

4871 TELEVISION SOUND aud/tv
f son *m* de télévision
e sonido *m* de televisión
i suono *m* di televisione
n televisiegeluid *n*
d Fernsehton *m*

4872 TELEVISION STANDARD SIGNAL tv
A signal which conforms to the TV transmission standards.
f signal *m* étalon de télévision
e señal *f* normal de televisión, señal *f* patrón de televisión
i segnale *m* campione di televisione
n standaardtelevisiesignaal *n*
d Normfernsehsignal *n*

4873 TELEVISION STANDARDS tv
f normes *pl* de télévision
e normas *pl* de televisión
i norme *pl* di televisione
n televisienormen *pl*
d Fernsehnormen *pl*

4874 TELEVISION STATION LINK tv
f chaîne *f* de stations émettrices de télévision
e cadena *f* de estaciones de televisión
i ponte *m* radio per televisione
n televisiekoppelnet *n*
d Fernsehbrücke *f*, Fernsehrelaisstrecke *f*

4875 TELEVISION SYSTEM tv
A system in accordance with a set of specifications which completely defines the radiated TV signal and method of picture production and permits the reproduction of the picture.
f système *m* de télévision
e sistema *m* de televisión
i sistema *m* di televisione
n televisiesysteem *n*
d Fernsehsystem *n*

4876 TELEVISION SYSTEM CONVERTER tv
Equipment for converting the video signal from one TV system into the corresponding signal for another system.
f convertisseur *m* de norme
e convertidor *m* de sistemas de televisión
i convertitore *m* di norma televisiva, convertitore *m* di sistema televisivo
n standaardomzetter
d Normwandler *m*

TELEVISION TEST CARD
see: TELEVISION CHART

4877 TELEVISION TOWER aea/tv
f tour *f* de télévision
e torre *f* de televisión
i torre *f* di televisione
n televisiemast
d Fernsehturm *m*

4878 TELEVISION TRANSMITTER tv
f émetteur *m* de télévision, transmetteur *m* de télévision
e emisor *m* de televisión, transmisor *m* de televisión
i emettitore *m* di televisione, trasmettitore *m* di televisione
n televisiezender
d Fernsehsender *m*

4879 TELEVISION TRANSMITTING STATION tv
f station *f* émettrice de télévision
e estación *f* emisora de televisión
i stazione *f* emettitrice di televisione
n televisiezendstation *n*
d Fernsehsendestation *f*

4880 TELEVISION WAVEFORM tv
Composite waveform of TV transmission resulting from the joint modulation of the carrier by picture signals and synchronizing pulses.
f forme *m* d'onde de télévision
e forma *f* de onda de televisión
i forma *f* d'onda di televisione
n televisiegolfvorm
d Fernsehwellenform *f*

4881 TELSTAR tv
The first of the experimental, low-altitude, active satellites which was launched into orbit on 10 July 1962.
f telstar *m*
e telstar *m*
i telstar *m*
n telstar
d Telstar *m*

4882 TEMPORAL AND SPATIAL FACTORS ct
f facteurs *pl* de temps et d'espace
e factores *pl* de tiempo y de espacio
i fattori *pl* di tempo e di spazio
n tijds- en ruimtefactoren *pl*
d Zeit- und Raumfaktoren *pl*

4883 TERRESTRIAL CONNECTING NETWORKS tv
f réseaux *pl* de connexion terrestres
e redes *pl* de conexión terrestres
i reti *pl* di connessione terrestri
n grondverbindingsnetten *pl*
d Erdeverbindungsnetze *pl*

TERRESTRIAL MAGNETIC FIELD
see: EARTH'S MAGNETIC FIELD

4884 TEST CARD,
TEST CHART (GB),
TEST PATTERN (US) ge
Chart displaying geometrical patterns and fine detail which is transmitted at specified times for checking purposes.
f image *f* test,
mire *f*
e carta *f* de ajuste,
imagen *f* de prueba
i figura *f* di prova
n toetsbeeld *n*
d Testbild *n*

4885 TEST FILM tv
f film *m* d'essai
e película *f* de prueba
i pellicola *f* di prova
n proeffilm
d Probefilm *m*,
Testfilm *m*

TEST KIT
see: ACCESSORY KIT

4886 TEST LINE cpl
f ligne *f* d'essai,
ligne *f* test
e línea *f* de prueba
i linea *f* di prova
n toetslijn
d Prüfleitung *f*

TEST MICROPHONE
see: PROBE MICROPHONE

4887 TEST SHOT tv
f bout *m* d'essai,
essai *m* image
e toma *f* de prueba
i ripresa *f* di prova
n proefopname
d kalte Probe *f*,
Probeaufnahme *f*

TEST SIGNAL
see: MEASURING SIGNAL

TEST SIGNAL GENERATOR
see: MEASURING SIGNAL GENERATOR

TEST STRIP
see: HAND TEST

4888 THEATRE ACOUSTICS aud
f acoustique *f* du théâtre
e acústica *f* del teatro
i acustica *f* del teatro
n theaterakoestiek
d Theaterakustik *f*

4889 THEATRE NETWORK TELEVISION tv
A TV program(me) transmitted from a theatre.
f émission *f* théâtrale de télévision,
théâtre *m* télévisé
e televisión *f* en la red teatral
i emissione *f* di televisione dal teatro
n televisie-uitzending uit een theater
d Fernsehübertragung *f* aus einem Theater,
Theaterübertragung *f*

4890 THERMOPLASTIC RECORDING aud/vr
A method of electron beam recording of information in picture or other form on a deformable plastic film.
f enregistrement *m* thermoplastique
e grabación *f* termoplástica,
registro *m* termoplástico
i registrazione *f* termoplastica
n thermoplastische opneming
d thermoplastische Aufzeichnung *f*

4891 THERMOPLASTIC RECORDING TAPE aud/vr
f bande à surface thermoplastique
e cinta *f* de superficie termoplástica
i nastro *m* a superficie termoplastica
n band met thermoplastisch oppervlak
d Band *n* mit thermoplastischer Oberfläche

4892 THREE-DIMENSIONAL ANIMATION ani
f animation *f* à trois dimensions
e animación *f* tridimensional
i animazione *f* tridimensionale
n driedimensionale tekenfilm
d dreidimensionaler Zeichentrick *m*,
dreidimensionaler Zeichentrickfilm *m*

4893 THREE-ELEMENT AERIAL,
THREE-ELEMENT ANTENNA aea
Aerial (antenna) consisting of a directing dipole and a receiving and a directing dipole.
f antenne *f* à trois éléments
e antena *f* de tres elementos
i antenna *f* a tre elementi
n drie-elementenantenne
d Dreielementenantenne *f*

4894 THREE-GUN COLO(U)R PICTURE TUBE,
THREE-GUN COLO(U)R TUBE crt
f tube *m* image couleur à trois canons électroniques,
tube *m* trichrome
e tubo *m* imagen de colores de tres cañones electrónicos,
tubo *m* tricolor
i tubo *m* immagine a colori con tre cannoni elettronici,
tubo *m* tricolore
n beeldkleurenbuis met drie elektronenkanonnen,
driekanonsbeeldbuis

d Dreifarbenröhre *f*,
Dreistrahlfarbfernsehröhre *f*,
Dreistrahlröhre *f*

4895 THREE-TUBE CAMERA, crt/ctv
THREE-TUBE COLO(U)R CAMERA
A camera with three image orthicon tubes in a relay lens configuration.
f caméra *f* à trois tubes couleur
e cámara *f* de tres tubos de colores
i camera *f* a tre tubi a colori
n driebuizencamera
d Dreiröhrenfarbfernsehkamera *f*,
Dreiröhrenkamera *f*

4896 THREE-TUBE TELECINE, ctv
THREE-TUBE VIDICON TELECINE CAMERA
f télécinéma *f* à trois tubes
e telecinema *f* de tres tubos
i telecinema *f* a tre tubi
n filmaftaster met drie buizen
d Dreiröhrenfilmabtaster *m*

THRESHOLD OF AUDIBILITY
see: AUDIBILITY THRESHOLD

THRESHOLD OF LUMINESCENCE
see: LUMINESCENCE THRESHOLD

THRESHOLD TUBE (GB)
see: CLIPPER TUBE

4897 THRESHOLD VOLTAGE cpl
f tension *f* de seuil
e tensión *f* de umbral
i tensione *f* di soglia
n drempelspanning
d Schwellenspannung *f*

4898 TIGHT ALIGNMENT cpl
In TV it is compulsory to operate with staggered radio-frequency circuits, which makes it necessary to have one frequency governing the radio- and/or the intermediate-frequency amplifiers in the sound channel.
f alignement *m* serré
e sintonía *f* a frecuencia piloto
i sintonia *f* a frequenza pilota
n instelling van de doorlaatkromme
d Abgleich *m* der Durchlasskurve

4899 TIGHT FRAMING tv
f cadrage *m* serré
e encuadro *m* apretado
i inquadratura *f* stretta
n enge omlijsting
d enge Umrahmung *f*

4900 TILT dis
Upward tilt of the horizontal scanning of a picture tube caused by distortion of the TV waveform.
f déclivité *f*,
dénivellation *f*
e caída *f*
i avvallamento *m* dell'impulso
n impulsvervorming,
sprongverval *n*
d Dachschräge *pl*,
Impulsabflächung *f*

4901 TILT tv
Pivotal movement of the camera in a vertical plane.
f panoramique *m* vertical
e panoramización *f* vertical
i inclinazione *f* della macchina
n verticale zwenking
d Senkrechtschwenk *m*

4902 TILT-AND-BEND SHADING dis
Adjusted compensation for inequality of effectiveness of a mosaic in translating an image of an external screen into a video signal.
f correction *f* de défauts de mosaïque
e corrección *f* de faltas de mosaico
i correzione *f* di difetti di mosaico
n correctie van mozaïekfouten
d Korrektur *f* von Mosaikfehlern

4903 TILT MIXER dis
f correcteur *m* de déformation de ligne
e corrector *m* de deformación de línea
i correttore *m* di deformazione di linea
n lijnvervormingscompensator
d Störsignalkompensator *m*,
Zeilenverzerrungskompensator *m*

4904 TILTABLE VIEWFINDER opt
f viseur *m* orientable
e visor *m* basculable
i mirino *m* orientativo
n verstelbare zoeker
d schwenkbarer Sucher *m*

TILTING (US)
see: PAN DOWN

4905 TIME-ADJACENT FIELDS tv
Fields which follow one another in time but which do not necessarily contain similar information.
f trames *pl* consécutives dans le temps
e campos *pl* consecutivos en el tiempo
i trame *pl* consecutive nel tempo
n opeenvolgende rasters *pl* in chronologische volgorde
d zeitlich unmittelbar aufeinanderfolgende Teilbilder *pl*

4906 TIME AND CONTROL CODE, vr
TIME AND LOCATION CODE
f code *m* temporel de commande
e código *m* de la pista de órdenes
i codice *m* di tempo
n tijd- en stuurcode
d Merkspurkode *m*

4907 TIME AXIS, cpl
TIMEBASE LINE
f axe *m* des temps
e eje *m* de tiempo
i asse *m* di tempo
n tijdas
d Zeitachse *f*

4908 TIMEBASE cpl
The waveform of the current pulses used to deflect the electron beam in a cathode-ray tube during scanning, which is a function of time.
f base *f* de temps
e base *f* de tiempo
i base *f* di tempi
n tijdbasis
d X-Achse *f*,
Zeitbasis *f*

4909 TIMEBASE ERROR CORRECTION cpl
f correcteur *m* d'erreur de base de temps
e corrector *m* de error de base de tiempo
i correttore *m* d'errore di base di tempi
n tijdbasisfoutcorrectie
d Zeitfehlerausgleicher *m*

4910 TIMEBASE FREQUENCY cpl
f fréquence *f* de la base de temps
e frecuencia *f* de la base de tiempo
i frequenza *f* della base di tempi
n tijdbasisfrequentie
d Kippfrequenz *f*

TIMEBASE GENERATOR
see: SWEEP GENERATOR

4911 TIMEBASE PRINT cpl
f plaque *f* base de temps
e placa *f* de base de tiempo
i placca *f* di base di tempi
n tijdbasisprint
d Zeitbasisleiterplatte *f*

TIMEBASE VOLTAGE
see: SWEEP VOLTAGE

4912 TIME-CONTROLLED MASTER SWITCHER cpl stu
f commutateur *m* principal à commande chronométrique
e conmutador *m* principal de mando cronométrico
i commutatore *m* principale a comando cronometrico
n hoofdschakelaar met chronometerbesturing
d Hauptschalter *m* mit Chronometersteuerung

4913 TIME DELAY aud
f temps *m* de transit
e tiempo *m* de tránsito
i tempo *m* di transito
n looptijd
d Laufzeit *f*

4914 TIME EQUALIZER dis
f correcteur *m* d'écho
e corrector *m* de eco
i correttore *m* d'eco
n echocorrector
d Echoentzerrer *m*

4915 TIME FUNCTION ge/tv
Formulation of the time dependency of a quantity.
f fonction *f* du temps
e función *f* del tiempo
i funzione *f* del tempo
n tijdsfunctie
d Zeitfunktion *f*

4916 TIME INTEGRAL ge/tv
An integral which contains a time function as integrands.
f intégrale *f* de temps
e integral *f* de tiempo
i integrale *f* di tempo
n tijdsintegraal
d Zeitintegral *n*

4917 TIME LAPSE SHOOTING fi/tv
f tournage *m* d'un accéléré
e toma *f* de vistas acelerada
i ripresa *f* cinematografica accelerata
n versnelde opname
d Zeitrafferverfahren *n*

4918 TIME MULTIPLEX ctv
Colo(u)r TV system characterized by a successive transmission of pulse samples of each of several signals.
f transmission *f* successive de signaux
e transmisión *f* sucesiva de señales
i trasmissione *f* successiva di segnali
n tijdmultiplexsysteem *n*
d Zeitfolgeverfahren *n*,
Zeitmultiplexsystem *n*

4919 TIME OF ARRIVAL DIFFERENCE aud
f différence *f* de temps d'arrivée
e diferencia *f* de tiempo de llegada
i differenza *f* di tempo d'arrivo
n verschil *n* in aankomsttijd
d Ankunftszeitdifferenz *f*

4920 TIME PATTERN tv
A TV picture-tube presentation of horizontal and vertical lines or dot rows generated by two stable frequency sources operating at multiples of the line and field frequencies.
f diagramme *m* horaire
e diagrama *m* horario
i diagramma *m* orario
n tijdprent
d Zeitplan *m*

TIMING IN ANIMATION
see: ANIMATION TIMING

4921 TIMING PULSE cpl
f impulsion *f* de marquage
e impulso *m* de tiempo
i impulso *m* di tempo
n tijdimpuls
d Taktimpuls *m*,
Zeitimpuls *m*

4922 TIMING PULSE GENERATOR cpl
f générateur *m* d'impulsions de marquage
e generador *m* de impulsos de tiempo
i generatore *m* d'impulsi di tempo
n tijdimpulsgenerator
d Taktimpulsgeber *m*,
Zeitimpulsgeber *m*

4923 TINGE ct
A trace of added colo(u)r.
f teinte *f* de couleur additionnée
e reflejo *m* de color adicionado
i tinta *f* di colore aggiunto
n zweem van toegevoegde kleur
d Anhauch *m*

TINT
see: BROKEN WHITE

4924 TIP vr
Component part of the magnetic head.
f pointe *f*
e punta *f*
i punta *f*
n tip
d Spitze *f*

4925 TIP AND GUIDE POSITION vr
The position of the mechanical parts of the head with respect to the tape.
f position *f* des pointes et du guide-bande
e posición *f* de las puntas y de la guía-cinta
i posizione *f* delle punte e del guida-nastro
n tip- en geleiderpositie
d Spitzen- und Führerlage *f*

4926 TIP ENGAGEMENT, TIP PENETRATION vr
f pénétration *f* des têtes
e penetración *f* de cabeza
i penetrazione *f* delle testine
n tippenetratie
d Kopfbandkontakt *m*

4927 TIP PROJECTION vr
f dépassement *m* des têtes
e avance *m* de cabeza
i avanzamento *m* di testina, sporgenza *f* delle testine
n kopvoorsprong
d Kopfüberstand *m*, Kopfvorsprung *m*

TITLE
see: MAIN TITLE

4928 TITLE INSERT tv
f carton *m* de générique
e cartón *m* de título
i inserto *m* di titolo
n titelinlas
d Titelkarton *n*

4929 TONAL QUALITY, TONAL RENDERING ctv/tv
f qualité *f* sonore
e cualidad *f* sonora
i qualità *f* sonora
n geluidskwaliteit
d Tonqualität *f*

4930 TONE ct
A slight variant of a colo(u)r.
f nuance *f* de couleur
e tinte *f*
i tinta *f*
n kleurnuance
d Farbton *m*, Färbung *f*

4931 TONE CONTROL aud/tv
f commande *f* de tonalité sonore, réglage *m* de tonalité sonore
e regulación *f* de tonalidad sonora
i regolazione *f* di tonalità sonora
n geluidstoonregelaar, toonregelaar
d Klangblende *f*, Toneinsteller *m*

4932 TONE CONTROL ctv/opt
f réglage *m* de nuance de couleur
e regulación *f* de tinte
i regolazione *f* di tinta
n kleurnuanceregeling
d Farbtonregelung *f*

4933 TONE KEY rep/tv
f touche *f* de tonalité
e tecla *f* de tonalidad
i pulsante *f* di tonalità
n klanktoets
d Tontaste *f*

4934 TONE REPRODUCTION ctv/opt
f rendu *m* des valeurs, reproduction *f* de la couleur
e reproducción *f* del color
i riproduzione *f* del colore
n kleurtoonweergave
d Farbtonwiedergabe *f*

4935 TONE SOURCE aud
f source *f* sonore
e fuente *f* sonora
i sorgente *f* sonora
n geluidsbron
d Schallquelle *f*, Tonquelle *f*

4936 TOP LINEARITY CONTROL rec/rep/tv
f réglage *m* de la résistance de crête
e regulación *f* de la resistencia de cresta
i regolazione *f* della resistenza di cresta
n regeling van de piekweerstand
d Regelung *f* des Spitzenwiderstandes

TOP SHOT
see: HIGH-ANGLE SHOT

4937 TOROIDAL COIL cpl
f bobine *f* toroïdale
e bobina *f* toroidal
i bobina *f* toroidale
n toroïdspoel
d Ringspule *f*

4938 TOROIDAL ELECTRON GUN crt
A cathode-ray tube so constructed as to generate a hollow beam.
f canon *m* électronique à faisceau toroïdal
e cañón *m* electrónico de haz toroidal
i cannone *m* elettronico di fascio toroidale
n elektronenkanon *n* met ringbundel
d Elektronenkanone *f* mit Ringbündel

4939 TOTAL DISTORTION dis
The total harmonic distortion of an audio-frequency waveform expressed as a percentage of the fundamental frequency.

f distorsion *f* totale
e distorsión *f* total
i distorsione *f* totale
n totale vervorming
d Summenklirrfaktor *m*

4940 TOTAL INTERNAL REFLECTION opt
f réflexion *f* interne totale
e reflexión *f* interna total
i riflessione *f* interna totale
n totale inwendige reflectie
d Gesamtinnenreflexion *f*

TRACE
see: FORWARD STROKE

TRACE
see: LINE

TRACE INTERVAL
see: FORWARD-STROKE INTERVAL

4941 TRACK aud/vr
A path for recording one channel of information on a magnetic tape or other magnetic recording medium.
f piste *f*
e pista *f*
i pista *f*
n spoor *n*
d Spur *f*

4942 TRACK ADHESION vr
f adhésion *f* de la piste
e adhesión *f* de la pista
i adesione *f* della pista
n spooradhesie
d Spurhaftung *f*

4943 TRACK ADJUSTMENT vr
f ajustage *m* de piste
e ajuste *m* de pista
i aggiustaggio *m* di pista
n spoorinstelling
d Spureinstellung *f*

TRACK CUT
see: DOLLY OUT

4944 TRACK ELEMENT vr
f élément *m* de piste
e elemento *m* de pista
i elemento *m* di pista
n spoorelement *n*
d Spurelement *n*

4945 TRACK ERROR vr
f erreur *f* de piste
e error *m* de pista
i errore *m* di pista
n spoorfout
d Spurfehler *m*

4946 TRACK GEOMETRY vr
Geometry in such a way that the line synchronizing pulses are aligned on adjacent recorded tracks.
f géométrie *f* de la piste
e geometría *f* de la pista
i geometria *f* della pista
n spoorgeometrie
d Spurgeometrie *f*

TRACK IN
see: DOLLY IN

4947 TRACK LAYING vr
The process of editing and synchronizing the sound tracks to accompany a videotape recording.
f préparation *f* au montage des bandes à mixer
e preparación *f* de las cintas sonoras miscibles para el montaje
i preparazione *f* dei nastri sonori miscibili per il montaggio
n gereedmaken *n* van de te mengen geluidsbanden voor de montage
d Vorbereitung *f* der zu mischenden Tonbänder für den Schnitt

4948 TRACK PITCH vr
f pas *m* de la piste
e paso *m* de la pista
i passo *m* della pista
n spoorspoed
d Spurteilung *f*

4949 TRACK PLACEMENT, TRACK POSITION vr
f emplacement *m* de la piste
e posición *f* de la pista
i posizione *f* della pista
n spoorplaats
d Spurlage *f*

4950 TRACK RELAY vr
f relais *m* de piste
e relé *m* de pista
i relè *m* di pista
n spoorrelais *n*
d Spurrelais *n*

4951 TRACK SELECTION vr
f sélection *f* de piste
e selección *f* de pista
i selezione *f* di pista
n spoorkeuze
d Spurwahl *f*

4952 TRACK SELECTOR, TRACK SWITCH vr
f sélecteur *m* de piste
e selector *m* de pista
i selettore *m* di pista
n spoorkeuzeschakelaar
d Spurschalter *m*

4953 TRACK WIDTH vr
f largeur *f* de piste
e ancho *m* de pista
i larghezza *f* di pista
n spoorbreedte
d Spurbreite *f*

4954 TRACKING tv
f commande *f* dynamique de la luminosité
e mando *m* dinámico de la luminosidad
i comando *m* dinamico della luminosità

n dynamische helderheidsregeling
d dynamische Hellesteuerung *f*

4955 TRACKING vr
f centrage *m* de piste
e centraje *m* de pista
i centratura *f* di pista
n sporing
d Spurhaltung *f*

4956 TRACKING CONTROL tv/vr
Used to secure optimum picture quality on playback.
f régleur *m* à la reproduction
e regulador *m* a la reproducción
i regolatore *m* alla riproduzione
n beeldkwaliteitscontrole
d Bildqualitätskontrolle *f*

4957 TRAFFIC CONTROL BY TELEVISION tv
f réglage *m* de la circulation par télévision
e control *m* del tráfico por televisión
i controllo *m* del traffico per televisione
n verkeersregeling door middel van televisie
d Verkehrsregelung *f* mittels Fernsehen

4958 TRAILER dis
A bright streak at the right of a dark area in a TV picture, or a dark area or streak at the right of a bright part.
f strie *f* lumineuse
e estría *f* luminosa
i striscia *f* luminosa
n lichtstreep
d Lichtstreifen *m*

4959 TRAILER fi/tv
f bande *f* annonce,
bande *f* de lancement
e cortometraje *m* de presentación
i presentazione *f* del prossimo film
n trailer,
vooraankondiging
d Trailer *m*

4960 TRAILER vr
f amorce *f* de fin
e cola *f* de fin
i coda *f* di fine
n eindband
d Endband *n*

4961 TRAILING EDGE cpl/ge
The major portion of the decay of a pulse.
f flanc *m* arrière
e flanco *m* posterior
i fronte *f* posteriore
n achterflank
d Hinterflanke *f*

4962 TRAJECTORY cpl
f trajectoire *f*
e trayectoria *f*
i traiettoria *f*
n baan, weg
d Bahn *f*

TRANSCODER
see: COLO(U)R TRANSCODER

TRANSCONDUCTANCE (US)
see: MUTUAL CONDUCTANCE

4963 TRANSCRIBE (TO) aud/vr
To record a program(me) by means of electric transcription or magnetic tape for future rebroadcasting.
f transcrire v,
transférer v
e transcribir v
i trascrivere v
n afschrijven v,
overschrijven v
d vom Band abschreiben v,
vom Band übertragen v

4964 TRANSCRIPTION vr
f transcription *f*
e transcripción *f*
i trascrizione *f*
n afschrift *n*,
transcriptie
d Abschreiben *n*,
Übertragung *f*

4965 TRANSFER CHARACTERISTIC aud/vr
Of magnetic recording, a graph of the magnetic induction in magnetic tape plotted against magnetizing force.
f caractéristique *f* de transfert
e característica *f* de transferencia
i caratteristica *f* di trasferimento
n overdrachtskromme
d Übertragungscharakteristik *f*,
Übertragungskonstante *f*

4966 TRANSFER FUNCTION ge
f fonction *f* de transfert
e función *f* de transferencia
i funzione *f* di trasferimento
n overdrachtsfunctie
d Übertragungsfunktion *f*

4967 TRANSFER RATIO crt
The relation between TV camera tube illumination and the corresponding signal current.
f rapport *m* de transfert
e relación *f* de transferencia
i rapporto *m* di trasferimento
n overdrachtsverhouding
d Übertragungsverhältnis *n*

4968 TRANSFORMATION OF COLO(U)R MIXTURE DATA ctv
Colo(u)r mixture date for one set of primaries can be used to compute the colo(u)r mixture data for any other set of primaries for the same observer.
f calcul *m* d'un changement de systèmes colorimétriques
e cálculo *m* de cambio de sistemas colorimétricos
i calcolo *m* di scambio di sistemi colorimetrici
n transformatie van kleurmengselgegevens
d Transformation *f* von Farbmischwerten

4969 TRANSIENT DISTORTION dis
A type of distortion arising only when

there is a rapid fluctuation in frequency and/or amplitude of the stimulus.
f distorsion *f* due aux phénomènes transitoires
e distorsion *f* debida a los fenómenos transitorios
i distorsione *f* dovuta ai fenomeni transitori
n vervorming door voorbijgaande verschijnselen
d Verzerrung *f* durch Ein- und Ausschwingvorgänge

TRANSIENT RESPONSE
see: SURGE CHARACTERISTIC

TRANSIENT TIME
see: BUILDING-UP TIME

4970 TRANSIENTS cpl
f phénomènes *pl* transitoires
e fenómenos *pl* transitorios
i fenomeni *pl* transitori, transitori *pl*
n voorbijgaande verschijnselen *pl*
d Ausschwingvorgänge *pl*, Einschwingvorgänge *pl*

4971 TRANSISTOR QUARTET cpl
f quartette *m* de transistors
e cuarteto *m* de transistores
i quartetto *m* di transistori
n transistorkwartet *n*
d Transistorenquartett *n*

4972 TRANSLATOR, TRANSLATOR STATION tv
A satellite TV station that takes the vision and the corresponding sound signals by direct radio transmission from a main station and rebroadcasts them on another channel without first demodulating and modulating them.
f émetteur *m* relais, réémetteur *m*
e reemisor *m*
i riemettitore *m*
n heruitzender
d Umsetzer *m*

TRANSLATOR
see: FREQUENCY TRANSLATOR

4973 TRANSMISSION ge
Conveying electrical energy over a distance by wires in the form of pictorial information.
f transmission *f*
e transmisión *f*
i trasmissione *f*
n transmissie, zending
d Sendung *f*, Übertragung *f*

4974 TRANSMISSION BAND cpl
f bande *f* de transmission
e banda *f* de transmisión
i banda *f* di frequenze trasmesse, banda *f* di trasmissione
n transmissieband
d Durchlassbereich *m*, Übertragungsbereich *m*

TRANSMISSION GAIN
see: GAIN

TRANSMISSION GAMMA
see: CONTRAST TRANSFER FUNCTION

4975 TRANSMISSION LEVEL cpl
f niveau *m* de transmission
e nivel *m* de transmisión
i livello *m* di trasmissione
n transmissieniveau *n*
d relativer Leistungspegel *m*, Übertragungspegel *m*

TRANSMISSION LINE AMPLIFIER
see: DISTRIBUTED AMPLIFIER

TRANSMISSION MONITOR (GB)
see: MASTER MONITOR

4976 TRANSMISSION OF COLO(U)R CHROMINANCE CARRIERS ctv
f transmission *f* de sousporteuses en télévision couleur
e transmisión *f* de subportadoras en televisión en colores
i trasmissione *f* di sottoportanti in televisione a colori
n hulpdraaggolfoverdracht in kleurentelevisie
d Hilfsträgerübertragung *f* im Farbfernsehen

TRANSMISSION OF LIGHT
see: LIGHT TRANSMISSION

4977 TRANSMISSION PRIMARIES ctv
The set of three primaries, either physical or non-physical, so chosen that each corresponds to one of the three independent signals contained in the colo(u)r picture signal.
f primaires *pl* de la transmission
e primarios *pl* de la transmisión
i primari *pl* della trasmissione
n primaire kleuren *pl* bij de uitzending
d Farbwertsignale *pl*

4978 TRANSMISSION STANDARDS tv
f normes *pl* de transmission
e normas *pl* de transmisión
i norme *pl* di trasmissione
n transmissienormen *pl*
d Übertragungsnormen *pl*

4979 TRANSMISSION TYPES tv
f types *pl* de transmission
e tipos *pl* de transmisión
i tipi *pl* di trasmissione
n transmissiesoorten *pl*
d Übertragungsarten *pl*

4980 TRANSMIT (TO) tv
f émettre v, transmettre v
e emitir v, transmitir v

i emettere v,
trasmettere v
n zenden v
d aussenden v,
senden v

4981 TRANSMITTANCE ge
Ratio of transmitted to incident flux.
f coefficient *m* de transmission,
transmittance *f*
e coeficiente *m* de transmisión,
transmitancia *f*
i coefficiente *m* di trasmissione
n doorlaatbaarheid
d Durchlassgrad *m*

4982 TRANSMITTED-LIGHT VIEWING SYSTEM tv
f système *m* de vision à lumière transmise
e sistema *m* de visión con luz transmitida
i sistema *m* di visione con luce trasmessa
n observatiesysteem *n* met doorgelaten licht
d Betrachtungssystem *n* mit durchgelassenem Licht

4983 TRANSMITTER cpl
f émetteur *m*,
transmetteur *m*
e emisor *m*,
transmisor *m*
i emettitore *m*,
trasmettitore *m*
n zender
d Sender *m*

4984 TRANSMITTER AND RECEIVER EQUIPMENT tv
f appareillage *m* pour la transmission et la réception
e equipo *m* para la transmisión y la recepción
i apparecchiatura *f* per la trasmissione e la ricezione
n zend-ontvangapparatuur
d Sende-Empfangsgeräte *pl*

4985 TRANSMITTER POWER tv
f puissance *f* de l'émetteur
e potencia *f* del emisor
i potenza *f* dell'emettitore
n zendervermogen *n*
d Senderleistung *f*

4986 TRANSMITTER SITE rep/tv
f situation *f* géographique de l'émetteur
e ubicación *f* del emisor
i sito *m* geografico dell'emettitore
n ligging van de zender
d Senderstandort *m*

4987 TRANSMITTING POWER aea
f puissance *f* d'émission
e potencia *f* de emisión
i potenza *f* d'emissione
n zendvermogen *n*
d Sendeenergie *f*,
Sendeleistung *f*

4988 TRANSMITTING STATION tv
The location at which a transmitter, transmitting antenna, and associated transmitting equipment of a TV station are grouped.
f station *f* d'émission
e estación *f* de emisión
i stazione *f* d'emissione
n zendstation *n*
d Sendestation *f*,
Sendestelle *f*

4989 TRANSPARENT SCREEN crt
f écran *m* transparent
e pantalla *f* transparente
i schermo *m* trasparente
n doorzichtig scherm *n*
d Durchsichtschirm *m*

4990 TRANSPONDER tv
A transmitter-receiver, used in active satellites, which transmits signals automatically but only when interrogated.
f émetteur-récepteur *m* asservi par impulsions,
transpondeur *m*
e contestador *m* del satélite,
respondedor *m*
i emettitore-ricevittore *m* azionado per impulsi
n zender-ontvanger van een satelliet
d Transponder *m*

4991 TRANSPORT LOCKING CATCH vr
f levier *m* de verrouillage du transport de bande
e enclavador *m* de transporte de cinta
i leva *f* di bloccaggio del trasporto di nastro
n bandtransportgrendel,
bandtransportvergrendeling
d Bandtransportverriegelung *f*,
Verriegelungsgriff *m*

4992 TRANSPORT MECHANISM vr
f mécanisme *m* de transport
e mecanismo *m* de transporte
i meccanismo *m* di trasporto
n transportmechanisme *n*
d Transportmechanismus *m*

TRANSVERSAL MAGNETIC FIELD
see: LATERAL MAGNETIC FIELD

4993 TRANSVERSAL RECORDING vr
f enregistrement *m* transversal
e grabación *f* transversal
i registrazione *f* trasversale
n transversale opname
d Querspuraufzeichnung *f*,
Querspurverfahren *n*

4994 TRANSVERSE HEAD vr
f tête *f* transversale
e cabeza *f* transversal
i testina *f* trasversale
n dwarskop
d Transversalkopf *m*

4995 TRANSVERSE MAGNETIZATION aud/vr
In magnetic reading the magnetization which is perpendicular to the moving direction of the recording medium.
f magnétisation *f* transversale
e magnetización *f* transversal
i magnetizzazione *f* trasversale
n dwarsmagnetisering
d Quermagnetisierung *f*

4996 TRANSVERSE SCAN vr
f analyse *f* transversale
e exploración *f* transversal
i analisi *f* trasversale
n dwarsaftasting
d Querabtastung *f*

TRANSVERSE-TRACK TELEVISION TAPE RECORDER
see: QUADRUPLEX VIDEOTAPE RECORDER

TRAPEZIUM DISTORTION
see: KEYSTONE DISTORTION

4997 TRAPEZIUM EFFECT dis
Phenomenon in which deflecting voltage applied to the deflector plates of a cathode-ray tube is unbalanced with respect to the anode.
f effet *m* trapézoïdal
e efecto *m* trapezoidal
i effetto *m* trapezoidale
n trapeze-effect *n*
d Trapezeffekt *m*

4998 TRAPEZOIDAL DEFLECTION crt
f déviation *f* trapézoïdale
e desviación *f* trapezoidal
i deviazione *f* trapezoidale
n trapezevormige afbuiging
d trapezförmige Ablenkung *f*

4999 TRAVEL SHOT, TRAVEL(L)ING SHOT tv
f prise *f* de vues en mouvement
e toma *f* en movimiento
i carrellata *f*,
ripresa *f* a seguimento
n volgopname
d Folgeaufnahme *f*

5000 TRAVEL(L)ING MAN fi/tv
f machiniste *m* dolly,
travellingman *m*
e maquinista *m* de carro portacámara
i macchinista *m* di carrello di telecamera
n camerawagenbestuurder
d Dollyfahrer *m*

TRAVEL(L)ING SHOT
see: DOLLYING SHOT

5001 TRAVELING WAVE ANTENNA (US), TRAVELLING WAVE AERIAL (GB) aea
An aerial (antenna) in which the signal is gradually attenuated.
f antenne *f* à ondes progressives
e antena *f* de ondas progresivas
i antenna *f* ad onde progressive
n lopende-golfantenne
d Antenne *f* mit fortschreitenden Wellen

5002 TRAVEL(L)ING WAVE AMPLIFIER cpl
f amplificateur *m* à ondes progressives
e amplificador *m* de ondas progresivas
i amplificatore *m* ad onde progressive
n lopende-golfversterker
d Wanderfeldverstärker *m*,
Wanderwellenverstärker *m*

TRIAD
see: PHOSPHOR CLUSTER

TRIAL PERFORMANCE
see: AUDITION

5003 TRIANGULAR NOISE dis
f bruit *m* triangulaire
e ruido *m* triangular
i rumore *m* triangolare
n driehoeksruis *n*
d Dreiecksrauschen *n*

5004 TRI-BARREL ELECTRON GUN crt
An electron gun as a component part of three-colo(u)r TV tubes emitting three streams of electrons for the three primary colo(u)rs.
f canon *m* électronique à trois faisceaux
e cañón *m* electrónico de tres haces
i cannone *m* elettronico a tre fasci
n elektronenkanon *n* met drie bundels
d Dreistrahlelektronenkanone *f*

5005 TRICHROMAT ct
One who possesses trichromatic vision.
f trichromate *m*
e tricrómato *m*
i tricromato *m*
n trichromaat
d Trichromat *m*

5006 TRICHROMATIC CAMERA ctv
A colo(u)r TV camera in which the luminance signal is produced by matrixing the three tristimulus signals.
f caméra *f* trichromatique
e cámara *f* tricromática
i camera *f* tricromatica
n driekleurencamera,
trichromatische camera
d Dreifarbenkamera *f*,
trichromatische Kamera *f*

5007 TRICHROMATIC RESPONSE ct
f courbes *pl* d'égalisation chromatiques
e curvas *pl* de igualación cromáticas
i curve *pl* d'uguagliamento cromatiche
n kleurmengingskrommen *pl*
d Farbmischkurven *pl*

5008 TRICHROMATIC SYSTEM ct
Any system of colo(u)r specification based on the possibility of matching colo(u)rs by the additive mixture of three suitably chosen standard stimuli.

f système *m* trichromatique,
trichromie *f*
e sistema *m* tricromático
i sistema *m* tricromatico
n driekleurensysteem *n*
d Dreifarbensystem *n*

5009 TRICHROMATIC UNITS ct
Relative units of stimulus quantity applicable to stimuli of any colo(u)r and such that the quantity of any stimulus, when expressed in these units, is equal to the sum of the tristimulus values.
f unités *pl* trichromatiques
e unidades *pl* tricromáticas
i unità *pl* tricromatiche
n trichromatische eenheden *pl*
d trichromatische Einheiten *pl*

5010 TRICK BUTTON ani
f bouton *m* de surimpression,
bouton *m* d'effet
e tecla *f* de efecto
i pulsante *m* d'effetto
n tructoets
d Tricktaste *f*

5011 TRICK EFFECTS ani/fi/tv
f trucages *pl*
e efectos *pl* de trucos de toma
i effetti *pl* di trucchi
n truceffecten *pl*
d Trickeffekte *pl*

5012 TRICOLO(U)R TUBE crt
f tube *m* tricolore
e tubo *m* tricolor
i tubo *m* tricolore
n driekleurenbuis
d Dreifarbenbildröhre *f*,
Dreifarbenröhre *f*,
Dreistrahlröhre *f*

TRICONOSCOPE (US)
see: SET OF COLOUR IMAGE TUBES

5013 TRIGGER, cpl
TRIGGER PULSE
f déclencheur *m*,
impulsion *f* de déclenchement
e impulso *m* de disparo
i impulso *m* di scatto
n trekkerimpuls
d Steuerimpuls *m*,
Triggerimpuls *m*

5014 TRIMMING FILTER ctv
An optical filter providing an adjustment to the spectral characteristics of a colo(u)r TV camera.
f filtre *m* d'ajustage
e filtro *m* de ajuste
i filtro *m* d'aggiustaggio
n vereffeningsfilter *n*
d Abgleichfilter *n*

TRIMS
see: CLIP

5015 TRINOSCOPE ctv
A colo(u)r display device in which the colo(u)r picture is formed by optically combining the output from three separate display devices.
f trinoscope *m*
e trinoscopio *m*
i trinoscopio *m*
n trinoscoop
d Trinoskop *n*

5016 TRIPLEXER aea
A filter device to enable three TV bands to use a common VHF downlead from the aerial (antenna) system.
f filtre *m* triplexeur
e filtro *m* triplexador
i filtro *m* per tre bande di frequenza
n triplexer
d Dreibandantennenweiche *f*

5017 TRIPOD opt
f trépied *m*
e trípode *m*
i cavalletto *m*, treppiede *m*
n driepoot
d Dreifuss *m*

TRISTIMULUS SIGNALS
see: RGB SIGNALS

TRISTIMULUS SPECIFICATIONS,
TRISTIMULUS VALUES
see: COLO(U)R MIXTURE DATA

TRISTIMULUS VALUES MONITOR
see: PRIMARY SIGNAL MONITOR

TRITANOMALOUS VISION
see: PARTIAL TRITANOPIA

5018 TRITANOPE ct
One who possesses tritanopia.
f tritanope *m*
e tritanope *m*
i tritanope *m*
n tritanoop
d Tritanop *m*

TROUBLE HUNTING (US),
TROUBLE SHOOTING (US)
see: FAULT FINDING

TRUCK
see: DOLLY

5019 TRUNCATED PICTURE dis
f image *f* tronquée
e imagen *f* truncada
i immagine *f* troncata
n afgeknot beeld *n*
d abgehacktes Bild *n*

TUBE AXIS
see: AXIS OF THE TUBE

5020 TUNER cpl
f dispositif *m* d'accord
e sintonizador *m*

i sintonizzatore *m*
n afsteminrichting
d Abstimmapparat *m*

5021 TUNER DRUM tv
f tambour *m* sélecteur de canaux
e tambor *m* selector de canales
i tamburo *m* selettore di canali
n kanalenkiezertrommel
d Kanalwählertrommel *f*

5022 TUNING rep/tv
f syntonisation *f*
e sintonía *f*,
sintonización *f*
i sintonia *f*,
sintonizzazione *f*
n afstemming
d Abstimmung *f*

5023 TURNSTILE AERIAL, aea
TURNSTILE ANTENNA
f antenne *f* croisée,
antenne *f* en tourniquet,
doublet *m* en croix
e antena *f* en torniquete
i antenna *f* ad arganello
n draaiboomantenne,
tourniquetantenne
d Drehkreuzantenne *f*,
Kreuzdipol *m*

5024 TURRET, opt/tv
TURRET LENS
f tourelle *f* de lentilles
e revólver *m* de lentes
i torretta *f* girevole
n objectiefrevolver,
revolverlens
d Linsenkranz *m*

5025 TURRET TUNER cpl/tv
A tuner in which the coil assemblies are carried in a rotating turret bringing the required set into contact with the rest of the active circuit.
f sélecteur *m* de canaux à tourelle
e revólver *m* de selectores de canales
i selettore *m* di canali a torretta girevole
n kanalenkiezerrevolver
d induktiver Kanalwähler *m*,
Kanalschalter *m*,
Kanalwählerkranz *m*

TVI
see: TELEVISION INTERFERENCE

5026 TWELVE-DIRECTION MIXING tv
UNIT
f pupitre *m* de mélange pour douze directions
e pupitre *m* de mezclado para doce direcciones
i tavolo *m* di mescolatura per dodici direzioni
n menglessenaar voor twaalf richtingen
d Mischpult *n* für zwölf Richtungen

5027 TWIN-CATHODE RAY BEAM
f faisceau *m* jumelé de rayons cathodiques
e haz *m* gemelo de rayos catódicos
i fascio *m* doppio di raggi catodici
n tweelingelektronenbundel
d Zwillingselektronenstrahl *m*

5028 TWIN-INTERLACED rec/rep/tv
SCANNING
Process by which adjacent lines in the scanned image belong to alternate fields, the odd field and the even field.
f analyse *f* entrelacée double
e exploración *f* entrelazada doble
i analisi *f* a doppio interlacciamento alle trame
n geïnterlinieerde aftasting met even en oneven raster
d Abtastung *f* nach dem Doppelzeilensprungverfahren

5029 TWIN-LENS FILM rec/rep/tv
SCANNER,
TWIN-LENS TELECINE
f télécinéma *m* à lentilles jumelées
e telecinema *m* de objetivos gemelos
i telecinema *m* a lenti gemelle
n filmaftaster met tweelinglens
d zweiäugiger Filmabtaster *m*

TWINNING
see: PAIRING

5030 TWO-COLO(U)R DIRECT crt
VIEW STORAGE TUBE
f tube *m* mémoire bicolore à vision directe
e tubo *m* memoria bicolor de visión directa
i tubo *m* memoria bicolore a visione diretta
n tweekleuren-directzichtgeheugenbuis
d Zweifarben-Direktsichtspeicherröhre *f*

5031 TWO-FIELD PICTURE tv
The TV picture made up of an odd and an even field which fields are interlaced to form the complete picture.
f image *f* complète de télévision formée par deux trames
e imagen *f* televisiva formada por dos campos
i immagine *f* completa televisiva ottenuta con due trame
n beeld *n* gevormd door twee geïnterlinieerde rasters
d Bild *n* gebildet durch zwei Teilbilder im Zeilensprungverfahren

5032 TWO-SHOT opt
f plan *m* américain
e plano *m* americano
i piano *m* americano
n tweeshot
d amerikanische Einstellung *f*,
Zweiereinstellung *f*

5033 TWO-SIDED MOSAIC PICKUP crt
TUBE
f tube *m* image à mosaïque à double couche
e tubo *m* imagen de mosaico de dos caras
i tubo *m* immagine a mosaico a due facce
n beeldbuis met tweezijdig mozaïek
d doppelseitige Bildabtaströhre *f*

5034 TWO-TUBE CAMERA crt/rec/tv
f caméra *f* à deux tubes
e cámara *f* de dos tubos
i camera *f* a due tubi
n tweebuizencamera
d Zweiröhrenkamera *f*

5035 TWO-WAY TELEVISION tv
f télévision dans les deux sens
e televisión *f* en los dos sentidos
i televisione *f* nei due sensi
n duplexverkeer *n* in televisie
d Gegensehen *n*

U

U CENTER(RE)
see: COLO(U)R CENTER(RE)

U MODULATOR
see: B-Y MODULATOR

U SIGNAL
see: B-Y SIGNAL

5036 UCS DIAGRAM, UNIFORM CHROMATICITY-SCALE DIAGRAM ct
Chromaticity diagram in which the co-ordinate scales are chosen with the intention of making equal intervals represent approximately equal steps of discrimination for colo(u)rs of the same luminance at all parts of the diagram.
f diagramme *m* à chromaticité uniforme
e diagrama *m* de cromaticidad uniforme
i diagramma *m* di cromaticità uniforme
n UCS diagram *n*
d UCS-Farbdreieck *n*

5037 UHF, ULTRAHIGH FREQUENCY cpl
f UHF *f*, ultrahaute fréquence *f*
e frecuencia *f* ultraelevada, UHF *f*, ultraalta frecuencia *f*
i iperfrequenza *f*, UHF *f*
n UHF, ultra hoge frequentie
d UHF *f*, Ultrahochfrequenz *f*, ultrahohe Frequenz *f*

5038 UHF BANDPASS FILTER cpl
f filtre *m* passe-bande UHF
e filtro *m* pasabanda UHF
i filtro *m* passabanda UHF
n UHF-bandfilter *n*
d UHF-Bandfilter *n*

5039 UHF CONVERTER cpl
An electronic circuit that converts UHF signals to a lower frequency to permit reception on a VHF-receiver.
f convertisseur *m* de fréquence UHF
e convertidor *m* de frecuencia UHF
i convertitore *m* di frequenza UHF
n UHF-frequentietransformator
d UHF-Frequenzwandler *m*

5040 UHF PREAMPLIFIER STAGE cpl
f étage *m* préamplificateur UHF
e etapa *f* preamplificadora UHF
i stadio *m* preamplificatore UHF
n UHF-voorversterkertrap
d UHF-Vorverstärkerstufe *f*

5041 UHF TUNER cpl
Consists of two transistors in the common-base configuration, the first acting as an RF amplifier and the second as a self-oscillatory mixer.
f syntonisateur *m* UHF
e sintonizador *m* UHF
i sintonizzatore *m* UHF
n UHF-afsteminrichting
d UHF-Abstimmvorrichtung *f*

5042 UHF VISION TRANSMITTER rec/tv
f transmetteur *m* image UHF
e transmisor *m* imagen UHF
i trasmettitore *m* immagine UHF
n UHF-beeldzender
d UHF-Bildsender *m*

ULTOR
see: SECOND ANODE

5043 ULTRASONIC CLEANING vr
The cleaning of magnetic tape by means of ultrasonic vibrations.
f nettoyage *m* par ultrason
e limpieza *f* por vibración ultrasónica
i pulitura *f* per vibrazione ultrasonica
n reiniging door middel van ultrasone trillingen
d Reinigung *f* mittels Ultraschallwellen

5044 ULTRAWHITE, WHITER THAN WHITE rep/tv
Increase in the luminance associated with white due to an overshoot in the magnitude of the picture signal, for example, during ringing.
f dépassement *m* d'amplitude, ultrablanc *m*
e ultrablanco *m*
i ultrabianco *m*
n ultrawit *n*
d Überschwingen *n* der Leuchtdichte, Ultraweiss *n*

5045 ULTRAWHITE REGION, WHITER THAN WHITE REGION rep/tv
f zone *f* de l'ultrablanc
e región *f* del ultrablanco
i zona *f* dell'ultrabianco
n ultrawitgebied *n*
d Ultraweissgebiet *n*

5046 UMBRELLA-TYPE AERIAL (GB), UMBRELLA ANTENNA (US) aea
f antenne *f* avec panneau réflecteur
e antena *f* en forma de paraguas
i antenna *f* in forma d'ombrello
n parapluantenne
d Gitterantenne *f*, Schirmantenne *f*

5047 UNATTENDED STATION tv
f station *f* nonsurveillée

e estación *f* desatendida
i stazione *f* non sorvegliata
n onbemand station *n*
d unbemannte Station *f*

5048 UNBLANKING CIRCUIT tv
f circuit *m* d'annulation de la suppression
e circuito *m* de cancelación de la supresión de la traza
i circuito *m* di cancellazione della soppressione dell'andata
n opheffingscircuit *n* van de onderdrukking van de heenslag
d Austastungsauslösungskreis *m*, Hinlaufhellsteuerungskreis *m*

5049 UNBLANKING OF FORWARD SWEEP crt
f annulation *f* de la suppression de l'aller
e cancelación *f* de la supresión de la traza
i cancellazione *f* di soppressione dell'andata
n opheffing van de onderdrukking van de heenslag
d Hinlaufhellsteuerung *f*

5050 UNBLANKING PULSES cpl
f impulsions *pl* d'annulation
e impulsos *pl* de cancelación
i impulsi *pl* di cancellazione
n opheffingsimpulsen *pl*
d Helltastimpulse *pl*

5051 UNCOATED vr
f sans couche magnétique
e sin capa magnética
i senza strato magnetico
n zonder magnetische laag
d unbeschichtet adj

5052 UNCONSCIOUS REACTION, UNCONSCIOUS RESPONSE ct
f réaction *f* inconsciente
e reacción *f* inconsciente
i reazione *f* inconscia
n onbewuste reactie
d unbewusste Reaktion *f*

5053 UNDERLAP dis
Diminished width of a TV picture when the length of the scanning lines is reduced by a defect in the line timebase.
f rétrécissement *m* de l'image
e estrechamiento *m* de la imagen
i restrizione *f* dell'immagine
n beeldversmalling
d Bildverschmälerung *f*

5054 UNDERMODULATION cpl
A condition in which the volume of a program(me) is too low for the efficient operation of a transmitter.
f sousmodulation *f*
e submodulación *f*
i sottomodulazione *f*
n ondermodulatie
d ungenügende Aussteuerung *f*

5055 UNDERRUN tv
The finishing of a program(me) before the scheduled time has been completed.
f terminaison *f* du programme en moins du temps prévu
e terminación *f* del programa dentro del tiempo previsto
i terminazione *f* del programma entro il tempo previsto
n beëindiging van het programma binnen de vastgestelde tijd
d Sendezeitunterschreitung *f*

5056 UNDERSHOOT, UNDERSWING (GB), UNDERTHROW DISTORTION dis
f dépassement *m* balistique en sens négatif, sousoscillation *f*
e hipomodulación *f*
i sottoelongazione *f*
n doorschieten *n* in negatieve richting
d Überschwingen *n* in negative Richtung, Unterschwung *m*

5057 UNDERWATER TELEVISION tv
f télévision *f* sousmarine
e televisión *f* subacuática, televisión *f* submarina
i televisione *f* subacquea
n onderwatertelevisie
d Unterwasserfernsehen *n*

5058 UNDERWATER TELEVISION CAMERA rec/tv
f caméra *f* de télévision sousmarine
e cámara *f* de televisión subacuática
i telecamera *f* per riprese subacquee
n onderwatertelevisiecamera
d Unterwasserfernsehkamera *f*

5059 UNDISTORTED PICTURE tv
f image *f* sans distorsion
e imagen *f* no distorsionada
i immagine *f* non distorta
n onvertekend beeld *n*
d unverzeichnetes Bild *n*

5060 UNIDIRECTIONAL MAGNETIC FIELD cpl
f champ *m* magnétique unidirectionnel
e campo *m* magnético unidireccional
i campo *m* magnetico unidirezionale
n eenzijdig gericht magneetveld *n*
d einseitig gerichtetes Magnetfeld *n*

UNIFORM CHROMATICITY-SCALE DIAGRAM
see: UCS DIAGRAM

5061 UNIFORM COLO(U)R SPACE ct
Colo(u)r space in which the distance between any two colo(u)r points is intended to represent a measure of the perceived difference between the corresponding colo(u)rs.
f espace *m* chromatique uniforme
e espacio *m* cromático uniforme
i spazio *m* cromatico uniforme
n uniforme kleurenruimte
d einförmiger Farbraum *m*

5062 UNISELECTOR tv
Used in TV as part of line selection equipment in e.g. master control.
f carrousel *m*, rotacteur *m*

e selector *m* rotatorio,
selector *m* unidireccional
i selettore *m* rotativo
n draaikiezer
d Drehwähler *m*

UNIT FUNCTION RESPONSE (GB)
see: SURGE CHARACTERISTIC

5063 UNMARRIED PRINT vr
Pair of spools one carrying the optical recording and the other the sound recording of a film, the synchronization taking place during projection.
f enregistrements *pl* séparés
e grabaciones *pl* separadas,
registros *pl* separados
i registrazioni *pl* separate
n gescheiden opnamen *pl*
d getrennte Aufnahmen *pl*

5064 UNRELATED PERCEIVED COLO(U)R ct
Colo(u)r perceived to belong to an area with completely dark surroundings.
f couleur *f* perçue en ambiance noire
e color *m* percibido en ambiente negro
i colore *m* percepito in ambiente nero
n in zwarte omgeving waargenomen kleur
d in schwarzer Umgebung wahrgenommene Farbe *f*

5065 UNRESTORED TELEVISION RECEIVER rep/tv
A TV receiver with partial or no d.c. restoration.
f téléviseur *m* sans restauration ou à restauration partielle de la composante continue
e televisor *m* sin restauración o con restauración parcial de la componente continua
i televisore *m* senza restaurazione o con restaurazione parziale della componente continua
n televisieontvanger zonder herstel of met slechts gedeeltelijk herstel van de werkzame nulcomponent
d Fernsehempfänger *m* ohne Wiederherstellung oder mit nur teilweise Wiederherstellung des Schwarzpegels

5066 UNSHARP IMAGE tv
f image *f* floue
e imagen *f* borrosa
i immagine *f* sfocata
n onscherp beeld *n*
d unscharfes Bild *n*

5067 UNSTEADINESS OF THE PICTURE fi/tv
f manque *m* de fixité
e inestabilidad *f* de imagen
i instabilità *f* d'immagine
n beeldinstabiliteit
d Bildinstabilität *f*

5068 UNWANTED REFLECTIONS opt
f réflexions *pl* inutiles
e reflexiones *pl* inútiles
i riflessioni *pl* inutili
n onnuttige reflecties *pl*
d Nebenreflexionen *pl*

5069 UNWANTED SIGNAL FREQUENCY dis
f fréquence *f* de signal inutile
e frecuencia *f* de señal inútil
i frequenza *f* di segnale inutile
n frequentie van het onnuttige signaal
d Nebensignalfrequenz *f*

5070 UNWANTED SIGNALS dis
f signaux *pl* inutiles
e señales *pl* inútiles
i segnali *pl* inutili
n onnuttige signalen *pl*
d Nebensignale *pl*

5071 UNWANTED SOUND dis
f son *m* inutile
e sonido *m* inútil
i suono *m* inutile
n bijgeluiden *pl*,
onnuttig geluid *n*
d Nebenton *m*

5072 UNWEIGHTED SIGNAL-TO-NOISE RATIO dis
f rapport *m* à la tension induite
e distancia *f* de tensión psofométrica
i distanza *f* di tensione esterna
n uitwendige-spanningsafstand
d Fremdspannungsabstand *m*

5073 UPSTREAM vr
f partie *f* en amont
e ---
i a monte
n stroomopwaarts
d gegen Bandlaufrichtung

5074 UPWARDS CONVERSION tv
Standards conversion to a higher standard.
f conversion *f* montante
e conversión *f* ascendente
i conversione *f* verso l'alto
n opwaartse omzetting
d Aufwärtswandlung *f*

V

5075 VACUUM TAPE GUIDE vr
f guide *m* à dépression,
guide-bande *m* à vide
e guía-cinta *f* de vacío
i guida *f* concava di nastro,
guida-nastro *m* a vuoto
n schoen,
vacuümbandgeleider
d Bandführungssegment *n*,
Vakuumbandführer *m*

5076 VAN ALLEN BELTS,
VAN ALLEN RADIATION BELTS tv
Two toroidal belts of radiation which encircle the earth at high altitude which may cause damage to TV satellite component parts.
f ceintures *pl* de Van Allen
e cinturones *pl* de Van Allen
i cinture *pl* di Van Allen
n stralingsgordels *pl* van Van Allen
d Van Allen-Gürtel *pl*

5077 VARIABLE APERTURE SHUTTER fi/tv
f obturateur *m* à ouverture variable
e obturador *m* de abertura variable
i otturatore *m* ad apertura variabile
n sluiter met veranderlijke diafragmaopening
d Verschluss *m* mit veränderlicher Blendenöffnung

5078 VARIABLE CORRECTION UNIT,
VCU aud
f filtre *m* de tonalité
e filtro *m* de tonalidad
i filtro *m* di tonalité
n toonhoogtefilter *n*
d Tonblende *f*

5079 VARIABLE FOCUS LENS,
ZOOM LENS opt
Compound lens mounted to many TV cameras which has a continuously variable range of focal length.
f objectif *m* à distance focale variable
e lente *f* con enfoque ajustable,
objetivo *m* ajustable
i lente *f* con fuoco variabile
n zoomlens
d Gummilinse *f*,
Variaoptik *f*

5080 VARIABLE GAIN DEVICE cpl
f dispositif *m* à amplification variable
e dispositivo *m* de amplificación variable
i dispositivo *m* d'amplificazione variabile
n inrichting voor veranderlijke versterking
d veränderlicher Verstärker *m*

5081 VARIABLE QUADRICORRELATOR ctv
f quadricorrélateur *m* à amplification variable
e cuadricorrelador *m* de amplificación variable
i quadricorrelatore *m* ad amplificazione variabile
n quadricorrelator met veranderlijke versterking
d Quadrikorrelator *m* mit veränderlicher Verstärkung

5082 VARIABLE RATIO
QUADRICORRELATOR ctv
f quadricorrélateur *m* à rapport variable
e cuadricorrelador *m* de relación variable
i quadricorrelatore *m* a rapporto variabile
n quadricorrelator met veranderlijke overdracht
d Quadrikorrelator *m* mit veränderlicher Übertragung

5083 VARIABLE-SPEED SCANNING tv
Scanning method in which the velocity of the scanning beam may be varied.
f analyse *f* à vitesse variable
e exploración *f* de velocidad variable
i esplorazione *f* a velocità variabile
n aftasting met veranderlijke snelheid
d Abtastung *f* mit veränderlicher Geschwindigkeit

5084 VARIABLE VIDEO ATTENUATOR tv
f potentiomètre *m* gain signal vidéo
e potenciómetro *m* de la ganancia de la señal video
i potenziometro *m* del guadagno del segnale video
n versterkingsregelaar in het videosignaal
d BA-Regler *m*

5085 VARIABLES OF PERCEIVED
COLO(U)R ct
Hue, saturation and brightness.
f variables *pl* d'une couleur perçue
e variables *pl* de un color percibido
i variabili *pl* d'un colore percepito
n variabelen *pl* van een waargenomen kleur
d Variabelen *pl* einer wahrgenommenen Farbe

5086 VARIETY STUDIO stu
f studio *m* de variété
e estudio *m* de variedades
i studio *m* per varietà
n studio voor amusementsprogramma's
d Unterhaltungsstudio *n*

VECTOR DIAGRAM OF THE CHROMINANCE SIGNAL
see: CHROMINANCE SIGNAL VECTOR DIAGRAM

5087 VECTORSCOPE ctv
A monitor for the carrier chrominance signal which displays the phase and amplitude.
f vectorscope *m*
e vectorscopio *m*

i vettoroscopio *m*
n vectorscoop
d Vektorskop *n*

5088 VELOCITY CONTROL SERVO cpl
Mechanism for remotely controlling the velocity or change of speed.
f régulateur *m* de vitesse asservi
e servorregulador *m* de velocidad
i servoregolatore *m* di velocità
n snelheidsservoregelaar
d Geschwindigkeitsservoregler *m*

5089 VELOCITY ERRORS vr
f erreurs *pl* de vitesse
e errores *pl* de velocidad
i errori *pl* di velocità
n snelheidsfouten *pl*
d Geschwindigkeitsfehler *pl*

5090 VELOCITY MODULATED ELECTRON BEAM crt
f faisceau *m* électronique modulé en vitesse
e haz *m* electrónico modulado en velocidad
i fascio *m* elettronico modulato in velocità
n in snelheid gemoduleerde elektronenbundel
d geschwindigkeitsgesteuerter Elektronenstrahl *m*

5091 VELOCITY MODULATION cpl
Method of modulating the electron beam of a picture tube in which the luminance of the scanning spot is varied by modulating the velocity of the electron beam, the beam current remaining constant.
f modulation *f* de la vitesse du faisceau
e modulación *f* de la velocidad del haz
i modulazione *f* della velocità del fascio
n modulatie van de bundelsnelheid, snelheidsmodulatie
d Geschwindigkeitsmodulation *f*, Geschwindigkeitssteuerung *f*, Modulation *f* der Strahlgeschwindigkeit

5092 VELOCITY MODULATION TELEVISION SYSTEM cpl
f système *m* de télévision à modulation de vitesse du faisceau
e sistema *m* de televisión de modulación de velocidad del haz
i sistema *m* di televisione a modulazione di velocità del fascio
n televisiesysteem *n* met snelheidsmodulatie van de bundel
d Fernsehverfahren *n* mit Liniensteuerung

5093 VENETIAN BLIND EFFECT dis
f effet *m* de persiennes
e efecto *m* de persianas
i effetto *m* di persiane
n jaloezie-effect *n*
d Jalousieeffekt *m*

VERTICAL AMPLITUDE
see: FIELD AMPLITUDE

VERTICAL AMPLITUDE CONTROL
see: FIELD AMPLITUDE CONTROL

5094 VERTICAL AND HORIZONTAL CORRECTION vr
Correction carried out in video recording when there is a slight time difference between the field sync pulses and the external reference field sync pulses.
f correction *f* verticale et horizontale
e corrección *f* vertical y horizontal
i correzione *f* verticale e orizzontale
n verticale en horizontale correctie
d vertikale und horizontale Korrektur *f*

5095 VERTICAL APERTURE CORRECTION tv
f correction *f* verticale de l'aperture
e corrección *f* vertical de la abertura
i correzione *f* verticale dell'apertura
n verticale apertuurcorrectie
d VAR *f*, vertikale Aperturkorrektur *f*

5096 VERTICAL AXIS, Y AXIS crt
f axe *m* verticale, axe *m* Y
e eje *m* vertical, eje *m* Y
i asse *m* verticale, asse *m* Y
n verticale as, Y-as
d vertikale Achse *f*, Y-Achse *f*

5097 VERTICAL BAR GENERATOR, VERTICAL BAR OSCILLATOR tv
f générateur *m* de barres verticales
e generador *m* de barras verticales
i generatore *m* di barre verticali
n verticale-balkengenerator
d Vertikalbalkengenerator *m*

5098 VERTICAL BLACK BAND tv
f trait *m* noir vertical
e trazo *m* negro vertical
i tratto *m* nero verticale
n verticale zwarte streep
d vertikaler schwarzer Strich *m*

VERTICAL BLANKING
see: FIELD BLANKING

VERTICAL BLANKING INTERVAL
see: FIELD BLANKING INTERVAL

VERTICAL BLANKING PULSE
see: FIELD BLANKING PULSE

5099 VERTICAL CENT(E)RING, VERTICAL POSITION crt
f position *f* verticale
e posición *f* vertical
i posizione *f* verticale
n verticale stand
d Y-Lage *f*

VERTICAL CENT(E)RING CONTROL
see: FIELD CENT(E)RING CONTROL

VERTICAL CONVERGENCE
see: FIELD CONVERGENCE

VERTICAL CONVERGENCE AMPLITUDE CONTROL
see: FIELD CONVERGENCE AMPLITUDE CONTROL

VERTICAL CONVERGENCE CONTROL
see: FIELD CONVERGENCE CONTROL

VERTICAL CONVERGENCE SHAPE CONTROL
see: FIELD CONVERGENCE SHAPE CONTROL

VERTICAL CONVERGENCE TILT CONTROL
see: FIELD CONVERGENCE TILT CONTROL

VERTICAL CONVERGENCE YOKE
see: FIELD CONVERGENCE YOKE

5100 VERTICAL DEFINITION, VERTICAL RESOLUTION — tv
f définition *f* verticale
e definición *f* vertical
i definizione *f* verticale
n verticale definitie
d Senkrechtauflösung *f*, Vertikalauflösung *f*

VERTICAL DEFLECTION
see: FIELD DEFLECTION

VERTICAL DEFLECTION ELECTRODES
see: FIELD DEFLECTION ELECTRODES

VERTICAL DEFLECTION OSCILLATOR, VERTICAL OSCILLATOR
see: FIELD DEFLECTION OSCILLATOR

VERTICAL DYNAMIC FOCUS
see: FIELD DYNAMIC FOCUS

VERTICAL FLYBACK
see: FIELD FLYBACK

VERTICAL FLYBACK PERIOD
see: FIELD FLYBACK PERIOD

VERTICAL FREQUENCY
see: FIELD FREQUENCY

VERTICAL FREQUENCY CONTROL
see: FIELD FREQUENCY CONTROL

VERTICAL FREQUENCY LOCKING
see: FIELD FREQUENCY LOCKING

VERTICAL HEIGHT
see: FIELD HEIGHT

VERTICAL HOLD CONTROL, VERTICAL SYNC CONTROL
see: FIELD HOLD CONTROL

VERTICAL HUNTING
see: BOUNCING

5101 VERTICAL INTERVAL TEST SIGNALS, VIT SIGNALS — cpl
Signals which sometimes are introduced during the field suppression period.
f signaux *pl* d'essai entre trames
e señales *pl* de prueba entre cuadros
i segnali *pl* di prova tra trame
n interrastersignalen *pl*
d Testsignale *pl* zwischen aufeinanderfolgenden Teilbildern

5102 VERTICAL LINEARITY CONTROL — rep/tv
A linearity control that permits narrowing or expanding the height of the image on the upper half of the screen of a TV picture tube.
f réglage *m* de la linéarité de trame
e regulación *f* de la linealidad de cuadro
i regolazione *f* della linearità di trama
n rasterlineariteitsregeling
d Teilbildlinearitätsregelung *f*

VERTICAL OUTPUT TRANSFORMER
see: FIELD OUTPUT TRANSFORMER

5103 VERTICAL POLARIZATION — cpl/dis
f polarisation *f* verticale
e polarización *f* vertical
i polarizzazione *f* verticale
n verticale polarisatie
d vertikale Polarisation *f*

5104 VERTICAL POSITIONING CONTROL — tv
f réglage *m* de la position verticale
e regulación *f* de la posición vertical
i regolazione *f* della posizione verticale
n beeldstandregeling
d Bildhöhenverschiebung *f*, Y-Lageregelung *f*

VERTICAL POSITIONING CONTROL
see: FRAME-AMPLITUDE CONTROL

VERTICAL SCANNING
see: FIELD SCANNING

5105 VERTICAL SHADING — dis
A type of raster shading occurring when the brightness differences appear from top to bottom.
f canevas *m* à luminosité inégale verticale
e cuadrículo *m* de luminosidad dispareja vertical
i quadro *m* rigato a luminosità disuguale verticale
n verticale ongelijke rasterhelderheid
d vertikale ungleiche Rasterhelligkeit *f*

VERTICAL SLIP
see: FRAME SLIP

VERTICAL SWEEP
see: FIELD SWEEP

VERTICAL SYNC PULSE
see: FIELD SYNC PULSE

VERTICAL SYNCHRONIZATION
see: FIELD SYNCHRONIZATION

5106 VERTICAL WHITE BAND tv
f trait *m* blanc vertical
e trazo *m* blanco vertical
i tratto *m* bianco verticale
n verticale witte streep
d vertikaler weisser Strich *m*

5107 VERY HIGH FREQUENCY, VHF cpl
f très haute fréquence *f*, VHF *f*
e muy alta frecuencia *f*, VHF *f*
i altissima frequenza *f*, VHF *f*
n VHF, zeer hoge frequentie
d sehr hohe Frequenz *f*, VHF *f*

5108 VESTIGIAL SIDEBAND cpl
f bande *f* latérale restante
e banda *f* lateral residual, banda *f* lateral vestigial
i banda *f* laterale residua, banda *f* laterale vestigiale
n semi-eenzijband
d Restseitenband *n*

VESTIGIAL SIDEBAND AERIAL (ANTENNA) FILTER
see: FILTER FOR TELEVISION TRANSMITTER

5109 VESTIGIAL SIDEBAND FILTER cpl
f filtre *m* de bande latérale restante
e filtro *m* de banda lateral residual
i filtro *m* di banda laterale residua
n semi-eenzijbandfilter *n*
d Restseitenbandfilter *n*

5110 VESTIGIAL SIDEBAND TRANSMISSION col/tv
System of transmission chiefly used in TV in which either the lower or upper sideband is transmitted normally and the complementary sideband is almost completely suppressed by filter after modulation.
f transmission *f* à bande latérale restante
e transmisión *f* de bande lateral residual
i trasmissione *f* a banda laterale residua
n semi-eenzijbandtransmissie
d Restseitenbandübertragung *f*

5111 VHF AERIAL ARRAY (GB), VHF ANTENNA ARRAY (US) aea
f antenne *f* VHF
e sistema *m* de antena VHF
i sistema *m* d'antenna VHF
n VHF-antennestelsel *n*
d Antennenanordnung *f* für UKW

5112 VHF BAND cpl
f bande *f* de fréquence VHF
e banda *f* de frecuencia VHF
i banda *f* di frequenza VHF
n VHF-frequentieband
d VHF-Frequenzband *n*

5113 VHF BANDPASS FILTER cpl
f filtre *m* passe-bande VHF
e filtro *m* pasabanda VHF
i filtro *m* passabanda VHF
n VHF-bandfilter *n*
d VHF-Bandfilter *n*

5114 VHF CHANNEL cpl
f canal *m* VHF
e canal *m* VHF
i canale *m* VHF
n VHF- kanaal *n*
d VHF-Kanal *m*

5115 VHF LINK cpl
f voie *f* de transmission VHF
e vía *f* de transmisión VHF
i via *f* di trasmissione VHF
n VHF-transmissieweg
d VHF-Übertragungsweg *m*

5116 VHF PREAMPLIFIER STAGE cpl
f étage *m* préamplificateur VHF
e etapa *f* preamplificadora VHF
i stadio *m* preamplificatore VHF
n VHF-voorversterkertrap
d VHF-Vorverstärkerstufe *f*

5117 VHF TRANSMISSION cpl
f transmission *f* VHF
e transmisión *f* VHF
i trasmissione *f* VHF
n VHF-overdracht
d VHF-Übertragung *f*

5118 VHF TUNER tv
f dispositif *m* d accord VHF
e sintonizador *m* VHF
i sintonizzatore *m* VHF
n VHF-kanalenkiezer
d VHF-Kanalenwähler *m*

5119 VI-METER, VOLUME INDICATOR aud
An electrical speech level meter which is specified in respect of its electrical and dynamic characteristics and of the method by which readings are taken.
f vumètre *m*
e indicador *m* de volumen, volúmetro *m*
i VU-metro *m*
n niveaumeter
d Aussteuerungsmesser *m*, VU-Meter *n*

5120 VIDEO ADJUSTMENT tv/vr
f compensation *f* du signal d'image
e compensación *f* de la señal de imagen
i compensazione *f* del segnale d'immagine
n beeldsignaalcompensatie
d Bildsignalausgleich *m*

5121 VIDEO AMPLIFICATION tv
f amplification *f* vidéo

e amplificación *f* video
i amplificazione *f* video
n videoversterking
d Videoverstärkung *f*

5122 VIDEO AMPLIFIER cpl/tv
A wideband amplifier capable of amplifying video frequencies in TV.
f amplificateur *m* vidéo
e amplificador *m* video
i amplificatore *m* video
n videoversterker
d Videoverstärker *m*

5123 VIDEO AMPLIFIER AND BLANKING SIGNAL, VIDEO SIGNAL WITH BLANKING tv
f signal *m* vision à suppression
e señal *f* de imagen con supresión
i segnale *m* d'immagine con soppressione
n beeldsignaal *n* met onderdrukking
d BA-Signal *n*, Bildaustastsignal *n*

5124 VIDEO AMPLITUDE cpl/tv
f amplitude *f* vidéo
e amplitud *f* video
i ampiezza *f* video
n videoamplitude
d Videoamplitude *f*

5125 VIDEOCASSETTE RECORDING, VIDEOTAPE CASSETTE RECORDING vr
f magnétoscopie *f* à cassettes
e magnetoscopía *f* de casetas, videorregistro *m* de casetas
i magnetoscopia *f* a cassette, videoregistrazione *f* a cassette
n videocassetteopname
d MAZ-Kassettenaufzeichnung *f*, Videokassettenaufzeichnung *f*

5126 VIDEO CHANNEL rec/rep/tv
f canal *m* vidéo
e canal *m* video
i canale *m* video
n videokanaal *n*
d Videokanal *m*

5127 VIDEO CIRCUIT tv
A broadband circuit carrying intelligence that could become visible.
f circuit *m* vidéo
e circuito *m* video
i circuito *m* video
n videocircuit *n*
d Videokreis *m*

5128 VIDEO COIL cpl
f bobine *f* vidéo
e bobina *f* video
i bobina *f* video
n videospoel
d Videospule *f*

VIDEO CONTROL ROOM
see: CAMERA CONTROL ROOM

5129 VIDEO CONVERTER crt
f convertisseur *m* vidéo
e convertidor *m* video
i convertitore *m* video
n videoconverter
d Videokonvertor *m*

5130 VIDEO DEMODULATOR cpl vr
f démodulateur *m* vidéo
e desmodulador *m* video
i demodulatore *m* video
n videodemodulator
d Videodemodulator *m*

5131 VIDEO DETECTION cpl
f détection *f* vidéo
e detección *f* video
i rivelazione *f* video
n videogelijkrichting
d Videogleichrichtung *f*

5132 VIDEO DETECTOR rec/rep/tv
Crystal or thermionic diode following the I.F. stages of a TV receiver to demodulate the video signal.
f détecteur *m* vidéo
e detector *m* video
i rivelatore *m* video
n videodetector
d Videodetektor *m*

5133 VIDEO DISK APPARATUS vr
f appareil *m* à disque vidéo
e aparato *m* de disco video
i apparecchio *m* a disco video
n videoplatenspeler
d Videoplattenspieler *m*

5134 VIDEO DRUM APPARATUS vr
f appareil *m* vidéo à tambour
e aparato *m* video de tambor
i apparecchio *m* video a tamburo
n videotrommelapparaat *n*
d Videotrommelgerät *n*

5135 VIDEO FREQUENCY rec/rep/tv
A frequency component of the video signal.
f fréquence *f* vidéo, vidéofréquence *f*
e frecuencia *f* video, videofrecuencia *f*
i frequenza *f* video, videofrequenza *f*
n videofrequentie
d Videofrequenz *f*

5136 VIDEO-FREQUENCY AMPLIFIER tv
An amplifier capable of handling the entire range of frequencies that comprise a periodic visual presentation in TV.
f amplificateur *m* à vidéofréquence
e amplificador *m* de videofrecuencia
i amplificatore *m* a videofrequenza
n videofrequentieversterker
d Videofrequenzverstärker *m*

5137 VIDEO-FREQUENCY BAND tv
f bande *f* de vidéofréquences
e banda *f* de videofrecuencias
i banda *f* di videofrequenze, banda *f* video
n videobandbreedte
d Videofrequenzband *n*

5138 VIDEO-FREQUENCY CHROMINANCE COMPONENTS ctv
f signaux *pl* de chrominance dans le domaine de vidéofréquence
e señales *pl* de crominancia en el campo de videofrecuencia
i segnali *pl* di crominanza nel campo di videofrequenza
n chrominantiesignalen *pl* in het video-frequentiegebied
d Chrominanzsignale *pl* im Videofrequenz-bereich

5139 VIDEO HEAD vr
f tête *f* vidéo
e cabeza *f* video
i testina *f* video
n videokop
d Videokopf *m*

5140 VIDEO HEAD ASSEMBLY vr
A plug-in unit including the video head drum, slip rings and brushes, head drum motor, vacuum tape guide, magnetic tachometer and the control track head.
f bloc *m* de têtes vidéo
e conjunto *m* cabeza video
i complesso *m* testina video
n videokopsamenstel *n*
d Videokopfaggregat *n*

5141 VIDEO HEAD WHEEL vr
f roue *f* de tête de vidéo
e rueda *f* de cabeza video
i ruota *f* di testina video
n videokoprad *n*
d Videokopfrad *n*

VIDEO INFORMATION
see: IMAGE INFORMATION

5142 VIDEO INSERTION tv
f insertion *f* vidéo
e inserción *f* video
i inserzione *f* video
n video-invoeging
d Videoeinblendung *f*

5143 VIDEO INTEGRATION tv
A method of utilizing the redundancy of repetitive signals to improve the output signal-to-noise ratio, by summing the successive video signals.
f intégration *f* des signaux vidéo
e integración *f* de las señales video
i integrazione *f* dei segnali video
n bundeling van de videosignalen
d Bündelung *f* der Videosignale

5144 VIDEO INTERCONNECTION POINT cpl/vr
f point *m* de jonction de lignes de télévision locales
e punto *m* de interconexión de líneas de televisión locales
i punto *m* di giunzione di linee di televisione locali
n aansluitpunt *n* voor locale televisie-leidingen
d Übergabepunkt *m* an Fernsehleitungen

5145 VIDEO LEVEL INDICATOR vr
f indicateur *m* du niveau vidéo
e indicador *m* del nivel video
i indicatore *m* del livello video
n beeldniveau-aanwijzer
d Bildpegelanzeiger *m*

5146 VIDEO LIMITER tv
f limiteur *m* vidéo
e limitador *m* video
i limitatore *m* video
n videobegrenzer
d Videobegrenzer *m*

5147 VIDEO LONG-PLAY SYSTEM vr
f système *m* du vidéodisque LP
e sistema *m* del videodisco de larga duración,
sistema *m* del videodisco LP
i sistema *m* del videodisco LP
n kleurenbeeldplaatsysteem *n*,
videolangspeelsysteem *n*
d Videolangspielplattensystem *n*

5148 VIDEO LONG-PLAYING RECORD vr
f disque *m* vidéo longue durée
e disco *m* video de larga duración
i disco *m* video a lunga durata
n videolangspeelplaat
d Videolangspielplatte *f*

5149 VIDEO MATRIX vr
f grille *f* de distribution compensée de signaux vidéo
e rejilla *f* de distribución compensada de señales video
i griglia *f* di distribuzione compensata di segnali video
n videomatrix
d Filterkreuzschiene *f*,
Videoverteiler *m*

5150 VIDEO MIXER, VIDEO MONITORING AND MIXING DESK tv
A mixer used to combine the output signals of two or more TV cameras.
f pupitre *m* de mélange image,
pupitre *m* régie image
e pupitre *m* de control video
i tavolo *m* di regia video
n beeldmenglessenaar
d Bildmischpult *n*

5151 VIDEO MODULATOR tv
f modulateur *m* vidéo
e modulador *m* video
i modulatore *m* video
n videomodulator
d Videomodulator *m*

5152 VIDEO OUTPUT, VIDEO OUTPUT SIGNAL tv
f signal *m* de sortie vidéo
e señal *f* de salida video
i segnale *m* d'uscita video
n video-uitgangssignaal *n*
d Videoausgangssignal *n*

5153 VIDEO OUTPUT STAGE tv
f étage *m* de sortie vidéo
e etapa *f* de salida video
i stadio *m* d'uscita video
n video-eindtrap
d Videoendstufe *f*

VIDEORECORDING
see: MAGNETIC PICTURE TRACING

5154 VIDEORECORDING EQUIPMENT, VIDEOTAPE RECORDING EQUIPMENT, VTR EQUIPMENT tv
f magnétoscope *m*
e equipo *m* de videorregistro, magnetoscopio *m*
i apparecchiatura *f* di videoregistrazione, magnetoscopio *m*
n videobandopnameapparatuur, video-opnameapparatuur
d Magnetaufzeichnungsanlage *f*, MAZ-Anlage *f*, Videoaufzeichnungsanlage *f*

5155 VIDEO REPEATER tv
A special type of repeater used in video transmissions on telephone, balanced and coaxial pairs.
f répéteur *m* vidéo
e repetidor *m* video
i ripetitore *m* video
n videoversterker
d Videoverstärker *m*

5156 VIDEO SHEET APPARATUS vr
f appareil *m* vidéo à feuille magnétique
e aparato *m* video de hoja magnética
i apparecchio *m* video a foglio magnetico
n videofoelieapparaat *n*
d Videofoliengerät *n*

5157 VIDEO SIGNAL tv
The combined picture and synchronizing signals.
f signal *m* d'image complet
e señal *f* de video
i segnale *m* di video composito
n videosignaal *n*
d BAS *n*, Signalgemisch *n*, Videosignal *n*

VIDEO SIGNAL BLACK-OUT
see: BLACK-OUT OF VIDEO SIGNAL

5158 VIDEO SIGNAL MEAN LEVEL tv
f valeur *f* moyenne du signal image
e valor *m* medio de la señal imagen
i valore *m* medio del segnale immagine
n gemiddelde waarde van het videosignaal
d Bildsignalmittelwert *m*

5159 VIDEO SIGNAL MONITOR, WAVEFORM MONITOR tv
f moniteur *m* du signal d'image complet
e monitor *m* de la señal de video
i monitore *m* del segnale di video composito
n beeld- en profielmonitor
d Bild- und Wellenformmonitor *m*

5160 VIDEO SIGNAL SEGMENTATION tv/vr
f segmentation *f* du signal vidéo
e división *f* de la señal video en segmentos
i segmentazione *f* del segnale video
n segmentatie van het videosignaal
d Segmentierung *f* des Videosignals

VIDEO SIGNAL WITH BLANKING
see: VIDEO AMPLIFIER AND BLANKING SIGNAL

5161 VIDEO SOCKET vr
Used for connection of a TV set or monitor to a video recorder.
f connexion *f* vidéo
e conexión *f* video
i connessione *f* video
n videoaansluiting
d Videoanschluss *m*

5162 VIDEO SWITCHER cpl
Switch-selection device enabling a number of cameras, telecine machines or other video sources to be mixed.
f commutateur *m* vidéo
e conmutador *m* video
i commutatore *m* video
n videoschakelaar
d Videoschalter *m*

5163 VIDEOTAPE tv
f bande *f* image, bande *f* vidéo
e cinta *f* imagen, cinta *f* video
i nastro *m* immagine, nastro *m* video
n beeldband, videoband
d Bildband *n*, Videoband *n*

5164 VIDEOTAPE ADAPTER vr
f adaptateur *m* magnétoscope
e adaptador *m* para magnetoscopio
i adattatore *m* per videoregistrazione, videoadattatore *m* per sistema VTR
n videobandaanpasinrichting
d MAZ-Anpassglied *n*, Videobandanpassung *f*

5165 VIDEOTAPE APPARATUS vr
f appareil *m* à bande vidéo
e aparato *m* de cinta video
i apparecchio *m* a nastro video
n videobandapparaat *n*
d Videobandgerät *n*

5166 VIDEOTAPE COLO(U)R INFORMATION DECODING vr
f décodage *m* de l'information de bande vidéo en couleur
e decodificación *f* de la información de cinta video en colores
i decodificazione *f* dell'informazione di nastro video in colori
n decodering van de informatie in kleurenbeeldband
d Dekodierung *f* der information des Farbbildbandes

5167 VIDEOTAPE DUBBING, VIDEOTAPE RERECORDING vr
f repiquage *m*
e doblaje *m*, regrabación *f*
i trascrizione *f* di registrazioni sonore
n overspelen *n*
d Überspielung *f*, Umschnitt *m*

VIDEOTAPE RECORDING
see: MAGNETIC PICTURE TRACING

VIDEOTAPE RECORDING EQUIPMENT
see: VIDEORECORDING EQUIPMENT

5168 VIDEOTAPE REPRODUCER, VIDEOTAPE REPRODUCTION APPARATUS vr
f magnétoscope *m* de lecture, vidéoscope *m*
e magnetoscopio *m* de reproducción, videoscopio *m*
i riproduttore *m* video magnetico, videoscopio *m*
n beeldbandapparaat *n*, videorecorder voor weergave
d Bildbandgerät *n*, Videobandwiedergabegerät *n*

5169 VIDEOTAPE SPLICER vr
f colleuse *f* pour bande image
e empalmadora *f* para cinta imagen
i incollatrice *f* per nastro immagine
n beeldbandlasapparaat *n*
d Bildbandklebepresse *f*

5170 VIDEO TECHNIQUE cpl
f technique *f* vidéo
e técnica *f* video
i tecnica *f* video
n videotechniek
d Videofrequenztechnik *f*

5171 VIDEO TELEPHONE tv
f téléphone *m* visuel, vidéotéléphone *m*
e visioteléfono *m*
i visiotelefono *m*
n visiotelefoon
d Fernsehtelephon *n*

5172 VIDEO TELEPHONY tv
f vidéotéléphonie *f*
e visiotelefonía *f*
i visiotelefonia *f*
n visiotelefonie
d Fernsehtelephonie *f*

5173 VIDEO TEST SIGNAL GENERATOR tv
f générateur *m* de signal vidéo d'essai
e generador *m* de señal video de prueba
i generatore *m* di segnale video di prova
n videomeetsignaalgenerator
d Videoprüfsignalgeber *m*

VIDEO TIMEBASE
see: IMAGE OUTPUT

5174 VIDEO TRACK vr
f piste *f* image, piste *f* vidéo
e pista *f* imagen, pista *f* video
i pista *f* immagine, pista *f* video
n beeldspoor *n*, videospoor *n*
d Bildspur *f*, Videospur *f*

5175 VIDEO TRANSMITTER tv
f émetteur *m* d'images
e emisor *m* de imágenes
i emettitore *m* d'immagini
n beeldzender
d Bildsender *m*

5176 VIDEO TRANSMITTER OUTPUT, VIDEO TRANSMITTER OUTPUT SIGNAL tv
f signal *m* de sortie de l'émetteur d'images
e señal *f* de salida del emisor de imágenes
i segnale *m* d'uscita dell'emettitore d'immagini
n beeldzenderuitgangssignaal *n*
d Bildsenderausgangssignal *n*

VIDEOTRON
see: MONOSCOPE

5177 VIDEO VOLTAGE cpl
f tension *f* vidéo
e tensión *f* video
i tensione *f* video
n videospanning
d Videospannung *f*

5178 VIDICON crt
Basic type of camera tube making use of the photoconductive effect.
f vidicon *m*
e vidicón *m*
i vidicon *m*
n vidicon *n*
d Vidikon *n*

5179 VIDICON FILM SCANNER crt/rec/tv
f télécinéma *f* à vidicon
e telecinema *f* de vidicón
i telecinema *f* a vidicon
n vidiconfilmaftaster
d Vidikonfilmabtaster *m*

5180 VIEWER tv
f spectateur *m*, téléspectateur *m*
e espectador *m* de televisión, telespectador *m*
i spettatore *m*, telespettatore *m*
n kijker, televisiekijker
d Fernsehzuschauer *m*, Zuschauer *m*

VIEWFINDER
see: FINDER

VIEWFINDER MONITOR
see: FINDER MONITOR

VIEWFINDER TUBE
see: FOCUSING MICROPHONE

5181 VIEWING ANGLE opt
f angle *m* visuel
e ángulo *m* de visión,
aspecto *m*
i angolo *m* visivo
n gezichtshoek
d Gesichtswinkel *m*

5182 VIEWING DISTANCE rep/tv
f distance *f* de vision
e distancia *f* visual
i distanza *f* visiva
n kijkafstand
d Betrachtungsabstand *m*

5183 VIEWING RATIO rep/tv
f distance *f* optimale de vision
e distancia *f* óptima de visión
i distanza *f* visiva ottima
n optimale kijkafstand
d Betrachtungsverhältnis *n*

VIEWING ROOM
see: REVIEW ROOM

5184 VIEWING SCREEN crt
f écran *m*
e pantalla *f*
i schermo *m*
n beeldscherm *n*
d Bildschirm *m*

5185 VIRTUAL FOCUS opt
f foyer *m* virtuel
e enfoque *m* virtual
i fuoco *m* virtuale
n virtueel brandpunt *n*
d virtueller Brennpunkt *m*

VIRTUAL IMAGE
see: AERIAL IMAGE

5186 VISIBLE IMAGE opt
f image *f* visible
e imagen *f* visible
i immagine *f* visibile
n zichtbaar beeld *n*
d sichtbares Bild *n*

5187 VISIBLE RADIATION ct
f rayonnement *m* visible
e radiación *f* visible
i radiazione *f* visibile
n zichtbare straling
d sichtbare Strahlung *f*

5188 VISION ge
The action of seeing with the bodily eye, the exercise of the normal faculty of sight, or the faculty itself.
f faculté *f* visuelle,
vision *f*,
vue *f*
e facultad *f* visiva,
vista *f*
i facoltà *f* visiva,
vista *f*
n gezichtsvermogen *n*,
zien *n*
d Sehen *n*,
Sehvermögen *n*

5189 VISION BANDWIDTH tv
f largeur *f* de bande image
e ancho *m* de banda imagen
i larghezza *f* di banda immagine
n beeldbandbreedte
d Bildbandbreite *f*

5190 VISION CARRIER cpl
f porteuse *f* image
e portadora *f* imagen
i portante *f* immagine
n beelddraaggolf
d Bildträger *m*

5191 VISION CARRIER FREQUENCY, tv
VISION FREQUENCY
The frequency of the carrier wave of the vision transmitter as distinct from that of the sound transmitter.
f fréquence *f* porteuse image
e frecuencia *f* portadora imagen
i frequenza *f* portante immagine
n beelddraaggolffrequentie
d Bildträgerfrequenz *f*

VISION CARRIER SPACING
see: ADJACENT PICTURE CARRIER SPACING

5192 VISION CHANNEL tv
The circuits of a TV receiver devoted to detection and amplification of the video signal and reproduction of picture.
f canal *m* image
e canal *m* imagen
i canale *m* immagine
n beeldkanaal *n*
d Bildkanal *m*

5193 VISION CIRCUIT cpl
f circuit *m* image
e circuito *m* imagen
i circuito *m* immagine
n beeldcircuit *n*
d Bildkreis *m*

5194 VISION CONTROL cpl
f régie *f* image
e control *m* imagen
i regia *f* immagine
n beeldregie
d Bildregie *f*

5195 VISION CONTROL ENGINEER, tv
VISION SUPERVISOR
f technicien *m* image
e técnico *m* imagen
i tecnico *m* immagine
n beeldtechnicus
d Bildingenieur *m*,
Bildtechnikus *m*

VISION CONTROL ROOM
see: CAMERA CONTROL ROOM

VISION CROSSTALK
see: CROSS-FIRE

5196 VISION FREQUENCY tv
f fréquence *f* image
e frecuencia *f* imagen
i frequenza *f* immagine
n beeldfrequentie
d Bildfrequenz *f*

5197 VISION FREQUENCY RANGE cpl
f gamme *f* de fréquences image
e gama *f* de frecuencias imagen
i gamma *f* di frequenze immagine
n beeldfrequentiegebied *n*
d Bildfrequenzbereich *m*

5198 VISION FREQUENCY SIGNAL cpl
f signal *m* de fréquence image
e señal *f* de frecuencia imagen
i segnale *m* di frequenza immagine
n beeldfrequentiesignaal *n*
d Bildfrequenzsignal *n*,
hochfrequentes Fernsehsignal *n*

5199 VISION MIXER,
VISION MIXER CONTROL PANEL tv
Equipment installed in a TV studio for fading one picture out and another picture in under guidance from a group of monitor tubes which display the separate scenes.
f pupitre *m* régie image
e pupitre *m* de control imagen
i banco *m* di regia immagine
n beeldmenglessenaar
d Bildmischpult *n*

5200 VISION MIXTURE OPERATOR tv
f opérateur *m* de mélange image
e operador *m* de mezclado imagen
i operatore *m* di mescolanza immagine
n beeldmenger
d Bildmischer *m*

5201 VISION MODULATION tv
f modulation *f* image
e modulación *f* imagen
i modulazione *f* immagine
n beeldmodulatie
d Bildmodulation *f*

5202 VISION MODULATOR cpl
f modulateur *m* image
e modulador *m* imagen
i modulatore *m* immagine
n beeldmodulator
d Bildmodulator *m*

VISION SIGNAL (GB)
see: TELEVISION SIGNAL

5203 VISION SWITCHER cpl
f commutateur *m* image
e conmutador *m* imagen
i commutatore *m* immagine
n beeldomschakelaar
d Bildschalter *m*

5204 VISION SWITCHING MATRIX tv
f matrice *f* commutatrice image
e matriz *f* conmutadora imagen
i matrice *f* commutatrice immagine
n beeldschakelmatrix
d Bildschaltmatrize *f*

5205 VISION TRANSMITTER,
VISUAL TRANSMITTER tv
f émetteur *m* image
e emisor *m* imagen
i emettitore *m* immagine
n beeldzender
d Bildsender *m*

5206 VISION TRANSMITTER
MONITORING EQUIPMENT tv
f ensemble *m* de contrôle de l'émission
e equipo *m* de control de la emisión
i insieme *m* di controllo dell'emissione
n beeldzendercontroleapparatuur
d Bildsendermessgestell *n*

5207 VISION TRANSMITTER OUTPUT,
VISION TRANSMITTER POWER,
VISUAL TRANSMITTER POWER tv
The peak power output when transmitting a standard TV signal.
f puissance *f* d'émission image
e potencia *f* de emisión imagen
i potenza *f* d'emissione immagine
n beeldzendvermogen *n*
d Bildsendeleistung *f*

VISTA SHOT
see: LONG SHOT

VISUAL ACUTENESS
see: ACUITY OF THE EYE

5208 VISUAL PRINCIPLES IN
TELEVISION tv
f principes *pl* de vision en télévision
e principios *pl* de visión en televisión
i principi *pl* di visione in televisione
n visuele grondslagen *pl* in televisie
d visuelle Grundlagen *pl* beim Fernsehen

VISUAL SENSORY CHARACTERISTICS
see: SENSORY CHARACTERISTICS

VIT SIGNALS
see: VERTICAL INTERVAL TEST SIGNALS

VOICE COIL (US)
see: MOVING COIL

5209 VOLTAGE ADAPTER ge
f sélecteur *m* de tension
e selector *m* de tensión
i cambiotensioni *m*,
commutatore *m* di tensione
n spanningskiezer
d Spannungsumschalter *m*,
Spannungswähler *m*

5210 VOLTAGE GAIN ge
The ratio of the output and input voltages under specified conditions of impedance termination.
f facteur *m* d'amplification de tension
e factor *m* de amplificación de tensión
i fattore *m* d'amplificazione di tensione
n spanningsversterkingsfactor
d Spannungsverstärkungsfaktor *m*

5211 VOLUME aud
f volume *m* acoustique
e volumen *m*
i volume *m*,
volume *m* sonoro
n geluidssterkte
d Tonpegel *m*

VOLUME COMPRESSION
see: COMPRESSION

VOLUME COMPRESSOR
see: COMPRESSOR

VOLUME CONTROL
see: GAIN CONTROL

VOLUME CONTROL
see: LOUDNESS CONTROL

VOLUME INDICATOR
see: LEVEL INDICATOR

VOLUME INDICATOR
see: VI-METER

VOLUME LEVEL
see: ELECTRICAL SPEECH LEVEL

VOLUME RANGE
see: DYNAMIC RANGE

5212 VOLUME RANGE CONTROL cpl
f réglage *m* de la dynamique
e regulación *f* de la dinámica
i regolazione *f* della dinamica
n dynamiekregeling
d Aussteuerungsregelung *f*

VTR
see: MAGNETIC PICTURE TRACING

VTR EQUIPMENT
see: VIDEORECORDING EQUIPMENT

WALKIE-LOOKIE
see: PORTABLE TELEVISION RECEIVER

WALKIE-TALKIE
see: PORTABLE TELEVISION TRANSMITTER

5213 WALL ANODE crt
f anode *f* sur la paroi
e ánodo *m* sobre la pared
i anodo *m* sulla parete
n wandanode
d Wandanode *f*

WALLMAN AMPLIFIER
see: CASCODE AMPLIFIER

5214 WANTED SIGNAL cpl
f signal *m* utile
e señal *f* útil
i segnale *m* utile
n nuttig signaal *n*
d Nutzsignal *n*

5215 WARM COLO(U)RS ct
f couleurs *pl* chaudes
e colores *pl* cálidos
i colori *pl* caldi
n warme kleuren *pl*
d warme Farben *pl*

5216 WARM-UP TIME, WARMING-UP TIME crt
The time interval between application of power to a system and the instant at which the system is stabilized and ready to perform its intended function.
f temps *m* de chauffage
e tiempo *m* de calentamiento
i tempo *m* di riscaldamento
n opwarmtijd
d Anheizzeit *f*

5217 WARNING SIGNAL tv
f signal *m* d'alarme
e señal *f* de alarma
i segnale *m* d'allarme
n waarschuwingssignaal *n*
d Warnsignal *n*

5218 WARNING SYSTEM tv
Closed circuit TV system for warning purposes.
f système *m* d'alarme
e sistema *m* de alarma
i sistema *m* d'allarme
n waarschuwingssysteem *n*
d Warnsystem *n*

5219 WASH DISSOLVE fi/tv
f volet *m* ondulé
e fundido *m* ondulado
i dissolvenza *f* ondulata
n golvende overgang
d Wellenblende *f*

WATERED-SILK
see: MOIRÉ

5220 WAVEFORM COMPONENTS cpl
The set of signals which make up the complete TV waveform.
f composantes *pl* du signal de télévision
e componentes *pl* de la señal de televisión
i componenti *pl* del segnale di televisione
n componenten *pl* van het televisiesignaal
d Komponenten *pl* des Fernsehsignals

5221 WAVEFORM DISTORTION cpl
f distorsion *f* de la forme d'onde
e distorsión *f* de la forma de onda
i distorsione *f* della forma d'onda
n golfvormvervorming
d Wellenformverzerrung *f*

5222 WAVEFORM MEASUREMENTS cpl/tv
f mesures *pl* de la forme d'onde
e medidas *pl* de la forma de onda
i misure *pl* della forma d'onda
n golfvormmetingen *pl*
d Wellenformmessungen *pl*

WAVEFORM MONITOR
see: VIDEO SIGNAL MONITOR

5223 WAVEFORM MONITOR, WAVEFORM OSCILLOSCOPE tv
Cathode-ray oscilloscope for displaying TV signals, in which the X or horizontal axis represents time and the Y or vertical axis represents the amplitude of the signal.
f oscilloscope *m* de contrôle de télévision
e monitor *m* de forma de onda
i monitore *m* di forma d'onda
n profielmonitor, videosignaaloscilloscoop
d Kontrolloszilloskop *n*

5224 WAVEFORM RESPONSE cpl
f réponse *f* de forme d'onde
e respuesta *f* de forma de onda
i risposta *f* di forma d'onda
n golfvormresponsie
d Wellenformfrequenzgang *m*

5225 WAVELENGTH ge
f longueur *f* d'onde
e longitud *f* de onda
i lunghezza *f* d'onda
n golflengte
d Wellenlänge *f*

5226 WAVELENGTH OF LIGHT ct
f longueur *f* d'onde de la lumière
e longitud *f* de onda de la luz

i lunghezza *f* d'onda della luce
n golflengte van het licht
d Wellenlänge *f* des Lichtes

5227 WAVELENGTH OF SOUND aud
f longueur *f* d'onde du son
e longitud *f* de onda del sonido
i lunghezza *f* d'onda del suono
n golflengte van het geluid
d Wellenlänge *f* des Schalls

5228 WAVE MOTION aud/ct
f mouvement *m* ondulatoire
e movimiento *m* de las ondas elásticas
i movimento *m* ondulatorio
n golfbeweging
d Wellenbewegung *f*

5229 WAVE RANGE SWITCH, WAVEBAND SWITCH cpl
f commutateur *m* de gammes d'ondes
e conmutador *m* de gamas de ondas, conmutador *m* de gamas de ondas
i commutatore *m* di lunghezza d'onda
n golflengteschakelaar
d Wellenlängenschalter *m*

5230 WAVE RANGE SWITCHING, WAVEBAND SWITCHING cpl
f commutation *f* de gammes d'ondes
e conmutación *f* de gamas de ondas
i commutazione *f* di lunghezza d'onda
n golflengtekeuze
d Wellenlängenschaltung *f*

5231 WEAK PICTURE tv
f image *f* douce, image *f* peu contrastée
e imagen *f* sin contraste
i immagine *f* senza contrasto
n contrastarm beeld *n*
d weiches Bild *n*

5232 WEAK SIGNAL rep/tv
f signal *m* faible
e señal *f* débil
i segnale *m* debole
n zwak signaal *n*
d schwaches Signal *n*

5233 WEAKER ct
A difference apparently due to the presence of less colo(u)r than in the original sample.
f plus flou
e más flojo
i più sbiadito
n flauwer
d matter

5234 WEAR LIFE OF TAPE vr
f limite *f* de résistance à l'usure de la bande magnétique
e límite *m* de resistencia al desgaste de la cinta magnética
i limite *m* di resistenza all'usura del nastro magnetico
n slijtgrens van een magneetband
d Verschleisswiderstandsgrenze *f* eines Magnetbandes

5235 WEAVE dis
f manque *m* de fixité latérale
e inestabilidad *f* lateral
i instabilità *f* laterale
n laterale instabiliteit
d laterale Instabilität *f*

5236 WEDGE opt
f coin *m*
e cuña *f*
i cuneo *m*
n wig
d Keil *m*

5237 WEIGHTED SIGNAL-TO-NOISE-RATIO dis
f rapport *m* signal/bruit subjectif
e relación *f* señal/ruido subjetiva
i rapporto *m* segnale/rumore soggettivo
n zichtbare signaal/ruisverhouding
d visueller Störabstand *m*

WHIP PAN
see: SWISH PAN

5238 WHITE ct
f blanc *m*
e blanco *m*
i bianco *m*
n wit *n*
d Weiss *n*

5239 WHITE ADJUSTMENT svs
f ajustage *m* du blanc
e ajuste *m* del blanco
i aggiustaggio *m* del bianco
n witinstelling
d Weiss-Einstellung *f*

5240 WHITE AFTER BLACK dis/rep/tv
Picture defect characterized by the appearance of a white streak after a black element.
f blanc *m* après le noir
e blanco *m* detrás del negro
i bianco *m* dopo il nero
n wit *n* na het zwart
d Überschwinger *m*, Weiss *n* hinter Schwarz

5241 WHITE BALANCE tv
f balance *f* du blanc
e equilibrio *m* del blanco
i equilibrio *m* del bianco
n balancering op wit, witbalans
d Weissbalance *f*

5242 WHITE BALANCE STABILITY tv
f stabilité *f* de la balance du blanc
e estabilidad *f* del equilibrio del blanco
i stabilità *f* dell'equilibrio del bianco
n stabiliteit van de balancering op wit
d Stabilität *f* der Weissbalance

5243 WHITE BAR tv
A number of lines of peak white used as a reference or test signal.
f barre *f* blanche
e barra *f* blanca

i barra *f* bianca
n witte balk
d weisser Balken *m*

5244 WHITE CLIPPER cpl
Provided to prevent unwanted signal excursions above peak-white.
f écrêteur *m* du blanc,
limiteur *m* du blanc
e limitador *m* del blanco,
recortador *m* del blanco
i limitatore *m* del bianco,
limitatore *m* del valore di soglia del bianco
n drempelwaardebegrenzer van het wit,
witbegrenzer
d Schwellenwertbegrenzer *m* für Weiss,
Weisswertbegrenzer *m*

5245 WHITE CLIPPING tv
f limitation *f* du blanc
e limitación *f* del blanco
i limitazione *f* del bianco
n witbegrenzing
d Weisswertbegrenzung *f*

5246 WHITE COMPRESSION tv
f écrasement *m* du blanc,
tassement *m* du blanc
e compresión *f* del blanco
i compressione *f* del bianco
n compressie van het wit
d Weissstauchung *f*,
Weisswertstauchung *f*

5247 WHITE CONTENT ct
The subjectively estimated amount of whiteness seen in the visual sensation arising from a surface colo(u)r.
f contenu *m* en blanc
e contenido *m* en blanco
i contenuto *m* in bianco
n witgehalte *n*
d Weissanteil *m*

5248 WHITE CRUSHING dis
f distorsion *f* de la crête du blanc
e distorsión *f* de la cresta del blanco
i distorsione *f* della cresta del bianco
n vervorming van het helderste wit
d verzerrtes Maximum *n* an Weiss

5249 WHITE EDGING dis
f bord *m* blanc
e borde *m* blanco
i bordo *m* bianco
n witte rand
d Vorläufer *m*,
Vorplastik *f*

5250 WHITE EXPANSION tv
f expansion *f* du blanc
e expansión *f* del blanco
i espansione *f* del bianco
n witexpansie
d Weissexpansion *f*

5251 WHITE HALO dis
f halo *m* blanc
e halo *m* blanco
i alone *m* bianco
n witte halo,
witte overstraling
d weisser Halo *m*,
weisser Lichthof *m*

5252 WHITE LEVEL tv
In the colo(u)r video signal, the maximum permissible level of the luminance signal.
f niveau *m* du blanc
e nivel *m* del blanco
i livello *m* del bianco
n witniveau *n*
d Weisspegel *m*,
Weisswert *m*

5253 WHITE-LEVEL INSTABILITY tv
f instabilité *f* du niveau du blanc
e inestabilidad *f* del nivel del blanco
i instabilità *f* del livello del bianco
n instabiliteit van het witniveau
d Instabilität *f* des Weisspegels

WHITE LEVEL RASTER,
WHITE RASTER
see: PEAK-WHITE RASTER

5254 WHITE LIGHT ct
f lumière *f* blanche
e luz *f* blanca
i luce *f* bianca
n wit licht *n*
d weisses Licht *n*

5255 WHITE MODIFYING RELAY rep/tv
f relais *m* modificateur de blanc
e relé *m* modificador de la intensidad de blanco
i relè *m* modificatore dell'intensità di bianco
n de intensiteit van wit veranderend relais *n*
d die Intensität von Weiss änderndes Relais *n*

WHITE NOISE
see: FLAT RANDOM NOISE

5256 WHITE OBJECT ct
An object which reflects all wavelengths of light with substantially high efficiency and with considerable diffusion.
f objet *m* blanc
e objeto *m* blanco
i oggetto *m* bianco
n wit voorwerp *n*
d weisser Gegenstand *m*,
weisses Ding *n*

5257 WHITE-PATTERN SIGNAL svs
f signal *m* du canevas blanc
e señal *f* de la cuadrícula blanca
i segnale *m* del quadro rigato bianco
n witrastersignaal *n*
d Weissrastersignal *n*

WHITE PEAK
see: PEAK WHITE

WHITE POINT
see: ACHROMATIC POINT

WHITE REFERENCE
see: NORMALIZING WHITE

5258 WHITE SATURATION tv
f saturation *f* du niveau de blanc
e saturación *f* del nivel de blanco
i saturazione *f* del livello di bianco
n witniveauverzadiging
d Weisspegelsättigung *f*

WHITE-TO-BLACK AMPLITUDE RANGE
see: BLACK-TO-WHITE AMPLITUDE RANGE

WHITE-TO-BLACK FREQUENCY SWING
see: BLACK-TO-WHITE FREQUENCY SWING

5259 WHITER ct
A difference apparently due to the presence of more white than in the original sample.
f plus blanc
e más blanco
i più bianco
n witter
d weisser

WHITER THAN WHITE
see: ULTRAWHITE

WHITER THAN WHITE REGION
see: ULTRAWHITE REGION

5260 WIDE-ANGLE DEFLECTION crt
f déviation *f* à grand angle
e desviación *f* granangular
i deviazione *f* grandangolare
n groothoekige afbuiging
d weitwinklige Ablenkung *f*

5261 WIDE-ANGLE LENS opt
An optical lens having a large angular field.
f objectif *m* à grand angle, objectif *m* grandangulaire
e objetivo *m* granangular
i obiettivo *m* grandangolare
n groothoekobjectief *n*
d Weitwinkelobjektiv *n*

WIDEBAND AMPLIFIER
see: ALL-BAND AMPLIFIER

WIDEBAND AXIS
see: I-AXIS

5262 WIDEBAND CHROMINANCE DECODING ctv
f décodage *m* de chrominance à large bande
e descodificación *f* de crominancia de ancha banda
i decodificazione *f* di crominanza a larga banda
n brede-bandchrominantiedecodering
d Breitbandchrominanzdekodierung *f*

5263 WIDEBAND CHROMINANCE RECEIVER ctv
In the NTSC system, a receiver in which all the information contained in the I signal is utilized.
f récepteur *m* à décodage de chrominance à large bande
e receptor *m* de descodificación de ancha banda
i ricevitore *m* a decodificazione a larga banda
n ontvanger met brede-band-chrominantie-decodering
d Empfänger *m* mit Breitbandchrominanz-dekodierung

5264 WIDEBAND DIPOLE aea
f dipôle *m* à large bande
e dipolo *m* de ancha banda
i dipolo *m* a larga banda
n brede-banddipool
d Breitbanddipol *m*

5265 WIDEBAND INTERPOLATION tv
f interpolation *f* par filtre à large bande
e interpolación *f* por filtro de ancha banda
i interpolazione *f* per filtro a larga banda
n interpolatie door brede-bandfilter
d Interpolation *f* mit Breitbandfilter

5266 WIDE-CUT FILTER ct/ctv
A colo(u)r filter that transmits an extensive spectral band so that the resultant colo(u)r is less saturated.
f filtre *m* chromatique à large bande
e filtro *m* cromático de ancha banda
i filtro *m* cromatico a larga banda
n brede-bandkleurenfilter *n*
d Breitbandfarbfilter *n*

5267 WIDE-LONG SHOT opt
f plan *m* de grand ensemble
e plano *m* más legano
i campo *m* lunghissimo
n ruimtotaalshot
d Weiteinstellung *f*

5268 WIDE-MESH SPACING crt
f gaze *f* à larges mailles
e gasa *f* de anchas mallas
i garza *f* a larghe maglie
n gaas *n* met wijde mazen
d weitmaschige Gaze *f*

5269 WIDTH rep/tv
The horizontal dimension of a TV picture.
f largeur *f*
e anchura *f*
i larghezza *f*
n breedte
d Breite *f*

5270 WIDTH CHOKE cpl
f bobine *f* d'ajustage de la largeur de l'image
e bobina *f* de ajuste de la anchura de la imagen
i bobina *f* d'aggiustaggio della larghezza dell'immagine
n instelspoel voor de beeldbreedte
d Bildbreiteeinstellspule *f*

WIDTH CONTROL
see: HORIZONTAL DEFLECTION CONTROL

5271 WIG-MAKER fi/tv
f perruquier *m*,
posticheur *m*
e peluquero *m*
i parrucchiere *m*
n pruikenmaker
d Perückenmacher *m*

WILD TRACK
see: NON-SYNC SOUND TRACK

5272 WINDING vr
Process of transferring film or tape from one reel to another.
f bobinage *m*,
enroulement *m*
e arrollamiento *m*,
bobinado *m*
i avvolgimento *m*
n wikkelen *n*
d Wickeln *n*

WINDOW
see: DOUBLE LIMITER

5273 WINDOW FRAME AERIAL, WINDOW FRAME ANTENNE aea
f antenne *f* de vitre
e antena *f* de ventana
i antenna *f* di finestra
n vensterantenne
d Fensterantenne *f*

5274 WINKING dis
f papillotement *m* partiel
e centelleo *m* parcial
i sfarfallamento *m* parziale,
sfarfallio *m* parziale
n deelflikkering
d teilweises Flickern *n*

5275 WIPE rep/tv
A method of replacing one TV picture by another, in which the boundary between the two pictures is caused to move progressively.
f commutation *f* par volet,
découvrement *m*,
fondu *m* effacé
e conmutación *f* por cortinillas
i commutazione *f* a tendina
n trucbeeldovergang
d rollender Schnitt *m*

WIPING
see: ERASING

WIPING CURRENT
see: erasing current

WIPING HEAD
see: ERASING HEAD

WIRE BROADCASTING (US)
see: LINE BROADCASTING

5276 WIRED SYSTEM tv
Method of distributing sound and or TV signals by means of cables.
f filodiffusion *f*,
filovision *f*
e hilodifusión *f*,
hilovisión *f*
i filodiffusione *f*,
filovisione *f*
n draadomroep,
draadtelevisie
d Drahtfernsehen *n*,
Drahtfunk *m*

5277 WIRED TELEVISION RECEIVER rep/tv
f téléviseur *m* à filovision
e televisor *m* de hilovisión
i televisore *m* a filovisione
n draadtelevisieontvanger
d Drahtfernsehempfänger *m*

5278 WITHOUT CONTRAST tv
f sans contraste
e sin contraste
i senza contrasto
n **contrastarm adj**
d **flau adj**

5279 WOBBULATION cpl
f vobulation *f*
e vobulación *f*
i vobulazione *f*
n wobbelen *n*
d Wobbelung *f*,
Wobblen *n*

5280 WOBBULATOR tv
Frequency-modulated signal generator, the frequency of which is varied periodically by a fixed amount at constant amplitude above and below a central frequency.
f générateur *m* vobulé
e generador *m* de barrido,
vobulador *m*
i vobulatore *m*
n wobbelgenerator
d Wobbelgenerator *m*,
Wobbelmesssender *m*,
Wobbler *m*

WOMP (US)
see: FLARE

WORKING DIAGRAM
see: DYNAMIC DEMONSTRATOR

5281 WOW AND FLUTTER dis/vr
f scintillation et pleurage
e ululación y trémolo
i miagolio e trillo
n hoge en lage jank,
toonhoogtevariaties *pl*
d Tonhöhenschwankungen *pl*

5282 WRITING AND READING TECHNIQUE tv
f technique *f* d'écrire et de lire

e técnica *f* de escribir y de leer
i tecnica *f* di scrivere e di leggere
n schrijf- en leestechniek
d Schreibe- und Lesetechnik *f*

5283 WRITING BEAM crt
f faisceau *m* écrivant
e haz *m* escribente
i fascio *m* scrivente
n schrijvende straal
d Schreibstrahl *m*

5284 X AXIS crt
f axe *m* des X
e eje *m* de tiempo,
eje *m* X
i asse *m* di tempi,
asse *m* X
n tijdas,
X-as
d X-Achse *f*,
Zeitachse *f*

5285 X DEFLECTION crt
f déviation *f* X
e desviación *f* X
i deflessione *f* X,
deviazione *f* X
n X-afbuiging
d X-Ablenkung *f*,
Zeitablenkung *f*

X PLATES
see: HORIZONTAL DEFLECTION ELECTRODES

5286 X-Z AXES ctv
f axes *pl* X-Z
e ejes *pl* X-Z
i assi *pl* X-Z
n X-Z-assen *pl*
d X-Z-Achsen *pl*

5287 X-Z CIRCUITS ctv
f circuits *pl* X-Z
e circuitos *pl* X-Z
i circuiti *pl* X-Z
n X-Z-ketens *pl*
d X-Z-Kreise *pl*

5288 X-Z DETECTION ctv
f démodulation *f* X-Z
e desmodulación *f* X-Z
i demodulazione *f* X-Z
n X-Z-demodulatie
d X-Z-Demodulation *f*

5289 X-Z MATRIX ctv
f matrice *f* X-Z
e matriz *f* X-Z
i matrice *f* X-Z
n X-Z-matrix
d X-Z-Matrize *f*

Y AXIS
see: VERTICAL AXIS

5290 Y DEFLECTION crt
f déviation *f* Y
e desviación *f* Y
i deflessione *f* Y,
deviazione *f* Y
n Y-afbuiging
d Y-Ablenkung *f*

Y LEVEL
see: BLACK LEVEL

Y PLATES
see: FIELD DEFLECTION ELECTRODES

Y SIGNAL
see: LUMINANCE SIGNAL

5291 Y TO M CONVERTER ctv
A circuit for changing the monochrome or luminance signal (Y) to a form (M) suitable for direct decoding in a single gun display device.
f convertisseur *m* Y à M
e convertidor *m* Y hacia M
i convertitore *m* Y a M
n Y naar M omzetter
d Y-M-Umsetzer *m*

5292 YAGI AERIAL, aea
YAGI ANTENNA
f antenne *f* Yagi
e antena *f* Yagi
i antenna *f* Yagi
n yagiantenne
d Yagi-Antenne *f*

YELLOW
see: MINUS BLUE

YELLOW SPOT
see: MACULA LUTEA

Z

5293 Z AXIS crt
f axe *m* des Z
e eje *m* Z
i asse *m* Z
n Z-as
d Z-Achse *f*

5294 Z AXIS MODULATION crt
f modulation *f* dans l'axe des Z
e modulación *f* en el eje Z
i modulazione *f* nell'asse Z
n modulatie in de Z-as
d Steuerung *f* in der Z-Achse, Z-Achsensteuerung *f*

5295 Z DISTORTION dis
f distorsion *f* en Z
e distorsión *f* en Z
i distorsione *f* in Z
n Z-vervorming
d Z-Verzeichnung *f*

5296 ZEBRA COLO(U)R TUBE crt
Single-gun picture tube which has a screen covered with narrow vertical stripes of red, freen and blue phosphors alternately.
f tube *m* zèbre
e tubo *m* cebra
i tubo *m* zebra
n zebrabuis
d Zebraröhre *f*

5297 ZENER VOLTAGE cpl
f tension *f* Zener
e tensión *f* Zener
i tensione *f* Zener
n zenerspanning
d Zenerspannung *f*

5298 ZERO ADJUSTMENT cpl
f mise *f* à zéro
e ajuste *m* a cero, puesta *f* a cero
i messa *f* a zero
n nulinstelling
d Nulleinstellung *f*

5299 ZERO BEAT cpl/dis
f battement *m* nul
e batido *m* cero
i battimento *m* zero
n nulzweving
d Nullschwebung *f*, Schwebungslücke *f*

5300 ZERO CARRIER cpl
f porteuse *f* zéro
e portadora *f* cero
i portante *f* zero
n nuldraaggolf
d Trägernull *n*

5301 ZERO LEVEL cpl
f niveau *m* zéro
e nivel *m* cero
i livello *m* zero
n nulniveau *n*
d Nullpegel *m*

5302 ZERO LUMINANCE ct
f luminance *f* zéro
e luminancia *f* cero
i luminanza *f* zero
n nulluminantie
d Nullhelle *f*

ZERO LUMINANCE PLANE
see: ALYCHNE

5303 ZERO SIGNAL tv
f signal *m* zéro
e señal *f* cero
i segnale *m* zero
n nulsignaal *n*
d Nullsignal *n*

5304 ZERO SUBCARRIER CHROMATICITY ctv
The chromaticity which is intended to be displayed when the subcarrier amplitude is zero.
f couleur *f* résultant de l'élimination de la sousporteuse
e color *m* resultante de la eliminación de la subportadora
i colore *m* dovuto all'eliminazione della sottoportante
n kleur ontstaan door eliminatie van de hulpdraaggolf
d Farbe *f* entstanden durch Wegfall des Hilfsträgers

5305 ZERO-VOLT ADJUSTMENT svs
f mise *f* à tension zéro
e ajuste *m* a tensión cero
i aggiustaggio *m* a tensione zero
n nulvoltinstelling
d Nullspannungseinstellung *f*

ZIP PAN
see: SWISH PAN

ZONE TELEVISION
see: RECTILINEAR SCANNING

ZOOM (US)
see: CAMERA SHIFTING

5306 ZOOM AWAY (TO), ZOOM OUT (TO) opt/tv
f ouvrir v
e aumentar v la distancia focal
i aumentare v la distanza focale
n zoomlens openen v
d Gummilinse aufziehen v

5307 ZOOM-AWAY SHOT, ZOOM-OUT SHOT rec
f prise *f* de vues à augmentation de la distance focale
e toma *f* con aumento de la distancia focal
i ripresa *f* con aumento della distanza focale
n zoomlensopname met toenemende brandpuntsafstand
d Gummilinsenaufnahme *f* mit zunehmender Brennweite

5308 ZOOM CAMERA tv
f caméra *f* avec objectif à distance focale variable
e cámara *f* con lente de enfoque ajustable
i camera *f* con obiettivo a lunghezza focale variabile
n zoomlenscamera
d Kamera *f* mit Gummilinse

5309 ZOOM IN (TO) opt/tv
f fermer v,
serrer v
e diminuir v la distancia focal
i diminuire v la distanza focale
n zoomlens sluiten v
d Gummilinse zuziehen v

5310 ZOOM-IN SHOT tv
f prise *f* de vues à diminution de la distance focale
e toma *f* con disminución de la distancia focal
i carrellata *f* rapida in avanti
n zoomlensopname met afnemende brandpuntsafstand
d Gummilinsenaufnahme *f* mit abnehmender Brennweite

ZOOM LENS
see: VARIABLE FOCUS LENS

FRANÇAIS

F

ESPAÑOL

ITALIANO

NEDERLANDS

DEUTSCH